AF783178

Advances in IMAGING AND ELECTRON PHYSICS

VOLUME 172

Neutron and X-ray Microscopy - Part 1

EDITOR-IN-CHIEF

PETER W. HAWKES
CEMES-CNRS
Toulouse, France

Advances in IMAGING AND ELECTRON PHYSICS

VOLUME **172**

Neutron and X-ray Microscopy - Part 1

By

JAY THEODORE CREMER, Jr.

Chief Scientist, Adelphi Technology, Inc.

ELSEVIER

AMSTERDAM • BOSTON • HEIDELBERG • LONDON
NEW YORK • OXFORD • PARIS • SAN DIEGO
SAN FRANCISCO • SINGAPORE • SYDNEY • TOKYO
Academic Press is an imprint of Elsevier

Cover photo credit:
Jay Theodore Cremer, Jr.
Introduction to Neutron and X-ray Optics
Advances in Imaging and Electron Physics (2012) **172**, p. 58.

Academic Press is an imprint of Elsevier
Radarweg 29, PO Box 211, 1000 AE Amsterdam, The Netherlands
32 Jamestown Road, London NW17BY, UK
225 Wyman Street, Waltham, MA 02451, USA
525 B Street, Suite 1900, San Diego, CA 92101-4495, USA

First edition 2012

Copyright © 2012 Elsevier Inc. All rights reserved.

No part of this publication may be reproduced, stored in a retrieval system or transmitted in any form or by any means electronic, mechanical, photocopying, recording or otherwise without the prior written permission of the publisher.

Permissions may be sought directly from Elsevier's Science & Technology Rights Department in Oxford, UK: phone (+44) (0) 1865 843830; fax (+44) (0) 1865 853333; email: permissions@elsevier.com. Alternatively you can submit your request online by visiting the Elsevier web site at http://www.elsevier.com/locate/permissions, and selecting *Obtaining permission to use Elsevier material*.

Notice
No responsibility is assumed by the publisher for any injury and/or damage to persons or property as a matter of products liability, negligence or otherwise, or from any use or operation of any methods, products, instructions or ideas contained in the material herein. Because of rapid advances in the medical sciences, in particular, independent verification of diagnoses and drug dosages should be made.

Library of Congress Cataloging-in-Publication Data
A catalog record for this book is available from the Library of Congress

British Library Cataloguing in Publication Data
A catalogue record for this book is available from the British Library

ISBN: 978-0-12-394422-1
ISSN: 1076-5670

For information on all Academic Press publications
visit our Web site at *store.elsevier.com*

Working together to grow
libraries in developing countries

www.elsevier.com | www.bookaid.org | www.sabre.org

ELSEVIER BOOK AID International Sabre Foundation

Contents

Foreword *xi*
Preface *xv*
Contributor *xvii*
Future Contributions *xix*

1. Introduction to Neutron and X-ray Optics **1**

1. Compound Refractive Lenses for Neutrons and X-rays: Background and Theory 2
2. Experiments with Neutron Compound Refractive Lenses, Magnetic Lenses, and Microscopes 44
3. Experiments with X-ray Compound Refractive Lenses and Microscopes 100
4. Thermal Neutron Radiography, Coded Aperture, and Phase-Contrast Imaging 199
5. Experiments in Fast Neutron Radiography 237
6. A Brief History of Neutron Optics Before Compound Refractive Neutron Lenses 272
7. Brief History and Overview of Diffractive and Reflective X-ray Optics and Microscopy 282

References 312

2. Compound Refractive Lenses and Prisms **335**

1. The Compound Refractive Lens 336
2. Numerical Aperture and Thin Lenses 338
3. Biconcave Parabolic Lens: Path Length, Focal Length, and Absorption Aperture Radius 339
4. Biconcave Spherical Lens: Path Length, Focal Length, and Absorption Aperture Radius 346
5. Parabolic Aperture Radius in Spherical Biconcave CRLs 349
6. Matrix Representation of X-ray and Neutron Optics with Paraxial Approximation 350
7. Thin Lens Matrix and Ray Angle Deviation in a Thin Lens 353
8. Average CRL Transmission for X-rays or Neutrons 358
9. CRL Intensity Gain Including Attenuation 361

10. CRL Transverse and Axial Magnification 368
11. CRL Depth of Field and Depth of Focus 371
12. Modulation Transfer Function for CRL Resolution Determination 373
13. Calculation of CRL Modulation Transfer Function by Line Profile Measurements of Knife-Edge Images 377
14. CRL Field of View 380
15. Thick Lens CRLs 382
16. CRL Surface Roughness and Lens Alignment 390
17. Compound Refractive Prisms: X-ray and Neutron Deflection by a Single Prism 398
18. Deflection of X-rays and Neutrons in a Compound Refractive Prism 402
19. Calculation of X-ray or Neutron Absorption Aperture Depth and Transmission in a Compound Refractive Prism 404
20. Use of a Compound Refractive Prism to Offset Neutron Gravity Droop in Small-Angle Scattering 406
21. Curvature and Radius of Curvature of Lens Surface and Wave Surface 409
22. Measurement of Spherical and Parabolic Lens Curvature 411
23. Wave Vergence and Wave Vergence Change by Lens Surface 413
24. Refractive Index, Snell's Law, Huygens' Principle, and Fermat's Principle of Least Time 414
25. The Paraxial Approximation and the Fundamental Paraxial Equation 417
26. The Gaussian Lens Equation for Thin Lenses 418
27. Thick Lens Power and Focal Length, and Wave Vergence Change in Thick Lenses 420
28. Wave Vergence Change and Power and Focal Length of a Separated Lens Doublet 424

References 426

3. Geometric Neutron and X-ray Optics – Aberrations 429

1. Chromatic Aberration in Spherical and Parabolic Biconcave Lenses 430
2. Spherical Aberration in Spherical Biconcave Lenses and Total Object Resolution 434
3. CRL Image and Object Resolution: Spherical and Chromatic Aberration 438
4. CRL Aberrations: Astigmatism, Curvature of Field, Coma, and Distortion 444
5. Parabolic and Spherical Waves Converge to Different Focal Points by CRL 450
6. Monochromatic and Chromatic Aberrations Occur in Parabolic and Spherical Lenses 453
7. Aberrations from an Imperfect Spherical Lens Surface 456
8. Derivation of the Five Seidel Monochromatic Aberrations of a Spherical Biconcave Lens 461

9. Relation of the Optical Path Length Difference to its Associated Aberration 470
10. Spherical Aberration 473
11. Coma 475
12. Coma: The Abbe Sine Condition and the Coddington Shape Factor 477
13. Curvature of Field and the Petzval Condition 479
14. Astigmatism 481
15. Combined Astigmatism and Curvature of Field 483
16. Astigmatism: Sturm's Interval and Remedy by Upstream Aperture Stop 485
17. Distortion 487
18. Images without and with Distortion 490
19. Prevention of Distortion and the Petzval Condition 492
20. Distortion in Compound Refractive Lens Imaging with Synchrotron X-ray Sources 494
References 495

4. X-ray Optics 497

1. Damped Resonance for Driven Atomic Electron Oscillation and X-ray Emission 498
2. The Complex Atomic Scatter Factor for a Single Element 502
3. The Complex Atomic Scatter Factor for a Mixture or Compound 503
4. Maxwell's Equations Yield the Electromagnetic Vector Wave Equation in Material Media 505
5. The Electromagnetic Vector Wave Equation Contains the Refractive Index 506
6. The Complex Refractive Index and the Complex Atomic Scatter Factor 508
7. The Complex Refractive Index—Decrement, Absorption Index, and Linear Attenuation 510
8. X-ray Complex Index of Refraction—Kramers–Kronig Relation 513
9. X-ray Rayleigh Scatter Total Cross Section and the Complex Atomic Scatter Factor 514
10. Differential Cross Section for the Rayleigh Scatter and Complex Atomic Scatter Factor 517
11. Rayleigh and Thomson X-ray Scatters are Coherent, Elastic Scatter 519
12. Compton Scatter of X-rays is Incoherent and Inelastic 521
13. X-ray Attenuation by Photoelectric Absorption 525
14. X-ray Absorptive Attenuation by Pair Production 527
15. Collision Losses of Charged Particle Passage in Matter 528
16. Bremsstrahlung from Charged Particle Passage in Matter 536
17. Forward-Directed Bremsstrahlung from Relativistic Electrons Undergoing Acceleration 538

18. X-ray Wave Packets—Phase Velocity and Group Velocity, Spatial and Spectral Widths 541
19. X-ray Wave Packets—Dispersion and Group Velocity, and Relation to Real Part of Complex Atomic Scatter Factor 544
20. X-rays in Material Medium—Phase and Group Velocity and Pulse Broadening 548
21. Reflection and Transmission of X-rays at a Planar Interface of Two Media 550
22. Fraction of Transmitted and Reflected Polarized X-rays with an Electric Field Component Parallel to the Planar Surface Interface 554
23. Fraction of Transmitted and Reflected Polarized X-rays with a Magnetic Field Component Parallel to the Planar Surface Interface 556
24. X-ray Brewster Angle Occurs Only for Electric Field Parallel to Plane of Incidence 557

References 559

5. Neutron Optics 561

1. Neutron Phase and Group Velocity 561
2. Derivation of the Schrödinger Wave Equation for Neutron Wave 566
3. Derivation of the Schrödinger Wave Equation by the Electromagnetic Wave Equation Analogy 568
4. Derivation of the Schrödinger Wave Equation by Assumption of the Neutron Plane Wave Solution 572
5. Operator form of the Schrödinger Wave Equation for a Neutron Wave 574
6. Reflection and Transmission of Neutrons at Media Interfaces 576
7. Neutron Reflectometry 582
8. Measurement of the Complex Refractive Index via Refraction 588
9. X-ray and Neutron Interferometry 590
10. Interferometry and the Fizeau Effect 592
11. Pendellösung Oscillations and Anomalous Absorption in Perfect Crystals 596
12. Measurement of Refractive Index via Interferometry 601
13. The Maxwell–Boltzmann Distribution for the Reactor Source of Thermal Neutrons 603

References 606

6. X-ray and Neutron Optics 607

1. Derivation of the Index of Refraction for X-rays and Neutrons 607
2. Types of X-ray and Neutron Lenses and Significance of Delta/Mu Ratio 612
3. Coherent, Incoherent, Elastic, and Inelastic Scatter of X-rays and Neutrons 615
4. X-ray and Neutron Attenuation Comparison 619
5. Useful Formulas for X-rays and Neutrons 622
6. X-ray Multilayer Mirrors 624
7. Neutron Multilayer Mirrors 636

8. Capillary Optics and Kumakhov X-ray and Neutron Lenses 641
9. Bright-field, Dark-field, and Phase-Contrast Microscopy 646
References 653

Contents of Volumes 151–171 **655**

Index **663**

Foreword

The theory, experiments, and applications are presented in a detailed, tedious, and sometimes redundant manner in an attempt to provide neutron and X-ray optics and microscopy to a wide international readership of varying backgrounds, specialties, and educations in which English may be a second language. This book attempts to provide rapid assimilation of the presented topics that include neutron and X-ray scatter, refraction, diffraction, and reflection and their potential application.

The broad topics of neutron and X-ray optics covered in this AIEP publication can be explained in part by the author's background. Following undergraduate studies at M.I.T. in physics, chemistry, biology, engineering, and applied math, the author ended up with a S.B. in Electrical Engineering under the guidance of Prof. M. Dresselhaus. His S.B. thesis was directed by M.I.T. Prof. A.J. Grodzinsky on the electrical property measurements of real and artificial skin.

Dr. Cremer then received the M.S. and PhD in Electrical Engineering degrees at the University of Maryland, where he worked at the Charged Particle Beam Laboratory under the mentorship of Prof. W.W. Destler (thesis advisor), Prof. M. P. Reiser, and Prof. M.J. Rhee. The author's PhD thesis was on the acceleration of laser-produced ions and collective ion acceleration for inertial confinement fusion applications.

The author then obtained a postdoctoral traineeship at the University of California at San Francisco (UCSF) in Dept. of Biochemistry and Biophysics under the mentorship of Prof. R.M. Stroud to study electrically-mediated gene transfer and electroporation for application in membrane bound protein structure-function research.

From the X-ray protein crystallography lab of UCSF Prof. Stroud, the author was launched into coherent X-ray sources and optics in a postdoctoral position in X-ray microscopy and micropole undulator research at the Stanford Synchrotron Radiation Laboratory (SSRL) and the Institute of Theoretical Science at the University of Oregon under the mentorship of Dr. R.O. Tatchyn and Prof. P. Csonka. Also several years later, the author was a visiting scientist at SSRL performing multilayer X-ray optics research and development under the mentorship of Dr. R.O. Tatchyn.

After graduate school, the author has held positions in several aerospace and technology-based industries. The author was introduced to

Adelphi Technology, Inc. near Stanford University in 1988 by Dr. Tatchyn. The name Adelphi, which is a pseudo-acronym for advanced electron physics, reflects the company's origins from the research group of Prof. R.H. Pantell at Stanford University, where the Adelphi founder and president, Dr. M.A. Piestrup, received his PhD, and performed postgraduate research in electron physics and free electron lasers.

Founded in 1986, and under the vision and direction of Dr. M.A. Piestrup, Adelphi Technology and its collaborators have performed research and development in a wide variety of topics in X-ray and neutron sources and optics. Prof. Pantell has been a consultant and mentor for Adelphi Technology over the many years since its founding, and many of the Stanford PhD students and postdoctoral researchers of Prof. Pantell have worked or consulted at Adelphi Technology.

Starting in 1996 the author began full-time work at Adelphi Technology in X-ray sources and optics, which eventually expanded into neutron sources and optics. In particular, the 1996 Nature paper by Snigerv *et al.* on compound refractive lenses for X-rays inspired the author to pursue compound refractive X-ray lenses with Adelphi Technology. This was followed by compound refractive neutron lenses, along with fast and thermal neutron radiography, and other approaches such as dark-field and phase contrast imaging with thermal neutrons and X-rays.

Prior to 2006 Adelphi Technology had been focused on X-ray microscopy and the development of novel X-ray sources including a compact betatron for mammography and angiography. However, in the last 6 years Adelphi Technology has focused on the development and manufacture of thermal and fast neutron generators for a variety of applications including boron neutron capture therapy for cancer treatment.

The author's pursuit of X-ray and neutron optics and experimental work was strongly enabled by the mentorship of Dr. M.A. Piestrup of Adelphi Technology, Prof. R.H. Pantell of Stanford University, and Dr. D.G. Boyers of Phifer Smith Corp., as well as the collaboration of my Adelphi Technology colleagues Dr. H.R. Beguiristain, Dr. H. Park, Dr. C.K. Gary, Dr. L.W. Lombardo, Dr. D.L. Williams, D. Doggett, J. Harris, M.J. Fuller, J. H. Vainionpaa, and M. Apodaca. The author also acknowledges the valued scientific influence of N.B. Orr of the Science Foundation for Public Policy.

Adelphi Technology has benefited immensely from numerous collaborations with the U.S. National Laboratories and universities, and international collaborations with university and National Laboratory researchers outside the United States, as indicated by the many joint publications. Adelphi Technology was enabled to perform research and development in neutron and X-ray sources and optics due to the crucial funding and support provided by the United States Small Business Innovative Research (SBIR) and Small Business Technology Transfer (STTR) programs.

Finally, the author wishes to express his thanks to the editor, P.W. Hawkes, for providing the opportunity to write this work, and for his review of numerous drafts, as well as to the excellent Elsevier staff.

Preface

From time to time, I receive a text for the Advances that is long enough to fill an entire volume but for the first time, the series includes a text that will occupy at least two volumes. Neutron and X-ray Microscopy by J.T. Cremer began as a regular chapter but has grown considerably in an attempt to cover the subject thoroughly, in a way that will be of use to newcomers as well as to those familiar with these subjects.

This first volume opens with a long introduction to the subject and, in particular, to compound refractive lenses. Chapter 2 examines these in depth and leads on naturally to the presentation of geometric neutron and X-ray optics including aberrations in Chapter 3. The following two chapters go into X-ray optics and neutron optics in greater detail. Further details are collected in the chapter that concludes this volume.

I am confident that this meticulous study of an important form of imaging will be welcomed and I thank the author for all the trouble he has taken with this ambitious project.

Peter W. Hawkes

Contributor

Dr. Jay Theodore (Ted) Cremer, Jr. **1–654**
Chief Scientist, Adelphi Technology, Inc., 2003 East Bayshore Rd, Redwood City, CA 94063

Future Contributions

S. Ando
Gradient operators and edge and corner detection

D. Batchelor
Soft X-ray microscopy

E. Bayro Corrochano
Quaternion wavelet transforms

C. Beeli
Structure and microscopy of quasicrystals

C. Bobisch and R. Möller
Ballistic electron microscopy

F. Bociort
Saddle-point methods in lens design

A. Cornejo Rodriguez and F. Granados Agustin
Ronchigram quantification

C. Dwyer
Atomic-scale chemical mapping in the scanning transmission electron microscope

M. El-Gomati
Quantitative scanning electron microscopy

M. Ferroni
Transmission microscopy in the scanning electron microscope

R.G. Forbes
Liquid metal ion sources

H. Gelderblom
Helmut Ruska

A. Gölzhäuser
Recent advances in electron holography with point sources

D.E. Greenfield, M.A. Monastyrskiy and M. Ya. Schelev
Photoemission high-speed tubes for the investigation of ultra-short events

J. Grotemeyer
Time-of-flight mass spectrometry

M. Haschke
Micro-XRF excitation in the scanning electron microscope

A. Heuer and D. Hovis
Laser scanning confocal microscopy

M.I. Herrera
The development of electron microscopy in Spain

R. Herring and B. McMorran
Electron vortex beams

D. Ioanoviciu and K. Tsuno
Wien filters

M.S. Isaacson
Early STEM development

A.A. Ischenko and P.M. Weber
Structure and dynamics of free molecules and condensed matter

K. Ishizuka
Contrast transfer and crystal images

A. Jacobo
Intracavity type II second-harmonic generation for image processing

M. Jourlin
Logarithmic methods for image classification

L. Kipp
Photon sieves

T. Kohashi
Spin-polarized scanning electron microscopy

O.L. Krivanek
Aberration-corrected STEM

B. Lencová
Modern developments in electron optical calculations

H. Lichte
New developments in electron holography

A. Lisowska
Smoothlets and wedgelets

P. Maragos
Representations of morphological image operators and comparisons with linear operators

S. Marshall and P. Murray
The hit-or-miss transform

M. Matsuya
Calculation of aberration coefficients using Lie algebra

P.P. Mondal and A. Diaspro
Point-spread function engineering for super-resolution single- and multi-photon fluorescence microscopy

J.A. Monsoriu
Fractal zone plates

L. Muray
Miniature electron optics and applications

N. Neskovic
Rainbow lenses

M.A. O'Keefe
Electron image simulation

V.P. Oleshko
Electron tweezers and potential applications

D. Paganin, T. Gureyev and K. Pavlov
Intensity-linear methods in inverse imaging

N. Papamarkos and A. Kesidis
The inverse Hough transform

J.L. Prince, Y. Bai and X. Han
Adaptive octree grids

P. Rocca and M. Donelli
Imaging of dielectric objects

J. Rodenburg
Lensless imaging

A.T. Ruiz, A.J. den Dekker and W. van den Broeck
System and control theory in electron microscopy

S. Sangwine
Color image processing

K. Schultheiß
Phase plates for the electron microscope

R. Shimizu, T. Ikuta and Y. Takai
Defocus image modulation processing in real time

T. Soma
Focus-deflection systems and their applications

P. Sussner and M.E. Valle
Fuzzy morphological associative memories

M. Treacy
Speckle and disordered materials

M. Yacaman, A. Ponce-Pedraza, S. Meija and L. Francis
High-resolution transmission electron microscopy of catalysts

E.M. Yakushev and S. Bimurzaev
Theory and computation of electron mirrors

Chapter 1

Introduction to Neutron and X-ray Optics

Jay Theodore Cremer, Jr.*

Contents

1. Compound Refractive Lenses for Neutrons and X-rays: Background and Theory 2
 1.1. The Compound Refractive Lens 2
 1.2. The Neutron and X-ray Complex Refractive Index 4
 1.3. Biconcave Parabolic and Spherical Lenses: Transmission and Aperture 10
 1.4. CRL Resolution Comparison with Pinhole 13
 1.5. Optical System Analysis with Kirchhoff Equation: Mathematical Expression of Huygen's Principle 13
 1.6. Neutron and X-ray Reflection 17
 1.7. Light Microscopy Archetypes for Neutron and X-ray Microscopy 19
 1.8. Neutron and X-ray Microscopes: Amplitude-Contrast Microscopes and Phase-Contrast Microscopes 22
 1.9. Neutron and X-ray Temporal and Longitudinal Spatial Coherence Length, and Transverse Spatial Coherence Length 26
 1.10. Neutron and X-ray Fresnel Zone Plates 29
 1.11. Neutron and X-ray Achromat Lenses Using Fresnel Zone Plate and Compound Refractive Lens 31
 1.12. Neutron and X-ray Fresnel and Compound Fresnel Lenses 34
 1.13. Neutron and X-ray Monochromator with Compound Refractive Lens and Pinhole 38
 1.14. Comparison of Neutron, X-ray, and Light Source Intensities and Images 39
2. Experiments with Neutron Compound Refractive Lenses, Magnetic Lenses, and Microscopes 44

* Chief Scientist, Adelphi Technology, Inc.

Advances in Imaging and Electron Physics, Volume 172, ISSN 1076-5670, DOI: 10.1016/B978-0-12-394422-1.00001-3.
Copyright © 2012 Elsevier Inc. All rights reserved.

3. Experiments with X-ray Compound Refractive Lenses and Microscopes 100
4. Thermal Neutron Radiography, Coded Aperture, and Phase-Contrast Imaging 199
5. Experiments in Fast Neutron Radiography 237
6. A Brief History of Neutron Optics Before Compound Refractive Neutron Lenses 272
7. Brief History and Overview of Diffractive and Reflective X-ray Optics and Microscopy 282
References 312

1. COMPOUND REFRACTIVE LENSES FOR NEUTRONS AND X-RAYS: BACKGROUND AND THEORY

1.1. The Compound Refractive Lens

The X-ray or neutron compound refractive lens CRL consists of N biconcave lenses with $2N$ refractive interfaces. This enhances refraction and reduces focal lengths by a factor of $1/N$ as a function of X-ray or neutron wavelength λ. The refraction of X-rays and neutrons has been long established. However, only recently have CRLs been applied to the focusing of X-rays (Snigerev *et al.*, 1996) and cold neutrons (Choi *et al.*, 2000; Eskildsen *et al.*, 1998). In visible light optics it is well established that the focal length $f_{crl} = f/N$ of a stack of N lenses, which form a CRL, is $1/N$ times shorter than the focal length f of a single spherical biconcave lens, where (Hecht and Zajac, 1974)

$$f_{crl} = \frac{f}{N}. \tag{1}$$

A multilayer mirror is a periodic structure with repeating reflecting layers A and B. The multilayer is composed of N period's layer A with refractive index n_A and layer B with refractive index n_B, which present $2N$ reflecting layers to incident light, X-rays, or neutrons. The cumulative sharing of a tiny reflection from a multitude of N interfaces allows for a sum of N amplitudes to be summed constructively in particular directions. This allows for an N-fold increase in reflectivity if the ratio n_A/n_B is fixed, and for a fixed number of periods N, the reflectivity increases when n_A/n_B increases (see Born and Wolf, 2002). Similarly, the cooperation of a multitude of $2N$ weakly biconcave lens refractive interfaces reduces the focal length of a CRL to $1/2N$ of that of a single lens (Figure 1).

The X-ray and neutron CRLs obeys the lensmaker's equation, where given the object distance r_0 and image distance r_i, the reciprocal of the focal length is equal to the sum of the reciprocals of the object and image

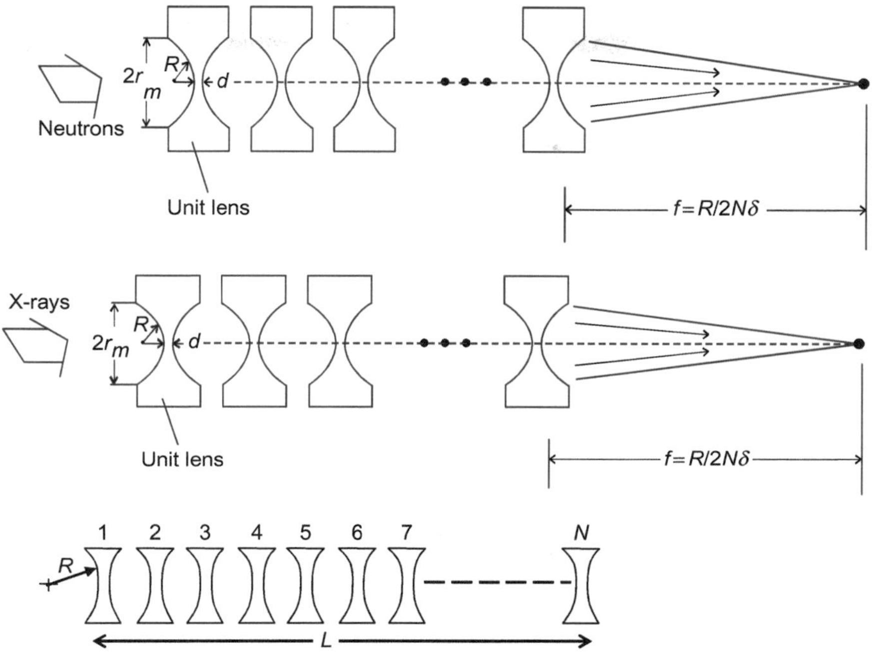

FIGURE 1 The stacked biconcave lenses that form a neutron compound refractive lens (CRL) or an X-ray CRL. Lenses fabricated from certain materials, such as beryllium, amorphous Teflon, aluminum, and copper, can focus both X-rays and neutrons. The length of the CRL is L, and R is the radius of curvature for spherical biconcave lenses and the on-axis radius of curvature for parabolic biconcave lenses. The radius r_m is the lens's physical aperture radius. The N-lens CRL focal length $f = R/2N\delta$ is N-fold shorter than a single-lens focal length $R/2\delta$. The biconcave lenses have a center minimum thickness d.

distance, where

$$\frac{1}{f_{crl}} = \frac{1}{r_0} + \frac{1}{r_i}. \tag{2}$$

Given axial propagation of a beam of X-rays or neutrons, the magnification M or ratio of image transverse dimension y relative to the object's transverse dimension x is the familiar magnification formula from visible light lens optics. The X-ray or neutron image—like an optical light image—is inverted relative to the object, and hence the minus one in the transverse magnification formula (Figure 2):

$$M = -\frac{r_i}{r_0} = -\frac{y}{x}. \tag{3}$$

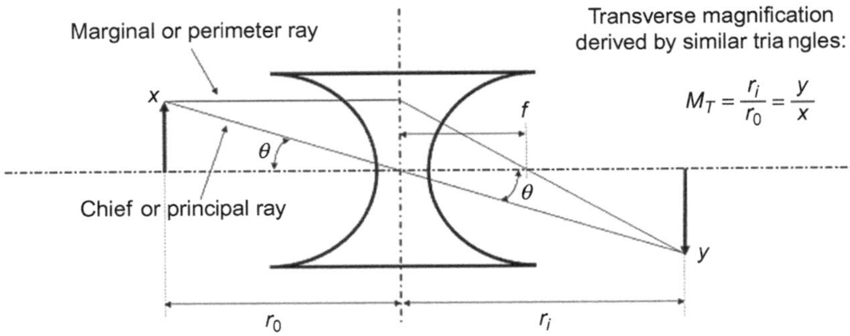

FIGURE 2 The magnification of the lens is obtained by similar triangles formed by the chief ray. The chief (or principal) ray passes from an object point through the center point of the lens to the inverted conjugate image point. The marginal or perimeter ray passes from an object point through the outermost aperture point or perimeter of the lens and then the marginal ray refracted by the lens to the conjugate image point.

1.2. The Neutron and X-ray Complex Refractive Index

The extremely small refractive decrement δ of lens materials for X-rays and neutrons, and the requirement of a very small radius of curvature for a single X-ray or neutron lens discouraged the development of neutron optics for X-rays and thermal neutrons. Long-wavelength X-rays below a few kiloelectron volts, which can be refracted by a single lens to focal lengths of less than 1 meter, are not practical because of the strong absorption. Very cold and ultracold neutron beams, which could be used with single neutron lenses to produce focal lengths of several meters, were not readily available, and if available, had very low neutron fluxes. The above CRL equations are derived or examined in more detail in the next few sections.

Given the lens material's complex refractive index n with refractive decrement δ and linear attenuation μ for X-rays or neutrons, an X-ray or neutron CRL consists of a stack of N biconcave, spherical or parabolic lenses with radius of curvature R, with focal length f_{crl} for neutrons of wavelength λ, where

$$f_{crl} = \frac{f}{N} = \frac{1}{N}\frac{R}{2\delta}. \tag{4}$$

The refractive index n with refractive decrement δ and linear μ applies to X-rays and neutrons, although their values as a function of wavelength λ differ for neutrons and X-rays, where

$$n = 1 - \delta + i\frac{\lambda}{4\pi}\mu. \tag{5}$$

The refractive index n is the ratio of the wave number of the X-ray or neutron in the material medium k, divided by the wave number in a vacuum (field-free region) k_0, where

$$n = \frac{k}{k_0}. \tag{6}$$

The ranges of refractive decrement δ for materials are similar for X-rays and thermal neutrons in the range of typical X-ray wavelengths of interest—namely, 0.5 to 2 Å—and in the range of typical thermal and cold neutron wavelengths—namely, 0.5 to 20 Å. For most materials the attenuation of thermal and cold neutrons in the range of 0.5 to 20 Å is much less than X-rays in the range of 0.5 to 2 Å.

For neutrons the decrement δ is due to the strong nuclear force interactions and is expressed by the neutron scatter length b for a material of one isotope. For X-rays the decrement δ is due to the photon interaction with electrons and can be expressed by the X-ray scatter length b for one element (see the top diagram in Figure 3). The Ångstrom-range neutron wavelength is much larger than the dimensions of the femtometer dimensions of the nucleus, whereas only long-wavelength very soft X-rays and vacuum ultraviolet radiation greater than 100 Å are much larger than the 1- to 10-Å diameter of the atomic electron cloud. Hence, plane waves of neutrons are scattered equally in all directions (isotropic) into spherical outgoing waves by the pointlike nucleus, and only the very long-wavelength vacuum ultraviolet and very soft X-rays ($1/\text{fm}^2$), which dwarf the atomic electron cloud, are scattered into outgoing spherical waves. The soft X-rays between the 100- to 1-Å range and especially the hard X-rays in the 1- to 0.1-Å range are on the order of the atomic electron cloud diameter. These shorter-wavelength, incident X-ray plane waves are non-isotropically scattered into outgoing waves, which do not have spherical outgoing wave fronts as do the isotropic scattered incident neutron plane waves.

Given the nuclei number density ρ of a material composed of a single isotope of atomic weight A and atomic number Z, and scatter length b for neutrons or X-rays, the neutron or X-ray decrement for neutron or X-rays of wavelength λ is

$$\delta = \frac{\lambda^2 b \rho}{2\pi}. \tag{7}$$

For the nuclear scatter of neutrons, the neutron scatter length b is experimentally determined for each isotope, whereas for the atomic electron cloud scatter of X-rays, the X-ray scatter length b_{xr} is the product of the atomic number of an atom Z multiplied by the classical radius of

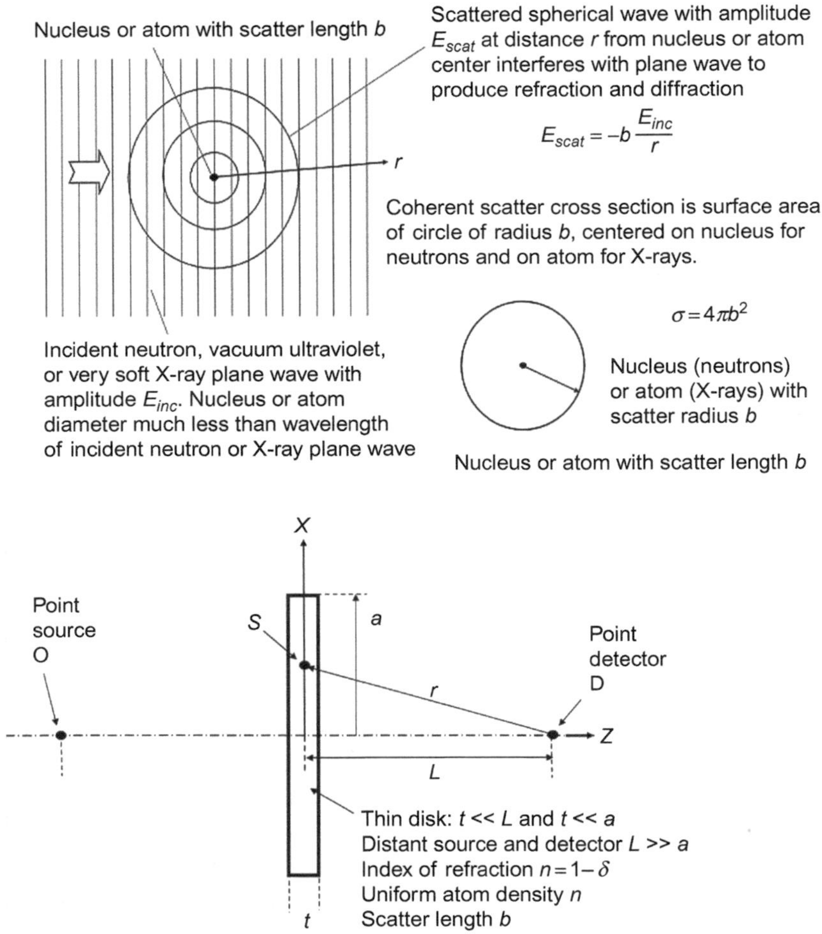

FIGURE 3 (Top) The scatter of a neutron plane wave from the pointlike nucleus into an outgoing spherical wave. This diagram applies to vacuum ultraviolet or soft X-ray plane wave, which has a wavelength much larger than the diameter of the atomic electron cloud. (Bottom) The refractive index δ for X-rays or neutrons for a material in the form of a thin disk is derived in the text.

the electron r_e, where $b_{xr} = Zr_e$ (see the National Institute for Standards and Technology [NIST] Center for Neutron Research website for its Scattering Length Density Calculator; http://www.ncnr.nist.gov/resources/sldcalc.html). The neutron and X-ray scatter lengths b are measured constants that are independent of the respective neutron and X-ray wavelengths. The decrement δ of the refractive index $n = 1 - \delta$ contains the wavelength λ and scatter unit (atom or nuclei) number density ρ dependence.

The bottom diagram in Figure 3 shows a thin disk of thickness t and radius R that can be used to derive the expression for the refractive decrement δ for X-rays or neutrons. A point source O on the z-axis emits neutrons or X-rays, which pass through a thin disk of material of radius a and thickness t and are scattered by the nuclei (neutrons) or atomic electron cloud (the nuclei or electron cloud scatter points are denoted by S). The material disk is centered on the z-axis, and the summed amplitude of the X-rays or neutrons is measured by a point detector D on the z-axis at downstream axial distance L. The phase shift $\Delta\phi$ between the incident X-ray or neutron waves with wave vector $\mathbf{k}_0$ and the coherently forward-scattered X-ray or neutron waves with wave vector $\mathbf{k}$ are denoted by

$$\Delta\phi = \left(k - k_0\right) \cdot t = k_0\,(n-1) \cdot t = -\frac{2\pi}{\lambda_0} t\delta. \tag{8}$$

The phase shift $\Delta\phi$ results from the sum of the scattered wave amplitudes from the disk material scattering centers, atomic electron cloud (X-rays), or nuclei (neutrons). The number Q of scattering centers is given by the volume of the disk $V = \pi a^2 t$ multiplied by the number density of the scatter centers n (cm^{-3}), where

$$Q = nV = n\pi a^2 t. \tag{9}$$

Given an incident wave amplitude $E_0 = 1$, the 180° phase-shifted amplitude of the scattered wave E at distance r from a single scatter center with scatter length b is

$$E = -E_0 \frac{b}{r}. \tag{10}$$

A single scatter center in the disk is modeled as a point. For x-rays the scatter center is the electron cloud of an atom, and for neutrons the scatter center is a nucleus. Each scatter center S of the disk material is at radial distance x from the disk center, and the distance r from scatter point S to the point detector D is

$$r = \sqrt{x^2 + L^2} \cong L\left(1 + \frac{1}{2}\frac{x^2}{L^2}\right) = L + \frac{1}{2}\frac{x^2}{L}. \tag{11}$$

The change of distance Δr between disk scatter point S and detector point D as the disk scatter point S moves from the disk center $x = 0$ to disk perimeter $x = a$ is

$$\Delta r = \frac{1}{2}\frac{a^2}{L}. \tag{12}$$

One picks the radius a of the disk as the first Fresnel zone for the incident wavelength λ_0

$$\frac{1}{2}\frac{a^2}{L} = \frac{\lambda_0}{2}. \tag{13}$$

The summed scatter amplitudes E at the detector point D from all the disk scatter points S are averaged over the π radian phase shift, which occurs in the first Fresnel zone between disk center $x = 0$ and the disk perimeter radius $x = a$, where (Squires, 1996)

$$\frac{E}{\pi} \cong -\frac{Q}{\pi}\frac{b}{L} = -\frac{1}{\pi}\left(\pi \cdot a^2 t \frac{b}{L} n\right) = -\lambda_0 tbn. \tag{14}$$

The phase shift $\Delta\phi$ of the scattered X-ray or neutron wave amplitude E at detector point D and the incident unity amplitude $E_0 = 1$ is then

$$\Delta\phi = \tan^{-1}\left(\frac{E}{E_0}\right) \cong \frac{E}{E_0} = E = -\lambda_0 tbn. \tag{15}$$

Equating the two expressions for $\Delta\phi$ gives

$$\Delta\phi = -\frac{2\pi}{\lambda_0} t\delta = -\lambda_0 tbn. \tag{16}$$

One solves for the neutron or X-ray refractive decrement δ of the disk material so that

$$\delta = \frac{\lambda_0^2 bn}{2\pi}. \tag{17}$$

The scatter length density bn has units of $1/\text{fm}^2$, and is the product of the X-ray scatter length b (fm/atom) or neutron scatter length b (fm/nucleus) multiplied by the atom number density (atom/fm^3) or nucleus number density n (nuclei/fm^3).

Unlike X-rays, the neutron refractive index $n = 1 - \delta$ and decrement δ are significantly dependent on gravity due to the neutron mass, dipole magnetic fields arising from unpaired atomic electrons in materials and vacuum space–applied magnetic fields. The slightly uneven positive and negative charge distribution within the charge neutral neutron and the neutron magnetic dipole moment together cause the neutron to interact weakly with the atomic electric fields, which leads to a very small refractive decrement contribution δ_e due to the atomic electric fields that is usually ignored. Assuming neutron wavelengths decrements much less than 1, one can obtain the refractive index of a neutron passing through a material composed of isotope nuclei with nuclear force decrement contribution δ_n, which have unpaired atomic electrons causing the atoms to be ferromagnetic with decrement contribution δ_n, so that the total neutron decrement δ for the ferromagnetic material is

$$\delta = \delta_n + \delta_f. \tag{18}$$

As another example, a neutron passing through a non-magnetic gas of one isotope with nuclear force decrement contribution δ_n, with an applied magnetic field **B** with decrement contribution δ_B, and accounting for the gravitational acceleration **g** acting on the neutron and characterized by decrement contribution δ_g, where

$$\delta = \delta_n + \delta_B + \delta_g. \tag{19}$$

For CRL lens material comprised of M isotopes, the linear attenuation coefficient μ of neutrons or X-rays in CRL is the sum of the M products of the ith isotope nuclei number density σ_i multiplied by the total neutron attenuation cross section σ_i, where

$$\mu = \sum_{i=1}^{M} n_i \sigma_i. \tag{20}$$

With X-rays the linear attenuation is based on the M elements and the ith element's atom number comprising the lens material. Given the lens material mass density ρ, Avogadro's number N_A, the ith isotope atomic weight A_i and weight fraction w_i, the ith isotope's nuclei number density n_i is then

$$n_i = \frac{w_i \rho}{A_i} N_A. \tag{21}$$

The neutron the decrement δ for the CRL material, given the neutron scattering length b_i for each ith isotope, is then

$$\delta = \frac{\lambda^2}{2\pi} \sum_{i=1}^{M} n_i b_i. \tag{22}$$

We now consider the sources of neutron attenuation in a neutron CRL. For neutrons the coherent elastic cross section σ_{coh}^{e} of the CRL lens material gives rise to refraction and the decrement δ. The absorption cross section

$\sigma_{abs}(\lambda)$ is due to thermal neutron capture by the nuclei comprising the lens material and increases linearly with neutron wavelength λ, where

$$\sigma_{abs}(\lambda) \propto \lambda. \tag{23}$$

The neutron wavelength-dependent attenuating scatter cross section includes (1) the coherent inelastic scatter $\sigma^{i}_{coh}(\lambda)$ due to collective or coherent thermal vibrations of lattice atoms in the lens material that result in phonon exchange between the incident neutrons and lens nuclei, (2) the incoherent elastic scatter $\sigma^{e}_{inc}(\lambda)$ due to small angle scatter, and (3) incoherent inelastic scatter $\sigma^{i}_{inc}(\lambda)$ due to random thermal vibrations of individual lens atoms. The unwanted coherent, elastic scatter due to diffraction of the neutrons from the polycrystals or single crystals, which comprise the CRL lenses, acts as another source of neutron attenuation by the CRL. The total attenuation cross section $\sigma_{att}(\lambda)$ of the ith isotope nuclei that comprise the lens material is then

$$\sigma_{att}(\lambda) = \sigma_{abs}(\lambda) + \sigma^{i}_{coh}(\lambda) + \sigma^{e}_{inc}(\lambda) + \sigma^{i}_{inc}(\lambda) + \sigma_{dif}(\lambda). \tag{24}$$

Next we consider the source of attenuation of X-rays in an X-ray CRL. With X-rays the decrement δ is based on the M elements and ith element atom number that make up the lens material. For X-rays below 25 keV, the linear attenuation in the lens material is due primarily to photoelectric absorption cross section $\sigma_{ph}(\lambda)$, which is much larger than the coherent, Rayleigh scatter cross section $\sigma_{ray}(\lambda)$, which causes refraction and gives rise to decrement δ. The wavelength-dependent X-ray photoelectric absorption cross section $\sigma_{ph}(\lambda)$ gives rise to the X-ray linear attenuation coefficient $\mu(\lambda)$ of the CRL material for X-rays. The coherent Rayleigh scatter cross section $\sigma_{ray}(\lambda)$ is much larger than the incoherent, Compton scatter cross section $\sigma_{c}(\lambda)$, which contributes to X-ray attenuation, where for X-ray energies less than 25 keV (or wavelengths λ larger than 0.5 Å),

$$\sigma_{ph}(\lambda) \gg \sigma_{ray}(\lambda) \gg \sigma_{c}(\lambda) \quad \lambda > 0.5\,\text{Å}. \tag{25}$$

1.3. Biconcave Parabolic and Spherical Lenses: Transmission and Aperture

The CRL consists of N biconcave lenses, which can be parabolic or spherical. The minimum thickness d of a biconcave, spherical lens is at its center, and thus the minimum N-lens CRL material thickness through which pass incident X-rays or neutrons is Nd. Given an incident X-ray electric field amplitude E_0 or a neutron wave function amplitude E_0, the transmitted X-ray or neutron wave amplitude E through the center of the CRL is

$$E = E_0 \exp(-Nd\mu). \tag{26}$$

The transmitted intensity I is the squared magnitude of the transmitted amplitude E, where

$$I = |E|^2 = \left|E_0 \exp(-Nd\mu)\right|^2 = |E_0|^2 \exp(-2Nd\mu). \tag{27}$$

The X-ray or neutron intensity transmission fraction T through the center of the CRL is

$$T = \frac{I}{I_0} = \left(\frac{E}{E_0}\right)^2 = \exp(-2Nd\mu). \tag{28}$$

The top of Figure 4 shows the absorption aperture radius r_a and physical aperture radius r_m of a single, biconcave, spherical or biconcave lens. The bottom diagram shows the absorption aperture radius r_a of an N-lens CRL composed of N biconcave spherical or parabolic lenses. The single lens and CRL are assumed to be thin lenses in which the rays change direction, but not radial position, within the lens or CRL.

The CRL acts as a cylinder of length l and effective aperture diameter $2r_e$, with limited angular acceptance $2r_e/l$ and numerical aperture r_e/f. If $f \ll l$, a neutron ray passing through a thin lens CRL changes direction, but not its radial distance r from the CRL center axis, provided each individual lens has a radius of curvature R much larger than its radial-dependent thickness $t(r)$,

$$t\,(r) \ll R \quad 0 < r < r_e. \tag{29}$$

The thickness $t(r)$ of each biconcave lens increases with r from its minimum center thickness d. For an N-lens CRL, the neutron amplitude transmission fraction $T(r)$ decreases with radial distance r from its center axis value, $T(0)$, where

$$T\,(r) = \exp(-tN\mu) \tag{30}$$

and

$$T(0) = \exp(-N\mu d). \tag{31}$$

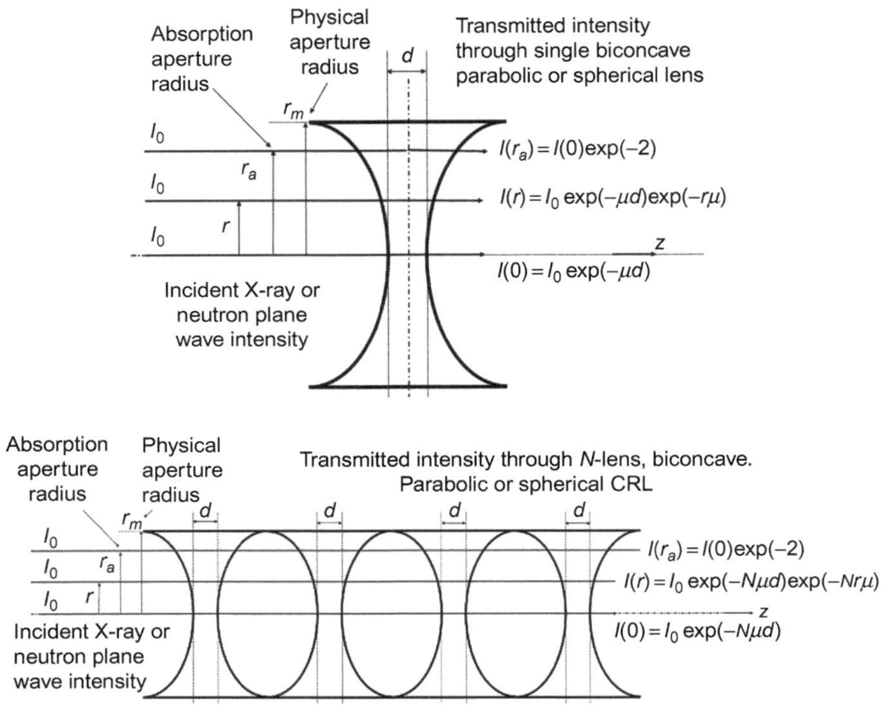

FIGURE 4 (Top) The absorption aperture radius r_a of a single, biconcave spherical or parabolic lens, and the lens physical aperture radius r_m. (Bottom) The absorption aperture radius r_a of a biconcave spherical or parabolic N-lens CRL.

Absorption can limit the CRL effective aperture radius r_e, which is the minimum of the CRL physical aperture radius r_m and absorption aperture radius r_a, where

$$r_e = MIN\,(r_m, r_a)\,. \tag{32}$$

The transmission of rays of neurons or X-rays passing through a parabolic biconcave lens at the absorption aperture radius r_a is $1/e^2$ of their transmission through the CRL center axis, where

$$T(r_a) = \frac{T(0)}{e^2}. \tag{33}$$

Given (1) typical focal lengths f_{crl} of 10 to 250 cm and (2) the relative small effective aperture radius r_e, which typically for X-rays is less 1.0 mm and for neutrons is less than 2.5 cm, the numerical aperture NA of an X-ray or neutron CRL in vacuum or air is very small, where typically

$$NA = \frac{r_e}{f_{crl}} < 0.01. \tag{34}$$

1.4. CRL Resolution Comparison with Pinhole

We now compare the Rayleigh, diffraction-limited object resolution σ_{df} of a biconcave parabolic N-lens CRL with the object resolution σ_{pin} of a pinhole camera. A CRL with effective aperture radius r_e produces better resolution than a pinhole of the same radius r_e. The pinhole transmits the same number of neutrons or X-rays to the detector at image distance r_i; however, the pinhole discerns specimen details with a lower object resolution σ_{pin} in which the object resolution produced by a pinhole imaging is simply the pinhole diameter $2r_e$, where

$$\sigma_{pin} = 2r_e \tag{35}$$

and

$$\sigma_{df} = 0.61\frac{r_0}{r_e}\lambda. \tag{36}$$

The biconcave, parabolic N-lens CRL with resolution σ_{crl} can resolve finer object details than the pinhole as long as

$$\frac{\sigma_{crl}}{\sigma_{pin}} = \frac{0.3\lambda r_0}{r_e} < 1. \tag{37}$$

Conversely, a pinhole of radius r_{pin} produces the same object resolution as a biconcave, parabolic N-lens CRL with larger effective aperture radius $r_e \gg r_{pin}$. However, the much smaller diameter pinhole reduces the image intensity by a factor $(r_{pin}/r_e)^2$ compared with that of the CRL.

1.5. Optical System Analysis with Kirchhoff Equation: Mathematical Expression of Huygen's Principle

The Kirchhoff equation is a mathematical representation of Huygen's principle and is a convenient tool to analyze the performance of an X-ray or neutron CRL; the Kirchhoff equation accounts for both monochromatic and chromatic aberrations. We consider an X-ray or neutron wave with wave vector $\mathbf{k}$, wavelength λ, and wave number $k = 2\pi/\lambda$, which arrives at the image plane after emission or transmission by an object and subsequent refraction by a CRL. We consider the one-dimensional (1D) case in which at a particular point along a line of image points, the amplitude of the arriving neutron or X-ray wave has wavelength-dependent amplitude $E_y(\lambda, y, l, x)$. For neutrons $E_y(\lambda, y, l, x)$ is the neutron wave function amplitude, and for X-rays $E_y(\lambda, y, l, x)$ is the electric field amplitude.

Measured from the center optical axis is the radial distance y of image point I, the radial distance l of lens point L on the CRL principal plane (center plane), and the radial distance x of source point O. Amplitude $E_y(\lambda, y, l, x)$ is determined by the total ray path length L_T, and the resulting phase shift φ_T of the local wave front. The normal to the wave front at a particular point on the wave from is the wave vector **k**, which is represented by the ray of the neutron or X-ray that propagates along a single ray path. Phase shift φ_T is given by

$$\varphi_T = 2\pi \frac{L_T}{\lambda}. \tag{38}$$

The wave amplitude $E_y(\lambda, y, l, x)$ at image position y is the product of the amplitude magnitude and the exponential phase factor, where $r \cong z$

$$E_y(\lambda, y, l, x) = \frac{E_s(\lambda, x, l)}{r} \exp[-i\varphi_T(\lambda, y, l, x)]. \tag{39}$$

The distance r is between object point O and image point I, and distance r is approximated by the axial distance z between the object and image planes.

For a noncoherent source of X-rays or neutrons, the thin-lens CRL 1D image or radial image intensity profile I_y is obtained by determining the intensity at each image point I. The intensity of each image point I is the sum of the intensities from each object point O. The intensity contribution by a single object point O to a single image point I is the integral sum of the intensity contributions from all the paths that pass through the lens. That is, each point L on the lens represents a particular path OLI from a single object point O to lens point L, and then from lens point L to image point I. The image intensity at image point I from conjugate object point O is the squared magnitude of the integral sum of the amplitude contributions $E_y(\lambda, y, l, x)$ contributions from all possible ray paths O-L-I through all the lens point L. The summed amplitude at y is

$$I_y(\lambda, y) = \int_{object} dx \left| \int_{lens} E_y(\lambda, y, l, x)\, dl \right|^2. \tag{40}$$

For a coherent source of X-rays, then the image intensity distribution is the squared magnitude of the resulting amplitude after both integration over the lens and over the coherent source, where

$$I_y(\lambda, y) = \left| \int\limits_{object} dx \int\limits_{lens} E_y(\lambda, y, l, x)\, dl \right|^2 . \tag{41}$$

For coherent X-ray sources, see papers by Arthur and Tatchyn (2000) and Schulte-Schrepping (2000). Chapter 25 (in an upcoming volume) provides the detailed expressions for X-ray electric field amplitude or neutron wave function amplitude $E_y(\lambda, y, l, x)$ and intensities. We consider the simple case of a 1D biconcave, parabolic, N-lens CRL with focal length f, which images a line source object of incoherent neutrons or X-rays. The object is a line source of neutrons or X-rays with a Gaussian intensity distribution, which has standard deviation w_0. The 1D biconcave parabolic N-lens CRL images the line source object on the image plane as a line of image points with image intensity distribution $I_y(\lambda, y)$, which is given by the 1D Kirchhoff equation (Pantell *et al.*, 2001):

$$I_y(\lambda, y) = \left(\frac{E_0}{r_i + r_0}\right)^2 \int\limits_{obj} dx \exp\left(-\frac{x^2}{w_0^2}\right) \times \left| \int\limits_{lens} \exp\left[-i\frac{2\pi}{\lambda}\left(\frac{(x-l)^2}{2r_0} - \frac{l^2}{2f} + \frac{(l-y)^2}{2r_i}\right)\right] \exp\left(-\frac{l^2\mu}{4f\delta}\right) dl \right|^2 . \tag{42}$$

The exponential term in the integrand for integration over the object is the Gaussian object intensity distribution with w_0 as its full width half maximum (FWHM). The integration over the lens is along a line lying in the thin lens CRL center (principal) plane. The exponential with imaginary arguments is the phase factor that accounts for the three phase shifts encountered by a wave vector **k**, which propagates from object point O to a lens point L and then to image point I. The first exponential phase factor is the phase shift for wave vector **k** or ray propagation from the object point O at line coordinate x to the lens point L at line coordinate l.

The second exponential phase factor is the phase shift for wave vector **k** propagation through the CRL with focal length $f$at lens point L situated at line coordinate l. The third exponential phase factor is the phase shift for the wave vector **k**, which propagates from the lens point L at line coordinate l to the image point I at line coordinate y.

At the image plane the gain in image plane intensity can be calculated using a CRL in place of only an aperture at the CRL position. The aperture diameter is equal to the CRL effective aperture diameter $2r_e$. The 1D intensity gain G_{1D} at the image plane for CRL intensity $I_y^{crl}(y,\lambda)$, compared with the case of no CRL aperture only with intensity $I_y^0(\lambda,y)$, is

$$G_{1D}(\lambda,y) = \frac{I_y^{crl}}{I_y^0}. \tag{43}$$

One substitutes $I_y^{crl} = I_y^{par}$ or $I_y^{crl} = I_y^{sph}$ for a parabolic or spherical lens CRL. For an azimuthally symmetric lens the two-dimensional (2D) gain is the square of the 1D gain; that is,

$$G_{2D}(\lambda,y) = \left[G_{1D}(\lambda,y)\right]^2 = \left[\frac{I_y^{CRL}(\lambda,y)}{I_y^0(\lambda,y)}\right]^2. \tag{44}$$

The resolution can be determined by considering a specific object point (e.g., the center object point at $x = 0$) and determining the image plane intensity distribution $I_y(\lambda,y)$ from this single object point O; that is,

$$I_y(\lambda,y) = \left|\int_{lens} E_y(\lambda,y,l,x=0)\,dl\right|^2. \tag{45}$$

The FWHM of the image plane distribution along transverse position y yields the image resolution and division by the CRL transverse magnification determines the object resolution.

Alternatively, one can determine the field of view (FOV) of the CRL by considering a specific image point (e.g., the center image point at $y = 0$). One can determine the object intensity distribution along the transverse radial line x that contributes to the intensity of the image point at $y = 0$; that is,

$$I_y(\lambda,x) = \left|\int_{lens} E_y(\lambda,y=0,l,x)\,dl\right|^2. \tag{46}$$

From the Kirchhoff equation, which accounts for both monochromatic and chromatic aberrations, the CRL point spread function (PSF) and the modulation transfer function (MTF) can be calculated. The MTF is a quantitative and standard measure of image quality and resolution and describes the ability of a lens or system to transfer object contrast to the image. The resolution of a microscope generally means the capability to distinguish two point objects of about equal intensity. In the Rayleigh criterion the PSF center maximum of a second point is placed at the zero intensity position of the PSF of the first point. The Rayleigh resolution is the reciprocal of the spatial frequency s or period $1/s$ corresponding to MTF = 0.09. However, the period $1/s$ at MTF = 0.05 is a common resolution limit.

1.6. Neutron and X-ray Reflection

Reflection of X-rays or neutrons occurs at grazing incidence angle θ_i with respect to a reflecting surface (Figure 5). The X-rays or neutrons are fully reflected at angles less than the critical angle θ_c, measured relative to the surface tangent plane at the point of incidence. For X-ray or neutron incidence angles larger than the critical angle $\theta > \theta_c$, partial reflection and partial refracted transmission occur at the interface between two medium. At each point along an X-ray or neutron incident wave front is a ray that is perpendicular to the wave front and point in the direction of the X-ray or neutron wave vectors $\mathbf{k}_i$. The magnitude $k_i = |\mathbf{k}_i|$ of the incident wave vector is the wave number $k_i = 2\pi/\lambda_i$ or the reciprocal of the incident X-ray or neutron wavelength λ_i.

The reflected wave front rays are depicted by the X-ray or neutron reflection wave vectors $\mathbf{k}_r$ with associated reflection angle θ_r and wave number for $k_r = 2\pi/\lambda_r$ reflected wavelength λ_r, and the angle of reflection is equal to the angle of incidence θ_i, and the magnitude of the incident

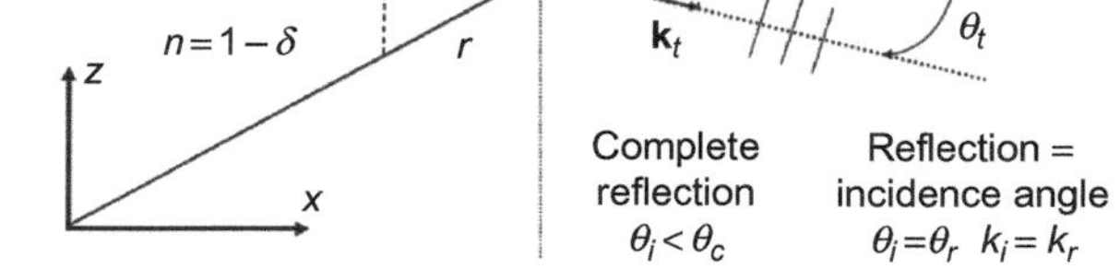

FIGURE 5 X-ray and neutron plane wave reflection from a surface follows the same rules of reflection of light in which there is partial reflection and refraction (transmission). The electric field amplitude for X-rays and the wave function amplitude for neutron waves play the role of the electric field amplitude of a light wave. For grazing incident angles θ_i less than the critical angle θ_c, there is complete reflection and no transmission. The electric field is transverse to the direction of the electromagnetic wave.

and reflected wave vectors are equal (i.e., the incident and reflected waves have the same wavelength $\lambda_i = \lambda_r$), where

$$\frac{2\pi}{\lambda_i} = k_i = |\mathbf{k}_i| = |\mathbf{k}_r| = k_r = \frac{2\pi}{\lambda_r} \quad \text{and} \quad \theta_i = \theta_r. \tag{47}$$

The refracted or transmitted wave front rays are depicted by the refracted or transmitted X-ray or neutron wave vectors $\mathbf{k}_t$ and associated refraction or transmission angle θ_t. Here, the incidence θ_i, reflection θ_r, and refracted θ_t angles are measured relative to the tangent plane of the surface at the point of ray incidence. Alternatively, the angles can be specified relative to the normal vector to the tangent plane of the surface at the point of ray incidence. The wave vector $\mathbf{k}_t$, wave number k_t, and wavelength λ_t of the refracted or transmitted wave differ from the incident and reflected waves, where

$$\frac{2\pi}{\lambda_i} = k_i = |\mathbf{k}_i| \neq |\mathbf{k}_t| = k_t = \frac{2\pi}{\lambda_t} \quad \text{and} \quad \theta_i \neq \theta_t. \tag{48}$$

The surface roughness causes a decrease in reflectivity of X-rays by a reflecting surface. In terms of the root mean squared (RMS) mirror surface roughness σ_{sur} (μm), the reflectivity of an ideal, smooth surface mirror R (derived above), the incident X-ray wave number k_i (μm^{-1}), and the refracted (transmitted) wave number k_t $\left(\mu\text{m}^{-1}\right)$, the reflectivity R_{sur} of a non-ideal X-ray mirror with surface roughness is given in section 4.2 the Lawrence Berkeley National Laboratory X-Ray Data Booklet (Lawrence Berkeley National Laboratory, Center for X-Ray Optics and Advanced Light Source, LBNL Pub-490 Rev. 3; http://xdb.lbl.gov/xdb-new.pdf, 2009), where

$$R_{sur} = R \cdot \exp\left(-k_i k_t \sigma_{sur}^2\right) = R \cdot \exp\left(-4\pi^2 \frac{\sigma_{sur}^2}{\lambda_i \lambda_t}\right). \tag{49}$$

See papers by Henke *et al.* (1993), Savage *et al.* (1991), Windt *et al.* (1994), and Lodha *et al.* (1998) for more details. The above equation for the reflectivity R_{sur} of a mirror with surface roughness for X-rays can be applied to neutron mirrors with surface roughness characterized by RMS surface deviation σ_{sur} (μm). Also, see chapters by Harvey (chapter 44), Takacs (chapter 46), Mildner (chapter 63), and Gubarev and Ramsey (chapter 64) in the *Handbook of Optics* (Bass *et al.*, 2010), and Daillant and Gibaud (2009).

1.7. Light Microscopy Archetypes for Neutron and X-ray Microscopy

The development of neutron and X-ray optics in large part follows the path of the development of light optics. The development of X-ray and neutron microscopy mirrors the evolution and development of light microscopy in which the bright-field, amplitude-contrast microscope developed by the Dutch spectacle-makers Hans Janssen and his son, Zacharias Janssen in 1590, and by Galileo in 1609 and Cornelius Drebbel in 1619 was followed by a published paper in 1665 by Robert Hooke on dark-field microscopy. These developers based their work on sample-scattered light, although J.J. Lister is credited with developing the achromatic microscope and promoting dark-field microscopy in 1830.

In dark-field microscopy, the object specimen is illuminated with a hollow cone of light that does not enter the objective lens unless the light is scattered by the specimen into the objective lens. That is, the dark-field microscope removes the zero-order illumination of the sample in which light is transmitted through the sample undeviated. However, sample-scattered light or the higher orders of diffracted (deviated) light are scattered into the objective lens and imaged.

The left diagram of Figure 6 shows a modern Abbe condenser/ objective lens bright-field microscope with oil immersion. Oil immersion is the immersing of both the objective lens and the specimen in a transparent oil of high refractive index. This increases the numerical aperture of the objective lens. The right diagram shows a modern transmitting dark field microscope.

The development of even illumination of the microscope specimens by Köhler (1893) followed dark-field microscopy. Figure 7 shows a diagram of Köhler illumination in which two light paths from the source through the sample to the image plane are established. In the first path (the illuminating light path), the set of rays emitted from each single source point is made to diverge and pass through every point in the sample plane (providing uniform illumination of the sample). These rays are then converged and diverged again to terminate at each point in the image plane. In the second path (the image-forming path), a ray from every source point is made to converge to a single sample point and then these rays are diverged and converged again to a single point in the image plane.

The weak X-ray and neutron sources and the large attenuation of the X-rays and neutrons in CRLs render the use of two or more CRLs in a microscope difficult or impractical. The smaller collection area of a CRL compared with a mirror for X-rays or neutrons dictates the use of a mirror or capillary array optic as the condenser lens and a CRL as the objective lens. Köhler illumination was devised to prevent superimposition of an image of the source filament on the specimen image. Following

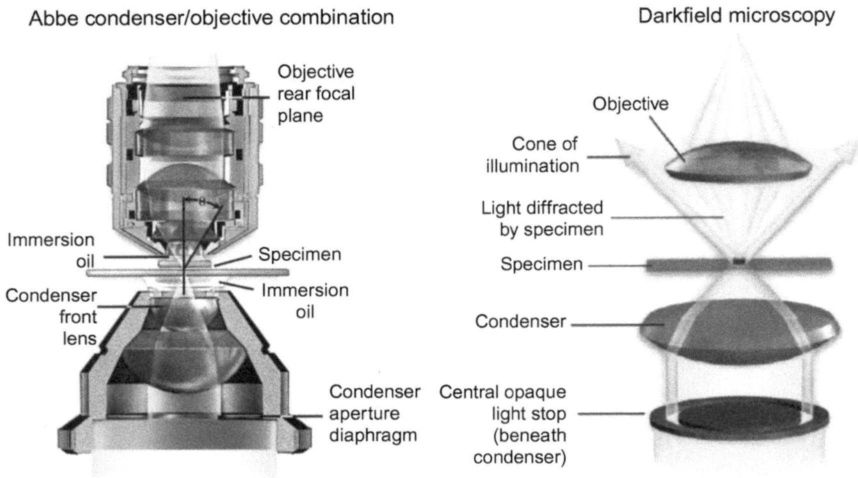

FIGURE 6 (left) A modern bright-field microscope with the Abbe condenser/objective combination, invented by Ernst Karl Abbe of University of Jena in 1870. This combination includes oil immersion of the specimen and objective lens. Immersion objective lenses were suggested as early as 1678 by Hooke and then by Sir D. Brewster in 1812, and were first made by Giovanni Amici of the University of Modena in 1840. Carl Zeiss introduced Abbe's water immersion objectives in 1872 and Abbe's oil immersion objectives in 1877. Ernst Abbe, Otto Schott (glass chemist pioneer and borosilicate glass inventor), and Carl Zeiss (innovative lens and microscope maker) pioneered the development of the modern light microscope. Zeiss founded Carl Zeiss AG in Jena in 1846 and was joined later by Abbe in 1872. Then in 1884 Schott joined with Abbe and Zeiss to found Schott and Associates Glass Technology Laboratory. August Köhler, who invented uniform illumination of specimens, joined Carl Zeiss in 1900 (see Molecular Expressions website, http://micro.magnet.fsu.edu/optics/timeline/people/index.html). The Optical Museum in Jena, Germany, is a scientific-technological museum displaying optical instruments from eight centuries. (Right) A modern transmission dark-field light microscope. In dark-field microscopy only sample-scattered light can reach the objective lens and be focused onto the detector. Without a sample the detector plane is dark. This figure is from *Optical Microscopy* by M. Abramowitz and M.W. Davison (see the Florida State University website, Optical Microscopy, http://micro.magnet.fsu.edu/primer/pdfs/microscopy.pdf). (See Color Insert.)

the development of light microscopy, Köhler illumination serves as a reminder of the possibility of the occurrence of a superimposed image of the source and the specimen X-ray and neutron microscope setup.

The synergy between the University of Jena and the Carl Zeiss Company is a good example of academic-industrial partnerships in which the right set of circumstances and people meet to produce great achievements in society. For example, Abbe introduced the 8-hour workday to the employees of Carl Zeiss, which was an improvement over the

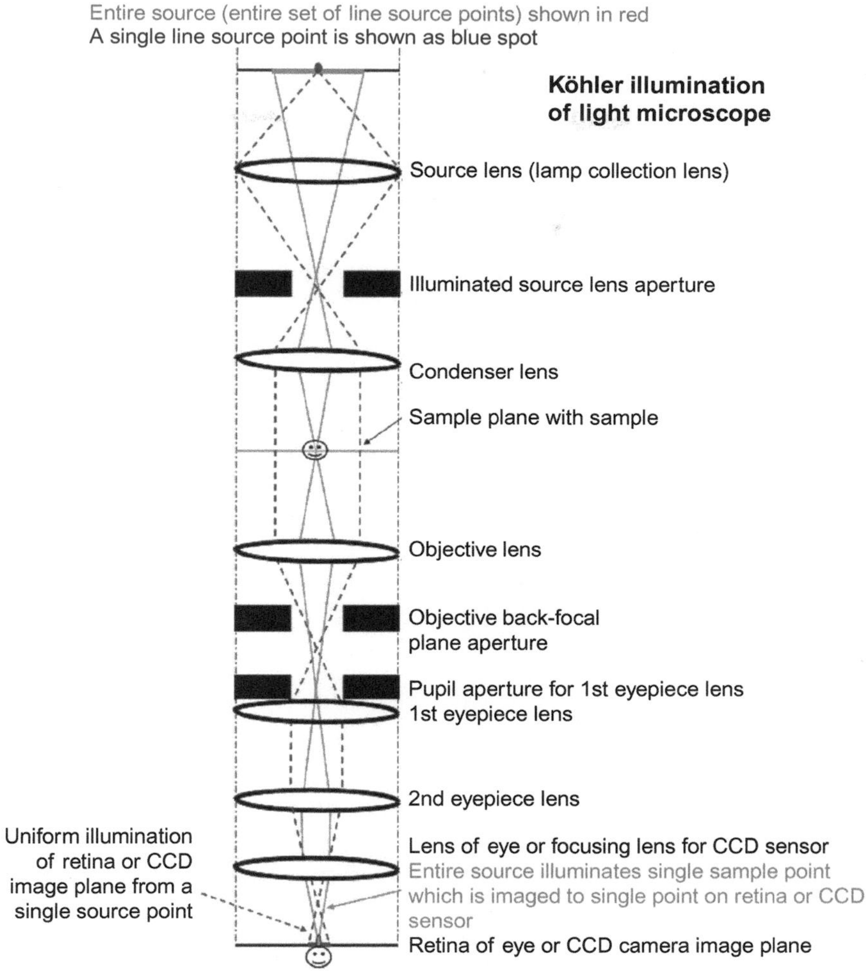

FIGURE 7 Köhler illumination for light microscope sets up to two light paths: (1) rays from each source point illuminate entire sample plane and (2) rays from all source points focus to single sample points. CCD, charge-coupled device. (See Color Insert.)

14-hour workdays prevalent in those times. The University of Jena–Carl Zeiss Company relationship was similar to the relationships between the University of California at San Francisco and Genentech; Stanford University with Varian; Hewlett-Packard, Google; California Institute of Technology with Beckman-Coulter; Massachusetts Institute of Technology with Bose, EG&G, and Thermo Electron; Oxford University and Oxford Instruments; Karlsruhe University and Bruker Physik AG; Leiden University and Crucell; Cambridge University and Plastic Logic, to name a few.

The-bright-field and dark-field microscopes had long been established before the phase-contrast microscope was invented by Dutch physicist Frits Zernike in 1934. In phase-contrast microscopy the light transmitted by the specimen without scatter and the light scattered by the specimen are separated into two light paths via an annular light source, which is matched to a disk of material with a thinner annulus region that phase-shifts the light by 90° or –90° relative to light that passes through the disk inside the annulus (Figure 8). The annular light source is focused onto the sample such that the specimen-transmitted (non-scattered light) passes through the annulus and acquires a 90° or –90° phase shift relative to the specimen-scattered light that passes inside the annulus. The light scattered by a thin specimen acquires a 90° phase shift due to the thin sample and recombines with the sample-transmitted light, which acquired a 90° or –90° phase in passage through the annulus relative to the specimen-scattered light passed though the disk inside the annulus. In the 90° phase shift of specimen-transmitted light in the annulus, both the specimen-scattered and specimen-transmitted light have the same 90° phase shift and their amplitudes constructively interfere and sum. In the –90° phase case, the specimen-transmitted and specimen-scattered light have a relative phase shift of 180° that causes destructive interference or subtraction of amplitudes.

1.8. Neutron and X-ray Microscopes: Amplitude-Contrast Microscopes and Phase-Contrast Microscopes

The amplitude-contrast neutron microscope at the top of Figure 9 uses a condenser lens CRL to focus X-rays or neutrons through a small aperture at the condenser lens focal length so that only narrow-band neutrons or X-rays are focused on the object specimen (see Adelphi Technology U.S. patents by Beguiristain *et al.*, 2004 and Piestrup *et al.*, 2001). A second downstream CRL acts as an objective lens and focuses the neutrons or X-rays onto a detector to form an amplitude-contrast image.

An amplitude-contrast neutron or X-ray microscope can be converted into a phase-contrast microscope that images objects using phase contrast between X-ray or neutron waves (see Adelphi Technology U.S. patent by Piestrup *et al.*, 2004). The Zernike phase-contrast microscope separates the direct zero-order light from the diffracted light at the rear focal plane of the objective. The conversion from an amplitude-contrast to a phase-contrast instrument is done by adding an annular condenser optic, which is formed by placing an annular aperture just downstream of the compound refractive condenser lens (see the bottom diagram in Figure 9). A phase plate is placed in the conjugate plane of the annular condenser lens, which illuminates the object specimen. If there were no diffracted neutron rays from the specimen, an image of the annular diaphragm would form. The phase plate applies a 90° phase shift to the zero-order X-ray or neutron rays

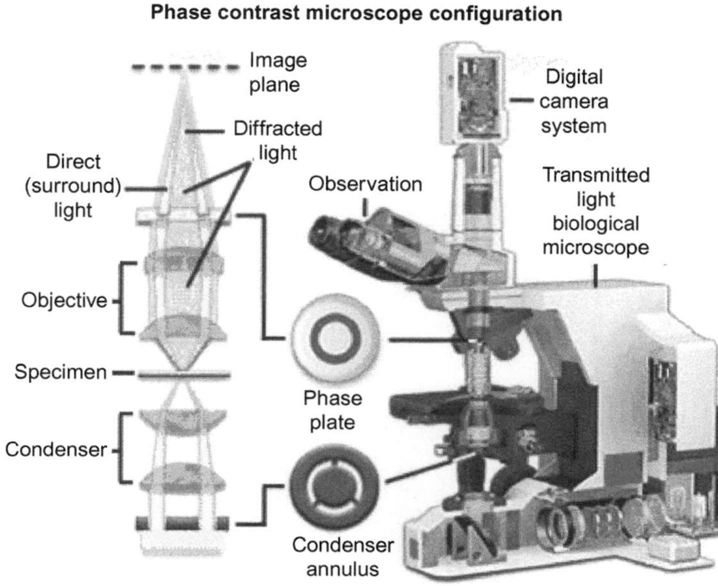

FIGURE 8 A diagram of a phase-contrast light microscope. In this microscope a single annular beam of light is focused onto the sample. Some of the light is transmitted through the sample without interaction. The rest of the light interacts with the sample by scatter. The phase-contrast microscope separates the sample-deviated light via the annular aperture ring, which undergoes a 90° phase shift by thin specimens, and the zero-order, nondiffracted transmitted light. Downstream of the objective lens is a phase plate with an annular indentation or region of less thickness, which is aligned and matches the source annular aperture. The specimen-diffracted light passes inside the inner radius of the annular indentation of the phase plate, whereas the specimen-transmitted light passes through the annular indentation in the phase plate. The phase-plate annular indentation produces a 90° phase shift of the specimen-transmitted light relative to the specimen-diffracted light that passes inside the annular indentation. The light, specimen-transmitted light, and the specimen-diffracted light then arrive at the detector plane with the same 90° phase shift. The phase-shift ring thus causes the deviated and nondeviated light to have the same phase, which allows constructive interference. Alternatively, the phase plate annulus can provide a −90° phase shift of the specimen-transmitted light relative to the specimen-diffracted light passing inside the annular inner radius. This causes a 180° phase shift between the specimen-transmitted and diffracted light with destructive interference or amplitude subtraction in the image plane. The figure is from "Introduction to Phase Contrast Microscopy" on the website of the Nikon MicroscopyU (http://www.microscopyu.com/articles/phasecontrast/phasemicroscopy.html), written by Douglas B. Murphy, Ron Oldfield, Stanley Schwartz, and Michael W. Davidson. (See Color Insert.)

coming from the object with respect to the rest of the rays deflected by the object. The thickness and material of the phase plate determine the phase shift introduced by the phase ring.

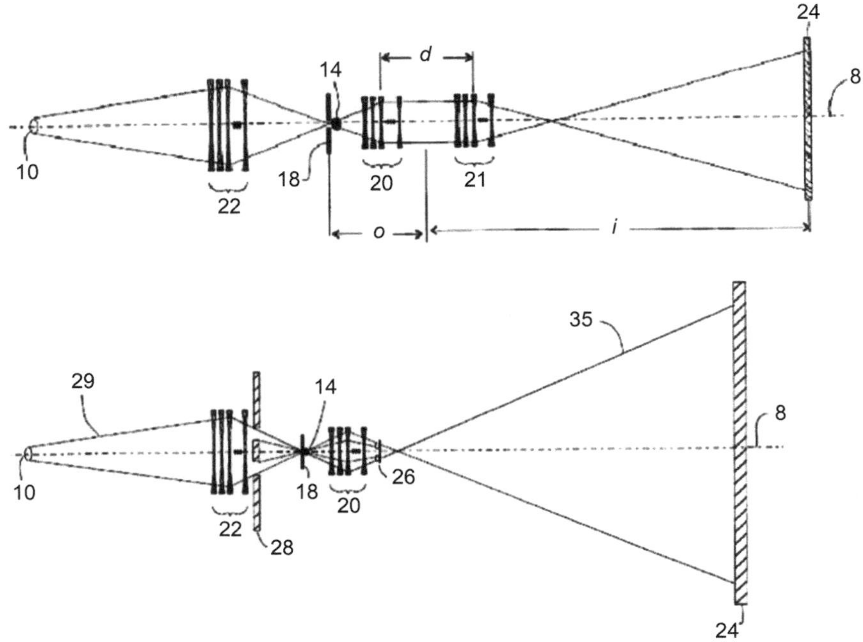

FIGURE 9 (Top) An amplitude-contrast X-ray or neutron microscope. A condenser CRL lens and aperture act as a monochromator, which focuses narrow-band neutrons or X-rays through a pinhole onto a sample. Downstream at object distance $r_0 = 0$ is a second CRL, which is the objective lens that produces an amplitude-contrast image on the detector at image distance $r_i = i$. (Bottom) A phase-contrast X-ray or neutron microscope. A condenser CRL lens with adjacent annular ring aperture focuses narrow-band neutrons or X-rays onto a sample. Downstream at object distance $r_0 = 0$ is a second CRL objective lens and downstream annular phase shifting plate that produces a phase-contrast image on the detector at image distance $r_i = i$. See text for details (Piestrup *et al.*, 2004, 2001; Beguiristain *et al.*, 2004).

The incidence of X-rays on the structures in the specimen leads to the absorption of some X-rays or neutrons, the passage of some X-rays or neutrons nondeviated or nondiffracted, and diffraction of some X-rays or neutrons. The image of the specimen is the result of (1) absorption contrast resulting from the omission of incident X-rays or neutrons at the detector due to absorption by the object specimen and (2) phase contrast resulting from the pattern created by the wave interference of phase-shifted, diffracted waves with the nondiffracted waves that did not undergo phase shifting and uniformly illuminate the detector.

In a thick absorptive sample not only is a pattern produced at the detector by absorption of incident X-rays on the specimen leading to absorption contrast, but there is also an overlay pattern resulting from phase contrast. The X-rays refracted by the thick sample undergo typically a $\lambda/2$ (180°) phase shift with respect to the nondiffracted waves. The nondiffracted waves uniformly illuminate the detector and the interference pattern with the diffracted X-rays results in the overlying phase-contrast image of the specimen.

In a thin sample, however, absorption contrast is not only weak, but if the sample is sufficiently thin, the scattered neutrons or X-rays are phase-shifted only by $\lambda/4$ (90°) and thus no overlying phase-contrast image is produced. To use the phase information carried by the scattered X-rays to produce a visible intensity modulation (i.e., contrast) in the image plane, one the nondiffracted neutron or X-ray waves must somehow be shifted by $\lambda/4$ (90°) or $3\lambda/4$ (270°). In the phase-contrast microscope this is accomplished with a phase-shifting optical element just downstream of the objective lens CRL (see item 26 in the bottom diagram of Figure 9).

The phase-shifting element is matched in its geometry to the distribution of nondiffracted neutrons or X-rays. The phase-shifting element has an annular ring, which corresponds to the ring-shaped hollow cone of X-rays focused by the condenser optic onto the specimen. The hollow cone of focused, nondiffracted neutrons or X-rays diverges from the specimen to the objective lens. The objective lens CRL then refracts and directs these nondiffracted rays to the phase-shifting ring. The phase-shifting ring adds $\lambda/4$ (90°) or $3\lambda/4$ (270°) phase shift due to these nondiffracted neutrons or X-rays.

Meanwhile, the focused hollow cone of neutrons or X-rays, which is diffracted by specimen, then undergoes a $\lambda/4$ (90°) phase shift due to the specimen. The diffracted neutrons or X-rays are scattered out of the hollow cone, which contains the nondiffracted X-rays. The diffracted neutrons or X-rays diverge to the objective lens and are incident on a different part of the objective lens than the nondiffracted rays. Due to their position on a different portion of the CRL objective lens, the diffracted neutrons or X-rays are refracted and directed through the center hole or alternatively around the phase-shifting ring. In summary, the diffracted neutrons or X-rays are phase-shifted by $\lambda/4$ due to the specimen and then pass to the detector from the specimen with no further phase shifting.

The diffracted and the nondiffracted rays each have acquired a $\lambda/4$ phase shift and are thus in phase and undergo constructive interference. As a result, the diffracted and nondiffracted neutrons or X-rays interfere constructively at the detector with summation of their amplitude and thereby produce an interference pattern. This interference pattern is a bright phase-contrast image of the specimen. Alternatively, the diffracted neutrons or X-rays with the $\lambda/4$ phase shift interfere destructively with the nondiffracted neutrons or X-rays, which have acquired a $3\lambda/4$ phase shift or equivalently a $-\lambda/4$ phase shift due to passage through the phase-shift ring. The diffracted and nondiffracted neutrons of X-rays at the detector now have a $\lambda/2$ phase-shift difference and thus undergo destructive interference. The resulting destructive interference between the nondiffracted and specimen diffracted neutrons or X-rays produces a dark, phase-contrast image of the specimen.

In addition to the phase shifting of the nondiffracted neutron or X-rays, the phase ring also attenuates the nondiffracted neutrons or X-rays. Since the nondiffracted neutrons or X-rays typically dominate over the diffracted neutrons or X-rays, the attenuation helps equate the amplitudes of the nondiffracted and diffracted neutrons or X-rays. The phase ring's elemental composition, density, and thickness can be chosen to produce a quarter-wavelength $\lambda/4$ or three-quarter-wavelength $3\lambda/4$ phase shift as well as attenuating the nondiffracted neutrons or X-rays. This yields very high–phase-contrast (40% to 50%) images at the detector with monochromatic or polychromatic neutrons or X-rays. The past development of light microscopy has guided the recent development of neutron and X-ray microscopy.

The condenser optic of the phase-contrast X-ray microscope must illuminate the specimen, ideally with defocused parallel neutron or X-ray wave fronts. Consider the specific case for X-rays. The omnidirectional X-rays emitted by the anode of the X-ray tube are collected by a collector optic (mirror or CRL) and are made parallel. These parallel neutrons or X-rays from the collector optic pass through an annular aperture, which is aligned with the annular phase-shifting ring, and onto the condenser optic, which directs the defocused parallel X-rays as a hollow cone onto the specimen. The parallel neutron or X-ray wave front illuminating the specimen has a high spatial or lateral coherence. This allows wave amplitudes to be highly correlated between different points transverse to the direction of X-ray propagation. That is, the phase at any point on the wave front may be fluctuating randomly with time, but its phase angle difference with other points on the wave front remains constant. Hence, points of the specimen illuminated by an X-ray wave can lead to a fixed interference pattern at the detector.

1.9. Neutron and X-ray Temporal and Longitudinal Spatial Coherence Length, and Transverse Spatial Coherence Length

We now consider the temporal and spatial coherence of a wave train with a temporal coherence length τ_{coh}. The top diagram of Figure 10 depicts the temporal coherence length τ_{coh} and its associated longitudinal spatial coherence length L_{coh}. The longitudinal spatial coherence length L_{coh} is related to the temporal coherence length τ_{coh} by the wave-phase velocity v_{ph} of the center wavelength λ and frequency f of an X-ray or neutron wave train, where

$$L_{coh} = v_{ph}\tau_{coh}. \qquad (50)$$

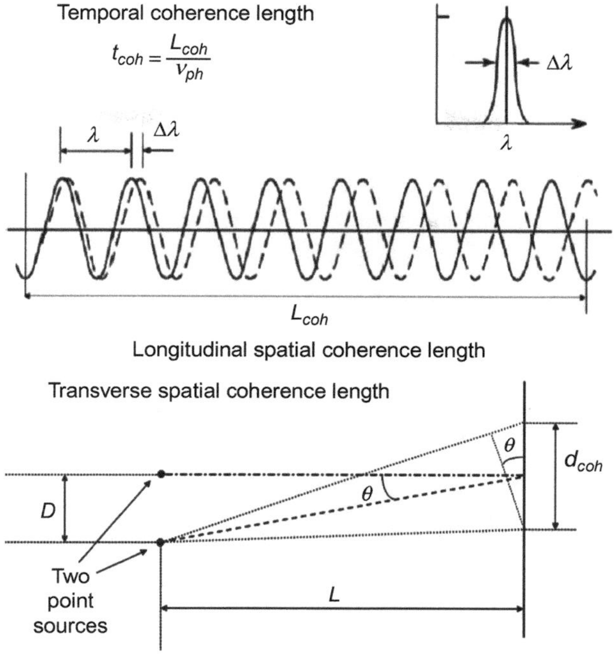

FIGURE 10 (Top) A diagram for temporal coherence length τ_{coh} and longitudinal spatial coherence length L_{coh}, which are related by the phase velocity v_{ph} of the center wavelength λ and frequency f. (Bottom) A diagram for the transverse spatial coherence length d_{coh} (Lauter-Pasyuk, 2007).

The temporal width τ_{coh} of the wave train is related to the frequency width Δf of the train by

$$\Delta f = \frac{1}{\tau_{coh}}. \tag{51}$$

The center frequency f is related to the center wavelength λ of the wave train with wave train wavelength width $\Delta\lambda$ and wave train frequency width Δf by

$$\frac{\Delta f}{f} = \frac{|\Delta\lambda|}{\lambda}. \tag{52}$$

The phase velocity v_{ph} of the center of the wave train wavelength λ and frequency f is

$$f\lambda = v_{ph}. \tag{53}$$

Substitution for v_{ph} of Eq. 53 and τ_{coh} of Eq. 51 with use of Eq. 52 in the expression for the longitudinal spatial coherence length L_1 of Eq. 50 gives

$$L_{coh} = \frac{\lambda^2}{\Delta\lambda}. \tag{54}$$

Now consider a finite-source diameter D set on the z-axis and an observation plane at axial distance L as shown in the bottom diagram of Figure 10. At the observation plane the lateral coherence length d_{coh} is the transverse distance interval about the z-axis, which subtends polar angle interval $\Delta\theta$. To be coherent over the transverse spatial coherence length d_{coh} requires the wave to not vary more than 2π radians (i.e., one wavelength λ), where

$$\lambda = d_{coh} \sin(\Delta\theta) \cong d_{coh}\Delta\theta \tag{55}$$

and

$$\Delta\theta = \frac{D}{L}. \tag{56}$$

The transverse spatial coherence length d_{coh} is then

$$d_{coh} = \frac{\lambda}{\Delta\theta} = \frac{L}{D}\lambda. \tag{57}$$

The synchrotron X-ray source diameter D_{xry} is much smaller than the neutron source diameter D_{neu}, where

$$D_{xry} \ll D_{neu}. \tag{58}$$

As a result, the X-ray transverse spatial coherence length d_{coh}^{xry} is much larger than the neutron transverse coherence length d_{coh}^{neu}, where

$$D_{coh}^{xry} \gg D_{coh}^{neu}. \tag{59}$$

Both X-ray and neutron sources have relatively large distances L between the source and the observation plane in which the observation plane could be the object plane or entrance plane of a CRL. Assuming typical distances $L = 20$ m and a synchrotron X-ray source size of 0.02 cm, a center wavelength $\lambda = 2.0$ Å, and a typical neutron source diameter 2.0 cm, then

$$d_{coh}^{xry} = \frac{2 \times 10^4}{0.02} \cdot 2 \times 10^{-8} = 200\,\mu\text{m} \tag{60}$$

$$d_{coh}^{neu} = \frac{2 \times 10^4}{2.0} \cdot 2 \times 10^{-8} = 2\,\mu\text{m} \tag{61}$$

1.10. Neutron and X-ray Fresnel Zone Plates

Neutrons can also be focused by diffraction using a FZP (Figure 11). The radius of the annular zones varies so that the path length difference between every other zone is a single wavelength λ. The alternating zones can be either opaque or composed of a given thickness of material that causes a 180° phase shift (a "phase-reversal" FZP). Neutrons or X-rays, which are incident on the zone plate perpendicular to its surface, are then diffracted from the zones and add constructively at the focal point. For a given design focal length f, the required radius r_n of the nth zone of M zones comprising the FZP is given by

$$r_n^2 + f^2 = \left(f + \frac{n\lambda}{2}\right)^2. \tag{62}$$

The FZP focal length f is much larger than the zone plate diameter, which is the number of half-wavelength zones of the zone plate multiplied by the intended half-wavelength, where

$$f \gg \frac{n\lambda}{2}. \tag{63}$$

As a result, the radius r_n of the nth zone becomes

$$r_n \cong \sqrt{n\lambda f}. \tag{64}$$

Setting the radius of the first zone r_1 to l, with $r_1 = l$, and assuming the FZP focal length f is much greater than the X-ray or neutron wavelength λ,

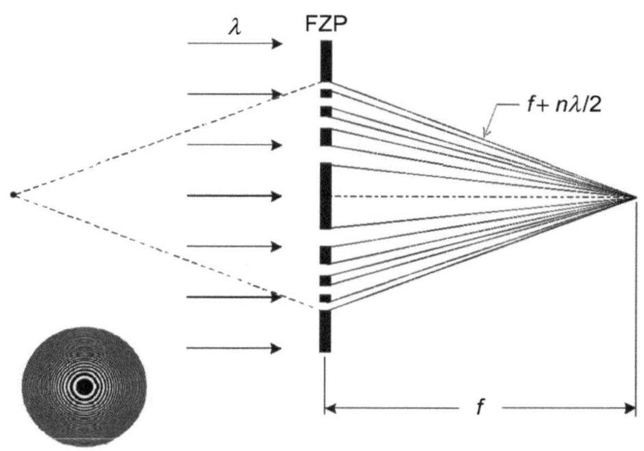

FIGURE 11 A schematic diagram of a Fresnel zone plate (FZP). Incident X-rays or neutrons are diffracted by the radial zones. The size of the zones varies so that the diffracted radiation is focused to the axis at a distance *f*.

with $f \gg \lambda$, then gives

$$f = \frac{l^2}{\lambda}. \quad f \gg \lambda \tag{65}$$

Hence, the FZP focal length f is determined by the size of the first zone radius $r_1 = l$. The aperture diameter D of a zone plate is simply the diameter of the last zone $n = M$, where

$$D = 2r_M = 2\sqrt{M}l, \tag{66}$$

Typically the manufacturing limit for FZPs is the width of the outer zone; that is,

$$r_M - r_{M-1} = \Delta r. \tag{67}$$

For a given FZP focal length f, the first zone radius $r_1 = l$ is then determined, so that Δr limits the total number of zones that can be produced, and thus determines the aperture diameter D. With the above results the limiting aperture diameter D can be calculated from Δr, where

$$r_M^2 - r_{M-1}^2 = \lambda f = r_M^2 - (r_M - \Delta r)^2 = 2r_M \Delta r - (\Delta r)^2 \approx 2r_M \Delta r = D\Delta r \tag{68}$$

or

$$D = \frac{f\lambda}{\Delta r}. \tag{69}$$

If we assume a minimum zone spacing $\Delta r = 1.0\,\mu\text{m}$ and focal length $f = 10\,\text{m}$ for neutron wavelength $\lambda = 10$ Å, then the aperture diameter of the neutron FZP is D = 1.0 cm, which is similar to the apertures of neutron CRLs.

Concentric gold or nickel rings on a silicon dioxide substrate were used by Sir Lawrence Bragg to produce some of the first usable X-rays images. The practical use of soft X-ray imaging via diffractive optics with an FZP imaging a sample onto a soft X-ray CCD sensor occurred in 1974 at the Georg-August University in Göttingen (see Schmahl and Rudolph, 1969, and Niemann *et al.*, 1974). This was followed in 1988 by the development of the scanning soft X-ray microscope in which the sample is scanned past a focused spot and the image is electronically produced one pixel at a time at SUNY Stony Brook and the Brookhaven National Laboratory (see Rarback *et al.*, 1988).

Neutron Fresnel zone plates (NFZPs) were first fabricated and tested by Kearney *et al.* (1980). The NFZPs have typically been phase plates, rather than opaque, due to the low absorption of neutrons in micron thicknesses of most materials. FZPs have applications in coded aperture imaging and holography. Mertz and Young (1961) first proposed shadow-casting (total absorption) as a viable option for high-energy photon imaging, particularly for X-ray astronomy. Zone plate shadow-casting is a technique complementary to the Leith–Upatnieks (1964) modification of Gabor's interferometric method (transmission holography) for producing holograms. The Leith–Upatnieks hologram—or offset-reference hologram—uses a separate and distinct reference wave rather than object transmittance as in the Gabor hologram in which the reference wave is introduced at an offset angle, which spatially separates the virtual and real twin images (Gabor, 1949).

In neutron holography, an FZP with alternating zones of total absorption and total transmission was used as a reticle to cast geometrical shadows. An object is placed in a cold neutron beam from a research reactor. The object scatters the neutrons through an FZP, with alternate zones of gadolinium and aluminum, which produces a hologram of the object on an X-ray film. The image is constructed by placing the hologram, linearly reduced in size, in a converging laser beam, which produces a real image of the original object (Beynon and Pink, 1980). A zone plate, which has a sinusoidal transmissivity distribution, is called a *Gabor zone plate.*

1.11. Neutron and X-ray Achromat Lenses Using Fresnel Zone Plate and Compound Refractive Lens

We consider now two methods to fabricate achromat lenses for X-ray or neutrons. The first approach to achromat lenses uses multiple optical elements separated by finite distances. The individual lenses can be made of the same material and method (e.g., both CRLs or both FZPs). The typical system is two lenses separated by a distance d, where f_1 and f_2 are the focal lengths of the lenses, and

$$d = \frac{1}{2}(f_1 + f_2). \tag{70}$$

Unfortunately, with such a method it is not possible to correct both lateral and longitudinal aberration. While the focal length is invariant, giving constant magnification, the locations of the principal planes in the system move with the wavelength.

A second method of making an achromat requires two component lenses with different dispersions. This method is generally referred to as a *crown and flint* since it was first produced using two types of glass (crown and flint) that had different dispersions for visible light. This design provides both lateral and longitudinal chromatic correction. Figure 12b shows the design of an optical crown and flint achromat; one lens is positive and

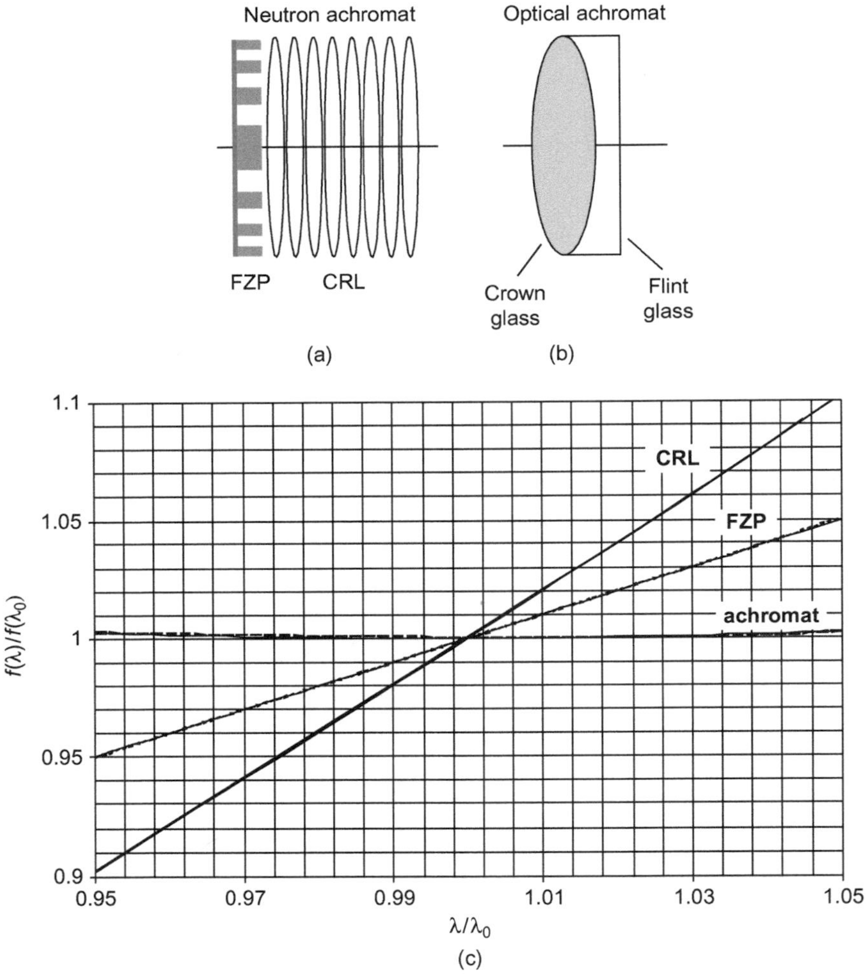

FIGURE 12 Achromats. (a) Neutron or X-ray achromatic lens made from positive (focusing) Fresnel zone plate (FZP) and negative compound refractive lens (CRL). (b) An optical achromat is made from a positive lens made of crown glass and a negative lens made of flint glass. (c) Plot of the Kirchhoff equation calculation: wavelength dependence of focal length over 10% bandwidth for a CRL, FZP, and achromat made from a CRL and FZP with $f_{CRL} = -2f_{FZP}$.

the other negative. Figure 12a shows a proposed CRL and FZP achromat. FZPs, regardless of their design, all have a dispersion relation is expressed by the $1/\lambda$ dependence of the focal length f, where

$$f = \frac{l^2}{\lambda}. \quad f \gg \lambda. \tag{71}$$

The dispersion of refractive, biconcave spherical or parabolic lenses is expressed by the $1/\lambda^2$ focal length dependence, where

$$f = \frac{R}{2N\delta} = \frac{\pi R}{N\rho b}\frac{1}{\lambda^2}. \tag{72}$$

Thus, since CRLs and FZPs have different dispersions, an achromat can be designed from their combination analogous to the design of crown and flint achromats for visible optics.

The general design of the neutron or X-ray achromat proceeds as follows. The total focal length of the achromat is given by

$$\frac{1}{f_{total}} = \frac{1}{f_{CRL}} + \frac{1}{f_{FZP}}. \tag{73}$$

This assumes the lenses are collocated. Since the CRL has a finite thickness, the length of the CRL must also be treated but it should not affect the ability to create an achromat, since the chromatic correction is due to the compensating dispersions of the two lenses rather than their spacing. Locating the FZP midway in the CRL would eliminate any problem with translating principal planes if necessary. The power P of a lens is the reciprocal of its focal length f, where the total power P_{tot} of the combined FZP and CRL, the CRL power P_{CRL}, and FZP power P_{FZP} are given by

$$P_{tot} = \frac{1}{f_{tot}} \quad P_{CRL} = \frac{1}{f_{CRL}} \quad P_{FZP} = \frac{1}{f_{FZP}}. \tag{74}$$

The focusing power of the combined optics P_{tot} can be written from the individual focal lengths of Eqs. 71 and 72, with substitution of Eq. 74 into Eq. 73, gives

$$P_{total} = P_{CRL} + P_{FZP} = \frac{2N\delta_{CRL}(\lambda)}{R_{CRL}} + \frac{\lambda}{l_{FZP}^2}. \tag{75}$$

The above focal power P_{tot} is scaled to the design wavelength λ_0, where

$$P_{tot} = \frac{2N\delta_{CRL}(\lambda_0)}{R_{CRL}}\frac{\lambda^2}{\lambda_0^2} + \frac{\lambda_0}{l_{FZP}^2}\frac{\lambda}{\lambda_0} = P_{CRL}(\lambda_0)\frac{\lambda^2}{\lambda_0^2} + P_{FZP}(\lambda_0)\frac{\lambda}{\lambda_0}. \tag{76}$$

To form an achromat, the derivative of the total focal power is equal to zero at the design wavelength λ_0, where

$$\frac{d}{d\lambda}P_{tot}\Big|_{\lambda_0} = 0 = 2P_{CRL}(\lambda_0)\frac{\lambda}{\lambda_0^2} + P_{FZP}(\lambda_0)\frac{1}{\lambda_0}\Big|_{\lambda_0} = 2P_{CRL}(\lambda_0) + P_{FZP}(\lambda_0) \tag{77}$$

or

$$2P_{CRL}(\lambda_0) = -P_{FZP}(\lambda_0) \tag{78}$$

and

$$P_{total}(\lambda_0) = -P_{CRL}(\lambda_0) = \frac{1}{2}P_{FZP}(\lambda_0). \tag{79}$$

The focal length could also be set equal at two different wavelengths to design the achromat. Via the Kirchhoff equation, Figure 12c plots the calculated wavelength dependence of focal length over 10% bandwidth for a CRL, FZP, and achromat made from a CRL and FZP with $f_{CRL} = -2f_{FZP}$. Considerable improvement is possible with the combination of an FZP and CRL according to this design.

1.12. Neutron and X-ray Fresnel and Compound Fresnel Lenses

One means of overcoming the limited absorption aperture of CRLs is to use a Fresnel lens shape. A refractive Fresnel lens is constructed with stepped setbacks of many divided annular segments. That is, the Fresnel lens contains a series of concentric grooves of prismatic or curved shape, which focus light just as an ordinary lens. The Fresnel lens is based on geometric optics rather than the diffractive optics of the Fresnel and Gabor zone plates. Invented by Augustin-Jean Fresnel in 1822, Fresnel lenses were first used in Courdouan lighthouse in France in 1823 and are used extensively today in automobile headlamps, projectors, and ophthalmic applications.

Transmission loss through the lens may be reduced with a Fresnel CRL as shown in the left diagram of Figure 13. The diagram shows how a parabolic refractive lens is conceptually converted to a refractive Fresnel lens. Because the lens surface has the same curvature as the conventional CRL, the focal length is unaltered but the loss is less, thereby permitting an increase in lens aperture. However, the Fresnel CRL is much more difficult to fabricate, the step size between zones is critical, and the alignment of individual lenses is important. Only the material-vacuum (air) interfaces of the Fresnel segments or "steps" are useful for deflecting and focusing X-rays.

The absorbing segment behind the refractive segment is of no use and results in increased X-ray absorption. Removing material in steps does not interfere with the ability of the lens to refract the X-rays. Indeed, the gain of the lens increases with the number of steps. Since (1) the Fresnel lens can be thin and (2) the Fresnel lens surface follows a parabolic shape in steps, the refractive Fresnel lens can achieve a larger clear aperture. Fresnel lenses are used for visible optics when a thin lens is needed.

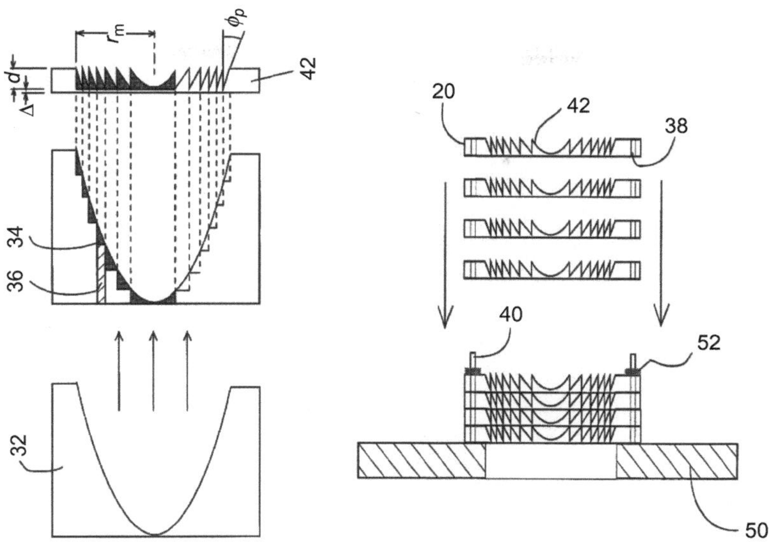

FIGURE 13 (Left) A diagram showing the derivation of a Fresnel lens from a thick plano-concave lens. (Right) The stacking and precise alignment of *N* Fresnel lenses to form a compound Fresnel lens (CFL), which can be used as a large-aperture condenser lens to focus X-rays or neutrons onto a sample, which is then imaged with a CRL (Piestrup *et al.*, 2001).

The relatively large transverse, spatial coherence length d_{coh} for X-rays ($\sim$100 μm) can cover several annular lens zones of the Fresnel lens. As a result, a fixed phase relationship can exist between portions of the wave front refracted by the inner lens zones, which are covered by the transverse coherence length of the wave front. To take advantage of the large X-ray transverse coherence length, the removed lens material from a single plano-concave, parabolic lens to form the Fresnel lens should be exactly an integer number n of wavelengths $n\lambda$ or corresponding integer number $n2\pi$ of 2π phase shifts. The inner annular lens zones contained within the transverse coherence radius of the X-ray wave front provide coherent focusing.

In the case of neutrons, the neutron sources provide very small, transverse coherence length d_{coh} of only a few microns, which is smaller than the annular lens zone widths. In this case, the precise removal of lens material equal to an integer number of wavelengths $n\lambda$ is not necessary. The refracted portions of the waves from each nth lens annulus of the M lens annuli is coherent within that zone with summing of amplitudes, and then the amplitude magnitude is squared to obtain the intensity for the nth lens annulus. However, the phase relation between the adjoining lens zones is random and incoherent. Hence, the focused intensities from the M lens annuli sum to produce the total intensity, which is refracted by the stack of N Fresnel lenses to the observer or point detector.

For neutrons, the removal of the absorbing material of a parabolic lens to produce a Fresnel lens reduces the neutron absorption, but the coherence is lost across the refracted wave front because each annular lens zone is incoherent with the other annular lens zones. That is, there is no interference pattern between the neutron waves refracted by the M annular zones of the single Fresnel lens or the stack of N Fresnel lenses in the CFL. Thus the neutrons are focused incoherently on the Fresnel lens in the same manner as the incoherent light focusing by the lighthouse Fresnel lenses.

Given the object distance r_0 from the object to the Fresnel lens of diameter D, the diffraction limited resolution d_r of the Fresnel lens is given by

$$d_r = 0.61\frac{r_0\lambda}{D}. \tag{80}$$

For typical neutron imaging conditions (e.g., object distance $r_0 = 5\,\text{m}$, wavelength $\lambda = 10$ Å, and Fresnel lens diameter $D = 1.0\,\text{cm}$), the diffraction-limited resolution of the d_r is on the order of tens of nanometers. However, resolution typically is limited by aberrations, including chromatic, gravity, the surface quality of the lens, and most typically, the resolution of the detector.

In order to have phase addition from all the zones it is necessary for the phase shift through all the lens elements between adjacent zones of the Fresnel CRL to be an integer n multiplied by 2π. If the phase-shift error between zones is required to less than a quarter-wavelength $\lambda/4$, then the standard deviation σ_s for the step height of an individual lens element in the axial z-direction is

$$\sigma_s \leq \frac{\lambda}{4\delta\sqrt{N}}. \tag{81}$$

This result assumes the variation in error is random from lens element to lens element. The required roughness σ_s becomes more demanding as the number N of lenses increases. An N-lens compound Fresnel lens (CFL) has approximately the same focal length as a CRL with given radius of curvature R of the spherical surface or the on-axis radius of a curvature R of parabolic surface. The focal length f_{CFL} of the N-lens parabolic or spherical plano-concave lens–based CFL is then approximately

$$f_{CFL} = \frac{R}{N\delta} = \frac{2\pi R}{N\rho b}\frac{1}{\lambda^2}. \tag{82}$$

Since cold and thermal neutron sources are inherently broadband, these chromatic aberrations generally pose the first limit to achievable resolution.

The z-axis is assumed to be the center optical axis for a stack of N Fresnel lenses. Each Fresnel lens is mounted on a support structure of thickness Δ (Figure 13). For a CFL composed of parabolic shaped surfaces, the groove surface of a single lens element is described by the equation for a parabola in which the on-axis radius of curvature R is pointed in the z-direction. Consider a single parabolic, concave-surfaced Fresnel lens of the N lenses in the CFL. Given radial distance r from the center z-axis, the axial height h_z of the $(n+1)^{\text{th}}$ groove surface or Fresnel lens annulus is (Freeman and Hull, 2003)

$$h_z = \frac{r^2 - r_n^2}{2R} = \frac{r^2 - r_n^2}{2Nf_{CFL}\delta} = \frac{\left(r^2 - r_n^2\right)\pi}{Nf_{CFL}\rho b\lambda^2} \leq d. \tag{83}$$

The thickness d of a single Fresnel lens increases quadratically with increasing aperture radius r.

The Fresnel lens was developed to increase the aperture area of the lens without increasing the thickness of the lens. That is, the refractive interface surface of the lens was preserved, whereas interior lens material was removed to reduce attenuation and the weight and size of the lens. A large-area condenser lens for X-rays or neutrons can be fabricated by using a single Fresnel lens with a very long focal length or a stack of precision-aligned Fresnel lenses to form a CFL, which reduces the focal length by N-fold (see the right diagram in Figure 13).

Surrounding each curved lens surface in the Fresnel lens are the flat cylindrical surfaces or *relief surfaces,* which are parallel to the optical z-axis. Small-angle X-ray and neutron scatter and reflection can occur from these relief surfaces. The small-angle scatter and reflection of X-rays can be reduced by adding surface roughness to the relief surface; in the case of neutrons, a thin neutron-absorbing gadolinium layer can be added to the relief surface. However, the fabrication of a single X-ray or neutron Fresnel lens is technically difficult. The treatment of the relief surfaces to suppress small-angle scatter and reflection is even more difficult.

Such treatment of the relief surfaces may not be necessary with the use of a well-collimated beam of X-rays or neutrons that propagates parallel to the optical axis. The precise stacking of Fresnel lenses to form an X-ray or neutron CFL is particularly difficult because of the very small radial width of the outer lens zones. An achromat composed of two CRLs is technically more feasible, although the aperture diameter of the CRLs will be smaller than that of the CFLs.

The left diagram of Figure 14 shows an X-ray or neutron achromat lens formed by two identical CFLs, each of focal length f, and separated by distance d equal to the focal length, where $d = f$. The right diagram shows an X-ray or neutron telescope formed from a convex lens and a concave lens. The convex lens can be fabricated from a CFL composed of a stack of N plano-convex Fresnel lenses. Similarly, the concave lens can

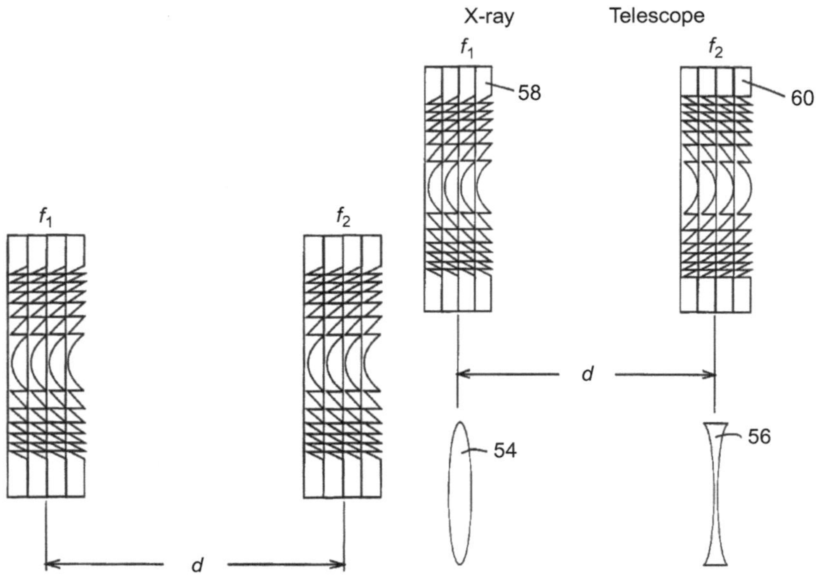

FIGURE 14 (Left) A diagram of an achromat formed from two identical compound Fresnel lenses (CFLs), each of focal length *f*, which are separated by distance $d = f$ to reduce transverse chromatic aberration. Technically more feasible is the use of two identical compound refractive lenses (CRLs) in place of the pair of CFLs. The CFL provides more aperture diameter than the CRL; however, the CFL is much more difficult to fabricate and align. (Right) A CFL composed of *N* plano-convex Fresnel lenses is paired with a CFS composed of *N* plano-concave Fresnel lenses to form an X-ray or neutron telescope (Piestrup *et al.*, 2001).

be fabricated from a CFL composed of a stack of *N* plano-concave Fresnel lenses. However, the fabrication and stacking in precision alignment is very difficult for both the CFL achromat and CFL telescope.

1.13. Neutron and X-ray Monochromator with Compound Refractive Lens and Pinhole

An X-ray or neutron monochromator can be fabricated with a CRL and a pinhole aperture in the CRL focal plane (Figure 15). The pinhole aperture is set to transmit the focused neutrons or X-rays. For a monochromatic source, the neutrons or X-rays focus to monochromatic aberration and diffraction-limited spot diameter. As the bandwidth of the neutron or X-ray source is increased, chromatic aberration leads to an increasing focal spot diameter in the focal plane of the center wavelength. The pinhole diameter *d* can be set to transmit a specified narrow bandwidth of X-rays or neutrons in which the pinhole aperture is placed in the focal plane

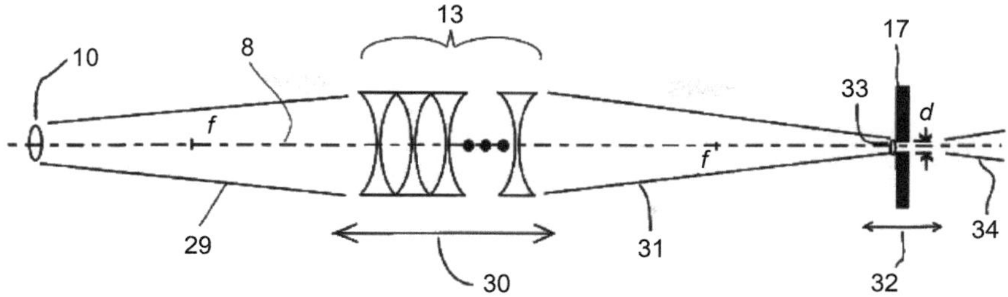

FIGURE 15 An X-ray or neutron monochromator that uses a CRL (13) and a pinhole (33) with aperture diameter *d* in the focal plane of the CRL, which is set for the center transmission wavelength of interest. The CRL focuses a narrow bandwidth of X-rays or neutrons from a wide band source (10) of X-ray or neutrons (Piestrup *et al.*, 2001, 2004, and Beguiristain *et al.*, 2004).

for the center wavelength. As the pinhole diameter d is decreased to the diffraction-limited focal spot diameter, the transmitted X-ray or neutron intensity decreases but the bandwidth decreases as well.

1.14. Comparison of Neutron, X-ray, and Light Source Intensities and Images

Strong visible light sources and modest laser sources are much brighter than X-rays sources, and typical X-ray sources are far brighter than reactor and spallation neutron sources. The photon or neutron brightness B ($\# \, \mathrm{mm}^{-2} \cdot \mathrm{mr}^{-2} \cdot \mathrm{sec}^{-1}$) is the number of photons or neutrons N (#) emitted per unit time interval Δt (sec) from source unit area ΔA_{sou} (nm^2) into a unit solid angle $\Delta\Omega$ (mr^2), where mr is milliradian units, and

$$B = \frac{N}{\Delta t \Delta A_{\mathrm{sou}} \Delta\Omega}. \tag{84}$$

The photon or neutron spectral brightness B_{sp} is photon or neutron brightness per fractional angular frequency bandwidth $\Delta w/w$ or per fractional wavelength bandwidth $\Delta\lambda/\lambda$, where

$$B_{sp} \equiv \frac{B}{\Delta w/w} \equiv \frac{N}{\Delta t \Delta A_{sou} \Delta\Omega \left(\Delta w/w\right)} \tag{85}$$

or

$$B_{sp} \equiv \frac{B}{\Delta\lambda/\lambda} = \frac{N}{\Delta t \Delta A_{sou} \Delta\Omega \left(\Delta\lambda/\lambda\right)}. \tag{86}$$

Here w is the center angular frequency of frequency bandwidth Δw (rad) and λ is the corresponding center wavelength of the wavelength bandwidth $\Delta\lambda$(Å) corresponding to Δw. Angular frequency w (rad $\cdot$ sec^{-1}) is related to temporal frequency f(sec^{-1}) by $w = 2\pi f$.

Flux Φ (# cm$^{-2}\cdot$ sec^{-1}) is the number of photons or neutrons N (#) that cross a unit area ΔS (cm^2) per unit time interval Δt (sec), where

$$\Phi = \frac{N}{\Delta t \Delta S}. \tag{87}$$

Spectral irradiance I (# of photons or neutrons per sq. cm. per sec per rad) is the number of photons or neutrons N (#) crossing a unit area ΔS (cm^2) per unit time interval Δt (sec), per unit photon or neutron radial frequency interval Δw (rad) or wavelength interval $\Delta\lambda$ (Å), where

$$I \equiv \frac{N}{\Delta t \Delta S \Delta w} \tag{88}$$

or

$$I \equiv \frac{N}{\Delta t \Delta S \Delta \lambda}. \tag{89}$$

The photon or neutron irradiance I(W cm^{-2} $\cdot$ sec^{-1} $\cdot$ rad^{-1}) can also be expressed by power P(W) per unit area ΔS (cm^2) per frequency Δw (rad) or wavelength $\Delta\lambda$ (Å) interval, where

$$I \equiv \frac{P}{\Delta S \Delta \lambda}. \tag{90}$$

The neutron or X-ray fluence φ(# cm^{-2}) is the number of photons or neutrons N(#) that cross a unit area ΔS (cm^2), where

$$\varphi = \frac{N}{\Delta S}. \tag{91}$$

Compared with visible light optics, the development and application of X-ray and neutron optics has been slow and difficult due to the lack of high brightness and spectral brightness sources. X-ray sources with large spectral brightness B_{sp} require large and expensive synchrotron rings. Neutron sources with high fluxes Φ such as reactors and spallation neutron sources are large and expensive. Presently there are no neutron sources with large spectral brightness B_{sp} comparable to the X-ray or laser light sources. That is, in visible light optics or photonics the easy availability of inexpensive and very compact, high brightness as well as high

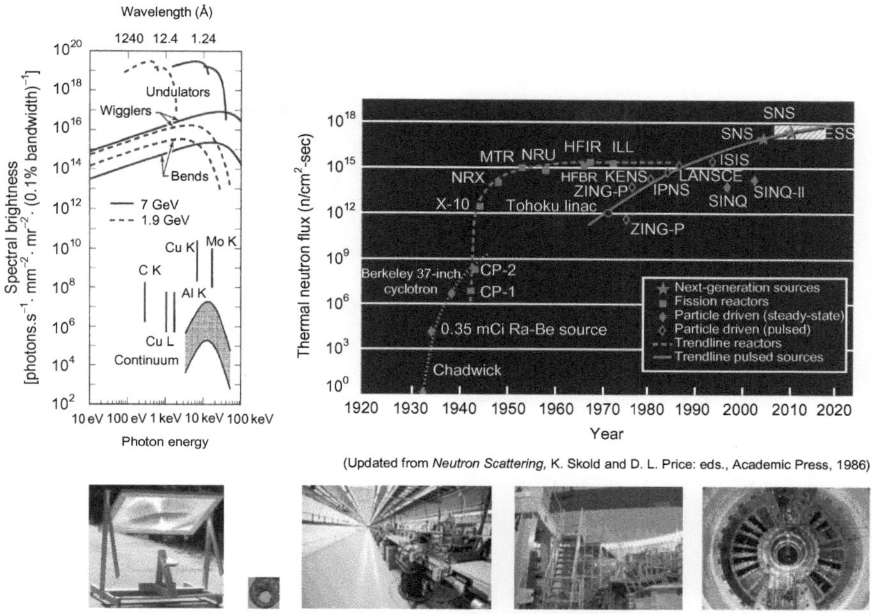

FIGURE 16 (Top left) The spectral brightness B_{sp} for several synchrotron and conventional X-ray sources (from Kim, 2009). (Top right) The thermal neutron flux Φ from a wide variety of sources over time (from Skold and Price, 1986). (Bottom) The 1.0-m Fresnel lens solar furnace and the ~6.25-mm hole produced by focused sunlight on a U.S. copper penny shown in the adjoining photo (from Science Toys website, mhttp://scitoys.com/). The three photos to the right of the penny are (from left to right) the electron accelerator of the Linear Coherent Light Source at Stanford Linear Accelerator Center, the research area above the 58-MW reactor at Institut Laue Langevin in Grenoble, and the target structure for the Spallation Neutron Source at Oak Ridge National Laboratory.

spectral brightness sources of photons and the lack of radioactivity give visible light optics a big advantage over X-ray and neutron optics. The left plot of Figure 16 is the spectral brightness of several synchrotron and conventional X-ray sources. The right plot is the cold neutron flux as a function of neutron wavelength at the NIST Center for Neutron Research on the SANS NG-7 beam line.

On a sunny day the sun can deliver a peak of $\sim 1.0\,\text{kW/m}^2$ of visible, infrared, and ultraviolet light. Approximating the average photon energy as 1.5 eV, this averages to solar photon flux $4.2 \times 10^{17}\text{ph/cm}^2/\text{sec}$ (see the bottom plot in Figure 17 of solar spectral distribution at Earth's surface). With an inexpensive, flat, thin plastic $1.0\,\text{m} \times 1.0\,\text{m}$ Fresnel lens

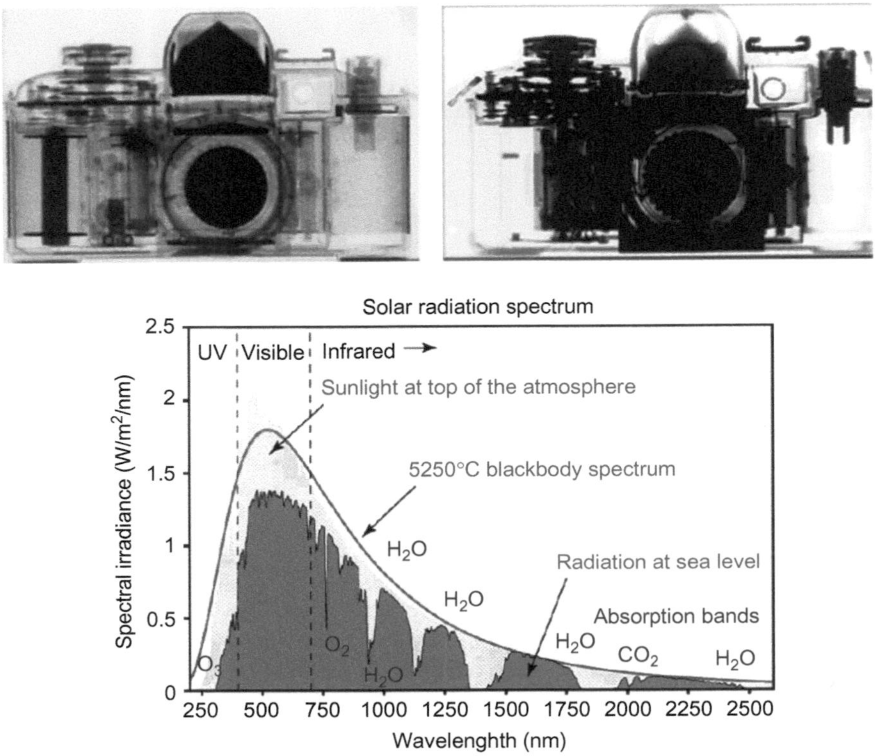

FIGURE 17 (top) Neutron radiographic image of a camera (right) X-ray radiographic of an image of a camera (Lehmann, 2000). As can be seen higher contrast can be obtained of the camera's low-Z materials with the thermal neutrons. (bottom) Plot of solar spectral flux per reaching Earth given in units of watts per square meter per nanometer wavelength interval. (From Wikipedia File:Solar Spectrum.png.)

from a 40-inch projection television set (shown in the photo at the bottom of Figure 16), 1.0 kW of sunlight can be focused to a 0.25-inch spot diameter (1.25×10^{-4} m^2), as demonstrated by the copper U.S. penny (adjoining photo), and continuously deliver photon flux of approximately 1.1×10^{22} ph/cm^2/ sec of average energy 1.5 eV.

In 2005 the Vulcan laser, which is located in the Central Laser Facility of Rutherford Appleton Laboratory in Oxfordshire, England, delivered a peak, pulsed focused intensity of 10^{21} W/cm^2/ sec of 1.16-eV infrared photons, which corresponds to a peak photon flux of 8.6×10^{39} ph/cm^2/ sec. The relatively low flux from X-ray and neutron sources compared to ordinary visible light from the sun or a laser, and the relatively strong attenuation of X-rays and neutrons by the X-ray and neutron optical components together produce technical challenges for X-ray and neutron optics.

The three bottom-right photos in Figure 16 (from left to right) are from the X-ray laser source (Linac Coherent Light Source) at Stanford Linear Accelerator Center, the 58-MW reactor neutron source of the Institut Laue-Lan in Grenoble, France, and the spallation neutron source (SNS) of the Oak Ridge National Laboratory. The three photos on the right show the massive scale required to produce the X-ray brightness and neutron fluxes, which are comparable to or less than the sunlight spectrum collected by the Fresnel lens solar collector (bottom-left photo). The raising and herding of X-rays and neutrons is a difficult and expensive task compared with raising and herding visible and infrared photons. These organized scientific endeavors are certainly far more demanding, but less hazardous, than the raising and herding of sheep in the rough and violent times of thousand of years ago.

The fusion reactor or sun in the sky provides photons for the solar furnace free of charge, but the production of 10-keV X-rays and 0.025-eV neutrons (mean lifetime of $\sim$15 minutes) requires the raising and herding of the workhorse electrons, which directly produce the X-rays and indirectly produce the neutrons with the essential ingredient—electricity. The thin plastic Fresnel lens is a representative descendant of the eighteenth- and nineteenth-century master craftsmanship combined with experimental research and development, which played an essential role in the development of science, technology, medicine, and industry that shaped society yesterday and today.

Figure 17 compares radiographic images of a camera with thermal neutrons (top-leftimage) and with X-rays (top-right image). Hydrogen strongly scatters thermal neutrons and acts as a neutron contrast agent, whereas high-Z elements act as a contrast agent for X-rays. Thermal neutrons interact with nuclei and magnetic fields of unpaired atomic electrons, X-rays interact with the atomic electron cloud in an ionizing fashion, and visible light interacts with the atomic electron cloud in a non-ionizing manner. The modalities of light, X-ray, and neutron imaging can be performed on the same object to produce three different views of the same object. The three different images can be fused to provide more detailed information of the object-image fusion. In addition, the object can be imaged with electron microscopy, terahertz waves, radio waves, sound waves, and so on to provide more information.

The bottom plot in Figure 17 is the visible light spectral distribution from the sun incident at the top of the Earth's atmosphere and at sea level. This light spectral distribution determines the life processes on Earth and is the spectral distribution that the camera is designed to refract and image.

2. EXPERIMENTS WITH NEUTRON COMPOUND REFRACTIVE LENSES, MAGNETIC LENSES, AND MICROSCOPES

Before the development of capillary and refractive lens optics, neutron optics was confined to the reflection of neutrons with mirrors (Sears, 1989) and the guiding and collecting of neutrons via diffraction using crystals (see Mikula *et al.*, 1986, and Majkrzak and Wood, 1992). The guiding and focusing of neutrons with capillary optics was reported in 1992 by the publications of Kumakhov and Sharov (1992) and Chen *et al.* (1992). Four year later, *Nature* published a paper on X-ray refraction by X-ray compound refractive lenses (CRLs) by Snigerev *et al.* (1996), which inspired M.R. Eskildsen *et al.* (1998) at Risø in Denmark to use compound refractive neutron lenses to focus neutrons and improve the performance of small-angle neutron scatter (SANS). Four years after the original publication, Choi *et al.* (2000) of the U.S. National Institute for Standards and Technology [NIST] used compound refractive neutron lenses to replace the focusing collimation slits by neutron CRLs and thus improve small angle-scatter resolution.R Neutron lenses can achieve a smaller center spot in a SANS beam than a set of collimating holes whose radii decrease toward the sample. In 1998 at the SANS beam line in the cold neutron guide hall at the Risø DR3 reactor, Ekildsen *et al.* (1998) used a stack of 30 biconcave magnesium fluoride (MgF_2) CRLs to focus cold neutrons. The $N = 30$ synthetic MgF_2 lenses were biconcave, and each lens each had an $R = 25$ mm radius of curvature and a $2r_m = 25$ mm physical aperture diameter with center thickness $d = 1.0$ mm (top photo in Figure 18). The MgF_2 biconcave spherical lenses are commercially available and were obtained from International Scientific Products (Irvington, NY, USA) by the Risø group. With $\delta = 8.9 \times 10^{-5}$ at neutron wavelength $\lambda = 100$ Å, the calculated focal length is $f = 5.2$ m. The smaller center spot of nonscattered neutrons, which was provided by the CRL, allowed the measurement of small-angle scatter points, which without a CRL are covered by the intense background of neutrons transmitted through the sample without scatter.

In the Risø experiment, the particular wavelength λ was selected from a range of 9–20-Å neutrons with bandwidth $\Delta\lambda/\lambda = 0.09$ using a mechanical velocity selector in which the MgF_2 CRL achieved gains $G > 15$ with focal lengths ranging from $f = 1.0$ m to $f = 6.0$ m. The average overall CRL transmission was 77% at neutron wavelength $\lambda = 9.5$ Å and 50% at $\lambda = 16.0$ Å. The neutron area detector was composed of an 128 × 128 array of pixels, each with a 5 × 5 mm area. The angular beam divergence was very anisotropic. In the horizontal direction, the beam divergence was constant at $\Delta\theta_{hor} = 20$ arcmin, and in the vertical direction the beam divergence was wavelength dependent, where $\lambda_{\theta \mathrm{ver}}$ (arcmin) = 12 λ (Å). The plot and focal spot image in the bottom diagram of Figure 18 show that the CRL

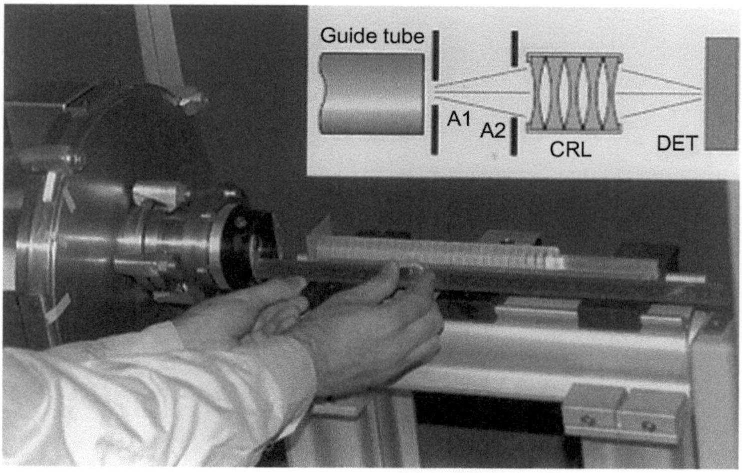

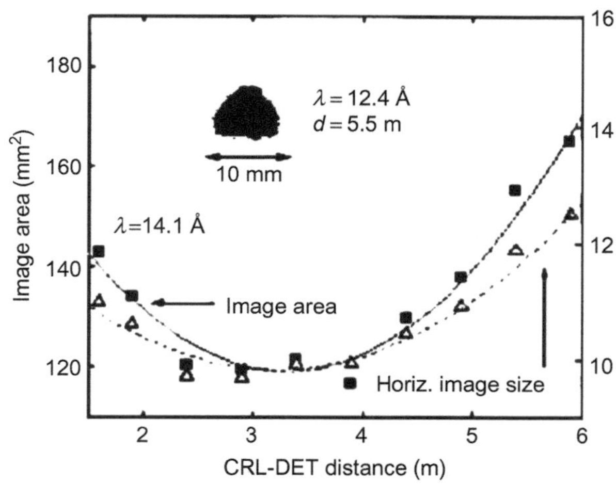

FIGURE 18 (Top) A photograph with an inset of the setup and the Risø SANS beam line with the positioned stack of $N = 30$ biconcave, 25-mm physical aperture diameter MgF_2 lenses. The incident 2.0-mm diameter aperture A1 was positioned 5.6 m from the CRL in an evacuated neutron guide, and the 25-mm aperture A2 was placed at the CRL entrance to act as the CRL-limiting aperture. (Bottom) Plot of the image size dependence on the distance (CRL-DET) from the MgF_2 CRL to the detector (DET) for $\lambda = 14.1$ Å neutrons. The horizontal full width half maximum (FWHM) image size is represented by triangles, and the total FWHM image area is denoted by squares. The inset is a 12.4-Å neutron image of an 8.3-mm triangular source aperture for a 5.5-m CRL to detector distance (Ekildsen *et al.*, 1998).

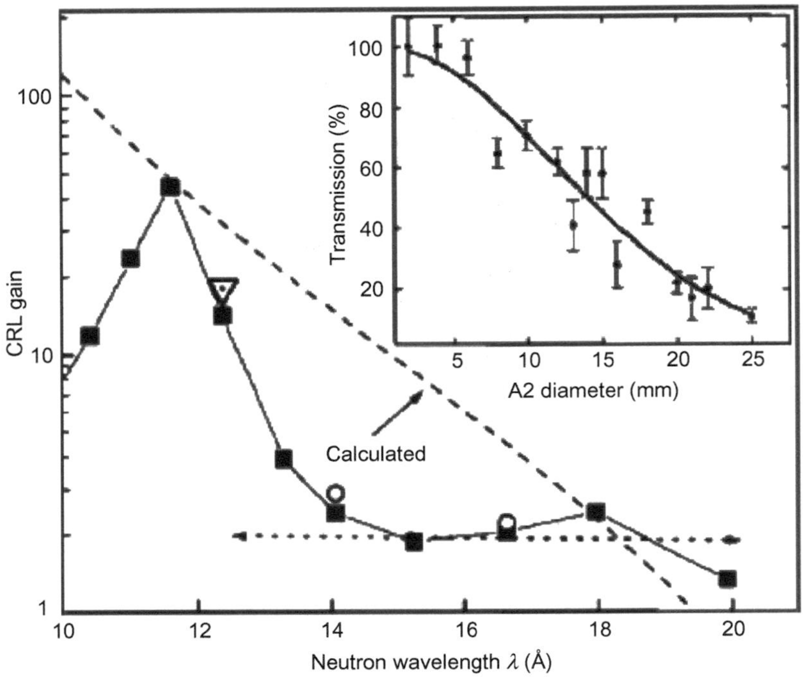

FIGURE 19 Plot of the CRL gain G dependence on neutron wavelength λ. The CRL intensity gain G of the triangle source aperture object was evaluated at the wavelength, which is indicated by the inverted open triangle plot point. The dashed line shows the result of a simple calculation of gain G, which included the effective aperture radius r_e, the minimum absorption along the lens centerline $T_0 = \exp(-N\mu d)$, chromatic aberration $f(\lambda)$, and demagnification $M = r_i/r_0 < 1$. The dotted line indicates the gain $G \geq 2$ over more than an octave in wavelength. The CRL-to-detector distance was limited to a maximum of 6.0 m, which led to an artifact of a measured peak gain at 11.5 Å. The gain G increases with decreasing wavelength λ, as indicated by the diagonal dashed line. The inset is a plot of total CRL transmission $T(r_{A2})$ versus diameter $2r_{A2}$ of aperture $A2$, which was used to determine the effective lens aperture radius r_e at wavelength $\lambda = 14.1$ Å. The CRL transmission T was normalized to the same configuration with no CRL, and the transmission T is plotted as a function of the aperture $A2$ diameter $2r_{A2}$ (Ekildsen *et al.*, 1998).

acted as a $\pi \cdot r_{A1}^2$ lens by producing an inverted image of the object source; the experimentally measured image size was a good fit to the calculated demagnified image size base on lens geometric optics. Some chromatic aberration smearing of the image occurred due to the 9% neutron source bandwidth.

Figure 19 plots the CRL gain G dependence on neutron wavelength λ. The CRL gain G is defined as the intensity at the CRL focus relative to intensity produced by pinhole collimation for pinhole apertures $A1$ with diameter $2r_{A1} = 2.0\,\text{mm}$ (solid squares) and $2r_{A1} = 4.0\,\text{mm}$ (open circles). The gain was estimated by multiplication of the image area by the total attenuation and then dividing by the area $\pi \cdot r_{A1}^2$ of the pinhole aperture $A1$. Excess attenuation occurs at the lens perimeter due to the biconcave shape of the lens, and thus the effective aperture $2r_e$ became less than the CRL-defining aperture A2, which was set to the physical aperture diameter $2r_m$ of the CRL lenses. Detector smearing was not included in the gain calculation and was attributed to the measured gain being less than the calculated gain. The CRL-to-detector distance was limited to a maximum of 6.0 m, which led to an artifact of a peak gain at 11.5 Å. The gain G increases with decreasing wavelength λ, as indicated by the diagonal dashed line in Figure 19. The effective lens aperture diameter $2r_e$ was determined by the absorption aperture radius r_a. Via Gaussian fitting, the effective aperture diameter was $2r_e = 15.7\,\text{mm}$ at wavelength $\lambda = 9.5$ Å and was $2r_e = 11.9\,\text{mm}$ at $\lambda = 11.4$ Å.

Ekildsen *et al.* (1998) noted the effect of gravity on the neutron mass and the effect of magnetic fields on the neutron magnetic dipole moment. Relative to mirrors, a CRL is more compact, does not require high surface quality, is easier to align, and does not alter the beam path. However, as they noted, the CRLs suffer from chromatic aberration but mirrors are achromatic; however, the mirror's critical angle for total reflection is small and wavelength dependent. They also noted the CRLs were more flexible and efficient for focusing a neutron beam to a focal spot than the Kumakhov lens, which is based on internal reflecting light pipes that are tapered to produce a focal spot. Finally, they noted the CRL could be used (1) as a microprobe to guide and focus neutron beams into local crystallographic and magnetic domains, (2) to reduce neutron beam divergence and enhance the resolution of Bragg reflection rocking curves, and (3) to magnify the pattern of scattered neutrons from a sample to cover more pixels of the large-area detectors, which have low spatial resolution due to the large pixel size (~3- to 5-mm dimensions).

Figure 20 is a table from the Ekildsen *et al.* (1998) *Nature* paper. The table lists the neutron refractive (coherent scatter length b_c) and absorption (thermal neutron capture cross section σ_a at room temperature 293°K) properties of candidate elements for neutron lenses. The refractive properties of an isotope can be succinctly described by the ratio b_c/σ_α (293° K) of the coherent scatter length b_c (independent of wavelength and temperature) divided by the thermal capture absorption cross section σ_a at 293°K. That is, one assumes the nonrelativistic neutron velocity v has a Maxwell–Boltzmann distribution $\Phi(v)$, where

$$\Phi(v) \propto v^3 \left(\frac{3k_B T}{m_n}\right). \tag{92}$$

Candidate elements for CRLs

Element/isotope	Atomic weight	b_c (fm)	σ_a (barns)	b_c/σ_a (fm^{-1})
O	15.99	5.8	1.9×10^{-4}	310
C	12.01	6.6	3.5×10^{-3}	19
Be†	9.01	7.8	7.6×10^{-3}	10
Pb*	208.0	9.5	4.8×10^{-3}	8.0
F†	18.99	5.6	9.6×10^{-3}	5.8
Zr*	90.0	6.4	1.1×10^{-2}	5.3
Pb*	206.0	9.2	3.0×10^{-2}	3.1
Bi†	208.98	8.5	3.4×10^{-2}	2.5
H*†	2.0	6.7	5.2×10^{-4}	2.1
Zr*	94.0	8.2	5.0×10^{-2}	1.6
Mg†	24.3	5.4	6.3×10^{-2}	0.86
Mo*	94.0	6.8	1.5×10^{-2}	0.85
Mo*	92.0	6.9	1.9×10^{-2}	0.68
Sr*	88.0	7.1	5.8×10^{-2}	0.43
N*†	15.0	6.4	2.4×10^{-5}	0.34
Tl*†	205.0	9.5	1.0×10^{-1}	0.24

A complete list of elements and isotopes useful for neutron lenses, with data for 2,200 m s^{-1} (1.8 Å) thermal neutrons. Nuclei with $b_c > 5$ fm, $\sigma_a < 0.1$ barn and natural abundance >5% (for atomic weight >40) and resultant ratio $b_c/\sigma_a > 0.1$ are included.

* For these elements, the final column has been adjusted for the relative natural abundance of different isotopes.

† Materials with an incoherent scattering length $b_i > 0.1$ fm, which affects use with polarized neutrons, or in cases where diffuse scattering is important.

FIGURE 20 The table from the Ekildsen *et al.* (1998) *Nature* paper listing the neutron refractive (coherent scatter length b_c) and absorption (thermal neutron capture cross section σ_a at room temperature 293°K) properties of candidate elements for neutron lenses. The table omits aluminum lenses $b/\sigma_a = 14.9$, which produced the first cold neutron microscope images (Beguiristain *et al.*, 2002), and copper lenses $b/\sigma_a = 2.0$, which produced the first thermal neutron images (Cremer *et al.*, 2007a).

The neutrons with rest mass m_n are assumed to be thermalized at the room temperature of 293°K in the Maxwell–Boltzmann distribution. The thermal neutrons are nonrelativistic and the neutron wave group velocity v_g is equal to the neutron particle velocity v, which is related to the neutron wavelength λ by the de Broglie relationship, where given Planck's constant h,

$$v = v_g = \frac{h}{m_n \lambda}. \tag{93}$$

The neutron kinetic energy of the neutrons E is related to the neutron particle or group wave velocity by

$$E = \frac{1}{2} m_n v^2 = \frac{3}{2} k_B T. \tag{94}$$

The following are useful formulas that relate the wavelength λ, velocity v, kinetic energy E, and equilibrium temperature $T$of the neutrons, where (Squires, 1996)

$$\lambda(\text{Å}) = \frac{30.81}{\sqrt{T(°\text{K})}} = \frac{3.956}{v(\text{km/s})} = \frac{9.045}{\sqrt{E(\text{meV})}} \tag{95}$$

The neutron coherent scatter length b_c, incoherent scatter length b_{inc}, absorption cross section σ_a, as well as the coherent σ_{coh} and incoherent scatter σ_{inc} cross sections for a large set of isotopes are given at the NIST website (http://www.ncnr.nist.gov/resources/n-lengths). The thermal neutron capture cross section σ_a linearly increases with decreasing temperature T, where

$$\sigma_a \propto \frac{1}{v} \tag{96}$$

The use of CRLs to improve SANS performance was adopted by the NIST Center for Neutron Research NG7 30-meter SANS instrument (Figure 21). The pinhole collimation was replaced by lens focusing of the neutrons. The top plot of Figure 22 shows the increased intensity on the sample as a function of scatter neutron wave vector magnitude Q. The bottom plot shows a photo of the CRL composed of a variable number N of spherical biconcave MgF_2 lenses with $2r_e = 25$-mm aperture diameter, biconcave and radius of curvature $R = 25$ mm. The number N of lenses varies depending on the required neutron wavelength λ and focal distance f. The bottom photo of Figure 22 shows the setup for 8 and 28 lenses, respectively, focusing 18.1 Å and 8.44 Å. For $Q \leq 0.0038\,A^{-1}$, the CRL performed better than the pinhole collimation by delivering more focused intensity on the sample.

The first refractive imaging of neutrons occurred at Institut Laue-Lan (ILL) in Grenoble, France, in 2002 with an aluminum compound refractive neutron lens as a collaboration between Adelphi Technology, Inc. and I. Anderson and C. Dewhurst of ILL (see Beguiristain *et al.*, 2002). The aluminum CRL consisted of a stack of $N = 105$ biconcave aluminum lenses with an $R = 12.7$ mm radius of curvature, which produced an $M = 35$ magnified image of a steel spring at 214-μm resolution with $\lambda = 20$ Å. The neutron image of the entire spring was produced by combining 400 exposures of small, partial views of the spring taken at 500-μm intervals horizontally and vertically with 2-minute exposure times. This was the first neutron image by a refractive lens neutron microscope (Figures 23 and 24).

This proof-of-principle experiment demonstrated a neutron microscope could be fabricated from CRLs in a manner similar to ordinary lenses in an optical microscope. The two-dimensional (2D) neutron detector pixel size limited the neutron microscope resolution to 214 μm,

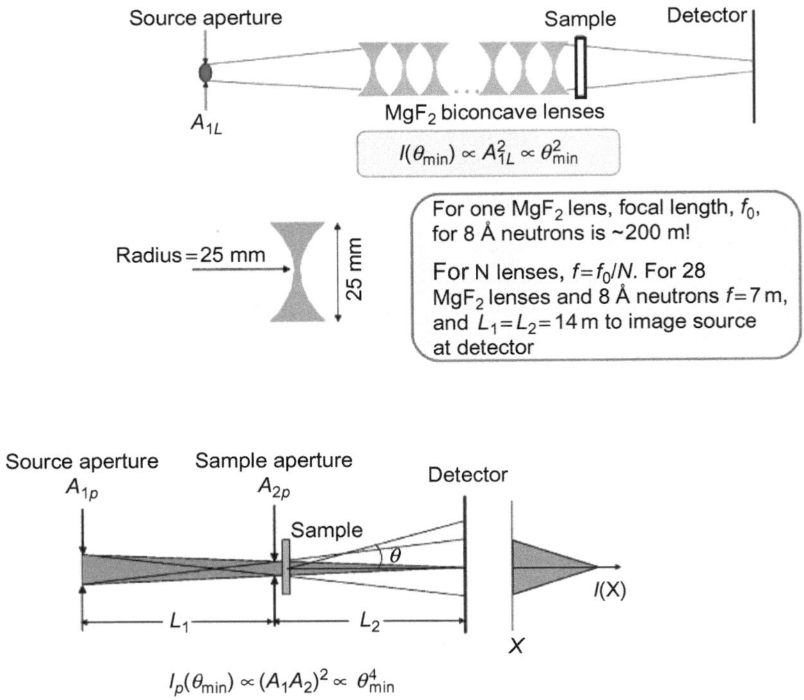

FIGURE 21 (Top) A focusing refractive optics for SANS from the NIST Center for Neutron Research NG7 30-m SANS instrument. The diagram shows the use of a CRL with small-angle neutron scatter. With an MgF_2 CRL to image the source aperture at the detector, the intensity variation on the sample varies as θ_{min}^2, which is nearly independent of the sample size contained within the lens aperture diameter. (Bottom) The commonly used "pinhole" collimation for SANS. The intensity on the sample varies with the minimum resolvable scattering angle as θ_{min}^4.

which was estimated to be larger than the CRL microscope lens aberrations. The neutron images were highly magnified ×35, and similar to images from a light microscope, possessed field of view (FOV), depth of field, and improved spatial resolution via magnification. This experiment demonstrated neutron microscopy techniques could become a valuable complement and extension of the already mature imaging techniques with light, X-rays, and electrons. The experiment also demonstrated the need for stronger and higher-quality neutron sources and higher-resolution neutron imaging detectors (preferably real-time).

The neutron microscopy experiment was performed on the D22 beam line at ILL, which is a 58-MW reactor neutron source that produces a thermal neutron flux of 1.5×10^{15}n/cm^2/s in its moderator region. The

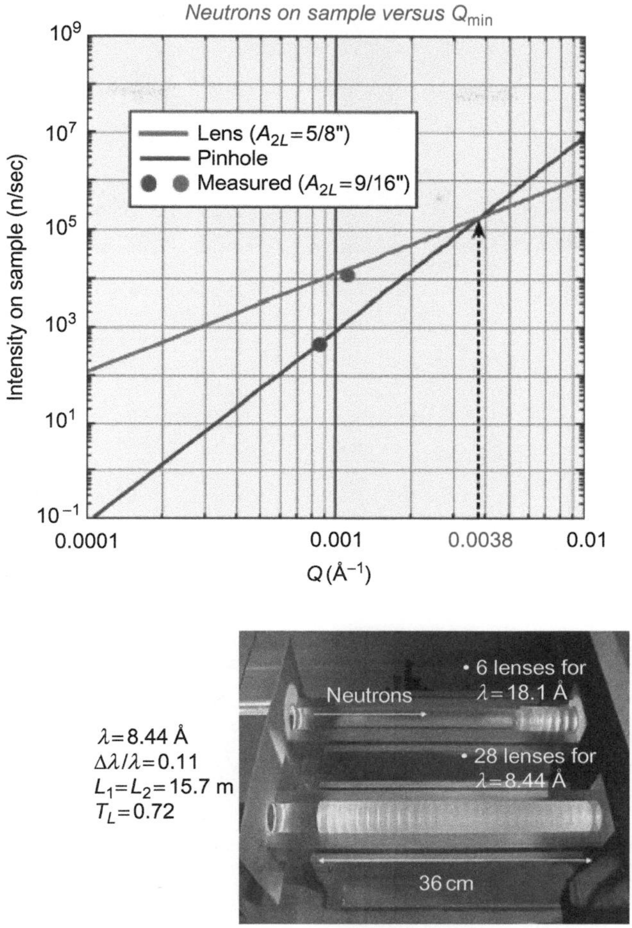

FIGURE 22 (Top) Plot of focused intensity versus the scattered neutron vector Q(Å^{-1}) comparing the pinhole and CRL performance. For $Q \leq 0.0038$, the lenses provided more intensity than the pinhole collimation; at $Q = 0.001$, the CRL provides 10-fold more intensity than the pinhole collimation (Choi *et al.*, 2000). (See Color Insert.)

D22 beam line is a small-angle scatter beam line that uses a mechanical velocity selector with a 10% bandwidth, which can select neutrons of wavelengths ranging from 4.5 Å to 40 Å. The quasi-monochromatic beam illuminate an object place in front (upstream) of the CRL. The D22 imaging detector was a position-sensitive multidetector composed of an array 128 × 128 pixels, each 7.5 mm × 7.5 mm in area. The pixels were arranged in a square matrix covering an area of 960 mm × 960 mm, which can be positioned at distance ranging from 1.3 to 18 m downstream of the CRL.

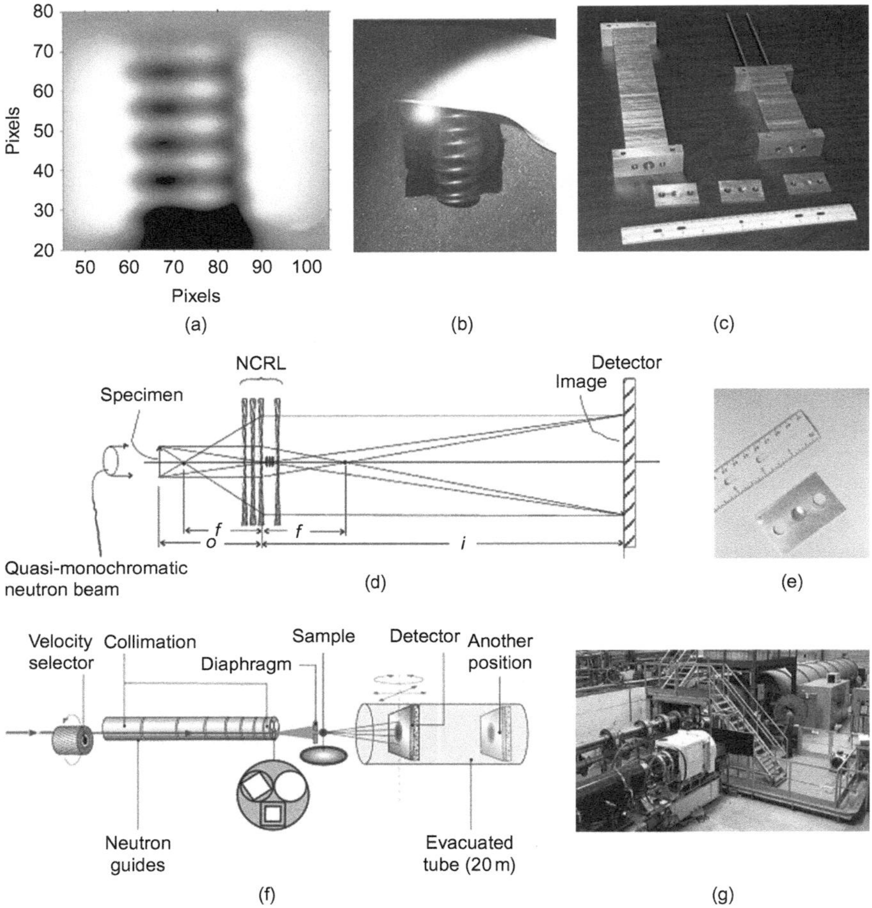

FIGURE 23 (a) Aluminum CRL image (×35) at 214-μm resolution of an entire spring at $\lambda = 20$ Å produced by combining 400 exposures of small partial views of spring at 500-μm intervals horizontally and vertically with 2-minute exposure times. This was the first neutron image produced by a refractive lens neutron microscope. The aluminum CRL had $N = 105$ biconcave aluminum lenses with radius of curvature $R = 12.7$ mm and focal length $f = 46$ cm at 20 Å. (b) Photo of spring. (c) Photo of aluminum CRL and lenses. (d) Neutron microscope setup at ILL beam line D22 (from Beguiristain *et al.*, 2002). (d) Lens scale. (f) Diagram and (g) photo of D22 (ILL beam line D22). (See Color Insert.)

The permanent configuration of the D22 beam line required setting the object distance to remain fixed at a slightly greater distance than the CRL focal length. An image of high magnification is obtained at the detector plane due to the long range of motion of the detector. The object r_0 and image r_i distances and the CRL focal length f for a particular neutron wavelength λ then satisfied the lensmaker's equation $f^{-1} = r_0^{-1} + r_i^{-1}$,

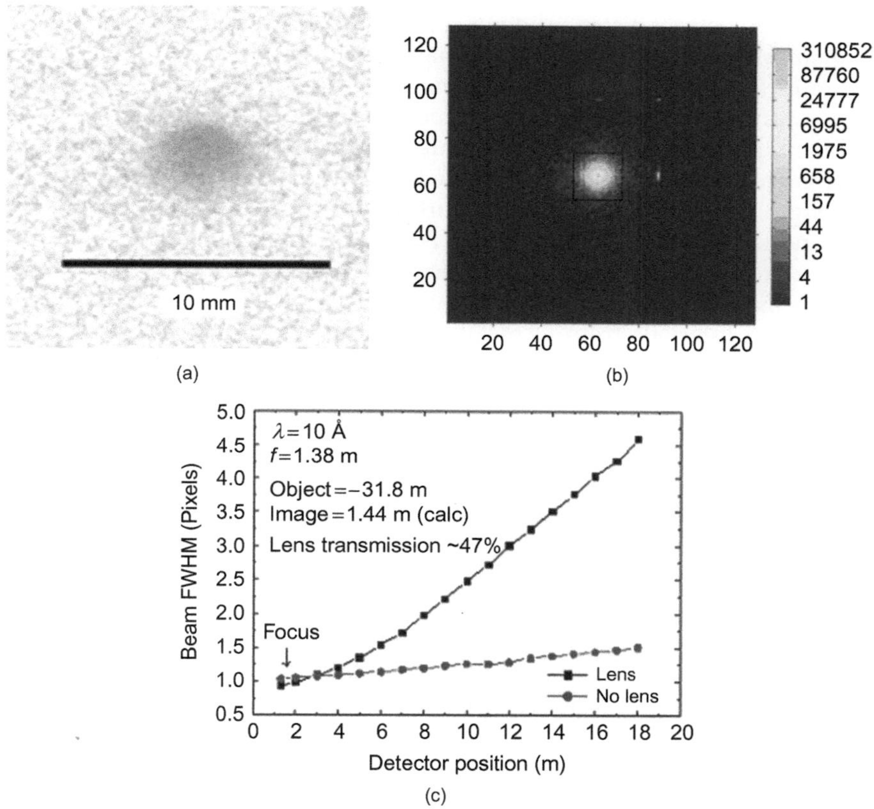

FIGURE 24 (a) and (b) Neutron demagnified image of the neutron source achieved an intensity gain $G = 10$ with the Adelphi aluminum biconcave, spherical lens CRL set with focal length $f = 1.5$ m set for the focused $\lambda = 10$-Å neutrons. The image is from a location past the focal point of the CRL and thus is defocused. (c) The actual source size is given by the black square in the neutron image. Such defocused images were used to measure focal lengths of the neutron CRL at different wavelengths. (d) Plot of the spot size measurement as a function of distance from the prototype lens showing the shape of the defocusing beam as it traverses the focal distance, the image distance, and beyond. Also shown is the natural divergence shape of the collimated neutron beam from a source located at 31.8 m in front of the lens (Beguiristain *et al.*, 2002). (See Color Insert.)

yielding a magnified image of magnification $M = r_i/r_0$. Real inverted images with magnifications of up to $M = 35$ when the object is illuminated with 20-Å neutrons in which the CRL has focal length $f = 46$ cm. The detector array allowed digitization and processing of the captured images. However, the detector pixel size limited the simple neutron microscope resolution to 214 μm and $M = 35$ magnification.

The long object distance between the neutron source and object caused the simple neutron microscope to have a small 2.5-mm FOV at $M = 35$ magnification. Within the field of view (FOV) of the microscope a one-hole cadmium mesh with 400-μm micron holes with a 2.5-mm pitch was imaged. Also imaged was one square of plastic mesh with 400-μm filaments and 2.0-mm pitch. Also imaged was a stainless steel spring 16 mm in length, 6 mm in diameter with a thread pitch of ~2 mm and spring wire thickness 0.7 mm. With the limited FOV of the microscope, the neutron image of the spring required stitching 400 two-minute exposures of the spring. Each image was obtained during a raster scan of the CRL taken at 500-μm intervals from –5 mm to +55 mm in the vertical direction to produce a 1 × 1-cm image.

The FOV can be much larger if the specimen is placed closer to the output of a neutron wave guide as was shown in the collaboration between Adelphi Technology, Inc. and the Center for Neutron Research at NIST on the small-angle scatter NG-7 beam line (see Cremer *et al.*, 2004). An aluminum CRL was used to form a simple neutron microscope. The following three conclusions were made as a result of this experiment. First, the spherical aberration of the lenses does not distort the image and resolution is not greatly reduced by spherical aberration. Second, the wide 11% bandwidth of the source and the CRL chromatic aberration does not affect the resolution as would be expected from simple analysis. And third, large FOVs are possible with CRLs. Thus the refractive neutron lenses control the propagation of neutrons just as refractive optics controls the propagation of light. This experiment demonstrated neutron lenses can be used for direct imaging of microscopic and macroscopic objects.

The top diagram of Figure 25 shows the layout of the NG-7 SANS beam line and the bottom plot shows the neutron flux intensity $\Phi(\mathrm{n/cm^2 sec})$ at the sample as a function of wavelength λ. NG-7 SANS instrument at NIST has a position-sensitive multi-element ^{3}He detector composed of an array of 128 × 128 pixels, each 5 mm × 5 mm in area. The pixels are arranged in a square matrix covering an area 640 × 640 mm, which could be positioned from 1 meter to 17 m downstream of the CRL. The end of the NG-7 guide tube acted as the neutron source and determined the size and distribution of the emitted neutrons. The guide tube conducted neutrons from the NIST reactor after they were monochromatized to an 11% bandwidth full width half maximum (FWHM). The rectangular guide tube is coated internally with a thin layer of nickel, which reflected down the guide the neutrons whose angle of reflection was less than the critical angle of reflection $\theta_c = \sqrt{2\delta}$ or approximately 17 mrad for nickel at 10 Å. The imaged objects were exposed to a nominal intensity of $3 \times 10^4\,\mathrm{n/cm^2/s}$ at 10 Å for 5-minute exposure times. Samples were placed inside a two window SiO_2 cell.

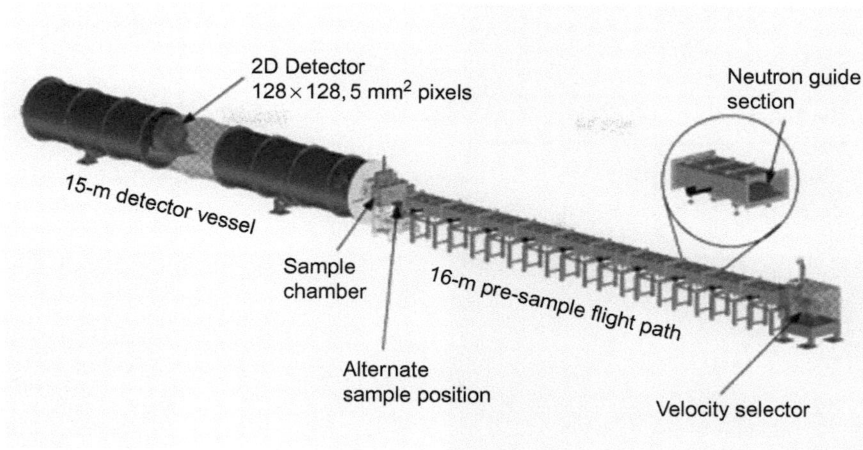

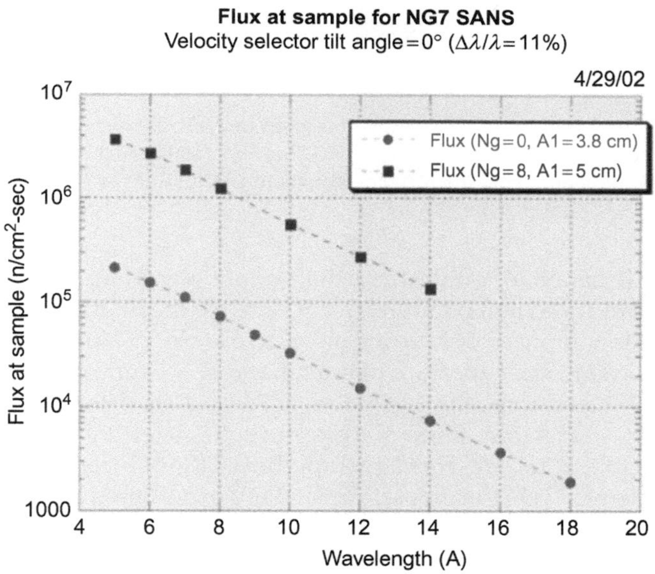

FIGURE 25 (Top) A layout of the NIST Center for Neutron Research NG-7 SANS 30-m beam line. (Bottom) A plot of the NG-7 beam line neutron flux (n/cm^2-sec) at the sample versus wavelength (Å). The upper curve is the flux at sample using the maximum number of guides (Ng = 8) with a 5.0-cm source aperture diameter. The lower curve is the flux at sample with no guides (Ng = 0) in the pre-sample collimation section with a 3.8-cm diameter source aperture. The mechanical neutron velocity selector can select a center wavelength λ from 5–20 Å with bandwidth $\Delta\lambda/\lambda = 11\%$. (From the NIST NCNR website for NG-7.)

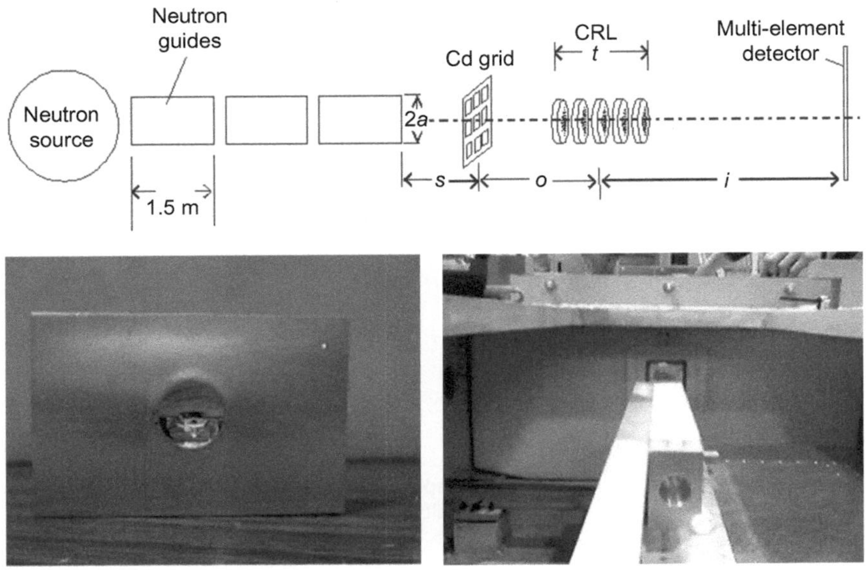

FIGURE 26 (Top) Schematic of experimental setup for neutron microscope experiments at NIST NG-7 SANS. (Bottom left) Photo of a single aluminum biconcave, spherical lens. (Bottom right) The 100-lens aluminum CRL (Cremer *et al.*, 2004). (See Color Insert.)

Using all the NG-7 guides and the sample placed at 35 cm from the guide exit produced a maximum 4.1-cm FOV for the neutron microscope. The CRL was composed of $N = 100$ biconcave lenses of aluminum with an $r_m = 12.1$ mm physical aperture diameter and $R = 5$ mm radius of curvature. For 10-Å neutrons, the 30-cm long CRL had an absorption aperture diameter $2r_a = 11.4$ mm, parabolic aperture diameter $2r_p = 2.0$ mm, and a focal length $f = 1.62$ m. This large-aperture CRL of biconcave spherical aluminum lenses was used to perform an FOV experiment and an imaging experiment in the bottom of Figure 26 shows photos of a single biconcave, aluminum spherical lens and a the CRL formed from 100 aluminum lenses.

Before the imaging experiment, an FOV experiment was performed with the large-aperture aluminum lens CRL. A diagram of the setup for the FOV experiment is shown at the top of Figure 26. Figure 27 shows the photo of the FOV test object and the neutron images for three source to object distances ($s = 35$ cm, $s = 3.0$ m, and $s = 6.0$ m) for which the calculated and measured CRL FOVs are shown in the table. For 5-minute exposure times, the FOV increased from 1 imaged square hole (8 mm × 8 mm) to 9 imaged square holes as the 9 square-hole cadmium test object was moved from a 600-cm to 35-cm source-to-object distance from the neutron guide exit.

Measured and Calculated values of the FOV

Source-to-object Distance, s	Calculated FOV	Measured FOV
0.35 m	39.3 mm	41.4 mm
3.0 m	20.8 mm	20.5 mm
6.0 m	15.1 mm	11 - 16 mm

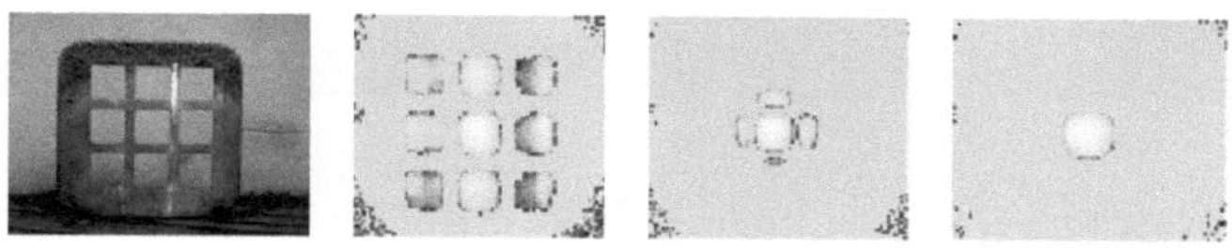

(a) (b) (c) (d)

FIGURE 27 (Top) The table shows the calculated and measured FOV for three source-to-object distances *s*. Below the table are the corresponding three neutron images. (a) The nine square-hole cadmium FOV test object and (b)-(d) the right three neutron images taken with 5 minute exposures. (b) The nine-hole neutron image was recorded at source-to-object distance $s = 35$ cm. (c) The five-hole neutron image was taken at source-to-object distance $s = 3$ m, and (d) the neutron image was taken at source-to-object $s = 6$ m (Cremer *et al.*, 2004). (See Color Insert.)

The 35-cm source-to-object distance, which gave the largest FOV, was then used in the imaging experiments in which the samples were placed 35 cm downstream of the neutron source guide exit. The placement of the CRL was set at $r_0 = 1.8$ m object distance and the CRL image distance $r_i = 16.7$ m was set for focal length $f = 1.62$ m at the neutron center wavelength setting of 10 Å within the 10% bandwidth of the chopper. This placement of samples yielded sample images with magnification $M = 9.4$. The aluminum CRL had a 44-mm FOV.

A preserved rat's hind foot was chosen as the biological object, which was magnified by $M = 9.4$ with $\lambda = 10$-Å neutrons ($\Delta\lambda/\lambda = 10\%$ bandwidth) in 5-minute exposure times (Figure 28). The 30-cm long CRL was composed of $N = 100$ biconcave lenses of aluminum with $r_m = 12.1$ mm physical aperture diameter and $R = 5$ mm radius of curvature with absorption aperture diameter $2r_a = 11.4$ mm, parabolic aperture diameter $2r_p = 2.0$ mm, and focal length $f = 1.62$ m for the 10-Å neutrons. The CRL object resolution was intrinsically 1.1 μm, but the large 5-mm pixel size of the ^{3}He detector yielded a detector-limited resolution of 0.5 mm for this simple neutron microscope. A 5-minute source image exposure was taken without the sample, rat paw, and then a 5-minute exposure with the object rat paw. The source image was subtracted from the object's image. Thus, two images each with 5-minute exposures at 10 Å were taken: images with

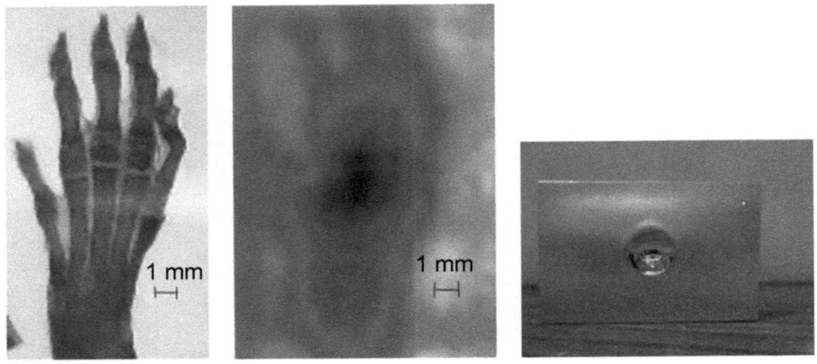

(a) (b) (c)

FIGURE 28 (a) Photo of rat paw. (b) The $M = 9.4$ magnified, 5-minute, $\lambda = 10$-Å wavelength neutron image of the rat paw. (c) One of the aluminum biconcave, spherical lenses of the $N = 100$-lens CRL. The rat paw photo in (a) and its neutron image (b) appeared on the front cover of the July 19, 2004, issue of *Applied Physics Letters* (Cremer *et al.*, 2004). (See Color Insert.)

and without the specimen (foot or grid). Given the source image intensity I_0 without the sample, and the image intensity I with the source and rat paw, then the processed image intensity of the rat paw ΔI is given by

$$\Delta I = I - KI_0. \tag{97}$$

Here the $K = 0.5$ was the arbitrary constant that provided the best image.

Another image of the rat paw was obtained with an 85-lens aluminum (Al) CRL; the magnification for the rat paw image was $M = 9.4$ using $\lambda =$ 5.0-Å wavelength neutrons (with $\Delta\lambda/\lambda = 10\%$ bandwidth) (Figure 29). The false grey linear scale image of the rat paw was taken with a 5-minute exposure from which was subtracted an 80% background-only image, also taken at a 5-minute exposure. A 1.3-cm scale set was imposed in the specimen (or object) plane. The rat paw was 35 cm downstream of the source. The rat paw was at 1.6-m object distance from the Al CRL center. The Al CRL center is 15 m upstream of the detector. The individual spherical biconcave lenses of the Al CRL had a radius of curvature $R = 1.98$ mm, a physical aperture diameter $2r_e = 3.9$ mm, a minimum wall thickness $d = 100\,\mu$m, and an absorption aperture diameter of $2r_a = 4.5$ mm with an FOV = 2.3 cm. This smaller, 3.9-mm physical aperture diameter aluminum CRL had less spherical and chromatic aberration than the larger 12.1-mm physical aperture diameter aluminum CRL.

Again in collaboration with the Center for Neutron Research at NIST, Adelphi Technology performed imaging experiments with hydrogen-rich polymers and biological specimens on the NG-7 SANS beam line (see Cremer *et al.*, 2005). For hydrogen-rich objects 98% to 99% of the neutron attenuation is by incoherent scatter and the remaining 1% to 2% is due to neutron absorption by hydrogen atoms. Hydrogen acts as a contrast agent

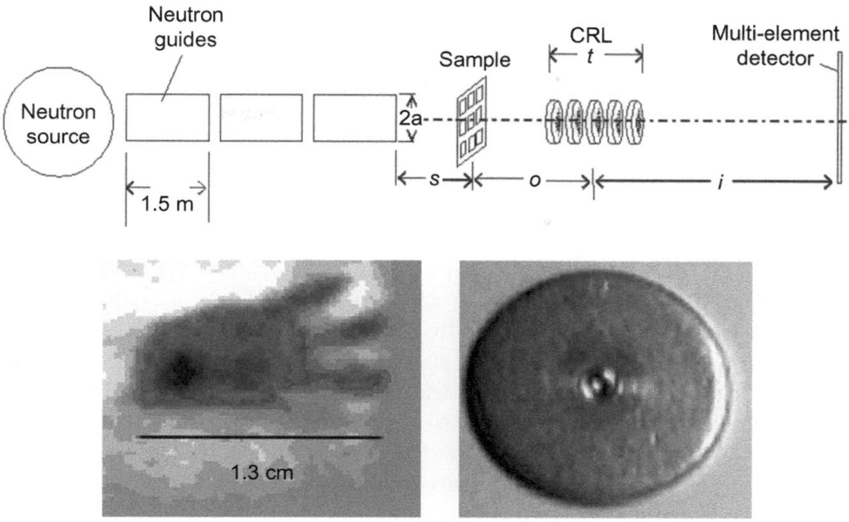

FIGURE 29 (Top) Imaging experiment carried out at the NG-7 beam line. Neutron guides are removed to change the distance to guide system output, which acts as an effective source. The source-to-object distance $s = 35$ cm provided a 4.1 cm FOV object distance $o \equiv r_0 = 1.8$ m and image distance $i \equiv r_i = 16.7$ m provided a magnification $M = 9.4$. (Bottom left) Neutron image of a rat paw. The 85-lens aluminum CRL imaged the rat paw with a ×9.4 magnification using 5.0-Å neutrons (10% bandwidth). The false grey linear scale image of the rat paw is taken with a 5-minute exposure from which is subtracted an 80% background-only image, also taken at a 5-minute exposure. (Right) Photo of a single biconcave, spherical aluminum lens of the AL CRL (Cremer *et al.*, 2004).

in hydrogen-containing objects. This experiment pursued the selection of the best lens materials for the CRL. Desirable materials for a CRL are those with (1) a large coherent scatter cross section to achieve large decrements in index of refraction and (2) small absorption and incoherent cross sections to minimize attenuation of the beam through the CRL.

Ideally, the material should be of uniform density so that all the strong coherent scattering remains in the forward "null" scattering direction. Materials of high purity (to avoid formation of second-phase precipitates) and free from voids are required. At shorter wavelengths, Bragg coherent scattering can also cause significant attenuation. Amorphous materials (e.g., fluoropolymer resin [Teflon, DuPont] in amorphous form) or single crystals (e.g., MgF_2) are those for which the resultant coherent scattering length can be large, the incoherent scattering and absorption cross sections are minimal, and the Bragg scattering can be eliminated. The motion of atoms as the neutron is transmitted through the material also causes coherent phonon scattering. Materials with high atomic bond energies or those cooled cryogenically can reduce the amount of attenuation by phonons.

This experiment was performed to further develop the possibility of neutron microscopes using CRLs, which may provide new imaging procedures that outperform conventional methods. Since biological material is composed of hydrocarbons, radiographs have extensive blurring and very poor resolution due to scattered neutrons reaching the detector. A neutron microscope, using a CRL with its tight collimation, images the paraxial trajectory neutrons that are transmitted through the sample. Unlike the CRL materials, objects to be imaged are ideally composed of materials with large or a large variation of neutron absorption and/or incoherent scattering cross sections. The CRL can provide higher-resolution images than radiographs of materials with large neutron absorption and/or incoherent scattering cross sections (inappropriate material for CRL lenses), such as those found in hydrogen-rich biological specimens, petroleum products, and fluid flow processes. The small acceptance angle of the CRL and distance of the CRL from the object allows the CRL to receive and transmit the object-transmitted neutrons rather than the object-scattered neutrons.

A MgF_2 CRL was composed of $N = 100$ MgF_2 biconcave, spherical lenses with $R = 8.0$-mm radius of curvature, 8.0-mm physical aperture diameter, 1.0-mm center (minimum) thickness, and 3.14-mm edge (maximum) thickness (Figure 30). Each MgF_2 CRL was fabricated from a single crystal by ISP Optics (Irvington, New York; http://www.ispoptics.com). The MgF_2 lenses were fabricated so the crystal axis was set parallel the CRL optical axis to avoid diffraction of the incident neutrons. The 31.5-cm long MgF_2 lens produced $M = 22.5$ magnified images of hydrogen-rich grids polyethylene and polypropylene, and biological specimens at 8.5 Å in 5-minute exposure times.

At 8.5 Å the MgF_2 readily transmits the neutrons in which the 10.2-mm absorption aperture diameter exceeds the 8.0-mm physical aperture

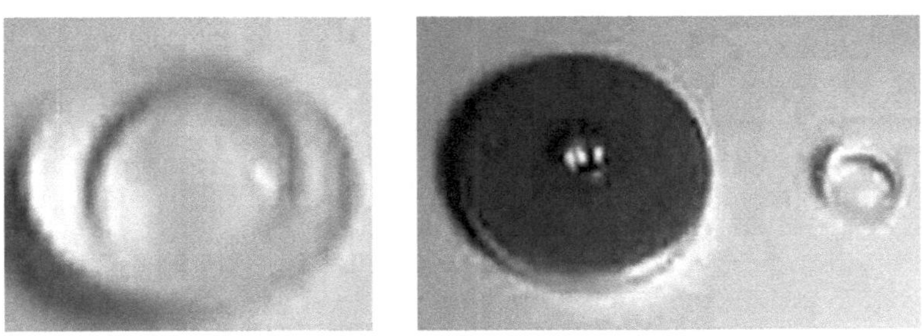

FIGURE 30 (Left) Photograph of biconcave, spherical MgF_2 lens with an 8.0-mm aperture diameter, 3.14-mm maximum thickness, 1.0-mm minimum thickness, and 8.0-mm radius of curvature. (Right) The MgF_2 lens is set next to the aluminum lens. The aluminum lens has a 3.9-mm aperture, 1.9-mm radius of curvature 2.54-cm diameter, and 0.1-mm minimum thickness (0.1 mm) and maximum 3.0 mm maximum thickness (Cremer *et al.*, 2005).

TABLE 1 Dimensions and material of the 3 sample plastic meshes

Mesh No.	Material	Hole dimension in mesh	Filament width	Filament thickness	Transmission fraction
#1	Polypropylene	0.76 mm × 0.94 mm	0.63 mm	0.91 mm	0.44
#2	Polyethylene	1.52 mm × 1.52 mm	0.51 mm	1.14 mm	0.36
#3	Polypropylene	0.76 mm × 0.94 mm	0.36 mm	0.36 mm	0.72

diameter. However, at 8.5 Å the 100-spherical biconcave lens MgF_2CRL has spherical aberration because of the small 1.2-mm parabolic aperture diameter. The 69.2-cm thin lens focal length for the MgF_2 CRL at 8.5 Å corresponds to a 74.7-cm thick lens focal length. The thick lens focal length formula is used to calculate the focal length of a long neutron or X-ray CRL, whose length is smaller but significant compared to an ideal thin lens focal length (see Pantell *et al.*, 2003).

Three plastic meshes of fine and coarse polyethylene and polypropylene grids, as well as biological specimens (preserved scorpion and leaf), were used as the objects in an imaging experiment with cold neutrons. The plastic meshes should not exhibit blurring from small-angle scatter or double Bragg scattering. Sample objects imaged by the 100-lens MgF_2 at $M = 22.5$ magnification at $\lambda = 8.5$ Å included polypropylene and polyethylene mesh ranging from filament width and thickness 0.36 mm ×0.36 mm and square-hole dimension 0.76 mm ×0.94 mm to filament width and thickness 0.63 × 0.91 and square hole dimension 0.76 mm × 0.94 mm (see Table 1 and Figure 31). The imaged objects were exposed for 5-minutes to a $\Phi = 3 \times 10^4$ n/cm^2/s nominal flux of $\lambda = 8.5$-Å neutrons.

Using the same experimental arrangement as for the plastic grids, a preserved scorpion and leaf at $\lambda = 8.5$ Å were imaged at a 5-minute exposure time. Figure 32a shows a photo of the scorpion and a leaf with a superimposed 10-mm scale. Image (b) shows the neutron image of the scorpion and leaf with a CRL 24-mm FOV. The MgF_2 CRL neutron image of the scorpion and leaf is approximately 200 mm × 200 mm at the detector, but the 4.4-mm scale corresponds to their actual size in the object plane. The dark areas in the leaf and scorpion are presumed to be water in this linear scale, false-color neutron image. Note the leaf dominates the image and hides the left claw of the scorpion.

Figure 32c shows a stitched neutron image of the scorpion in a different measurement with the leaf removed, the scorpion reoriented, and the neutron wavelength changed to 10 Å. The limited 24-mm CRL FOV required stitching two linear-scale neutron images of the scorpion as

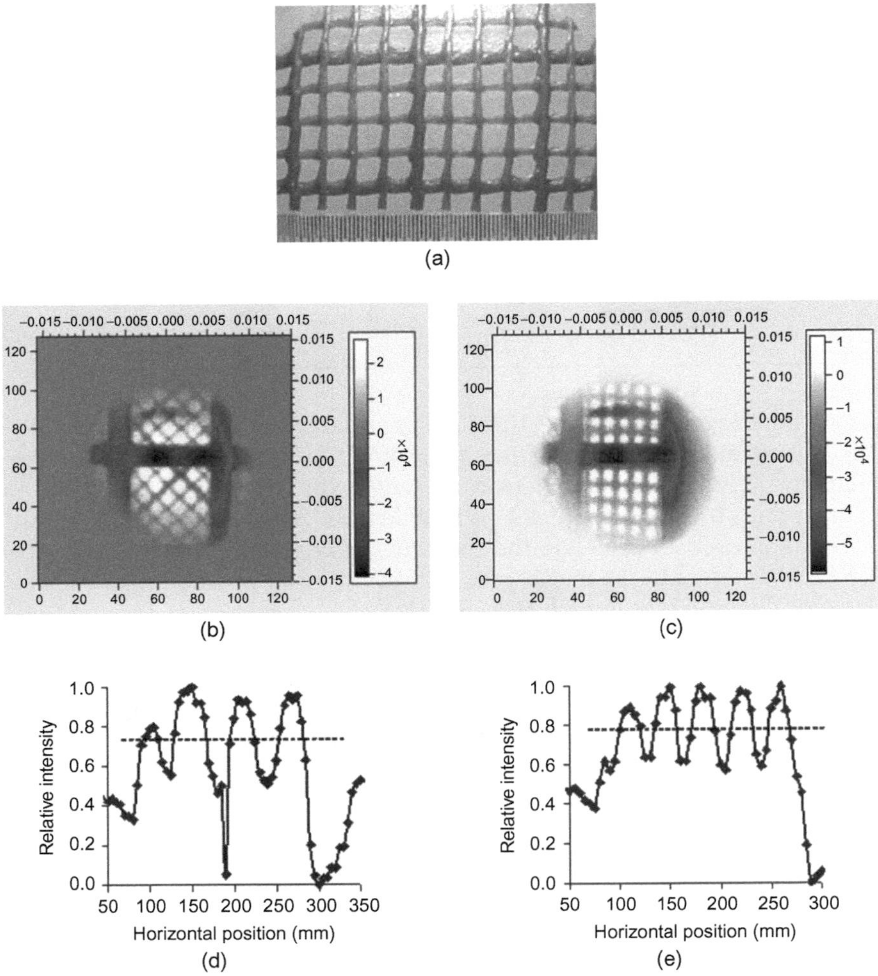

FIGURE 31 Results of neutron imaging with the 100-lens MgF_2 CRL neutron images at 8.5 Å and 5-minute exposure time with magnification $M = 22.5$. (a) Coarse green polyethylene diamond mesh grid (mesh #2 in Table 1) whose neutron image with 60% background subtraction is shown in (b) and intensity profile in (c). (b) A neutron image with 80% background subtraction of a coarse white polypropylene square grid (mesh #1 in Table 1) is shown with (e) its intensity profile (Cremer *et al.*, 2005). (See Color Insert.)

shown in Figure 32a and is displayed with a superimposed 4.4-mm scale. The scorpion neutron image shows a variation in density, indicating some areas of the scorpion contain more hydrogen than others. The lengths of the object specimens in Figures 31 and 32 are found by dividing their corresponding neutron detector image dimensions (200 mm $\times$ 200 mm) by the $M = 22.5$ magnification.

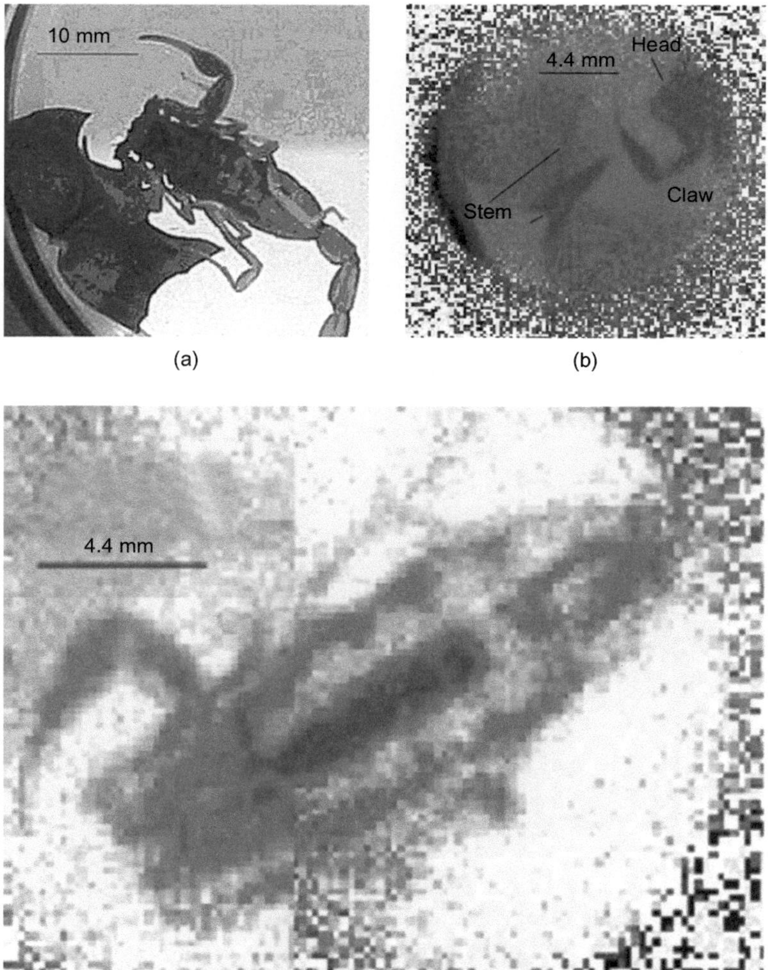

FIGURE 32 (a) A photo of a scorpion and leaf. (b) The 5-minute neutron exposure at $\lambda = 8.5$ Å with the MgF_2 CRL at magnification $M = 22.5$. (c) Two stitched, 5-minute, $\lambda = 8.5$-Å neutron images of the scorpion at $M = 22.5$ (Cremer *et al.*, 2005). (See Color Insert.)

The imaging of hydrogen-rich materials with a neutron microscope was continued by Adelphi Technology in collaboration with the University of California at Davis McClellan Nuclear Radiation Center (MNRC) (see Cremer *et al.*, 2007a). Thermal neutron imaging experiments were conducted in Bay 4 of the 2 MW MNRC nuclear reactor facility. A 40° C Maxwell–Boltzmann distributed source of thermal neutrons with 1.1-Å peak wavelength and 74% bandwidth FWHM was assumed for the MNRC

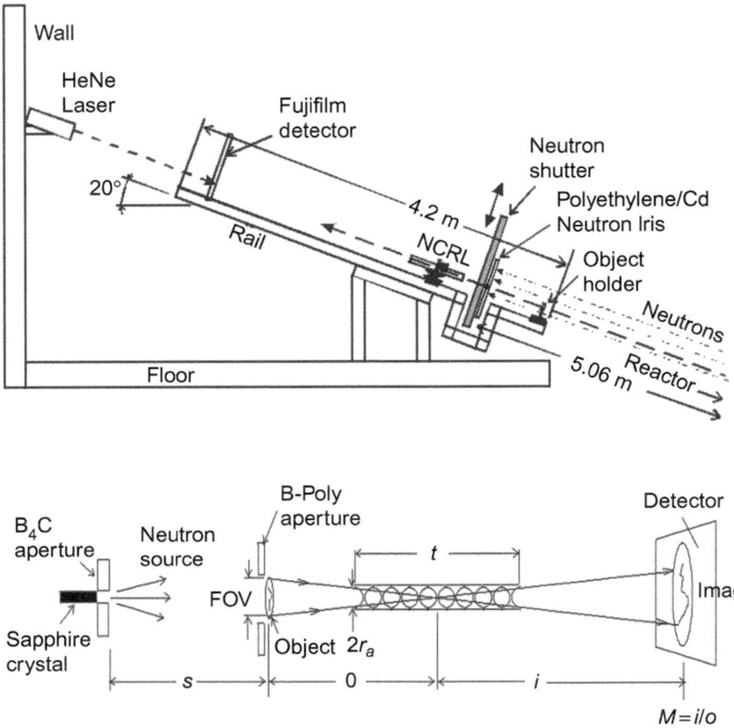

FIGURE 33 (Top) The physical setup of the thermal neutron microscope in Bay 4 of MNRC. (Bottom) Schematic of thermal neutron microscope with source *s*, object *o*, and image *i* distances. See text for details (Cremer *et al.*, 2007a).

Bay 4 beam line. Bay 4 provided about a 6.5-meter length from the massive concrete sliding door admitting the neutrons to the most downstream placement of the imaging plate detector at the beam stop wall.

The thermal neutrons emerge from the reactor water core, entering the Bay 4 beam tube through a graphite (moderator) end plug, and pass through a 28-cm long sapphire crystal, which improves the thermal content by reducing the relative number of higher-energy neutrons (Figure 33). The neutrons emerge from a 3.2-cm square source aperture in a boron carbide plate and travel a distance $s = 4.0\,\text{m}$ through a collimation tube to the sample position. The sample is mounted in the aperture of a 2.54-cm thick plate of enriched lithium polyethylene. The 1.25-cm aperture is matched to the CRL FOV. This aperture reduces unwanted neutrons peripheral to the CRL FOV, which can scatter and decrease the image contrast and resolution. At the sample is a collimated beam (6.3-mrad divergence angle) with thermal neutron flux $1.7 \times 10^6\ \text{n/cm}^2/\text{s}$.

Depending on the desired wavelength, the object distance r_0 from the sample to the CRL center plane (thin lens principal plane) is 1–2 m, and the image distance r_i of the CRL center plane to the imaging detector is 1–4 m. The thin lens focal length f_0 for the spherical lens CRL was determined by the biconcave spherical lens radius of curvature R, number of lenses N, and lens material decrement $\delta(\lambda)$ corresponding to neutron wavelength λ. One approximates the thin lens, spherical-biconcave, N-lens focal length f_0 by the thin lens, parabolic-biconcave, N-lens focal length, so that

$$f_0 = \frac{R}{2N\delta}. \tag{98}$$

The thick lens focal length f is obtained from the thin lens focal length f_0 for a given neutron wavelength λ and CRL length L, where

$$f = f_0 \frac{\sqrt{\frac{L}{f_0}}}{\sin\sqrt{\frac{L}{f_0}}}. \tag{99}$$

The CRL center plane is the single principal plane for the thin lens approximation of the thick lens CRL. However, the center plane of the actual thick lens CRL is between the thick lens left and right principal planes.

To set up the microscope, the object r_0 and image r_i distances were established using the calculated CRL's thick focal length f and the lensmaker's equation for a particular wavelength (e.g., 2.6 Å), which was expected in the 40°C Maxwell–Boltzmann MNRC thermal neutron spectrum, where the lensmaker's equation is

$$\frac{1}{f} = \frac{1}{r_0} + \frac{1}{r_i}. \tag{100}$$

The FujiFilm detector (FujiFilm Medical Systems USA, Inc., Stamford, CT), the CRL, and the sample objects were positioned using the calculated values of the object r_0 and image r_i distances. Images were obtained using a FujiFilm neutron imaging plate, which is an integrating detector with high detective quantum efficiency, 50-μm pixels, and 5 orders of dynamic range (FujiFilm, 2012). The FujiFilm plates not only have high resolution but also can range from small to large areas. The thermoluminescent FujiFilm imaging plate in the experiment was approximately 24 cm × 30 cm in area. The sensitivity and contrast of the time-integrated FujiFilm plate was better than the real-time stacked images produced by a thermal neutron charge-coupled device camera–based imaging system, which was also used.

The two CRLs used in the simple thermal neutron microscope were biconcave, spherical CRLs of beryllium and copper. The two CRLs consist of a stack of biconcave, spherical lenses that are fabricated with inexpensive compression-molding techniques. The copper CRL has $N = 120$

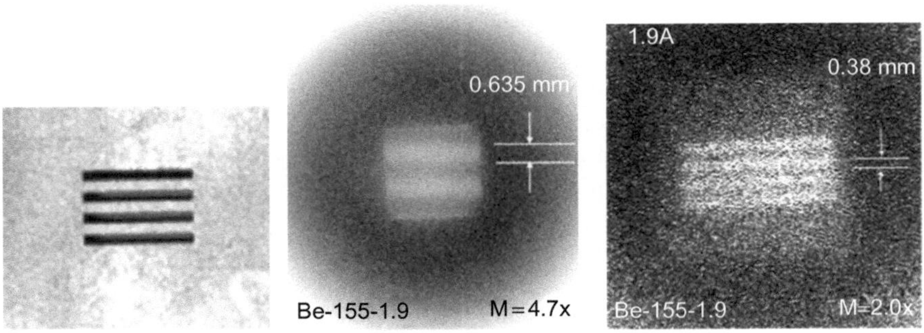

FIGURE 34 (a) Photo and (b) neutron images of 0.635-mm wide cadmium slits (1.3-mm period) at $M = 4.7$ and $\lambda = 2.6$ Å for 1-hour exposure, taken with the beryllium CRL. Also taken with the beryllium CRL is a neutron image (c) of 0.4-mm wide cadmium slits (0.8-mm period) at $M = 2.0$ and wavelength $\lambda = 1.9$ Å for 1-hour exposure (Cremer *et al.*, 2007a).

lenses, each with radius of curvature $R = 2.0$ mm, physical aperture diameter $2r_m = 2.3$ mm, and center thickness $d = 76$ μm. The beryllium CRL has $N = 155$ lenses, each with radius of curvature $R = 1.9$ mm, physical aperture diameter $2r_m = 2.3$ mm, and center thickness $d = 40$ μm. The unit lenses were stacked and supported inside a cylinder that was supported and aligned using conventional laser/optical mounts.

Thermal neutron imaging results were obtained with the beryllium and copper CRLs. Figure 34 shows a neutron image of 0.64-mm wide cadmium slits at ×4.7 magnification and 1-hour exposure time produced by the beryllium CRL, with focal length $f = 60$ cm at $\lambda = 2.6$ Å. The modulation transfer function (MTF) for the beryllium and copper CRLs was measured using periodic arrays in neutron-absorbing cadmium. The 5 periodic cadmium slit arrays had spatial periods of 0.5, 0.8, 1.3, 2.5, and 5.1 mm in which the slit widths were equal to the spacing between the slits. The cadmium slit array acted as a periodic square wave structure with an object contrast $C_0(s) = 1$. Figure 34 shows a photo (image a) and neutron image (image b) of 0.635-mm wide cadmium slits (1.3-mm period) at $M = 4.7$ and $\lambda = 2.6$ Å for 1-hour exposure, taken with the beryllium CRL. Figure 34c is a neutron image of the 0.4-mm wide cadmium slits (0.8-mm period) at $M = 2.0$ and $\lambda = 1.9$ Å for 1 hour exposure taken with the beryllium CRL.

The measured MTF (s) for five spatial periods $1/s$ was obtained from the measured image contrast $C_i(s)$. Intensity cuts perpendicular to the slit pattern were done with line-scan software. The calculated and measured MTF plots for the beryllium and copper CRL are shown in Figure 35. Figure 35a shows calculated MTF curves and measured MTF points for the copper CRL at 2.6 Å and Figure 35b the beryllium CRL at 1.9 Å, each at

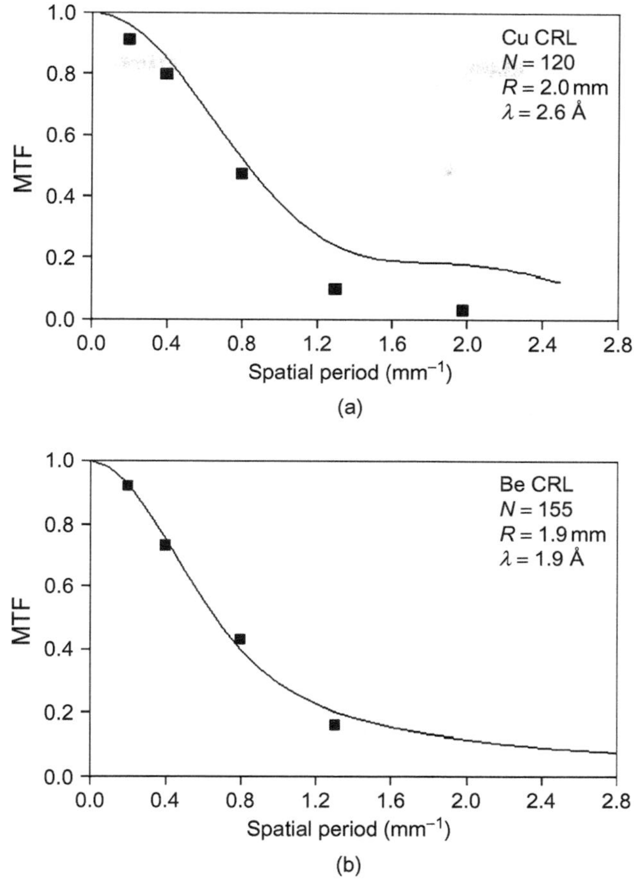

FIGURE 35 (a) Plots of the calculated MTF of beryllium CRL for 1.9 Å and four measured MTF values. (b) Plots are calculated for MTF of copper CRL for 2.6 Å and five measured MTF values (Cremer *et al.*, 2007a).

$M = 2$ and 1-hour exposure with a 6.1×10^9 n/cm^2 neutron fluence. Given MTF $= 0.05$ s, the calculated and measured resolutions for the beryllium CRL at $\lambda = 1.9$ Å are both 0.28 mm. For the copper CRL at $\lambda = 2.6$ Å the calculated resolution is 0.31 mm and the measured resolution is 0.56 mm.

Spherical biconcave lens CRLs have slightly better resolution than parabolic lens CRLs for wide-bandwidth neutron sources, such as thermal Maxwell–Boltzmann-distributed neutron sources. Also, the geometric optics–derived object resolution σ_{tot} yields larger values for MTF-measured and calculated object resolutions. Kirchhoff analysis shows that chromatic aberration is ameliorated by spherical aberration and absorption, resulting in better resolution than expected by σ_{tot} because the interactions among absorption, spherical, and chromatic aberration are not included in the geometric optics resolution result σ_{tot}.

The geometric optics–derived object resolution $\sigma_{tot} = 1.1$ mm yielded a larger value for both the copper and beryllium CRLs than the above MTF-measured and calculated object resolutions. Kirchhoff analysis showed that chromatic aberration is ameliorated by spherical aberration and absorption, resulting in better resolution than expected by σ_{tot} because the interactions among absorption, spherical, and chromatic aberration are not included in the geometric optics resolution result σ_{tot}. With increasing radial distance r from the CRL center axis, the CRL thickness $t_s(r)$ increases, and the transmission fraction $T(r)$ decreases for longer wavelength incident rays because $\mu(\lambda) \propto \lambda$, where

$$T(r) \propto \exp(-t_s(r)N\mu(\lambda)). \tag{101}$$

The longer-wavelength neutrons focus closer to the CRL than those of shorter wavelengths, given the CRL focal length f_{sp} is inversely proportional to the squared wavelength λ^2, where

$$f_{sp} \propto \frac{1}{\lambda^2}. \tag{102}$$

However, with increasing radial distance r from the CRL center axis, the focal lengths f_{sp} of the shorter-wavelength rays are reduced by spherical aberration, where

$$f_{sp} \propto \sqrt{R^2 - r^2} = R\sqrt{1 - (r/R)^2}. \tag{103}$$

Decrease of the CRL focal length deviation about the CRL mean focal length yields a finer CRL resolution.

The copper and beryllium CRLs were compared with their equivalent pinhole-aperture cameras by using cadmium irises with diameters equal to the CRL's physical aperture ($2r_m = 2.3$ mm). A one-aperture or simple pinhole camera was fabricated by the placement of a single cadmium iris at the center of the empty CRL lens holder. The two-aperture case or cylinder formed a "thick iris" that was the same length as the matching CRL and was formed from two cadmium irises mounted on each end of an empty cylindrical CRL. The hole radius of the cadmium irises was set equal to the CRL effective aperture radius r_e for the chosen neutron wavelength λ and the corresponding thick lens CRL focal length f.

The chosen neutron wavelength λ and corresponding thick lens focal length f and selection of the desired magnification M then determined the

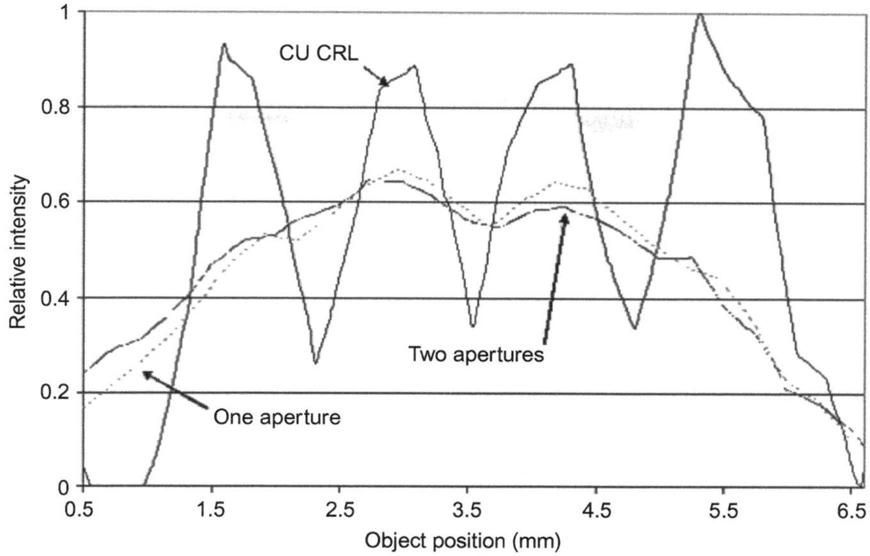

FIGURE 36 For $\lambda = 2.6$ Å and 1-hour exposure times, one-pinhole camera, two-pinhole camera (cylinder), and copper CRL are compared, each with 2.3-mm aperture. The plot shows normalized intensity versus cut width across the slit array image (Cremer *et al.*, 2007a). (See Color Insert.)

object r_0 and image r_i distances. Images from the CRL microscope and the equal-aperture–sized simple pinhole camera and two-pinhole (cylinder) camera were obtained using identical exposure times and identical image and identical object distances. The copper and beryllium CRLs were compared with their equivalent pinhole aperture cameras by use of cadmium irises with diameters equal to the CRL's physical aperture ($2r_m = 2.3$ mm). A one-aperture or simple pinhole camera was constructed by the placement of a single cadmium iris at the center of the empty CRL lens holder. In addition, a two-aperture camera was constructed by two cadmium irises mounted on each end of an empty cylindrical CRL lens holder of the same length as the copper or beryllium CRL. This two-aperture case or cylinder formed a "thick iris" that was the same length as the matching CRL. Images from the CRL microscope and the equal-aperture–size simple pinhole camera and two-pinhole (cylinder) camera were obtained using identical 1-hour exposure times and image and object distances at $\lambda = 2.6$ Å. Figure 36 compares the measured intensity profiles from the copper CRL and its equivalent one- and two-pinhole cameras and shows that copper CRL image contrast and resolution are superior.

The imaged cadmium slits demonstrated the copper and beryllium CRLs were acting as refractors of neutrons and not simply as pinhole neutron cameras. For thermal wavelengths the CRL apertures are small, such that their absorption-limited aperture diameters $2r_a < 2r_m$ approach the expected CRL object resolution σ_{tot}. At $\lambda = 2.6$ Å and $M = 2$, the copper and beryllium CRLs have effective aperture diameter $2r_e = 2r_a = 1.0\,\text{mm}$ and resolution $\sigma_{tot} = 1.1\,\text{mm}$. A pinhole camera with the same aperture diameter $2r_e = 1.0\,\text{mm}$ has calculated object resolution $\sigma_{pin} = 1.5\,\text{mm}$ for $M = 2$, where

$$\sigma_{pin} = \frac{2r_e\,(M+1)}{M} \tag{104}$$

The above resolutions of 1.1 mm to 1.5 mm are coarser than the range of MTF-calculated and measured-resolution values of 0.28 mm to 0.56 mm.

A CRL used with a pulsed neutron source and a synchronized time-of-flight (TOF) imaging detector can improve resolution by causing the chromatic aberration to be dependent on the frame acquisition time of the detector rather than the CRL. In a neutron pulse the neutrons arrive with steadily decreasing velocities and increasing wavelengths during the frame acquisition time. A short acquisition time Δt can result in a sufficiently small bandwidth $\Delta\lambda/\lambda = 0.025$, where $\Delta\lambda \propto \Delta t$—for example, the beryllium spherical lens CRL resolution could increase by 23-fold to $\sigma_{tot} = 47\,\mu\text{m}$. Spherical aberration is removed by a parabolic lens CRL.

Adelphi Technology performed imaging experiments with the CRLs using neutron contrast agents of gadolinium-based compounds in Bay 4 at MNRC (unpublished results, Cremer *et al.*, 2007b). The top of Figure 37 shows copper CRL neutron images of boron nitride that was pressed between polyethylene sheets with $M = 1.2$ at $\lambda = 30$ Å for a 30-minute exposure. The bottom schematic and neutron image in Figure 37 show a copper CRL neutron image with 30-minute exposure and $M = 2$ at wavelength $\lambda = 2.6$ Å of gadolinium-based solution, gadoteridol (ProHance, a magnetic resonance imaging contrast agent) that is placed in holes of varying diameter drilled into an aluminum plate. Both 30-minute images with the copper CRL delivered fluence $3 \times 10^9\,\text{n/cm}^2$ to the samples. Gadolinium-based solutions are also commonly used as a contrast agent in thermal neutron radiography.

The top of Figure 38 shows a schematic and a beryllium CRL image of boron nitride that was placed between polyethylene sheets and imaged at $\lambda = 3$ Å for a 1.5-hour exposure with $M = 1.2$. The bottom photo shows the green polyethylene mesh #2 (see Table 1) and a beryllium neutron image taken in 1.5 hours at $M = 1.2$ at $\lambda = 2.1$ Å. Both beryllium CRL images of 1.5-hour exposures delivered fluence $10^{10}\,\text{n/cm}^2$ to the samples. Finally, in the leftmost image in Figure 39 is a photo of a rat paw. The middle image is a neutron image obtained with the beryllium CRL with $M = 2.1$ at $\lambda = 1.8$ Å and 30-minute exposure. The rat paw was then

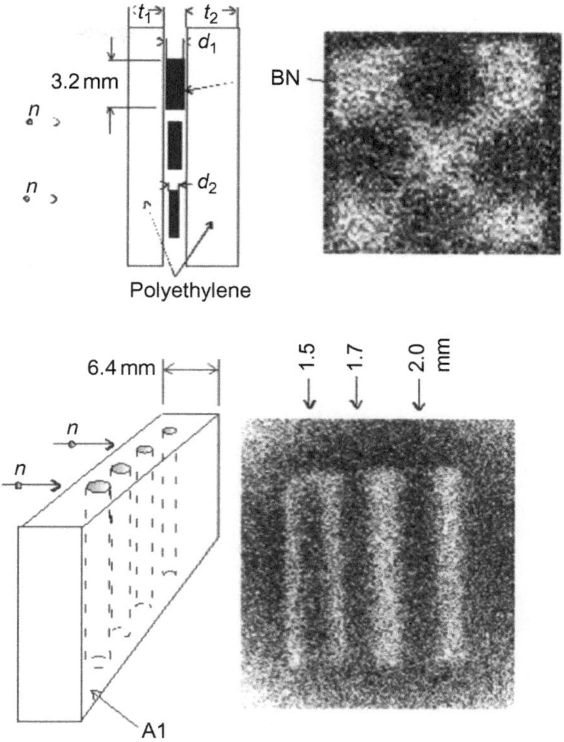

FIGURE 37 (Top) Copper CRL magnified $M = 1.2$ image of boron nitride (BN) inside polyethylene with $3 \times 10^9 n/cm^2$ at 3 Å which required 30-minute exposure. (Bottom) Copper CRL magnified $M = 2$ image at 2.6 Å of a gadolinium-based solution (ProHance) in holes drilled in the aluminum plate. This image required an accumulated fluence of 3×10^9 n/cm^2, which required a 30-minute exposure (Cremer *et al.*, 2007b).

injected with ProHance to act as a thermal neutron contrast agent and then imaged again for 30 minutes with the beryllium CRL (right image).

Figure 40 shows photos of the MNRC Bay 4 setup for the neutron microscopy experiments. The photos show the beryllium and copper CRLs mounted in regular Newport Corporation (Irvine, California) gimbal optical mounts, which are set on an optical aluminum rail from Thorlabs (Thornton, New Jersey). The object specimens are mounted on a sliding stage, which allows the object distance to be adjusted. The object distance is determined by the particular focal length and image distance corresponding to a desired wavelength and magnification. The imaging magnification and wavelength is limited by the 7.0 m available in Bay 4 from the white, enriched-lithium polyethylene plate with 1.25-cm source aperture, which is adjacent to the massive concrete sliding door that

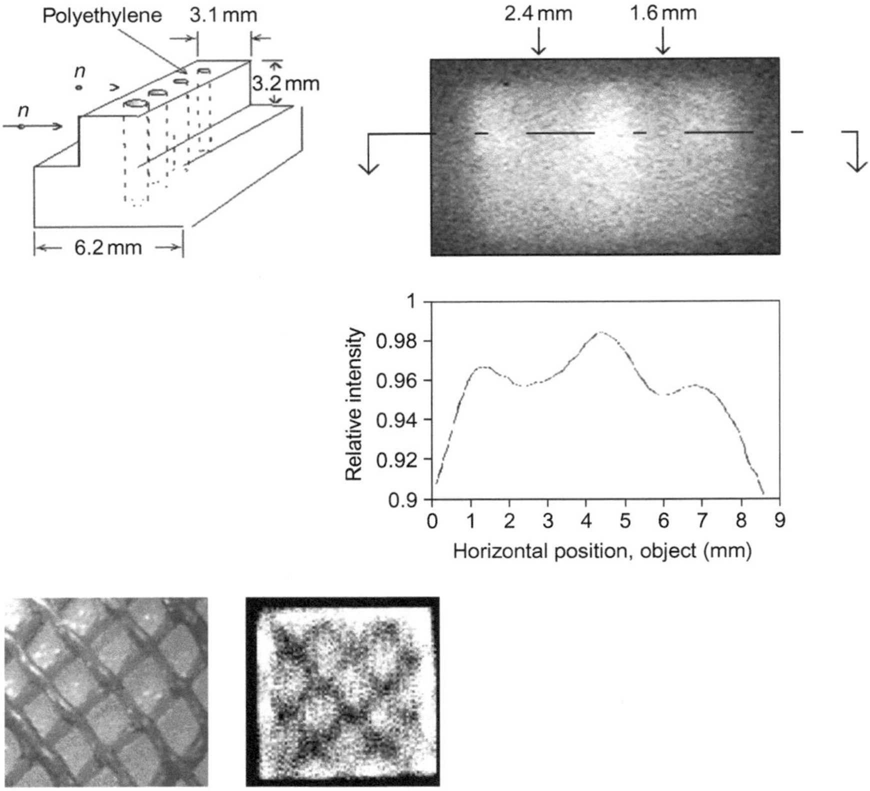

FIGURE 38 (Top) Beryllium CRL magnified $M = 1.2$ image of boron nitride inside polyethylene with 10^{10} n/cm^2 neutron fluence on the sample at wavelength 3 Å, which required a 1.5-hour exposure. (Bottom left) Photo of polyethylene mesh #2 (see Table 1). (Bottom right) Beryllium CRL image of mesh at $M = 1.2$ magnification at wavelength 2.1 Å, which required a 1.5-hour exposure with resulting neutron fluence 10^{10} n/cm^2 on the sample (Cremer *et al.*, 2007b).

admits neutrons into Bay 4 and the downstream beam stop with alignment laser and mount for the FujiFilm imaging detector. The sliding stages of the CRL and object specimen mount allow easy setting of the object distance, and the FujiFilm imaging plate is placed on the beam stop at maximum distance downstream of the CRL to allow maximum magnification. A photo of the rat paw is also shown in the object specimen mount. A 2.5-cm thick plate of enriched lithium polyethylene with a 1.25-cm aperture admits neutrons onto the sample and CRL and acts as the source stop aperture.

Figure 41 shows a variety of neutron lenses that Adelphi tested after the initial experiments neutron microscope experiments at ILL in

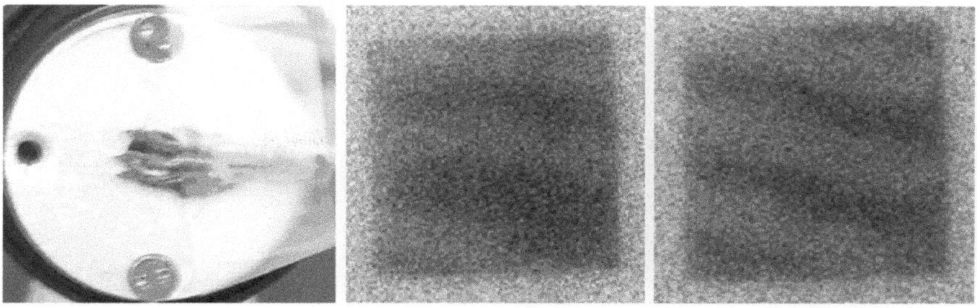

FIGURE 39 (Left) A photo of a rat paw and (middle) a neutron image obtained with the beryllium CRL with magnification $M = 2.1$ at wavelength $\lambda = 1.8$ Å for 30 minutes. The rat paw was injected with ProHance to act as a thermal neutron contrast agent and then was imaged again with the beryllium CRL with a 30-minute exposure (center and right images) (Cremer *et al.*, 2007b).

2002. The 2.5-cm in diameter lenses are simply stacked in the black polyoxymethylene (Delrin $(CH_2O)_n$, DuPont) cylindrical holder shown at the bottom of the figure. The smaller MgF_2 lenses can also be mounted in a black Delrin holder with the inner bore diameter matched to physical aperture diameter of the lens. The black Delrin holder is designed to fit into a standard gimbal optical mount to facilitate lens alignment. The 2.5-cm diameter disks of beryllium or copper, which have embossed spherical biconcave lenses at their centers, are stacked in a black Delrin cylinder. At each end of the stack of lenses is a cadmium stop aperture with a diameter that matches the physical aperture diameter of the lens. The MgF_2 biconcave, spherical lenses can be purchased commercially. The coin lenses are debossed by inexpensive compression-molding techniques, whereas the larger beryllium and Teflon lenses are precision machined and polished.

The large Teflon and beryllium lenses were not successful in imaging neutrons. However, the small radius of curvature $R = 1.9$ mm beryllium lenses, which were machined and then compression molded into smooth surface biconcave, spherical lenses were successful. The ability to compress the small beryllium lenses at relatively high mechanical pressure with an inexpensive, commercially available press appears to reduce the neutron scatter and diffraction by the metals due to compaction of the polycrystals. The Teflon lenses did not focus neutrons, apparently because of the polycrystalline nature of the Teflon. However, sufficiently amorphous Teflon is a candidate for neutron lenses. Teflon contains only fluorine and carbon molecules, which refract neutrons with minimal attenuation. The new amorphous fluoropolymer resin (Teflon AF 1600, DuPont) already has applications as a visible optics lens because of its transparency and excellent surface quality. The typical molding temperatures of 240°C to 275°C for Teflon AF 1600 allow it to be

FIGURE 40 Neutron microscope imaging experiments at MNRC Bay 4. (Top left) A photo of the beryllium CRL in a regular optical gimbal mounted on an aluminum optical rail from Thorlabs. (Top right) A photo of copper CRL in the gimbal mount. (Bottom left) The rat paw object specimen mounted. (Bottom middle) View of the CRL and mounted object specimen. (Bottom right) Beam stop, alignment laser, and mount for the FujiFilm imaging plate. The object and image distance are set for the thick lens focal length corresponding to the desired wavelength and magnification, which is limited by the 7-m linear distance from the massive sliding concrete door entrance and the downstream beam stop. At the foot of the rail is a white, 2.5-cm thick plate of an enriched lithium-polyethylene with a center 1.25-cm diameter stop aperture to limit the source neutron illumination of the object specimen, CRL, and FujiFilm imaging detector. The CRL has cadmium stop apertures at its entrance and exit matched to the CRL physical aperture diameter (Cremer *et al.*, 2007b). (See Color Insert.)

compression-molded into plano-concave parabolic neutron lenses using the existing compression molding systems. Figure 41 shows a plano-concave, parabolic, amorphous Teflon lens with on-axis radius of curvature $R = 0.186\,\text{mm}$ and $d = 125\,\mu\text{m}$ minimum wall thickness with physical aperture diameter $2r_m = 3.0\,\text{mm}$, as well as a lens holder, gimbal mount, and adjustable lens stage for a CRL of $N = 78$ amorphous Teflon lenses.

A broadband spallation neutron source (Adelphia Technology) was used to measure the characteristics of two CRLs, an aluminum CRL and a magnesium fluoride, in collaboration with R. P. Hjelm on the low-Q diffractometer (LQD) beam line at the Los Alamos Neutron Center (LANSCE; see Park *et al.*, 2006). This allowed testing of CRLs using a broadband pulsed neutron source and its associated TOF detector for the first time. Prior to this experiment, neutron CRLS had been tested only

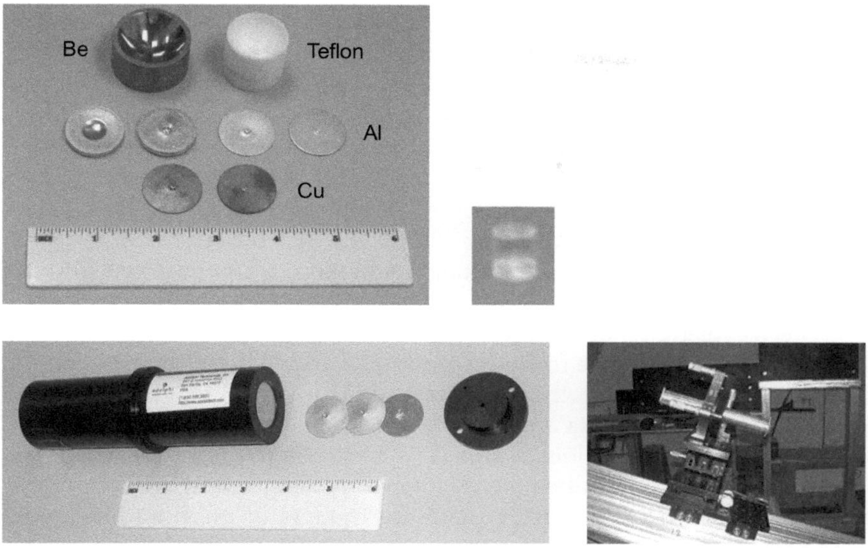

FIGURE 41 (Top) A variety of neutron lenses fabricated and tested by Adelphi Technology since 2002. Disks of 2.5-cm diameter metal, such as aluminum, copper, and beryllium, are embossed with a biconcave, spherical lens at the disk center are fabricated via inexpensive compression-molding techniques. The beryllium and Teflon lenses were precision machined and polished. (Bottom) A photo of the Delrin lens holder, which fits in a standard optical gimbal mount. Cadmium stop apertures with center aperture diameter matching the physical aperture diameter the lenses are placed at the input and output ends are. The top right photo shows a plano-concave parabolic, amorphous Teflon lens with an on-axis radius of curvature of $R = 0.186$ mm, minimum center thickness $d = 125\ \mu m$, and physical aperture diameter $2r_m = 3.0$ mm. The bottom right photo is the lens holder in its gimbal mount and adjustable lens stage with 78 amorphous Teflon lenses (Cremer *et al.*, 2007b). (See Color Insert.)

with reactor-based neutron sources with monochromatized cold neutron beams. The LANSCE source produces a pulsed "pink" (broadband-width) beam of neutrons.

The LANSCE experiment demonstrated the regions of the neutron spectrum where the neutron CRL operation was possible. The TOF detector showed the regions of the spectrum where Bragg diffraction of the microcrystal structure prevented CRL neutron transmission. This experiment showed the LQD beam line can determine the neutron wavelength range of operation of the CRL. The aluminum CRL transmission was good from 5–15 Å with maximum transmission at 8 Å. The MgF_2 CRL transmission was good from 9–15 Å with maximum transmission at 13 Å. Furthermore, the LQD beam line can guide design, fabrication, and testing efforts to select CRL materials, which minimizes microcrystal interference of neutron transmission.

Neutrons are generated by an 800-MeV proton beam by spallation. The resulting spectrum of hot neutrons is moderated to a cold spectrum by a liquid hydrogen moderator at 19°K. The pulse of neutrons emitted from the moderator has a root mean squared duration of approximately 300 μs. From the cold moderator the thermalized neutrons passed through a chopper system to provide a spectrum of wavelengths from 1.5 Å to 15.7 Å. The neutron imaging detector was a position-sensitive He-3 detector consisting of an array of 128 × 128 pixels, each 5 mm × 5 mm in area. A 37-mm diameter neutron-blocking beam stop can be inserted into the detector center.

The wavelengths of the individual neutrons are encoded by the TOF over the 12.7-m distance from the moderator to the detector in which the usable neutron wavelength band corresponds to a TOF range between 4.8 and 50 ms. The TOF data are collected as a histogram with a logarithmic binning scheme with a bin width Δt such that $\Delta t/t = 0.016$, where t is TOF. The TOF t determines the neutron wavelength and Δt determines the neutron bandwidth, which occurs for moderator pulse widths below ~18.8 ms ($\lambda \cong 5.9$ Å) and for histogram bin widths above 18.8 ms.

The aluminum CRL consisted of a stack of $N = 198$ biconcave, spherical, single-polycrystal aluminum lenses with $R = 2.0$ mm radius of curvature and physical aperture diameter $2r_m = 3.9$ mm with $d = 0.1$ mm center, minimum wall thickness. The MgF_2 CRL consisted of a stack of $N = 98$ biconcave, spherical, single-crystal MgF_2 lenses with $R = 8.0$ mm radius of curvature and physical aperture diameter $2r_m = 8.0$ mm with $d = 1.0$ mm center, minimum wall thickness. The unit lenses were stacked and supported inside a cylinder, which was supported and aligned using conventional laser/optical mounts. The single MgF_2 lenses were cut from a single crystal and were stacked with the crystal axis aligned with the center axis of the CRL to avoid diffraction of the incident neutrons.

The experimental determination of the operational wavelengths of the aluminum and MgF_2 CRLs was performed as follows (Figure 42). The CRL was placed in the beam line with its entrance aperture 8.4 m from the source in which the CRL was expected to transmit and refract neutrons around the 37-mm diameter beam stop composed of a mixture of lithium and lead. The beam stop is used to prevent the main neutron beam from striking the detector's center pixels and damaging the detector. An 8.3-mm diameter cadmium iris was located 50 cm upstream from the CRL and a 15.5-mm cadmium iris was located 420 cm upstream from the CRL. The two irises served to collimate the neutron beam to a triangular intensity cross section with a full width of 37 mm.

At various wavelengths the lensmaker's equation is satisfied and the image of the source occurs somewhere downstream of the CRL, either in front of or behind the detector. For long-wavelength neutrons the neutrons are strongly refracted and the source image will form in front of

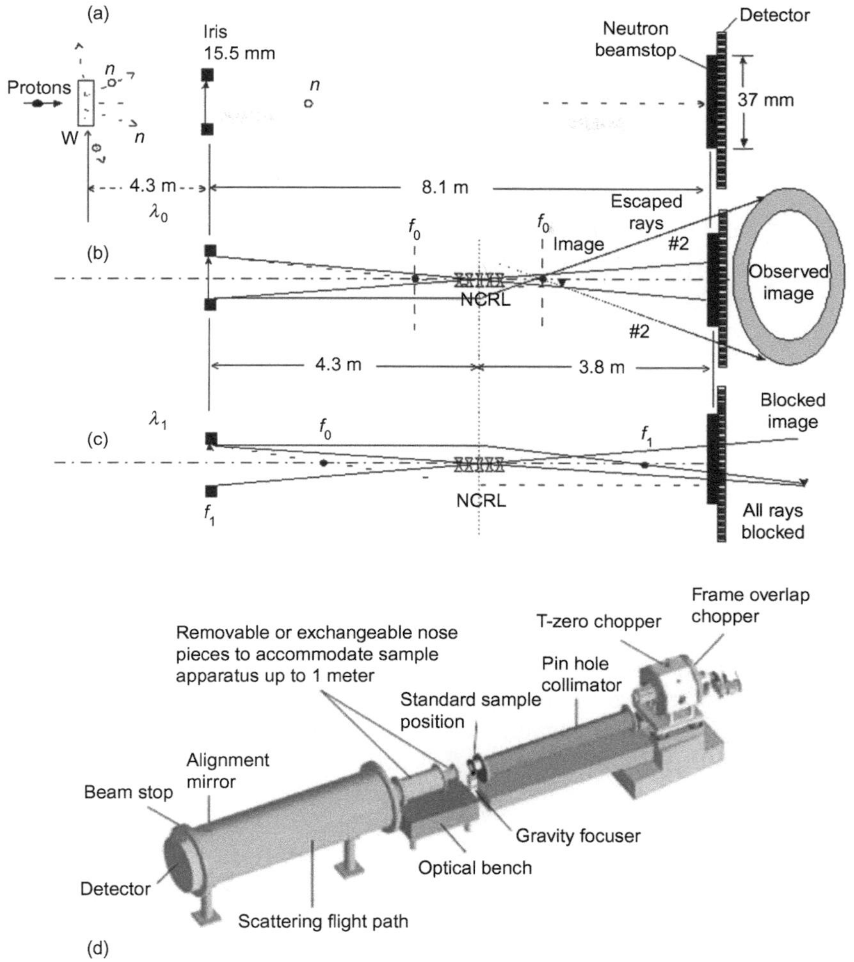

FIGURE 42 (a) The LANSCE LQD beam line with a 37-mm diameter beam stop in front of the detector. (b) A CRL is placed in the beam line and images the source in front of the detector. This causes some diverging rays to miss the beam stop. (c) The source is imaged at a shorter wavelength λ_1 into the back of the detector, such that the beam stop blocks all neutrons. (d) Schematic of the LANSCE LQD beam line (Park *et al.*, 2006).

the detector in which the neutron beam diverges sufficiently to fall on the detector outside the center 37-mm beam stop diameter. For shorter-wavelength neutrons the refraction is less and the source will be imaged behind the detector such that these neutrons strike and are absorbed by the center 37-mm beam stop. Thus, the CRL focal length f and the source iris object distance r_0 determine whether the detector image distance r_i

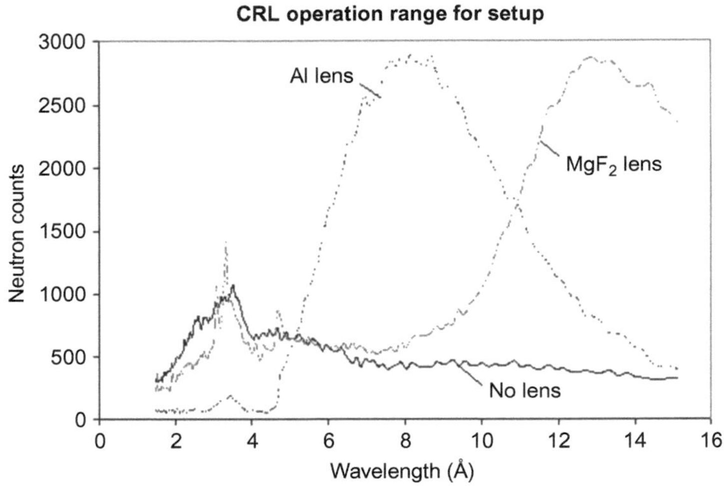

FIGURE 43 The observed operational wavelengths for the aluminum and MgF_2 CRLs with $r_i = 3.7$ m image distance and $r_0 = 8.8$ m object distance compared with the case when no CRL is present (Park *et al.*, 2006). (See Color Insert.)

receives neutron flux. That is, sufficient CRL refraction of the neutron rays causes the rays to hit the detector by falling outside the detector 37-mm diameter center beam stop. The detector intensity from the aluminum CRL began increasing at 5 Å and peaked at about 8 Å, whereas the detector intensity from the MgF_2 CRL began increasing at 9 Å and peaked at ~13 Å (Figure 43).

The transmission spectrum of the aluminum and MgF_2 CRLs was experimentally determined as follows. Using a TOF neutron detector, the neutron transmission through the two CRLs was measured and compared with the case of no CRL lens. The 37-mm diameter neutron-blocking beam stop in the detector center was removed and a 0.5-mm diameter cadmium aperture was placed 0.6 m upstream of the detector. The 0.5-mm diameter cadmium aperture limited the neutron intensity and beam size that arrive at the detector. In addition, cadmium irises with diameters equal to the physical aperture diameter $2r_e$ of the CRLs were placed on either end of the CRL. Hence, to reach the detector, the neutrons must pass through the physical aperture diameter $2r_e$ of the CRLs.

The detector measurement of the spectrum was performed with (1) no CRL, (2) with the aluminum CRL, and (3) with the MgF_2 CRL. The measured spectrum without a CRL has a peak ~3.3–3.7 Å beyond which the spectral intensity decreases exponentially with increase in wavelength. The spectra with the aluminum CRL and MgF_2 CRL had multiple dips or absorption bands due to Bragg diffraction. The polycrystal lenses of the aluminum CRL and the single crystal lenses of MgF_2 CRL caused Bragg

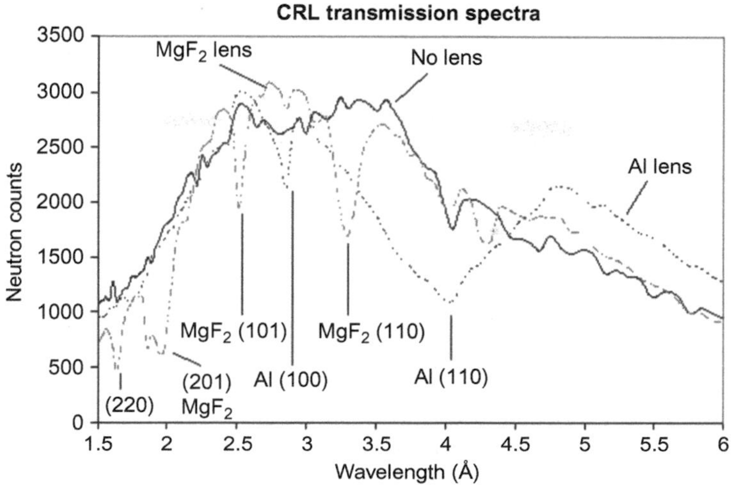

FIGURE 44 Plot of the observed spectra without the CRL and through the aluminum (Al) and MgF_2 CRLs with $r_i = 3.7$ m image distance and $r_0 = 8.8$ m object distance. The beam stop is removed in the transmission mode, and a 0.5-mm–upstream cadmium aperture is placed 0.6 m upstream of the CRL limits the beam at the detector (Park *et al.*, 2006). (See Color Insert.)

diffraction as a function of the neutron wavelength λ and the distance d between adjacent and parallel crystal diffraction planes, and the Bragg diffraction angle θ. Given integer n, the diffraction condition is given by the Bragg equation, where

$$n\lambda = 2d \sin\theta. \tag{105}$$

The detector received a narrow beam in which the Bragg angle was confined to an angle of approximately 90°, where $\theta \cong \pi/2$.

The corresponding planes for Bragg diffraction were calculated and found to match the experimentally measured decreases in detector intensity (Figure 44). The aluminum CRL lenses are polycrystals with a face-centered cubic lattice and lattice parameter $a = 4.05$ Å. Decreases in detector intensity were measured at 2.86 Å and 4.05 Å. Given $\theta = \pi/2$ and $n = 2$, the plane-to-plane separation distance is one wavelength, where $d = \lambda$. For the (100) aluminum planes, $d = a = 4.05$ Å and for the (110) planes $d = a/\sqrt{2}$ or $d = 2.86$ Å. The MgF_2 RL lenses are single tetragonal crystals with lattice unit cell constants $a = 4.621$ Å and $c = 3.053$ Å. The CRL transmission measurements through the MgF_2 CRL showed decreases of detector intensity at neutron wavelengths 1.9 Å, 2.05 Å, and 3.25 Å, which corresponded, respectively, to the MgF_2 diffraction planes (201), (101), and (110).

The application of neutron refractive optics, such as material prisms for small-angle scatter, was presented by Forgan and Cubitt (1998), and Shinohara *et al.* (2006) presented experimental results with fabricated arrays of stacked MgF_2 prisms. Each row of N prisms forms a compound refractive prism (CRP) whose refractive separation of neutron wavelengths is N times that of a single prism. Shimizu (2002) fabricated a neutron monochromator with a three dimensional array of prisms consisting of of parallel rows of MgF_2 prisms and perfluoropolymers. The MgF_2 prisms were right-angle isosceles triangle prisms 1 mm × 1 mm, and a CRP composed of 49 layers of prisms with layer-to-layer spacing of 4 mm. Shimizu and the Neutron Optics and Physics (NOP) group at The High Energy Accelerator Research Organization (KEK) in Japan also fabricated a stack of 44 Fresnel lenses of vitreous silica with each lens 45 mm in diameter, 21 mm thick, and grooved to a constant depth of 0.8 mm. The stacked Fresnel CRL had 65% transmission for $\lambda = 15$-Å neutrons (Figures 45 and 46).

Neutrons possess spin and an associated magnetic dipole moment, and the force exerted between the magnetic dipole moment and a magnetic field gradient may be used to control the neutrons' trajectory. By such means, a variety of neutron magnetic, optical elements can be fabricated, such as lenses, prisms, mirrors, field gradient devices, and spin flippers (see Williams, 1988). Polarized neutrons have also been produced by refraction with magnetic fields in vacuum (or air) in the shape of prisms and cylindrical lenses.

Just *et al.* (1973) performed neutron refraction experiments with magnetic fields shaped into prism and cylinders and measured the focusing and polarizing character of the one-dimensional, cylindrical magnetic lenses with a double-crystal spectrometer. They obtained prism-shaped magnetic fields by taper-cutting the cylindrical iron-core pieces of a split electromagnet into triangle-shaped pole pieces (Figure 47). The neutron deflection achieved by a single 60° prism-shaped material or magnetic field is small and inadequate to efficiently spectrally disperse short-wavelength neutron beams. However, a row of N identical prisms or lenses (material or magnetic), which forms, respectively, a CRP or CRL, increases the deflection angle and dispersion by N-fold.

Shimizu (2002) presented experimental results with fabricated multipole magnets for producing focused and polarized neutrons (see also Shimizu *et al.*, 2010). Shimizu produced focused and polarized 14.4-Å neutrons with a 2D quadrupole magnetic lens and with a permanent magnet sextupole magnet (Figure 48). The quadrupole and sextupole magnets are used to produce focused, polarized neutrons as depicted by the diagram at the bottom of Figure 48. In 2004 the NOP group presented results of their fabrication and testing of a superconducting sextupole magnet with a 46.8-mm bore (Shimizu *et al.*, 2004). The Shimizu group has also fabricated and tested a 2D permanent, sextupole magnet with 25-mm bore

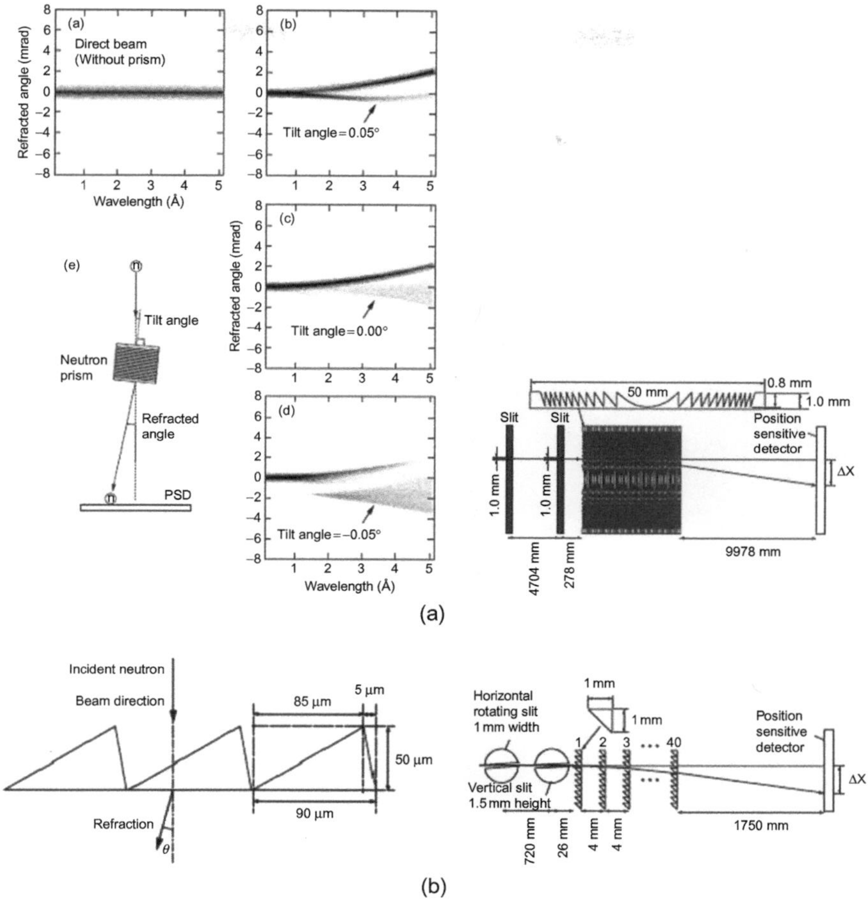

FIGURE 45 Compound refractive prisms and lenses of magnesium fluoride and perfluoropolymers designed, fabricated, and tested by the Neutron Optics and Physics group at KEK in Japan (Shimizu, 2002; Shinohara *et al.*, 2006).

diameter, as well and a pulsed quadruple magnet to produce focused polarized neutrons (Figure 49).

Five years later Littrell *et al.* (2007) fabricated and tested a one-dimensional (1D) compound refractive magnetic lens that used 99 pairs of cylindrical magnets with a radius of curvature $R = 1.27\,\text{cm}$ to produce cylinder-shaped $B = 1.0\,\text{T}$ magnetic fields in air, which focused the polarized cold neutrons. The neutron focusing was accomplished with a row of 1D cylindrical magnetic fields (radius $R = 1.27\,\text{cm}$) at the ASTERIX beam line at LANSCE by Littrell *et al.* (2007) (Figure 50). Figure 51 plots the measured intensity is plotted for the focused spin-down neutrons (spin anti-parallel to B) and defocused spin-up neutrons (spin parallel to B).

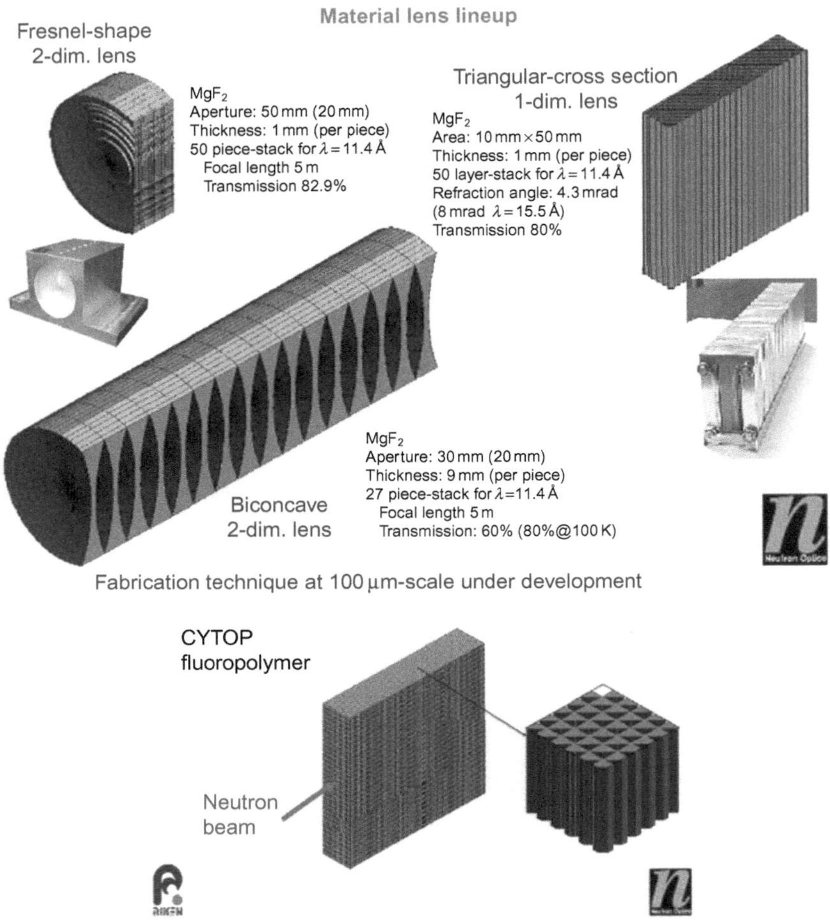

FIGURE 46 The neutron material lenses and prisms fabricated by the Neutron Optics and Physics group at KEK (from slide presentation, Shimizu 2004).

The detector measured vertical intensity profile is shown in false color versus neutron wavelength. The $L = 157.5$-cm long compound magnetic refractive lens (CMRL) has a thin lens focal length of 48 cm and a thick lens focal length of $f = 101$ cm at $\lambda = 12.6$ Å with principal plane distance $d = 138$ cm from the end of the magnet.

Figure 52 shows how a novel periodic magnetic field (PMF) optic acts as a prism, lens, and polarizer for neutrons and particles with a magnetic dipole moment. The PMF has a 2D field in the axial direction of neutron propagation (Figure 52). The PMF alternating magnetic field polarity

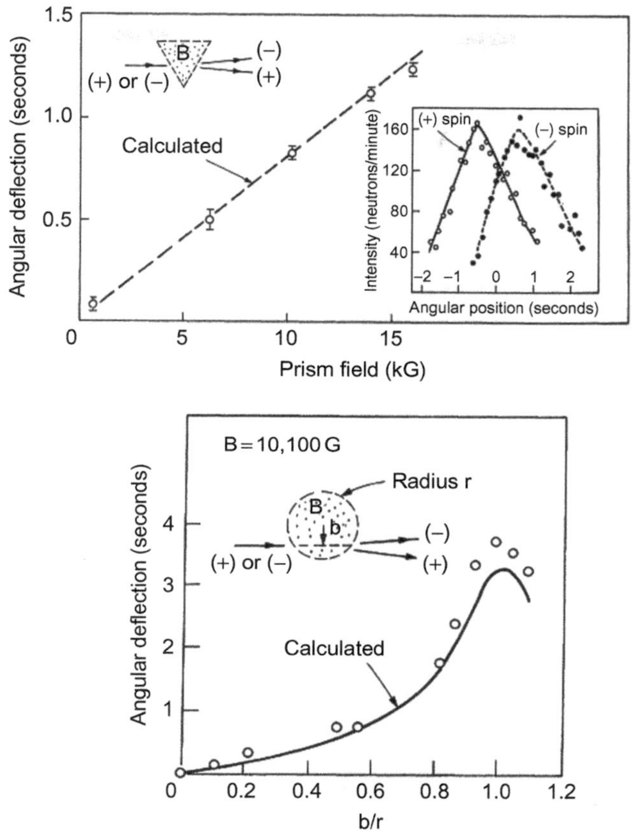

FIGURE 47 (Top) The plot shows the measured and calculated angular deflection of neutrons as a function of the prism-shaped magnetic field flux and neutron spin polarization; see inset plot. (Bottom) Plots of angular deflection between the spin-up and spin-down neutrons as a function of radial distance from the center of the cylindrical lens of radius *r*, which is expressed by the radius fraction *b*/*r* (Just *et al.*, 1973).

provides strong gradients that cause separation of neutrons by wavelength axially and by spin state transversely. The spin-up neutrons exit the PMF with their magnetic spins aligned parallel to the PMF magnetic field and are deflected upward and line-focused at a fixed vertical height, proportional to the PMF period, at a downstream focal distance that increases with neutron energy. The spin-down neutrons in the PMF are deflected downward, unfocused toward the PMF pole faces (see Cremer *et al.*, 2010).

The PMF was tested at the thermal neutron beam line (Bay 4) at the MNRC reactor for its ability to refract, deflect, and separate neutrons vertically by spin state and axially by wavelength. Figure 52 shows

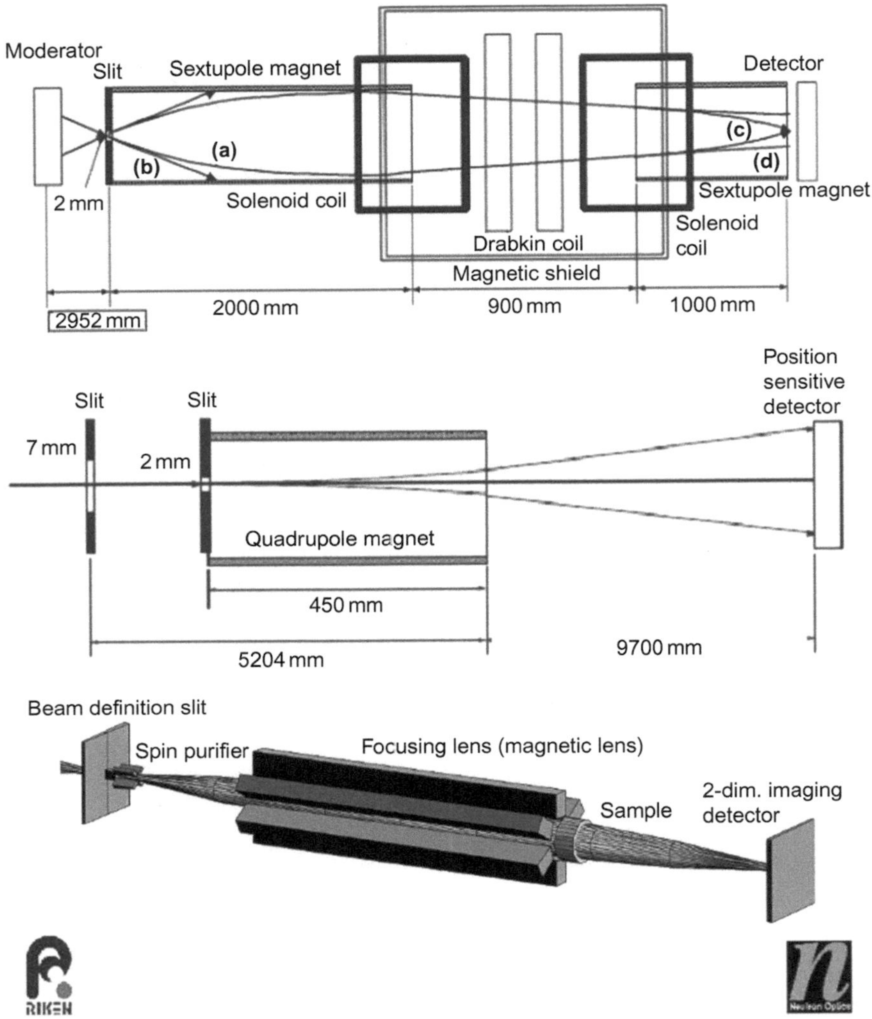

FIGURE 48 (Top) Experimental setup for measuring magnetic defocusing effect using two sextupole magnets and a Drabkin coil (from Shimizu, 2002). (Middle) Schematic of a quadrupole magnet used to measure magnetic refraction effect (from Shimizu, 2002). (Bottom) Diagram showing the use of a quadrupole or sextupole lens to provide focused polarized neutrons (From Shimizu, 2004).

the profile of magnetic field lines of a single period of the PMF and a schematic of the PMF with its alternating magnetic poles separated by aluminum spacers. The PMF provides equal and opposite vertical deflection of neutron spin states and axial separation of neutron wavelengths.

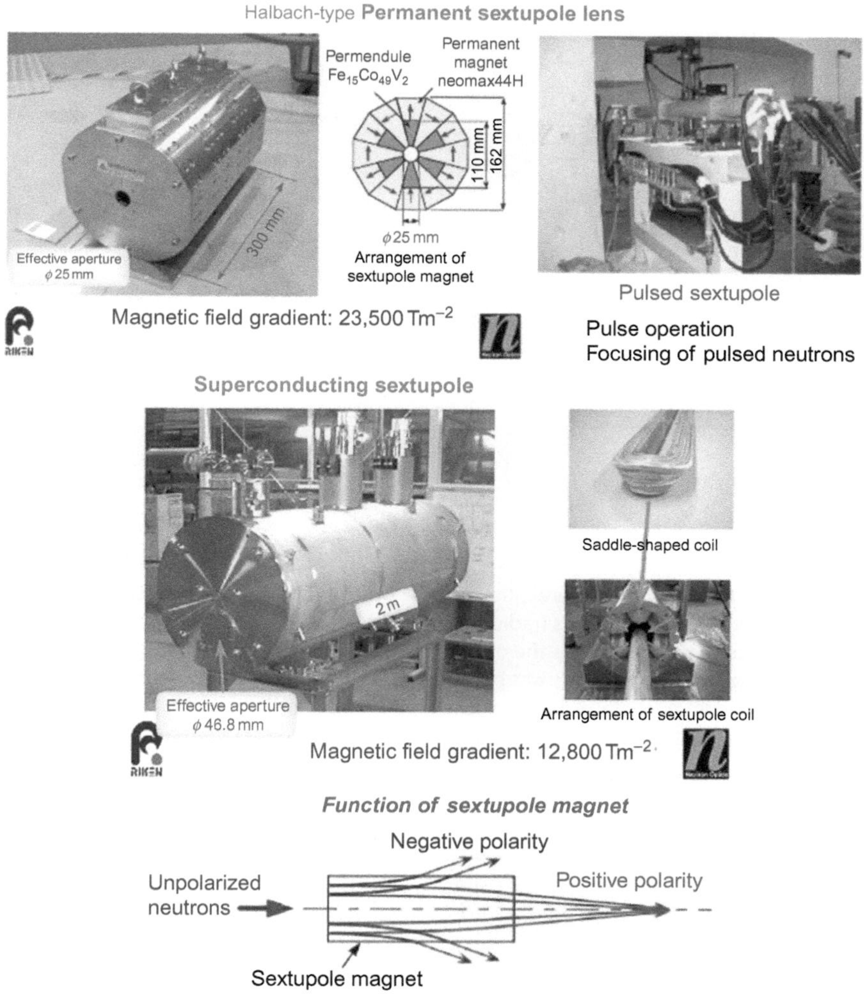

FIGURE 49 Photos and diagrams of superconducting, Halbach, and pulsed magnetic sextupoles designed, fabricated, and tested by the Neutron Optics and Physics (NOP) (Shimizu, 2004). (See Color Insert.)

The higher energy state, spin-up neutrons ($\mu_{||}$ aligned anti-parallel to **B**) are deflected upward from the PMF magnet surface. The pair of collimation slits reduces the initial neutron beam divergence to less than the deflection angle acquired by the neutrons in the PMF. The PMF provides 1D horizontal line-focusing of neutrons along an axial line parallel and above the PMF surface in which the focal distance measured from the PMF exit increases proportionally with neutron energy.

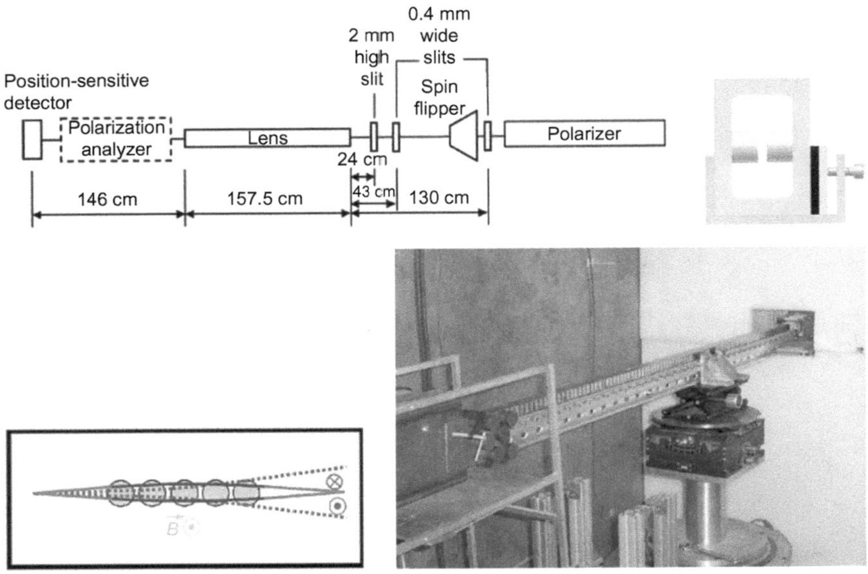

FIGURE 50 Setup schematic and photo of the neutron focusing experiment with row of 1D cylindrical magnetic fields (radius $R = 1.27$ cm) at the ASTERIX beam line at LANSCE (Littrell *et al.*, 2007). As the other two schematics show, a row of cylinder-shaped magnetic fields was produced by pairs of opposite-polarity cylindrical magnets. The commercially available and inexpensive but high-quality neodymium (NdFeB) permanent magnets provide surface fields exceeding 1.0 T (Littrell *et al.*, 2007).

The bottom-left photo in Figure 52 shows the front end of the 2.63-m long, 230-period, 460-magnet pole PMF with the second collimation slit at its entrance. Each magnet is 3.175 mm in axial width and is separated by 2.54-mm axial width aluminum spacers. Each magnet period of $p = 11.43$ mm is composed of two magnets and two aluminum spacers set in the following order: N-surface pole magnet, aluminum spacer, S-surface pole magnet, aluminum spacer. The bottom-right photo in Figure 52 shows the MNRC Bay 4 beam line with the PMF in the foreground; in the background is the 20°-inclined Bosch table on which is mounted the FujiFilm imaging plate.

Figure 53 shows the measured data in the PMF experiment is presented in the top-left and top-right plots. The top-left plot shows the splitting of the neutron beam in the presence of the PMF field at horizontal (x) positions of 42.5 mm. The direction of the splitting was in the y-direction. To a good approximation, the background intensity was observed to be linearly dependent upon y. This allowed background subtraction by assuming a straight-line fit to intensity across the y-direction of the FOV.

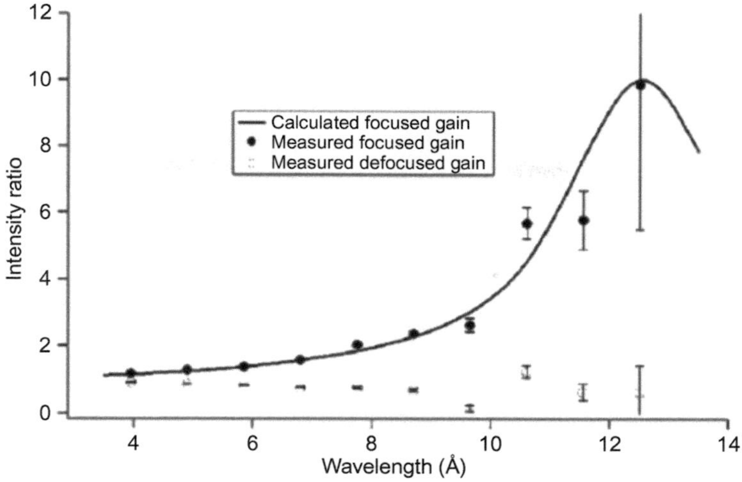

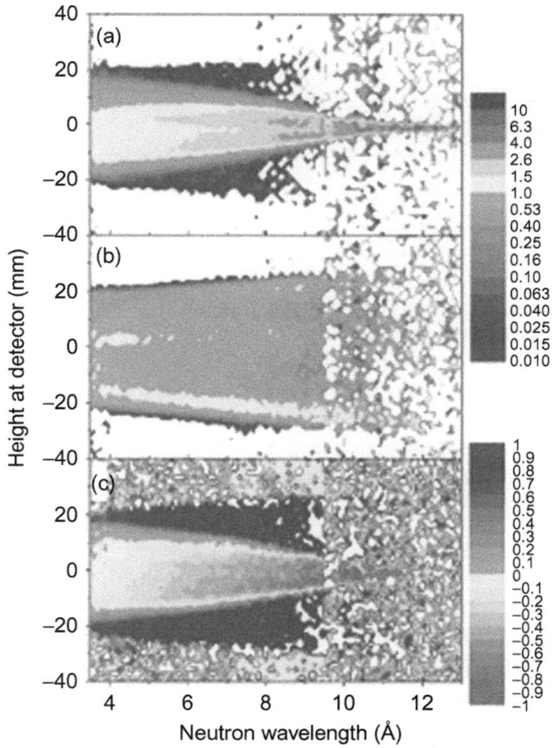

FIGURE 51 (Top) Plot of the focused spin-down neutrons (spin anti-parallel to B) and defocused spin-up neutrons (spin parallel to B). The detector-obtained vertical intensity profile is shown in false color versus neutron wavelength. This $L = 157.5$ cm -long compound magnetic refractive lens has a thin lens focal length of 48 cm and a thick lens focal length of $f = 101$ cm at $\lambda = 12.6$ Å with principal plane distance $d = 183$ cm from the magnet end (Littrell *et al.*, 2007).

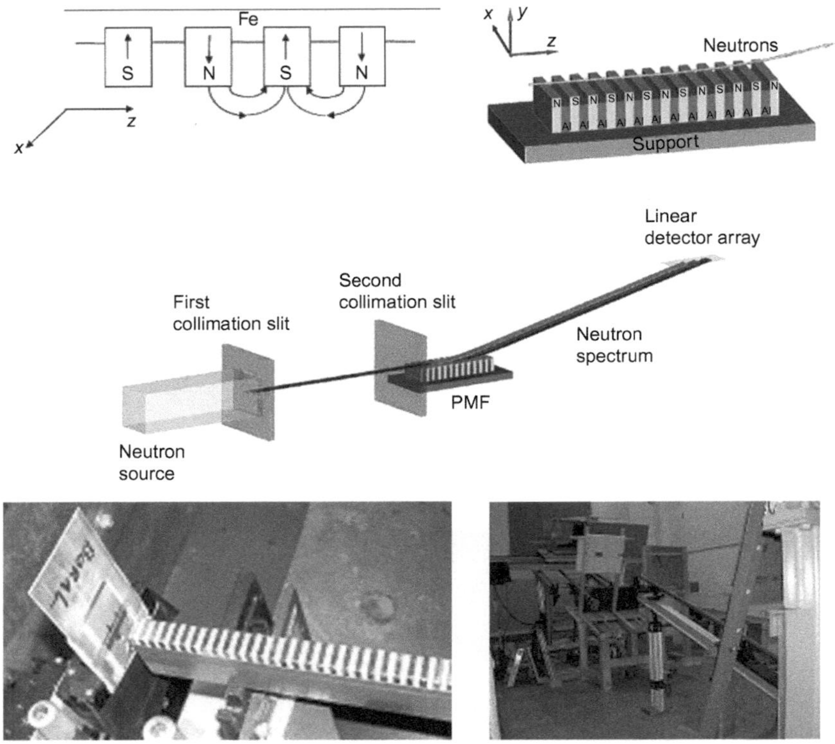

FIGURE 52 (Top and middle) Schematics of the PMF magnetic field, the magnet arrangement, and the setup of the PMF with collimation slits to provide focused polarized neutrons whose energy increases with distance from the PMF entrance. The bottom-left photo shows the second collimation slit at the entrance of the PMF. The first collimation slit is upstream and placed on the aluminum face plate of the Bay 4 collimation tube. The bottom-right photo shows the PMF setup in Bay 4; see text for details (Cremer *et al.*, 2010). (See Color Insert.)

Following background subtraction, the best fit with two Gaussian distributions was calculated. The top-right plot shows the vertical (y) position of the peaks as a function of the horizontal (x) position where the scan was taken, which clearly indicates the neutron beam breaks into two components. The top component of the beam, corresponding to spin-up neutrons, is deflected upward by 1.2 mm and the lower component of the beam, corresponding to spin-down neutrons, is deflected downward by 0.8 mm relative to the center position of the nondeflected high-energy neutrons.

The bottom plots of Figure 53 shows PMF line-focusing of a strip neutron beam downstream of the PMF exit. (The derivation of the trajectory

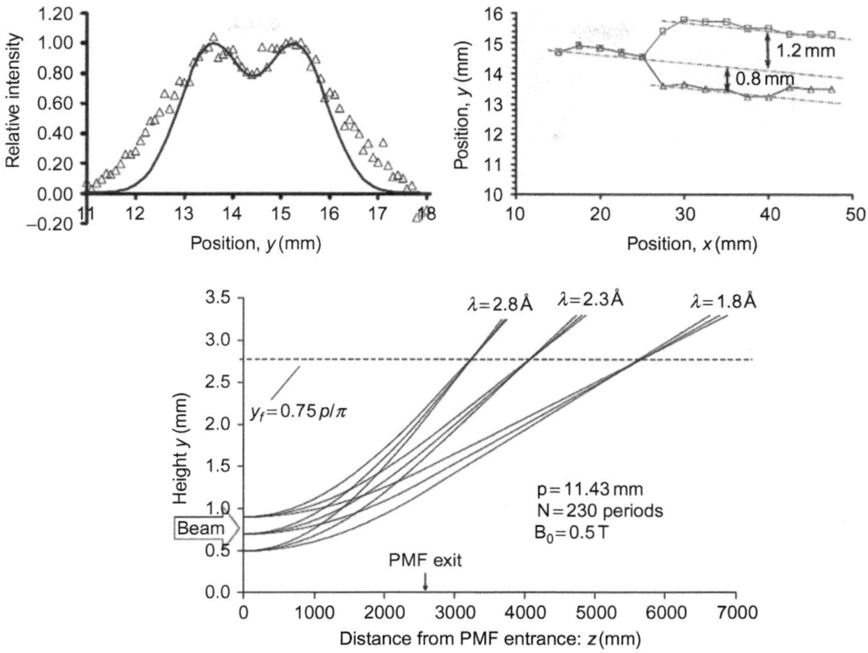

FIGURE 53 (Top-left) The measured vertical (y) neutron relative intensity profile at horizontal (x) position 42.5 mm from the left side shows the vertical splitting of the neutron beam into two peaks at 13.5 mm and 15.3 mm. (Top-right) Plot of the vertical (y) position of the peaks as a function of the horizontal (x) position. (Bottom) Trajectory plots for 1.8-Å 2.3-Å, and 2.8-Åspin-up neutrons in a horizontal strip beam that passes through the PMF and line-focuses downstream of the PMF exit. For each wavelength λ, neutrons with zero initial divergence angle are centered at initial height $y_0 = 0.7$ mm, with lower and upper beam boundaries at $y_0 = 0.5$ mm and 0.9 mm. The PMF parameters are $N = 230$ periods, period length $p = 11.43$ mm, and surface magnetic field amplitude $B_0 = 0.5$ T (Cremer *et al.*, 2010).

equations for these plots is derived in Chapter 31 in an upcoming volume.) The strip neutron beam is composed of neutrons of three wavelengths $\lambda = 1.8$ Å, 2.3 Å, and 2.8 Å. The strip beam is produced by the pair of collimation slits just upstream of the PMF entrance. For each wavelength, three spin-up neutrons with a zero initial y-velocity, $v_T(0) = 0$, start from three different initial positions, $y_0 = 0.5$ mm, 0.7 mm, and 0.9 mm, which define the bottom, center, and top of the strip beam. In the bottom figure, the PMF has $N = 230$ periods, with equal pole and gap thickness ($z_r = 0.9$), and period $p = 11.43$ mm, with an overall length $l_1 = 2629$ mm, and surface magnetic field amplitude $B_0 = 0.5$ T. Spin-up neutrons are focused along a line parallel to the PMF surface at critical height $y_f = 2.73$ mm, shown by the dashed horizontal line. The shorter-wavelength 1.8-Å neutrons are line-focused farther out at distance 5.5 m downstream of the PMF entrance, whereas neutrons of the longest-wavelength 2.8-Å focus in closer at 3.2 m. The spin-up neutrons transition from an upward curved trajectory, which is caused by the magnetic field gradient in the 2.63-m long PMF, to a straight-line trajectory extending from its exit through the field-free drift space to a linear detector.

One possible application of the PMF is spectral splitting of a white neutron beam for use in residual stress analysis. Consider the PMF spectrometer and monochromator configurations in Figure 54. In either case the white (or wide-band) neutron beam is collimated by two narrow horizontal slits placed before the PMF. The PMF can provide a line-focused beam of aligned (polarized), spin-up neutrons, given a sufficiently high flux neutron source. The horizontal band of spin-up neutrons exits the PMF with the neutrons' longitudinal magnetic moment component $\mu_{||}$ aligned anti-parallel to the direction of the PMF magnetic field lines at the PMF exit. As the neutrons propagate away from the PMF, their $\mu_{||}$ remain pointed in a fixed direction anti-parallel to the direction of their last encountered PMF magnetic lines. The fixed direction of the neutrons' $\mu_{||}$ at the PMF exit assumes the direction and amplitude of the last encountered PMF magnetic field lines are well defined and dominant over the fringing magnetic fields that extend into the drift space beyond the PMF exit. A very thin, cryocooled superconducting, niobium foil could be placed at the PMF exit to better define the PMF exit magnetic field direction and amplitude and prevent the undesired outward fringing of PMF magnetic field lines via the Meissner effect.

In the case of the PMF spectrometer shown at the top of Figure 54, the diffracted neutrons from the sample pass through the PMF, which causes the spin-up neutrons to be deflected vertically upward. The spin-down neutrons are downward deflected and strike the PMF. The upward-deflected spin-up and polarized neutrons are line-focused at points along a parallel axial line above the PMF surface at axial distances from the PMF exit that increase proportionally with neutron energy. From the maximum axial distance downstream of the PMF, a horizontal slit detector can be translated axially toward the PMF to measure the energy spectrum of the diffracted neutrons from the sample. Alternatively, without the need for translation, a long fixed, 1D axial position-sensitive detector, which is set y_f to $2y_f$ critical heights ($y_f = 0.75p/\pi$) above the PMF surface plane, could measure the energy spectrum of sample diffracted neutrons.

In the case of the PMF monochromator at the bottom of Figure 54, neutrons first pass through the PMF to produce a narrow-band, spin-up, polarized (aligned) neutron beam, which then propagates to the sample. Upward-deflected, spin-up, and polarized neutrons are line-focused at

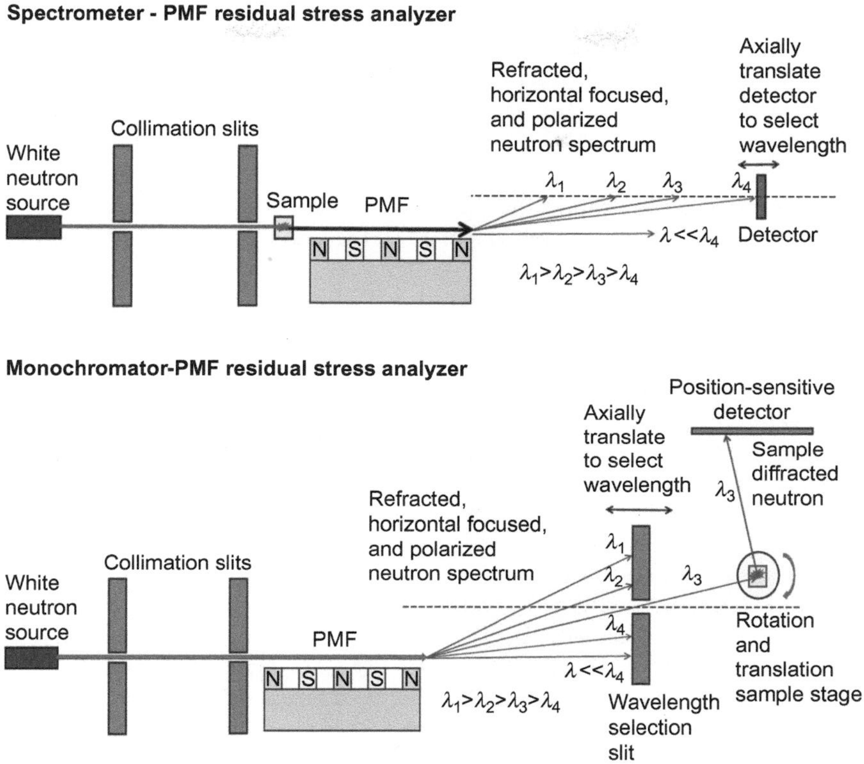

FIGURE 54 (Top) A PMF is used as a spectrometer to perform residual stress analysis with a neutron beam that passes through a sample to the PMF. The PMF line-focuses the upward-deflected, spin-up neutrons along a horizontal line parallel to the PMF surface at axial focal distances downstream of the PMF exit that are proportional to the neutron energy. The energy spectrum of the spin-up neutrons from the sample is then measured by an axial-translatable detector. (Bottom) A PMF acts as a monochromator to produce a narrow bandwidth of spin-up, polarized (aligned) neutrons that illuminate a sample, which then diffracts these neutrons to a fixed, position-sensitive detector (Cremer *et al.*, 2010).

points along an axial line y_f to $2y_f$ critical heights above the PMF surface. A horizontal slit in a cadmium sheet is translated axially toward the PMF, which intercepts line-focused neutrons with wavelengths that increase as the scanning slit approaches the PMF. A narrow bandwidth of line-focused, spin-up, and polarized neutrons passes through the slit and strikes the sample position at a particular height that depends on the drift space distance between the scanning slit and sample. Thus, the sample (e.g., machined, cast, or welded part) should be placed on a small translation stage to position it at the center of the beam.

Using motorized translation stages results in a system that can be tuned to different wavelengths. The sample rotation stage allows the monochromatized, polarized neutrons to interrogate different crystal planes of the sample lattice. These admitted neutrons are then diffracted by the sample lattice planes to a stationary position-sensitive detector. This detector records the intensity of the sample-diffracted neutrons of wavelength as a function of the diffraction angle, which is measured relative to the line from the PMF exit through the wavelength-selection slit to the sample.

In the above two embodiments of a PMF residual stress instrument, the diffracted neutron intensity as a function of angle and wavelength can be compared with a "calibration" unstressed sample. The sample can then be subjected to compression, torsion, tension, heat, and so on to measure residual stress as a function of applied environment and thus provide structural information and feedback.

Imaging of very cold neutrons (VCNs) with refractive material lenses (copper and aluminum CRL), magnetic lenses, and magnetic prisms was conducted in August-September 2007 at ILL in a collaboration between Adelphi Technology (Ted Cremer) with Argonne National Laboratory, (IPNS) Intense Pulse Neutron Source (Jack Carpenter, Jyotsana Lal, Ed Lang), ILL (Peter Geltenbort) and Hirohiko Shimizu of KEK, Japan, and Katsuya Hirota, Kenji Mishima of RIKEN, Japan. The experiments were carried out on the VCN/UCN (very cold neutron/ultracold neutron) beam line Pf2 at ILL. Figures 55 through 58 are rearranged diagrams, photos, and neutron images from a slide presentation of the experiment and results prepared by Jyotsana Lal.

Figure 55 shows a diagram of the Pf2 ILL beam line, a plot of its neutron spectrum, and photos showing of the white-domed ILL reactor next to the white annular European Synchrotron Radiation Facility (ESRF) synchrotron ring, as well as a photo of Level D (top floor of reactor), which contains the Pf2 hutch. The beam line hutches are mounted on the steel-plate floor that covers the pool of water above the reactor core. The ILL research reactor produces the most intense continuous neutron flux (1.5×10^{15} n/cm^2/s) in the world, although it has a thermal power of just 58 MW (http://www.ill.eu/). The great advantage of VCN refraction is the need for only (1) a few refractive material lenses with relatively thin, center minimum walls to minimize absorption or (2) just a few magnetic field lenses or prisms. The disadvantage of VCN and UCN sources is the increased neutron absorption cross section in lens materials and the relatively low available beam fluxes.

The high intensity of the Pf2 UCN/VCN beam line is due to a combination of several design features (see the top-left diagram in Figure 55). A beam of VCNs is extracted from the ILL's vertical cold source by an optimized neutron guide system that dips into the liquid deuterium. The

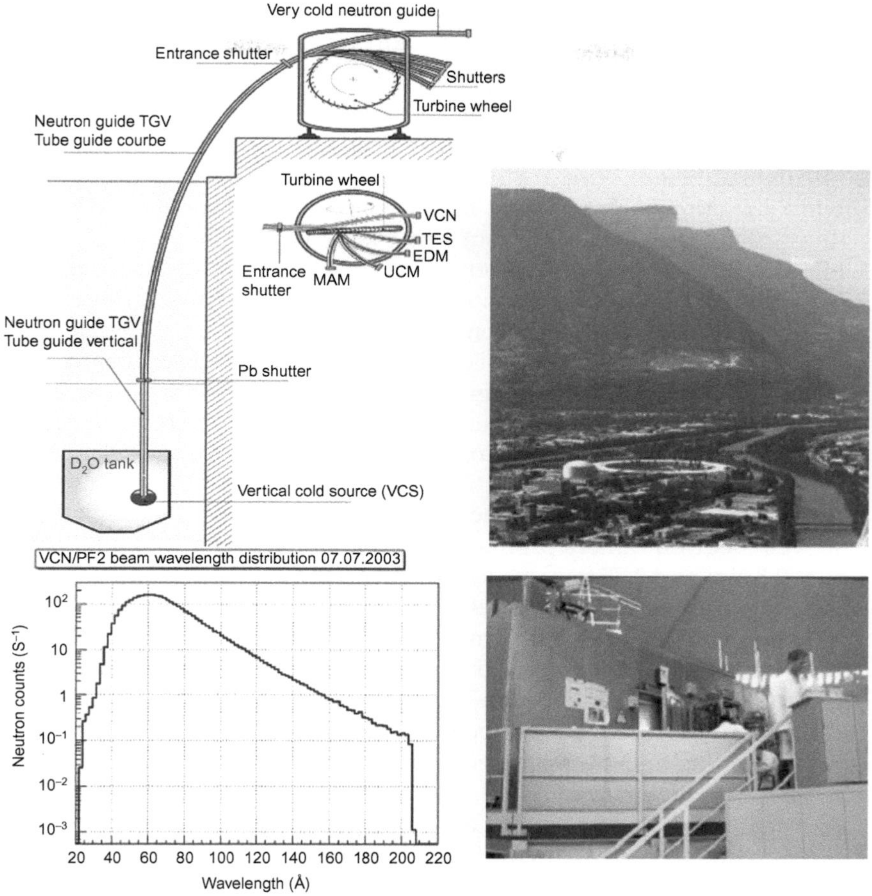

FIGURE 55 (Top left) ILL Pf2 very cold beam line diagram. (Top right) White-domed ILL reactor next to the white annular ESRF in Grenoble. (Bottom left) Wavelength spectrum of Pf2 with peak at $\sim\lambda = 62$ Å. (Bottom right) Ed Lang standing under the reactor dome at ILL Reactor Hall Level D on the UCN/VCN Pf2 hutch platform (Lal *et al.*, 2007). (See Color Insert.)

UCNs are produced "externally" by a Doppler shift device, the Steyerl neutron turbine. Part of the beam can be used for experiments where VCNs are required. The bottom end of the lower straight vertical section of the Pf2 neutron guide dips into the deuterium of the vertical cold source. The upper part of the Pf2 neutron guide is curved with 12.8-m length and 13-m radius of curvature. The neutron mirrors of the Pf2 guide are high-quality nickel surfaces. The Pf2 guide transmits a square beam (7 cm × 7 cm) with wavelengths from $\lambda = 20$ Å to $\lambda = 400$ Å to Level D of the ILL reactor.

The beam enters the turbine vessel and is split into two beams. One beam bypasses the turbine wheel and is the source of the VCNs. The other beam is Doppler-shifted into the UCN region by action of a neutron turbine, which consists of a 1700-mm diameter wheel. On the wheel's outer radial portion are mounted 690 cylindrically shaped blades. Each blade is 160 mm in height with 158° arc and 65-mm radius of curvature. The blade's reflecting surfaces are of high-quality nickel. The receding speed of the nickel-coated blades is about 25 m/s at rotational frequency 230 rpm, which Doppler-shifts the neutrons into the low-velocity range from 0 to 15 m/s. The Doppler shifting action of the neutron-striking blades causes an enlargement of the beam from 7 cm × 3.4 cm to 8 cm × 16 cm cross section and causes the beam to acquire a wide divergence angle range.

The top right image in Figure 56 shows the real-time, position-sensitive, neutron-imaging camera developed by K. Hirota *et al.* (2005) of RIKEN and the NOP group at KEK in Japan (http://nop.kek.jp/). A 1-mm diameter pinhole was set 79 cm downstream of the ILL Pf2 VCN guide and at downstream distances of 35 cm and 60 cm from the pinhole. The NOP neutron camera produced neutron images with FWHM diameters of 2.5 mm and 3.75 mm, respectively (top-left image in Figure 56) with a chopper in place. The bottom-left photos show the collaborators at work.

Without the chopper the 1-mm diameter pinhole was imaged with a time-integrated, neutron-imaging plate (thermoluminescent) at downstream distances of 39 cm and 67 cm. The polychromatic neutrons from the Pf2 waveguide may have been incident at different angles on the pinhole, leading to multiple images at different wavelengths and/or reflection angles in the neutron guide. Nine images of the pinhole are seen at downstream distances of 39 cm and 67 cm with respective 8.7-mm and 13.06-mm horizontal widths for three holes in a row (bottom-right image in Figure 56). Each of the nine pinhole images has a diameter consistent with the FWHM diameters of 2.5 mm and 3.75 that were measured by the NOP neutron camera .

Refractive imaging of the 1-mm diameter pinhole was done with a CRL composed of $N = 4$ copper biconcave, spherical lenses. Each copper lens had a radius of curvature $R = 1.98$ mm, center, minimum wall thickness $d = 20$ μm, perimeter maximum thickness 0.81 mm, and physical aperture diameter $2r_{\mathrm{m}} = 2.3$ mm. With $\lambda = 62$ Å neutrons, the four-lens copper CRL had an absorption aperture of diameter $2r_a = 1.1$ mm, parabolic aperture diameter $2r_p = 0.6$ mm, focal length $f = 6.19$ cm, 76% center peak transmission $T(0) = 0.76$, and 33% average transmission $T_{avg} = 0.33$. A single copper lens is shown in the middle of Figure 57.

The copper lens CRL imaging was done without the chopper, and the polychromatic beam from the wave guide had multiple points of incidence through the 1-mm diameter pinhole object. The top left of Figure 57 (top left) is a diagram of the CRL imaging setup with lensholder and four

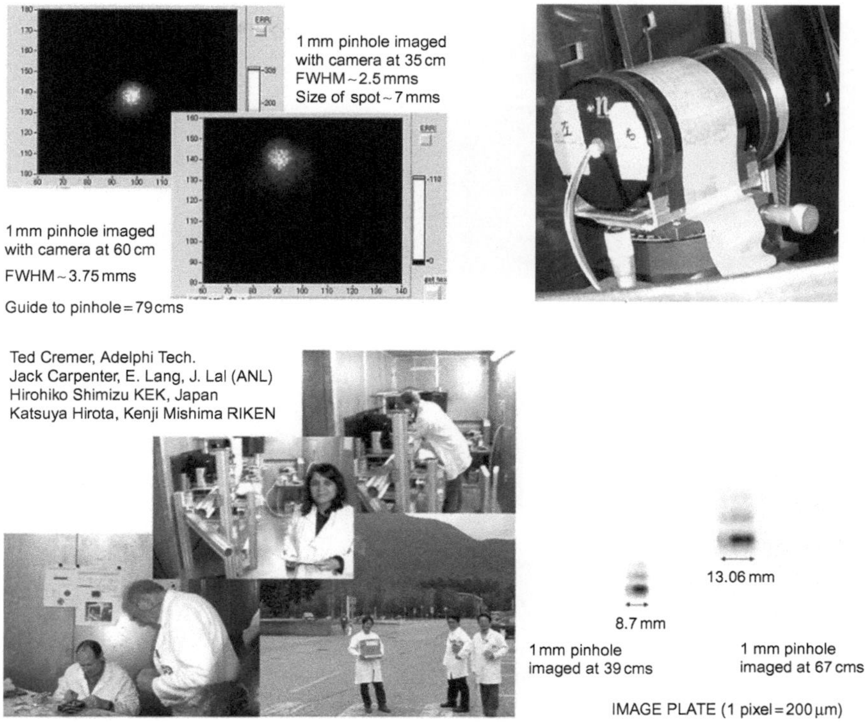

FIGURE 56 (Top left) The neutron images of 1-mm diameter pinhole shown at left were made with the KEK NOP at ILL pf2 with a chopper in place. Without the chopper the polychromatic neutrons beams with various reflection angles pass through the pinhole at different incident angles producing multiple (nine) images of the pinhole. (Top right) Photo of the NOP neutron camera (Bottom left) Photos of the author (Ted Cremer) seated fabricating magnetic lens, Jack Carpenter examining a lens, Jyotsana Lal in the beam hutch holding imaging plate, Ed Lang in the beam hutch doing tedious alignment (unfortunately for Ed, snap-together, Lego-like modules for assembling beam line experiments are not yet available). (Bottom right photo) Kenji Mishima (left), Hirohiko Shimizu (center), and Katsuya Hirota (right) holding boxes with the NOP neutron camera and compound refractive prisms. (Bottom far right) Multiple neutron images of the pinhole (Lal *et al.*, 2007). (See Color Insert.)

copper lenses. The top right images show the 1-mm diameter pinhole by the four-lens copper CRL at three magnifications, $M = 3.4$, $M = 8.74$, $M = 17.27$, set by the CRL-to-pinhole object distance and the CRL-to-neutron image plate detector (thermoluminescent) for center wavelength $\lambda = 62$ Å of the wideband VCN beam.

In addition, an off-axis image of the pinhole (without the chopper) was taken with the neutron image plate in which the four-lens copper CRL was set at an off-axis angle from the beam alignment axis. The magnification

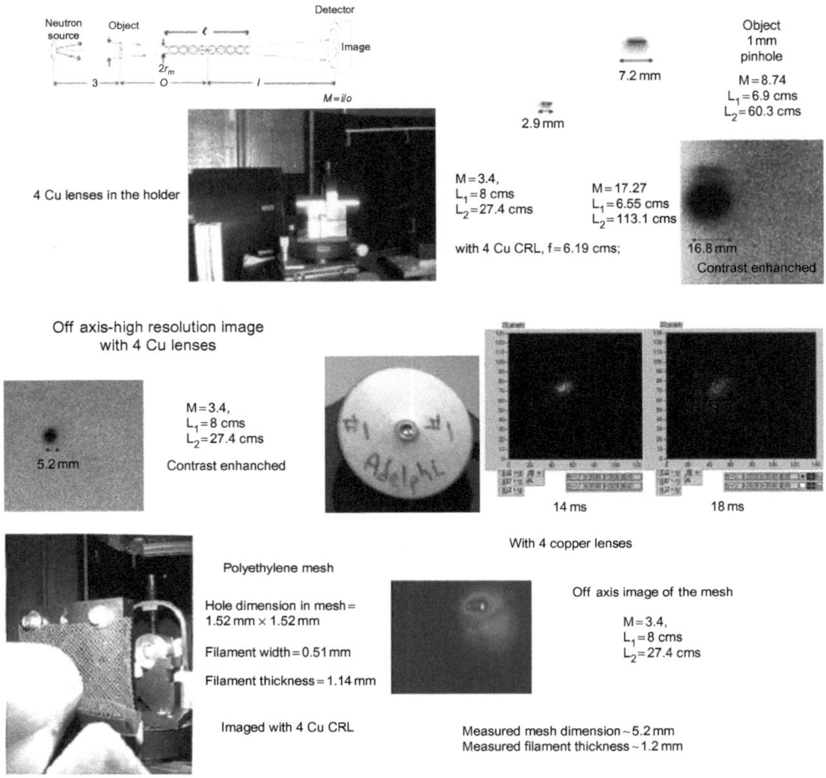

FIGURE 57 (Top left) Setup of the refractive imaging experiment at center wavelength $\lambda = 62$ Å of wideband VCN beam. (Top right) Images of 1.0-mm pinhole at three magnifications by copper CRL with $N = 4$ biconcave, spherical lenses. (Middle left) Photo of off-axis copper CRL set to magnify pinhole by ×3.4. (Middle) Photo of copper biconcave, spherical lens with $R = 1.98$ mm, $2r_m = 2.3$ mm, and $d = 20$ μm. (Middle right) NOP neutron camera images of neutrons with 14-ms and 18-ms arrival times corresponding to wavelengths 50 Å and 61 Å. See text for details (Lal *et al.*, 2007). (See Color Insert.)

$M = 3.4$ for the center wavelength $\lambda = 62$ Å of the wideband VCN beam was set by placement of the CRL along an off-axis direction from the beam alignment axis. The 1-mm pinhole was set at object distance $L_1 = 8$ cm and the neutron image plate detector was set at image distance $L_2 = 27.4$ cm. in the middle-left image in Figure 57 shows the off-axis image of the pinhole.

With the chopper in place, an $M = 3.4$ magnified image of the 1-mm pinhole object was obtained with NOP neutron camera, which combines a position-sensitive sensor and neutron TOF capability. The NOP camera recorded a set of images in which each image is composed of neutrons that arrive within a narrow arrival time window. The arrival time t is measured relative to the initial $t = 0$ (time fiducial), when a burst of neutrons passes through the narrow radial directed slit of the rotating chopper wheel. Each center time of arrival t of the narrow arrival time window Δt corresponds to a neutron velocity v within a narrow velocity range Δv. Given by the length of flight L from chopper to detector, where $v = L/t$, the neutron center velocity corresponds to de Broglie wavelength $\lambda = h/m_n v$.

The distance between the chopper (600 rpm) and TOF NOP neutron camera was 1375 mm. The minimum 3.25-ms detectable TOF between chopper and NOP detector set the cutoff neutron wavelength to 100 Å for obtaining images. The NOP neutron camera acquired neutron images of the 1-mm pinhole at different TOFs of 6, 10, 14, and 18 ms. These TOFs corresponded to neutron velocities and wavelengths 27, 38, 50, and 61 Å. The neutron images for chopper–to-detector TOFs 14 ms and 14 ms, which corresponded to wavelengths 50 Å and 61 Å, are shown in the middle right of Figure 57.

The four-lens copper CRL was set off-axis and imaged a coarse green polyethylene diamond mesh grid at magnification $M = 3.4$ at center wavelength $\lambda = 62$ Å of the polychromatic beam. The green mesh had seven holes per centimer, which were 1.52 mm × 1.52 mm square holes defined by filaments of 0.51-mm width and 1.14-mm thickness (bottom left of Figure 57). The small FOV of the off-axis CRL imaged a single hole of the green mesh onto the neutron image plate detector shown in the bottom right image).

A magnetic CRP, magnetic cylindrical lens CRL, and some novel magnetic neutron CRLs were also tested (Figure 58). Due to lack of preparation time the novel magnetic CRLs developed by the author were based on intuition and left to trial and error testing (Thomas Edison approach) rather than the more modern approach of Charles Proteus Steinmetz (founder of the General Electric Research Laboratory) using detailed calculations based on sound rigorous reasoning and understanding of the involved physical processes. Trial and error and accidental discoveries (e.g., the discovery of Teflon) are a time-honored tradition in science and engineering, but usually only the researcher who has performed painstaking and detailed preparation is in a position to have the accident happen and recognize the significance of the accident. The compound magnetic prism and cylindrical magnetic lens were similar to the magnetic cylindrical neutron lens and magnetic neutron prism fabricated and tested by Just *et al.* (1973).

The sphering magnetic lens consisted of a spherical magnet place on each side of a brass nut. Drilled through the brass washer was a hole to allow axial propagation of the neutron beam between the spherical north and south poles surfaces (top image in Figure 58). A row of such nuts with axial drilled holes for neutron passage and the pair of spherical magnets separated by each washer formed a magnetic CRL. The sphering magnetic

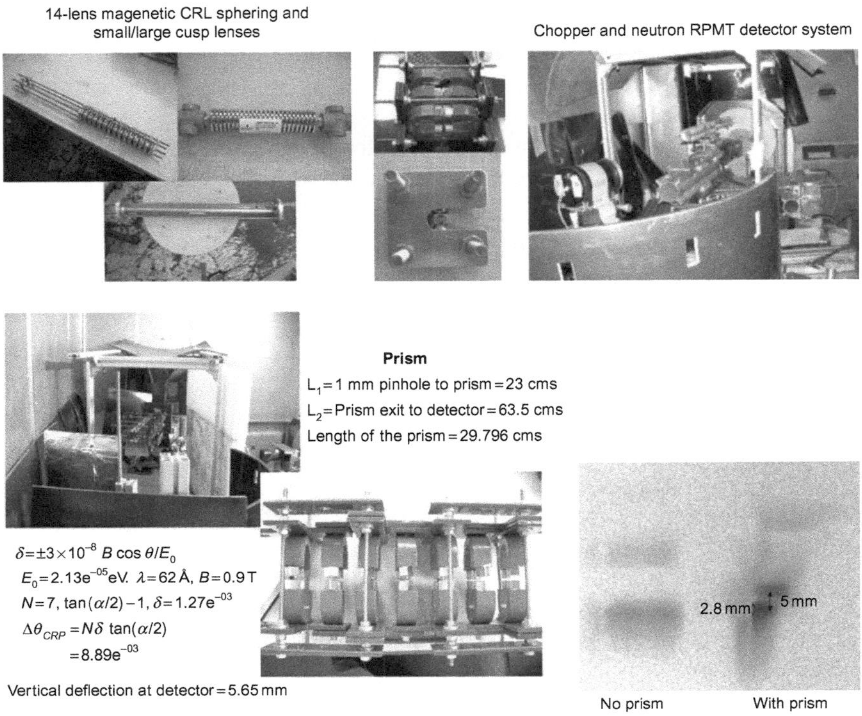

FIGURE 58 (Top) Photos of magnetic field lenses using a spherical, convex magnetic field (sphering lens) and cusp, radial-directed magnetic fields formed between faces of annular magnets of same polarity (e.g., N-N or S-S). Also shown is a C magnet with diamond-shaped magnet pole faces of opposite polarity, which forms a three-prism CRP. The bottom, cylindrical magnet pole faces of opposite polarity form a three-cylinder lens CRL. The top-right image shows the sphering lens mounted in the Pf2 beam line. (Middle Left) Photo of 7-prism CRP mounted in the beam line (Bottom Middle) A photo of the 7-prism CRP. (Bottom left) calculation of the 5.65-cm separation distance between spin-up (+) and spin-down (−) compared to the measured spin-up neutron (+) deflection of 2.8 cm from the center axis (the bottom dark band; bottom right), which implies a measured 5.6-cm separation distance between spin-up (+) and spin-down (−) neutrons. (Bottom Right) Multiple neutron source images without CRP (left) and with CRP (right). CRP caused upward deflection of spin-up neutrons, forming dark band above bottom dark band. Downward deflection of spin-down neutrons caused downward smearing. (Lal *et al.*, 2007).

lens produced a spherical, convex field between the two spherical NdFeB magnets, which were readily available from K&J Magnetics of Jamison, Pennsylvania, USA. (http://www.kjmagnetics.com).

The top middle photos in Figure 58 are close-ups of one sphering spherical, convex, magnetic lens (shown in the lower photo) and the top-right photo is a sphering magnetic CRL setup in the Pf2 beam line for testing. The spherical magnets naturally form a closed ring of spheres, which provides a return path of the spherical magnetic field across the gap of the brass nut. The sphering CRL was composed of a stack of spherical magnet rings. Each ring had a brass nut inserted between two of the spherical magnets to form a gap between the opposite-polarity spherical magnet poles. The axial-directed neutron beam was propagated between the spherical magnet pole faces through the axial holes drilled in the brass nut. That is, the neutron beam crossed the transverse, spherical magnetic field formed between two spherical magnets separated by the brass nut. The spin-up (+) neutrons are defocused and spin-down (−) neutrons are focused by the magnetic field gradient of spherical magnetic field in which the magnitude of the gradient of the spherical magnetic field decreases with the increase of radial distance from the center axis. In addition, the incident neutron beam propagates along the center axis, which passes through the line of drilled holes in the brass nuts.

The cusp magnetic lens shown (top of Figure 58) was composed of a row of annular magnets, which were magnetized in the axial z-direction of the neutron beam propagation. The facing pairs of pole surfaces had the same magnet polarity, thereby causing magnetic field repulsion along the axial direction. The magnetic field lines were thus repelled in the radial direction between the pairs of magnet surfaces with the same polarity. The neutron beam propagated in the axial direction and encountered the transverse radial magnetic field gradient at a gap between the two same-polarity (N-N or S-S) annular magnetic surfaces.

The decrease in the gradient magnitude in the radial inward direction causes radial inward deflection of the spin-up (+) neutrons, and the radial outward deflection of the spin down (−) neutrons as the radial magnetic field alternates between radial outward to radial inward from one gap to the next gap. The photos show a cusp magnetic lens with the annular magnets outside a neutron propagation pipe and with the annular magnets inside a neutron propagation pipe hidden from view. The results of the sphering and cusp magnetic lenses were inconclusive due to crude calculations and crude fabrication by the author. In addition, the Pf2 beam line was not optimized for the material and magnetic CRL imaging experiment in which alignment and multiple sources of VCNs emerging from the guide and late application of the chopper were problematic.

The red C magnets with silver NdFeB magnet pole faces in Figure 58 were diamond-shaped, rectangular magnets. The rectangular NdFeB magnets were rotated 45° to form prism-shaped pole faces along the top C-gaps, thus forming a CRP. Along the bottom C gaps, the C magnets had cylindrical magnets that acted as cylindrical magnetic lenses to thus form a CRL composed of a row of cylindrical magnetic lenses. The bottom-left photos in Figure 58 show the CRP and the beam line with the CRP installed for testing with the neutron imaging plate. A close-up photo shows the three-prism and three-cylindrical lens, red C-magnet in the upper photo (top-middle) photos.

The bottom images in Figure 58 show the neutron multiple source images without the CRP (left) and with the CRP. The prism caused an upward deflection of spin-up (+) neutrons, which forms the dark band 2.8 mm above the bottom dark band; the downward deflection of the spin-down (−) neutrons produced the downward smearing in the images. The calculated 5.65-cm separation distance between spin-up (+) and spin-down (−) is compared to the measured spin-up neutron (+) deflection of 2.8 cm from the center axis (bottom dark band). This implies a measured 5.6-cm separation distance between spin-up (+) and spin-down (−) neutrons, which is close to the calculated 5.65-cm separation.

3. EXPERIMENTS WITH X-RAY COMPOUND REFRACTIVE LENSES AND MICROSCOPES

Refractive lenses have been used for focusing and imaging of light due to the relatively large refractive index and low absorption that glass lenses present to visible light. Refractive lenses for X-rays were considered impractical due to the large absorption and very small refractive index, which materials present to X-rays and result in a single X-ray lens that has a very long focal length and very small aperture size, and strongly attenuates X-rays, see Michette (1991). As early as 1948, Kirkpatrick and Baez (1948) and Kirkpatrick (1949) considered refractive X-ray lenses impractical because of the very small real part or decrement $\delta$of the refractive index $n = 1 - \delta - i\beta$, which causes minimal refraction and long focal lengths, and the large imaginary component β, which causes strong photoelectric absorption and Compton scattering of the X-rays (blurring).

Suehiro *et al.* (1991) proposed biconcave refractive X-ray lenses, which had 50- to 100-m focal lengths. Two years later, Yang (1993) discussed the feasibility of fabricating refractive Fresnel X-ray lenses on a flat sheet with electron-beams. In 1983 (or about 10 years before Yang's paper), Ceglio *et al.* (1983) actually fabricated a Fresnel phase plate for 1.5-keV X-rays that had blazed zones to increase the diffraction efficiency. Blazed Fresnel phase plates, which could be fabricated, had flat tilted surfaces like the reflection grating with triangle-shaped grooves in which the tilt angle is constant for all zones, rather than curved surfaces of the traditional optical Fresnel lens in which the tilt angle is different for each zone. The technology required for fabricating an X-ray Fresnel lens is therefore far more difficult to design and presented more technical challenges than that of a blazed Fresnel phase plate.

The invention of the X-ray refractive lens by Tomie in 1993 was initiated by his consideration of a blazed zone plate (Tomie, 2010).

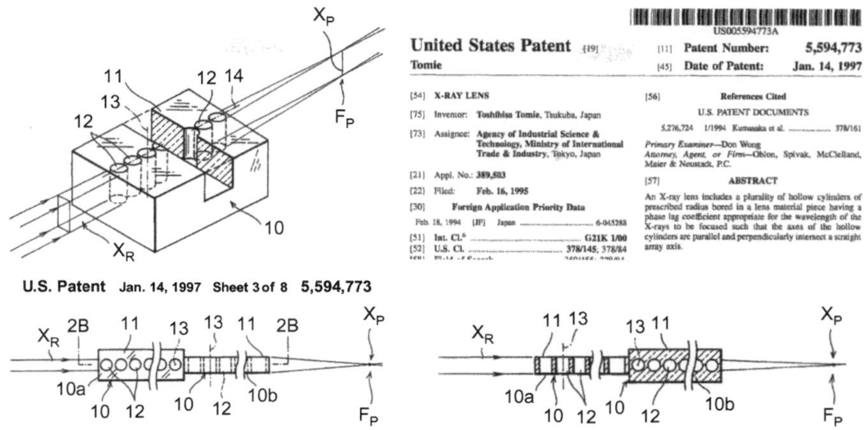

FIGURE 59 (Top) A 1D X-ray lens formed from a row of closely spaced cylindrical holes with minimally thin walls separating the holes. (Bottom) The top and side drawings of two crossed 1D X-ray lenses forming a 2D X-ray lens. The lower and upper rows of hemi-spherical lens can be merged to form the 2D compound refractive X-ray lens composed of a row of spherical biconcave lenses (Tomie, 1997).

This led Tomie from the concept of stacking thin films with debossed plano-concave lenses with two-dimensional (2D) focusing to drilling a line of closely spaced cylindrical holes, which act as a line of biconcave cylindrical lenses with one-dimensional (1D) focusing (Figure 59). In 1994 Tomie received a Japanese patent for the X-ray lens (Tomie, 1994) and in 1995 a German patent (Tomie, 1995). In February 1995, Tomie filed a patent with the U.S. patent office for the X-ray lens, which was granted in 1997 (see Tomie, 1997). Figure 59 shows the 1D embodiment of the Tomie X-ray lens from the U.S. patent. The Tomie X-ray compound refractive lens (CRL) is a row of drilled cylindrical holes; the 2D embodiment consists of two sets of drilled cylindrical holes in which the first and second set of cylindrical holes were perpendicular (see Figure 59).

In the U.S. patent Tomie also presented a 2D X-ray lens formed from the merged lower and upper rows of hemispherical holes, thus forming a CRL composed of a row of biconcave spherical lenses (Figure 60). Tomie realized the fabrication of the X-ray refractive lens can tolerate surface roughness because of the extremely small phase retardation caused by the refractive decrement δ of the considered X-ray lens materials. The most important issue was fabricating thin walls between unit lenses to avoid strong absorption of the X-rays through a stack of the unit X-ray lenses (Figure 61).

Tomie presented his X-ray lens at the 43rd Spring Meeting of the Japan Society of Applied Physics and Related Societies, Osaka, Japan, in March

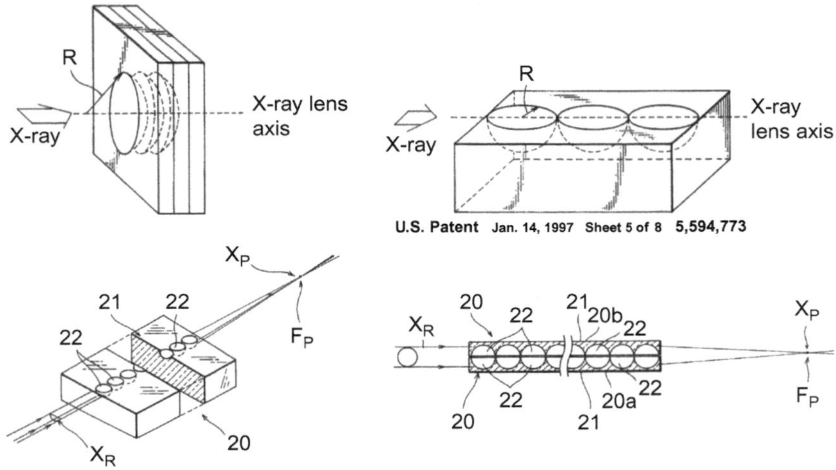

FIGURE 60 The four drawings are 2D, spherical biconcave lenses that form a compound refractive X-ray lens. The lower and upper rows of hemispherical lenses can be merged to form biconcave spherical lenses, which thus form the 2D compound refractive X-ray lens. The lower and upper rows of hemispherical lens can be merged to form the 2D compound refractive X-ray lens composed of a row of spherical biconcave lenses (Tomie, 1997).

1996. Later, in April 1996, Tomie presented his X-ray lens paper, which was accepted as a poster paper at the XRM 1996 International Conference on X-ray Microscopy and Spectroscopy in Wurzburg, Germany (August 19–23, 1996). Figure 62 shows one page from the program of the XRM 96 conference, which lists Tomie's X-ray lens paper in the same session with Snigerev (see Tomie, 2010). Snigerev at the time was developing linear Bragg–Fresnel diffracting lenses for hard X-ray focusing, imaging, and high-resolution diffraction techniques (see Snigerev *et al.*, 1995, 1997a).

The compound refractive X-ray lens invented by Tomie required a high-brightness synchrotron X-ray source for experimental demonstration, such as the European Synchrotron Radiation Facility (ESRF) in Grenoble, France; the Advanced Photon Source (APS) at the Argonne National Laboratory in Lamont, Illinois, or the Stanford Synchrotron Radiation Lightsource (SSRL) at the Stanford Linear Accelerator Center in Stanford, California, in the United States; or Super Photon Ring, 8 GeV (SPing-8) in Himeji, Japan. However, Tomie was not able to experimentally demonstrate his CRL in Japan because SPring-8, which started operation in 1997, was not available yet when Tomie disclosed the compound refractive X-ray lens during the 1996 meetings in Tokyo and Wurzburg.

The first demonstration of X-ray focusing by means of compound refractive X-ray lenses was left to Snigirev *et al.*, who in November

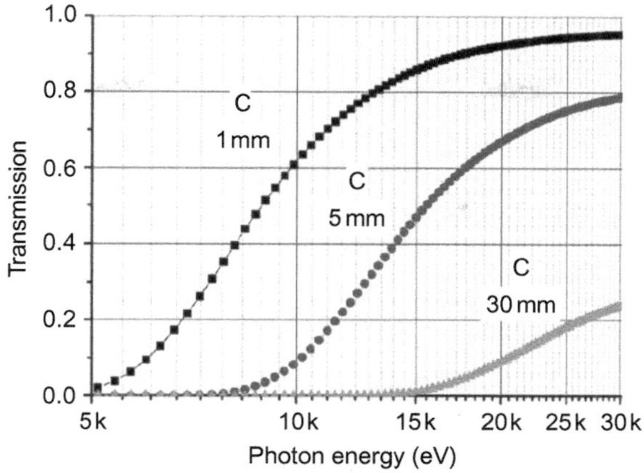

Table 1
Some design examples of X-ray lenses made of carbon.

Element	Radius of holes (mm)	Number of holes	Total length of lens (mm)	Focused X-ray wavelength (nm)	Focal length (mm)	Effective lens diameter (mm)	Focusing angle (mrad)	Microbeam diameter (μm)
c	0.5	50	50	0.1	1650	0.23	0.14	0.7
c	0.25	100	50	0.1	410	0.17	0.41	0.24
c	0.25	200	100	0.1	210	0.12	0.57	0.17
c	0.05	100	40	0.1	84	0.75	0.9	0.11
c	0.4	500	400	0.03	740	0.28	0.38	0.08

FIGURE 61 (Top) Plot of the transmission spectra for carbon with thickness of 1 mm, 5 mm, and 30 mm. X-ray CRLs made of carbon have a transmission fraction of 60% for 10-keV X-rays, which propagate along the center axis of the CRL, which has a material thickness of 1 mm along the center line. The CRL center line passes through the center minimum wall thickness of the row of biconcave or plano-concave lenses that form the CRL. The transmission of X-rays along the CRL centerline, which passes through 5 mm of lens material, drops to 10%. Therefore, the space between unit lenses (the lens center minimum wall thickness) should be smaller than $d = 20\,\mu\text{m}$ when the number of unit lenses is $N = 100$. (Bottom) Calculated parameters of candidate carbon CRLs (Tomie, 2010).

1996 published their results of their experimental demonstration X-ray focusing. Snigerev *et al.* (1996) used a 1D compound refractive X-ray lens in the form of a row of closely spaced cylindrical holes drilled in aluminum. Figure 63 shows the experimental results of their X-ray focusing experiments. The table in Figure 64 from their paper shows calculated parameters for X-ray CRLs of aluminum and boron. Their 19-mm long aluminum-copper (4% wt copper) CRL consisted of $N = 30$ closely spaced holes. Each hole had a 0.6-mm diameter ($R = 300$-μm radius of curvature) with a minimum wall thickness $d = 25\,\mu\text{m}$. On the D5 X-ray beam line at ESRF in Grenoble, France, in 1996, Snigerev *et al.* focused a 150-μm diameter beam of 14-keV X-rays (selected from an Si-111 monochromator) to an 8.0-μm line focus width within a focal length $f = 1.8\,\text{m}$. The 14-keV X-ray line focus width of 8.0 μm, which was measured by scanning with

XRM96 X-Ray Microscopy '96 held in Wurtsburg, Germany

Program; poster

Schneider, G., Maser, J.	**Zone Plates as Imaging Optics in High Diffraction Orders described by Coupled Wave Theory**	P 17
Snigirev, A., Snigireva, I., Souvorov, A., Kohn, V., Aristova, E., Hartman, Ya., Yunkin, V.	**Hard x-ray focusing by curved linear Bragg-Fresnel lens**	P 18
Snigireva, I., Souvorov, A., Legrand, F., Raven, C., Snigirev, A., Hartman, Ya., Yunkin, V.	**Composed Bragg-Fresnel lens for high resolution diffraction techniques**	P 19
Souvorov, A., Snigireva, I., Snigirev, A., Hartman, Ya., Aristova, E.	**Enhancement of the BFL light gathering power by modifying crystal reflectivity**	P 20
Tomie, T.	**Refractive x-ray lens for focusing hard x-rays**	P 21
Troussel, P., Bac, S., Olliivier, M., Pichet, M., Jalinaud, T., Le Breton, J.P., Sauneuf, R., M. Idir, B. Vidal	**Phase Fresnel zone plate focusing and application to x-ray imaging**	P 22

FIGURE 62 A page from the 1996 program of the International Conference on X-Ray Microscopy and Spectroscopy (Wurzburg, August 19–23, 1996). Tomie's paper on the X-ray lens was accepted as a poster paper. In 1996, people were theoretically studying Bragg–Fresnel optics, which is based on diffraction for focusing hard X-ray focusing, as seen in the program (Tomie, 2010).

a pinhole, presented a demagnification factor or gain $G = 20$. The demagnification factor of gain in this 1D X-ray focusing case is the CRL object distance r_0 divided by the image distance r_i, where $G = r_0/r_i$. Their measured gain of 20 with the compound refractive X-ray lens was threefold larger than the gain produced by a similar-sized 8-μm line width produced by slits. In their November 1996 *Nature* paper, Snigerev *et al.* noted that with proper material and geometry a CRL could be fabricated and used for thermal neutron applications. Snigerev *et al.*'s results inspired Eskildsen *et al.* (1998) at Risø in Denmark to use compound refractive neutron lenses to improve the performance of small-angle neutron scattering (SANS). In 2002 the neutron CRL for neutron imaging was designed, fabricated, and tested by Adelphi Technology in collaboration with I. Anderson and C. Dewhurst at the 58-MW neutron reactor source at Institut Laue-Langevin (neighboring ESRF in Grenoble) on the D22 SANS instrument (see Beguiristain *et al.*, 2002).

After the aluminum CRL, Snigerev *et al.* followed up with focusing experiments of CRLs composed of beryllium, pyrographite, aluminum, and polytetrafluoroethylene (PTFE) (see Snigerev *et al.*, 1997b, and Elleaume, 1998). Snigerev *et al.* (1997b) found the pyrographite CRL

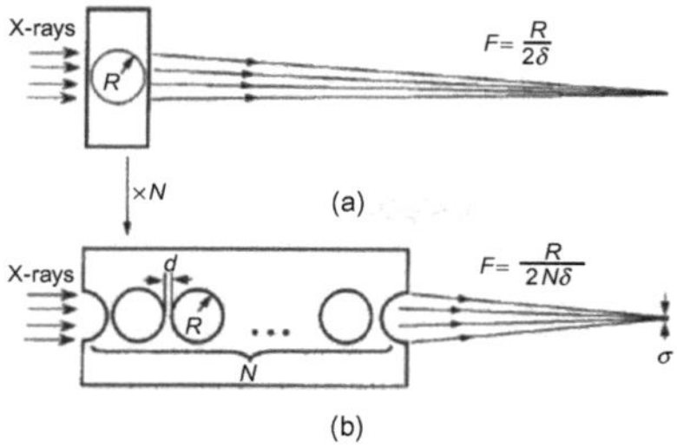

FIG. 1 Schematic diagram showing the principles of X-ray focusing by a compound refractive lens (CRL). As $(1-\delta)$ is smaller than 1 (where δ is the decrement of the refractive index), a collecting lens for X-rays must have a concave shape. *a*, A simple concave lens fabricated as a cylindrical hole in the material. *b*, A CRL consisting of a number (N) of cylindrical holes placed close together in a row along the optical axis, focuses the X-rays at a distance that is N times shorter compared to a single lens. R is the radius of the holes, d is the spacing between the holes, λ is the X-ray wavelength, and F is the focal distance for a parallel input beam.

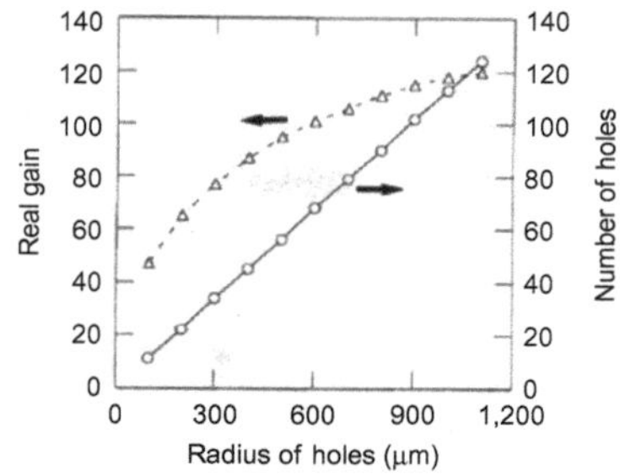

FIG. 2 Calculated number of holes N and real gain g versus radius of the holes R for a boron lens at fixed 1-m focal distance (10-keV X-rays). The number of holes increases linearly with increasing hole radius. Significant gain can be obtained with relatively few holes.

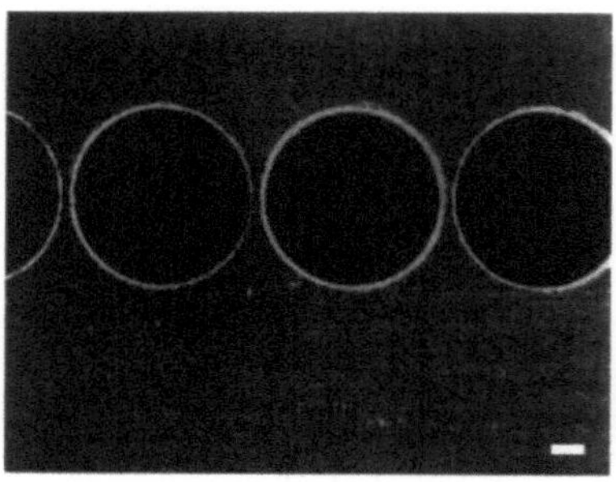

(a)

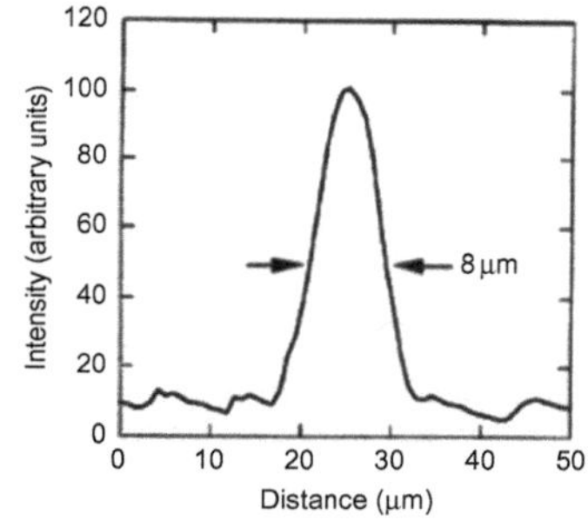

(b)

FIG. 3 *a*, Scanning electron microscope image of the tested compound refractive lens (CRL). 30 cylindrical holes of 300-μm radius were produced in an Al–Cu alloy (with 4 wt% Cu) plate using a computer-controlled drilling machine. Spacing d, the minimum alloy thickness between the holes, was ~25 μm. The length of the CRL was 19 mm. *b*, Measured profile of the focused X-ray beam taken at 1.8 m from the CRL. The width of the focal line is ~8 μm, which almost corresponds to the source size of 150 μm reduced by a demagnification factor $r_0/r_i = 20$.

FIGURE 63 The first experimental results of focusing X-rays with a CRL (Snigirev *et al.*, 1996).

showed very good focusing, but the small-angle scatter from the pyrographite produced a very strong background and the Teflon did not provide good focusing due to the rough hole surfaces. However, the aluminum and beryllium CRLs did produce good focusing. The beryllium CRL, which had $N = 69$ cylindrical holes or lenses, each with radius of curvature $R = 0.5$ mm and minimum wall thickness $d = 100$ μm, produced a focused line width of 2.5-μm full width half maximum (FWHM) at focal distance $f = 0.9$ m for 9-keV X-rays (Figure 65).

They also fabricated a 2D beryllium CRL with 37 horizontal cylindrical holes (lenses) alternated 37 vertical holes (lenses), each hole was 1-mm diameter (radius of curvature $R = 0.5$ mm) with minimum wall thickness $d = 50$ μm (Figure 66). With 9.2-keV X-rays the 2D beryllium CRL produced a focal spot of 6.8-μm vertical full width half maximum (FWHM)

TABLE 1 Calculated CRL parameters for boron and aluminium

Lens material	Energy (keV)	Number of holes	Effective aperture (μm)	Ideal gain	Real gain	CRL length (mm)
Boron	5	13	251	254	95	13
	10	55	211	359	93	55
	15	125	191	440	89	126
	20	222	177	508	84	224
	25	345	168	568	81	348
	30	501	160	622	78	506
	35	682	154	672	75	689
	40	892	149	719	73	901
Aluminium	5	11	85	22	0.2	11
	10	45	116	83	3	45
	20	184	163	336	18	186
	25	289	168	568	26	292
	30	417	160	622	31	421
	35	568	154	672	34	573
	40	743	149	718	37	750

Calculations were made under conditions typical for the ESRF beamlines: source size, 100 μm; source-to-lens distance, 50 m. Lens parameters were as follows: spacing between holes $d + 10\,\mu m$, hole radius $R = 500\,\mu m$, focal distance $F = 1$ m. The ideal gain (G) in the intensity at the focus for a point source was calculated according to ref. 4. The real gain (g) takes into account the finite source size and the attenuation of the X-rays owing to the absorption in the material between the holes[5]. The calculated focal spot size is $< 1\,\mu m$ in all cases.

FIGURE 64 Tabulation of calculated parameters for CRLs composed of carbon and boron (Snigirev *et al.*, 1996).

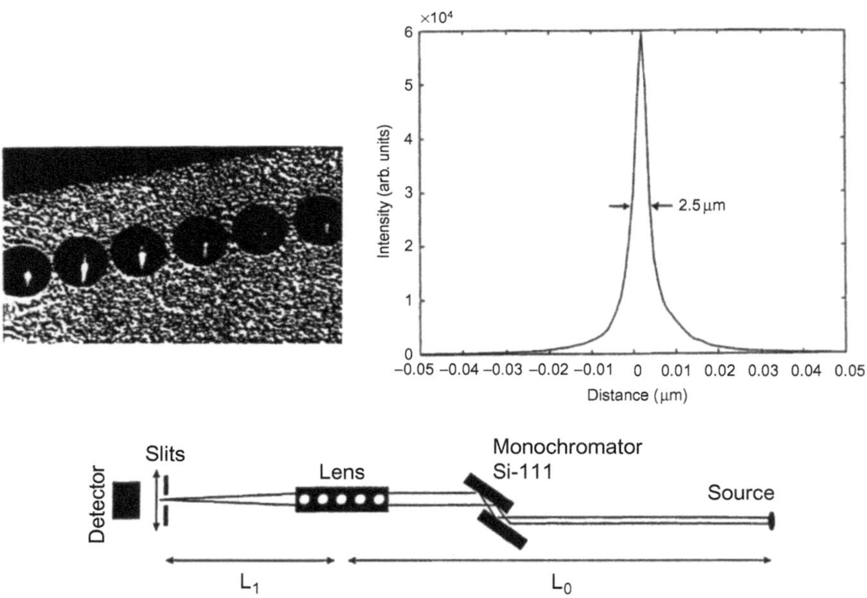

FIGURE 65 A layout of the beryllium CRL focusing experiment performed by Snigerev *et al.* at ESRF and a photo of a beryllium CRL, which has $N = 69$ cylindrical lenses, each with radius of curvature $R = 0.5$ mm and minimum wall thickness $d = 100\,\mu m$. The beryllium CRL is plotted as the measured intensity distribution of the focused line with minimum 2.5-μm FWHM line width at the focal distance $f = 0.9$ m produced by the beryllium CRL for 9-keV X-rays (Snigerev *et al.*, 1997b).

Material	Energy, keV	Focus FWHM, μm	Gain
Beryllium	12.6	3.4	5.7
Beryllium	9.2	6.8/17.5	13.6
Beryllium	9	2.5	2
Beryllium	12.5	4.2	10
Pyrographite	16.4	4	13.9
Aluminium(0.5%Mg)	18.8	3.5	1.3
Aluminium(3%Mg)	27.1	4.5	1.3
Teflon	16.7	5.2/18	

Table 2: Experimentally measured FWHM of the focal spot and gain form CRLs from low-Z materials. In the case of two-dimensional focusing the values for FWHM of the focal spot horizontally are given after a slash.

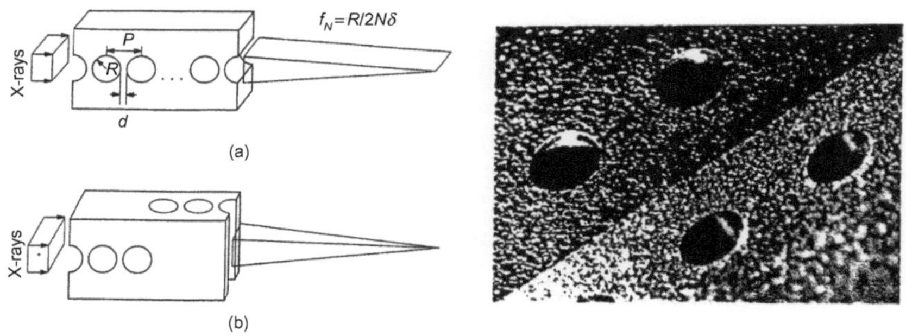

FIGURE 66 (Top) A table of the CRLs fabricated and tested at ESRF by Snigerev *et al.* (Bottom) A schematic of the 1D and 2D CRL composed of alternating horizontal and vertical cylindrical lenses (holes). The photo shows a 2D beryllium CRL with 37 horizontal lenses alternated 37 vertical lenses; each hole was 1-mm diameter (radius of curvature $R = 0.5$ mm) with minimum wall thickness $d = 50$ μm. With 9.2-keV X-rays the 2D beryllium CRL produced a focal spot of 6.8-μm vertical FWHM and 17.5-μm horizontal FWHM with a 2D intensity gain of 13.6 (Snigerev *et al.*, 1997b).

and 17.5-μm horizontal FWHM with a 2D intensity gain of 13.6. The detector was high-resolution charged-couple device (CCD) camera with 0.6-μm pixels and 14-bit dynamic range, which could take 100-ms to 1-minute exposure time.

Snigerev *et al.* (1997b) pointed out the following advantages of the CRL over other X-ray optical elements:

1. Relatively easy fabrication and low cost.
2. The CRL can be paired with an undulator without the need for a monochromator by selection of the correct harmonic frequency
3. Easy alignment.
4. The required surface quality of the CRL is much less than mirrors.
5. The instability of the CRL optical axis relative to the X-ray beam axis is not as critical as the case for mirrors.
6. The CRL can be easily cooled and has a high heat load capability.

7. The CRL and its ruggedness and microfocusing capabilities can be used with the powerful X-ray beams produced by the X-ray–free electron lasers.

They also concluded CRLs can be further improved by better drilling technology and parabola-shaped lenses.

Inspired by the 1996 *Nature* paper by Snigerev *et al.* (1996), Adelphi Technology embarked on the design, fabrication, and testing of X-ray CRLs. As its first step, Adelphi designed and fabricated 1D CRL X-ray lenses of aluminum, polymethyl methacrylate (PMMA) with tradename lucite ($C_5H_8O_2$) with mass density $1.2\,g/cm^3$, and polyethylene ($-CH_2-$) with mass density $0.96\,g/cm^3$ and at the SSRL beam line 2-3 measured the intensity profile of the line-focused X-rays produced by the cylindrical lens CRL (see Cremer *et al.*, 1999). The X-ray source was a synchrotron emitter whose source size in the vertical dimension was 445 μm, and the source to lens distance was 16.8 m. The SSRL beam line 2-3 has a double-crystal monochromator that was capable of producing X-rays from 2.4 keV to 30 keV with a 5×10^{-4} resolution. The approximate FWHM source size was $0.445 \times 1.74\,mm^2$, and the distance from the source to CRL was 16.81 m. The top of Figure 67 shows a diagram of the experimental configuration.

The X-ray beam size from this source was approximately $2 \times 20\,mm^2$ at the entrance to the experimental station; however, this size was reduced to be about $0.7 \times 22\,mm^2$ by a tantalum slit just upstream of the CRL position. The X-ray beam was further blocked in the x-dimension by an aluminum support and lens holder, which further reduced the X-ray beam spot entering the lens to $0.7 \times 3.2\,mm^2$. In this experiment, the source was imaged by the cylindrical lens CRLs only in the horizontal plane xz-plane, in which the smallest source dimension is 0.445 mm.

The important parameters for this type of focusing are the focal length and absorption, and for wavelengths shorter than 3 Å, low atomic number plastics are suitable. With the Lucite and polyethylene CRLs, Adelphi experimentally demonstrated 1D focusing for photon energies between 9 and 19.5 keV with focal lengths between 20 and 100 cm. For example, using 12-keV X-rays Adelphi achieved focal FWHM line widths down to 21 μm at a distance of only 20 cm from the CRL.

The Lucite CRL had $N = 50$ drilled cylindrical holes with 0.5-mm diameter ($R = 0.250$-μm radius of curvature) with $d = 50$-μm minimum wall thickness between adjoining holes, and for 9-keV X-rays, a calculated $2r_a = 0.284$-mm absorption aperture diameter, $2r_p = 150\,\mu m$ parabolic aperture diameter, 24% average transmission, $f = 87.1$-cm focal length, and gain $G = 1.5$.

The polyethylene CRL had $N = 200$ drilled cylindrical holes with 0.32-mm diameter ($R = 0.160$-μm radius of curvature) with $d = 75$-μm minimum wall thickness between adjoining holes, and for 19.5-keV X-rays,

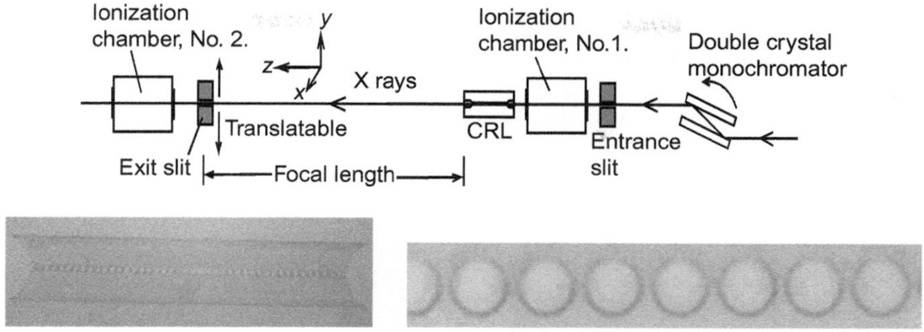

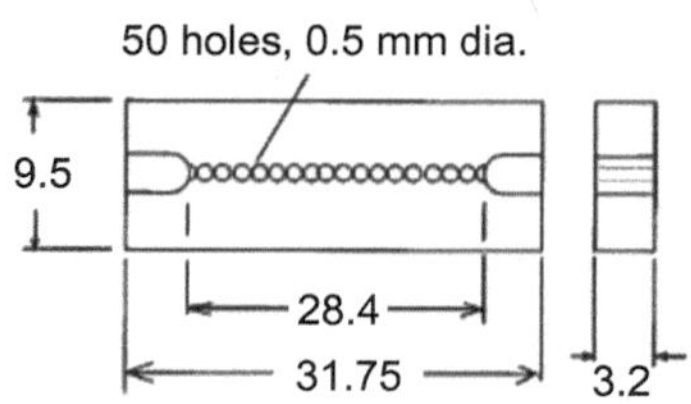

Photon energy (keV)	Calculated focal length (m)	Calculated image distance (m)	Measured image distance (m)	Measured image line-width (μm)	Source width from image line-width (μm)
12	0.35	0.35	0.33	21	1070
15	0.55	0.56	0.58	25	720
18	0.79	0.82	0.79	30	640
19.5	0.92	0.97	0.84	32.1	644

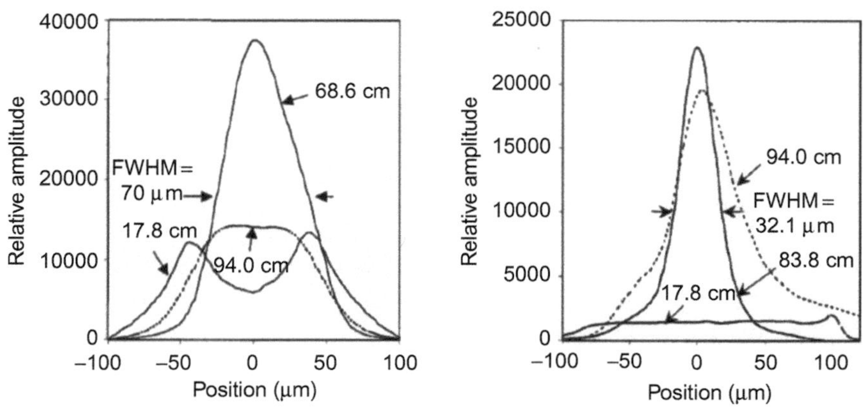

FIGURE 67 (Top) Adelphi Technology experiment at SSRL beam line 2-3 with images of Lucite CRL. (Middle) Lucite CRL drawing and tabulated CRL parameters. (Bottom left) Lucite $N = 50$ CRL beam profile plot at 9 keV with line focus width 70-μm FWHM at $f = 68.6$ cm. (Bottom right) Polyethylene $N = 200$ CRL beam profile at 19.5 keV with line focus width 32.1-μm FWHM at $f = 83.8$ cm. See text for details (Cremer *et al.*, 1999).

a calculated $2r_a = 0.298$-mm absorption aperture diameter, $2r_p = 50$-μm parabolic aperture diameter with 75% average transmission, $f = 93$-cm focal length, and gain $G = 3$.

Figure 67 shows a diagram of the experimental layout at beam line 2-3 at SSRL with images of the Lucite CRL and a diagram of the 50-hole Lucite CRL. The plot at the bottom left shows the measured relative amplitude of

9-keV focused photons with the 50-hole Lucite CRL for three lens to detector distances of 17.8 cm, 68.6 cm, and 94.0 cm, which shows the line focus occurs at 68.6 cm with a 70-μm FWHM. The plot at the bottom right shows the measured relative amplitude of 19.5-keV focused photons with the 200-hole polyethylene CRL for three lens to detector distances of 17.8 cm, 83.8 cm, and 94.0 cm, which shows the line focus occurs at 83.8 cm with a 32.1-μm FWHM. The table in the middle of the figure shows the calculated and measured focal lengths, image distances, and focal line widths for the 200-hole polyethylene CRL.

Lengeler *et al.* (1998) fabricated parabolic biconcave lenses of aluminum and beryllium and performed imaging experiments in which spherical aberration was eliminated, 2D focusing was introduced, and the variation of the number of lenses in the CRL produced focal lengths in the range of 0.5 to 2.0 m for photon energies from 6 to 60 keV (see Lengeler *et al.*, 1999a,b, 2001, and Schroer *et al.*, 2000a, 2001). The biconcave parabolic lenses of polycrystalline aluminum were designed and manufactured at the University of Technology in Aachen (RWTH Aachen University) and tested at ESRF in Grenoble on the ID22 beam line. Figure 68 shows the schematics of a single biconcave parabolic lens and the stack of lenses, as well as a plot of the measured and calculated parabolic lens profile and a table of the refractive index decrement δ and attenuation $\beta = \lambda \mu / 4\pi$ for 8- to 30-keV X-rays for aluminum and beryllium.

The aluminum and beryllium biconcave, parabolic lenses were fabricated at University of Technology in Aachen by a pressing or debossing technique using a pressing tool composed of two convex paraboloids with rotational symmetry that faced each other and were guided by a center ring. The 12-mm diameter aluminum or beryllium blank fits tightly in the centering ring, and then the blank is pressed simultaneously on both sides by the paraboloids in a similar manner to the production of coins. The physical aperture diameter of their lenses was about 1 mm, which matched the angular divergence of the undulator beams of the third-generation synchrotron radiation X-ray sources. As noted by Lengeler *et al.* (1999a), the 2D parabolic lens and aluminum and beryllium CRLs were well suited to withstand the heat of the intense X-ray beams produced by undulators and the X-ray–free electron laser developed 10 years later, the Linear Coherent Light Source (LCLS) at the Stanford Synchrotron Radiation Laboratory.

Yang (1993) has shown the ratio of the decrement δ and the absorption index $2\pi\beta$ is an excellent guide in determining the best lens material. The decrement δ of the X-ray refractive index $n = 1 - \delta - i\beta$ is proportional to the density of atomic electrons ρ and thus to the product of the CRL material atomic number Z and density ρ; that is, $\delta \propto \rho Z$. However, the X-ray absorption index varies as $\beta \propto Z^4 \rho$, and the ratio varies $\delta/\beta \propto Z^{-3}$. Hence, using the lowest atomic number material for the CRL lenses gives

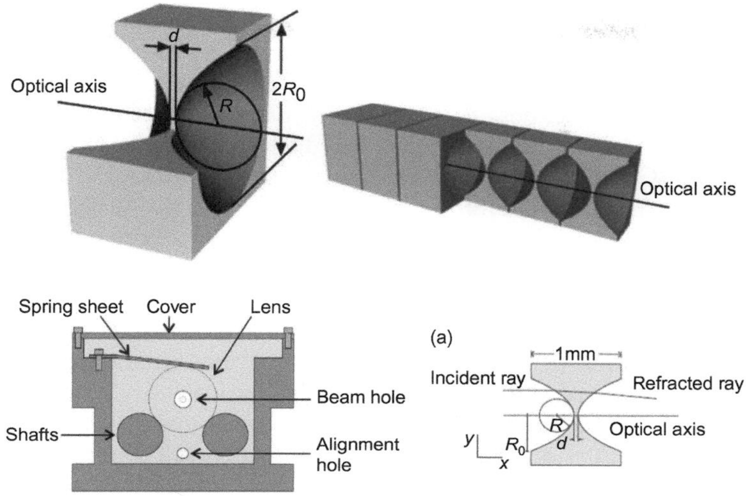

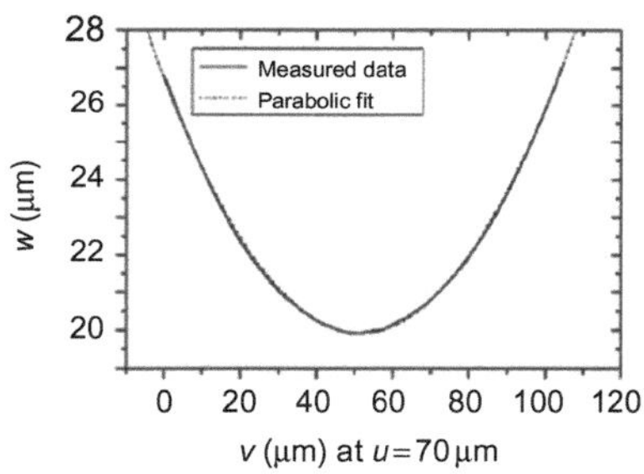

	Be		Al	
E(keV)	$\delta(10^{-6})$	$\beta(10^{-9})$	$\delta(10^{-6})$	$\beta(10^{-9})$
8	5.334	2.419	8.579	158.20
10	3.412	1.095	5.468	66.61
15	1.515	0.341	2.414	13.50
20	0.852	0.195	1.355	4.40
30	0.379	0.113	0.601	0.95

FIGURE 68 (Top) Diagrams of the biconcave parabolic lenses of polycrystalline aluminum or beryllium and the mounting structure that were designed and manufactured at the University of Technology in Aachen (Lengeler *et al.*, 2001). (Center, left) Schematic diagrams of the X-ray CRL and mount, and (Center Right) a schematic of the biconcave parabolic lens. (Bottom Left) A plot of the measured and calculated parabolic lens profile. (Lengeler *et al.*, 1999a). (Bottom Right) A table of the refractive index decrement δ and attenuation $\beta = \lambda\mu/4\pi$ for 8- to 30-keV X-rays for aluminum and beryllium (Lengeler *et al.*, 1999a).

the largest value of ratio of refraction to attenuation δ/β, which is the most refraction δ for the least attenuation β. Using a more sophisticated analysis, Yang (1993) showed lithium and beryllium would be the best choices for lens materials followed by carbon. Smither *et al.* (1997) also showed beryllium would be an excellent lens material, providing the largest possible lens aperture other than lithium.

The ESRF ID22 X-ray undulator source had a vertical dimension $d_v = 35\,\mu\text{m}$ and horizontal dimension $d_h = 700\,\mu\text{m}$, and Lengeler *et al.* (1999a) were able to obtain a demagnified image of the source with the biconcave parabolic aluminum lens CRL. The aluminum CRL had $N = 33$ biconcave parabolic lenses, each with radius of curvature $R = 200\,\mu m$, center minimum wall thickness $d = 16 \pm 2\,\mu m$, physical aperture diameter $d_m = 870\,\mu m$, and surface roughness $\sigma = 1.0\,\mu m$. For 15-keV X-rays the parabolic lens CRL had an absorption aperture diameter $2r_a = 154\,\mu m$, focal length $f = 1.26$ m, and overall transmission $T = 5.3 \times 10^{-3}$. The CRL X-ray transmission was limited due to Compton scattering, which causes blurring, and photo-absorption, which causes loss of image signal.

The 15-keV X-ray source had 1.6-μm FWHM vertical dimension and 14-μm FWHM horizontal dimension, and the parabolic aluminum CRL produced a vertical demagnification $M_v = 1/22$ and horizontal demagnification $M_h = 1/50$. Their measured elliptical spot size of 1.6 ± 0.3-μm vertical FWHM $\times 14.1 \pm 0.3$-μm horizontal FWHM was in good agreement with the calculated values of, respectively, $1.5 \pm 0.1\,\mu m$ and $14.3 \pm 1.4\,\mu m$. The detector has a point spread function of $1.2 \pm 0.1\,\mu m$. The theoretical focal spot intensity gain over pinhole imaging was $G_{th} = 200 \pm 40$, whereas the measured experimental gain was $G_{\exp} = 177 \pm 35$. Figure 69 shows a schematic of the focusing and imaging experiments performed at ID22, a configuration for imaging the undulator source, and a demagnified X-ray image of the source with 15-keV X-rays. The demagnification, along with losses due to CRL absorption, still lead to very large intensity gains in the CRL focal spot. Lengeler *et al.* (1999a) also note the CRL can act as a source of parallel X-rays by using the CRL to refract the diverging rays of a smaller, distant undulator source to parallel rays.

Figure 70 shows the configurations used for (1) imaging an object illuminated by the undulator source and (2) obtaining a diffraction-limited CRL focal spot, which is used to provide a transverse coherent illumination of an object. Producing a diffraction-limited CRL focal spot of X-rays with wavelength λ requires the CRL effective aperture diameter d_e in relation to the source vertical height s_v and undulator source to CRL distance r_s to be $d_e \leq \lambda r_s/s_v$. This result is obtained from the CRL focal spot vertical height $f_v = d_v f/r_s$ for focal length f, the vertical height o_v of object illumination by the CRL $o_v = d_e r_0/f$ for CRL focal spot to object distance r_0, and the expression for lateral coherence width $l_{coh} = \lambda r_0/f_v$.

For the 23.5-keV X-ray imaging experiments, the aluminum CRL had $N = 62$ biconcave parabolic lenses, each with radius of curvature $R = 200\,\mu m$, center minimum wall thickness $d = 16 \pm 2\,\mu m$, physical aperture diameter $d_m = 870\,\mu m$, and surface roughness $\sigma = 1.0\,\mu m$. For 23.5-keV X-rays, the parabolic lens CRL had an absorption-limited effective aperture diameter $2r_e = 211\,\mu m$, focal length $f = 1.65$ m, and aluminum linear attenuation $\mu = 5.81\,cm^{-1}$. The aluminum parabolic lens CRL produced a 23.5-keV X-ray image of a zone plate with 169 gold zones on an Si_3N_4 substrate with magnification $M = 12$ that clearly resolved the 0.3-μm

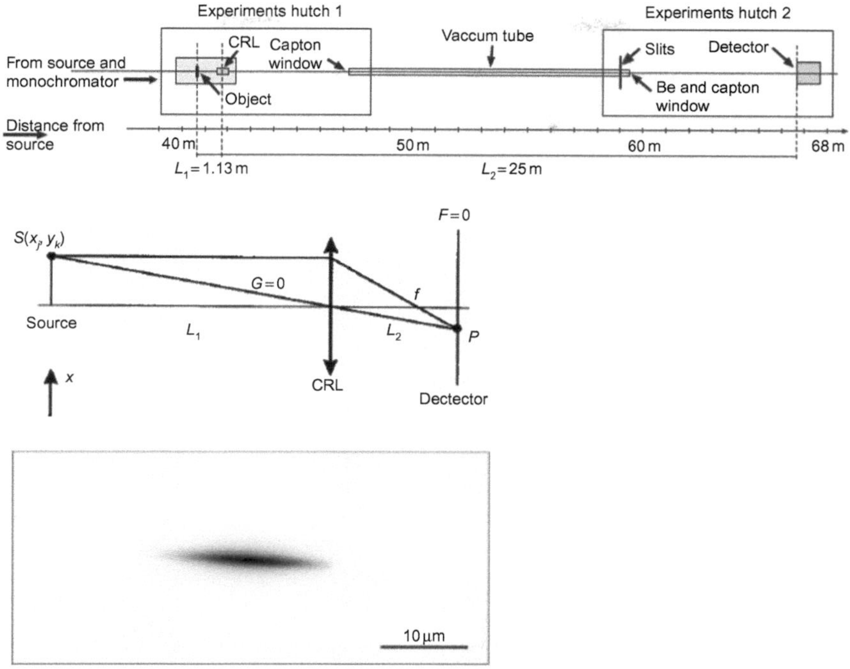

FIGURE 69 (Top) Experimental layout of the ID22 beam line undulator source at ESRF (Schroer *et al.*, 2000a). (Middle) Schematic for the configuration for obtaining demagnified X-ray images of the undulator source (Lengeler *et al.*, 1999a). (Bottom) A demagnified 15-keV X-ray image of the source produced by the aluminum CRL composed of 33 biconcave parabolic lenses (Lengeler *et al.*, 1999a).

width of the outer zone (see Figure 70). Figure 12 also shows a 23.3-keV X-ray image of an insect antenna produced by an aluminum biconcave, parabolic lens CRL with $N = 80$ lenses, and focal length $f = 1.218\,\text{m}$ with an absorption-limited effective aperture diameter $d_e = 180.6\,\mu\text{m}$ (see Lengeler *et al.*, 2001).

Beryllium lenses transmit X-rays better than aluminum lenses particularly at lower X-ray energies, and thus can be used to increase the flux of focused X-rays in microbeam X-ray probes for absorption and fluorescent spectroscopy as well as diffraction at energies as low as 2-keV. The focal spot size of the microbeam is limited by geometric optics of the source size, source to CRL distance, and CRL focal length, which remains unchanged independent of the choice of aluminum or beryllium CRLs. In full-field microscopy, parallel projection imaging or contact radiography a high-resolution X-ray camera is placed behind the sample in which the resolution is limited by the detector for thin samples. High-resolution

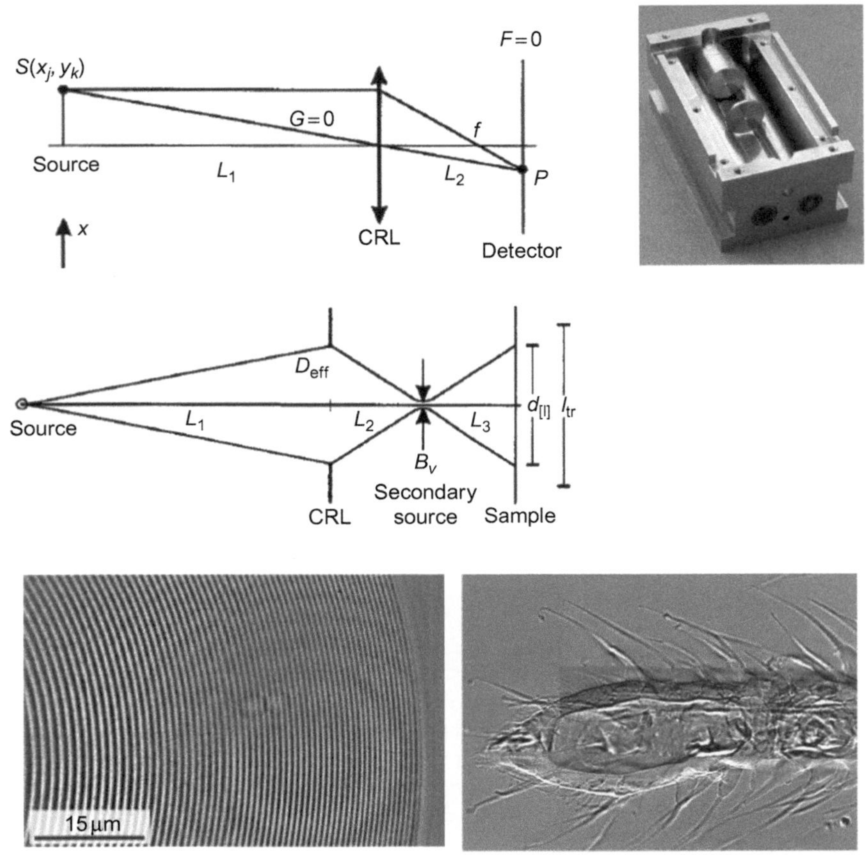

FIGURE 70 (Top left) Configurations for CRL imaging of an object illuminated by an undulator X-ray source (Lengeler *et al.*, 1999a), and (top right) a photo of the parabolic lens CRL (Lengeler *et al.*, 2001). (Middle) Configuration for producing a diffraction-limited spot by the CRL using an undulator source. The diffraction-limited spot then acts as a transverse coherent X-ray source that illuminates an object (Lengeler *et al.*, 1999a). (Bottom left) A 23.5-keV X-ray image of a zone plate with 169 gold zones on an Si_3N_4 substrate with magnification $M = 12$, which resolves the 0.3-μm wide outer zone (Lengeler *et al.*, 1999a). (Bottom right) An insect antenna is imaged at 23.3 keV with aluminum parabolic lens CRL with magnification $M = 17.6$ (Lengeler *et al.*, 2001).

detectors use a thin monocrystalline scintillator screen to convert X-rays into visible light which is transferred to a charge-coupled device (CCD) chip using a visible light microscope optic. The maximum resolution achieved by these detectors is limited to slightly below 1 μm due to X-ray scatter in the scintillator and the diffraction limit of the visible light optic. At the highest 1-μm resolution the efficiency of detection is reduced.

However, magnifying X-ray optics such as the CRL can achieve submicron resolution. Magnification of the object by an X-ray Fresnel zone plate, Bragg–Fresnel optic, or refractive X-ray lens presents object features of less than 1 μm, which spans a multitude of the pixels of the high-resolution detector, thereby achieving submicron object resolution. The biconcave parabolic lenses of aluminum and beryllium provide high-quality magnification due to the lack of spherical aberration Beryllium lenses are especially valuable in that beryllium has a significantly smaller X-ray attenuation cross section than aluminum especially at the lower X-ray energies.

The use of parabolic X-ray lens CRLs in tomography can produce the three-dimensional (3D) inner structure of opaque samples, which can be imaged with submicron resolution free of spherical aberration. For scanning microscopy techniques with hard X-rays, such as micro-fluorescence mapping or tomography, microdiffraction, micro-small angle scattering, or micro-absorption spectroscopy, an intensive hard X-ray microbeam is required, which can be generated by various hard X-ray optics. Parabolic refractive lenses composed of aluminum are routinely used to generate hard X-ray microbeams at several beam lines of the ESRF. The RWTH Aachen University and ESRF groups pursued the development of biconcave parabolic lenses of beryllium, having previously achieved success with the parabolic biconcave aluminum lenses. Schroer *et al.* (2002b) fabricated and tested a beryllium CRL compsed of biconcave parabolic lenses and obtained the first experimental results in both-full field microscopy and microbeam production with the new parabolic X-ray CRL at the undulator beam line ID22 at ESRF

The upper left plot in Figure 71 shows the parabolic profile measured by white light interferometry for biconcave parabolic lenses of beryllium, aluminum, and nickel in which the measured surface parabolic surface profile of these lenses closely matches the theoretical parabolic profile in which the beryllium parabolic lens had an on-axis radius of curvature $R =$ 184 μm. The minimum wall thickness of the biconcave parabolic lenses was measured by X-ray transmission as $d = 170\,\mu\text{m}$ and the lenses had a physical aperture diameter $2r_m = 800\,\mu\text{m}$. A hard X-ray microscope with a beryllium CRL composed of biconcave, parabolic lenses was fabricated by Schroer *et al.* (2002b) and tested on the undulator ID22 beam line at ESRF. The beryllium X-ray CRL was composed of $N = 91$ biconcave parabolic lenses with a focal length $f = 493$ mm at 12 keV.

Schroer *et al.* (2002b) obtained a demagnified image of the ID22 undulator X-ray source at 12 keV using the beryllium CRL, measured by an X-ray fluorescent knife edge. To obtain 12-keV X-rays without the higher harmonics, the undulator beam was directed to a silicon mirror at 15° grazing incidence angle. The reflected X-rays were incident on an Si-111 doublecrystal monochromator whose output was monochromatic 12-keV

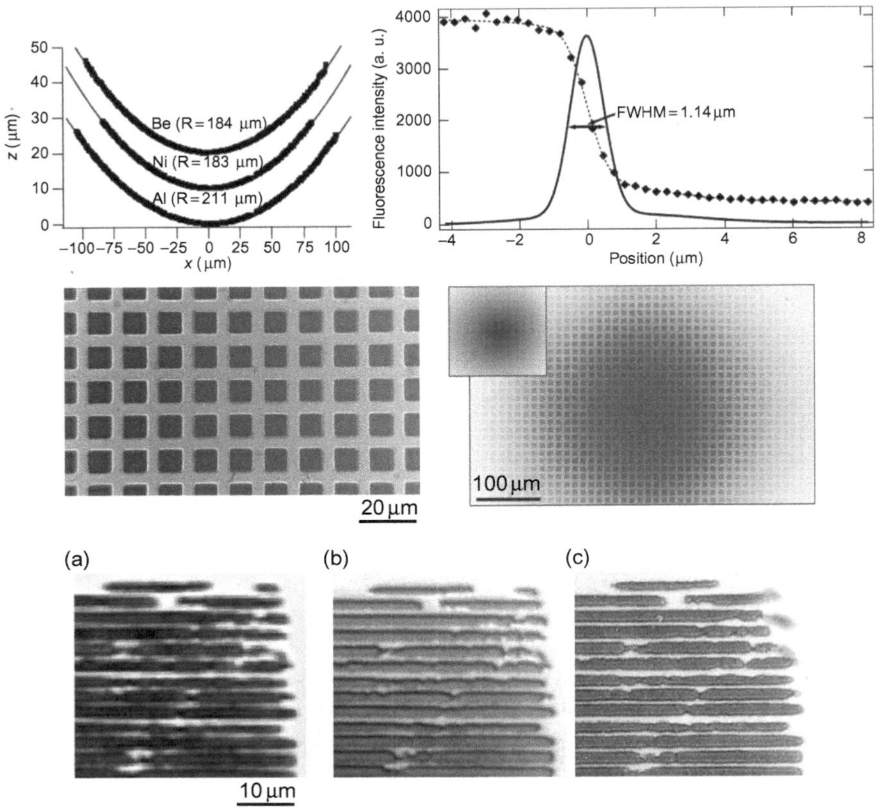

FIGURE 71 (Top left) Parabolic profiles of beryllium, aluminum, and nickel lenses measured by light interferometry. (Top right) The vertical scan of $M = 1/81$ demagnified image of undulator source by beryllium parabolic CRL at 12 keV. (Middle left) A close-up view of the refractive image of the 2000-mesh nickel grid by the $N = 120$ beryllium CRL at $M = 10$ with 24.9-keV X-rays. The right image shows the full $N = 120$ beryllium CRL image. The inset shows the image of the same nickel mesh by the $N = 120$ aluminum parabolic CRL at 24.9 KeV. (Bottom left) X-ray image (a) by contact radiography imaging of the gold bar test structure at 20 keV; the middle image (b) is by $N = 120$ aluminum parabolic CRLs with $M = 10$ at 24.9 keV, and the right image (c) is by $N = 91$ beryllium parabolic CRL with $M = 10$ at 12 keV (Schroer *et al.*, 2002b).

X-rays. The beryllium CRL was set at source to CRL; that is, object distance was $r_0 = 41.4$ m and the detector was set at image distance $r_i = 498$ mm to produce a demagnified image with magnification $M = 1/83$.

Schroer *et al.* (2002b) measured the vertical profile of the beryllium CRL image of the 12-keV source with a vertical knife edge scan using the X-ray fluorescent knife edge. Figure 71 plots the knife edge scan results for the demagnified image undulator source. The measured image profile data in the plot are indicated by diamonds, fitted by the sum of two error functions (indicated by the dashed line). The demagnified image profile of the undulator source is the derivative of measured intensity of the knife edge scan across the image spot (indicated by the solid line) The measured 1.14-μm FWHM of the vertical beam profile was slightly larger than the expected 0.73-μm vertical beam dimension, which was expected for an effective 60-μm vertical source dimension. The measured 11-μm FWHM horizontal beam dimension of the demagnified image corresponded to a 900-μm FWHM horizontal source dimension via magnification $M = 1/83$, which is set by the object $r_0 = 41.4\,\mathrm{m}$ and image $r_i = 498\,\mathrm{mm}$ distances.

Schroer *et al.* (2002b) performed imaging experiments at 12 keV with the beryllium parabolic CRL with a nickel mesh with 2000 mesh and a 12.7-μm pitch. The nickel mesh sample was positioned at object distance $r_0 = 545$-mm from the parabolic beryllium CRL, and the undulator X-ray source to sample nickel mesh was $r_s = 40.85\,\mathrm{m}$. The undulator X-ray had a large transverse spatial coherence, which produces a complicated phase-contrast image. A series of defocused images of the test structure can be mathematically processed to reconstruct the projection of the sample X-ray refractive index. A pure absorption contrast image of the sample can be acquired with incoherent X-rays, which then yields the projection of the sample X-ray linear attenuation coefficient.

To obtain an absorption contrast image a 1.0-mm thick B_4C diffuser was used in these beryllium CRL imaging experiments to reduce the transverse coherence of the X-rays incident on the gold test structure. The beryllium parabolic CRL imaged the nickel mesh onto a Kodak high-resolution X-ray film positioned at image distance $r_i = 5160\,\mathrm{mm}$ with a magnification $M = 10$ with 12-keV X-rays. The middle-left image in Figure 71 is a close-up view of a portion of the 12-keV nickel mesh image, and the middle-right image shows the complete image of the 12-keV nickel mesh produced by the parabolic beryllium CRL. The beryllium CRL had a 450-μm field of view (FOV) of the nickel mesh at 12 keV.

In Figure 71 the inset at the upper-left corner of the beryllium nickel mesh image is the X-ray image of the same nickel mesh, which was produced by an aluminum parabolic lens CRL. The aluminum parabolic CRL was composed of $N = 120$ biconcave parabolic lenses with a focal length $f = 1050\,\mathrm{mm}$ at 24.9 keV. The inset X-ray image is of the same nickel mesh taken with the aluminum parabolic CRL at 24.9 keV with a magnification $M = 21$ in which the aluminum CRL had a smaller 130-μm FOV than the beryllium CRL FOV of 450-μm. The 24.9-keV aluminum and 12-keV beryllium hard X-ray microscope images were nearly free of spherical aberration, which indicated the respective aluminum and beryllium objective CRLs were composed of parabolic lenses.

Schroer *et al.* (2002b) imaged a gold test structure with the beryllium parabolic CRL. The gold test structure was a set of horizontal gold bars with 2-μm, 1-μm, 1-μm widths, and 2-μm thickness. The gold test structure was set on a silicon wafer. The beryllium CRL again was composed of $N = 91$ biconcave parabolic lenses with $f = 493$-mm focal length for 12-keV X-rays, and the gold test structure was positioned at object distance $r_0 = 545$-mm from the parabolic beryllium CRL. The beryllium parabolic CRL imaged the test structure of parallel gold bars onto a high-resolution X-ray fast-readout, low-noise (FReLoN 2000) detector positioned at image distance $r_i = 5160$-mm. The undulator X-ray source to gold test structure position was $r_s = 41.4$ m. The FReLoN 2000 detector was developed by ESRF in which X-rays are converted by the detector into visible light via a 3.5-μm thick europium (Eu)-doped lutetium-aluminum-oxygen (LAO $Lu_3Al_5O_7$) single-crystal scintillator. The emitted visible light is imaged onto a low-noise CCD chip using a visible light optic. The detector resolution is slightly below 1.0 μm due to X-ray scatter in the scintillator.

The left-most X-ray image at the bottom right of Figure 71 is a contact 20-keV radiograph of the gold test structure in which the gold test structure was set 2.0 mm in front of the detector. In comparison, the bottom-middle X-ray image is a 24.9-keV refracted image with magnification $M = 10$, which is produced by the aluminum CRL with $N = 120$ biconcave parabolic lenses. The bottom-right X-ray image is a 12-keV refracted image with magnification $M = 10$, which is produced by the beryllium CRL with $N = 91$ biconcave parabolic lenses. The images of the gold test structure produced by the parabolic aluminum and beryllium CRLs were of higher resolution than the contact radiographic image, and the far-right beryllium CRL image had a significantly better resolution than the middle aluminum CRL image.

The large transverse coherence of X-rays from third-generation undulator sources allows phase-contrast imaging by slightly defocusing the sample. Similar to Gabor in-line holography, the contrast is produced by sample refraction effects. Weakly absorbing objects can be imaged with high contrast and a minimum of radiation damage with phase-contrast imaging in which an object can be imaged in the edge contrast or Fresnel regime by appropriate choice of the defocusing distance L_p. Schroer *et al.* (2000a) performed phase-contrast full-field microscopy experiments at ESRF.

In Figure 72, the sample is illuminated by a third-generation undulator source (top). The sample is slightly defocused by varying the object to image plane distance L_p. The middle X-ray images are phase-contrast micrographs (small rectangle on detector) of a tungsten borate-boron interface inside a boron fiber with tungsten borate core for object to image plane distances L_p of 1 mm, 50 mm, 70 mm, and 90 mm (left to right). The

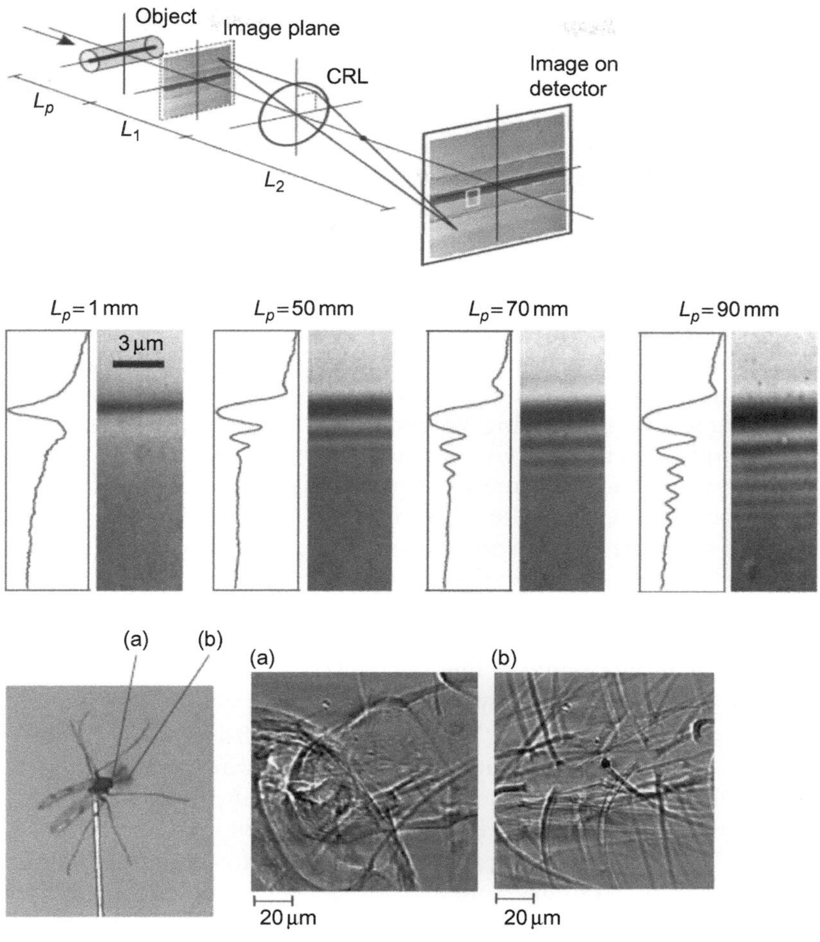

FIGURE 72 (Top) A diagram of the phase-contrast full-field microscope. The sample is illuminated by a third-generation undulator source. The sample is slightly defocused by varying the object to image plane distance L_p. (Middle) Phase-contrast micrographs (small rectangle on detector) of a tungsten borate-boron interface inside a boron fiber with tungsten borate core for object to image plane distances L_p of (from left to right) 1 mm, 50 mm, 70 mm, and 90 mm. (Bottom) Two 23.5-keV X-ray images of parts (a) and (b) of the insect in the photo at the left with $M = 12$ magnification using a CRL for phase-contrast imaging (Schroer *et al.*, 2000a).

bottom of Figure 72 shows a photo of the insect with locations (a) and (b) marked. The bottom two 23.5-keV X-ray phase-contrast images are of parts (a) and (b) in the insect photo. The images were acquired using a scintillator-based high-resolution CCD camera imaging system. The imaging system used by Schroer *et al.* at ESRF was a FReLoN 2000 detector in which a visible light microscope projects the light produced by a 3.5-μm thick europium-doped LAG scintillator onto a cooled, 2048 × 2048 pixel CCD camera in which the detector resolution is 1 μm.

Parabolic refractive X-ray lenses with short focal lengths can generate intensive hard X-ray microbeams with focal dimensions in the 100-nm range even at a short distance from a synchrotron radiation source. Schroer *et al.* (2003b) fabricated planar parabolic lenses made of silicon, which have focal lengths in the range of a few millimeters at hard X-ray energies. The lenses are fabricated using electron beam lithography and deep trench reactive ion etching. The lens pattern is written into a 30-nm chromium mask by electron-beam (e-beam) lithography and subsequent wet etching. Next the chromium mask transfers the patterns using reactive ion etching into an underlying 300-nm thick silicon dioxide layer. The silicon dioxide layer then serves as mask for the deep trench reactive ion etching process, and the lens structure is imprinted into the underlying bulk silicon. The depth of the lenses is larger than 20 mm. See Figure 73, which is a scanning electron micrograph of a planar array of the parabolic refractive X-ray lenses made of silicon.

Given synchrotron X-ray source object distance $r_0 = 42.1$ m the design of Schroer *et al.* (2003b) was composed of two crossed CRLs designed to focus the ^{third} harmonic of the ESRF ID22 undulator source of 25 keV in which the measured X-ray intensity incident on the lens was $I_0 = 4.3 \times 10^{12}$ photons/sec/mm^2. The first CRL was 8.4 mm long with $N = 100$ lenses for vertical focusing with on-axis radius of curvature $R = 2.15\,\mu$m, physical aperture diameter $2r_m = 30\,\mu$m, and image distance $r_i = 15.6$ mm, which produced an image vertical line width 210 ± 50 nm. The second CRL was 4.2 mm long with $N = 50$ lenses for horizontal focusing with on-axis radius of curvature $R = 2.0\,\mu$m, physical aperture diameter $2r_m = 18\,\mu$m, and image distance $r_i = 26.7$ mm, which produced an image horizontal line width 380 ± 90 nm. The two CRLs had a center minimum wall thickness $d = 3.0\,\mu$m.

These CRLs made tiny demagnified images of the undulator source. The CRL-produced image size of the microbeam was measured by scanning a gold knife edge across the beam horizontally and then vertically, and recording the fluorescence radiation in an energy-dispersive detector. Figure 73 shows the plotted results of the vertical and horizontal knife edge scans of a gold knife edge through the focused microbeams. Also shown are plots of the ideal diffraction-limited diameter (Airy disk for the smallest focal diameter), which is produced by CRLs composed of various materials as a function of the X-ray energy. The plots indicate the use of diamond as the lens material may produce microbeams with image sizes of 20 nm and smaller in the energy range from 10 to 100 keV.

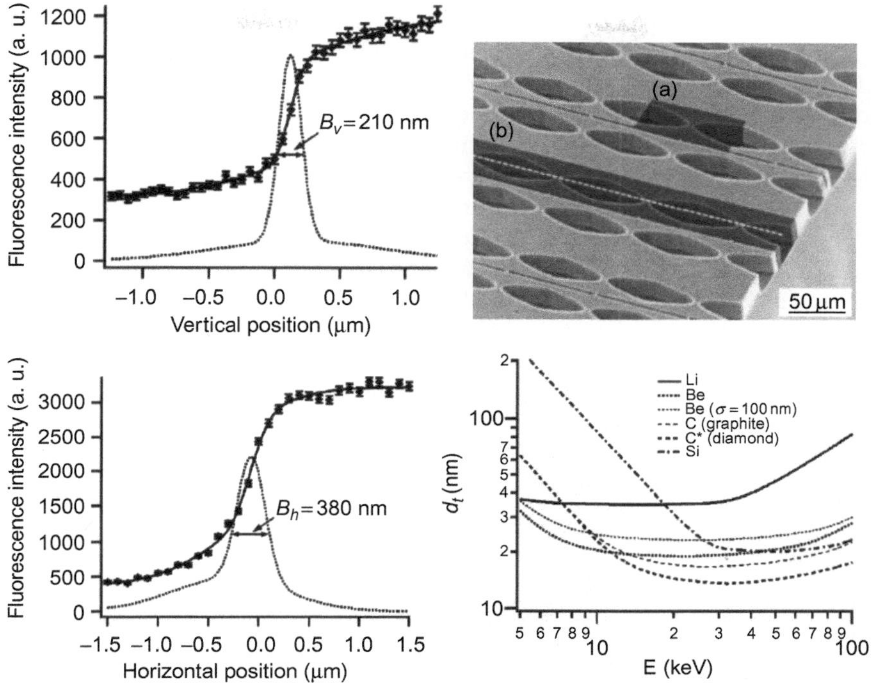

FIGURE 73 (Top right) A scanning electron micrograph of a planar array of parabolic refractive X-ray lenses made of silicon. The two plots at the left are the vertical and horizontal knife edge scans of a gold knife edge through the focused microbeams with focal distances of a few millimeters and focal spot dimensions of 380 nm × 210 nm for 25-keV synchrotron X-ray ray source at 42 m from the CRL. (Bottom right) The diffraction-limited Airy disk diameter for various CRL materials as a function of X-ray energy. Diamond CRLs produce the smallest Airy disk (Schroer *et al.*, 2003b).

Artemiev *et al.* (2006) fabricated parabolic planar CRLs made from glassy carbon by means of laser ablation. The 1D and 2D lenses were tested at the ESRF at the bending magnet beam line BM-5 in which the minimum experimentally achieved focal line width is 1.4 μm. Their glassy carbon CRLs are 1D biconcave parabolic lenses in which the use of two CRLs in a crossed geometry produces 2D focusing. The 1D glassy carbon CRLs were fabricated using a copper vapor laser with wavelengths of 510.6 nm and 578.2 nm, repetition rate of 16 kHz, pulse duration of 20 ns, and 10-W average power. The laser light was directed into an optical projection system, which had exchangeable objectives lenses with 8.2-mm, 13.9-mm, and 25-mm focal lengths. The CRLs were positioned with 2D mounting, which allowed minimal 0.8-μm steps over a 150-mm full range.

Artemiev *et al.* (2006) used two methods to fabricate the CRLs in glassy carbon. In the first (the direct method), which can be used only to produce profiles with a large curvature radius of the order of millimeters, the material is removed by focused laser beam in which the 30-μm laser beam. The laser beam is controlled by a mirror and burns deep cylindrical holes in a step-by-step manner, thereby creating a biconcave parabolic surface profile. In the second (projection method, shown in Figure 74), an optical demagnification of a mask of the desired parabolic profile is used in which optical lenses produce an image of the mask, which is two orders of magnitude smaller than the physical mask. The laser light is focused through the mask onto a glassy carbon block in which the glassy carbon block acts

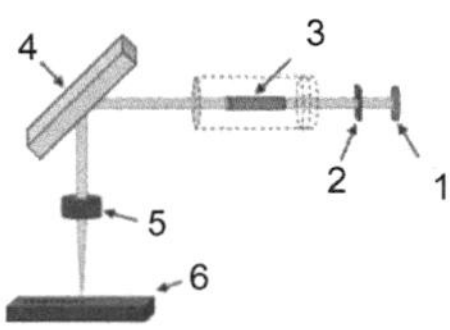

Lens	Number of biconcave elements N	Curvature radius R (μm)	Thin part of the biconcave element d (μm)	Relief depth h (μm)	Geometric aperture A (μm)	Manufacturing method
CRL-4	4	200	<100	∞1000	900	Direct
CRL-7	7	5	≤20	∞50	40	Projected
CRL-200	200	5	≤20	∞50	40	Projected

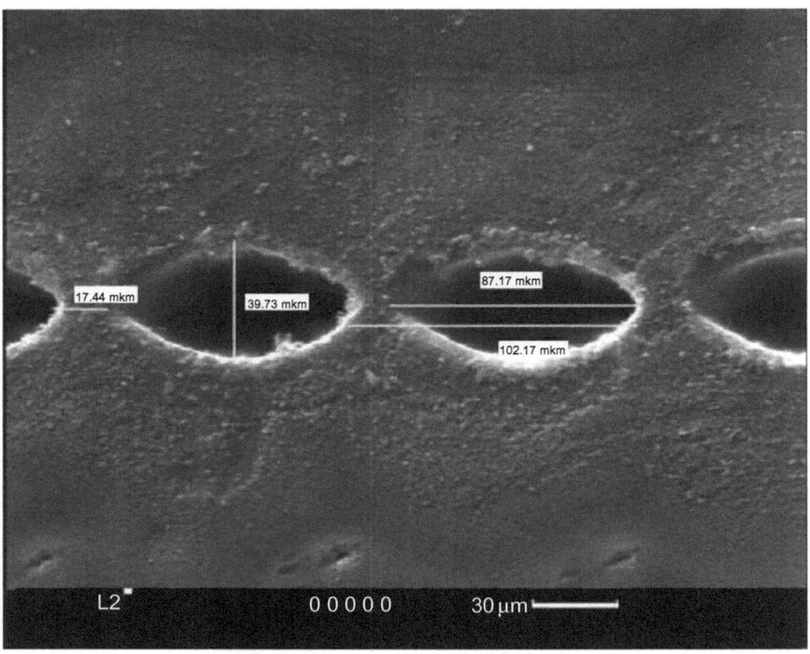

FIGURE 74 (Top) Mask projection scheme of X-ray CRL production using glassy carbon with (1) laser cavity spherical mirror, (2) mask, (3) laser active media, (4) flat mirror, (5) projection optic lens, and (6) glassy carbon block. (Bottom) Scanning electron micrograph of parabolic X-ray lens CRL-7 (Artemiev *et al.*, 2006).

as one of the laser cavity mirrors. The mask was fabricated from a thin metal sheet, and the mask aperture was shaped as the junction of two parabolas. The mask is placed inside the laser cavity formed by the copper vapor laser and glassy carbon block in which the aperture allowed the focused light to remove material from the exposed glassy carbon surface.

The placement of the mask and glassy carbon inside the laser cavity allowed high radiation density to be directed on the glassy carbon, which is needed for ablation. Placement inside the laser cavity also prevents laser damage of the mask because lasing takes place only through the mask gap. Figure 74 shows a schematic of the projection method, the tabulated parameters of the fabricated and tested 1D and crossed 2D CRLs, and an image of the glassy carbon CRL-7 with $N = 7$ lenses, on-axis radius of curvature $R = 5\,\mu m$, and physical (geometric) aperture diameter $2r_e = 40\,\mu m$.

Figure 75 shows the experimental setup for testing the glassy carbon CRLs at the bending magnet beam line BM-5 at ESRF; the tabulated calculated and measured focal line or spot dimensions; and measured focal distance and gain for 12 keV for the glassy carbon CRL-4 X-ray lens (1D focusing), 12.2 keV for the pair of crossed CRL-4 lenses (2D focusing), 20 keV for CRL-7 (1D focusing), and 25 keV for glassy carbon CRL-200 (1D-focusing). The figure also plots the focal size versus X-ray energy for the 1D CRL-4 X-ray lens and the result of the knife edge scan of the 1D CRL-200 line focus at 25 keV, which yielded a minimum FWHM focal width of 1.4 μm. The circles show the measured values of the knife edge intensity profile, and the solid line is a Gaussian fit of the derivative of the intensity profile, which thus yields the relative focal spot intensity. The inset is the FReLoN camera image of the line focus of 25-keV X-rays by the CRL-200, which had $N = 200$ lenses and on-axis radius of curvature $R = 5\,\mu m$ curvature radius and 40 micron physical aperture diameter.

Microbeam techniques perform element analysis of sample via microbeams of X-rays, electrons, and ions and are an important tool with many research, industrial, and medical applications. X-ray microbeams can be produced at synchrotron undulator sources and microfocus X-ray tubes. In conventional microprobe experiments, a microbeam scans a thin sample slice or a surface, which has the region of interest (ROI). ROIs inside the sample require sample preparation, which often includes shock-freezing and cryosections of the sample to yield sample slices with the ROI at the surface.

X-ray fluorescence element microtomography (XFEMT) allows interrogation of ROIs inside the sample and can be implemented at third-generation synchrotron sources or with microfocus X-ray tubes using CRLs. The sample is rotated to allow the tomographic slices to reconstruct a 3D image of the object, which includes the 3D element distribution that is enabled by the detection of the fluorescent signal emitted by sample

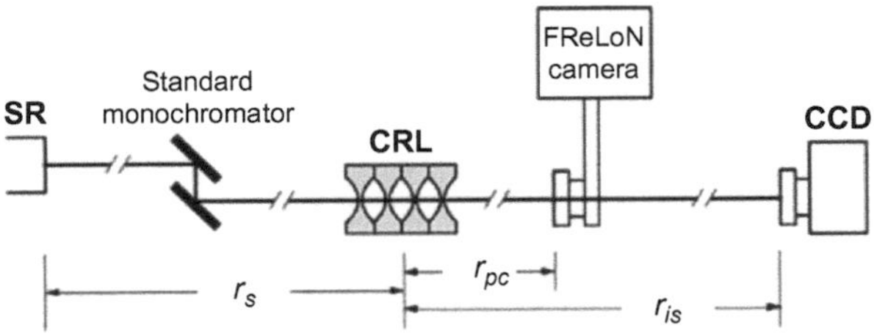

Lenses	Energy of radiation (keV)	Focus distance calculated (cm)	Focus size calculated (μm)	Focus size measured (μm)	Gain measured
CRL-4	12.2	1194	38	73	4.8
Two CRL-4 in crossed geometry	12.2	1194	38×163	73×292	6.4
CRL-7	20	47	0.9	2.7	3.0
CRL-200	25	2.8	0.05	1.4	3.3

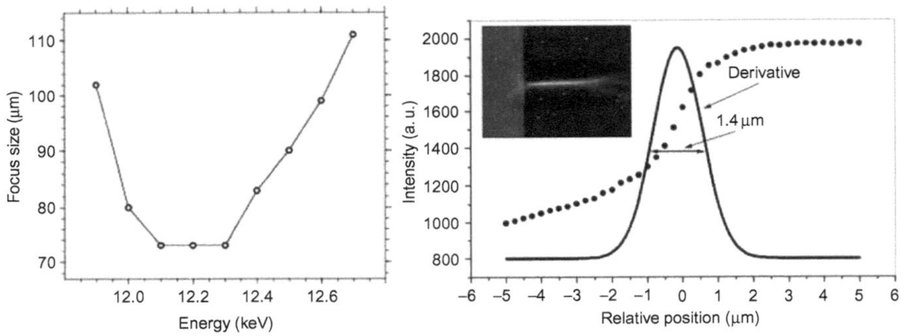

FIGURE 75 (Top left) Setup at bending magnet beam line BM-5 at ESRF. Tabulation of focal size, distance, and gain versus energy for tested CRLs. The table shows the measured plots of vertical focus (FWHM) for CRL-4 versus photon energy and knife edge measurement of CRL-200 focal spot relative intensity. (Center images) (a), Linear focus made by CRL-4 X-ray lens, and X-ray image (b) of 2D focus made by two CRL-4 X-ray lenses in crossed geometry. The CRL-4 lens has $N = 4$ lenses with on-axis radius of curvature $R = 200\,\mu m$ and physical aperture diameter $2r_m = 900\,\mu m$. See text for details (Artemiev *et al.*, 2006).

elements due to excitation by the monochromatic X-rays. Complementary to XFEMT is differential absorption contrast tomography, which is a powerful method for 3D elemental mapping. However, differential absorption contrast tomography requires sufficiently large pixel sizes and high concentrations of the elements of interest (see Bonse *et al.*, 1986, and 1989).

The fluorescent detection of elements can detect elements in very small concentrations. The combination of fluorescence detection and tomographic techniques has been applied to larger samples with millimeter-range resolutions (see Hogan *et al.*, 1991; Yuasa *et al.*, 1997 and Rust and Weigelt, 1998). With the recent advent of high-intensity beams at synchrotron sources and the continuing development of microfocus X-ray tubes, higher resolution, and more sensitive detectors, resolutions in the micron range are becoming feasible (see Simionovici *et al.*, 1999).

At ESRF Schroer *et al.* (2000b,c) applied CRLs with intense synchrotron X-ray microbeams to perform high-resolution tomographic imaging of samples and simultaneously provide a high-resolution 3D map of the elemental distribution in the sample via X-ray fluorescence (also see Simionovici *et al.*, 2000). Schroer *et al.* (2000b,c) conducted XFEMT experiments with the 19.5-keV monochromatic X-rays at the undulator X-ray source at the ESRF beam line ID18. The X-rays were focused onto the sample by an aluminum CRL with $N = 150$ biconcave, parabolic lenses, on-axis radius of curvature $R = 200\,\mu\text{m}$, and focal length $f = 45\,\text{cm}$ with source to lens distance $r_0 = 58\,\text{m}$, and CRL to detector distance $r_i = 46\,\text{cm}$. The focal spot dimensions were 1.4-μm vertical FWHM and 6-μm horizontal FWHM at 19.5 keV.

The top-left schematic in Figure 76 (from Schroer, 2000b) shows that the monochromatic X-rays from the undulator source are focused onto the sample, and a positive-intrinsic-negative (PIN) diode between CRL and sample (upstream of the sample) measures the incident X-ray flux, while the second PIN diode behind the sample (downstream of the sample) measures the transmitted radiation transmitted through the sample. The sample is scanned through the microbeam in both translation and rotation in which the measure microbeam flux exceeded 10^{10} ph/sat 19.5 keV. The X-ray fluorescence radiation emitted from the sample is recorded at each rotation or translation step by a high-purity germanium (HPGe) detector.

Figure 76 also shows the reconstruct distributions of several physiologically relevant ions inside the root of the mahogany plant (Swietenia macrophylla), and for comparison the Compton and Rayleigh scattering sources are reconstructed as well. The fluorescence microtomogram of the root of the mahogany plant was recorded by scanning the sample through the microbeam in 128 steps with a 6-μm step size. The step size was adjusted to the FWHM size of the microbeam in which the step size was chosen to minimize the total acquisition time with the expense of a slight undersampling. There were 151 horizontal scans (projections) each rotated in steps of 2.38° for 360° rotation. The fluorescence X-rays for chlorine, potassium, calcium, iron, and rubidium were simultaneously measured with 1.0-second acquisition times for each step together; in addition, the inelastic (Compton) and elastic (Rayleigh) X-rays were measured. An HPGe detector with 7-mm^2 aperture was positioned about 18 mm from the microbeam. The measured HpGe energy resolution at manganese $K\alpha$ line was 170 eV.

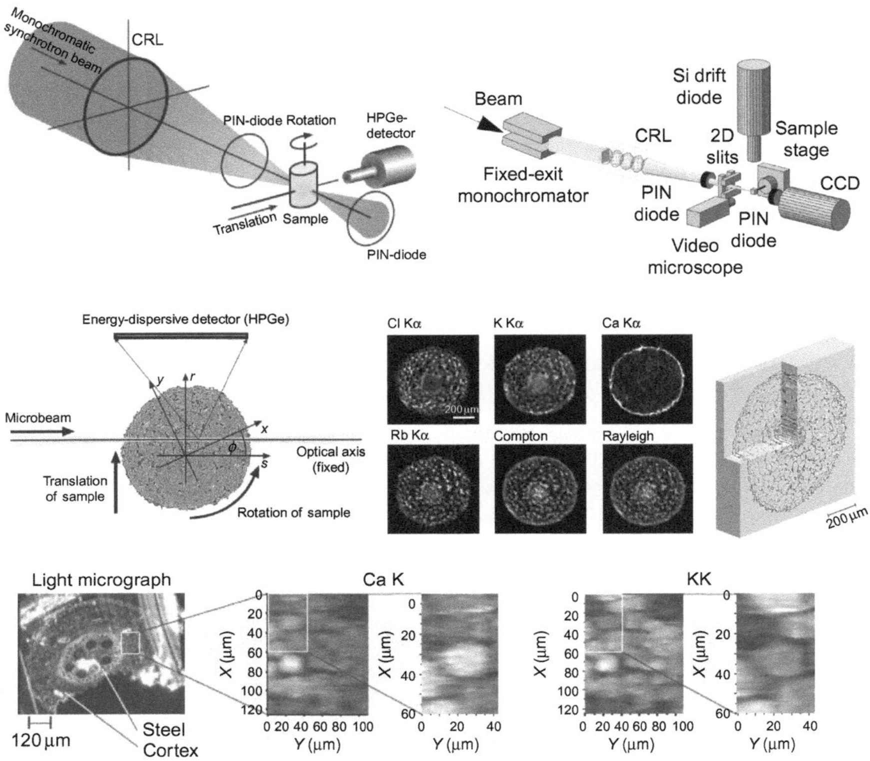

FIGURE 76 (Top left) Experimental setup of the fluorescence microtomography (XFEMT) apparatus on ESRF beam line ID22 (Schroer, 2000b). (Upper right) Tomography setup on ID22 (Simionovici *et al.*, 2000). (Middle left) XFEMT sample setup (Schroer, 2000b). (Middle) Compton and Rayleigh sample emission, and X-ray fluorescent emission from physiologically relevant ion distributions inside mahogany plant root (Schroer, 2000b). (Middle right) Phase-contrast X-ray microtomogram of mahogany root (Schroer *et al.*, 2000b). (Bottom) Maps of K and Ca in corn plant root at 15 keV focused by $N = 63$ Al, parabolic CRL to 1.1-μm vertical ×11.6-μm horizontal spot (Schroer *et al.*, 2000c).

The top-right schematic in Figure 76 (from Simionovici *et al.*, 2000) shows the setup for tomographic X-ray imaging by rotation and translation of the sample. The phase-contrast, X-ray micro-tomogram, reconstruction of the mohagany root is shown at bottom right in Figure 75. Schroer *et al.* (2000b) obtained the phase-contrast microtomogram at the tomography station 11 of beam line ID22 of the ESRF at 20 keV in which 1250 projections over equally spaced intervals in a 108° rotation were recorded using the FReLoN2000 high-resolution CCD camera. The effective pixel size was 1.4 μm, while the object to camera distance was

$r_i = 50\,\text{mm}$, and filtered backprojection was used to reconstruct the image from the 1250 projections. The phase-contrast microtomogram of a root of the mahogany plant showing the 3D reconstruction can be used to relate the element distribution obtained by the XFEMT to the structure of the root. The mahogany plant sample used by Schröer *et al.* (2000b) was hock frozen and freeze-dried root from *S. macrophylla*. The distributions of several physiologically relevant ions in a virtual section through the sample are shown; these distributions are used in the study the long-range ion transport in plants. By using XFEMT the difficult step of cryosectioning was not required.

Schroer *et al.* (2000c) also performed XFEMT on the root of a corn plant with an aluminum CRL composed of $N = 63$ biconcave lenses provided a 1.1 μm vertical × 11.6 μm horizontal focal spot or microbeam of 15-keV X-rays with an intensity gain $G = 20$. With the source to CRL distance $r_s =$ 41.3 m and a CRL to sample distance $r_i = 68\,\text{mm}$, the 15-keV microbeam scanned a corn plant root whose light microscope image (shown at the bottom left of Figure 76). The fluorescent emission of X-rays from the potassium and calcium in the corn plant are shown as reconstructed elemental distributions of potassium and calcium in the bottom two right figures.

Each figure shows the total area and a close-up of a portion of the scanned area of the corn plant. The reduction in the amount of required sample preparation is important, and the eventual goal is the acquisition of *in vivo* fluorescent microtomograms without the need for sample preparation. Real-time *in vivo* imaging allows one to truly see the structure and function of the ROIs in the sample. Only then can rapid progress be made in understanding basic phenomena, just as real-time, nonintrusive *in vivo* monitoring (and even control) of vital physiological functions can advance medicine and medical research.

Hard X-rays can probe the interior of opaque samples without destructive sample preparation in which X-ray absorption spectroscopy is performed, which allows researchers to determine the chemical state and local chemical environment of an elemental species in a variety of sample states. The large penetration depth of the hard X-rays, combined with absorption spectroscopy and tomographic imaging, allows determination of the chemical state distribution of sample elements. Projection tomograms of a sample are at a small number of hard X-ray energies around the absorption edge of the element of interest, such that X-ray near-edge spectroscopy X-ray absorption near-edge structure (XANES) can be combined with scanning microtomography. The microbeam technique allows reconstruction of the full XANES spectra at each location in a virtual sample section, and in the fluorescent mode the microbeam technique can be applied to the study of dilute systems. Schroer *et al.* (2003a) demonstrated that a quick scanning monochromator and data acquisition system can be combined with a scanning microprobe setup based on refractive X-ray lenses. This allows the acquisition of an element chemical state and allows analysis of the chemical reactions in a volume or identification of different chemical compounds inside a sample.

The absorption spectroscopy tomography experiment was performed by Schroer *et al.* (2003a) on the undulator at beam line 1-ID of the Advanced Photon Source (APS) in (Figure 77). The X-rays from a synchrotron radiation source are monochromatized with a channel-cut (Si-111) double-crystal monochromator and the monochromatic radiation is focused by a parabolic CRL onto the sample position, creating an intensive pencil beam with micron to submicron lateral dimensions. The incident flux I_0 and the sample-transmitted flux I_1 are measured with two ionization chambers, and the transmission through a reference material, which is placed downstream of the second ionization chamber, is recorded by a third ionization

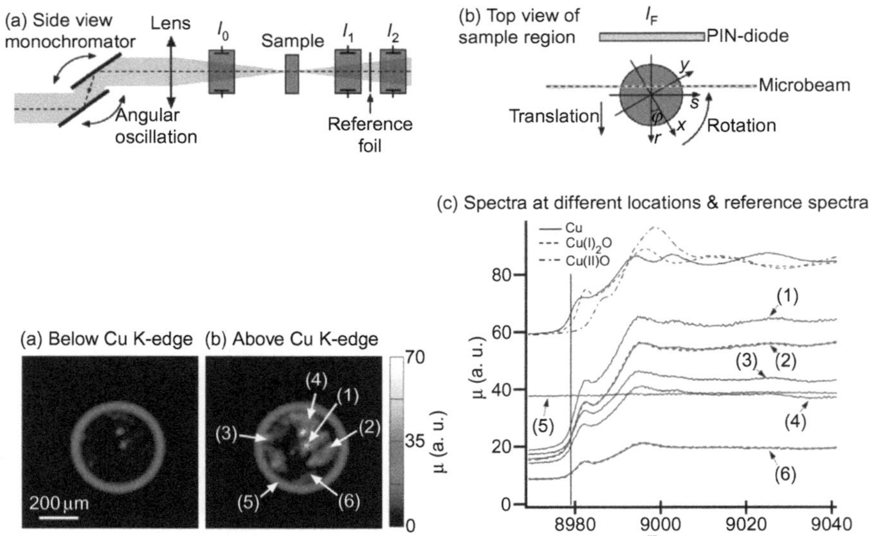

FIGURE 77 (Top) (a) A side view and (b) top view of the setup for absorption spectroscopy tomography experiment performed on the undulator at beam line 1-ID of the Advanced Photon Source. The coordinates $(x, y)$are fixed to the sample and are rotated with respect to the coordinates (r, s) by angle φ in which both coordinate systems (x, y) and (r, s) are translated with the sample; coordinate r specifies the microbeam position. (Bottom) A reconstructed tomogram of a capillary filled with a CuO/ZnO catalyst at (a) below the E = 8970 eV and (b) above the copper K edge E = 8995 eV. Image (c) is the reconstructed XANES spectra at different locations marked in (b) and (offset) reference spectra of metallic Cu, $Cu(I)_2O$, and Cu(II)O. The dashed red curves in (c) represent two fits using the reference spectra (Schroer *et al.*, 2003b).

chamber. In addition, a PIN diode records the sample fluorescent X-ray emission. The absorption spectroscopy tomogram is recorded by scanning the sample through the microbeam via translation and rotation. A single tomographic projection is acquired by translating the sample in the perpendicular direction r to the microbeam. At each position of the translation scan a full absorption spectrum is recorded both in transmission and fluorescence. After each projection, the sample is rotated by an integer fraction of 360°, and this process is repeated until the sample has completed one complete 360° rotation.

The purpose of the Schroer (2003) experiment was to analyze the structure and oxidation state in a 30% CuO/ZnO catalyst in a sample volume in which the sample was twice reduced and reoxidized by 4% H_2/He and 2% O_2/He at 200–300°C. After the final reduction the copper particles grown on the ZnO were re-oxidized at room temperature. A XANES tomogram was recorded at the copper K edge in which 101 projections with 90 translational steps of 10 μm each were recorded over 360°. The bottom tomograms in Figure 77 show the reconstructed attenuation coefficients μ, respectively, below and above the K edge of copper. The reconstructed spectra at the locations in image (b) are indicated in (c). The spectrum of the glass capillary (spectrum 5) has a weak energy dependence because the glass contains very little copper. At different locations in the reconstructed slice, the spectra show slight differences in addition to the varying edge heights.

The spatial distribution and concentration of trace elements in tissues are important indicators of tissue physiological and pathological conditions. The study of biological processes is greatly aided by nondestructive quantitative techniques with high spatial resolution and high sensitivity for the spatial distribution of elements and compounds in tissues and single cells. The technique of synchrotron-induced X-ray fluorescence (SXRF) microprobe is one such tool. Imaging of trace elements in single cells was demonstrated by Bohic *et al.* (2001) at ESRF by using synchrotron-induced X-ray fluorescence (SXRF) with hard X-rays. Monochromatic and "pink" excitations at 14 keV were used with CRLs resulting in 1.0 μm × 10 μm beam size. The experiment by Bohic *et al.* (2001) demonstrated SXRF can be applied to the microanalysis of freeze-dried cell in which the SXRF technique produces accurate quantitative imaging of the trace elements in cells. The SXRF analysis of single cultured cells can be applied to the study of the effects of various high-Zlabeled drugs on cells.

In this particular experiment, the human ovarian adenocarcinoma (IGROV1) cell line was used for this study in which the cells were grown directly on thin (0.2-μm) polyvinyl formal resin (Formvar; SPI Supplies/Structure Probe, Inc., West Chester, PA) film supports. The IGROV1 cells were incubated with complete culture medium, and the exponentially growing cells were exposed to 5 μm of the anticancer drug,

4′-iodo-4′-deoxydoxorubicin (IDX). Then the sample IGROV1 cell monolayers were rinsed, cryofixed into liquid nitrogen–chilled isopentane, and freeze-dried at 30°C. Figure 78 shows a schematic of the experimental setup used by Bohic *et al.* (2001) to fluorescently image trace elements in a single cell. The freeze-dried samples were then mounted on a scanning stage, which could be translated in both transverse vertical and horizontal directions, in which the CRL focused monochromatic or narrowband peak beam of X-rays onto the cells.

Photons from an aluminum parabolic CRL are focused onto the sample at a rate of N_0 (ph/s) with energy E_0. In the pink beam, an aluminum CRL with $N = 50$ biconcave parabolic lenses produce a flux of 5×10^{10} ph/s/μm^2 and a beam spot 10 μm horizontally × 1 μm vertically within a focal length $f = 713$ mm at 14 keV. A monochromatic beam

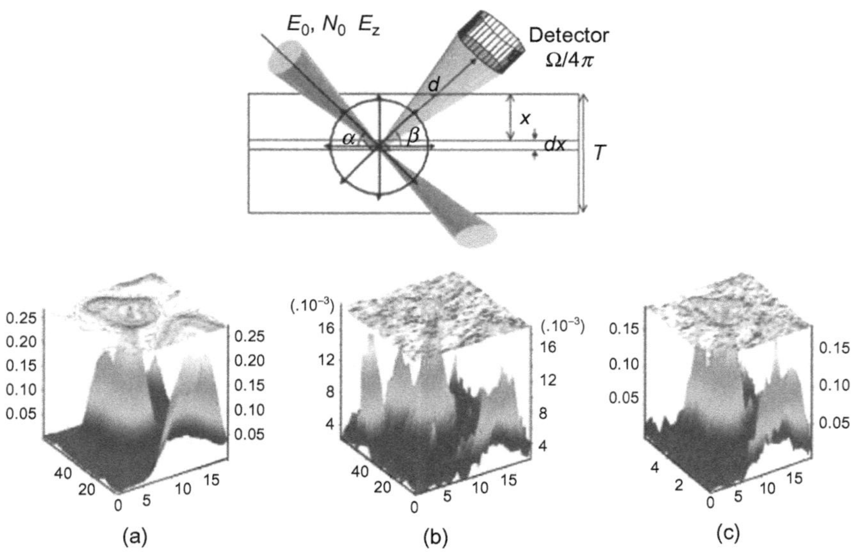

FIGURE 78 (Top) Schematic of the experimental setup. Photons of an aluminum parabolic CRL are focused onto the sample at a rate of N_0(ph/s) with energy E_0. The isotropic emission of fluorescent X-rays occurs from a layer of thickness *dx* at depth *x* within a sample of thickness *T*. The fluorescent X-rays detector is in the backscatter position with $\alpha = \beta = 45°$. (Bottom) The 2D elemental distribution of a freeze-dried cancer cell treated with 5 μM iodo-deoxydoxorubicin. A single cell was interrogated with a 14-keV polychromatic pink excitation in which the $K\alpha$ and $L\beta$ fluorescence line emission was measured. The sample surface was scanned in 2.5-second composed of transverse 1-μm vertical and 3-μm horizontal steps for a 2-hour scan time for area 60 μm × 60 μm. The surface mass values (μg/cm^2) along the *z*-axis are plotted versus *x*- and *y*- cell coordinates. The three plots show cell (a) potassium, (b) iron, and (c) iodine distribution (Bohic *et al.*, 2001).

focused by the CRL to the same spot size produced a flux, which is approximately about 10 times less than the peak beam. Isotropic emission of fluorescent X-rays occurs from a layer of thickness dx at depth x within a sample of thickness T, and the intensity of the isotropic fluorescent emission from an element with atomic number Z is a function of the local concentration of the element's atoms. The incident focused beam of X-rays from the CRL are incident at angle $\alpha = 45^\circ$ relative to the sample surface, and the fluorescence is detected in a solid angle of $\Omega/4\pi$ in which the detector is set at 90° from the incident X-ray beam. This places the detector at reflection angle $\beta = 45^\circ$ from the sample surface in a backscatter geometry. The bottom plots in Figure 78 are based on the fluorescent X-ray emission from the cell. The surface mass values ($\mu g/cm^2$) for potassium (bottom-left plot a), iron (bottom–middle plot b), and iodine (bottom-right plot c) distributions in the cell are plotted along the z-axis as a function of the transverse horizontal x and vertical y coordinates of the single IGROV1 cell containing the IDX anticancer drug.

Aluminum and beryllium parabolic compound refractive X-ray lenses can be made to image a synchrotron radiation source onto a sample with large de-magnification, which provides a dramatic increase in the focused intensity. The CRL can produce small-diameter, intense pencil beams that can be used for (scanning) microprobe experiments such as microdiffraction and microfluorescence. The aluminum lenses are most effective above 10 keV and are particularly suited for hard X-rays up to at least 60 keV. The pencil beam has micron range transverse dimension and can yield intensity gains of 100 to 1000. The small beam convergence in the spot (typically $\Delta k/k < 10^{-4}$) was ideal for microdiffraction and micro–small-angle microscattering (SAXS; see Schroer *et al.*, 2000c).

Schroer *et al.* (2000c) conducted microdiffraction experiments on the ESRF undulator ID22 beam line with a 16-keV monochromatic beam, which was focused into the center of rotation of a six-circle diffractometer (see the top of Figure 79 for setup). The signal from the Bragg reflection from a single crystal grain was recorded by a 2D gas-filled detector in which the typical sample crystal grain size was 20 to 50 μm. An aluminum CRL with $N = 40$ biconcave parabolic lenses produced a microbeam focus of 16-keV X-rays with 3 μm vertical and 12 μm horizontal dimensions at the $f = 1.18$m focal length with an intensity gain $G = 150$, which delivered 4.5×10^9 ph/s to the microspot. The source to CRL distance (i.e., object distance) was $r_0 = 45$ m, and the CRL to demagnified image of the X-ray source on the crystal (i.e., image distance) was $r_i = 1.21$ m.

The angular resolution of the incident focused X-ray beam was $\Delta k/k = 1.3 \times 10^{-4}$, which is 10-fold larger than the widths of the Bragg reflections. The line profiles of as many as 8 different Bragg reflections, (110), (210), (211), (220), (310), (322), (222), and (400) were measured for each individual crystal grain. The sample was set on a xyz-stage inside the

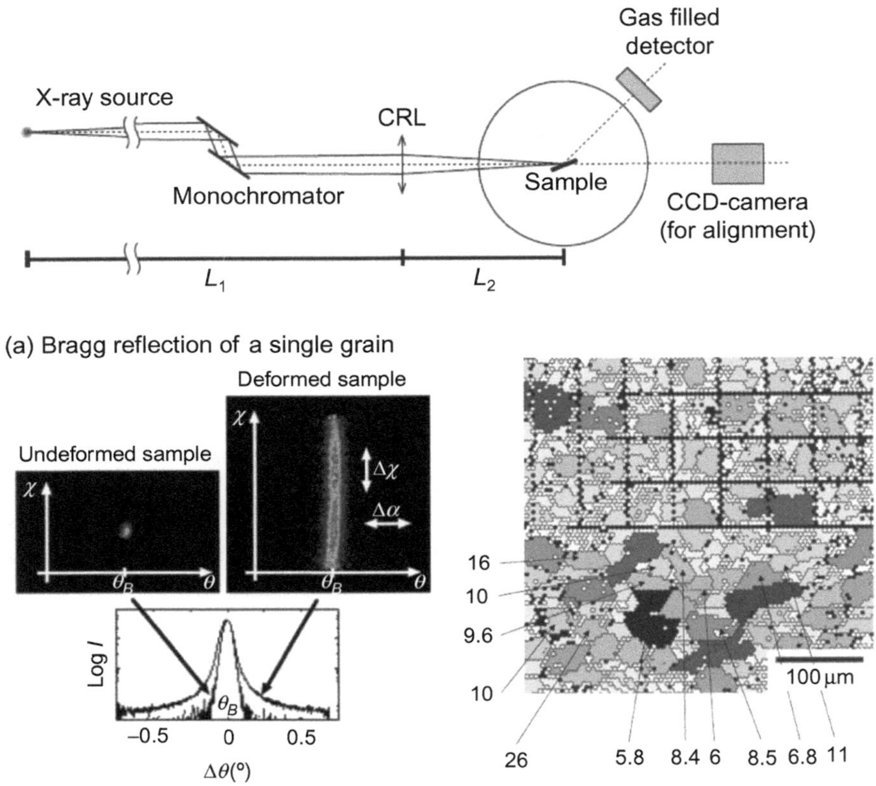

FIGURE 79 (Top) Experimental setup for microdiffraction experiment at ESRF ID22 undulator beam line. (Bottom) The lower-left image is the Bragg reflection and line profile of a single crystal grain. The lower-right image is the sample Electron backscatter diffraction (EBSD) image in which the dislocation densities are in arbitrary units and the dislocation density of a non-deformed (normal) grain is 0.16 (Schroer *et al.*, 2000c).

diffractometer, which allowed the different crystal grains to be translated to the center of rotation, and a CCD camera was used for alignment. The bottom of Figure 79 shows the results of the microdiffraction from single crystal grains by Schroer *et al.* (2000c). The lower-left image is the Bragg reflection and line profile of a single crystal grain. The lower-right image is the sample Electron backscatter diffraction (EBSD) image in which the dislocation densities are in arbitrary units and the dislocation density of a non-deformed (normal) grain is 0.16.

Tomography can be performed with an X-ray microscope but requires a sample stage that provides precise rotation and translation of the sample as demonstrated in the experiments of Schroer *et al.* at the undulator beam line ID22 at ESRF (see Schroer *et al.*, 2000a,b). Magnified tomography provides a means of producing magnified 3D absorption contrast images of the object with object-illuminating X-rays of low transverse coherence length, and phase-contrast images of the object with X-rays of large transverse coherence length with an undulator or free electron laser (FEL) X-ray source.

The tomogram of an object is reconstructed from a dataset composed of a set of X-ray micrographs recorded as a function of sample rotation. The sample is rotated with equal angular steps from 0 to 180° and an image is recorded at each angular position. The sample is moved out of the X-ray beam every few rotations to record a flat-field image in which the images are corrected for dark current and readout noise of the CCD detector. The reconstruction of the 3D image of an object in tomography requires each X-ray image in the dataset to be nondistorted and free of spherical aberration in which the depth of field must be larger than the sample thickness. High-fidelity tomographic reconstruction of the object requires the microscope resolution to be larger than the eccentricity of the rotation stage and the repeat accuracy of the translation stage.

Absorption contrast images with flat-field and dark-field corrections provide direct tomographic reconstruction of the sample in which the absorption contrast reconstruction is a spatial distribution of the sample linear attenuation coefficient. However, a phase-contrast–reconstructed image of the sample requires several defocused images of the object at each sample rotation angle in which the resulting phase-contrast image is a spatial distribution of the sample complex refractive index. The large transverse coherence of the ID22 undulator source must be reduced to obtain an absorption contrast tomogram, which can be accomplished by a 2.5-mm B4C diffuser placed in the beam.

The left image in Figure 80 shows a magnified tomogram of a fiber of reinforced concrete obtained by Schroer *et al.* (2002a). The tomogram required 250 X-ray images, each taken with a 50-second exposure time. Each X-ray image was recorded with 20.65-keV X-rays at magnification $M = 10.63$ using an aluminum CRL composed of $N = 42$ biconcave parabolic lenses with $f = 1892\,\text{mm}$. The fiber to the CRL object distance was $r_0 = 2070\,\text{mm}$ and the CRL to FReLoN 2000 detector image distance was $r_i = 22.03\,\text{m}$. Mechanical instability of the stage rotation and translation produce artifacts in the tomogram in which object features are all distorted in the same manner.

The right (magnified) tomogram of an AMD K6 microprocessor in Figure 80 was obtained by Schroer *et al.* (2002b) at 24.9 keV on the ID22 beam line by the reconstruction of 500 images taken at equal rotation angle intervals from 0 to 180° with 40-second exposure times. An aluminum CRL composed of $N = 120$ biconcave parabolic lenses with focal length $f = 1048\,\text{mm}$ produces the individual 24.9-keV X-ray micrographs. The AMD K6 chip fragment was at $r_0 = 1098\,\text{mm}$ object distance, and

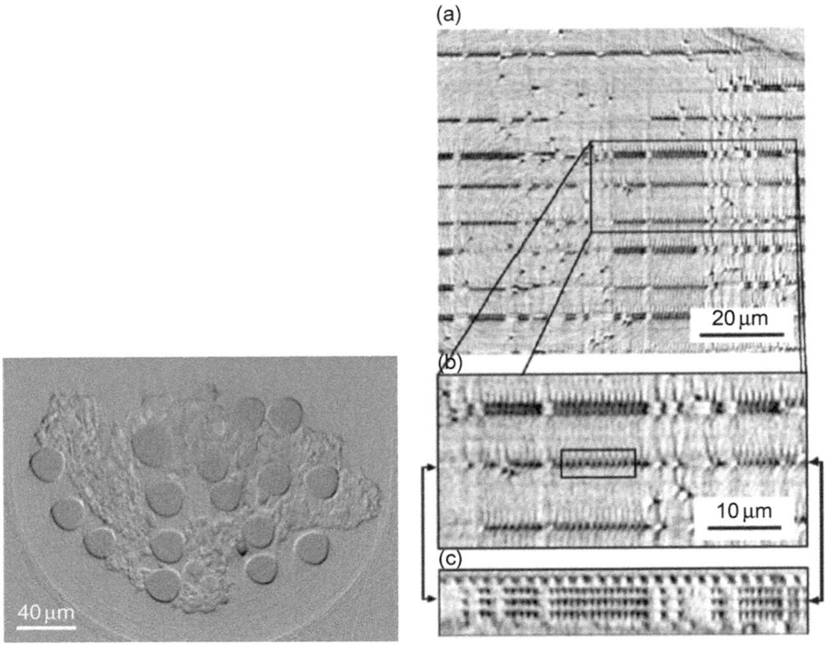

FIGURE 80 (Left) Tomographic slice through a fiber-reinforced concrete sample imaged with 20.65-keV X-rays using an aluminum CRL with $N = 42$ biconcave parabolic lenses with $f = 1892$ mm. The slice was reconstructed using 250 X-ray micrographs at magnification $M = 10.6$ (Schroer *et al.*, 2002a). (Right) (a) Horizontal tomogram slice through the second layer of tungsten plug vias in an AMD K6 microprocessor. (b) Magnification of the rectangular region shown in (a). (c) The vertical section through the processor along the line of plug vias marked by arrows in (b) is shown below the tomogram (Schroer *et al.*, 2002b).

the FReLoN 2000 detector was downstream of the CRL at image distance $r_i = 23.02\,\text{m}$, which resulted in a magnification $M = 20.9$. The magnification produced an effective pixel size of 131 nm compared with the 230-nm FWHM Airy disk of the aluminum CRL.

Schroer *et al.* (2002a) investigated the use of CRLs in lithography in which a CRL images a binary mask onto X-ray-sensitive resist in a reducing geometry. A lithography mask was illuminated by monochromatic undulator radiation at 25 keV. To avoid interference fringes at the edges between bright and dark regions in the mask, a 5-mm thick B_4C diffuser was placed 30 cm before the mask. An aluminum CRL with $N = 130$ biconcave parabolic lenses with on-axis radius of curvature $R = 209\,\mu\text{m}$ was set at object distance $r_0 = 3.62\,\text{m}$ downstream of the mask. The CRL imaged the mask pattern onto a downstream resist-covered silicon wafer

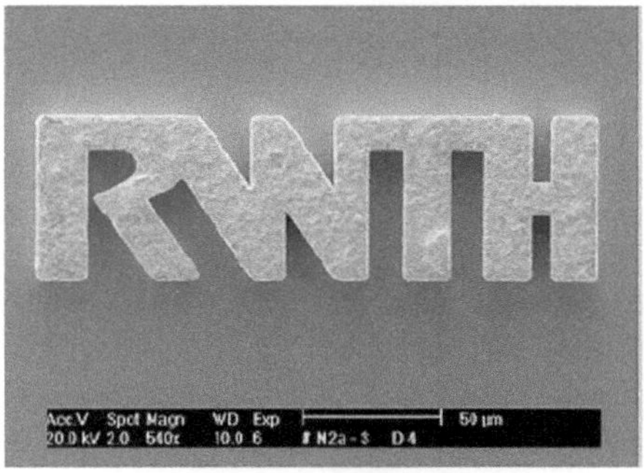

(a) SEM image of lithography mask

(b) Image on resist (reduction 2.8x)

FIGURE 81 (Top) Scanning electron micrograph 15-μm gold mask with acronym letters RWTH (for RWTH Aachen University, where many innovative compound refractive X-ray lenses have been fabricated). The mask is formed by electroplating gold onto a silicon wafer. (Bottom) Aluminum CRL ($N = 130$ biconcave parabolic lenses) demagnified ($M = 1/2.82$) image of the mask on the negative resist Novolac with 25-keV X-rays (Schroer *et al.*, 2002a).

at image distance $r_i = 1.284\,\text{m}$ in which the mask was demagnified by a factor of 2.82.

The top image in Figure 81 shows a scanning electron micrograph of the lithography test mask. The lithography test mask was formed by electroplating a 15-μm layer of gold onto a silicon wafer. The CRL then imaged the 25-keV X-rays passing through the mask onto the negative resist Novolac coated on a silicon wafer with magnification $M = 1/2.82$ (i.e., 2.8-fold demagnification), shown in the bottom image of Figure 40. Novolac is a phenol-formaldehyde resin named and invented by Leo Baekeland in 1907. The features of the mask are two large to demonstrate the resolution of the CRL-enabled X-ray lithography. However, this demonstration by Schroer *et al.* (2002a) shows the promise of CRLs for X-ray lithography.

In his Ph.D. dissertation, Cederström (2002) noted that curved surfaces require elaborate manufacturing techniques. He realized the curved surfaces could be replaced by planar surfaces arranged in a simple saw-tooth profile, which greatly simplified the process and cost of fabrication. Cederström presents in detail the multi-prism lens (MPL) for X-rays. Cederström also patented his multi-prism or alligator X-ray lens (see Cederström, 2003).

Cederström had fabricated the alligator lens (two sets of saw teeth or jaws joined at one vertex) or MPL from polymethyl methylacrylate (PMMA), which has average atomic number of 6 and good mechanical properties. The PMMA lens halves are supported on aluminum plates, which were joined and adjusted with micrometer screws by placing the setup under a microscope. The MPL is composed of identical sets of N prisms (teeth) of height y_t symmetrically placed above and below the center optical axis. An incident ray propagating in the axial z-direction, which passes through a PMMA isosceles triangle–shaped tooth, will be refracted transversely at both air-tooth interfaces by a small angle toward the center optical axis (Figure 82)

A ray at a larger transverse distance from the center optical axis will encounter more teeth than a ray at a smaller transverse distance. Hence, the number of tooth-air refraction interfaces encountered by a ray increases with increasing transverse distance of an incident ray from the center optical axis. That is, the number of teeth an X-ray passes through and the amount of lens material traversed by a ray increase with transverse distance y from the center axis. Given the slopes $\pm a$ of the two diagonal line, which form each isosceles triangle–shaped tooth of the N teeth of each jaw of the MPL, the rate of increase dt/dy of the ray axial path length t through lens material as a function of incident ray transverse distance y from the center optical axis is the equation of a straight line, where

$$\frac{dt}{dy} = 2ay. \qquad (106)$$

Integration of the area under the straight line gives a parabola, which is the area under the straight line as a function of transverse distance$y$. That is, the integration of the rate of increase dt/dy of the ray axial path length t through lens material for a ray incident at the MPL entrance, as a function of transverse distance y of the incident ray from the center axis, is the equation of a parabola, where

$$t = ay^2. \qquad (107)$$

The projection of the thickness of lens material traversed by a ray as a function of transverse y onto the y-axis forms the shape of a parabolic lens.

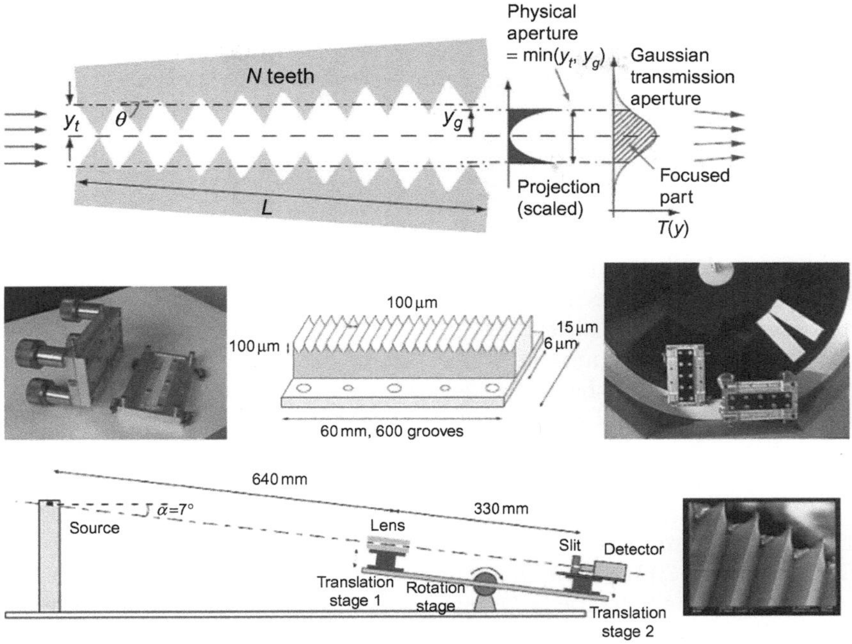

FIGURE 82 (Top) Schematic of MPL or alligator lens, which acts as a plano-concave parabolic lens. (Middle) Photo of the two jaws that form an MPL, a schematic of one jaw of the MPL composed of PMMA, and the famous vinyl record X-ray alligator lens. (Bottom) Experimental setup for focusing experiments with MPL and scanning electron micrograph of the PMMA isosceles triangle–shaped teeth of a jaw of the MPL (Cederström, 2002).

The MPL thus functions as a plano-concave parabolic lens. The saw-tooth profile is an example of redistributing the lens material in a manner that facilitates fabrication yet produces the parabolic-based increase of optical path lengths through the lens material that is necessary for refractive focusing to a single line focus. The focal length of the MPL can be varied by adjusting the gap between the two halves of the MPL.

Cederström *et al.* (2000) realized two sections cut from a vinyl long-playing (LP) record can form a refractive lens for hard X-rays without spherical aberration. Each jaw of the MPL resembles the saw-tooth surface of a vinyl LP record in which the vinyl has an approximate 0.18-mm pitch or spacing interval between adjacent triangle-shaped grooves. It appeared two saw-tooth sections cut from the vinyl record could be mounted on the aluminum jaws and thus form an alligator lens. However, it was not that straightforward. Their measurement of a vinyl LP record showed that the groove depth was only 25 μm and there was a radial layer of material between each groove (the grooves did not form a classic saw-tooth pattern. The radial layer of material separating the annular triangle-shaped grooves would strongly attenuate the X-rays.

The drawbacks of an actual sound-producing vinyl LP required Cederström to contact a record engraving company, which produced a silent vinyl LP that would not play opera. That is, the engraver produced a master with a 100-μm layer of lacquer on an aluminum disk and pressed a classic saw tooth pattern of 100-μm deep triangle-shaped grooves into the lacquer. With the master the engraver pressed a "silent" vinyl record, and from the vinyl record was cut two rectangular pieces, which were mounted on the two aluminum plates joined at one end, thus forming the vinyl record alligator lens (see Figure 82). The vinyl record alligator X-ray lens had a short focal length of $f = 22\,\mathrm{cm}$ for 23-keV X-ray. In the *Nature* paper, Cederström *et al.* (2000) noted the alligator X-ray lens showed good promise for use in small-scale line-focusing experiments with conventional X-ray tubes.

Pereira *et al.* (2001) fabricated and tested prototypes of Cederström's 1D saw-tooth refractive lens (or alligator lens) using lithium (see also Dufresne *et al.*, 2001, and Arms *et al.*, 2002). A single alligator lithium lens provides 1D line focusing; whereas 2D focusing requires crossed alligator lenses. The lithium saw-tooth refractive lens was tested by Pereira *et al.* on the 10-keV beam line 7ID at the APS in Argonne National Laboratories.

On beam line 71D the horizontal source size was 0.8 mm FWHM, and a white beam slit was set 26.5 m downstream from the source to collimate the beam to 1-mm horizontal and 0.5-mm vertical dimensions. A single jaw (half alligator lithium lens) or pair of hinged jaws open with small vertex angle 2θ (full alligator lens) is set at object distance $r_0 = 49.2\,\mathrm{m}$ downstream of the source to provide demagnified imaging of the source.

A second slit was set 75 mm upstream of the alligator half lens to act as an aperture stop to limit the incident 10-keV beam and 30-keV hardmonic harmonic to 0.7 mm, which is less than the $h = 0.75$-mm saw-tooth height in the single jaw under testing. In addition, the stop aperture eliminates the line-focus blurring, which is due to the X-ray focusing outside the parabolic lens-shape approximation for the single-jaw half lens. With two jaws (full lens) the aperture stop would be twice as wide to allow admit X-rays into both the lower and upper jaws and the line focus intensity would double.

X-ray images of the line focus of the single jaw (half lens) were taken by means of the CCD camera (Qmax 650 CCD18) imaged X-ray-induced fluorescent light emission from a 0.5-mm thick YAG:Ce–doped single-crystal scintillator. The imaging detector resolution was 0.02 mm or about 3 pixels FWHM. The results of the lithium alligator imaging experiments of Periera *et al.* (2001), Dufresne *et al.* (2001), and Arms *et al.* (2002) are shown in Figure 83.

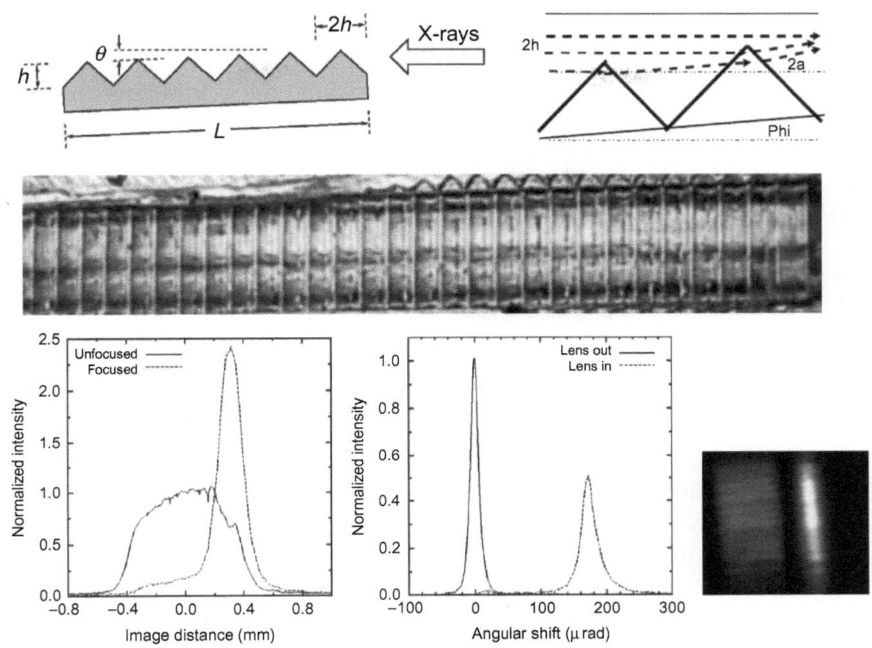

FIGURE 83 (Top) Schematic of the bottom jaw of a lithium alligator lens (Arms *et al.*, 2002) and refractive focusing by the saw teeth (Peirera *et al.*, 2001). (Middle) Photo of a jaw with lithium teeth (Peirera *et al.*, 2001). (Bottom left) Plot of the intensity profile across unfocused 10-keV X-ray beam and single-jaw lens focused beam (Arms *et al.*, 2002). (Bottom middle) Plot of a slice through the CCD detector image before (dotted curve) and after the lens is inserted into the 30-μm diameter X-ray beam (Arms *et al.*, 2002). (Bottom right) X-ray images of unfocused 10-keV X-rays and (right) line focus by bottom jaw of lithium lens (Peirera *et al.*, 2001).

The top of Figure 83 shows diagrams of a single jaw of the two jaws of the alligator lithium X-ray lens and a diagram of the refractive focusing of each jaw. The full alligator lens is composed of the upper inverted jaw and lower jaw joined at one end, forming an angle 2θ between the two jaws. Each tooth has height h, the width of one tooth or the pitch is $2h$, and the total lens length is L. Below the diagrams is a photograph of one jaw of the lithium lens, and the intensity profile is plotted across an unfocused X-ray beam and intensity profile with a single jaw of the alligator lens.

The bottom-left plot in Figure 83 shows the intensity profile across the unfocused 10-keV X-ray beam and the intensity profile across the line-focused beam imaged by a single jaw (half alligator) lens. The intensity with the single jaw of the alligator lens was 2.5-fold larger maximum than without the lens; however, the measured value was half of predicted factor of 5.2. The bottom-middle plots in Figure 83 shows a slice through the

CCD detector image before and after the lens is inserted into the 30-μm diameter X-ray beam. The plot shows the intensity profile of the beams with and without the CRL as a function of angular deflection downstream of the single-jaw alligator lens. The left curve is the intensity profile of the no-lens beam intensity profile, which is defined by the intrinsic beam divergence and upstream slit width. The left curve has a 14-μrad FWHM beam centered at about 0 μrad.

With the lithium lens jaw inserted in the X-ray beam, there appears the first harmonic of 10-keV photons with 25-μrad FWHM appears to be centered at about 175 μrad (see the right curve in the bottom-middle plot of Figure 83). A tiny curve of the third harmonic of 30-keV photons also appears centered at about 15 μrad, which is under the left curve.

Finally, the bottom right of Figure 83 shows an X-ray image of the unfocused beam of 10-keV X-rays (left) and an X-ray image (right) of the X-ray line focus, which is produced by a single jaw of the lithium alligator lens.

In 2004 Pereira *et al.* (2004) fabricated and tested a lithium CRL composed of $N = 80$, biconcave parabolic lenses with on-axis radius of curvature $R = 0.263\,\text{mm}$. The depth of the parabola on each side of the lithium coin lens was 0.5 mm, the lens physical aperture diameter was $2r_e = 1.0\,\text{mm}$, and the center minimum wall thickness of these biconcave parabolic lithium lenses ranged between $d = 50\,\mu\text{m}$ and $d = 100\,\mu\text{m}$. Each lithium coin lens was slightly thicker than 1.0 mm, resulting in a 100-mm length for the lithium CRL, and was mounted in a similar fashion to the mounting of parabolic aluminum and beryllium lenses by Lengeler *et al.* (1999a).

The lithium lenses were housed in a glass tube filled with helium gas in which X-rays are transmitted through two 127-μm thick beryllium windows. They found helium was as effective as vacuum but easier to protect with a helium atmosphere, and lithium lenses in a glass container filled with argon and a screw top remained clean for over 1 year. The biconcave parabolic lenses of lithium were fabricated by placing a disk of lithium in a precise fixture between parabolic dies. The pair of opposed parabolic dies were then pressed into the soft lithium metal in a manner similar to producing coins by debossing or pressing. The coin shape and manner of producing the coin lenses gives coin lenses their name.

Pereira *et al.* (2004) tested the lithium parabolic lens CRL at the APS on beam line 7-ID whose X-ray source is monochromatized by a liquid nitrogen–cooled double-crystal Si (111) monochromator, which is detuned by 50% to suppress the third harmonic. Upstream of the CRL by 120 mm is a pair of limiting slits (0.5 mm × 0.5 mm square aperture) that allowed the divergent X-ray beam to be contained within the 1-mm physical aperture diameter of the lithium parabolic lens CRL. Downstream of the CRL was an ionization chamber that measured the X-ray transmission through the lithium parabolic lens CRL. The CRL transmitted X-rays passed through the ionization chamber into a helium-filled flight path with 12-mm thick Kapton window foils. The X-rays then passed through the helium chamber and exited through the Kapton window to excite a cerium-doped YAG scintillator crystal. The scintillator-emitted visible light was imaged via a 5× microscope (Mitutoyo M Plan-APO long working distance objectives with X1 tube lens; Mitutoyo, Kawasaki, Japan) coupled with a CCD camera (Roper Scientific, Inc., CoolSNAP CF). The camera had 1392 × 1040 pixels that were 4.65 μm square. The fivefold magnification allowed the imaging system to have an FOV of 1.29 mm × 0.97 mm.

At 10.87 keV, the lithium parabolic CRL had 90% peak transmission with 49% average transmission and $f = 200$ cm. With the source to CRL object distance $r_0 = 49.2$ m and the detector at image distance $r_i = 2.25$ m, Pereira *et al.* (2004) obtained a 33-μm horizontal FWHM and a 17-μm vertical FWHM demagnified image of the X-ray source. The measured gain $G_{\text{exp}} = 40$ was about sixfold smaller the calculated (theoretical) gain $G_{th} = 234$.

Figure 84 (top left) shows the lithium parabolic CRL demagnified X-ray image of the source and the plot of the intensity profile in the x and y directions through the image spot. Figure 84 (bottom left) shows the CRL focused image of the 12.87-keV first harmonic and the unfocused higher X-rays harmonics from the source. The higher harmonics are imaged further downstream of the first harmonic and thus are effectively unfocused in the first-harmonic image plane.

If the pure, dull silver–colored lithium is put in contact with moist air, the lithium rapidly reacts with the water vapor to form a black coating of lithium hydroxide (LiOH) and its monohydrate ($LiOH{\cdot}H_2O$). However, lithium reacts more slowly with the nitrogen in the air to form lithium nitride (Li_3N), and a secondary reaction occurs between lithium oxide (*LiOH*) and the carbon dioxide in the air (CO_2) to form lithium carbonate (Li_2CO_3). Lithium is a soft metal that can be easily worked in a very cold room (which has little humidity), a dry room, or with a dry box in which the air has been completely dehumidified to a point, which is equivalent to the extreme cold condition of arctic air. The dry box can be filled with argon gas to eliminate the reaction of lithium with nitrogen. The reaction with nitrogen is rather slow, but the water vapor is rapid; hence, one can work in a dry room, which is dehumidified.

Lithia water springs in Lithia Springs, Georgia, and in Ashland, Oregon, promote the medicinal and beneficial value of spring water that contains lithium carbonate. The modern soft drink, 7 Up, is descended from the original product "Bib-Label Lithiated Lemon-Lime Soda" introduced 2 weeks before the 1929 market crash by the Howdy Corporation founded by C. L. Grigg, who changed the company name to the Seven-Up Company in 1936 (see Cadbury-Schweppes, 2005). However, lithium carbonate is an accepted drug by the medical community used to treat

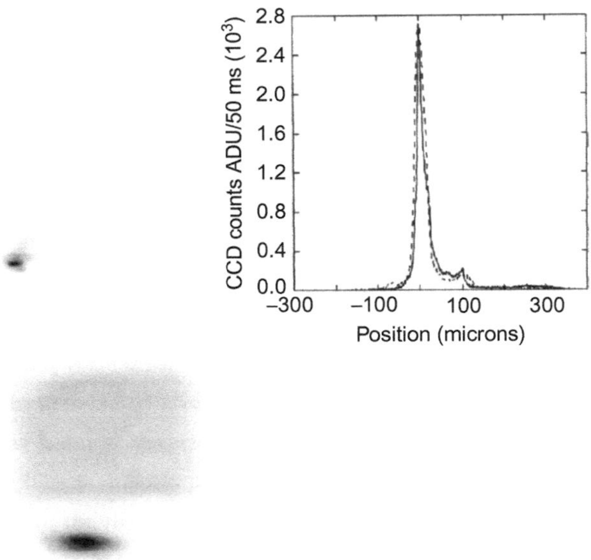

FIGURE 84 (Top left) The lithium parabolic lens CRL produced a demagnified X-ray image (inverted grey-scale) of the Argonne 7-ID beam line X-ray source at 10.87 keV, which required several aluminum foils to be placed in front of the CCD camera to reduce the intensity by a factor of 24, which could be tolerated by the detector. (Top right) The plot is the intensity profile of the demagnified source in the CRL image plane along the *x*-directed (dashed curve) and *y*-direction (solid curve). The detector pixel size was 93 μm, and the image horizontal and vertical FWHMs were, respectively, 33 μm and 17 μm. (Bottom left) The X-ray image of the demagnified source at 12.87-keV of the first-harmonic image plane and the unfocused higher harmonics in the first-harmonic image plane (Pereira *et al.*, 2004).

patients with mania and depression (bipolar disorder) (see Mitchell and Hadzi-Pavlovic, 2000). Lithium compounds such as lithium carbonate and lithium fluoride are modest moderators of fast neutrons and are good absorbers of thermal neutrons in which 99% of the thermal neutorn capture reactions emit no gammas (only tritium and an alpha particle from the 7.5% Li-6 neutron capture in naturally occurring lithium). Lithium thermal neutron shielding with its low gamma noise production is used with gamma-sensitive detectors. Pure lithium metal and lithium compounds can be obtained in large quantities at relatively inexpensive cost (as from Chemetall Foote Corporation, Kings Mountain, North Carolina, USA).

In the course of fabricating biconcave parabolic lithium lenses at Adelphi Technology, the author (J. T. Cremer) heated a 3-mm diameter glob of lithium (0.534 g/cm^3 mass density) to its 180° melting point on a hot plate and with a tiny spatula dropped the liquid glob into a Pyrex beaker filled with 1 liter of water. A small but somewhat terrifying fireball was propelled through the water from the lithium reaction gases. The lithium fireball moved to the wall of the beaker and cracked the Pyrex glass. The author had previously worked as postdoctoral fellow of P. Csonka and R. Tatchyn at Stanford Synchrotron Radiation Laboratory (SSRL) to develop a furnace to produce a tiny gas jet of lithium atoms and perform light spectroscopy on the jet. The atomic lithium gas was a potential lasing medium for pumping by synchrotron X-rays and a low photon energy laser (see Csonka, 1976, and Rozsnyai *et al.*, 1985). After an experiment (before coming up to air from vacuum), the lithium was slowly neutralized with ethyl alcohol.

A CRL composed of uniform, small, closely spaced air bubbles in epoxy in a precise alignment in a glass capillary tube was developed by Dudchik and Kolchevsky (1999). The bubble CRL is an inexpensive, practical, and robust compound refractive X-ray lens (see Dudchik *et al.*, 2000, and Dudchik and Kolchevsky, 2001). The five attributes of bubble lenses are as follows:

1. The lens surface quality is superb due to physicochemical hydrodynamics of bubble formation. The use of bubble arrays (bubble rafts) to represent crystal lattices (including grain boundaries, dislocations, and fault lines) was shown by Bragg and Nye (1947), whose work was presented in chapter 30 of the *Feynman Lectures on Physics* (Feynman *et al.*, 1964).
2. The capillary enclosure ensures the series of unit bubble lenses has excellent coaxial alignment.
3. The center minimum thickness of the lens can be extremely thin.
4. The radius of curvature of the bubble lens can be made extremely small.
5. The method of the formation of a line of closely spaced bubbles and the bubble material is inexpensive and very robust.

Figure 85 shows a diagram for the mechanism of producing bubble lenses in epoxy and a schematic of a spherical biconcave lens produced at the interface of two air bubbles in the epoxy taken from Dudchik and Kolchevsky (1999). Bubble lenses are produced by air from a compressor injected by an injector needle into a capillary filled with liquid epoxy. The needle is withdrawn at a uniform rate as the bubbles are blown, thereby producing a row of identical air bubbles that are uniformly spaced with center minimum wall thickness as small as $d = 10\,\mu m$.

After a row of air bubbles is blown into the liquid epoxy–filled capillary, ultraviolet (UV) light is applied to the UV light–sensitive epoxy ($C_{100}H_{200}O_{20}N$ and $1.02\,g/cm^3$ density) to cause it to harden. In dentistry UV-sensitive materials are used as fillings. Also shown are a photo of

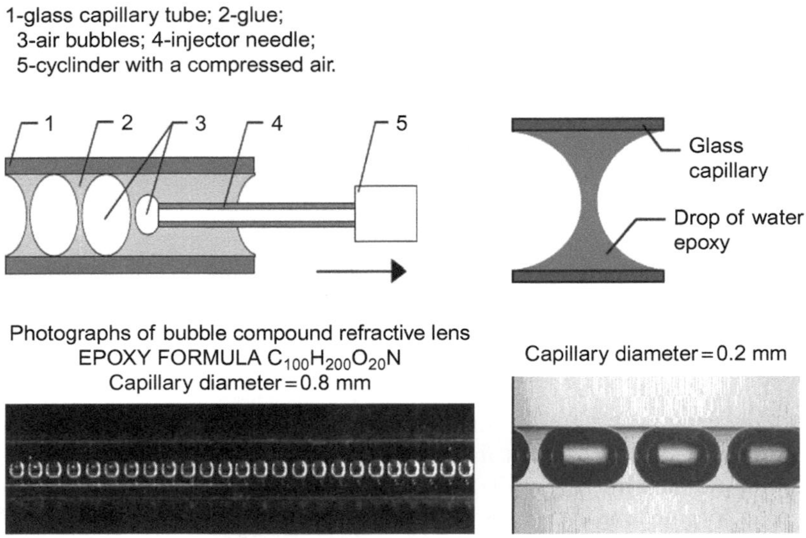

FIGURE 85 (Top left) The mechanism used to produce air bubbles. (Top right) Schematic of the spherical biconcave lens produced at the interface between two air bubbles in epoxy with UV light–sensitive fillings. The row of air bubbles thus produces a line of biconcave spherical lenses. (Bottom) The capillary inside diameter sets the effective aperture diameter of the bubble lenses. The photos show (left) a bubble compound refractive lens in 0.8-mm capillary and (right) a close-up photo of a bubble in a 0.2-mm diameter capillary (Dudchik and Kolchevsky, 1999).

a bubble CRL with a 0.8-mm capillary inside diameter, and a close-up photo of the bubble lenses in a 0.2-mm inside diameter capillary. The capillary inside diameter sets the physical aperture diameter $2r_e$ of the bubble lenses.

Adelphi Technology fabricated and tested a variety of two-dimensional (2D) CRL composed of spherical or parabolic biconcave lenses, which included lenses fabricated from Mylar (DuPont Teijin Films, Chester, VA) and Kapton (DuPont), beryllium, and lithium, as well as bubble lenses in epoxy. The following set of equations was used to characterize the CRLs of various materials under the conditions of both incoherent and coherent Gaussian beam X-ray sources. The description of these lenses and the experimental results are described after the brief presentation of the following basic equations. First, given the decrement δ and linear attenuation μ of the refractive index n with $\beta = \lambda\mu/4\pi$, the lens material refractive index n and the thin lens CRL focal length f are

$$n = 1 - \delta + i\beta = 1 - \delta + i\frac{\lambda\mu}{4\pi} \tag{108}$$

and

$$f = \frac{R}{2N\delta}. \tag{109}$$

The CRL absorption aperture radius r_a and, for a CRL with spherical biconcave lenses, the CRL parabolic aperture radius r_p are

$$r_a = \sqrt{\frac{2R}{\mu N}} = 2\sqrt{f\frac{\delta}{\mu}} \tag{110}$$

and

$$r_p = \left(4R^2\lambda r_i\right)^{1/4} = \left(\frac{2R^3\lambda}{\delta N}\right)^{1/4}. \tag{111}$$

The numerical aperture NA is a very compact way to describe the CRL performance because NA provides the ratio of the CRL effective aperture radius r_e divided by the CRL focal length f. For the CRL surrounded by air or vacuum, the numerical aperture NA is

$$NA \cong \frac{r_e}{f} = \frac{r_a}{f} = 2\sqrt{\frac{\delta}{f\mu}}. \tag{112}$$

For wavelength λ and CRL numerical aperture NA at λ (absorption-limited effective aperture diameter changes with wavelength), the diffraction-limited resolution σ (μm) is

$$\sigma = 0.61\frac{\lambda}{NA} = 0.61\frac{\lambda}{r_a}f = 0.3\lambda\sqrt{\frac{f\mu}{\delta}}. \tag{113}$$

Given object distance r_0, the FOV of each CRL of length L and absorption aperture radius r_a is given by

$$FOV = \frac{4r_a r_0}{L}. \tag{114}$$

The thick lens CRL focal length f_{thk} is expressed in terms of the CRL thin lens focal length f and CRL length L by

$$f_{thk} = f \cdot \frac{\sqrt{\frac{L}{f}}}{\sin\sqrt{\frac{L}{f}}}. \tag{115}$$

Consider an incoherent X-ray source area A_{sou} defined by vertical source width σ_v and horizontal source width σ_h, and a CRL with effective aperture area $A_{crl} = r_e^2\pi$ for effective aperture radius r_e, which is the minimum of the absorption aperture radius r_a and physical aperture radius r_m. The CRL has peak transmission T along the center optical axis, and the average transmission T_{avg} over the CRL effective aperture radius r_e, and the CRL produces magnification $M = r_i/r_0$. If the CRL image distance r_i is much less than the object distance r_0, which is the case with synchrotron X-ray sources, then the image distance r_i is approximately equal to the CRL focal length f, where

$$r_i \cong f \quad \text{and} \quad M < 1 \quad r_i \ll r_0. \tag{116}$$

The 2D gain of an N-lens CRL from geometric optics (ignoring diffraction effects and beam coherence) for the typical synchrotron ring, where $r_0 \gg r_i$ is then

$$G = \frac{r_e^2}{\sigma_v\sigma_h}\left(1+\frac{r_0}{r_i}\right)^2 T_{avg} \cong \left(\frac{r_0}{f}\right)^2 T_{avg}\frac{A_{crl}}{\sigma_v\sigma_h} \cong \frac{T_{avg}}{M^2}\frac{A_{crl}}{A_{sou}} \quad r_0 \gg r_i. \tag{117}$$

The on-axis peak transmission T along the center optical axis of the CRL is

$$T = \exp(-\mu N d). \tag{118}$$

Here d is the individual biconcave lens, center minimum thickness and μ is the lens material linear attenuation. The average transmission can be calculated and approximated by

$$T_{avg} \cong 0.43T = 0.43e^{-\mu N d}. \tag{119}$$

The transverse magnification $M = r_i/r_0$ can be measured to confirm its theoretical value as the image distance r_i divided by the object distance r_0 in which both are measured from the center plane of a thin lens CRL. For a thick lens CRL object distance r_0 is measured from the secondary principal plane, which is shifted from the CRL center plane toward the image plane. The thick lens image distance r_i and focal length f are measured from the primary principal plane, which is shifted from the center CRL plane toward the object plane.

The intensity at the axial spot of the image plane of a coherent source, which is produced by a Gaussian beam, is well known from laser optics (Figure 86). Combined with the Fresnel–Kirchhoff integral of the intensity at the same spot with the CRL in place, this yields the expression of the coherent CRL 2D gain G_{coh} for a coherent source (see Beguiristain *et al.*, 2000a,b, and Tatchyn *et al.*, 1997).

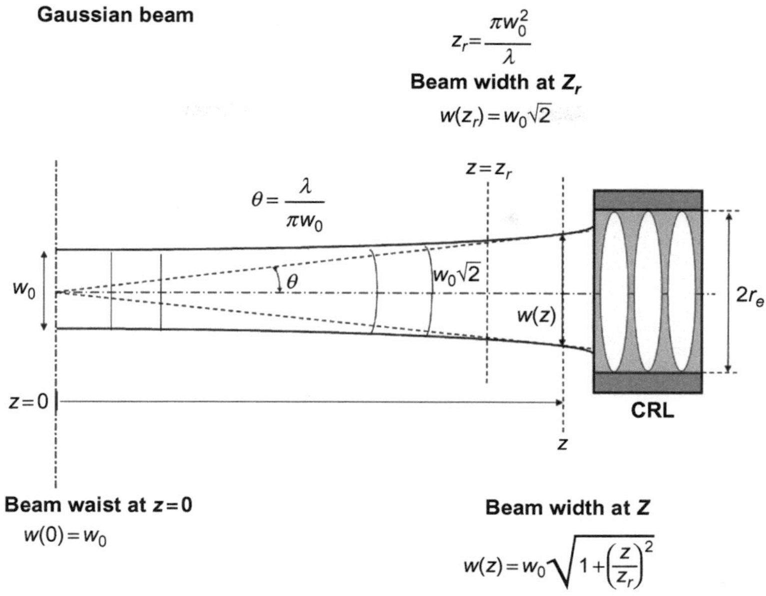

FIGURE 86 The propagation of a Gaussian beam in the $+z$ direction from its waist at the $z = 0$ plane. The standard deviation $w(z)$ of the beam diameter increases with distance z. The beam diverges out from the $z = 0$ beam waist plane with divergence angle θ. The divergence angle θ is inversely proportional to the minimum beam waist diameter w_0. Smaller beam waist diameters w_0 lead to larger divergence angles θ.

If the CRL effective aperture diameter $2r_e$ is larger than the Gaussian beam diameter $w(z)$, then the CRL 2D gain G_{coh} for a coherent source (Beguiristain *et al.*, 2000a) is

$$G_{coh} = \frac{\left[1 + \left(\frac{2(r_0+r_i)}{k\sigma^2}\right)^2\right]\exp(-\mu N d)}{\left[\frac{r_i\sqrt{b}}{k}\left(\frac{2}{b\sigma^2} + \frac{\mu}{2f_{coh}\delta}\right)\right]^2} \quad 2r_e \geq w(z). \tag{120}$$

Under coherent illumination, the CRL focal length f_{coh} is expressed in terms of the incoherent focal length f by

$$f_{coh} = \frac{br_0 f}{f - br_0}. \tag{121}$$

Here r_i is the image distance, r_0 is the object distance, and σ is the source diameter with

$$k = \frac{2\pi}{\lambda} \quad b = 1 + \frac{r_0^2}{z_r^2}. \tag{122}$$

The minimum beam size is related to the Rayleigh length z_r that is associated with the Gaussian beams of laser optics, and

$$z_r = \frac{\sigma^2 \pi}{\lambda}. \tag{123}$$

If the CRL effective aperture diameter $2r_e$ is less than the Gaussian beam diameter $w(z)$, then the CRL 2D gain G_{coh} for a coherent X-ray source (Beguiristain *et al.*, 2000a) is

$$G_{coh} = \frac{\left[1 + \left(\frac{2(r_0+r_i)}{k\sigma^2}\right)^2\right] \exp(-\mu N d)}{\left[\frac{r_i\sqrt{b}}{k}\left(\frac{2}{b\sigma^2} + \frac{\mu}{2f_{coh}\delta}\right)\right]^2} \cdot \left[erf\left(r_e \sqrt{\frac{k}{2}\left[\frac{1}{bz_r} + \frac{\mu}{2kf\delta}\right]}\right)\right]^4$$
$$\times\, 2r_e < w(z). \tag{124}$$

The 2D CRL gain G_{in} for an incoherent source (Beguiristain *et al.*, 2000a), which is found by solving the appropriate Fresnel–Kirchhoff integrals to account for diffraction and all aberrations, is given by

$$G_{in} = \left(\frac{r_0^2}{4\sigma^2} + fk^2 \frac{\delta}{\mu}\right)^{-1} \left(\frac{r_0 k}{\sigma}\right)^2 \left(\frac{\delta}{\mu}\right)^2. \tag{125}$$

The 2D biconcave parabolic and spherical coin lenses of the plastics (Mylar and Kapton), beryllium, and lithium (fabricated by Adelphi Technology), as well as bubble lenses in epoxy fabricated by both Y. Dudchik of the Institute of Applied Physics Problems at Belarus State University in Minsk, Belarus, and Adelphi Technology, were tested on the bending magnet beam line 2-3 at the Stanford Synchrotron Radiation Laboratory (SSRL). This beam line possessed a double-crystal Si(111) monochromator, which was capable of delivering X-rays from 2,400 to 30,000 eV with a 5×10^{-4} resolution. The SSRL beam line 2-3 X-ray source is ellipse-shaped with 0.4-mm vertical FWHM by 1.7-mm horizontal FWHM dimensions. The CRL was placed at 16.8 m distance downstream from the X-ray source.

The CRL was mounted in a goniometer head (gimbal mount) that could be manually tilted in two axes—pitch and yaw (see SSRL experimental setup at the top of Figures 87, 89, 90, and 91). The CRL could also be translated in the orthogonal (x- and y-) directions to the axial z-direction of the X-ray beam. These adjustments maximize the X-ray transmission though the CRL by aligning it with the beam. X-ray gas ionization detectors were placed after the entrance tantalum slit #1 and after the exit tantalum slit #2. The X-ray beam was profiled using the translatable tantalum exit slit #2. The tantalum exit slit #2 dimension was adjusted to below 25 μm by using a thin stainless steel shim. It is likely the exit tantalum slit #2 jaws were not ideally parallel at these small dimensions. Consequently, the exit tantalum

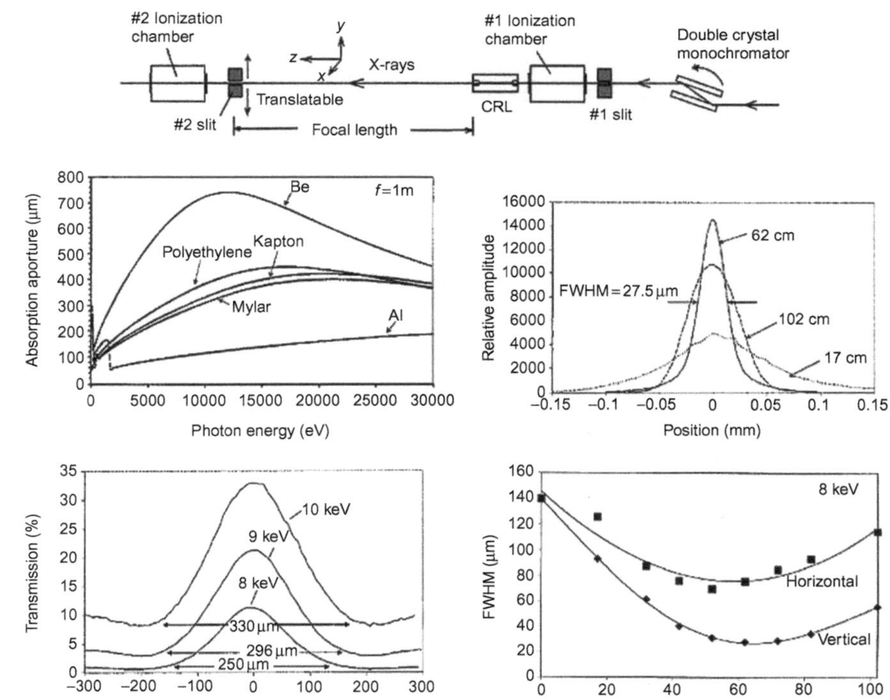

FIGURE 87 (Top) The SSRL BL 2-3 experiment setup. The elliptical X-ray source is 0.44-mm vertical × 1.7-mm horizontal and 16.8 m upstream of Mylar CRL ($N = 193$ biconcave spherical lenses) or Kapton CRL ($N = 330$ biconcave, parabolic lenses). (Middle left) Plots for CRL absorption aperture diameter $2r_a$ versus photon energy for plastic and beryllium CRLs with $f = 1$-m focal length. (Middle right) Plot for the intensity profile of the refracted beam at various distances downstream of Mylar M5 CRL for 8-keV X-rays (bottom left). The plot is for Mylar M5 CRL transmission $T(r)$ versus radial position r for 8-, 9-, and 10-keV X-rays. (Bottom right). Plot showing the FWHM horizontal and vertical beam widths versus the axial distance downstream of Mylar M5 CRL for 8-keV X-rays (Piestrup *et al.*, 2000).

TABLE I. Lens parameters.

Lens No.	Lens material	R (mm)	N	Min. lens thickness d (μm)	Photon energy (keV)	Calc. f (cm)	Calc. image distance (cm)	Meas. image dist (cm)	Phys. apert. (μm)	Calc. abs. apert. (μm)	Meas. abs. apert (μm)	Peak transmission (%)	Gain calc	Gain exp.
M5	Mylar	1.6	193	10	8	99	105	62	645	287	250	12	2.5	1.5
M5	Mylar	1.6	193	10	9	125	134	82	645	343	296	22	3.8	-
M5	Mylar	1.6	193	10	10	153	169	92	645	405	330	33	4.9	2.0
K5	Kapton	1.98	400	15.8	11	97	103	72	370	375	330	11	2.0	2.0
K5	Kapton	1.98	400	15.8	12	100	107	85	370	381	360	19	2.7	2.5
K7	Kapton	1.98	400	12.5	10	60	62	65	286	266	303	11	2.7	1.6
K7	Kapton	1.98	400	12.5	12	81	85	103	286	345	364	27	6.4	3.8
K6	Kapton	1.98	600	15.8	13	77	81	70	370	348	350	21	2.3	2.2
K6	Kapton	1.98	600	15.8	14	87	92	75	370	383	—	19	3.2	—
K11[a]	Kapton	1.35	330	12.5	9	48	49	47	350	230	262	8	2.9	1.7
K11[a]	Kapton	1.35	330	12.5	10	59	61	62	350	270	291	16	5.4	3.8
K11[a]	Kapton	1.35	330	12.5	11	69	72	72	350	310	310	24	8.0	5.0

[a]Parabolic lens.

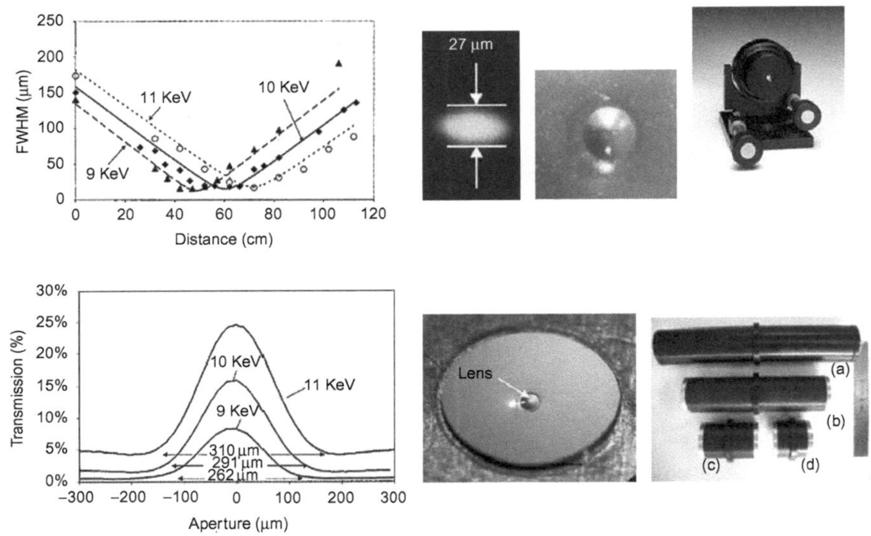

FIGURE 88 (Top) Parameters of Mylar and Kapton CRLs. (Middle left) Plot of the calculated and measured refracted vertical beam widths at 9, 10, and 11 keV produced by Kapton K11 CRL versus distance from the CRL. (Middle right) X-ray image of demagnified BL 2-3 X-ray source at 8 keV by Mylar M5 CRL and photos of the biconcave parabolic Kapton lens and lens holder. (Bottom left) Plot of the measured transmission $T(r)$ through Kapton K11 CRL as a function of radial position r for 9-, 10-, and 11-keV X-rays. (Bottom right) Photos of the biconcave spherical Mylar lens and lens holders of varying lengths. See text for details (Piestrup *et al.*, 2000). (See Color Insert.)

slit #2 width was minimally greater than 3 μm when the tantalum exit slit #2 jaws appeared to be entirely closed.

The ionization chamber #2 was set at a fixed position just downstream of the exit tantalum slit #2. Refracted X-rays from the CRL passed through ionization chamber #2, which measured the total X-ray power of the CRL-refracted X-rays. The relative intensity profiles of the refracted X-ray beam in the transverse, vertical y-direction, and the transverse horizontal

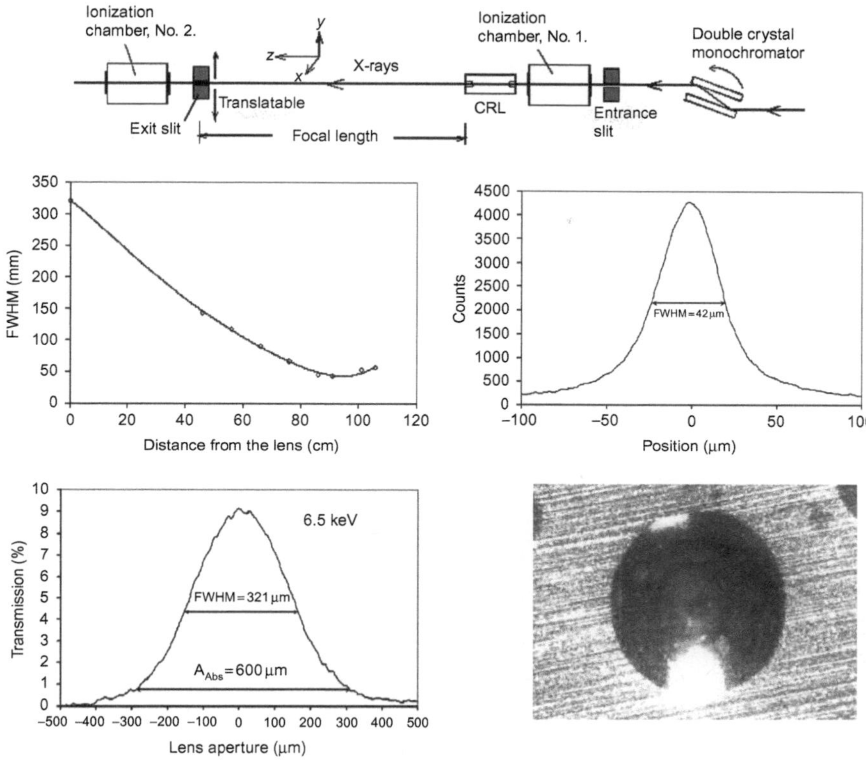

FIGURE 89 (Top) Experimental setup for the prototype beryllium CRL ($N = 160$, $2r_m = 2.29$ mm, $d = 40$ μm, and $2r_p = 2r_e = 0.6$ mm at 6.5 keV) set 16.8 m from the 0.44-mm vertical × 1.7-mm horizontal FWHM elliptical bending magnet X-ray source at SSRL BL 2-3. (Middle left) The plot of the vertical width of the refracted beam of 6.5-keV X-rays, which indicates the demagnified image of source has 42-μm FWHM width at $r_i = 93$ cm. (Middle right) Plot of the vertical intensity profile (number of counts measured by the detector vs. vertical position) of the demagnified image of the 6.5-keV X-ray source. (Bottom left) Plot of the transmission $T(r)$ through the beryllium CRL versus radial position r. (Bottom right) Photo of one side of the beryllium biconcave spherical lens. See text for details (Beguiristain *et al.*, 2002). (See Color Insert.)

x-direction at various axial z-distances downstream from the CRL were measured by translating the tantalum exit tantalum slit #2 across the ionization chamber #2.

The translating tantalum exit slits #2 were moved to a particular axial z-distance downstream from the CRL. At the set z-distance, the vertical exit slits #2 were translated in the horizontal x-direction to obtain the relative intensity of the refracted beam in the horizontal x-direction. Next, the horizontal exit tantalum slits #2 were translated in the vertical

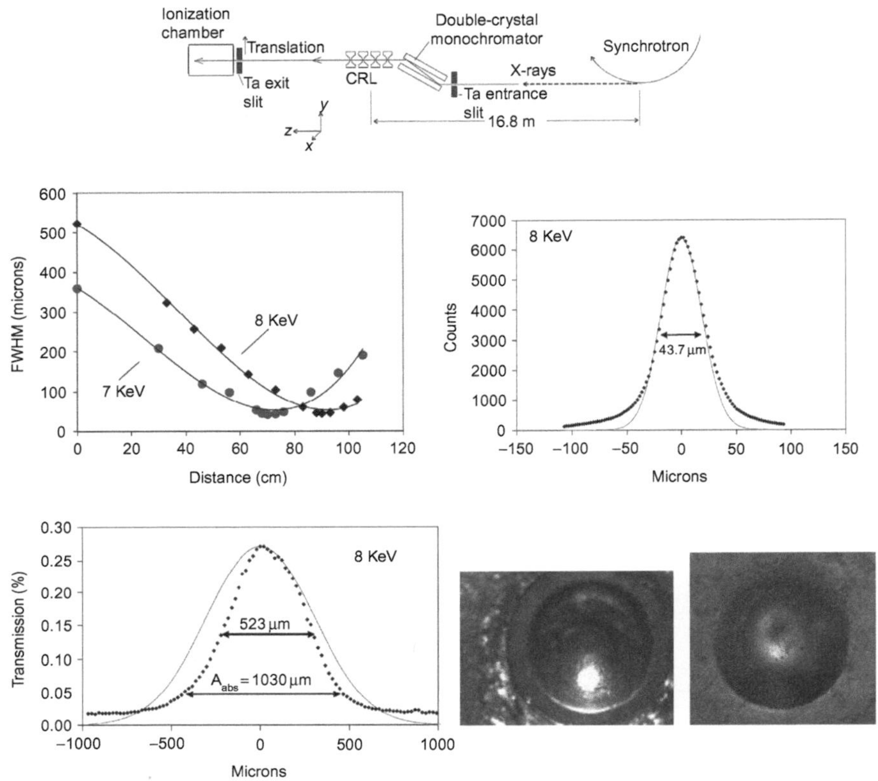

FIGURE 90 (Top) Lithium CRL testing at SSRL BL 2-3. (Middle left) Plot of the vertical widths of CRL-refracted 7.0-keV and 8.0-keV X-rays. (Middle right) Plot of the measured vertical intensity profile of the demagnified source image at 8 keV (dashed curve). Data are fitted with Gaussian curve (solid line). (Bottom left) Plot of the CRL transmission $T(r)$ of 8-keV X-rays versus radial position r. The plot shows $2r_a = 1030$-μm absorption and slightly larger $2r_m = 1200$-μm physical aperture diameters and 523-μm FWHM. The solid line shows the calculated transmission profile with $2r_a = 1200$-μm absorption aperture diameter. (Bottom right) Intel toy microscope photos of the lithium parabolic lens at 100× (Cremer *et al.*, 2003). (See Color Insert.)

y-direction to obtain the relative intensity of the refracted beam in the vertical y-direction. The X-rays could reach ionization chamber #2 only by passing through the scanning tantalum slits #2. Vertical and horizontal beam widths and relative beam intensity of the CRL-refracted X-ray beam were thus measured by scanning the horizontal and vertical exit tantalum slits #2, respectively, over the X-ray beam at each of a set of axial locations downstream from the CRL. The measured relative intensity profile of the beam in the vertical and horizontal directions and the resulting measured beams widths in the vertical and horizontal directions were plotted as a function of axial z-distance from the CRL.

The X-ray transmission through the CRL was also measured as a function of radial position across the CRL aperture. The transmission through the CRL as a function of the radial position in the CRL was obtained by translation of the CRL through a very narrow, square 25-μm × 25-μm X-ray beam, which was produced by a crossed pair of tantalum slits. The X-ray transmission fraction $T(r)$ as a function of radial position r in the CRL is plotted. The CRL transmission plot can be used to find the $1/e^2$ decrease in transmitted intensity $T(r = r_a)$ at absorption aperture radius r_a relative to the maximum transmission $T(r = 0)$ along the center CRL axis ($r = 0$). The measured width between the $1/e^2$ points on the Gaussian-shaped X-ray transmission $T(r)$ versus CRL radial position r yields the measured CRL absorption aperture diameter $2r_a$.

For X-rays the absorption aperture diameter $2r_a$ is usually smaller than the physical aperture diameter $2r_e$. If the CRL absorption aperture diameter $2r_a$ is less than or equal to the parabolic aperture diameter $2r_p$ of a CRL composed of biconcave spherical lenses, where $2r_a \leq 2r_p$, then refracted and imaged X-rays pass through only the center portion of the spherical lenses, which are almost identical in shape to parabolic lenses. Thus, by (1) setting a limiting aperture of the spherical lens CRL to equal the CRL parabolic aperture diameter $2r_p$, or (2) if the CRL absorption aperture diameter $2r_a$ equals or exceeds the CRL parabolic aperture diameter $2r_p$, then only the parabola-shaped portion of the CRL spherical lenses refracts and images the X-rays. A spherical lens CRL designed with effective aperture diameter $2r_e$ set equal to the parabolic aperture diameter $2r_p$, where $2r_e = 2r_p$, then produces X-ray images free of spherical aberration, but at the expense of reduced image intensity and reduced object FOV.

Adelphi Technology designed, fabricated, and tested 2D CRL composed of biconcave spherical lenses and biconcave parabolic lenses of Mylar ($C_5H_4O_2$ with density 1.39 g/cm^3) and Kapton ($C_{22}H_{10}N_2O_5$ with density 1.43 g/cm^3) (see Piestrup *et al.*, 2000, 2001, and Beguiristain *et al.*, 2000a,b). The X-ray optical constants were obtained from the Center for X-ray Optics at the Lawrence Berkeley National Laboratory. X-ray optical constants can also be obtained from the U.S. National Institute of Standards and Technology (NIST) using the XCOM database (see Berger *et al.*, 1998). The lenses were fabricated by compressing two opposed spherical balls into a thin disk of plastic film, which was held precisely in the fixture to produce the spherical biconcave lenses in 76.2-μm thick Mylar film and 25.4-μm thick Kapton films. The biconcave parabolic lenses in Kapton were produced by pressing a pair of opposed parabolic glass microlenses into the Kapton film. The 25.4-μm thick Kapton film was amorphous grade and was heated in the lensmaking fixture to improve the lens surface quality.

Adelphi measured the intensity profile and transmission of X-rays focused by a series of either spherical or parabolic lenses fabricated with Mylar or Kapton. The use of plastics can extend the range of operation of CRLs, increasing peak T and average $T_{avg} \cong 0.43\,\mathrm{T}$ transmission, increasing aperture the effective diameter $2r_e$, and decreasing focal length f. The number N of unit lenses range from 193 to 600 for each CRL. Two-dimensional focusing was obtained for photon energies 8–14 keV with imaging distances less than 1 m $r_i \leq 1.0\,\mathrm{m}$. For example, FWHM line widths down to 16 mm at an image distance $r_i = 47\,\mathrm{cm}$ were achieved with 9-keV X-rays. The effective aperture diameters $2r_e$ of the plastic refractive lenses were measured between 250 and 364 mm with peak transmissions T along the plastic CRL center axis of between 10% and 33% (see Table in Figure 88).

The top of Figure 87 shows the SSRL BL 2-3 setup for Mylar and Kapton biconcave spherical and parabolic lenses. The bending magnet X-ray source was ellipsis-shaped with 0.44-mm vertical × 1.7-mm horizontal dimensions, and the bending magnet X-ray source was located at object distance $r_0 = 16.81\,\mathrm{m}$ upstream from the plastic CRL.

The middle-left plot in Figure 87 shows the absorption aperture diameters $2r_e$ for CRLs composed of beryllium, polyethylene, Kapton, Mylar, and aluminum with focal length $f = 1.0\,\mathrm{m}$. The middle-right plot shows the measured vertical intensity profile of the refracted 8-keV X-ray beam for distances 17, 62, and 102 cm downstream from the Mylar M5 CRL. The Mylar M5 CRL was composed of $N = 193$ biconcave spherical lenses with $R = 1.6$-mm radius of curvature, $d = 10$-μm minimum wall thickness, $2r_e = 645$-μm physical aperture diameter, and measured absorption aperture diameter $2r_a = 250\,\mu\mathrm{m}$ with 8-keV X-rays. The Mylar M5 CRL demagnified the 8-keV X-rays to the vertical image width of 27.5-μm FWHM at image distance $r_i = 62\,\mathrm{cm}$ as shown in the plot. The demagnified image of the 8-keV source X-rays shows a measured 2D gain $G_{\mathrm{exp}} = 1.5$ compared with the theoretical gain $G_{th} = 2.5$.

The bottom left of Figure 87 plots the X-ray transmission $T(r)$ through the Mylar M5 CRL as a function of CRL radial position r for 8, 9, and 10 keV X-rays. The bottom-right plot shows the FWHM horizontal and vertical refracted beam profiles versus distance downstream of the Mylar M5 CRL for 8-keV X-rays.

The top of Figure 88 shows a table of the calculated and measured parameters of the Mylar and Kapton CRLs. The Kapton K11 CRL was composed of $N = 330$ biconcave parabolic Kapton lenses with on-axis radius of curvature $R = 1.35\,\mathrm{mm}$, minimum wall thickness $d = 12.5\,\mu\mathrm{m}$, physical aperture diameter $2r_m = 350\,\mu\mathrm{m}$, and measured absorption aperture diameter $2r_a = 262\,\mu\mathrm{m}$ at 9 keV.

The three images in Figure 88 (middle right) show a demagnified X-ray image (shown with arrows and taken with X-ray film) of the bending magnet source imaged by the parabolic Kapton K11 CRL; (center) a photo of a 2.54-cm diameter, 72.2-μm thick orange Kapton disk with the biconcave parabolic lens at the disk center; and (far right) a photo of the black garolite lens holder that held and aligned the stack of 25.4-mm diameter plastic disks with their center biconcave lenses. The aluminum endcaps on the black garolite lens holder had center holes, which acted as limiting X-ray apertures. The hole diameters of the aluminum endcaps were set equal to the physical aperture diameter $2r_e$ of the contained Mylar or Kapton CRL lenses. The black garolite lens holder was mounted on a normal optical gimbal mount, which allows adjustment of the CRL pitch and yaw. The gimbal mount is then mounted on an xy-translation stage for vertical and horizontal adjustment transverse to the axial, z-directed X-ray beam. The xy-stage with the Mylar or Kapton CRL was set just downstream from the first ionization chamber and input slits.

The bottom-middle photo in Figure 88 shows the 2.54-cm diameter, 25.4-μm thick Mylar disk with the biconcave spherical lens at the disk center. The bottom-right photo shows the set of garolite lens holders of different lengths, that held and aligned the varying number N of lenses required to produce a particular focal length.

The middle-left plot in Figure 88 shows the measured and calculated vertical beam widths of 9-, 10-, and 11-keV X-rays refracted by the Kapton K11 CRL (with biconcave parabolic lenses). From the measured plotted data, the 9-keV demagnified image has vertical 10-μm FWHM width at image distance $r_i = 47\,\text{cm}$. The experimental measured 2D gains in the demagnified images of the bending magnet source obtained with the Kapton K11 CRL were about 60% of the theoretical values. That is, at 9 keV the measured and theoretical gains were $G_{\text{exp}} = 1.7$ and $G_{th} = 2.9$, and at 11 keV the measured and theoretical gains were $G_{\text{exp}} = 5.0$ and $G_{th} = 8.0$. The theoretical FWHM vertical beam widths were calculated by solving the Fresnel–Kirchhoff integral for the X-ray CRL experimental setup, which includes effects of diffraction and the CRL Gaussian absorption profile. The bottom-left plot shows the measured transmission through the parabolic Kapton K11 CRL as a function of the CRL radial position for 9-, 10-, and 11-keV X-rays.

Adelphi fabricated beryllium CRL composed of biconcave spherical lenses and measured the vertical intensity profile and transmission of X-rays at the SSRL bending magnet beam line 2-3 (Beguiristain *et al.*, 2002, see and Pantell *et al.*, 2001). The use of beryllium extends the operation range of CRL by improving transmission, aperture size, and overall performance. The beryllium CRL consisted of $N = 160$ biconcave spherical unit lenses.

The beryllium coin lenses were fabricated from a 2.54-cm diameter, 0.762-mm thick disk of beryllium by first machining two hemispherical cavities into each side of the disk. The beryllium disk was then pressed between two opposed spherical tungsten carbide balls with the desired radius of curvature. The opposed tungsten carbide balls were pressed into the pair of premachined hemispherical cavities of the beryllium disk at as large a pressure as the thin beryllium disk could withstand without cracking.

The resulting beryllium biconcave spherical lens had a radius of curvature $R = 1.9$ mm, minimum (center) wall thickness $d = 40\,\mu$m, physical aperture diameter $2r_m = 2.29$ mm, and maximum thickness 0.762 mm. The beryllium lenses were stacked in a black garolite lens holder. The lens holder was held in a regular optical gimbal mount, which allowed pitch and yaw adjustment. The gimbal mount with CRL was mounted on an xy-translation stage to allow transverse translation of the CRL relative to the axial, z-directed X-ray beam. The xy-stage with the beryllium CRL was set just downstream from the first ionization chamber and input slit.

Two-dimensional focusing was obtained at 6.5 keV with a focal length $f = 93$ cm and an X-ray transmissions fraction $T = 0.09$ along the center CRL axis. The measured absorption CRL aperture diameter $2r_a = 0.6$ mm at 6.5 keV was also equal to the calculated CRL parabolic aperture diameter, where $2r_p = 2r_a = 0.6$ mm, and the resulting beryllium CRL effective aperture diameter was $2r_e = 0.6$ mm, where $r_e = MIN(r_m, r_a)$. Although the beryllium CRL was composed of biconcave spherical lenses, the absorption limited the CRL effective aperture diameter to the center diameter portion of the spherical lens surface, which is approximately parabola shaped, where $2r_p = 2r_e = 0.6$ mm. Thus, the beryllium CRL with spherical biconcave lenses acted as a CRL with biconcave parabolic lenses. This is because the beryllium CRL aperture was limited to the center portion of spherical lenses in which the spherical surface is almost identical to a parabolic surface.

The top of Figure 89 shows is the experimental setup of the beryllium spherical biconcave lens CRL performance tested on the SSRL 2-3 beam line. The bending magnet X-ray source was ellipsis-shaped with 0.44-mm vertical × 1.7-mm horizontal dimensions, and the bending magnet X-ray source was located at object distance $r_0 = 16.81$ m upstream from the CRL along the z-axis. The middle-left plot shows the measured vertical width of a refracted 6.5-keV X-ray. From this plot the demagnified image of the X-ray source occurred at image distance $r_i = 93$ cm with a vertical 42-μm FWHM; the measured 2D intensity gain $G_{\text{exp}} = 1.5$ was 25% of the calculated gain $G_{th} = 6.0$. The middle-right plot shows the vertical 42-μm FWHM of the relative intensity profile of the imaged source along the transverse vertical y-direction. The bottom-left plot shows the transmission fraction $T(r)$ of X-rays through the beryllium CRL as a function of radial distance r from the CRL center axis for 6.5-keV X-rays. The bottom-right photo shows one side of the spherical biconcave beryllium lens.

The possibility of 2D biconcave parabolic lenses of lithium, inspired by Pereira *et al.* (2001) and directly suggested to Adelphi by Periera, prompted Adelphi to fabricate a lithium CRL composed of biconcave parabolic lenses and measure the lithium CRL performance at the SSRL bending magnet beam line 2-3 (see Cremer *et al.*, 2003, 2004). The use of lithium extends the operation range of CRL by improving transmission, aperture size, and overall performance. The lithium CRL consisted of $N = 335$ biconcave parabolic lenses with an on-axis radius of curvature $R = 0.95\,\text{mm}$, center minimum wall thickness $d = 100\,\mu\text{m}$, and physical aperture diameter $2r_m = 1.2\,\text{mm}$. Two-dimensional focusing was obtained at 8.0 keV with a focal length $f = 95\,\text{cm}$. The effective aperture of the lithium CRL was measured as $2r_e = 1030\,\mu\text{m}$ with 27% on-axis (peak), transmission $T(0) = 0.27$, and a measured on-axis intensity gain $G_{\text{exp}} = 18.9$ (see the table at the top of Figure 91).

The individual biconcave parabolic lenses (coin lenses) of lithium are stacked to form a CRL capable of focusing in two dimensions and achieving a larger absorption aperture diameter (1.0 mm) and a higher 2D gain (18.9) than previous CRLs given the source size (<1 mm), source distance (<20 m), and CRL focal length (<1 m). This results from the larger δ/μ for lithium than other CRL materials below 20 keV, and especially below 10 keV with the δ/μ ratio of lithium twice that of beryllium and sixfold larger than carbon (Kapton). The lithium CRL was fabricated by a compression molding technique in a dry environment. The lithium is a soft metal and a 1.0-inch diameter punch pressed into the 0.020-inch thick lithium foil (obtained from Chemetall Foote) produced 2.54-cm diameter, 0.51-mm thick lithium disks.

The lithium disks were placed in the lens press in which the disks were pressed by opposing parabolic glass lenses. The lens press consisted of two pistons with glass parabolic lenses mounted on their center surface. Each piston fit inside a cylinder in a square aluminum block in which the piston could be slid to a fixed position. The two square aluminum blocks were joined by rods, which allowed the cylinders and pistons of the sliding aluminum blocks to be precisely aligned. In this manner, a lithium disk could be set on top of a piston with a parabolic lens recessed inside the cylinder. By fixing the pistons at set positions within their cylinders, the two blocks could be moved and pressed against each other. The two parabolic lenses pressed on opposite sides into the lithium disk by a set distance. This compression molding technique produced a biconcave lens with a minimum wall thickness $d = 100\,\mu\text{m}$ and a physical aperture diameter of about $2r_m = 1.2\,\text{mm}$.

Object Distance (m)	16.8	
Source Vertical FWHM Width (μm)	400	
Source Horizontal FWHM Width (μm)	1700	
Number of Lenses in CRL	335	
Parabolic Lens On-Axis Radius of Curvature (μm)	950	
Calculated Lens Minimum Wall Thickness (μm)	253	
Photon Energy (keV)	7.0	8.0
Photon Wavelength (Å)	1.77	1.55
Tabulated Real Atomic Scattering Factor[26]	3.0	3.0
Calculated Refractive Index Decrement	1.95×10^{-6}	1.50×10^{-6}
Tabulated Linear Attenuation Coefficient[26] (m^{-1})	23.4	15.7
Calculated Focal Length (m)	0.73	0.95
Calculated Image Distance (m)	0.76	1.00
Measured Image Distance (m)	0.70	0.90
Calculated Minimum Waist Diameter (μm)	18.1	23.8
Measured Minimum Waist Diameter (μm)	41.6	43.7
Measured Peak Transmission (%)	9.7	26.3
Calculated Attenuation Aperture Diameter (μm)	984	1200
Measured Attenuation Aperture Diameter (μm)	667	1030
Calculated 2D Gain	20.0	47.7
Measured 2D Gain	3.84	18.9

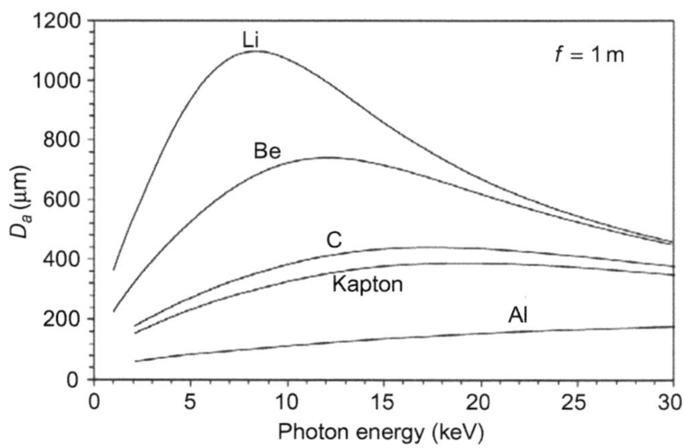

FIGURE 91 (Top) Tabulated parameters of the lithium CRL and measured and calculated results of the focusing experiments on SSRL beam line 2-3. (Bottom) Plot of the CRL absorption aperture diameter $2r_a$ versus photon energy for $f = 1.0$-m focal length for CRLs made of lithium, beryllium, Kapton, carbon, and alumimum. The $2r_a$ of lithium CRLs is almost twice as large at 8 keV than above 25 keV. See text for details (Cremer *et al.*, 2003).

The top of Figure 90 shows the SSRL beam 2-3 experimental setup. X-rays from a FWHM 400-μm × 1700-μm bending magnet source at SSRL beam line 2-3 traverse 16.8 m through a tantalum entrance slit and double-crystal monchromator to the CRL entrance. The focused monoenergetic X-rays from the CRL pass through a scanning tantalum exit slit into a gas ionization detector. The exit slit scan in the vertical and horizontal directions yields the respective vertical and horizontal intensity profiles in the CRL image plane with and without the CRL present in order to experimentally determine the CRL gain.

The middle-left plot in Figure 90 shows the transverse y-directed vertical profiles of 7.0-keV and 8.0-keV X-rays imaging along the axial z-direction of X-ray propagation. The minimum vertical FWHM of the image spots are 41.6 μm and 43.7 μm at image distances $r_i = 70$ cm and $r_i = 90$ cm, respectively, for 7-keV and 8-keV photon energies. The middle-right plot shows the vertical profile of the 8.0-keV X-ray image spot, which shows the 43.7-μm vertical FWHM produced by the parabolic lithium CRL at the image plane. The data are fitted with a solid line Gaussian distribution of 18.6-μm standard deviation and 43.7-μm FWHM about the center axis of the lithium CRL.

The bottom-left plot in Figure 90 shows the measured transmission $T(r)$ through the lithium CRL as a function of the radial position r relative to the center axis $r = 0$. The measured transmission shows the $2r_a = 1030$-μm absorption aperture diameter of the lens, as well as the 523-μm FWHM of the lens aperture. The lithium CRL absorption aperture diameter $2r_a = 1030$ μm is slightly less than its physical aperture diameter $2r_m = 1200$ μm. The solid line is the calculated transmission profile with an attenuation diameter of 1200 μm. The bottom-right photos show one side of the biconcave parabolic lithium lens debossed at the center of the lithium disk. The two photos were taken at 100× with an Intel toy microscope.

The table at the top of Figure 91 provides the calculated (theoretical) and measured parameters of the parabolic lithium CRL at 7 keV and 8 keV in the SSRL BL 2-3 experiments. The plot at the bottom shows the CRL absorption aperture diameter $2r_a$ versus photon energy for $f = 1.0$ m focal length for lenses made of lithium, beryllium, Kapton, carbon, and aluminum. The absorption aperture diameter $2r_a$ of the lithium CRLs is almost twice as large at 8 keV than above 25 keV.

Adelphi Technology collaborated with Y. Dudchik of the Institute of Applied Problems of the Institute of Applied Physics Problems at Belarus State University in Minsk, Belarus, in the fabrication and testing of short–focal length CRLs composed of microbubbles embedded in epoxy (see Dudchik *et al.*, 2004a,b). The bubbles were formed in epoxy inside glass capillaries. The interface between the bubbles formed 90 to 196 spherical biconcave microlenses, reducing the overall focal length inversely by the number of lenses. When compared with CRLs manufactured by other methods, the microbubble lenses have shorter focal lengths with higher transmissions and gain for moderate energy X-rays (e.g., 7–12 keV).

The beam line 2-3 at the SSRL was used to measure bubble CRL focal lengths f between 100 and 150 mm and absorption aperture diameters $2r_a$ between 90 and 120 μm. Transmission profiles $T(r)$ were measured giving, for example, a peak transmission $T = 0.27$ for a 126-mm focal length CRL at 8 keV. The focal spot sizes were also measured. For example, an elliptical focal spot of 5-μm vertical × 14-μm horizontal was measured as a result of an approximate 80-fold demagnification of the 0.44-μm vertical × 1.7-mm horizontal elliptical source of X-rays from the bending magnet. The measured gains in intensity over that of unfocused beam were between 9 and 26. Theoretical gain calculations, which include spherical aberrations, were in good agreement with measured results. These experiments demonstrated the microbubble technique is an effective approach for designing lenses in the 8- to 9-keV range with focal lengths f of less than 30 to 40 mm.

The top-left diagram in Figure 92 shows the experimental setup on SSRL BL 2-3 for the testing of the set of fabricated bubble lens CRLs. The top-right photo shows a the brass housing that contains the capillary with bubble lenses. The middle-left plot shows the vertical width of the 8-keV beam refracted by the bubble CRL lens 4-1 versus the distance downstream from the CRL. The middle-right plot shows the relative intensity profile across the 8-keV demagnified image of the bending magnet source produced by lens 4-1. The bottom-left plot is the 8-keV X-ray transmission profile $T(r)$ through the CRL lens 4-1 versus radial position r from the CRL center axis at $r = 0$. The bottom-right photo shows the glass capillary and contained bubble lenses.

The table at the top of Figure 93a shows the design specifications and measured and calculated parameters of various fabricated bubble lens CRLs tested on BL 2-3 at SSRL. The middle-left plot shows the vertical intensity profile through the demagnified image of the bending magnet source at 7 keV. The middle-right plot shows the 9-keV (bottom left) and 12-keV X-ray transmission profile $T(r)$ through the CRL lens 4-1 versus radial position r (see Dudchik *et al.*, 2004a).

The bottom-right photos of Figure 93a are among the first X-ray images obtained with the bubble lens CRL at SSRL beam line 2-3 by Adelphi Technology. The bottom-left X-ray image shows a compound eye (ommatidium) of a common fly taken in 5 seconds at 8 keV. The bottom-right X-ray image is of the antennae of a human body louse taken in 30 seconds at 7 keV. The visible area of each image (FOV) is approximately 100 μm. Preserved specimens of the common house fly and human body louse were obtained from Carolina Biological of North Carolina. The compound eye of the fly was removed with a scalpel, as was the head of the human body louse. The tiny specimens were then placed on the object mount and imaged with the bubble lens CRL 4-1 at 8 keV.

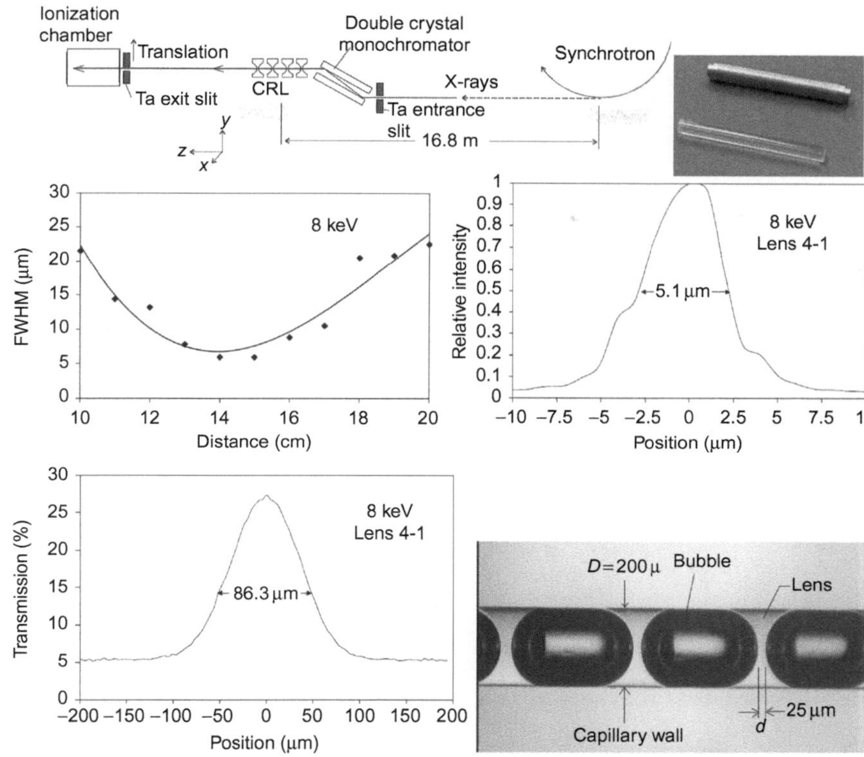

FIGURE 92 (Top) The setup for the bubble lens CRL experiment at SSRL bending magnet BL 2-3. (Top right) Photo of the brass cylinder housing for the glass capillary, which contains the bubble lenses. (Middle left) Plot of the vertical width of 8-keV X-rays refracted by bubble CRL lens 4-1. (Middle right) Vertical intensity profile through the demagnified 8-keV image of source by lens 4-1. (Bottom left) Transmission $T(r)$ through lens 4-1 versus radial position r. (Bottom right) Photo of the glass capillary with air bubbles in epoxy that form spherical biconcave bubble lenses. See text for details (Dudchik *et al.*, 2004a).

Using the same techniques and much of the same apparatus of the SSRL BL 2-3 experiment, Dudchik *et al.* (2004b) continued focusing experiments with the bubble lens CRLs at the APS on beam line 5BM-D-DND. The APS source had an x-directed (horizontal) width $\sigma_x = 23\,\mu\text{m}$ and a y-directed (vertical) width $\sigma_y = 97\,\mu\text{m}$, and the source was at object distance $r_0 = 58\,\text{m}$ upstream of the bubble lens CRL. The bubble CRL was mounted in a goniometer head (gimbal mount), which allowed adjustment of pitch (vertical θ_y) and yaw (horizontal θ_x). At APS the bubble lens CRLs could be tested at higher photon energies and with smaller X-ray source size. As in the SSRL experiment, limiting vertical and horizontal apertures in the form of tantalum slits were

Measured and Calculated Parameters of bubble CRLs for SSRL BL 2-3 source (440 μm vertical x 1700 μm horizontal FWHM) at 16.8 m distance from CRL.

Bubble Lens designation	1-1	3-1	3-4	3-3	4-1	4-1
Photon Energy	12	12	9	8	8	7
Number of lenses	90	196	103	93	102	102
R, Radius of Curvature (μm)	165	250	100	100	100	100
Observed Lens Minimum Wall Thickness (μm)	76	41	37	36	23	36
Visual Lens Minimum Wall Thickness (μm)	10	50	25	25	10	10
Calculated Refractive Index Decrement	1.74E-06	1.74E-06	3.09E-06	3.91E-06	3.91E-06	5.12E-06
Tabulated Linear Attenuation Coefficient (cm-1)	1.49	1.49	3.77	5.50	5.50	8.40
Calculated Focal Length (cm)	52.8	37	15.7	13.8	12.55	9.6
Calculated Image Distance (cm)	54.5	37.8	15.8	13.9	12.6	9.7
Measured image Distance (cm)	32	36	17.5	13	14	10
Calculated Vertical Minimum Waist Diameter (μm)	15.1	10.4	4.4	3.9	3.2	2.7
Measured Vertical Minimum Waist Diameter (μm)	12.8	12	3.9	4.8	2.7	4
Calculated Horizontal Minimum Waist Diameter (μm)	64.1	44.0	18.8	16.6	13.5	11.5
Measured Peak Transmission	36%	30%	24%	16%	27%	5%
Calculated Attenuation Aperture Diameter (μm)	314	262	143	125	119	96.7
Measured Attenuation Aperture Diameter (μm)	321	245	147	150	149	149
Calculated De-Magnification	0.032	0.022	0.009	0.008	0.007	0.006
Calculated 2D-Gain	16.6	20.0	25.6	16.9	28.9	6.0
Measured 2D-Gain	8.9	3.5	13.4	**	25.5	**

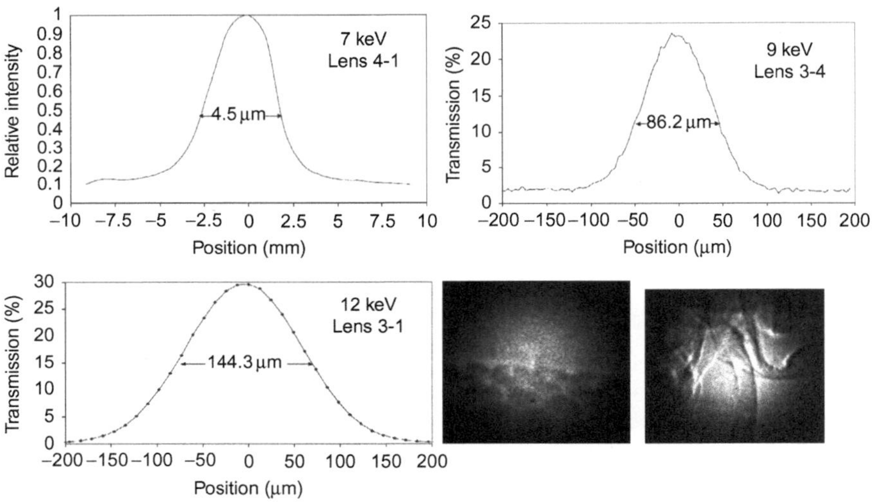

FIGURE 93A (Top) Tabulation of the measured and calculated parameters for fabricated bubble CRLs tested at bending magnet beam line 2-3 at SSRL. (Middle left) Vertical intensity profile of demagnified image of source at 7 keV produced by bubble CRL lens 4-1. (Middle right) Plot of the transmission $T(r)$ through lens 3-4 versus radial position r at 9 keV. (Bottom left) Plot of the transmission $T(r)$ through lens 3-1 versus radial position r at 12 keV (Dudchik *et al.*, 2004a). (Bottom right) Bubble CRL image of fly compound eye (left photo) and head louse antennae (right photo). (Adelphi Technology)

placed just upstream of the bubble lens CRL. The pair of crossed slits allowed only a 180-μm × 180-μm X-ray beam to enter the CRL. The CRL could then be translated across the beam along the CRL radial line to obtain the X-ray transmission $T(r)$ as a function of the CRL radius r, with center CRL axis at $r = 0$.

Also as with the SSRL experiment, the bubble CRL is set in a fixed position and X-rays are refracted by the CRL to the ionization detector. A scanning tantalum vertical slit is now placed downstream from the CRL at various axial distances downstream from the CRL. The X-rays must now pass through the scanning slit to reach the ionization detector. The downstream, y-directed (vertical) tantalum slit were set to 1.0 mm x-directed (horizontal) width, and the slits were moved to a particular axial distance z downstream of the CRL. The vertical slit was translated in the horizontal x-direction across the refracted X-ray beam. Thus, the refracted beam intensity profile along the horizontal x-direction was obtained for the axial distance z downstream of the CRL. The slits were then moved to a new axial distance z' downstream of the bubble lens CRL, and again the vertical slits were scanned in the horizontal direction to obtain a horizontal refracted beam intensity profile at distance z'. This was repeated for a set of axial distance downstream of the CRL.

The CRL 6-3-349 bubble lens CRL has $N = 349$ bubbles or lenses and since the capillary inside diameter is 200 μm, the bubble lens physical aperture diameter $2r_e = 200\,\mu\text{m}$ and the bubble radius of curvature $R = 100\,\mu\text{m}$ as shown in the table at the top of Figure 93b. The middle-right plot shows the vertical intensity profile of the demagnified APS X-ray source at beam line 5BM-D-DND produced by lens 6-3-349, which shows a 1.49-μm FWHM. The middle-left plot is the refracted vertical beam width versus axial distance z downstream distance from lens 6-3-349 for 13, 18, and 20 keV, which shows respective minimum waists at image distances $r_i = 13$ cm, $r_i = 21$ cm, and $r_i = 25$ cm. The bottom-left plot shows the transmission $T(r)$ through lens 6-3-349 versus radial position r at 18 keV, which shows a 112.6-μm FWHM. The bottom-right plot is the vertical intensity profile of the demagnified image of source with 20-μm FWHM at 15 keV by lens 6-2-300 and show the vertical intensity profile of X-ray beam with 169.5-μm FWHM, which is incident on CRL before refraction. The measured image intensity gain of lens 6-2-300 was $G_{\text{exp}} = 25$.

Lens designers can use Fresnel lenses to produce larger-aperture lens to overcome the absorption with increasing radius from the biconcave lens center. Diamond turning has been used for Fresnel lenses for the past 30 years with lens sizes progressively getting smaller. In this process a lens mold or "stamper" is manufactured by lathe turning (rotational cutting) using a V-shaped diamond-tipped tool to cut the Fresnel shape into the end of a brass cylinder. The stamper is then used to emboss or compress the Fresnel shape into plastics of a wide variety. However, the size and shape of the diamond tip limit the minimum feature size that can be achieved. In general, the finest zones can be no smaller than 2 to 5 μm wide and no edge can be steeper than 20°. In practice, this limits the size of the lens to at most five times that possible without a Fresnel design.

TABLE II. Measured and calculated parameters of bubble CRLs for APS.

Bubble lens designation	6-8-178	6-6-244	6-1-250	6-2-300	6-3-349	6-3-349
Photon energy	13	15	15	15	18	20
Number of lenses	178	244	250	300	349	349
R, radius of curvature (μm)	100	100	100	100	100	100
Calculated focal length (cm)	19	18.5	18	15	18.6	23
Calculated image distance (cm)	19.7	19.1	18.6	15.4	19.2	24.0
Measured image distance (cm)	20.6	19.1	19.1	17	20.8	25
Measured vertical min. waist dia. (μm)	2.7	1.35	2.55	1.24	1.50	2.10
Measured peak transmission	39%	54%	37%	25%	39%	46%
Calc. attenuation aperture dia. (μm)	198.44	214.8	211.9	193.5	242.85	288.96
Means. attenuation aperture dia. (μm)	171.5	176.8	182	177.5	191.42	208.93
Measured two-dimensional gain	25	11	...	25	...	...

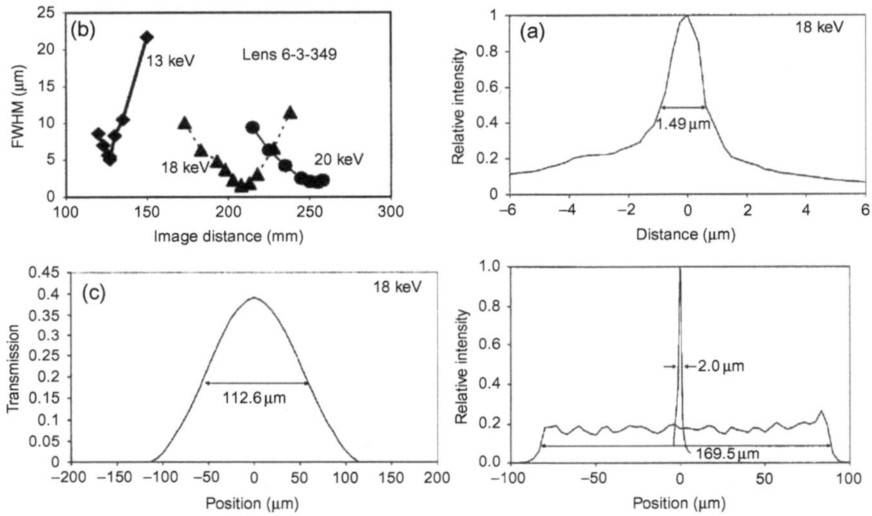

FIGURE 93B (Top) Tabulated specifications and measured and calculated parameters of the fabricated bubble lens CRLs tested at beam line 5BM-D-DND at APS. (Middle left) Plot of the refracted beam width at 13, 18, and 20 keV versus distance downstream of bubble CRL lens 6-3-349. (Middle right) The plot of vertical intensity profile of demagnified source at 18 keV by lens 6-3-349. (Bottom left) Transmission $T(r)$ through lens 6-3-349 verus radial position r at 18 keV. (Bottom right) Vertical intensity profile of demagnified image of source at 15 keV by lens 6-2-300 and vertical intensity profile of X-ray beam incident on CRL (Dudchik *et al.*, 2004b).

Adelphi Technology fabricated unit lenses using a diamond-turned compression tool in which the Fresnel shape was determined by a computer simulation program. A negative of the Fresnel shape was cut onto a $\frac{1}{8}$-inch diameter brass dowel pin. Adelphi used a simple compression technique similar to those used to fabricate spherical and parabolic biconcave lenses and an in-house lens-making jig aligned by dowel pins. The left photo in Figure 94 shows the thin film Fresnel lens. The diagram on the right shows the Fresnel lens preserving the surface curvature in discrete annular ring sections. Each annular ring contains only the portion of the lens thickness that contains the lens surface within the radial width

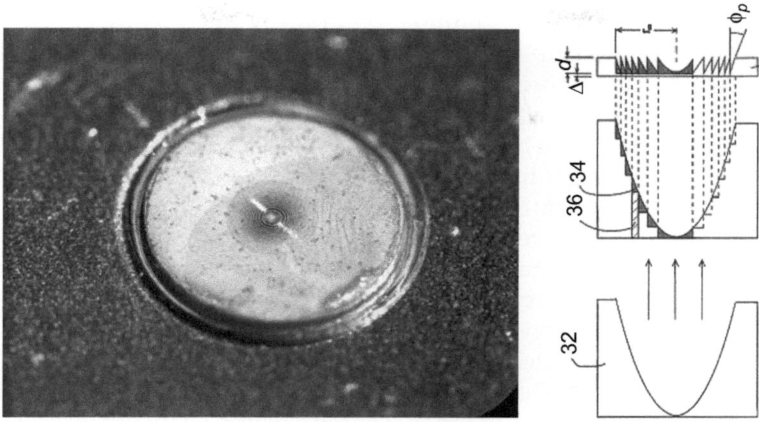

FIGURE 94 (Left) A freestanding 23-μm thick Fresnel lens 1.1 mm in diameter for X-ray focusing was fabricated by Adelphi Technology. Compare this unit lens with the ball lens in Figure 6, whose absorption aperture was 180 μm. (Right) Diagram showing the principle of the Fresnel lens (Piestrup *et al.*, 2001).

of the ring. The rest of the thickness of the lens is removed from the annular section of the lens. Ideally, the removed thickness of the lens is integer multiples of the wavelength λ of interest that corresponds to integer multiples of 2π phase shifts.

The Adelphi-fabricated Fresnel lens was 19-μm thick polyethylene (minimum thickness $\Delta \approx 5\,\mu m$) with a 1.2-mm diameter, which was a six-fold increase in diameter over that of an equivalent spherical biconcave lens. The manufacture of such a thin, freestanding Fresnel lens is important to limit X-ray absorption. For X-ray lenses, the refracted X-rays in a particular annular zone are coherent, but the X-rays refracted from the neighboring annular zones are incoherent because of the random thicknesses of the zones. The random thicknesses of the annular zones occur because of the inability of present-day technology to remove lens material in integer multiples of the wavelengths λ on the order of Ångstroms. Hence, the Fresnel lens is well suited as a condenser lens for focusing X-rays onto a sample. However, the coherent imaging of the sample should be done by a Fresnel zone plate or a CRL.

The CRL is limited in aperture diameter to less than 1.5 mm for beryllium lenses and less than 0.5 mm for plastic lenses. The use of CRLs with small X-ray sources, such as synchrotrons or microfocus X-rays tubes, produces significant intensity gains in demagnified images of the source and sizable object magnifications. Novel X-ray sources, such as channeling radiation, parametric X-rays, transition radiation, and bremsstrahlung, can be used with CRLs to produce an intense focused beam of X-rays that can be used for sample illumination in X-ray microscopy or sample fluorescence (see Piestrup *et al.*, 2001). These novel X-ray sources can be monochromatic, quasi-chromatic, or broadband. These sources can be relatively more efficient on a photon produced per electron basis. However, the novel sources lack the brightness of the synchrotron rings because the electron beam current is limited by the thermal capacity of the electron target and X-ray radiator.

The source size of novel X-ray radiators is determined by the diameter of the electron beam striking the radiator. The focusing of the electron beam on the radiator increases the radiator X-ray emission intensity, but the focusing of the electrons increases their range of incidence angles. And the increased range of electron incidence angles increases the range of X-ray emission angles. That is, the electron beam emittance is assumed constant in which a decrease in the area of the electron beam in the electron focal spot is offset by an increase in the electron divergence angles. Also, the thermal capacity of the radiator target limits the electron beam to that current which can be focused on the target without melting or vaporizing the radiator target.

Consider a relativistic electron beam produced by a linear accelerator (LINAC) or cyclic accelerator with moderate energy below 100 MeV. The electron beam is focused to a spot size less than 1 mm in diameter in which the resulting electron current that can be thermally tolerated by the radiator target for X-rays. The X-ray radiator target can be a crystal or foil. The resulting emission of X-rays from a focused beam of relativistic electrons, which strike a channeling crystal, parametric X-ray radiation (PXR) crystal, transition foil, or bremsstrahlung, then produces a narrow angular range of X-rays for each incident direction of converged electrons. The narrow angular range of X-ray emission from a single electron direction is convolved with the range of electron incident angles to determine the overall angular rang of emitted X-rays. The angular range of emitted X-rays is still relatively narrow for these novel X-ray sources (e.g., PXR, channeling radiation, bremsstrahlung, mirror-focused transition radiation), which allows the effective interception and focusing by the CRL despite a small effective CRL aperture diameter. Hence, the CRL can be used to focus X-rays from small intense, novel X-rays sources onto samples with resulting significant gains in intensity delivered from the novel X-ray source to the sample.

X-ray optics follows the lead of visible light optics. The use of smaller optical lenses to form large-area images with 1:1 imaging as well as magnified ($M > 1$) imaging was presented by R.H. Anderson (1979), who used two planar arrays of lenses aligned to a grid of holes placed between the two lens arrays. Anderson developed an unusually short, close-up imaging system of about 8.5 cm ($3\frac{3}{8}$ inch) long for document copiers, oscilloscope cameras, and computer peripheral hard-copy cameras (see

Anderson, 1979). He used arrays of simple lenses that were molded in rows and columns on two parallel plastic lens plates. Each lens on the first lens plate formed a demagnified, inverted intermediate image of a small area of the object in the space between lens plates. Each lens in the second plate erected and magnified an intermediate image and placed the resulting final image in partially overlapped registration with adjacent images, thereby forming an erect uniform composite image of the whole object.

Piestrup (2003, 2004) extended the concept of using two lens arrays aligned (with a set of aperture holes placed between the two lens arrays) to image objects larger than the FOV of the individual lenses in the array to X-rays using two arrays of CRLs separated by an aligned array of apertures (holes) in a plate that was opaque to X-rays (lead) or thermal neutrons (gadolinium or cadmium). The planar array of CRLs is actually a 3D array of lenses. A pair of planar arrays of X-ray or neutron CRLs can overcome the limitation of the individual small FOV of the X-ray or neutron CRL and allow X-ray or neutron imaging of large objects with 1:1 imaging with unit magnification ($M = 1$) (Figure 95). Furthermore, a pair of planar CRL arrays, one CRL array on each side of an aligned aperture array, can provide magnified imaging of large objects with $M > 1$ (Figure 96).

The two planar arrays of CRLs (3D array of X-ray or neutron lenses) can consist of bubble lenses or biconcave parabolic lenses fabricated by lithography or nanotechnology techniques. A single row of CRLs can be created in which the object or the CRL is scanned in a sequence of one row at a time (see the top of Figure 97). The image is formed by the set of scanned rows of the object. The use of a 1D single row of bubble lens or parabolic lens CRLs for the X-ray image scanning of large objects is relatively inexpensive and simple to fabricate compared with 2D arrays of CRLs. However, with improvements in technology a 3D array of lenses (planar array of CRLs), followed by an aligned aperture array and then followed by an aligned, second 3D array of lenses (planar array of CRLs), can form a imaging system for objects larger than the FOV of the individual CRLs. The lens array can thus provide 1:1 imaging of (1) the large object, (2) demagnified imaging of the large object ($M < 1$), or (3) magnified imaging of the large object ($M > 1$).

The pair of planar CRL arrays shown in the bottom of Figure 97 can be applied to X-ray and neutron imaging and detection of distant or near objects in the 1:1 or ($M = 1$) imaging or the magnified imaging mode ($M > 1$) or demagnified mode ($M < 1$). The bottom-left image in Figure 97 shows two arrays of CRLs, one array is on each side an aligned aperture array. This embodiment forms a reduction lens capable of imaging a demagnified circuit pattern onto a silicon wafer. The bottom-right image shows two arrays of CRLs straddling an aligned aperture array to form an image reduction lens capable of imaging a magnified image of breast tissue for cancer detection.

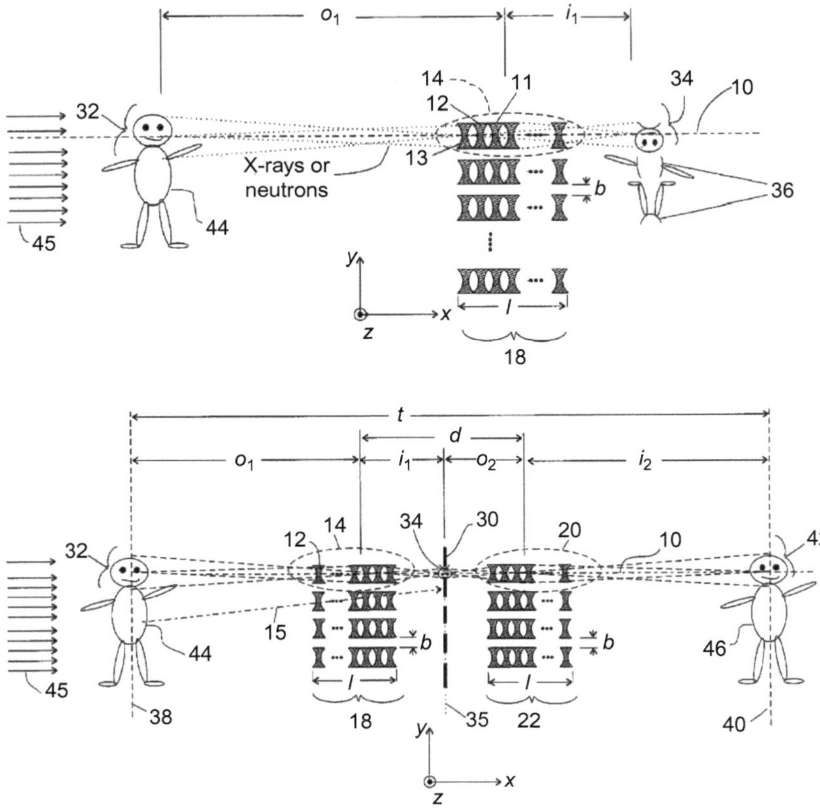

FIGURE 95 (Top) The side view of a single 3D lens array, which provides partial inverted images of an object using X rays. The unit lenses are in a cubic array. For a single 3D array, the final image is scrambled. (Bottom) The side view of a 3D lens array system, which provides a 1:1 image of an object using X-rays or neutrons (Piestrup, 2003, 2004).

CRLs are effective for collimating or focusing high-energy X-ray beams (50–100 keV) and can be used in conjunction with crystal optics in a variety of configurations to improve the intensity and narrow the bandwidth of synchrotron X-ray sources. Shastri *et al.* (2003) demonstrated this at the APS 1-ID undulator beam line. The collimating and focusing ability of the 1D aluminum CRL were exploited to improve the intensity in the vertical direction of the narrow-energy bandwidths delivered by the monochromator setup on this beam line. The undulator source was 600 μm and 800 μm in horizontal widths and 50 μm and 64 μm vertical widths, which corresponded to X-ray source energies of 67 keV and 81 keV.

The 1D CRLs were composed of a row of closely spaced (1.0-mm diameter) cylindrical holes in an aluminum bar and were fabricated by Adelphi

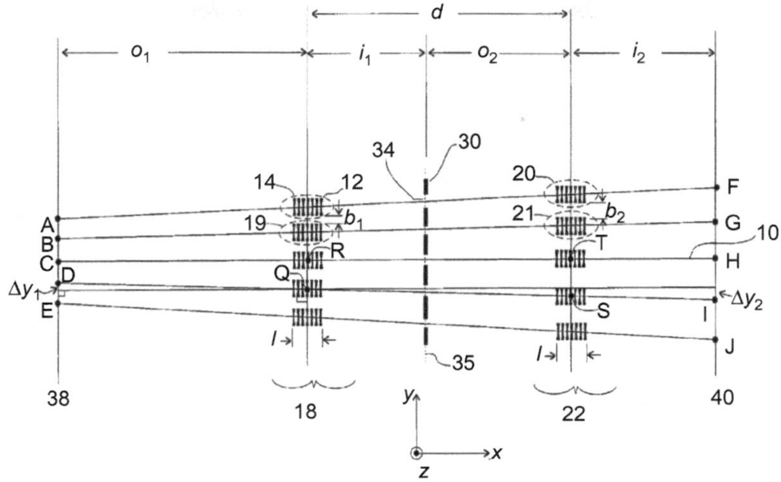

FIGURE 96 Shown is the side view of a pair of planar arrays of thin CRLs separated by an aligned grid of apertures in a plate opaque to X-rays (lead) or neutrons (cadmium). The setup produces a magnified image of object if the total CRL array lattice magnification is equal to the nodal magnification, which is produced by a lens pair of only the object portion that is covered by the particular CRL pair. Equivalently, the separate image portions will join like pieces in a puzzle to form the whole image if the size of each image portion has been magnified by the same amount as the spacing between the image portions (Piestrup, 2003, 2004, as taught by Anderson, 1979).

Technology for the X-ray line source at the APS 1-1D undulator beam line (Figure 98, top). The aluminum 1D CRLs had an $R = 0.5$-mm radius of curvature with the minimum wall thickness range of $d = 20$ to $50\,\mu\text{m}$, which caused 50% attenuation of the 67-keV and 81-keV X-rays used in the CRL collimation and focusing experiments. The aluminum 1D CRL for 67-keV X-rays had $N = 60$ holes or biconcave cylindrical lenses, whereas the aluminum 1D CRL for 81-keV X-rays required more lenses $N = 86$ to provide the same image distance $r_i = 24\,\text{m}$ to the end-station detector, given the CRL object distance $r_i = 36\,\text{m}$ from the source, where the source to image distance was $r_0 + r_i = 60\,\text{m}$.

The APS 1-1D undulator beam line had a pre-monochromator that was cryogenically cooled, bent double-Laue Si(111) crystal device that maintained the divergence and direction of the undulator source and produced X-rays with energy width of 0.1% or $\Delta E/E \sim 10^{-3}$. The 0.1% bandwidth of the bent Laue crystal pre-monochomator was sufficient for most experiments, which included pair-distribution function measurements, fluorescent spectroscopy, powder diffraction, material stress/texture

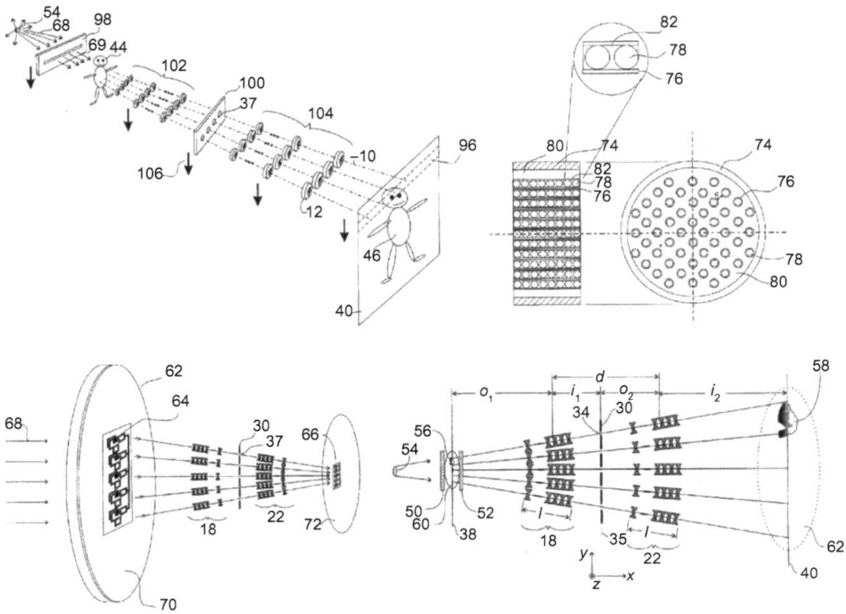

FIGURE 97 (Top left) Diagram of a 1D scanning system to image objects larger than the FOV of individual lenses. (Top right) A 2D array of bubble epoxy CRLs. A single row of bubble CRLs forms a 1D X-ray scanner and a matrix (rows and columns) of bubble CRLs forms 2D images. (Bottom left) Two arrays of CRLs straddling an aligned aperture array to form a reduction lens capable of imaging a demagnified circuit pattern onto a silicon wafer. (Bottom right) Two arrays of CRLs straddling an aligned aperture array to form an image-reduction lens capable of imaging a magnified image of breast tissue for cancer detection (Piestrup, 2003, 2004).

determination, small-angle scattering, and diffuse scattering. In one experiment, Shastri *et al.* (2003) tested the collimation ability of the CRL, which was placed after the bent Laue pre-monochromator. The top of Figure 98 shows the natural horizontal and vertical divergence of the X-rays emitted from the bent Laue pre-monochromator. The bottom shows the experimental results—namely, the horizontal and vertical beam profiles at the end-station detector at 67 keV and 81 keV—which are produced by placement of a collimating CRL downstream from the bent Laue pre-monochromator.

To obtain the 0.01% high resolution a second post-monochromator was placed downstream from the bent double-Laue crystal pre-monochromator. The second post-monochromator was a four-reflection, flat Si(111) crystal system that resembled two channel cuts in a dispersive arrangement. The high-resolution monochromator reduced the

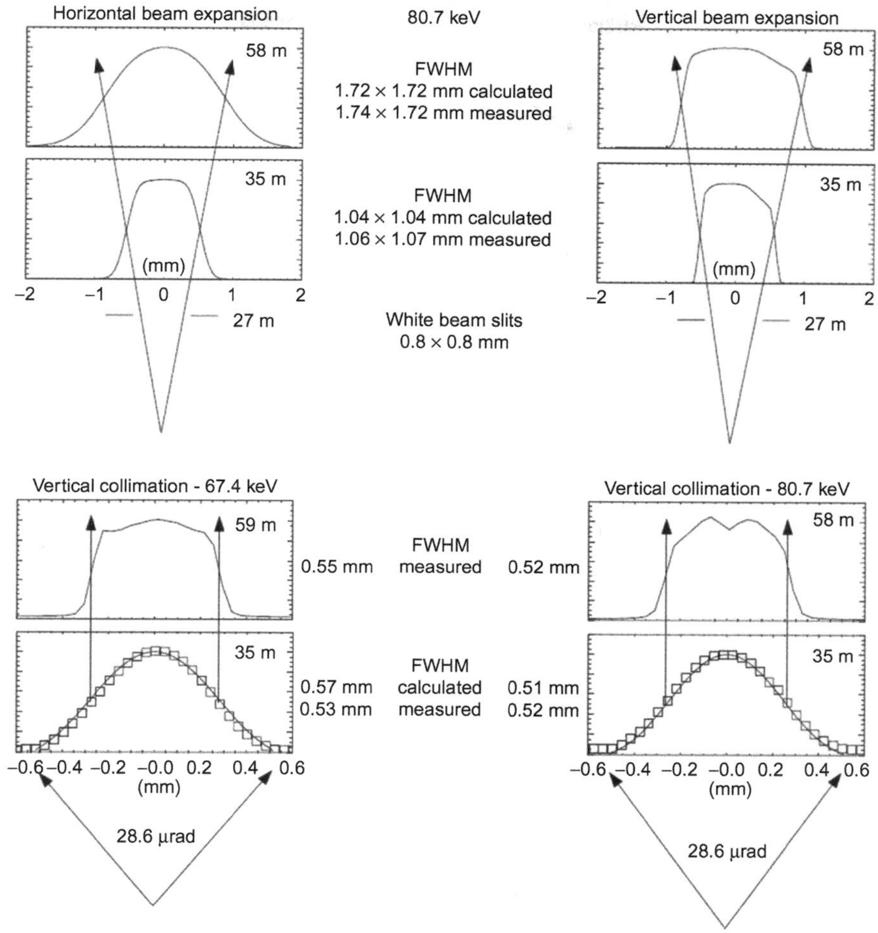

FIGURE 98 (Top left) The 81-keV horizontal profile and (right) vertical profile at distances 35 m and 58 m defined by 0.8-mm × 0.8-mm white beam slits set 27 m from the APS 1-1D undulator beam line. (Bottom left) Measured (squares) and calculated (solid line) vertical profile for 67 keV and 81 keV at distances 35 m (immediately after the CRL) from source (Bottom right) Measured (squares) and calculated (solid line) vertical beam profile for 67- and 81-keV X-rays at the end-station 58 m downstream from the source (Shastri *et al.*, 2003).

bandwidth to 0.01% or better ($\Delta E/E < 10^{-4}$), which is required for applications such as anomalous scatter with heavy elements, high-resolution stress/strain analysis measurements with line-shape analysis, atomic physics spectroscopy, and the excitation of nuclear resonances.

The use of the cylindrical lens CRLs enhanced the X-ray throughput by reducing the pre-monochromatized beam's vertical divergence from

29 μrad to a few μ rad. The CRL acted as a collimator to reduce the vertical divergence angle of X-rays that are emitted from the pre-monochromator. The collimation of the X-rays emitted from the pre-monochromator allowed more of these X-rays to fall within the narrow vertical acceptance angle of the post-monochromator. The resulting X-rays emitted from the post-monochromator and arriving at the detector had increased collimation and a fourfold throughput intensity increase. That is, the placement of the CRL between the pre-monochromator and post-monochromator increased the collimated 81-keV photon rate by fourfold to 10^{11} ph/s within bandwidth $\Delta E/E = 7 \times 10^{-5}$ in a 2-mm horizontal ×0.5-mm vertical beam width at the end-station detector. The measured fourfold throughput increase by the CRL also occurred for 67-keV X-rays. However, the fourfold increase of throughput X-ray intensity was half of the calculated eightfold value, which was attributed to imperfect collimation. The end-station detector was 60 m downstream from the 81-keV source, which was 0.6 mm horizontal × 0.05 mm vertical. The setup with the 1D aluminum CRL placed between the pre- and post-monochromators to increase throughput and collimation has been used for resonant powder diffraction studies. The diagram at the top of Figure 99 shows the experimental setup used by Shastri *et al.* (2003) with the $N = 86$ CRL with 81-keV X-rays set between the pre- and post-monochromators. The table at the bottom of the figure tabulates the measured horizontal and vertical beam widths, the beam flux, and the beam energy bandwidth at the locations labeled A, B, and C in the diagram.

Shastri *et al.* (2003) performed vertical line-focusing experiments with X-rays at 67 keV using a CRL with $N = 147$ holes of 1-mm diameter, and at 81 keV using a CRL with $N = 211$ holes of 1-mm diameter and minimum wall thickness range $d = 20$ to 50 μm. They placed the CRL 4 m downstream from the double-Laue monochromator in which the double-Laue monochromator was 32 m and the CRL was 36 m downstream from the source. The two 1D aluminum CRLs produced demagnified images of the source at the end-station 60 m downstream from the source with 1 : 0.7 demagnification (i.e., $M = 0.7$) The measured image vertical line focus profiles were 70 to 90 μm, which was twice the expected calculated vertical line widths and was attributed to cylindrical aberration in the CRLs. The removal of the CRL resulted in vertical line widths of 1.6 to 1.7 mm. The CRL reduced the vertical beam width from 1.6 to 1.7 mm to 67 μm FWHM for 67-keV X-rays and 89 μm for 81-keV X-rays, which was a 20-fold increase in focused intensity. Combined with the approximate average CRL transmission $T_{avg} = 0.3$, the overall gain increase by use of the CRL was measured as $\sim G = 6$. Figure 100 shows the vertical line-focusing experiment of Shastri *et al.* (2003).

The addition of a second $N = 60$ CRL after the high-resolution post-monochromator provides additional focusing for 67-keV X-rays in which

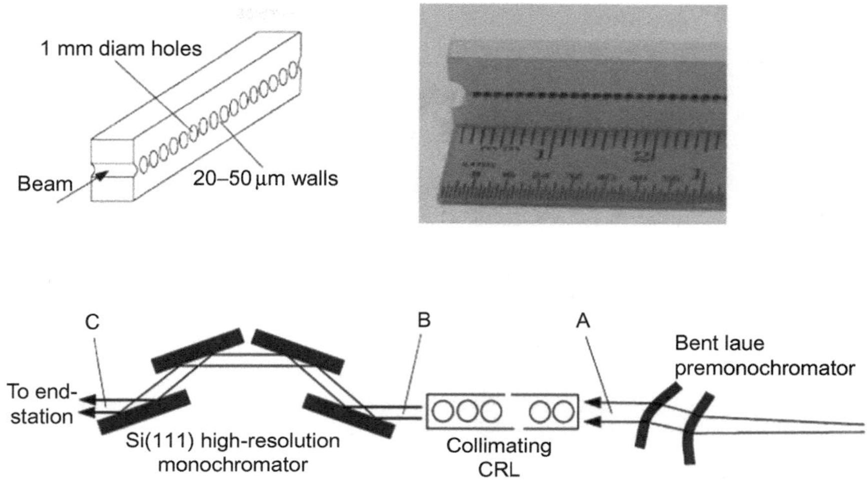

Location, distance from source	Beam size h × v FWHM (mm^2)	Flux (ph/s)	Energy width FWHM	
			ΔE (eV)	$\Delta E/E$
A 32 m	1 × 1	$I_0 = 4.0 \times 10^{12}$	127	1.6×10^{-3}
B 35 m	1 × 0.5	$I_0/2$	127	1.6×10^{-3}
C 36 m	1 × 0.5	$I_0/52 = 7.7 \times 10^{10}$	6	7.4×10^{-5}

FIGURE 99 (Top left) Diagram and (right) photo of 1D aluminum CRL in which CRLs with $N = 60$ and $N = 86$ biconcave cylindrical lenses collimated 67-keV and 81-keV X-rays. (Middle) Diagram of the CRL placed after the pre-monochromator to collimate X-ray beam and allow more X-rays to fall within the angle of acceptance of the Si(111) high-resolution monochromator at the APS 1-ID undulator beam line. The CRL increased throughput intensity of 0.01% narrowband output by fourfold. (Bottom) The table shows the measured horizontal and vertical beam size, beam intensity, energy spread, and bandwidth at indicated locations A, B, and C for 81-keV X-rays (Shastri *et al.*, 2003).

a 1:0.7 demagnified image of the source ($M = 0.7$) is produced at the end-station detector. The first $N = 60$ CRL was placed between the pre-monochromator and post-monochromator to collimate more 67-keV X-rays diverging from the pre-monochromator into the limited vertical acceptance angle of the post-monochromator. The second CRL (just after the post-monochromator) set the CRL object distance to $r_0 = 36$ m downstream from the source and the CRL image distance to $r_i = 24$ m upstream of the end-station detector (labeled by D in the diagram at the top of Figure 101).

The X-rays entering the second CRL were limited to a square beam 1 mm × 1 mm by a pair of vertical and horizontal slits acting as stop

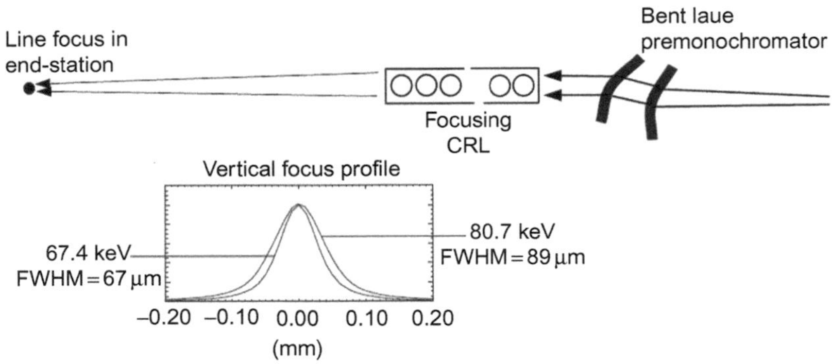

FIGURE 100 The experimental setup and results of vertical line-focusing experiments at the APS 1-1D undulator beam line at 67 and 81 keV using a CRL with $N = 147$ holes and a CRL with $N = 211$ holes (1-mm diameter) with source to image demagnification 1 : 0.7or $M = 0.7$. The 30% CRL transmission and reduction of vertical beam width from 0.5 mm resulted in a line-focused intensity gain $G = 6$ at the end-station detector 60 m from the source (Shastri *et al.*, 2003).

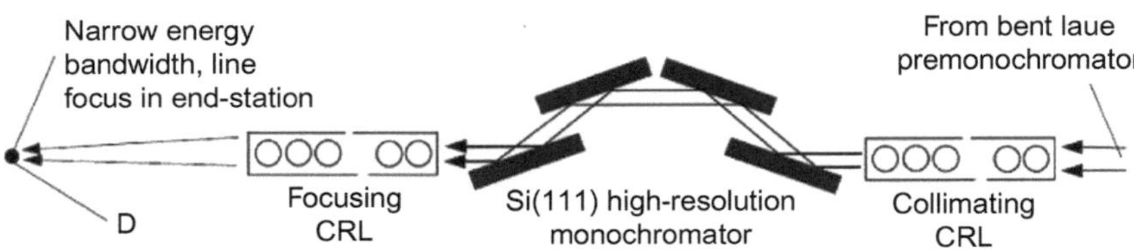

Location, distance from source	Beam size h × v FWHM (mm^2)	Flux (ph/s)	Energy width FWHM	
			ΔE (eV)	$\Delta E/E$
A 32 m	1 × 1	$I_0 = 4.0 \times 10^{12}$	90	1.3×10^{-3}
B 35 m	1 × 0.5	$I_0/1.9$	90	1.3×10^{-3}
C 36 m	1 × 0.5	$I_0/33.5 = 1.2 \times 10^{11}$	5	7×10^{-5}
D 60 m	1.6 × 0.067	$I_0/57.9 = 6.9 \times 10^{10}$	5	7×10^{-5}

FIGURE 101 (Top) The diagram shows the addition of a second $N = 60$ CRL just after the high-resolution post-monochromator, which set the CRL object distance to $r_0 = 36$ m downstream from the 67-keV source and the image distance $r_i = 24$ m upstream of the end-station detector (labeled D). The locations for labels A, B, and C are shown in Figure 99. (Bottom) The table shows the measured horizontal and vertical beam size, beam intensity, energy spread ,and bandwidth at indicated locations A, B, C, and D for 67-keV X-rays. The second CRL provides additional focusing for 67-keV X-rays in a 1 : 0.7 demagnified image of the source ($M = 0.7$) with a fourfold increase of intensity at the end-station detector (Shastri *et al.*, 2003).

apertures just upstream of the CRL and downstream from the post-monochromator at 36 m downstream from the source. The bottom of Figure 101 shows the measured horizontal and vertical beam widths, the beam flux, and the beam energy bandwidth at the indicated locations labeled A, B, and C in Figure 99 and by D in Figure 101. The addition of the second CRL transmitted only 58% of the 67-keV X-ray detector; however the 500-μm vertical width at the CRL was focused to 67 μm at the detector. Hence, overall the addition of the second CRL increased the intensity of X-rays delivered to the detector by fourfold.

Shastri *et al.* (2007) realized single-crystal silicon is a material well suited for refractive lenses operating at high X-ray energies above 50 keV. Single-crystal lens material minimizes small-angle scatter compared with polycrystalline lens materials. Motivated by this, they developed a silicon saw-tooth refractive lens, which provided vertical focusing and collimation of 50- to 100-keV high-energy X-rays (50–100 keV) at the APS 1-ID undulator beam line. The silicon saw-tooth lenses act as a 1D parabolic lens, providing high transmission and offering continuous tunability of photon energy and focal length. Their silicon saw-tooth lens, combined with the bent double-Laue monochromator of the 1-ID beam line, yielded both short– and long–focal length geometries, giving line foci with widths ranging from 2 μm to 25 μm for 81-keV X-rays.

Shastri *et al.* (2007) also designed and tested a compound focusing setup in which the radiation intercepted by a distant short–focal length saw-tooth lens was increased by having the saw-tooth lens receive a collimated beam from a nearer (upstream) saw-tooth lens. The collimation action of the silicon saw-tooth lenses was also used to increase the throughput of 50- to 80-keV X-rays in a small-angular–acceptance, high-energy–resolution post-monochromator placed downstream of the saw-tooth lens. Their saw tooth lenses were fabricated by subjecting single-crystal silicon to a crystallographically anisotropic chemical etching process that produced isosceles teeth, which were 100 to 200 mm in height depending on the device (see Ribbing *et al.*, 2003).

Said and Shastri (2010) reported a simple fabrication process readily available at the optics shop of a typical synchrotron radiation facility. The top diagram in Figure 102 shows a diagram of the silicon alligator lens, and the bottom photo shows the portion of one jaw of the silicon alligator lens. The fabrication process for a single-crystal silicon saw-tooth refractive lens used a dicing process via a thin diamond wheel. Said and Shastri tested the saw-tooth lens with 115-keV X-rays at the APS 1-ID undulator beam line. The silicon saw-tooth lens jaw produced a 17-μm line-focus width at the end-station detector located 58.5 m from the undulator X-rays source. A single jaw of the silicon saw tooth lens produced 2:1 demagnification of the source, which was measured with a 5-μm scanning slit (plotted at the bottom left). They measured the calculated (unconvolved) 17.4 -μm FWHM width for the vertical image profile.

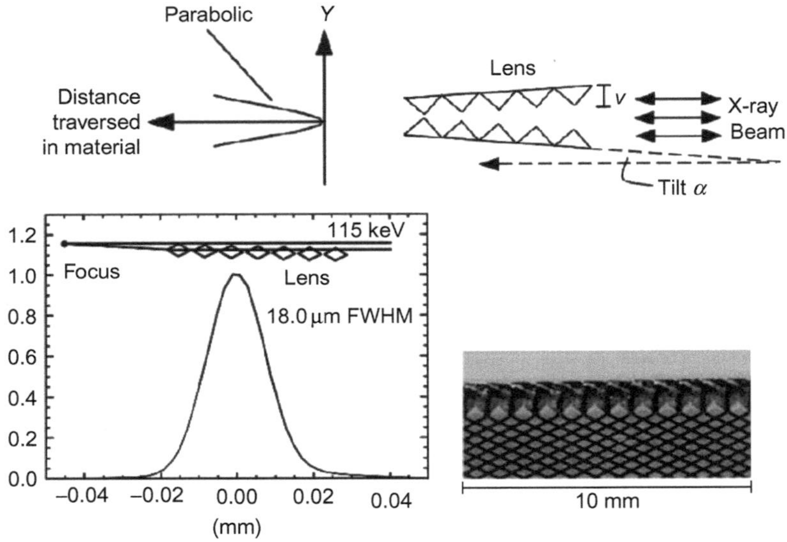

FIGURE 102 (Top) The diagram shows that the two opposite-facing saw-tooth lenses, which are tilted symmetrically about the beam axis, cause a parabolic increase in the optical path length with radial distance from beam axis, which is equivalent to a parabolic plano-concave lens. The spatial acceptance of each jaw is at most the tooth height v, which occurs for a sufficient saw-tooth pattern length $L > \upsilon/\sin\alpha$. The beam can enter from either end. (Bottom left) Plot showing a profile of the 115-keV line focus at 58.5 m from the APS 1-ID undulator beam line X-ray source by a single upright saw-tooth lens in a 2:1 demagnification setup, which was measured with a 5-μm slit. The expected (unconvolved) result for the width was 17.4 μm. (Bottom right) Photo of a segment within the actual 130-mm long saw-tooth device, which was produced by diamond saw cuts 2 mm deep in a 6-mm thick silicon plate, which gave the saw-tooth lens a 2-mm transverse beam acceptance (Said and Shastri, 2010).

The 2D CRL imaging of X-rays emitted for an X-pinch plasma source was performed by Gary *et al.*, (2004, 2005) using a bubble lens CRL. Gary *et al.* (2004) took diagnostic X-ray images of an X-pinch plasma X-ray source, which was driven by the XP pulser (100 ns, 500 kA) at Cornell University using an X-ray bubble CRL. The CRL obtained images X-pinch plasma source with approximately $\sigma = 2\,\mu\text{m}$ resolution (Figure 103, top left).

The bubble lens CRL lens consisted of 99 closely spaced air bubbles in an epoxy-filled glass capillary with inside diameter 200 μm, which formed $N = 100$ biconcave spherical lenses, each with radius of curvature $R = 100\,\mu\text{m}$ (one-half the 200-μm inside glass capillary tube). The minimum wall thickness of the biconcave spherical lens formed by two adjoining bubbles was $d = 28\,\mu\text{m}$. At 7 keV the bubble lens CRL focal length was $f = 13.5\,\text{cm}$

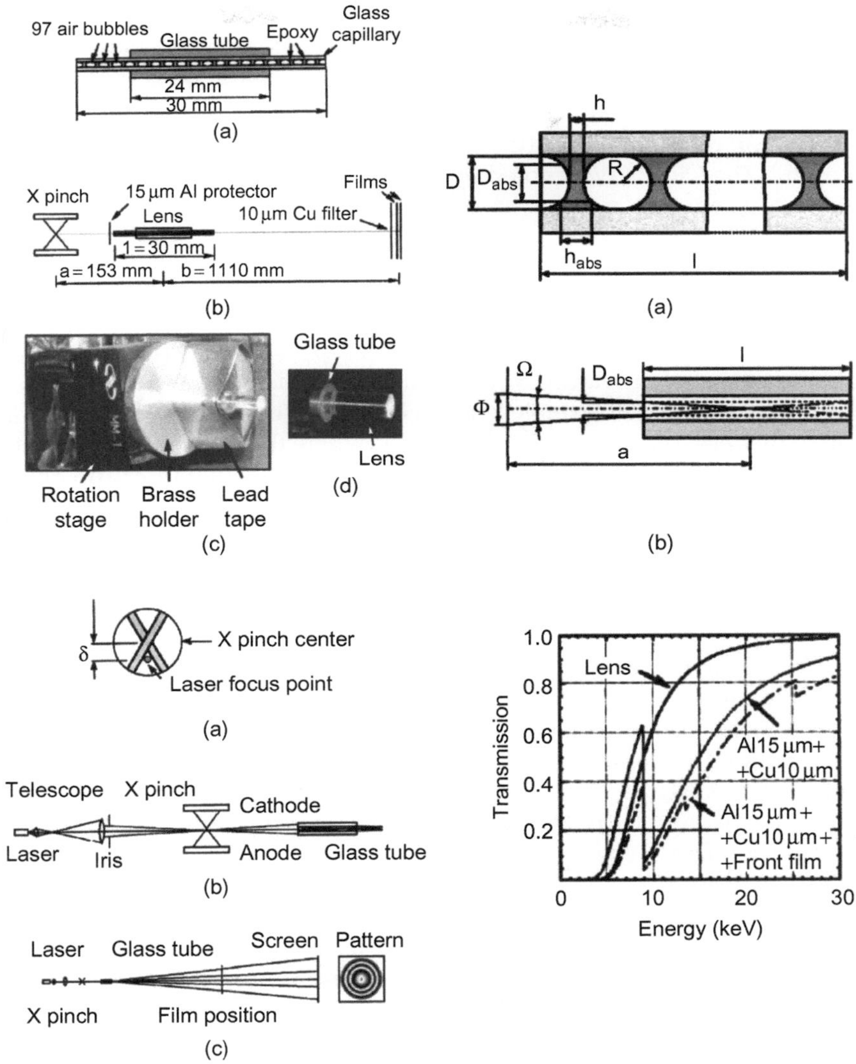

FIGURE 103 (Top left) Diagram of (a) the bubble epoxy lens CRL mounted glass tube (b) optical system for imaging X-pinch source, and (c and d) photos of the mounted CRL. (Top right) (a) Diagram of CRL indicating absorption aperture diameter $D_{abs} = 2r_a$ and (b) the CRL object FOV Φsubtended by angle Ω at object distance a. (Bottom left) (a) Position of the alignment laser focus spot relative to the X-pinch, (b) alignment configuration, and (c) alignment configuration showing the film position. (Bottom right) X-ray transmission versus X-ray energy through 15-μm protector film at CRL entrance, 10-μm copper light filter enclosing the X-ray film, and epoxy bubble lens CRL (Gary *et al.*, 2004).

with 20% on-axis transmission ($T = 0.2$) and absorption aperture diameter $D_{abs} = 2r_a = 112\,\mu\text{m}$. Compared with the CRL physical aperture diameter $2r_m = 200\,\mu\text{m}$, the effective CRL aperture diameter $2r_e = 112\,\mu\text{m}$ was limited by the absorption of 7.5-keV X-rays through the CRL.

The top right of Figure 103 shows a diagram of the bubble lens CRL of length l. The CRL FOV denoted by Φ was subtended by angle Ω at object distance a, where

$$\Phi = \frac{2D_{abs}a}{l}. \tag{126}$$

The CRL absorption aperture diameter $D_{abs} = 2r_a$ is expressed by the bubble radius of curvature R, minimum wall thickness $d = h$ between a pair of bubbles, and the single lens thickness h_{abs} (between two bubbles) at radial distance r from the CRL center axis, where the transmitted intensity fraction $T(r)$ decreases to e^{-2} of the transmission fraction of X-rays along the CRL center axis $T(0)$, where $T(r_a) = T(0)/e^2$ and

$$D_{abs} = 2\sqrt{R^2 - \left(R - \frac{h_{abs} - h}{2}\right)^2}. \tag{127}$$

The bottom-right plot shows the X-ray transmission fraction through the CRL versus X-ray energy.

A precise system for lens alignment with of 3- to 5-arcmin accuracy was required for obtaining X-ray images of four-wire X-pinches by the bubble lens CRL (see the bottom left of Figure 103). The XP pulser produced a hot X-pinch in the vicinity of the crossing point of the crossed wires. The plasma produced a Planck's radiation spectrum with temperature between 800 and 1000 eV. The bubble lens CRL was aligned with the anticipated hot plasma formation spot by a laser alignment setup. The laser was focused to focal spot diameter of less than the 20-μm diameter of the wires in which the laser light was completely blocked if the laser beam struck the center of the X crossing point of the 20-μm diameter wires. Hence, the laser was focused to a point 50 μm below the wire crossing point.

Laser light focused below the crossed wires diverged to a glass tube. The glass tube contained the removable glass rod, which was the support structure for the capillary of bubble lenses set along the rod center axis. The glass rod with the epoxy-filled capillary of air bubble lenses was removed from the glass tube. The divergent laser light from the focal spot below the crossed wires could enter the empty glass tube, reflect from the tube walls onto a detection screen, and produce a light interference pattern. The light interference pattern from the laser was extremely sensitive to the angular position of the glass tube. As a result, the accuracy of alignment relative to the laser focus position was better than 0.5 mrad.

After alignment of the glass tube, the glass rod containing the capillary of bubble lenses was slid back into the glass tube and covered by a 15-μm thick aluminum foil protector on the side facing the X-pinch. The film detector consisted of a sandwich of two Kodak BioMax 60 × 90-mm films. The X-ray films were protected from visible light exposure by enclosure in 10-μm thick copper foil. The copper K-edge at 8979 keV provided good X-ray transmission above the K-edge; the X-ray spectrum reaching the X-ray film was primarily determined by the 10-μm thick copper foil and secondarily by the 15-μm thick aluminum foil protector and the epoxy of the CRL. Figure 103 (bottom right) shows plots of the calculated X-ray transmission fraction versus X-ray energy plots for the 15-μm thick aluminum protection foil at the bubble lens CRL entrance and the light-stopping 10-μm copper foil enclosing the X-ray film and epoxy of the bubble lens CRL.

In the experiment by Gary *et al.* (2004), the X-pinch plasma sources that were imaged were 4-wire tungsten (17-μm diameter) and 4-wire nickel chromium (20-μm diameter). Pinhole camera X-ray images of the X-pinch plasma were recorded in addition to the bubble lens CRL imaging. The leftmost image (a) in Figure 104 shows the pinhole camera X-ray image. Image (b) shows the magnified $M = 7.5$ bubble lens CRL X-ray image; (c) is the detailed bubble lens CRL X-ray image at magnification ×7.5 of

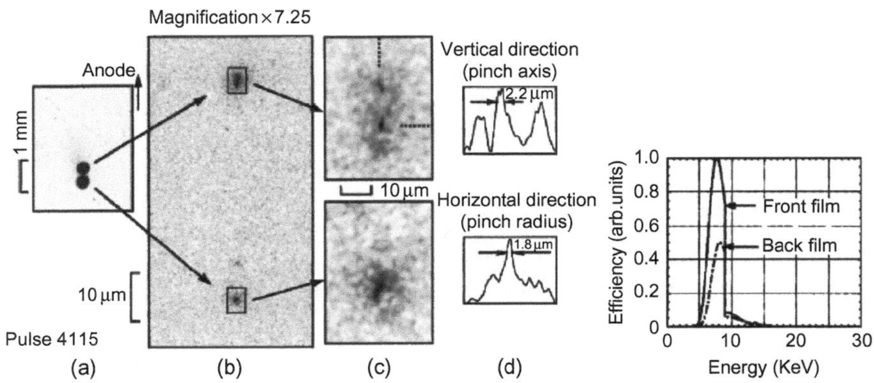

FIGURE 104 (Left) (a) The 0.15-mm diameter pinhole X-ray image of two NiCr plasma hot spots formed from the 4-wire (20-μm diameter) X-pinch for X-ray energies above 9 keV. (b) A magnified $M = 7.5$ X-ray image of the two NiCr hot spots produced by the bubble lens CRL with $N = 200$ biconcave spherical lenses with radius of curvature $R = 100$ μm, physical aperture diameter $2r_e = 200$ μm, center or minimum wall thickness $d = 28$ μm. For 7.5-keV X-rays the bubble lens CRL had focal length $f = 13.5$ cm with 20% on-axis transmission ($T = 0.2$) and absorption aperture diameter $2r_a = 112$ μm. (c) An enlargement of the bubble lens CRL - magnified X-ray image of the pair of NiCr hot spots. (d) Plots of the horizontal and vertical line scans of the upper hot spot image. (Right) Plot of the approximate spectral efficiency versus X-ray energy of the detector front and back Kodak BioMax films for the 800-eV X-pinch plasma (Gary *et al.*, 2004).

the two NiCr hot plasma spots; and (d) shows the horizontal and vertical line scan plots of the X-ray images of the upper hot spot. The right plot shows the efficiency of the 60 mm × 90-mm front and back Kodak BioMax films versus the X-ray energy for the 800-eV X-pinch plasma. The very fine structure of each NiCr hot spot is clearly seen in the image produced by the bubble lens CRL. The $\sigma = 2\,\mu m$ spatial resolution of the bubble lens X-ray image of the two NiCr hot spots was limited by the film scanner parameters.

Piestrup *et al.* (2005) designed, fabricated, and tested a simple X-ray microscope using an X-ray tube and a bubble lens CRL (air bubbles embedded in epoxy along a small capillary; see also Gary *et al.* (2005), and Dudchik *et al.* (2007). The small spherical bubbles of the bubble lens CRL are highly chromatic and have spherical aberration; however, with the wideband X-ray tube source, the bubble lens CRL produced $M = 4.5$ magnified images with good contrast and $\sigma = 6\,\mu m$ resolution by Piestrup *et al.* (2005). They reported the distortion, contrast, and resolution of the bubble lens CRL X-ray images obtained with the 8-keV characteristic line spectra of a standard X-ray tube was less than obtained with 8-keV X-rays of the SSRL bending magnet beam line BL 2-3; whereas, the FOV of the bubble lens CRL was larger with the X-ray tube source. The top of Figure 106 shows their comparison of the bubble lens CRL images of a 2000-grid mesh object with (1) the 8-keV characteristic line spectra X-rays from an 18-kV, 2.5-mA X-ray tube, and (2) the monochromatized 8-keV synchrotron X-rays of the SSRL bending magnet beam line 2-3.

The FOV of the bubble lens CRL varies depending on source radius a, source beam divergence angle θ_s, source to object distance s, object to CRL distance $o \equiv r_0$, the CRL length l, and the effective aperture radius r_e, which is the minimum of the CRL physical aperture radius r_m and CRL absorption aperture radius r_a, where

$$r_e = MIN\,(r_m, r_a) = MIN\left(r_m, 2\sqrt{\frac{f\delta}{\mu}}\right) = r_a \quad \text{Absorption limited.} \qquad (128)$$

The absorption limits the CRL effective aperture radius to the CRL absorption aperture radius $r_e = r_a$. From Figure 105 (top) the intrinsic FOV (FOV_{int}) of the bubble lens CRL determined by CRL pipe-like geometry for an assumed uniform source and object positioned at distance o from the CRL center plane is given by

$$FOV_{\text{int}} = \frac{4r_a o}{l} \quad \text{CRL pipe–limited object only.} \qquad (129)$$

The CRL FOV of an object at object distance o, which is illuminated by a source of diameter $2a$ at distance s upstream of the object and thus

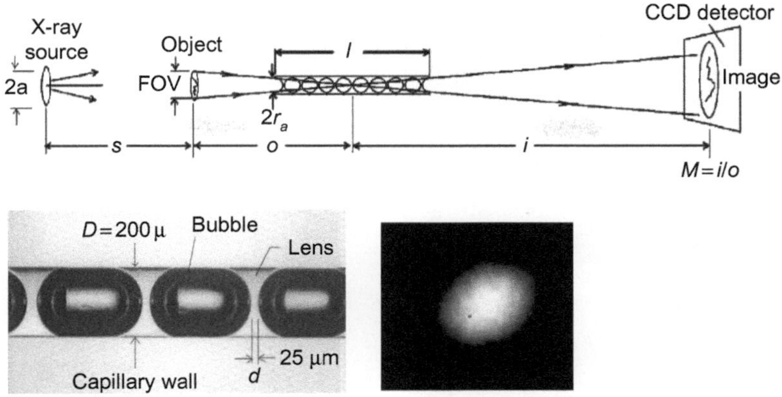

FIGURE 105 (Top) Shown is setup for X-ray imaging experiments with synchrotron X-ray source and tabletop X-ray microscope with an X-ray tube source. The SSRL bending magnet beam line 2-3 synchrotron X-ray source had a vertical FWHM of 0.44 mm and horizontal FWHM of 1.7 mm with source to object distance $s = 16.8$ m and X-ray monochromatized to 8 keV. Tabletop X-ray microscope experiments used a 18-kV, 2.5-mA X-ray tube with an elliptical source spot 0.5 mm × 1.5 mm on a copper anode target, which produced bremsstrahlung and 8-keV line spectra, further filtered by CRL. The $l = 39$ m-long bubble CRL had $N = 103$ biconcave, spherical lenses, each with $R = 100\,\mu$m radius of curvature and physical aperture diameter $2r_m = 200\,\mu$m. Measured at SSRL with 8-keV X-rays, bubble lens CRL peak, center-axis transmission $T(0) = 27\%$, absorption aperture diameter $2r_a = 149\,\mu$m, focal length $f = 13.5$ cm, and numerical aperture $NA = 10^{-4}$ (Piestrup *et al.*, 2005). (Bottom left) Photo of bubble CRL labeled with dimensions (Gary *et al.*, 2005). (Bottom right) The tabletop X-ray microscope image of a 50-μm circular spot on a copper anode X-ray tube source magnified threefold ($M = 3$) by bubble CRL with 1-minute image acquisition time (Gary *et al.*, 2005).

at distance $s + o$ upstream of the CRL center plane, is limited by either (1) the divergence angle θ_s of source-emitted X-rays or (2) the source diameter $2a$. Imaged rays from the source, which are not scattered or absorbed by the object, must pass straight from the source to the object and then to the CRL. An upper limit to the CRL (FOV) of the source-illuminated object is set by an FOV perimeter ray. The FOV perimeter ray extends from the CRL upper-perimeter point L at radial distance r_a (from the center optical axis) to the upper-perimeter object point O at radial distance $r_a + o\sin\theta$ to the perimeter source point S at radial distance a. The angle θ of the FOV perimeter ray with respect to the center optical axis is small such that

$$\theta = \sin^{-1}\frac{a - r_a}{s + o} \cong \frac{a - r_a}{s + o} \quad \text{Perimeter ray angle.} \tag{130}$$

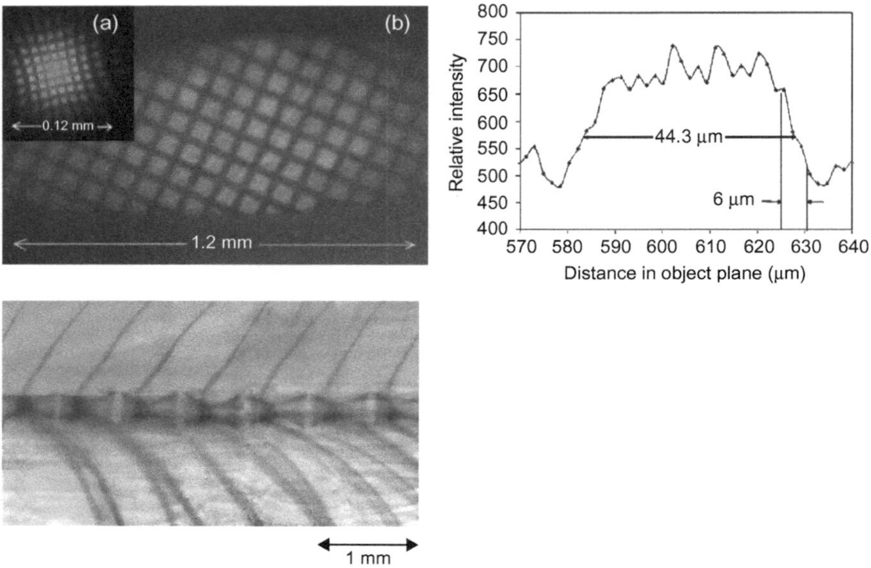

FIGURE 106 X-ray images with bubble CRL with $N = 78$ biconcave lenses, radius of curvature $R = 100\,\mu\text{m}$, physical aperture $2r_m = 200\,\mu\text{m}$, minimum wall thickness $d = 25\,\mu\text{m}$, and focal length $f = 17$ cm at 8 keV. (Top left, inset a) Synchrotron X-ray source with a bubble CRL produced a magnified ($M = 5.4$) image of gold 2000-mesh grid (2000 wires per inch) with 0.12-mm FOV. (Top right, b) X-ray tube source with a bubble CRL produced a magnified ($M = 4.5$) image of gold 400-mesh grid (400 wires per inch) with 0.5 mm × 1.2 mm FOV. (Top right) Plot of the 400-mesh grid using an X-ray tube source with measured 44.3-μm grid aperture width and 6-μm resolution at the grid edge. (Bottom) Magnified ($M = 4.5$) X-ray image of spinal column of Paracheirodon innesi (neon tetra fish) using a tabletop X-ray microscope composed of a bubble CRL and unfiltered X-ray tube (Piestrup *et al.*, 2005).

The diameter $2a$ of an isotropic X-ray source (uniform X-ray emission in all directions—at all angles over solid angle 4π ster) yields the limiting CRL FOV FOV_{dia} expression

$$FOV_{dia} \leq 2\left(r_a + o\frac{a - r_a}{s + o}\right) = 2r_a\left(\frac{s}{s+o}\right) + 2a\left(\frac{o}{s+o}\right) \tag{131}$$

Isotropic source size-limited.

If the source to object distance s is much larger than the object to CRL distance o (as is the case of synchrotron X-ray sources), then with size-limited sources and $s \gg o$,

$$FOV_{dia} \leq 2r_a \quad \text{Synchrotron source.} \tag{132}$$

If the source to object distance s is much smaller than the object to CRL distance o (as is the case with the tabletop X-ray tube source), then with size-limited sources and $s \ll o$,

$$FOV_{dia} \leq 2a \quad \text{X-ray tube source.} \tag{133}$$

Consider the case of a collimated X-ray source with X-rays emitted at divergence angle θ_s relative to the center optical axis. A perimeter ray from the CRL perimeter point L at radial distance a extends to the perimeter object point O at radial distance $r_a + o \sin\theta_s$ and then to the perimeter source point S at radial distance a is an angle θ relative to the center optical axis. The angle of the perimeter ray θ must be less than the source divergence angle θ_s, where

$$\theta_s \geq \theta \quad \text{Maximum perimeter ray angle for collimated source.} \tag{134}$$

Substitution for θ of Eq. 130 into Eq. 134 gives

$$\theta_s \geq \theta = \sin^{-1}\frac{a - r_a}{s + 0} \cong \frac{a - r_a}{s + 0}. \tag{135}$$

The perimeter ray divergence angle for a collimated source is assumed small, and hence the resulting CRL FOV of the source-illuminated object FOV_{div} is limited by the source divergence angle θ_s, where

$$FOV_{div} \leq 2r_a + 2o \sin\theta_s \cong 2r_a + 2o\theta_s \quad \text{Source divergence angle limited.} \tag{136}$$

If the source to object distance s is much larger than the object to CRL distance o (as is the case of an isotropic synchrotron X-ray source, where $s \gg o$), then the source angle θ_s is nearly zero at the object, where $\theta_s \ll r_a/o$. Hence, the CRL FOV is limited to its absorption aperture diameter $2r_a$, where

$$FOV_{dia} \leq 2r_a \quad \text{Synchrotron source with } \theta_s \ll \frac{r_a}{o}. \tag{137}$$

The divergence angle of tabletop X-ray tube sources is large, so the collimated source FOV expression does not apply. The source diameter $2a$ or the CRL absorption aperture diameter $2r_a$ limits the CRL FOV (FOV_{tube}) of the divergent X-rays from the X-ray tube source, where

$$FOV_{tube} \leq MIN\left(2a, \frac{4r_a o}{l}\right) \quad \text{X-ray tube source.} \tag{138}$$

The bubble lens CRL imaging experiments of Piestrup *et al.* (2005) with 8-keV X-rays (shown in Figure 105) consisted of two experiments. The first experiment was bubble lens CRL imaging with the SSRL bending magnet BL 2-3, and the second experiment was bubble lens CRL imaging with a tabletop X-ray microscope with an X-ray tube source. The SSRL X-ray source was elliptical with vertical FWHM 0.44 mm and horizontal 1.7mm FWHM located at source distance $s = 16.8$ m from the object. The X-rays from the bending magnet SSRL source were monochromatized to 8 keV, and the imaged object was a 2000-mesh Gilder gold grid with period 2000 gold wires per inch.

The tabletop X-ray source was an Oxford Instruments X-ray tube (model XTF5010A; Oxford Instruments, Abingdon, Oxfordshire, UK). The tube had a copper anode with an apparent X-ray elliptical source size of 0.5 mm × 0.5 mm. The X-ray tube voltage was 18 kV with 0.75-mA electron beam current, which resulted in standard bremsstrahlung X-ray emission and 8-keV characteristic line spectra emission with no tube filtering. The bubble lens CRL acted as an additional filter for the bremsstrahlung and 8-keV line-emitted X-rays, such that the object was effectively imaged by narrowband 8-keV X-rays. The imaged object for the tabletop X-ray microscope experiment with the X-ray tube was a 400-mesh Gilder gold grid with period 400 gold wires per inch.

The same bubble lens CRL and imaging detector were used for both the synchrotron X-ray source and X-ray tube imaging experiments. The bubble lens CRL consisted of a glass capillary of 200-μm inside diameter, which has 104 uniformly formed air bubbles in the epoxy filling ($C_{100}H_{200}O_2N$ and 1.08g/cm^3; see the bottom left of Figure 105). A biconcave spherical lenses is formed between each pair of adjacent bubbles such that the bubble lens in which the minimum wall thickness between a pair of air bubbles was 25 μm. The bubble lens CRL thus had $N = 103$ lenses with center minimum wall thickness $d = 25$ μm, and the 200-μm capillary determined the $R = 100$-μm radius of curvature and physical aperture diameter of the bubble lenses to be $2r_m = 200$ μm. The length of the bubble lens CRL was $l = 39$ mm. At SSRL beam line 2-3 the bubble lens parameters were measured at 8 keV, which included peak centerline transmission $T(0) = 27\%$, absorption aperture diameter $2r_a = 149$ μm, focal length $f = 13.5$ cm, and numerical aperture $NA = 10^{-4}$.

The detector was an X-ray Fast Digital Image (Photonic Science Limited, East Sussex, UK), which was a Peltier-cooled CCD sensor with a 1300 × 1300 array of pixels with 6.7-μm size, which was coupled to a light-emitting X-ray scintillator with a fiber optic. X-ray images of grids were obtained using either a synchrotron or X-ray tube source. The source "background" was first imaged without the object. Next, the object was imaged. And finally, the source background was subtracted from the imaged object. Thus, images were taken with the grid with imaged intensity I and without the grid with imaged intensity I_0. The processed image ΔI of the grid was then

$$\Delta I = I - KI_0. \tag{139}$$

Here K is a constant, which was varied to achieve the best contrast, where Piestrup *et al.* (2005) determined

$$K = 0.5. \tag{140}$$

In a completely different experiment, Gary *et al.* (2005) imaged a 50-μm diameter X-ray source spot on the copper anode of an X-ray tube with threefold magnification $M = 3$ and 60-second exposure time (see the bottom right of Figure 105). They used a bubble lens CRL with $N = 78$ biconcave spherical lenses. The motivation of Gary *et al.* (2005) was to demonstrate that the tabletop X-ray microscope can be used as an imaging diagnostic for X-ray tube sources. For example, in the X-ray image of the 50-μm X-ray spot on the copper anode of the X-ray tube, the illumination can be seen to be more uniform in the center of the spot than toward the edges. Gary *et al.* (2005) took a time sequence of images taken of the copper anode X-ray spot over several minutes, which its position drifted in time, allowing a direct measurement of the drift. This drift can be critical when the source is coupled to collection optics with a narrow acceptance angle in which the source could drift out of the acceptance area of the optic.

Gary *et al.* (2005) noted with direct imaging of the X-ray tube with a CRL or pinhole camera, the anode source spots are superior to test pattern images because the CRL or pinhole yields direct information on the position, shape, and intensity distribution over the source. The CRL can provide much better resolution than a pinhole and is less expensive and easier to use than other imaging X-ray optics. In addition, the X-ray flux through a CRL is much greater than that of a pinhole at resolutions finer than the absorption aperture of the CRL (typically on the order of 100 μm).

The imaging experiment with synchrotron X-rays at SSRL BL 2-3 used a gold 2000-mesh with 2000 wires per inch period. The source to object distance was $s = 16.8\,\text{m}$, the object to bubble lens CRL distance was $o \equiv r_0 = 16\,\text{cm}$, and the CRL to detector distance $i = r_i = 87\,\text{cm}$ provided a magnification $M = 5.4$. The SSRL-magnified image of the gold 2000-mesh obtained with 30-second exposure times is shown in the inset of Figure 107 (top left). The measured bubble lens CRL FOV of $FOV_{\text{exp}} = 120\,\mu\text{m}$ of the synchrotron source was close to the absorption aperture diameter $2r_a = 149\,\mu\text{m}$ for the calculated CRL FOV $FOV_{dia} \leq 2r_a$. The vertical divergence angle of radiation from a 3-GeV synchrotron was approximately

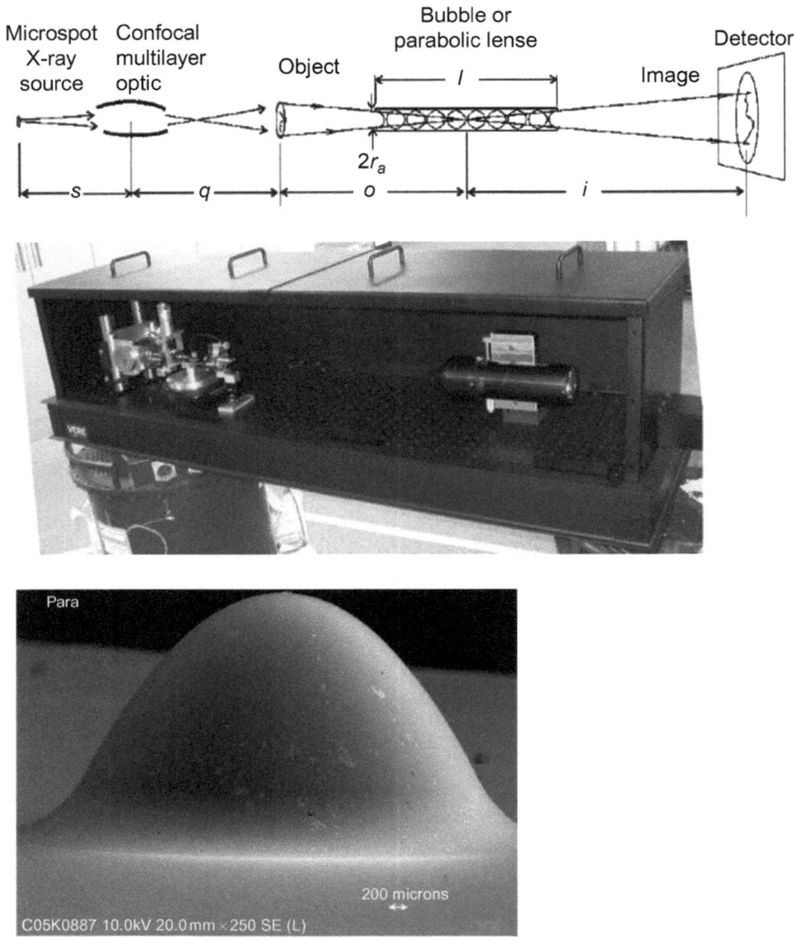

FIGURE 107 (top) Diagram of tabletop X-ray microscope with a multilayer optic to direct narrowband X-rays onto source, the object, a CRL with parabolic biconcave plastic lenses or biconcave, spherical bubble lenses, and Peltier-cooled CCD imaging sensor fiber-optic-coupled to X-ray scintillator. (middle) Photograph of tabletop X-ray microscope, X-ray tube source, confocal multilayer optic, rotation stage for sample and tomography, and CRL is at left whereas the X-ray camera is at right. (bottom) Scanning electron microscope (SEM) of an aluminum master, used to mold plastic parabolic CRLs with on-axis radius of curvature $R = 50\,\mu\text{m}$ (Gary *et al.*, 2007a,b).

0.17 mrad, and for CRL object distance $o = 16\,\text{cm}$, the source divergence limit on the CRL FOV was $FOV_{div} \leq 176\,\mu\text{m}$. The bubble lens image of the gold 2000-mesh object with the synchrotron X-rays was distorted.

For the X-ray-tube experiment, Piestrup *et al.* (2000) used the same bubble lens and detector. In this case, the source to object distance was only $s = 2\,\text{cm}$, the object to CRL distance was $r_0 \equiv o = 16.5\,\text{cm}$, and the CRL to detector distance was $r_i \equiv i = 74\,\text{cm}$ with resulting magnification $M = 4.5$. With the X-ray tube source, the resulting bubble lens CRL image of the 400-mesh grid with 400 wires per inch, which had $44.5\,\mu\text{m} \times 44.5$-μm square apertures and 19-μm wide bars, was acquired in a 10-minute exposure (top left of Figure 106).

The bubble lens CRL measured FOV increased from $FOV_{\text{exp}} = 0.12\,\text{mm}$ diameter with the synchrotron X-ray source to a measured FOV of $0.5\,\text{mm} \times 1.2\,\text{mm}$ with the X-ray tube source. The X-ray tube FOV was limited by X-ray tube source size in one dimension direction and by the CRL aspect ratio in the other dimension. A line profile of the bubble lens CRL image of the 400-mesh grid taken with the X-ray tube source is plotted at the top right of Figure 106. The plot shows the measured 44.3-μm FWHM grid aperture width from the X-ray tube bubble CRL image closely matched the optical measured 400-mesh grid aperture width of 44.5 μm. The resolution at the edge of the grid aperture was measured as 6 μm, which is the distance interval required for the change in contrast from 20% to 80%. However, the measured 6-μm resolution was far above the 190-μm diffraction-limited resolution σ (μm).

To demonstrate the ability of the X-ray tube microscope with bubble lens CRL to image biological materials, Piestrup *et al.* (2005) imaged the spinal column of *Paracheirodon innesi* (neon tetra aquarium fish) shown at the bottom of Figure 106. The image is a composite of six $1.9\,\text{mm} \times 1.9$-mm images that have been electronically stitched together. To obtain this image, they used a 600-W, $3\,\text{mm} \times 3$-mm focal spot X-ray tube with copper anode and a bubble lens CRL with $N = 78$ spherical biconcave lenses with radius of curvature $R = 100\,\mu\text{m}$ and focal length $f = 17\,\text{cm}$ at 8 keV. However, this CRL was transmitted better and had a smaller minimum wall thickness of $d = 12\,\mu\text{m}$.

The smaller fish ribs in the X-ray tube image in Figure 106 (bottom) are 20 μm in diameter, which demonstrated that the resolution of the tabletop bubble lens CRL X-ray microscope with X-ray tube source by Piestrup *et al.* (2005) produced a resolution equal to the best micro-computerized tomography machines in the 2005 market. Their tabletop X-ray microscope did not require filtering or condensing of the X-ray tube source.

Gary *et al.* (2007a,b) obtained X-ray images of grid meshes and biological material obtained using a microspot X-ray tube with a confocal multilayer optic and a 92-element parabolic CRL made of a plastic containing only hydrogen and carbon. These researchers also obtained (from images obtained from a bubble CRL with spherical biconcave lenses) a confocal multilayer optic and microspot X-ray source. The measured resolution using a 155-element parabolic CRL and a multilayer condenser with the microspot tube was $\sim 1.1\,\mu\text{m}$.

The top of Figure 107 shows a diagram of their tabletop X-ray microscope with microfocus X-ray tube, confocal multilayer optic, object, parabolic or bubble CRL, and imaging detector. Gary *et al.* (2007a,b) placed the sample on a rotation stage, which enabled regular X-ray imaging and X-ray tomography. The middle of the figure is a photo of their tabletop X-ray microscope as a portable tomographic imaging system. In the photo the X-ray tube source, confocal multilayer optic, rotation stage, and CRL are at the left and the X-ray camera is at the right. The drift space from the CRL to the camera was set to achieve a magnification $M = 4$ to $M = 10$, which depended on the configuration. The side panel shielding was removed for the picture. The system also includes a separate rack with the control system and power supplies. Gary *et al.* (2007a,b) found the best image quality using the multilayer condenser with a parabolic lens compared with images with a spherical lens and without the multilayer optics.

Gary *et al.* (2007a,b) fabricated two parabolic CRLs composed, respectively, of $N = 92$ and $N = 155$ biconcave parabolic lenses. They melted a plastic containing only carbon and hydrogen and molded it into lenses with aluminum parabolic shape masters. These low-atomic-number materials were chosen for the plastic because of their better X-ray–focusing ability compared with higher-atomic-number elements. The aluminum masters were micromachined and designed to have an on-axis radius of curvature $R = 50\,\mu\text{m}$. The bottom of Figure 107 shows a scanning electron microscope (SEM) of an aluminum master used to mold plastic parabolic CRLs. The equation for the parabolic shape of the master y (μm) with on-axis radius of curvature R (μm) was

$$y = \frac{R^2}{100}. \tag{141}$$

The molded polymer disks made with the aluminum master had a dimple in the center. The polymer in the dimple formed a biconcave parabolic lens. The disks were then closely stacked inside a stainless tube to form the parabolic CRL. The $N = 92$ CRL had 92 biconcave parabolic lenses with radius of curvature $R = 50\,\mu\text{m}$ with a $2r_a = 133\,\mu\text{m}$ absorption aperture diameter, focal length $f = 15.3$ cm, and numerical aperture $NA = 4.3 \times 10^{-4}$at 8 keV.

The tabletop microscope of Gary *et al.* (2007a,b) consisted of (1) a microspot X-ray tube or an area X-ray tube, (2) multilayer optics, (3) a CRL (parabolic coin CRL or spherical bubble CRL), and (4) an X-ray detector (shown at the top of Figure 107). The microspot X-ray tube was an Apogee X-ray tube (model XTG5011 from Oxford Instruments) with nominal 35-μm focal spot diameter and with a copper anode. The electron-tube voltage was set to 25 keV and a current of 1 mA, resulting in a standard bremsstrahlung and 8-keV characteristic line spectra from the microspot tube with no filtering. The area X-ray tube was a Brand X-ray copper anode X-ray tube (Brand X-Ray Tube Co., Inc., Addison, IL) with an apparent source size of 5 mm × 5 mm. The X-ray area tube was operated at 30 kV and 15 mA. The detector was a Photonic Sciences Ltd. X-ray Fast

Digital Imager. It consisted of a Peltier-cooled, CCD sensor with 1300 × 1300 pixels with square pixel dimensions 6.7 μm × 6.7 μm. The CCD sensor was coupled to the X-ray scintillator by a fiber optic. The multilayer optics was a Confocal Max-Flux Optics, CMF-12-38Cu6 (Osmic Inc., Auburn Hills, MI). The multilayer optic is served as an X-ray mirror that reflected only one wavelength (energy). Thus, the multilayer provided an ideal monochromatic spectrum for imaging with a CRL.

In addition to providing monochromatic X-rays, the multilayer optic was required to converge the X-rays through the full-sample FOV to the CRL aperture. The confocal multilayer optic, which was developed for X-ray diffraction studies, was designed to meet this condition. The confocal multilayer optic was designed to deliver the maximum possible X-ray flux to a 200-μm diameter spot with minimum divergence. With a 1 mm × 1-mm output aperture, the confocal multilayer optic produced the 200-μm diameter spot at a downstream distance of 38 cm. Importantly, the local divergence at each point of the output of the confocal multilayer optic was approximately 1 mrad, which was sufficient to fill the absorption-limited, effective aperture diameter $2r_e = 104\,\mu m$ CRL aperture for each point on the object. If the divergence angles were too small, then insufficient X-ray flux would be collected by the CRL to form an image. Images of objects were obtained using either a microspot or an area X-ray tube source.

The detector noise was subtracted and the source background was compensated. Thus, images were taken with the object (raw image) with intensity I and without the object (source image) with intensity I_0. An image without source also was taken to measure the detector noise I_N (detector noise). The processed image of the object I_p was then

$$I_p = \frac{I - I_N}{I_0 - I_N}. \tag{142}$$

Gary *et al.* (2007a,b) image processing was thus designed to eliminate any source nonuniformity and inherent detector noise in their imaging system.

Gary *et al.* (2007a,b) imaged wire grids, the vertebrae cartilage of a *P. innesi* (neon tetra aquarium fish), and Caribbean grass sponge to look at a lower-contrast object with fine features and examine the effects of imaging contrast agents. A 1000-mesh wire grid composed of 6-μm wires separated by 19 μm provided a strong qualitative presentation of the improved resolution of the imaging system. The top of Figure 108 shows the X-ray images of the 1000-mesh grid at 8 keV. The X-ray image (left, a) was taken with the $N = 103$ bubble lens CRL at $M = 4$ magnification and the unfiltered area X-ray tube at 250 W for a 30-minute exposure. The X-ray image (middle, b) was taken with the $N = 103$ bubble lens CRL at $M = 4$ magnification with the confocal multilayer optic and the microspot X-ray tube at 12.5 W for a 10-minute exposure. The X-ray image (right, c) was taken with a parabolic lens CRL at $M = 10$ magnification with the confocal multilayer optic and the microspot X-ray tube at 25 W for a 10-minute exposure. The parabolic CRL with microspot X-ray tube and confocal multilayer optic clearly provided the best images of the 1000-mesh grid.

Gary *et al.* (2007a,b) experimentally determined the resolution of the parabolic CRL with $N = 135$ biconcave parabolic lenses by a knife-edge measurement with 25-μm gold foil. With the microfocus X-ray tube

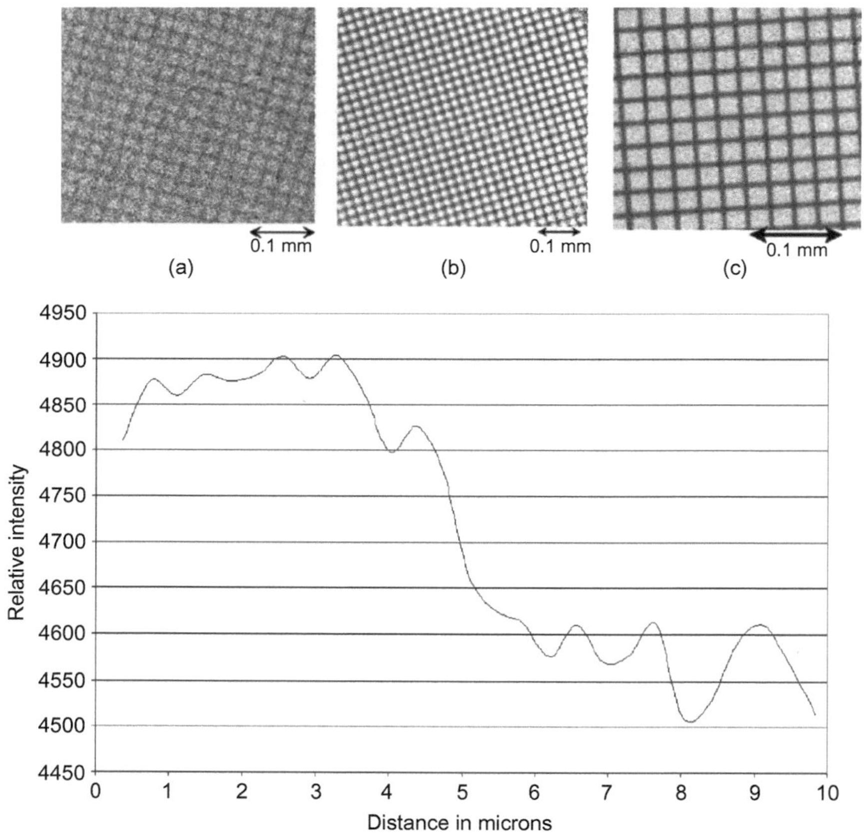

FIGURE 108 (Top) X-ray images of 1000-mesh grid at 8 keV. (Top left, a) Magnified $M = 4$ image with $N = 103$ bubble CRL and unfiltered area X-ray tube at 250 W for a 30-minute exposure. (Top middle. b) Magnified $M = 4$ image with $N = 103$ bubble lens CRL with confocal multilayer optic and microspot X-ray tube at 12.5 W for 10-minute exposure. (Right, c) Magnified $M = 10$ image with parabolic lens CRL with confocal multilayer optic and microspot X-ray tube at 25 W for 10-minute exposure. (Bottom) Plot of the line intensity profile of 25-μm thick gold foil knife-edge image taken with $N = 135$ parabolic CRL at $M = 18.4$ magnification. Measured $\sigma = 1.1$-μm resolution is the distance over which contrast changes from 20% to 80% (Gary *et al.*, 2007a,b).

and confocal multilayer mirror, the $N = 135$ parabolic CRL provided a $M = 18.4$ magnified image of the 25-μm gold foil boundary with air. They summed five line profiles of the gold foil knife-edge image (plotted at the bottom Figure 108). The measured resolution from the gold knife edge occurs over the $\sigma = 1.1$-μm distance in which occurs a contrast change from 20% to 8%. The measured $\sigma = 1.1\,\mu m$ was far above the diffraction-limited resolution $0.61\lambda/NA = 0.169\,\mu m$.

For magnification $M = 18.4$ and assuming two pixels are needed to resolve an object feature, given the detector's square pixels with side dimension $p = 6.7\,\mu m$, the minimum resolution due to the detector was $\sigma_{det} = 0.73\,\mu m$, where

$$\sigma_{det} = 2\frac{p}{M} \tag{143}$$

Gary *et al.* (2007a,b) noted that parabolic lenses and some form of filtering of the source are essential to improving image resolution from their attained $\sigma = 1.1\,\mu m$, which was an improvement over the 6-μm resolution attained by Piestrup *et al.* (2005) with a spherical bubble lens with an area X-ray tube source. However, resolutions better than $0.3\,\mu m$ were achieved with high-quality parabolic CRLs of aluminum with a synchrotron X-ray source in 1999 by Lengeler *et al.* (1999a,b).

Gary *et al.* (2007a,b) imaged the cartilage of the vertebrae of *P. innesi* (shown at the top of Figure 109) with different X-ray optical components. The width of the tetra fish vertebrae was 50 μm. The X-ray image (top left, a) was taken with the $N = 103$ bubble lens CRL at $M = 4$ magnification and the unfiltered area X-ray tube at 300 W for a 20-minute exposure. The X-ray image (top middle, b) was taken with the $N = 103$ bubble lens CRL at $M = 4$ magnification with the confocal multilayer optic and the microspot X-ray tube at 25 W for a 4-hour exposure. The X-ray image (top right, c) was taken with a $N = 92$ parabolic lens CRL at $M = 10$ magnification with the confocal multilayer optic and the microspot X-ray tube at 25 W for a 10-minute exposure. The parabolic CRL with microspot X-ray tube and confocal multilayer optic clearly provided the best image of the cartilage and vertebrae of the tetra fish spine.

Gary *et al.* (2007a,b) also imaged a Caribbean grass sponge with the $N = 92$ parabolic lens CRL using the confocal multilayer optic and the microfocus X-ray tube source at 25 W for a 10-minute exposure. Their purpose was to image an object with fine features at a lower contrast and then examine the effects of contrast agents injected into the object for improved image contrast. The X-ray images of the Caribbean grass sponge are shown in Figure 109 for the three conditions of (bottom left) a natural sponge, (middle left) a sponge with 1% iodine solution, and (bottom right) a sponge with Fenestra VC vascular contrast agent (Alerion Biomedical, Inc., San Diego, CA). Both the iodine and Fenestra VC were added by soaking and then wringing the sponge to remove the liquid, leaving only

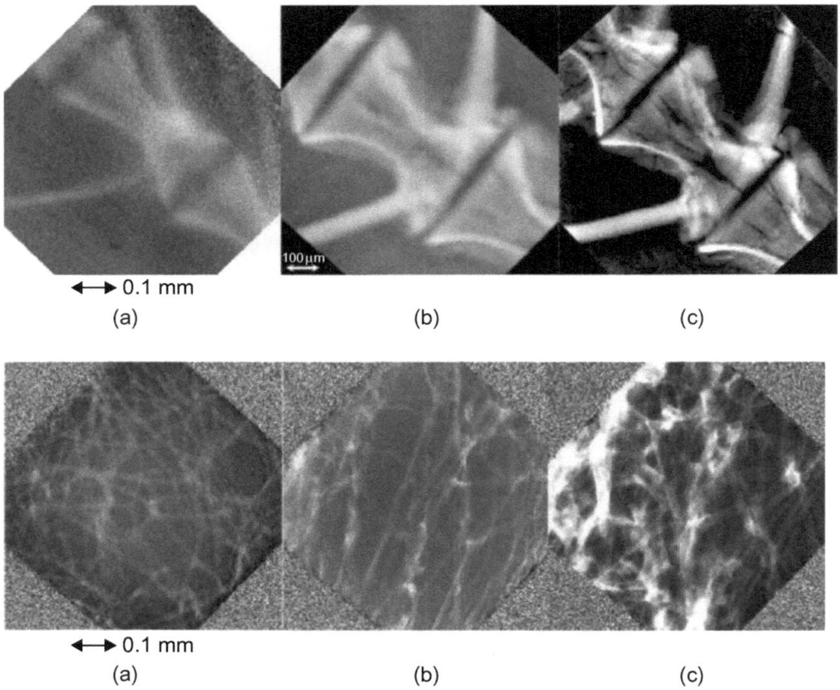

FIGURE 109 (Top) The width of the tetra fish vertebrae was 50 μm. (Top left, a) X-ray image obtained by $N = 103$ bubble CRL at $M = 4$ magnification and unfiltered area X-ray tube at 300 W for 20-minute exposure. (Top middle, b) X-ray image taken with $N = 103$ bubble CRL at $M = 4$ magnification with confocal multilayer optic and microspot X-ray tube at 25 W for a 4-hour exposure. (Top right, c) X-ray image taken with $N = 92$ parabolic CRL at $M = 10$ magnification with confocal multilayer optic and microspot X-ray tube at 25 W for a 10-minute exposure. (Bottom) Caribbean grass sponge imaged with $N = 92$ parabolic lens CRL using confocal multilayer optic and microfocus X-ray tube source at 25 W for a 10-minute exposure. (Bottom left) Natural sponge, (middle left) sponge with 1% iodine solution, and (bottom right) sponge with Fenestra VC vascular contrast agent (Gary *et al.*, 2007a).

a residual quantity in the sponge. While there is sufficient contrast to see the sponge without a contrast agent, the agents did increase the available signal. Interestingly, the Fenestra VC appears to thicken the strands of the sponge, which may be due to the size of the molecules used in the agent, though further study is needed to confirm this.

Gary *et al.* (2007b) obtained a tomographically reconstructed image of a 10-μm diameter wire in which scans were taken at 60 evenly spaced angles. Each scan image took 1 minute, for a total imaging time of 60 minutes (Figure 110).

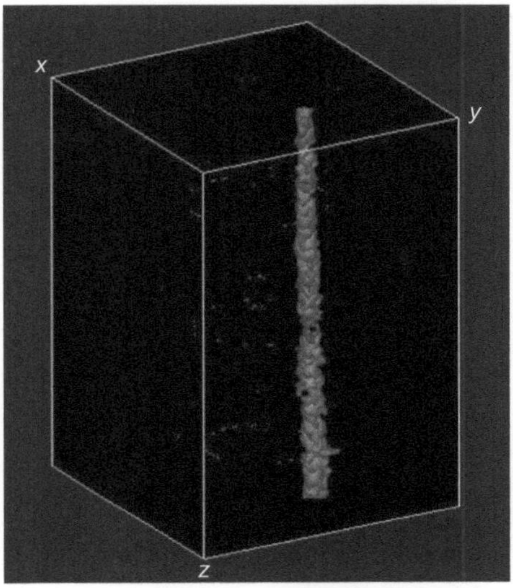

FIGURE 110 Tomographic reconstruction of 10-μm diameter gold wire. Scans were taken at 60 evenly spaced angles. Each scan image took 1 minute for a total imaging time of 60 minutes (Gary *et al.*, 2007b).

In order to provide a comparison with the Adelphi parabolic dies and method of biconcave parabolic lens fabrication just discussed, we now revisit the biconcave parabolic lenses and CRLs fabricated at the workshop of the II. Physikalisches Institut RWTH in Aachen, Germany. The images and schematic in Figure 111 are from the Ph.D. thesis of J. Tümmler (2000). The top left of Figure 111 shows a schematic of the RWTH Aachen biconcave parabolic lens, and the top right is a photo of the two dies, each with a parabolic debossing tip. The pressing tool consists of these two dies with a parabola on the top of each die and a ring to guide the two dies on one axis. Biconcave parabolic lenses of various materials (e.g., aluminum, magnesium, polycarbonate, boron carbide, lithium, and beryllium) can be manufactured in an inert argon atmosphere. In some cases, the two dies were heated, pressed into the lens material disk, and then allowed to cool to below 50°C before removal. In particular, dies were heated to 130°C for polycarbonate lenses and 450°C for magnesium. Figure 111 (middle) shows a light micrograph of a parabola die on the left and on the right is a micrograph view of a slice through a biconcave parabolic lens. Figure 111 (bottom left) is a photo of the CRL box at RWTH, and the bottom right shows an electron micrograph of one of the parabolic dies.

Dark-field microscopy uses oblique illumination to enhance contrast in specimens that are not imaged well under normal bright-field illumination conditions. After the direct X-rays have been blocked, X-rays passing through the object from oblique angles at all azimuths are diffracted,

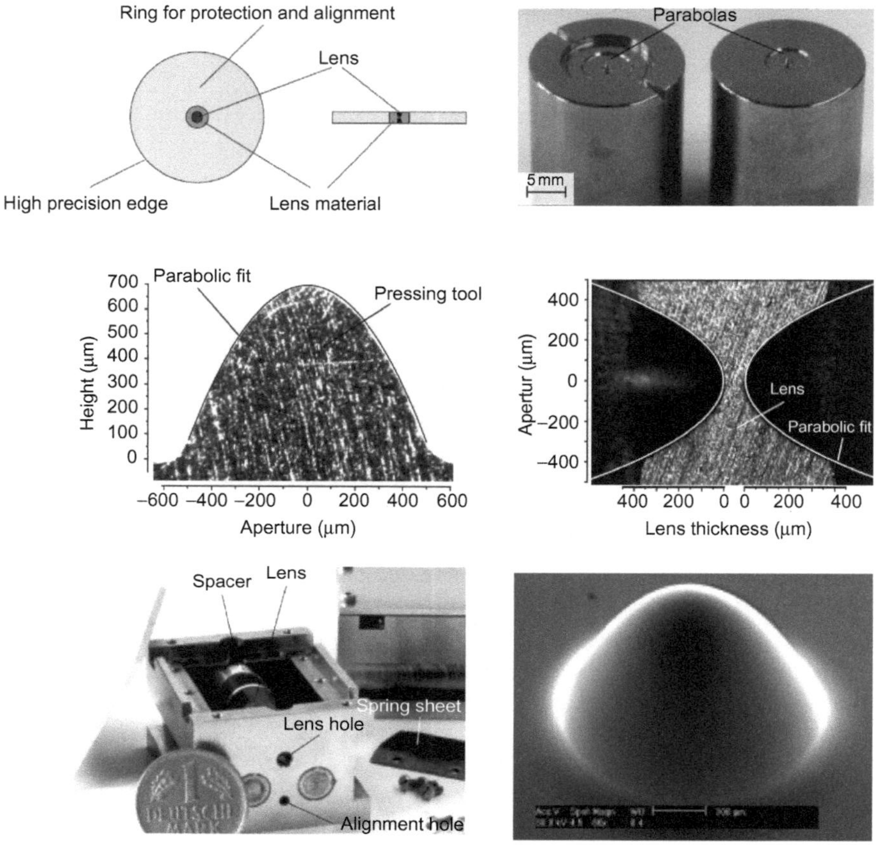

FIGURE 111 (Top left) Schematic of single parabolic lens. (Top right) Pressing tool and biconcave parabolic lenses of various materials (e.g., aluminum, magnesium, polycarbonate, and beryllium) were manufactured at workshop of the II. Physikalisches Institut RWTH (Aachen, Germany). The pressing tool consists of two dies with a parabola on top of each die and a ring to guide two dies on one axis. (Middle left) Light micrograph of parabola die and (right) micrograph view of a cut through the biconcave parabolic lens. (Bottom left) RWTH Aachen CRL. (Bottom right) Electron micrograph of a parabolic die. The schematic and images are from the Ph.D. thesis of J. Tümmler (2000) at RWTH.

refracted, and reflected to form an image of the object superimposed onto a dark background. Park *et al.* (2007) constructed a simple dark-field X-ray microscope using a bubble CRL as the objective lens and a tungsten wire to block the direct beam. X-rays (8 keV) from the SSRL source were used to illuminate a target, a gold grid with 6-μm bars and 19-μm spacing between bars.

The experimental results showed that dark-field imaging could be done with an X-ray source with a coherence length of several microns. The results also suggest that the dark-field microscope with a parabolic CRL and an opaque sphere or cross wire can be developed to view low-contrast objects with minimal distortion and less directional focusing. The CRL consists of a series of bubbles encased inside a glass capillary with the epoxy between two bubbles forming the biconcave lenses. The epoxy used in these experiments is $C_{100}H_{200}O_{20}N$ with density 1.08 g/cc.

The dark-field imaging experiments were performed using the wiggler on beam line 10-2 at SSRL. The source is a 30-pole, 1.45-T wiggler insertion device (ID) end-station. The X-ray beam was filtered using an Si(111) double-crystal monochromator. The usable X-ray energy range was ~5 keV to 20 keV. The beam size at the source is 14.3 μm (vertical FWHM) × 919 μm (horizontal FWHM), and the source was 21.8 m away from the input window at the hutch. The coherence length at the object for 8 keV was 236 μm (vertical) × 3.7 μm (horizontal), correspondingly. The unfocused wiggler beam was used throughout the experiments. The spot size of the unfocused beam at the hutch is 20 mm (vertical) × 20 mm (horizontal).

A CCD detector had a 1300 × 1300-pixel array, and each square pixel had dimensions 6.7 μm × 6.7 μm. The bubble lens CRL was composed of $N = 56$ biconcave unit lenses with radius of curvature $R = 100\,\mu m$. The top of Figure 112 shows the experimental setup for the dark-field experiment at SSRL beam line 10-2. The magnified bright-field shown in the middle of the figure and the three dark-field X-ray images at the bottom of a gold mesh object were obtained with 8-keV X-rays and a bubble lens CRL. The gold mesh object was a 1000-mesh gold grid with 1000 lines per inch, and the pitch of the grid bars was 25 μm with 19-μm gaps and 6-μm wide bars.

The gold mesh object was placed at $s = 21.9$ m from the beam line 10-2 wiggler source, and the bubble CRL was placed between the gold mesh and the detector. For 8-keV X-rays the bubble CRL focal length was $f = 24$ cm, and the fourfold magnification $M = 4$ of the 1000-mesh grid object required the object distance to be set to $o_1 = 30\,cm$, and the image distance to be set to $i_1 = 120$ cm. A 15-μm diameter tungsten wire was used as an X-ray block and was placed at the focal point of the bubble CRL. The bubble CRL had a dual function of (1) focusing the wiggler X-ray source (object 2) onto the X-ray–blocking 15-μm diameter tungsten wire, and (2) imaging the 1000-mesh gold grid onto the detector. For focusing the wiggler source onto the tungsten 15-μm wire, the image distance between the bubble CRL and the tungsten wire was $i_2 = 24$ cm, which was approximately equal to the bubble CRL focal length f. That is, the wiggler source to gold mesh object distance o_2 was much greater than the bubble CRL focal length f, where $o_2 \gg f$, and $i_2 \cong f = 24$ cm. This placement of the 15-μm tungsten wire blocked most of the 8-keV X-rays, and with the

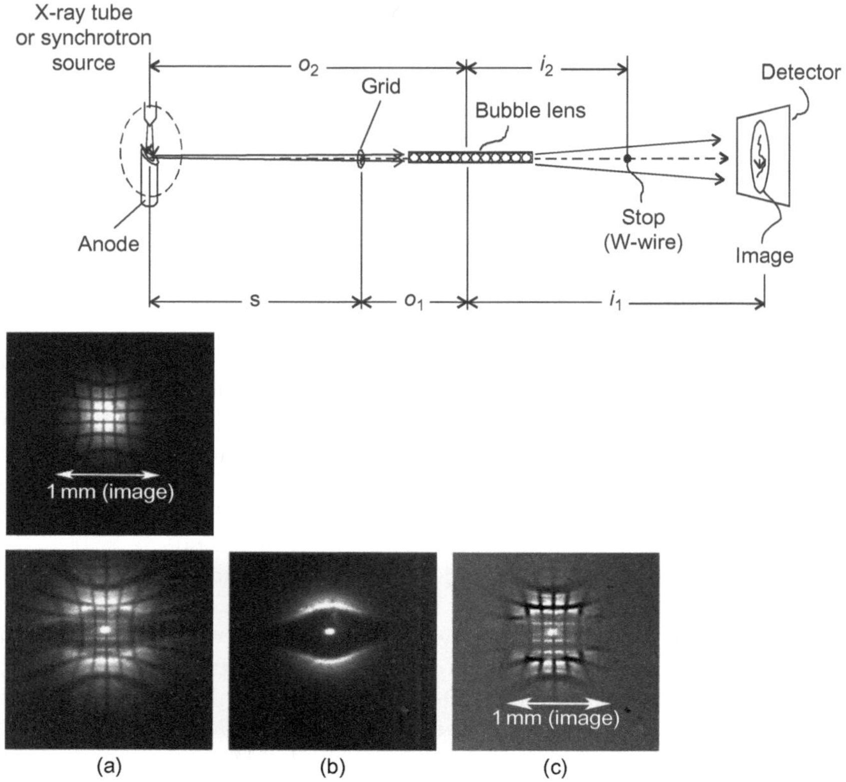

FIGURE 112 (Top) Setup for the dark-field imaging experiment at the SSRL wiggler beam line 10-2. (Middle) Bright-field image of gold 1000-mesh grid (1000 lines per inch) with 25-μm pitch, 19-μm gaps, and 6-μm bars. (Bottom left, a) Dark-field image of mesh object with 15-μm tungsten blocking wire at the CRL focal point, 6-minute exposure. (Bottom middle, b) Background image without mesh object, 4-minute exposure. (Bottom right, c) Background-subtracted image of mesh object. The 15-μm tungsten blocking wire was transverse to the axial *z*-directed wiggler X-ray beam (Park *et al.*, 2007).

direct X-rays blocked by the wire, very little X-ray intensity reached the detector. The gold mesh object placed upstream of the bubble CRL scattered X-rays, which missed the 15-μm tungsten blocking wire and thus reached the detector.

First, a bright-field image of the mesh was taken to compare with the dark-field image. In the bright-field image shown in the middle of Figure 112, the 15-μm tungsten wire stop was removed. The direct X-rays from the gold mesh object could then be imaged onto the detector. Next, the 15-μm tungsten wire stop was placed in the horizontal direction, transverse to the X-ray beam at the bubble CRL focal point.

A dark-field image of the gold 1000-mesh object was then obtained (shown at the bottom left of Figure 112). Then the gold mesh object was removed and a background-only dark-field image of the wiggler source was obtained (bottom middle). The dark-field background-only image of the wiggler source was then subtracted from the dark-field image of the gold 1000-mesh grid to obtain the dark-field background-subtracted image of the mesh object (shown at the bottom right of Figure 112).

The three dark-field images at the bottom of Figure 112 show pincushion distortion. Pincushion distortion occurs when an aperture is placed downstream of the lens, which causes the object magnification to increase with increasing radial distance of the rays from the center optical axis. However, this kind of aberration does not show up in the images of mesh objects taken with a spherical lens CRL using an area X-ray tube (see Piestrup *et al.*, 2005). Also, the pincushion distortion did not occur when imaging mesh objects with a microspot X-ray tube and a confocal multilayer optic (see Gary *et al.*, 2007a). The distortion may have been caused by the very small divergence angle of the SSRL wiggler source, whereas the tabletop X-ray microscopes used a microspot tube with a multilayer optic at 8 keV. The multilayer optic produced a sufficiently divergent source of X-rays in which the rays pass through the object (within the CRL FOV), and then those rays propagate into the CRL aperture.

The image size of the bright-field image in the middle of Figure 112 and the three dark-field X-ray images at the bottom were 300 pixels × 300 pixels, which corresponded to the 2.01 mm × 2.01-mm image size in the detector plane and to a mesh object FOV of ~170–180 μm square. In these bright-field and dark-field imaging experiments, the FOV was limited by the wiggler X-ray source size rather than the CRL acceptance angle.

The absorption-limited bubble CRL effective aperture diameter $2r_e = 168\,\mu m$ (where $2r_e = 2r_a$) closely matched the measured 170- to 180-μm FOV. The diffraction-limited resolution of the mesh object was $\sigma = 0.61\lambda/NA$ or $\sigma = 27\,\mu m$, and the detector-limited resolution was $\sigma_{det} = 3.35\,\mu m$, which was calculated by assumption of a two-pixel minimum for resolution of an object feature with 6.7-μm square pixels and bubble CRL magnification $M = 4$. The measured resolution (uncertainty) was larger than the calculated resolution, which was attributed to spherical aberration of the biconcave spherical lenses of the bubble CRL. The bright spot in the middle of the dark-field images at the bottom of Figure 112 was caused by focusing of the X-ray beam by the tungsten wire. The tungsten wire behaved as an X-ray–focusing element. In the visible spectrum, the bright spot in the middle is known as the Arago spot or Poisson spot. The phenomenon is associated with scattering at the periphery of a spherical (or a circular) object—or in this case, with a cylindrical object. Also, the spot size did not appear to change significantly as the diameter of tungsten beam-stop wire diameter was decreased from 75 μm to 15 μm.

A bright spot can also appear behind an opaque object immersed in a beam when the coherence length at the object is on the order or greater than the object dimension. When the coherence length at the wire exceeds or is comparable to the wire diameter, there can be constructive interference between the portions of the beam that are diffracted from the edges of the opaque object. This interesting effect with X-rays has been observed previously. Posey *et al.* (1998) reported imaging soft X-rays with an opaque sphere. They imaged a point X-ray source with chrome steel and Al_2O_3 spheres and observed the Arago spot in the middle of the shadow of the sphere.

The Institute of Microstructure Technology (IMT) at the Karlsruhe Institute of Technology (KIT) specializes in lithographic structuring (electron beam lithography, deep X-ray lithography, and ultraviolet lithography) and replication methods (hot embossing, thermoforming, and nanoimprinting). Impressive X-ray parabolic lens CRLs are fabricated at IMT of KIT (see their X-ray optics website at http://www.imt.kit.edu/english/236.php). Figure 113 shows some of the IMT CRLs.

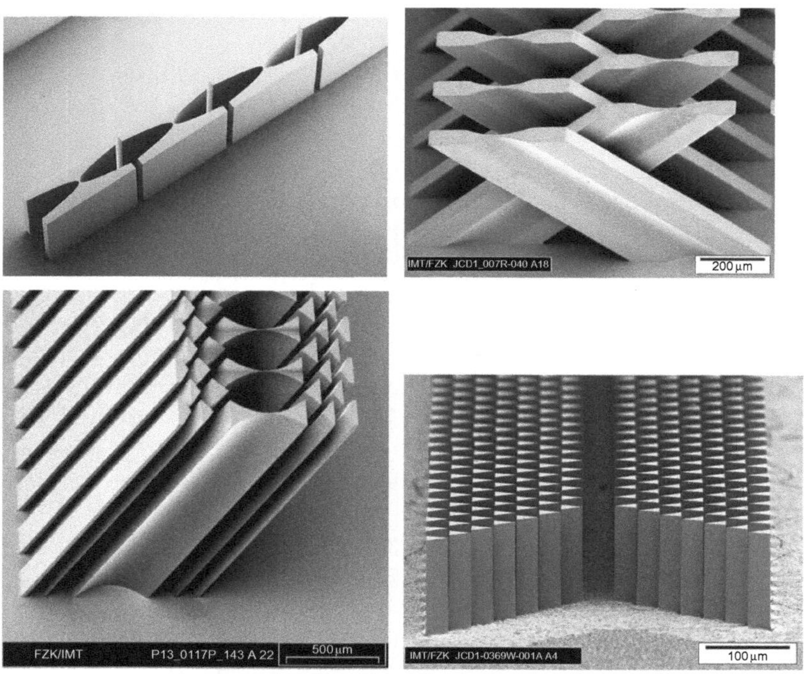

FIGURE 113 (Top left) The row of parabolic lenses. (Top right) The row of crossed parabolic lenses. (Bottom left) Mosaic lenses tilted by 45° . (Bottom right) Prismatic lens. From the website of Institute of Microstructure Technology at the Karlsruhe Institute of Technology, Germany (http://www.imt.kit.edu/english/236.php).

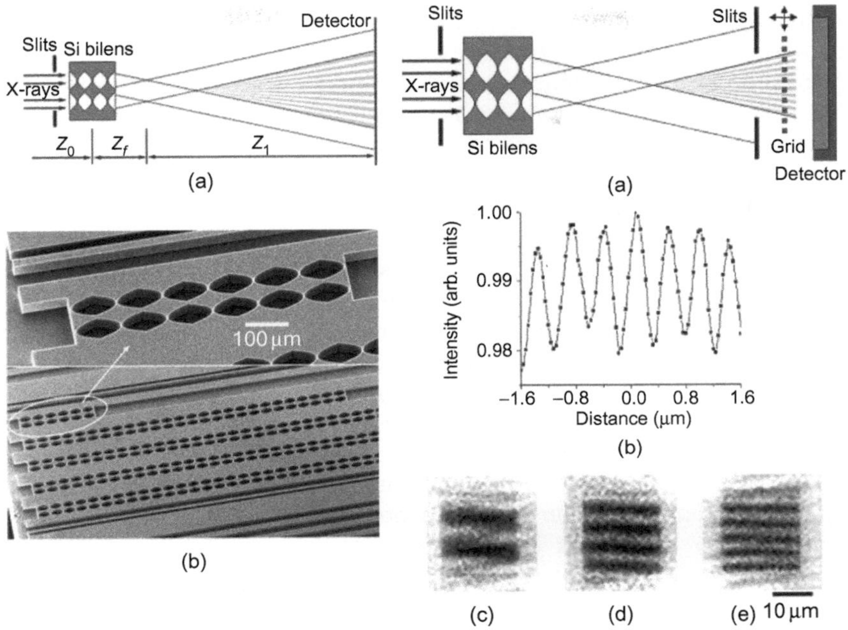

FIGURE 114 (Top left, a) Schematic of the X-ray bilens interferometer. (Top right, a) The diagram shows the setup for measurements of the interference pattern in the near field. (Bottom left, b) Images of five interferometers fabricated on the same silicon chip that produce X-ray energies from 10 to 50 keV. (Middle right, b) Plot showing intensity modulations recorded with a pin-diode by scanning a tantalum grid vertically. (Bottom right, c,d,e) X-ray moiré patterns taken by CCD X-ray camera at distances 280 mm, 288 mm, and 291 mm from the bilens. Schematics and images from the ESRF website: "An X-ray bilens nanointerferometer" (http://www.esrf.eu/news/spotlight/spotlight90; see Snigirev *et al.*, 2009).

Snigirev *et al.* (2009) devised, fabricated, and tested a hard X-ray interferometer that uses two parallel arrays of CRLs (bilens) (Figure 114). With incident coherent X-rays, the bilens produces two diffraction-limited beams that are mutually coherent. The overlapping beams produce an interference pattern with fringe spacings, which can vary over the approximate range of 10 nm to 10 μm. This X-ray bilens interferometer can create X-ray standing waves, which can be used to study nanostructures.

4. THERMAL NEUTRON RADIOGRAPHY, CODED APERTURE, AND PHASE-CONTRAST IMAGING

Thermal neutron radiography is a nondestructive imaging technique that uses highly collimated beams of thermal neutrons to provide contact radiographic images. Thermal neutron radiography has been used for quality control purposes in industries that require precision machining, such as aircraft turbines, structures, and wings. A dominant present-day application of thermal neutron radiography is testing the performance of fuel cells by imaging water flow in the cells *in situ*. Also, neutron radiography has been used in art and archeology to authenticate paintings and examine artifacts made of metal or stone. Thermal neutrons have many advantages for imaging over X-rays, electrons, and light, which include greater penetration depth for many materials, significant contrast variation between chemical elements and isotopes, and weaker radiation damage imposed on the irradiated samples.

There are (1) reactor sources of thermal neutrons for thermal neutron radiograph (e.g., the 20-MW FRM-II reactor facility at the Technical University of Munich in Garching, Germany, and (2) pulsed neutron sources (e.g., the spallation source at the Paul Scherrer Institute in Switzerland [SINQ]). The latter institute has provided thermal neutron transmission radiography for scientific and industrial applications at the NEUTRA station situated at the thermal beam port of SINQ (since 1997) and the more recently developed cold neutron radiography at the ICON station (since 2006). Thermal neutrons can also be generated by moderation of deuterium-deuterium (DD) and deuterium-tritium (DT) fusion fast neutron sources. Worldwide there are many research reactor sources of thermal neutrons for radiography and a smaller number of spallation neutron sources. The growing demand for nondestructive testing and imaging and hydrogen fuel cell development stimulated the growth of thermal neutron imaging facilities.

Thermal neutron radiography is based on the neutron interaction with the nucleus rather than the X-ray and visible light interaction with atomic electrons. The X-ray and gamma cross sections increase with increasing atomic number. This is not the case for thermal neutrons. Thermal neutron absorption and scatter cross sections vary with atomic weight, but not in the simple linear manner as X-rays and gammas. Thermal neutrons have high absorption (capture) cross sections in certain isotopes such as lithium-6, boron-10, chlorine-35, cadmium-113, mercury-196, uranium-235, silver-109, and many rare earth isotopes (particularly gadolinium-157).

Thermal neutron absorption and scatter is small in such biologically occurring elements as carbon, nitrogen, oxygen, sodium, phosphorous, calcium, and potassium. However, the neutron scatter cross section is large for hydrogen in organic and biological compounds as well as water. And compared to hydrogen, the thermal neutron absorption and scatter is small in structural and common component elements, such as magnesium, aluminum, silicon, chromium, iron, copper, zinc, zirconium, molybdenum, tin, bismuth, and lead.

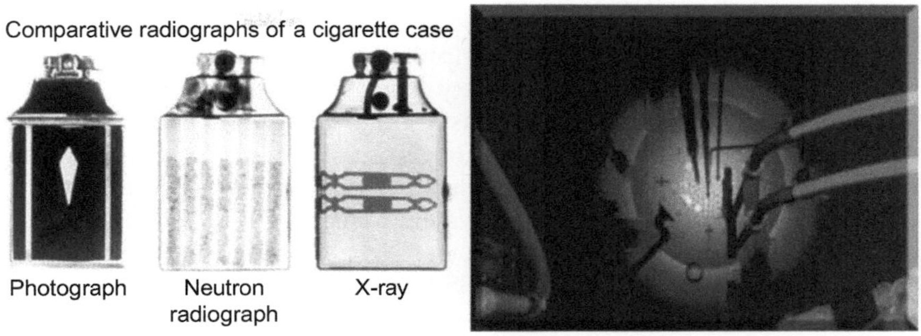

FIGURE 115 (Left) Comparison between the photo (left), neutron contact radiograph, and X-ray contact radiograph of a lighter in which the neutron radiograph indicates the lighter is full and ready to use. From the , University of California Davis McClellan Nuclear Radiation Center (MNRC) website (http://mnrc.ucdavis.edu). (Right) MNRC photo shows the MNRC TRIGA reactor core blue glow from the Cherenkov radiation "sonic boom or bow wave" of charged particles exceeding the phase velocity of light. (See Color Insert.)

Hydrogen has a 0.33 barn absorption cross section and 82 barn scatter cross section, but in comparison, deuterium has a 0.0056 barn absorption cross section and 8 barn scatter cross section. As a result, hydrogen-rich materials are excellent contrast agents for thermal neutrons. For example, one can determine the height of water in a lead standpipe by thermal (and fast) neutron radiography, which is difficult with X-ray or gamma-ray radiography (Figure 115.)

Standing above the 30-foot pool of water one receives less radiation than one would standing in a smoked-filled bar from secondhand smoke. The average 1.5–cigarette packs per day smoker receives on average 8000 millirem per year. The average radiation worker receives about a 300 millirem per year dose in addition to the average individual 360 millirem annual dose (mainly from natural sources and medical/dental X-rays) for a total dose rate of 660 millirem per year for the average radiation worker. The high smoker radiation dose rate is due to alpha-emitting polonium-210 in dust or moisture particles formed from radon-222, which diffuses from the earth. The polonium-210 becomes incorporated into smoking materials and is absorbed by the lung tracheobronchial tree through its thin, alpha-transmitting membrane, directly irradiating the radiation-sensitive bronchial tissue. Thick, relatively insensitive layers of protective tissue of the mouth and digestive tract layers absorb the alpha particles before they reach underlying radiation sensitive tissue, which can absorb the alphas (see Lamarsh, 1983).

Thermal neutron radiography has become a valuable nondestructive imaging and testing tool that has met standards, for example, set by the

ASTM International (ASTM). The ASTM, presently headquartered located in West Conshohocken (near Philadelphia), Pennsylvania, USA, was originally known as the American Society for Testing and Materials. ASTM, founded in 1898 by a group of scientists and engineers led by U.S. chemist C.B. Dudley, is an international standards organization that develops and publishes annually a voluntary consensus of technical standards for a wide range of materials, products, systems, and services.

Throughout the world standards were converged for thermal neutron radiography. Figure 116 shows a 1985 Euroatom working group publication. Euroatom started working on standards in 1979 and describes three calibration devices (Domanus, 1985). The first calibrator is the beam purity indicator (BPI), which checks the neutron beam gamma, thermal, and fast neutron components, and a modified BPI (BPI-F), which checks the nuclear fuel (see the top diagram of Figure 116.) The second calibrator is the sensitivity indicator, which determines the sensitivity of detail that is visible in the neutron radiograph (see bottom diagrams of Figure 116). The third calibrator is the Association Française de Normalisation (AFNOR) image quality indicator (IQI) to control the quality of neutron radiographs of non-radioactive items is shown in Figure 117.

In 1979, the Riso group noted, by use of the BPI (or BPI-F, F indicates fuel) and use of the sensitivity indicator (SI), the image quality of contrast and resolution in a neutron radiograph can be determined, which follows the method prescribed by ASTM E545 151. This ASTM standard was designed for neutron radiography using silver halide film and is limited only to the direct method of neutron radiography. AFNOR is the French national organization for standardization and its International Organization for Standardization member body.

Thermal neutron radiography was developed in the 1960–1980 period with the advent of power and research fission reactors. The technique was based on the pioneering work in thermal neutron radiography by H. Kallmann and E. Kuhn in the 1930s in Germany, and later by Berger (1964, 1971) of the Argonne National Laboratory, and Hawkesworth and Walker (1969) at the University of Birmingham in the United Kingdom, followed by publication of a handbook on neutron radiography by von der Hardt and Rottger (1981). In the same 1960–1980 time frame, fast neutron radiography developed with the publication of papers by Criscuolo and Polansky (1961), Polansky and Criscuolo (1964), Tochilin (1965), Wood (1967), Parks (1969), Berger (1969, 1970), and Richardson (1977).

The thermal neutron paper from the University of di Bologna in Italy by Casall *et al.* (1995) is more recent. In the 1990s the use of charged-coupled device (CCD) cameras as imaging detector for thermal neutrons became available due to the work of Kobayashi *et al.* (1990) of Rikkyo University in Japan, and Pleinert *et al.* (1997) at the Paul Scherrer Institute

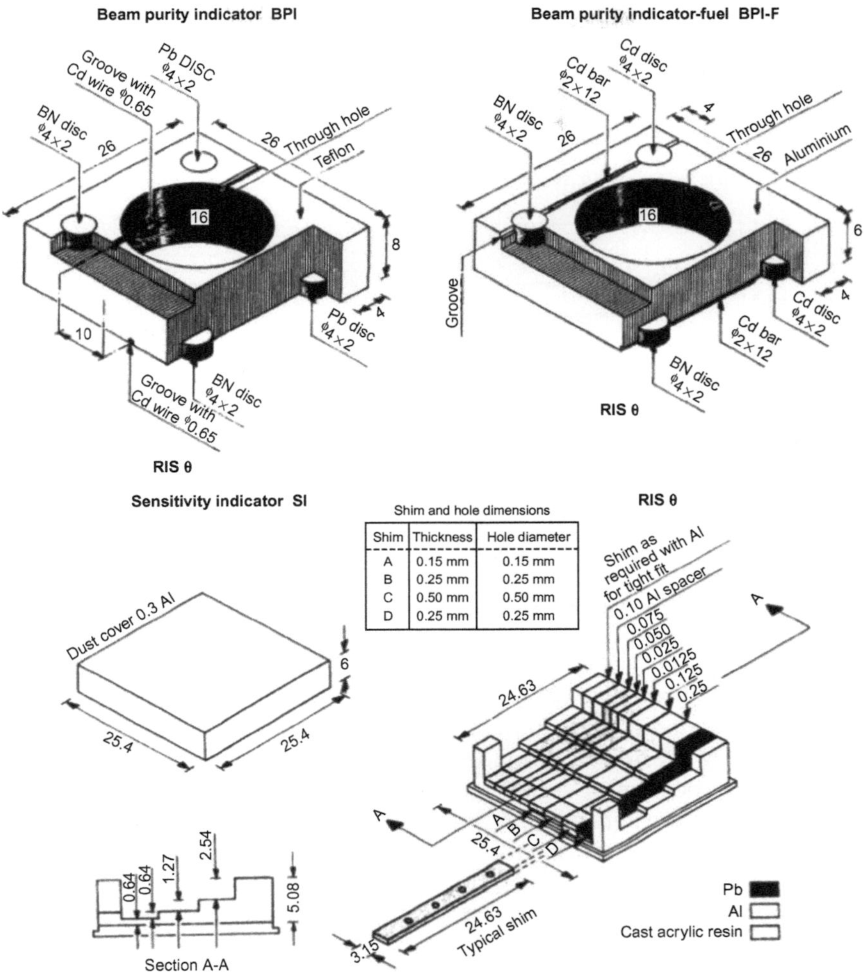

Shim	Thickness	Hole diameter
A	0.15 mm	0.15 mm
B	0.25 mm	0.25 mm
C	0.50 mm	0.50 mm
D	0.25 mm	0.25 mm

FIGURE 116 (Top) To check beam gamma, fast, and thermal neutron radiation components the left diagram is of the neutron beam purity indicator (BPI) and the right diagram is of modified beam purity indicator (BPI-F) for nuclear fuel. (Bottom) A diagram of the sensitivity indicator (SI) that measures image contrast and resolution quality (Domanus, 1985).

in Switzerland. More recent publications on thermal neutron radiography include the doctoral dissertation by Cao (2007) and the International Atomic Energy Agency report prepared by Lehmann (2008).

The distinction should be between contact radiography, projection radiography, refractive lens imaging, and pinhole camera imaging

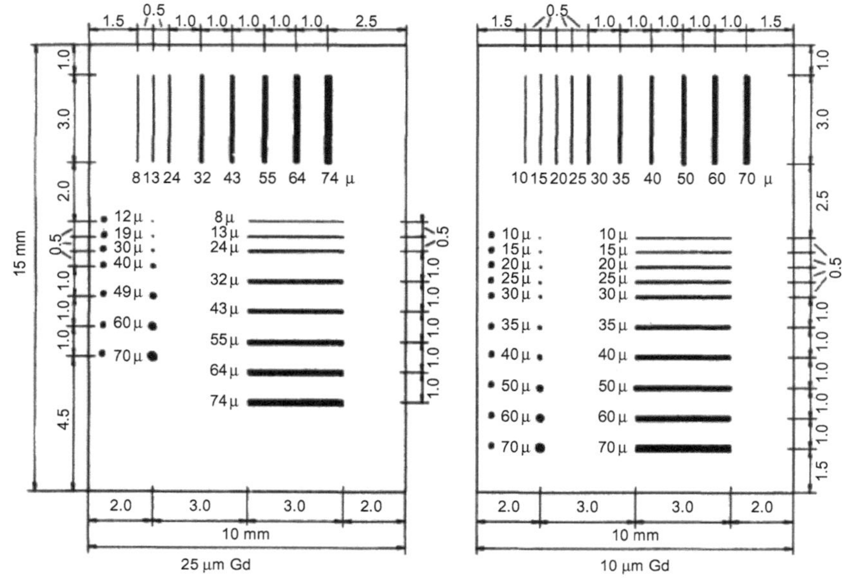

FIGURE 117 The AFNOR image quality indicator (IQI) used to control the quality of neutron radiographs of nonradioactive items (Domanus, 1985).

(Figures 118 through 121). The object resolution in projection radiography depends on the source diameter D, which requires a small but high-intensity source to provide high-resolution images within 5- to 30-minute exposure times. Microfocus X-ray tubes are intense X-ray sources with diameters smaller than 100 µm that are widely used in nondestructive imaging and inspection and specifically are used in ball grid array imaging in the semiconductor industry.

Point sources of fast neutron from the DT and DD reaction can potentially be used in projection radiography but intensity is still lacking. Similarly, the lack of sufficiently bright, small-diameter thermal neutron sources presently makes thermal neutron projection radiography impractical except at facilities with particularly intense thermal neutron sources. As a result, thermal and fast neutron radiography is presently limited to contact radiography. The projected image magnification M^{pr} of the object is found via similar triangles, where

$$M^{pr} = \frac{r_i + r_0}{r_0}. \tag{144}$$

The image resolution σ_i^{pr} is found by similar triangles as well, where

$$\sigma_i^{pr} = D\frac{r_i}{r_0}. \tag{145}$$

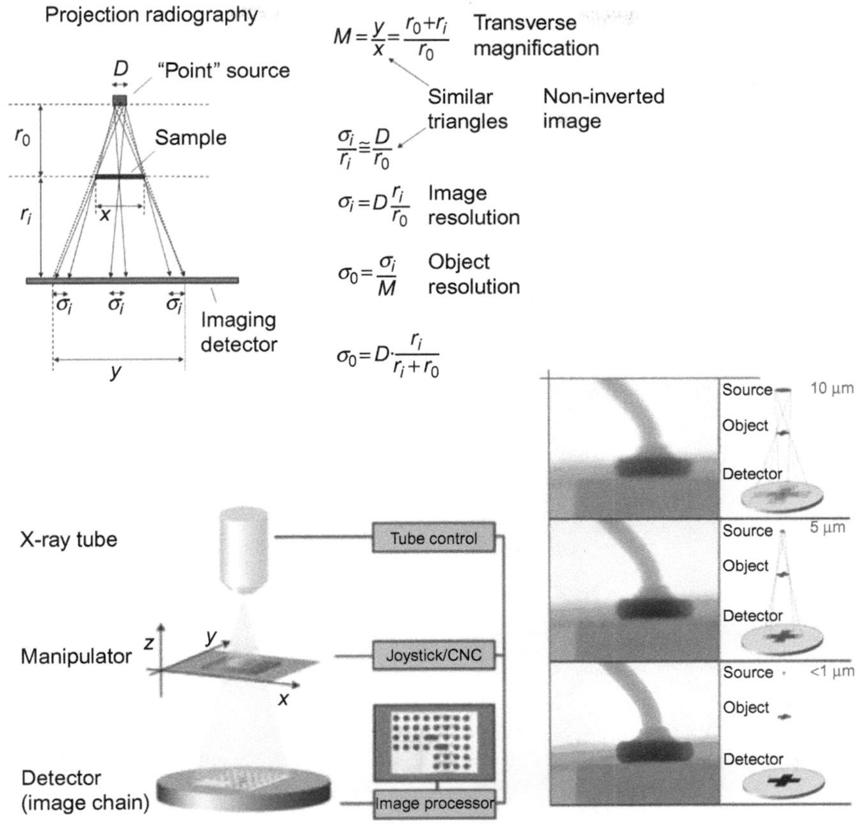

FIGURE 118 (Top) The diagram shows the setup, magnification, and image and object resolution with projection radiography. (Bottom) Projection radiography: X-ray inspection of ball grid arrays in the electronic industry (Roth and Lehman, 2006).

The object resolution σ_0^{pr} is obtained from the image resolution σ_i^{pr} by dividing by the magnification M, which yields object resolution σ_0^{pr} for projection radiography:

$$\sigma_0^{pr} = D\frac{r_i}{r_i + r_0}. \tag{146}$$

Projection radiography produces a magnified, but non-inverted image of the object. Figure 118 shows an example of X-ray projection radiography for ball grid array imaging in the semiconductor industry.

In contact radiography the object resolution depends on the divergence angle of the incident rays and thus requires small divergence angles θ. Ideally the rays are parallel or have a high degree of collimation with

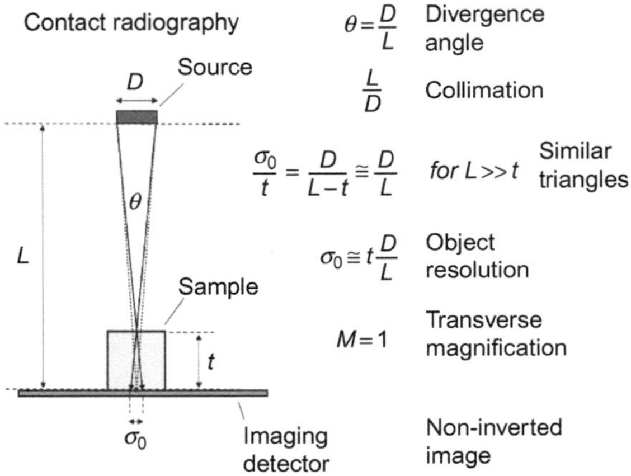

FIGURE 119 The diagram shows the setup and resolution with contact radiography.

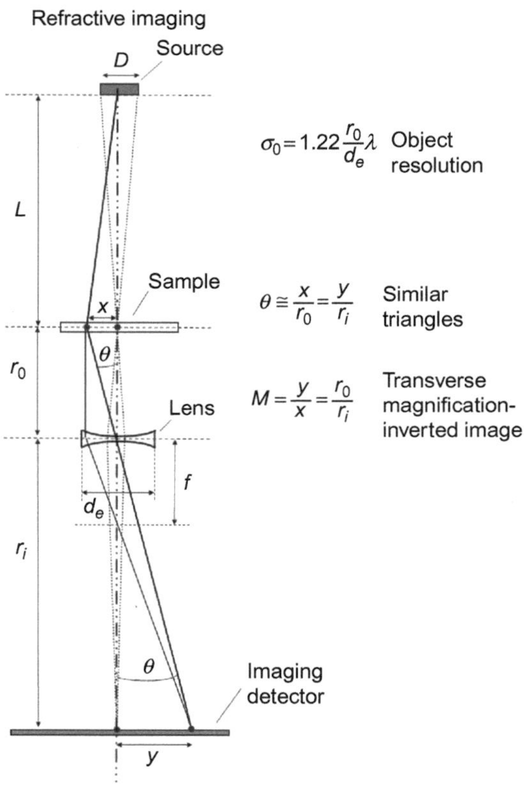

FIGURE 120 The diagram shows the setup, magnification, and resolution with refractive lens imaging.

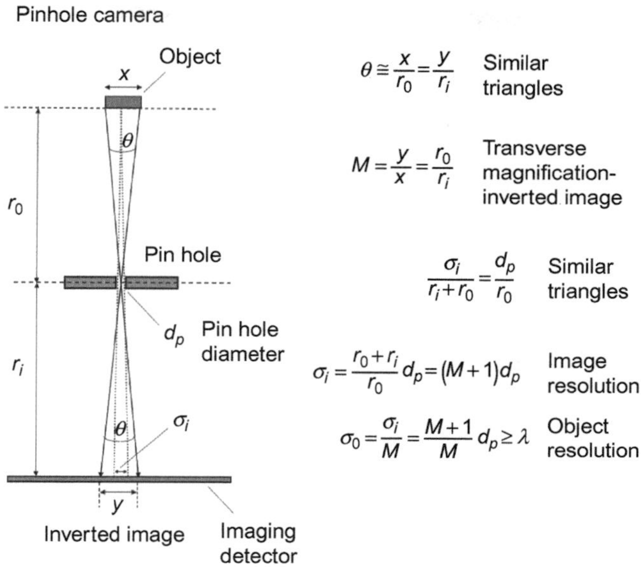

FIGURE 121 The diagram shows the setup, magnification, and resolution with a pinhole camera.

a large L/D ratio, where $\theta = D/L$ is the given source diameter and D is the source to the detector plane distance L. The contact radiograph object resolution σ_0^c depends on the object thickness t, where

$$\sigma_0^c = t\frac{D}{L}. \tag{147}$$

The contact radiograph image of the object placed adjacent to the detector is a non-inverted image with a transverse magnification $M^c$of unity,

$$M^c = 1. \tag{148}$$

Contact radiography is easily used with thermal and fast neutron sources and X-rays and gammas (see Figure 119). In particular, contact X-ray radiography is widely used in medicine, dentistry, and nondestructive testing and imaging and has been in use since Roentgen. In the past 30 years, thermal neutron radiography has become an important tool in nondestructive imaging and testing.

The quality factor $Q = 2.3$, the relatively short 1 cm or less penetration depth in hydrogen-rich material, and the thermal neutron capture gamma production and possible activation have hindered the use of thermal neutrons in medicine. However, the boron neutron capture therapy (BNCT) and gadolinium neutron capture therapy for brain and liver cancer and the use of the unique gamma signatures from thermal neutron capture are potential medical applications that are presently in the experimental stage.

Fast neutron contact radiography and projection radiography are relatively new and are evolving as commercial, compact, and intense sources of neutrons. Fast neutrons are being exploited in neutron stimulated emission computed tomography (NSECT) for noninvasive mapping of the concentration of any isotope, stable or radioactive, in any selected three-dimensional (3D) volume of the body.

NSECT uses a beam of fast neutrons that scatter inelastically from atomic nuclei in tissue, causing the excited state nuclei to emit characteristic gamma photons that are detected and identified using an energy-sensitive gamma detector. By measuring the energy and number of emitted gamma photons, the elemental composition of the target tissue can be determined. A practical application of NSECT is diagnosis of breast cancer by imaging the relative concentrations of trace elements in breast tissue and looking for subtle differences in chemical composition that are known to exist between benign and malignant tissue (see Kapadia *et al.*, 2008).

With refractive lens imaging the object resolution σ_0^{len} improves with increasing lens diameter d_e and one obtains an inverted image with transverse magnification M^{len}, given by the ratio of the object distance r_0 divided by the image distance r_i, where

$$M^{len} = \frac{r_i}{r_0} = \frac{y}{x}. \tag{149}$$

The object resolution for refractive lens imaging is given by the Rayleigh diffraction–limited resolution σ_0^{len}, where

$$\sigma_0^{len} = 1.22\lambda \frac{r_0}{d_e}. \tag{150}$$

Larger-aperture diameter d_e lenses not only provide better resolution, but they also collect more rays of light, X-rays, or neutrons, which produces larger intensity and better contrast in the image plane (see Figure 120.)

Refractive imaging is limited to sufficiently long wavelengths in which the refractive index is nonzero, which are limited to electromagnetic radiation below ~200 keV, including soft and hard X-rays, low-energy gammas; vacuum ultraviolet, visible light, infrared terahertz waves; microwaves; and radio waves. Also included in refractive imaging of neutrons are neutrons whose energies are below ~10 eV. The refractive index for epithermal neutrons and fast neutrons is near zero, and thus the imaging of epithermal and fast neutrons is limited to projection and contact radiography or pinhole camera.

In comparison, a pinhole camera also produces a magnified, inverted image with magnification M^{pin}, where

$$M^{pin} = \frac{r_i}{r_0} = \frac{y}{x}. \tag{151}$$

The pinhole camera object resolution σ_0^{pin} is limited by diffraction to the pinhole aperture diameter d_p, which must exceed the wavelength λ of the imaging light, X-rays, or neutrons, where

$$\sigma_0^{pin} = \frac{M+1}{M} d_p \geq \lambda. \tag{152}$$

Unlike the lens, the pinhole camera spot uncertainty diameter σ_0^{pin} increases (and object resolution decreases) as the pinhole aperture diameter d_p is increased (see Figure 121).

Pinhole imaging requires the material in which the pinhole is formed to be opaque to the radiation being imaged. However, pinhole camera imaging does not have chromatic aberration and monochromatic aberration and is commonly used to image high-energy X-rays, gammas, and neutrons, which are too energetic for refractive imaging or impractical for radiographic imaging. The resolution with a pinhole camera can reach the diffraction limit of the source wavelength. A useful image can be obtained with an extremely small-aperture, high-resolution pinhole camera if the source or object emitting or transmitting X-rays, gammas, or neutrons is sufficiently intense.

The use of an array of small-diameter pinholes or coded aperture imaging retains the high resolution due to the small pinhole diameter, yet increases the intensity of the image and the object field of view due to the multitude of pinholes (Figure 122). Each pinhole of the coded aperture array generates an image of the object on the detector in which all projections overlap. As a result, the detector produces a set of overlapped images of the object. From the overlapped image copies of the object and the set of known pinhole positions and aperture diameters, a single, high-fidelity image of the object is obtained.

Restoration of a single image of the object from the set of overlapping images can be done rapidly without creating artifacts if the pinholes, which are termed *coded apertures* or *masks,* are arranged in a particular manner. Each object point that transmits or emits X-rays, gammas, or fast neutrons casts a shadow of the mask on the detector. Hence, each object point is revealed not by a single bright spot, but rather by a pattern of spots. That is, the coded aperture mask encodes the signal from the object. Due to encoding, the coded aperture image data do not create a direct image, but rather an overlapped set of images, which must undergo image

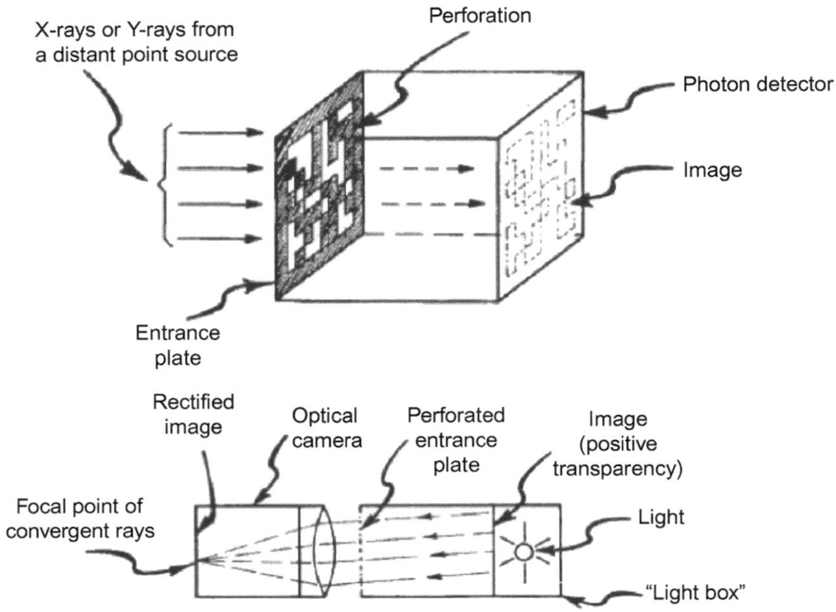

FIGURE 122 (Top) The diagram shows the coded aperture imaging camera that produces a set of overlapping images of each object point emitter or transmitter of X-rays or gammas. The overlapped image intensity distribution recorded by the position-sensitive detector is extracted by reversing the camera process. (Bottom) The extracted image is obtained by light illumination of a positive transparency of the recorded set of overlapping image. The light passes through the positive transparency though the coded aperture plate and is focused by a lens onto a detector, which records the rectified or single conjugate image of the object. Both figures are from Dicke (1968).

processing to distill the single image of the object (see Accorsi and Lanza, 2001; Caroli *et al.*, 1987; Dicke, 1968; Skinner, 1984; Ziock *et al.*, 2006).

Consider the form of the coded aperture camera suggested by Dicke (1968) (see Figure 122). The coded aperture camera has an opaque entrance plate of a few square centimeters or larger. The coded aperture array is a plate covered by a lattice of hundreds, thousands, or even millions of tiny squares, which form shaped patterns such as squares, triangles, and hexagons. However, half of the tiny squares are randomly open and transmit X-rays and gammas; the other half are opaque and do not transmit X-rays or gammas. A single object point, which acts as a single source point of X-ray or gammas, is an incident coded aperture array and casts a shadow on the position-sensitive detector (PSD).

The screen function $f(x, y)$ is defined over the coded aperture plate, and the center position coordinates of the tiny square opaque pieces and tiny square open holes of the plate are described by transverse Cartesian coordinates (x, y). That is, the screen function $f(x, y)$ characterizes the transmission or complete attenuation of X-rays of gammas through the coded aperture plate, where

$$f(x, y) = 1 \quad \text{open hole in opaque plate} \tag{153}$$

$$f(x, y) = 0 \quad \text{between holes in opaque plate.} \tag{154}$$

The image plane contains the PSD, and the overlapping images of the object form an intensity distribution $f_d(x_d, y_d)$ on the detector as a function of Cartesian transverse detector coordinates (x_d, y_d). If the object point is at infinity and given object transverse coordinates (x_0, y_0), then the overlapping images on the detector have an intensity distribution $f_d(x_d, y_d)$ given by

$$f_d\left(x_d, y_d\right) = f\left(x_d - x_0, y_d - y_0\right). \tag{155}$$

However, if the object point is not at infinity, scaling factors a_x and a_y are required and the image plane intensity distribution $f_i\left(x_i, y_i\right)$ becomes

$$f_d(x_d, y_d) = f(a_x(x_d - x_0), a_y(y_d - y_0)). \tag{156}$$

Given the object with coordinates (x_0, y_0) at some axial distance z from the coded plate aperture, the corresponding, extracted (processed or rectified) image distribution $g(x_i, y_i)$ is obtained by convolution of the coded aperture function $f(x, y)$ with the measured overlapped image distribution $f_d(x_d, y_d)$ at the detector. The convolution integral is over the area A_d of the detector with dimensions $(-d_x, d_x)$ and $(-d_y, d_y)$ in the respective x and y directions, where

$$g(x_i, y_i) = \int_{-d_x}^{d_x} dy_d \int_{-d_y}^{d_y} dy_d f_d(x_d, y_d) f(x_d - x_i, y_d - y_i). \tag{157}$$

The detector area A_d is

$$A_d = 4 d_x d_y. \tag{158}$$

Substitution in Eq. 157 of the expression of Eq. 155 for detector intensity distribution f_d in terms of the coded aperture plate function f of Eqs. 153–154 gives the convolution integral

$$g(x_i, y_i) = \int_{-d_x}^{d_x} dy_d \int_{-d_y}^{d_y} dy_d f(x_d - x_0, y_d - y_0) f(x_d - x_i, y_d - y_i). \tag{159}$$

The random distributions of the perforations (square holes) fill half of the coded aperture plate with $f = 1$ for the 50% plate area occupied by transmitting square holes, and $f = 0$ for the nontransmitting, opaque area between the holes. The maximum value of the convolution integral $g(x_i, y_i)$ is one half the detector area A_d when the transverse image position (x_i, y_i) is equal to the corresponding transverse object position (x_0, y_0), where the signal S is

$$S = g(x_i = x_0, y_i = y_0) \cong \frac{4d_x d_y}{2} = \frac{A_d}{2}. \tag{160}$$

If the transverse image position (x_i, y_i) is shifted from its conjugate position (x_0, y_0) of maximum image intensity $g(x_i, y_i)$ by one or more lattice positions, then the image position (x_i, y_i) is under an opaque square with 50% probability, where $f = 0$, or the image position (x_i, y_i) is under a square opening with 50% probability, where $f = 1$ with 50%. The image intensity distribution at non-object position coordinates forms the background intensity B, where

$$B = g(x_i \neq x_0, y_i \neq y_0) \cong \frac{1}{2}\frac{4d_x d_y}{2} = \frac{A_d}{4} = \frac{S}{2}. \tag{161}$$

For applications in X-ray or gamma-ray astronomy or medical imaging, the signal of the point source objects S exceeds the background B by two fold, where

$$\frac{S}{B} = 2 \tag{162}$$

The coded aperture plate has N squares. Half of the squares $N/2$ are opaque to X-rays, and the other half $N/2$ are open square holes that transmit X-rays. The point-to-point fluctuation in the background is determined by the smaller of (1) the $N/2$ square holes and (2) the total number of photons n counted by the detector. Consider the case of a weak source in which the counted photons n is much smaller than the number of square holes $N/2$ in the coded aperture plate, where

$$n < \frac{N}{2} \tag{163}$$

The root mean square (rms) background photon fluctuation Δn_{rms} for a background photon count $b = n/2$ at the detector plane is then

$$\Delta n_{rms} \cong \sqrt{\frac{1}{2}b} \cong \sqrt{\frac{n}{4}}. \tag{164}$$

With an off-peak $x_i \neq x_0$ and $y_i \neq y_0$ location on the detector for an image point of the single object (source) point, and a total of n detected photons from the single object (source) point, the probability P_f of a background fluctuation equal to a true image peak is

$$P_f = \left(\frac{1}{2}\right)^n. \tag{165}$$

The number of possible locations in the image plane of the point image of the point object is equal to the number of possible locations of the image point, which can be any one of the N square regions of the detector plane, which have a one-to-one correspondence to the N square regions of the coded aperture plate.

The probability P_{true} of no occurrence of a false peak anywhere in the N square regions of the detector plane converges in the limit of large N and sufficiently large n, where $N/2 > n$, to a decaying exponential, where

$$P_{true} = \left(1 - P_f\right)^N = \left[1 - \left(\frac{1}{2}\right)^n\right]^N \cong \exp\left(-\frac{N}{2^n}\right). \quad \frac{N}{2} > n \gg 1 \tag{166}$$

A 99% or better probability of no occurrence of a false peak with $P_{true} \geq 0.99$ requires the detection of $n_{\min}$ photons given by

$$n_{\min} \geq \frac{\log(100\,N)}{\log 2}. \tag{167}$$

For a coded aperture plate with $N = 10^6$ squares, half of which are opaque to X-rays and half of which are square holes that transmit X-rays, the minimum number of detected photons from a single distant object (source) point is only $n_{\min} = 26$ photons.

Adelphi Technology, in collaboration with the University of California Davis McClellan Nuclear Radiation Center (MNRC) performed thermal neutron imaging experiments on several preserved biological specimens from the Carolina Biological Supply Company in Bay 4 of MNRC. The MNRC was originally developed by the U.S, Air Force to detect low-level corrosion and hidden defects in aircraft structures using neutron radiography. In the intervening years, MNRC service has expanded to include computed tomography (3D neutron radiography), silicon doping, isotope production, neutron activation analysis, and radiation effects testing.

The MNRC reactor is a TRIGA (Training, Research, and Isotope Production General Atomic) reactor built in 1990; it is rated at 2 MW in steady state and can pulse to approximately 1000 MW for 20 ms. The TRIGA reactors, including the MNRC reactor, are passive reactors that require no action to maintain safety and are thus designed to be fail-safe. The imaging experiments were done in the Bay 4 beam line, and the samples imaged were preserved biological specimens and thin 1- and 2-mm slices of rat brain tissue for brain tumor rats injected with boron-based neutron capture therapy drug.

The MNRC Bay 4 beam line emerges from an underground reactor core at a 20° angle relative to the horizontal floor of Bay 4. The Maxwell–Boltzmann-distributed thermal neutrons pass from the reactor water core into the Bay 4 beam tube through a graphite (moderator) end plug and then pass through a 28-cm long sapphire crystal, which improves the thermal content by reducing the relative number of higher-energy neutrons and gammas.

Single-crystal sapphire (Al_2O_3) such as grades B1 (superior grade "hemlux") and C1 (standard optical grade "hemlite") are effective fast neutron filters because the transmission of neutrons at wavelengths less than 0.4 Å is less than 3% for 100 mm of thickness. It is also an effective filter for thermal neutrons with wavelengths less than 1 Å, since there is a great density of high-order reflections available to scatter the incident beam. For a 90-mm long single sapphire crystal filter the transmission is 0.8 for 1.2 to 2.4 Å wavelengths and 0.07 for epithermal neutrons. The sapphire crystals are aligned with the *c*-axis, parallel to the incident neutron beam.

From the sapphire crystal the neutrons emerge from a 3.2 cm by 3.2 cm square source aperture in a boron carbide shielding plate, and propagate 3.0 m through a collimation tube. The collimation tube has scraper structures along its walls that aid in the removal of divergent neutrons and suppress neutron scatter. At the exit of the beam tube is a massive sliding concrete door that opens to provide neutron exposure into Bay 4. The flux distribution $\Phi(\lambda)$ of the neutron beam is a wavelength-dependent Maxwell–Boltzmann distribution, given by

$$\Phi(\lambda) = \left(\frac{\lambda_p}{\lambda}\right)^5 \exp\left[2.5\left(1 - \frac{\lambda_p^2}{\lambda^2}\right)\right]. \tag{168}$$

The MNRC Bay 4 detectors used in the Adelphi Technology thermal neutron imaging experiments in Bay 4 included the time-integrated FujiFilm (FujiFilm Medical Systems USA, Inc., Stamford, CT) imaging plate and real-time imaging with a 2000-2003 vintage CCD camera–based imaging detector from NOVA Scientific (NOVA Scientific, Inc., Sturbridge, MA). The FCR XG-1NDT FujiFilm neutron imaging system uses an integrating PSD with high (18.2%) detective quantum efficiency (DQE), 50-μm pixels, and 5 orders of dynamic range and is based on the effect of photostimulated luminescence. One can obtain large area (23 cm × 25 cm) static, single shot images with the FujiFilm imaging plate (IP). The software package, Dynamix Image Share Viewer, was used to display the FujiFilm neutron images. The 2000-2003 vintage NOVA thermal neutron imaging system consists of a radiation-hard CCD camera (Thermo Fisher Scientific, Inc. Waltham, MA), which images the phosphor screen output of a borated microchannel plate (MCP).

Figure 123 is a photo of the experimental setup in Bay 4 of MNRC for thermal neutron radiography experiments performed by the author. The biological samples were exposed for 30 minutes to 3 hours to a neutron flux of approximately 10^6 n/cm^2/s with collimation $L/D = 270$. Figure 124 is a photo and a FujiFilm neutron image of a seahorse with background subtraction and pixel inversion. Pixel inversion is a valuable imaging tool to find features in faint images produced by low-intensity sources. In many cases, pixel inversion reveals features in an image that otherwise are not visible to the eye. The white background with dark dots or spots in astronomy images is an example of pixel inversion. In the case of thermal and fast neutron imaging with FujiFilm plates or CCD camera–based imaging of objects with weak sources, pixel inversion can reveal image features not visible to the eye.

Figure 125 shows a photo and FujiFilm neutron image of a rat paw and leg with background subtraction and no pixel inversion. Figure 126 shown a photo and a NOVA Scientific CCD neutron camera image of a rat paw and knee with 1800-frame summation and background subtraction. Based on the 2000-2010 period, the time integration FujiFilm imaging detector offers the advantage of larger image plates, better contrast, better resolution, and faster imaging time than CCD camera–based systems. However, the CCD camera–based systems have the advantage of real-time imaging, stacking frames for long exposures, background subtraction, and image processing.

The thermal neutrons are highly attenuated by hydrogen-rich materials in which 1-cm thick polyethylene or water is sufficient to stop a neutron flux of 10^6 n/cm^2/s (clearly seen in Figure 127), which shows the photo and FujiFilm neutron image of a preserved mouse.

Figure 128 shows a photo and a 3-hour FujiFilm image of a 2-mm thick coronal slice of rat brain with a brain cancer tumor (provided to Adelphi

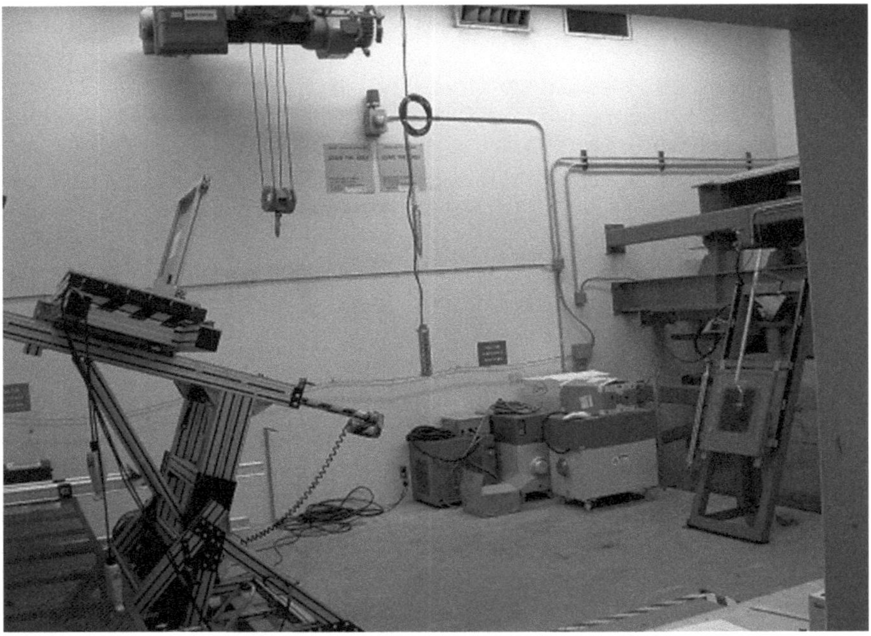

FIGURE 123 Photo of the experimental setup in Bay 4 of the 2-MW MNRC reactor for the thermal neutron radiography experiments performed by the author at Adelphi Technology in collaboration with the MNRC. The above thermal neutron imaging setup produced the thermal neutron radiographic images in Figures 124 to 128 in which the neutron beam collimation at the samples was $L/D = 270$ and the flux was approximately 10^6 n/cm^2/s (Cremer *et al.*, 2007).

Technology by Prof. R. Barth of Ohio State University Medical School). The brain slice was stored in formalin and rinsed with Hanks' balanced salt solution before neutron imaging. The brain tumor is in the right half of the coronal slice. The rat was raised and surgically tumorated (implanted) with cancer cells as part of research and development of boron-based drugs for BNCT. We attempted to image coronal slices of tumorated brain tissue with the typical concentrations of two boron-based drugs and delivery protocols; however, as expected, the low number density and moderate cross section of the ^{10}B did not provide sufficient absorption contrast to image. However, gadolinium-based neutron capture therapy drugs may have a higher likelihood of imaging due to the extremely large absorption cross section of gadolinium-157.

Figure 129 shows a graph of the resolution of major radiation imaging modalities for X-rays, and gamma, ultrasound, and microwaves compared with imaging modalities for thermal neutrons (from Cao, 2007). The CCD camera with MCP image intensifiers (light amplifiers), which

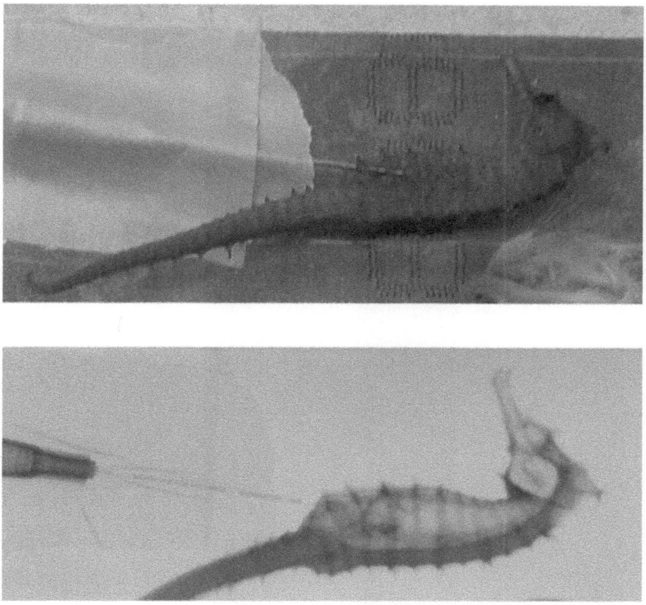

FIGURE 124 (Top) Photo of a complete seahorse. (Bottom) A 3-hour FujiFilm contact neutron image of the seahorse with subtraction of 2.16-hour background. The neutron image is pixel inverted to show more detail (Cremer *et al.*, 2007). (See Color Insert.)

view directly or indirectly (via a 45° mirror) the thermal neutron imaging detector, MCPs doped with ^{10}B, provide very high resolution compared with the other thermal neutron, X-ray, gamma, and microwave imaging modalities.

An MCP acts as a continuous-dynode electron multiplier. The MCP is an array of closely spaced small holes in a resistive material, such as glass, to which is applied a modest high voltage (e.g., ~1.5kV). This produces a large electric field across the typically thin 2-mm thickness of the MCP. An incident charged particle or energetic photon (X-ray or gamma) enters the MCP glass matrix and causes electrons to be ejected into the nearest microchannel, and these electrons cause further electron emission along the channel wall leading to an electron cascade in that channel. The hole array of the MCP consists of closely spaced holes or microchannels (e.g., 10μm) with typically 15-μm spacing between the center axis creating 5-μm walls. The holes through the MCP are parallel but are often set at a small angle (typically 8°) from the MCP surface normal.

Many MCP detectors are dual-MCP detectors, which have two MCPs, which are pressed together or separated by a small gap. The first MCP channels are typically at positive angle +8° from the surface normal. The

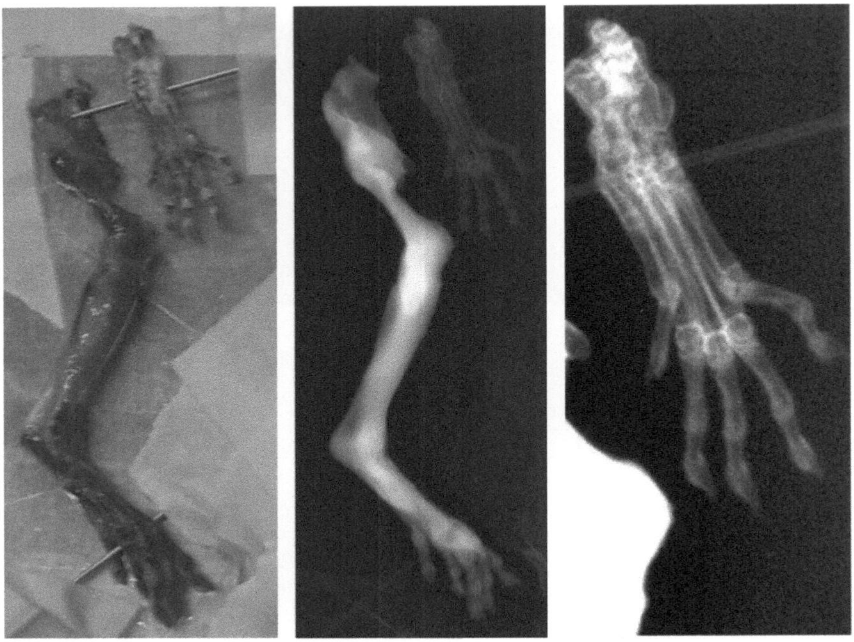

FIGURE 125 (a) A photo of a rat leg and paw. (b) A 3-hour FujiFilm contact neutron image of the rat leg and paw from which was subtracted a 2.16-hour background. (c) A 3-hour FujiFilm contact neutron image of the rat paw from which was subtracted a 2.16-hour background (Cremer *et al.*, 2007). (See Color Insert.)

upward-slanted holes of the first MCP are aligned to match the holes of the second MCP. The holes of the second MCP are slanted downward at negative angle -8° from the surface normal. The upward slanted holes of the first MCP and the downward slanted holes of the second MCP form a chevron shape ($\wedge$). The chevron arrangement of aligned holes of the two MCPs allows the electrons to exit the first MCP plate and start a cascade in the adjoining hole of the second MCP, but this reduces the ion feedback. As a result, the chevron dual MCP has much higher gain for a given voltage.

If the MCP is doped with ^{10}B, then the MCP becomes a high-resolution thermal neutron imaging detector, see Downing *et al.* (2000) and Tremsin *et al.* (2011). The top-left image in Figure 130 shows a diagram of an MCP plate doped with boron for the detection of thermal neutrons and a photo of an MCP neutron detector. A cascade of secondary electrons is produced in the microchannels due to the energetic He-4 and ^{7}Li due to neutron capture by the ^{10}B. The MCP is terminated with a light-emitting phosphor.

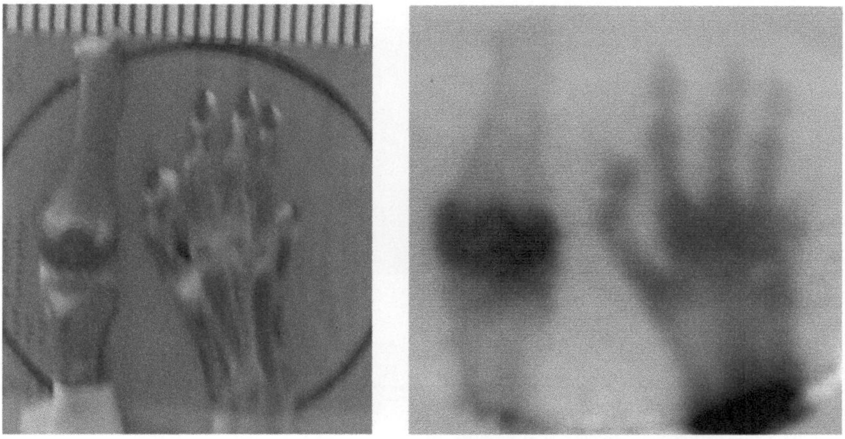

FIGURE 126 A photo (left) and the thermal neutron radiographic images (right) of a rat paw and knee joint using a 2002-2003 vintage NOVA Scientific thermal neutron imaging camera in MNRC Bay 4. The NOVA camera consists of a radiation-hard Thermo electron CCD camera that images the phosphor screen output of a borated microchannel plate. The neutron image was the summation of 1800 frames of the rat paw and leg object from which was subtracted an 1800-frame summed image of the background only (Cremer *et al.*, 2007). (See Color Insert.)

The light can be amplified by an ordinary (non-doped) MCP to provide a high-gain, low-noise amplification of light, which can be coupled by a tapered fiber optic to a radiation-hard CCD camera.

NOVA Scientific manufactures MCP neutron imaging detectors used with a CCD-based camera and also provides MCP plates as collimators (see the bottom-left image in Figure 130). The placement of a thin polyethylene film (0.1 mm to 1.0 mm, dictated by recoil proton-stopping distance for the incident fast neutron energy) over an input face of the MCP can act as a converter of fast neutrons to recoil protons. That is, a fast neutron undergoes elastic collision with a hydrogen atom in the polyethylene film. Within the proton-stopping distance of the recoil proton, which averages half the energy of the incident neutron. The recoil proton creates ionization in its path. The recoil proton and its produced secondary ions and electrons reach the nearest microchannels, where they cause electron cascades in those microchannels.

Phase-contrast imaging with neutrons and photons now is considered. Phase-contrast radiography provides edge enhancement of thin objects, which have a low absorption contrast but a large phase shift contrast due to large spatial gradients of object features and edges. The image formation occurs because the phase shift variations, which are induced by sharp changes in object thickness and/or refractive index, can cause

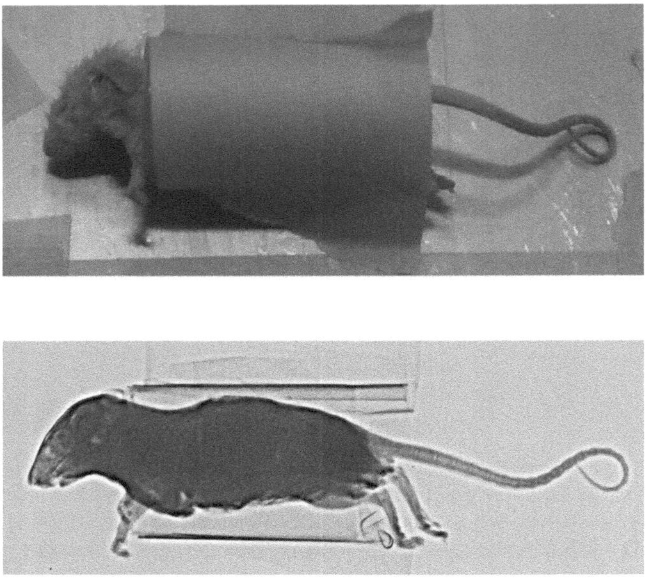

FIGURE 127 (Top) A photo of a complete mouse. (Bottom) A 1-hour FujiFilm contact neutron image of the mouse head; the neutron image is pixel inverted to show more detail. The thermal neutrons have less than about a 1.0-cm penetration in hydrogen-rich materials as the mouse radiography clearly indicates. Scatter by the hydrogen atoms rapidly attenuates the thermal neutrons even from a 2-MW reactor (Cremer *et al.*, 2007).

corresponding changes in the image intensity by use of X-ray or neutron beams with sufficiently large transverse coherence lengths.

The transverse coherence length l_{coh} of the X-ray or neutron beam must exceed transverse dimensions of the object edge or sample feature of interest. X-ray and neutron beams with large transverse coherence lengths can be created by small pinhole diameters d_{pin}, which are illuminated by high-intensity X-rays or neutrons of long wavelengths λ, and pinhole to object distance L, where

$$l_{coh} = \frac{L\lambda}{d_{pin}}. \tag{169}$$

Thermal and cold phase-contrast radiography was inspired by the pioneering work of Wilkins *et al.* (1996) in the development of X-ray phase-contrast radiography. Wilkins *et al.* described and demonstrated a simplified approach to X-ray phase-contrast imaging using a pointlike X-ray source (such as a microfocus S-ray tube or synchrotron-bending magnet source) with high spatial but almost no chromatic coherence

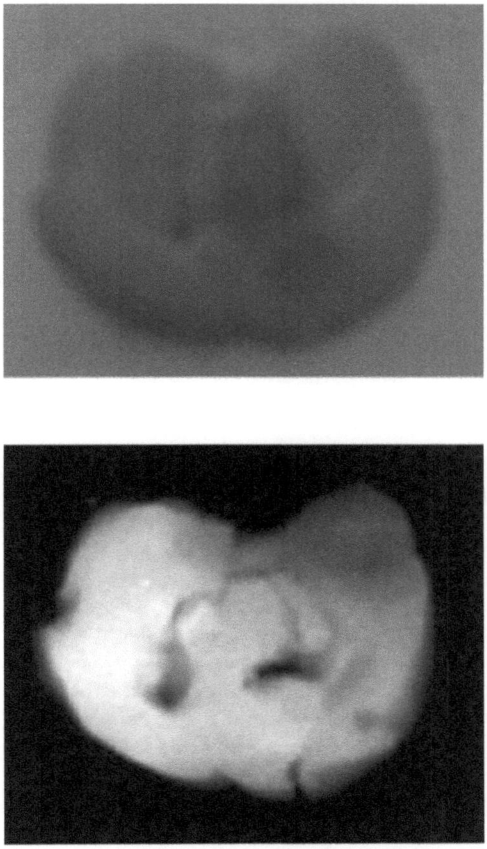

FIGURE 128 (Top) A photo of a 2-mm thick rat brain coronal slice with a brain cancer tumor (provided to Adelphi Technology by Prof. R. Barth of Ohio State University Medical School). The brain slice was stored in formalin and rinsed with Hanks' balanced salt solution before neutron imaging. (Bottom) A 3-hour contact thermal neutron radiograph of the 2-mm thick coronal brain slice with tumor. The tumor is in the right half of the coronal slice (Cremer *et al.*, 2007). (See Color Insert.)

(see also Gureyev and Wilkins, 1998a,b and Gureyev *et al.*, 2008). X-ray imaging based on phase contrast has greatly advanced via non-monochromatic synchrotron hard X-rays, which are progressing from microradiology and microtomography to the nanometer scale.

Improvements in synchrotron X-ray imaging with the goal of attaining high lateral resolutions down to 20 nm and time resolution down to 10 μs are of importance in the developing nanoscience and nanotechnology fields (see Weon *et al.*, 2006). In the meantime, Fezzaa and Wang (2008) have reported ultrafast, X-ray, phase-contrast imaging of the early

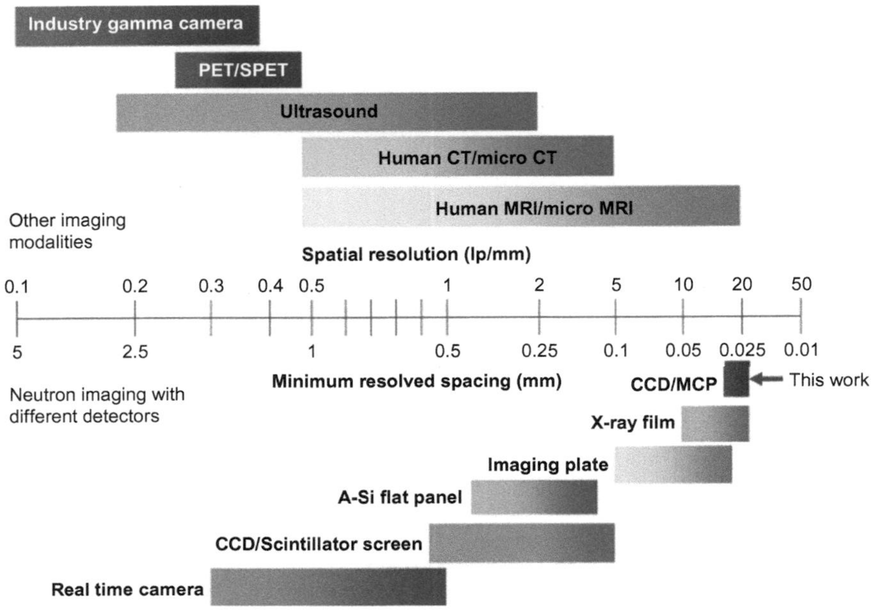

FIGURE 129 A graphical display of the resolution of major radiation imaging modalities for X-rays and gamma, ultrasound, and microwaves compared with imaging modalities for thermal neutrons. CT, computed tomography; MCP, microchannel plate; MRI, magnetic resonance imagine; PET, positron emission tomography; SPET, single photon emission computed tomography, more commonly denoted by SPECT (Cao, 2007). (See Color Insert.)

merging dynamics of two water drops in air, and Zoofan *et al.* (2006) demonstrated improvements in nondestructive evaluation of materials using a phase-contrast, microfocus X-ray system to image structural materials. Donnelley *et al.* (2011) used synchrotron, phase-contrast X-ray imaging to monitor surrogate fluid flow in live mouse air passages, attaining high spatial and temporal resolution.

X-ray phase-contrast radiography is advancing rapidly with the use of microfocus, polychromatic X-ray tubes and with the small monochromatic and polychromatic synchrotron X-ray sources. Microradiography was performed to obtain images in absorption-contrast and phase-contrast modes by changing the distance between the object and the detector (film). In the absorption-contrast mode, the object to detector distance was $R = 190\,\text{mm}$, and in the phase-contrast mode, the object to detector distance was a much longer $R = 500\,\text{mm}$.

Figure 131 shows a comparison of a conventional X-ray radiographic image (top left) and a phase-contrast radiographic image (top right). In both cases the same microfocus X-ray tube was set to 90 kV to image the bore hole in Al-2024 alloy in a 9-minute exposure with magnification

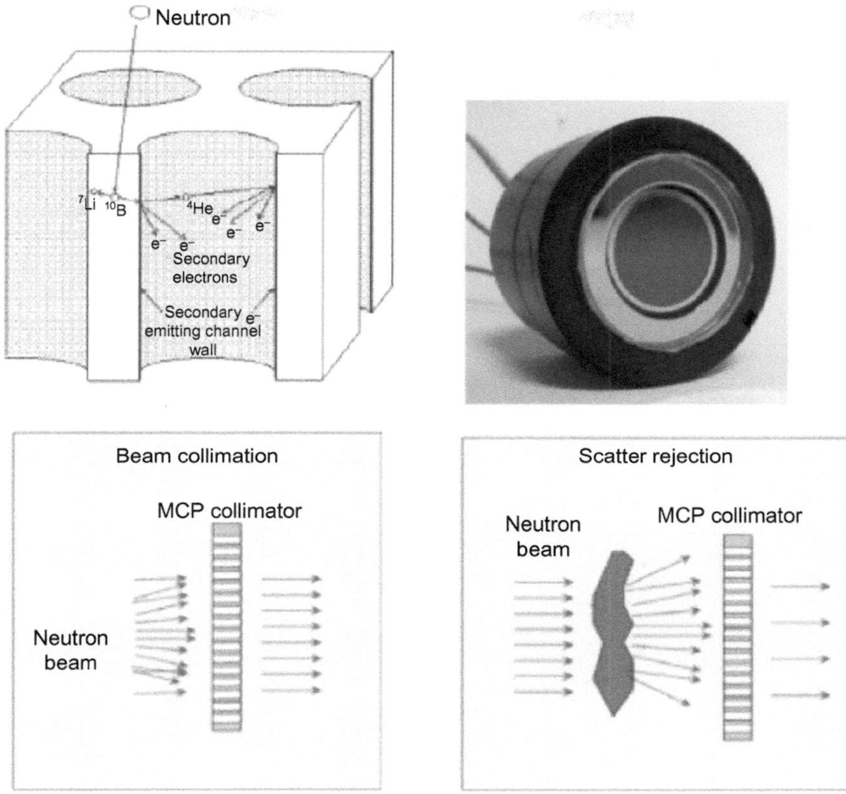

FIGURE 130 (Top left) Diagram of thermal neutron detection by a boron-loaded microchannel plate (from Cao, 2007). (Top right) Photo of an MCP neutron detector (from Cao, 2007). (bottom) The left diagram shows the use of an MCP as a collimator and thermal and cold neutron imaging detectors using a CCD camera, which images boron-loaded microchannel plates (from NOVA Scientific, Inc.). The right image shows scatter rejection of the neutron beam (from NOVA Scientific, website http://www.novascientific.com).

$M = 10$. The microfocus X-ray tube source was a Feinfocus 225-kV X-ray unit (Xylon International, Hamburg, Germany) with a 5-μm X-ray spot size. The X-ray detector was a high-resolution film or image intensifier integrated with a CCD camera for real-time microradiography and alignment of the sample before film microradiography. The ^{2024}Al sample contained a 340-μm diameter × 770-μm deep bore hole.

The first neutron phase-contrast imaging work was performed by Allman *et al.* (2000a). Their neutron phase-contrast imaging experiments followed from the successful X-ray phase-contrast imaging experiments reported by Wilkins *et al.* (1996). Allman *et al.* (2000a) performed their

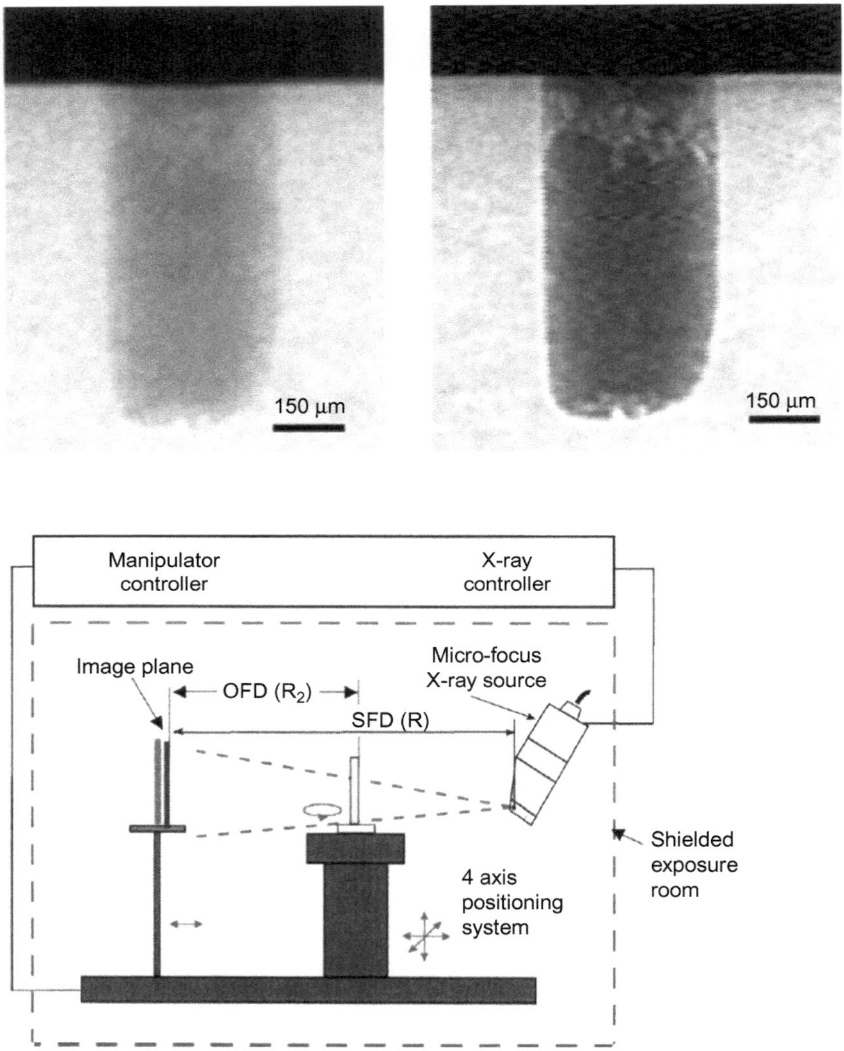

FIGURE 131 (Top left) Microfocus, X-ray absorption-contrast image of bore hole in a aluminum 2024 alloy with object-film (OFD) distance $R = 190$ mm. (Top right) X-ray phase-contrast image with object-detector distance $R = 500$ mm for ^{2024}Al. (Bottom) Schematic of experimental setup for in-line phase-contrast X-ray imaging. Microfocus X-ray tube voltage = 90 kV and X-ray projection magnification $M = 10$ (Zoofan *et al.*, 2006).

neutron phase-contrast imaging experiments on the National Institute for Standards and Technology [NIST] Center for Neutron Research NG-7 beam line with PG(002) crystal, which provided narrowband neutrons in the wavelength range $\lambda = 0.235$ nm to $\lambda = 0.475$ nm, and on the NG-0 beam line, which provided a broadband Maxwellian distribution centered at wavelength $\lambda = 0.432$ nm.

Figure 132 (top left) shows and compares conventional contact radiographic images (images a and c) with phase-contrast images of a lead sinker and a wasp (images b and c) obtained by Allman *et al.* (2000a). The top-middle and top-right) images show a reconstructed image of the lead sinker along a neutron beam direction and a plot of the measured (blue line) and computed (dotted line) phase variations of the along line AB of the lead sinker object (far-right top image). The line profile was obtained via Fourier processing of the data, and the predicted phase profile (dotted curve) was determined from the sample geometry and orientation. Their phase retrieval technique can be applied to a polychromatic beam and is not limited by low neutron fluxes.

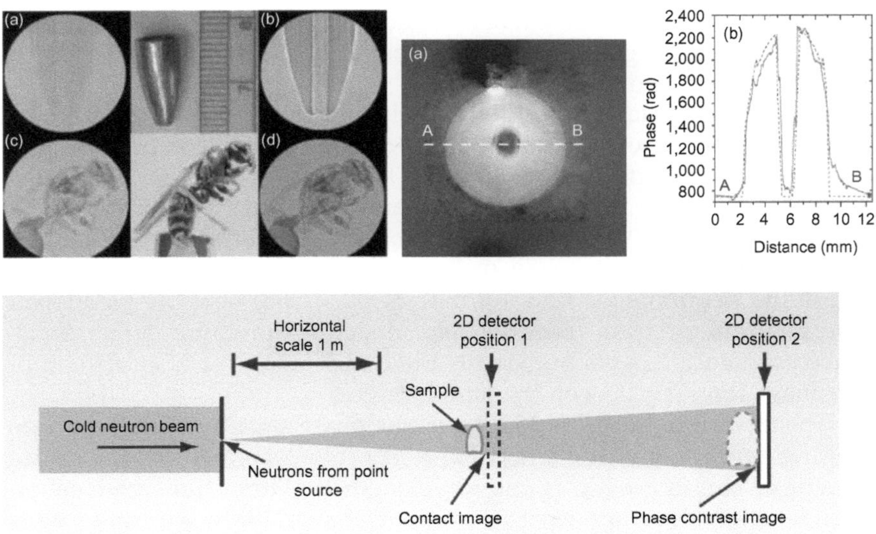

FIGURE 132 (Top left) Cold neutron images taken at the NIST NG-7 beam line with PG(002) crystal providing neutrons in the wavelength range $\lambda = 0.235$ nm to $\lambda = 0.475$ nm. The (a) contact and (b) phase-contrast images of a lead sinker are compared with the photo shown in the middle. The (c) contact and (d) phase-contrast images of a wasp are also compared; the photo is shown between the radiographic images (top middle and right). Also shown are the (a) reconstructed image of the lead sinker along the beam direction and (b) plot of the measured (blue line) and computed (dotted line) phase variations along line AB of the lead sinker object. The experimental layout shown at the bottom is for contact and phase-contrast radiography of neutrons at NIST. The diagram and images are from a 2000 NIST Research Highlights article reporting on the first phase-contrast imaging experiments with cold neutrons by Allman *et al.* (2000a) as reported in *Nature*.

The bottom of Figure 132 is a diagram of the setup used by Allman *et al.* (2000a). To perform conventional thermal neutron radiography the detector was in contact with the object, and in the phase-contrast radiography imaging the detector was moved downstream of the object. This phase-contrast image were taken with the CCD-based neutron imaging camera, which was placed 1.5 m to 1.8 m downstream of the samples. The CCD camera images were downloaded to a computer and the centroiding of the neutron events was determined by hardware processing, which achieved 60-μm as the best resolution.

Neutron and X-ray phase-contrast radiographic imaging together provide a very useful modality, which can be applied to the study of magnetic domain structures, interface boundaries, and density variations in multilayer thin films (see papers by Nugent *et al.*, 1996; Barty *et al.*, 1998; and Allman *et al.*, 2000a,b; and theoretical treatment by Teague, 1983; Gureyev *et al.*, 1995; and Paganin and Nugent, 1998). Phase-contrast imaging is based on Fresnel diffraction or imaging in the near field. Neutrons or X-rays originate from a small pinhole source, pass through a thin object, undergo phase shifting by the object, and are then imaged by a downstream detector (Figure 133).

Phase-contrast imaging is particularly effective for cold neutrons. As a result, high-flux, cold neutron, phase-contrast imaging experiments in Europe can be performed on the ANTARES beam line at FRM II in Munich, CONRAD at HMI in Berlin, and ICON at PSI in Switzerland (Kardjilov *et al.*, 2005). However, the cold moderator shifts the thermal neutron spectrum to longer wavelengths, which also leads to broadening of the neutron spectrum. Kardjilov *et al.* (2005) note the broadened cold neutron spectrum from the cold moderator produces larger transverse coherence lengths but entails the disadvantage of a smoothing of the phase-contrast effect on the sample edges.

The top image in Figure 133 shows the phase shift between the waves passing through points P1 and P2, which are separated by transverse coherence length $l = l_{coh}$, is one wavelength λ. Within transverse coherence length l, the wave front is coherent; that is, the wave front points within the transverse coherence length have fixed phase relationships with one another and can produce interference patterns. If the object were an opaque plate and points P1 and P2 were slits, then the two emergent waves would have the same phase, which would produce an interference pattern on the downstream detector. As shown in the bottom image of Figure 133, the object produces a phase shift in the incident wave, and the transverse coherence of the incident wave at the object allows the interfering, coherent waves at the detector to produce an amplitude-modulated image of the object.

The following is a derivation of the amplitude $\Psi(x)$ and intensity $I(x)$ of a phase-contrast image at the detector plane as a function of one

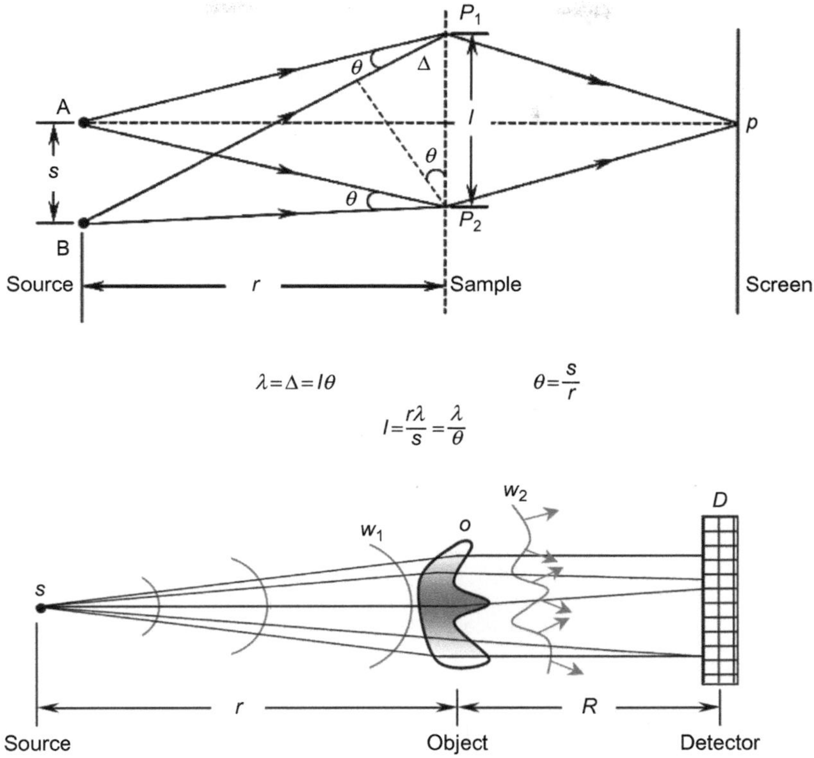

FIGURE 133 (Top) A diagram with the derivation of the transverse coherence length $l = r\lambda/s$. (Bottom) A diagram of phase-contrast imaging (Kardjilov, 2003).

transverse coordinate x. The same analysis is applied to obtain the amplitude $\Psi(y)$ and intensity $I(y)$ of a phase-contrast image at the detector plane as a function of the other transverse coordinate y. The z-axis (optical axis) extends from the center of a pinhole neutron or X-ray source of diameter d_{pin} at $z = 0$ to the object located in the object plane at axial distance $z = L$ (Figure 134).

The diagonal distance s extends from an off-axis point S in the source plane of the pinhole to an off-axis point O on the object. The object plane is modeled as a planar sheet with an aperture centered on the z-axis in which the planar sheet is opaque to X-rays and neutrons; a sheet of tungsten-gadolinium alloy is an example. The pinhole source point S is located at transverse distance x_s from center z-axis. For paraxial rays close to the center z-axis, the diagonal distance s from source point S to object point O is

$$s = \sqrt{(x_0 - x_s)^2 + L^2} \cong L + \frac{(x_0 - x_s)^2}{2L}. \qquad (170)$$

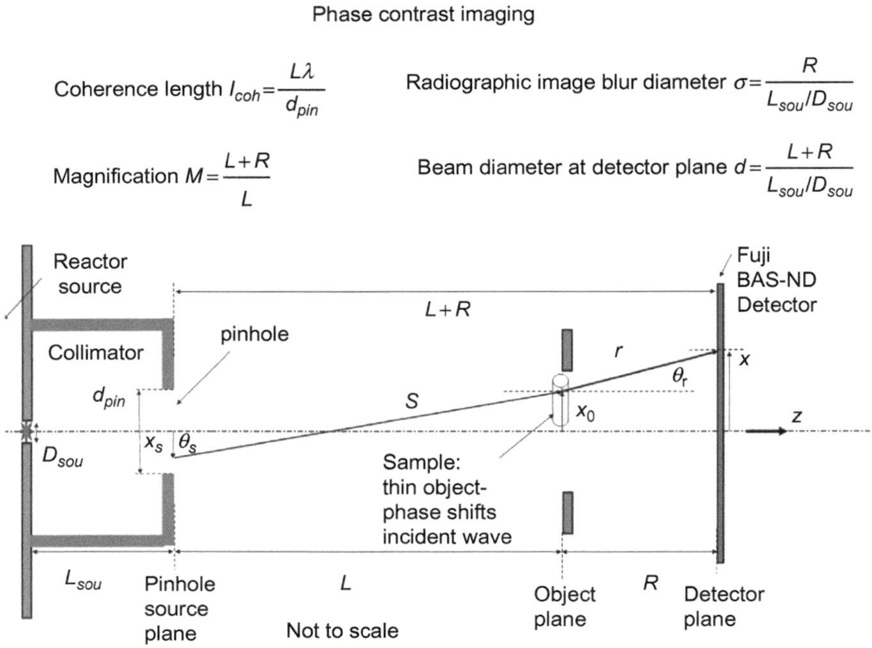

FIGURE 134 The layout for phase-contrast imaging for the derivations and phase-contrast imaging experiments of Kardjilov (2003) and Kardjilov *et al.* (2005).

The detector is centered on the z-axis at axial distance $z = L + R$ from the pinhole source plane, and the detector plane is at axial distance R from the object plane. The diagonal distance r is from off-axis object point O to an off-axis detector point P. A point O on the object at transverse distance x_0 from the center optical axis is at diagonal distance r from an off-axis detector point P at transverse distance x from the center z-axis. For paraxial rays, the diagonal distance from object point O to detector point P is

$$r = \sqrt{(x - x_0)^2 + R^2} \cong R + \frac{(x - x_0)^2}{2R}. \tag{171}$$

The detector wave amplitude for the 1D case is found via the Kirchhoff formula (see Kardjilov, 2003). The Kirchhoff equation expresses Huygen's equation and includes the direction cosine $\cos\theta_d$ of the ray from object point O to detector point D, which is at angle θ_d relative to the z-axis. The Helmholtz–Kirchhoff theorem yields the amplitude at detector point D given a source point S, which illuminates a tiny hole at point O on an opaque object, where for the 1D case

$$\Psi(x, z = L + R) = \frac{1}{4\pi} \int_{-\infty}^{\infty} (G\nabla\psi_0 - \psi_0\nabla G) \cdot \hat{\mathbf{z}} dx_0. \tag{172}$$

The above surface integral is on the object plane in which the spherical wave amplitude at the object point O is ψ_0, and the Green's function G is the image of a single object point O. In this 1D case, the above integral is a line integral over the transverse coordinate x_0 positions of the object point O. Here $\hat{\mathbf{z}}$ is the surface normal unit vector, which points outward in the negative z-direction from the object plane toward the source plane.

The Green's function G is the impulse response or image produced at the detector by an object point at transverse coordinate x_0. Vector $\mathbf{r}$ extends from object point O at transverse coordinate x_0 to detector point D at transverse coordinate x in which vector $\mathbf{r}$ is oriented at angle θ_r relative to the positive z-direction. If the source point O to detector point D distance is much larger than the neutron or X-ray wavelength λ, then

$$\hat{\mathbf{z}} \cdot \nabla G = \frac{\partial G}{\partial r} (\hat{\mathbf{z}} \cdot \hat{\mathbf{r}}) = \left(-i\frac{2\pi}{\lambda}G + \frac{1}{r}G\right) \cos\theta_r \cong -i\frac{2\pi}{\lambda} G \cos\theta_r \quad r \gg \lambda. \tag{173}$$

Relative to the positive z-axis, the unit vector $\hat{\mathbf{r}}$ points from object point O to detector point D at obtuse angle $\pi + \theta_r$, where

$$\hat{\mathbf{z}} \cdot \hat{\mathbf{r}} = \cos(\pi + \theta_r) = -\cos\theta_r. \tag{174}$$

The pinhole X-ray or neutron, spherical wave is emitted from a pinhole source point S at transverse source coordinate x_s, and is incident at object point O at transverse coordinate x_0. The spherical wave amplitude ψ_0 at the object point O is

$$\psi_0 = \frac{1}{s} \exp\left(i\frac{2\pi}{\lambda}s\right). \tag{175}$$

Vector $\mathbf{s}$ extends from to source point S at transverse coordinate x_s to object point O at transverse coordinate x_0, in which vector $\mathbf{s}$ is oriented at angle θ_s relative to the positive z-direction. If the source point S to object point O distance s is much larger than the neutron or X-ray wavelength λ, then

$$\hat{\mathbf{z}} \cdot \nabla\psi_0 = \frac{\partial\psi_0}{\partial s} (\hat{\mathbf{z}} \cdot \mathbf{s}) = \left(i\frac{2\pi}{\lambda}\psi_0 - \frac{1}{s}\psi_0\right) \cos\theta_s \cong i\frac{2\pi}{\lambda}\psi_0 \cos\theta_s. \quad s \gg \lambda \tag{176}$$

Relative to the positive z-axis the unit vector $\hat{\mathbf{s}}$ points from object point O to source point S at acute angle θ_s, where

$$\hat{\mathbf{z}} \cdot \hat{\mathbf{s}} = \cos\theta_s. \tag{177}$$

Substitution of Eqs. 173 and 176 into Eq. 172 yields the Kirchhoff equation for the wave amplitude $\Psi(x, L+R)$ at transverse position x in the detector plane at $z = L + R$, where

$$\Psi(x, L+R) = \frac{i}{2\lambda}\int_{-\infty}^{\infty} (G\psi_0 \cos\theta_s + \psi_0 G \cos\theta_r)\, dx_0. \tag{178}$$

At sufficient large distance s, the spherical wave from point S can be approximated as a plane at points of incidence O at the object plane, where $\theta_s \cong 0$ and at sufficient large distance r, the spherical wave emitted from object point O, which arrives at detector point D, can be approximated as a plane wave, where $\theta_r \cong 0$. A z-directed plane wave of unit amplitude $\Psi_0(x_0) = 1$ is assumed incident on the object, so the Kirchhoff equation becomes

$$\Psi(x, L+R) = \frac{i}{\lambda}\int_{-\infty}^{\infty} G dx_0 = \frac{i}{\lambda}\int_{-\infty}^{\infty} \frac{1}{r}\exp\left(i\frac{2\pi}{\lambda}r\right) dx_0. \tag{179}$$

Now consider the case of an object of material substance, which occupies the aperture in the opaque object plane. In this case, the object in the aperture of the object plane imposes a phase shift $\phi(x_0)$ on the incident wave amplitude ψ_0 as a function of the transverse distance x_0 from the center z-axis. That is, the object varies in thickness $t_z(x_0)$ along the z-direction as a function of transverse x-coordinate x_0 and induces phase shift $\phi(x_0)$ and amplitude change $T(x_0)$, where

$$T(x_0) = \exp[i\varphi(x_0)] = \int_{-\infty}^{\infty} \mathbf{\Phi}(u) \exp(i2\pi \cdot u x_0)\, du. \tag{180}$$

The Kirchhoff equation for a phase-shifting object in the aperture of the object plane becomes

$$\Psi(x, L+R) = \frac{i}{\lambda}\int_{-\infty}^{\infty} T(x_0)\, G dx_0 = \frac{i}{\lambda}\int_{-\infty}^{\infty} \frac{1}{r}\exp\left[i\left(\frac{2\pi}{\lambda}r(x_0) + \varphi(x_0)\right)\right] dx_0. \tag{181}$$

Substitution for r of Eq. 171 and $\phi(x_0)$ expression of Eq. 180 into Eq. 181 gives

$$\Psi(x) \cong \frac{i}{R\lambda}\exp\left(i\frac{2\pi}{\lambda}R\right)\int_{-\infty}^{\infty}\exp\left[i\frac{\pi}{R\lambda}(x-x_0)^2\right] \times \int_{-\infty}^{\infty}\Phi(u)\exp(i2\pi\cdot ux_0)\,dudx_0, \tag{182}$$

where

$$\frac{1}{r} \cong \frac{1}{R}. \tag{183}$$

The above integral is a convolution, where

$$\Psi(x) \cong \frac{i}{R\lambda}\exp\left(i\frac{2\pi}{\lambda}R\right)\int_{-\infty}^{\infty}\int_{-\infty}^{\infty}\Phi(u)f(x-x_0)\,g(x_0,u)\,dudx_0 \tag{184}$$

and

$$f(x-x_0) = \exp\left[i\frac{\pi}{R\lambda}(x-x_0)^2\right] \tag{185}$$

$$g(x_0,u) = \exp(i2\pi\cdot ux_0). \tag{186}$$

By definition, convolution allows swapping of $f(x_0)$ and $g(x_0,u)$, where

$$h(x) = \int_{-\infty}^{\infty} f(x-x_0)g(x_0,u)\,dx_0 = \int_{-\infty}^{\infty} f(x_0)g(x-x_0,u)dx_0. \tag{187}$$

Swapping the imaginary arguments of the two integrand exponentials yields

$$\Psi(x) \cong \frac{i}{R\lambda}\exp\left(i\frac{2\pi}{\lambda}R\right)\int_{-\infty}^{\infty}\int_{-\infty}^{\infty}\Phi(u)\exp\left(i\frac{\pi}{R\lambda}x_0^2\right) \exp[i2\pi\cdot u(x-x_0)]\,dudx_0, \tag{188}$$

where

$$\exp[i2\pi\cdot u(x-x_0)] = \exp(i2\pi\cdot ux)\exp(i2\pi\cdot ux_0) \tag{189}$$

and

$$\exp(i2\pi \cdot ux_0) = \cos(2\pi \cdot ux_0) + i\sin(2\pi \cdot ux_0)\,. \tag{190}$$

With substitution of Eqs. 189 and 190 into Eq. 188 the Kirchhoff integral becomes

$$\begin{aligned}\Psi(x) \cong \frac{i}{R\lambda}\exp\left(i\frac{2\pi}{\lambda}R\right)\int_{-\infty}^{\infty}\int_{-\infty}^{\infty}\Phi(u)\exp\left(i\frac{\pi}{R\lambda}x_0^2\right)\exp(-i2\pi \cdot ux)\\ \times\left[\cos\left(2\pi \cdot ux_0\right) + i\sin\left(2\pi \cdot ux_0\right)\right]du dx_0.\end{aligned} \tag{191}$$

The integral rule below can be applied to the above integral, where

$$\int_{-\infty}^{\infty}\exp\left(-a^2x_0^2\right)\cos\left(bx_0\right)dx_0 = \frac{\sqrt{\pi}}{a}\exp\left(-\frac{b^2}{4a^2}\right) \tag{192}$$

and

$$\int_{-\infty}^{\infty}\exp\left(-a^2x_0^2\right)\sin\left(bx_0\right)dx_0 = 0, \tag{193}$$

with

$$a^2 = -i\frac{\pi}{R\lambda} \quad b = 2\pi \cdot u. \tag{194}$$

The nonzero part of the Kirchhoff integral is then

$$\Psi(x) \cong \frac{i}{R\lambda}\exp\left(i\frac{2\pi}{\lambda}R\right)\int_{-\infty}^{\infty}\Phi(u)\exp\left(i\pi R\lambda u^2\right)\exp(i2\pi \cdot ux)\,du. \tag{195}$$

If the product of the distance R between the object and detector multiplied by the wavelength λ is much less than 1, where $R\lambda \ll 1$, then

$$\Psi(x) \cong \frac{i}{R\lambda}\exp\left(i\frac{2\pi}{\lambda}R\right)\int_{-\infty}^{\infty}\Phi(u)\left(1 + i\pi R\lambda u^2\right)\exp(i2\pi \cdot ux)\,du, \tag{196}$$

where

$$\frac{d^2}{dx^2}\exp(i2\pi \cdot ux) = -4\pi^2u^2\exp(i2\pi \cdot ux). \tag{197}$$

Application of the above derivative relation (used with Fourier transforms), the Kirchhoff integral for the amplitude at the detector $\Psi(x)$ becomes

$$\Psi(x) \cong \exp\left(i\frac{2\pi}{\lambda}R\right)\frac{i}{R\lambda}\left(1 - \frac{iR\lambda}{4\pi}\frac{d^2}{dx^2}\right)\int_{-\infty}^{\infty}\Phi(u)\exp(i2\pi\cdot ux)\,du. \quad (198)$$

By definition the

$$\exp[i\varphi(x)] = \int_{-\infty}^{\infty}\Phi(u)\exp(i2\pi\cdot ux)\,du. \quad (199)$$

Hence,

$$\Psi(x) \cong \exp\left(i\frac{2\pi}{\lambda}R\right)\left[\frac{i}{R\lambda}\exp[i\varphi(x)] + \frac{1}{4\pi}\frac{d^2}{dx^2}\left(\exp[i\varphi(x)]\right)\right], \quad (200)$$

where

$$\frac{d^2}{dx^2}\left(\exp[i\varphi(x)]\right) = \frac{d}{dx}\left(i\frac{d\varphi}{dx}\exp[i\varphi(x)]\right) = \left(i\frac{d^2\varphi}{dx^2} - \left(\frac{d\varphi}{dx}\right)^2\right)\exp[i\varphi(x)]. \quad (201)$$

Substitution into Eq. 200 for the derivative term of Eq. 201 yields the amplitude $\Psi(x)$ at transverse position x in the detector plane at $z = L + R$, where

$$\Psi(x) \cong \left[\left(\frac{1}{R\lambda} + \frac{1}{4\pi}\frac{d^2\varphi}{dx^2}\right)i - \frac{1}{4\pi}\left(\frac{d\varphi}{dx}\right)^2\right]\exp\left[i\left(\frac{2\pi}{\lambda}R + \varphi(x)\right)\right]. \quad (202)$$

The neutron or X-ray intensity $I(x)$ of the phase-contrast image at transverse position x in the detector plane $z = R + L$ is

$$I(x) = |\Psi(x)|^2 \cong \left(\frac{1}{R\lambda} + \frac{1}{4\pi}\frac{d^2\varphi}{dx^2}\right)^2 + \frac{1}{16\pi^2}\left(\frac{d\varphi}{dx}\right)^4. \quad (203)$$

Assume the reciprocal $1/R\lambda$ of the product of wavelength λ multiplied by the object plane to detector plane distance R is much larger than the object-induced spatial acceleration $d^2\phi/dx^2$ of the phase angle change of the wavefront along transverse direction x, where

$$\frac{1}{R\lambda} \gg \frac{1}{4\pi}\frac{d^2\varphi}{dx^2}. \quad (204)$$

Then one can apply the binomial approximation, where

$$I(x) \cong \frac{1}{R\lambda}\left(1+\frac{R\lambda}{2\pi}\frac{d^2\varphi}{dx^2}\right)+\frac{1}{16\pi^2}\left(\frac{d\varphi}{dx}\right)^4. \tag{205}$$

Then assume

$$\frac{1}{R\lambda} \gg \frac{1}{16\pi^2}\left(\frac{d\varphi}{dx}\right)^4. \tag{206}$$

Hence, the neutron or X-ray intensity $I(x)$ of the phase-contrast image at transverse position x in the detector plane $z = R + L$ is approximately

$$I(x) \cong \frac{1}{R\lambda}\left(1+\frac{R\lambda}{2\pi}\frac{d^2\varphi}{dx^2}\right). \tag{207}$$

For a fixed wavelength λ and object to detector plane distance R the constant multiplier $1/R\lambda$ can be dropped to obtain the effective variation of phase-contrast radiographic intensity $I(x)$ in the detector plane of an imaged object is

$$I(x) \cong 1+\frac{R\lambda}{2\pi}\frac{d^2\varphi}{dx^2}. \tag{208}$$

The phase shift $\phi(x)$ of the detected neutron or X-ray wave at the detector plane $z = L + R$ is equal to the phase shift of the wave by object just after the object plane $z = L + \varepsilon$, where $\varepsilon \ll L$ and

$$\frac{d^2\varphi}{dx^2}(x, z = L + R) = \frac{d^2\varphi}{dx^2}(x_0, z = L). \tag{209}$$

The transverse coordinate at the detector plane x is related to the transverse coordinate at the object plane x_0 via transverse magnification M, where as shown in Figure 132,

$$M = \frac{x}{x_0} = \frac{L+R}{L}. \tag{210}$$

The image blur diameter σ at the detector plane from an object at distance R from detector plane, given incident neutron or X-ray beam divergence angle θ at the object point is

$$\sigma = R\theta = R\frac{D_{sou}}{L_{sou}} = \frac{R}{(L_{sou}/D_{sou})}. \tag{211}$$

Here the source beam collimation from the reactor source is L_{sou}/D_{sou}, which illuminates the pinhole, is the reciprocal of the beam divergence angle θ. Given the pinhole diameter d_{pin} the intrinsic beam collimation L_{sou}/D_{sou} before the pinhole aperture and the distance $L+R$ between the pinhole and detector planes yields the pinhole image diameter d in the detector plane,

$$d = (L+R)\,\theta = \frac{L+R}{L_{sou}/D_{sou}}. \tag{212}$$

The phase shift $\phi(x)$ received by the neutron or X-ray wave front, which passes through the object is obtained from the refractive decrement $\delta\,(x_0, z_0)$, where given the nuclei number density $\rho_j(x_0, z_0)$ and coherent scatter length b_j^{coh} for each jth isotope of N isotopes, which comprises the sample object, the object refractive index is

$$n(x_0, z_0) = 1 - \delta(x_0, z_0) = 1 - \frac{\lambda}{2\pi}\sum_{j=1}^{N} b_j^{coh}\rho_j\,(x_0, z_0). \tag{213}$$

The phase shift $\phi(x_0)$ induced by the object due to object thickness $t(x_0)$ and refractive index $\delta(x_0, z_0)$ is a function of the axial z_0 and transverse x_0 object position coordinates within the object, where

$$\varphi(x_0) = -\frac{2\pi}{\lambda}\int_0^{t(x_0)} \delta(x_0, z_0)dz_0. \tag{214}$$

The negative sign appears because the phase shift without object with refractive index $n = 1$ is compared with phase shift with object and refractive index $n = 1 - \delta$.

We now present some of the experimental results of neutron phase-contrast imaging obtained by Kardjilov (2003). Also see the phase-contrast imaging results by Kardjilov *et al.* (2005). The conventional and phase-contrast imaging experiments were performed at the thermal neutron radiography station NEUTRA at the Paul Scherrer Institute in Switzerland. The imaging detector was a Fuji IP-ND with 100-μm best spatial resolution. The left image in Figure 135 shows a 90-minute exposure obtained by conventional thermal neutron radiography of a cogwheel; the right image is a 90-minute phase-contrast radiograph taken with $R = 120$ cm object to detector distance, $L = 7.0$ m pinhole source to object distance, and pinhole of diameter $d_{pin} = 0.5$ mm.

The combination of the phase-contrast effect and the magnification $M = (L+R)/L = 1.2$ provides a better resolution with phase-contrast

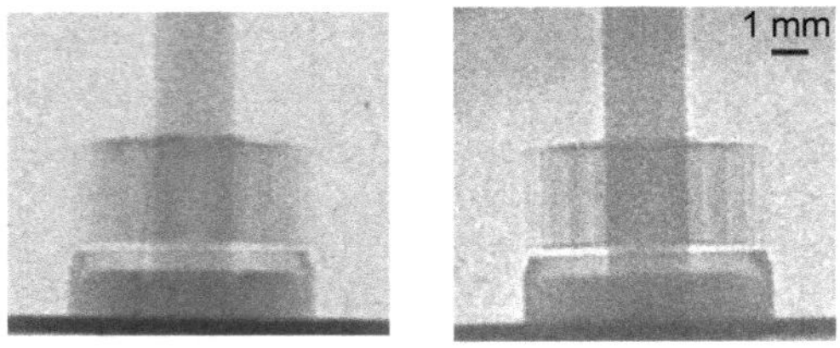

FIGURE 135 (Left) A 90-minute conventional thermal radiography of a cogwheel. (Right) A 90-minute phase-contrast radiography for $R = 120$ cm object to detector distance and $L = 7.0$ m pinhole source to object distance. This phase-contrast image was produced with a pinhole of diameter $d_{pin} = 0.5$ mm (Kardjilov, 2003).

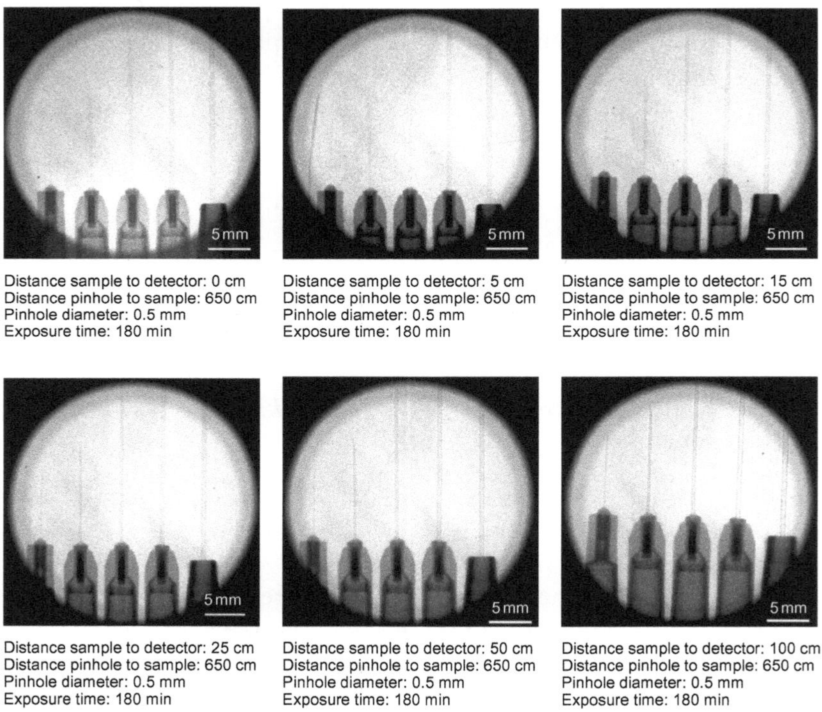

FIGURE 136 Shown are 180 minute exposure phase contrast images of a set of stainless steel capillaries (syringe needles) with varying diameters 0.3, 0.5, 0.6, 0.8 and 0.9 mm at different needle to detector distances R given fixed pinhole to object needle distance $L = 650cm$ and pinhole diameter $d_{pin} = 0.5mm$. (Kardjilov, 2003)

radiographic imaging than conventional radiography. The superposition of the secondary waves, which are emitted from coherent scattering centers in the sample, produces a magnified phase-contrast image that overwrites the radiographic blur of the object due to the neutron beam divergence angle $\theta = D_{sou}/L_{sou}$.

Figure 136 shows phase-contrast images by Kardjilov (2003) of hollow, stainless steel syringes at varying object to detector distances of $R = 0\,\text{cm}$ (top left), $R = 5\,\text{cm}$ (top middle), $R = 15\,\text{cm}$ (top right), $R = 25\,\text{cm}$ (bottom left), $R = 50\,\text{cm}$ (bottom middle), and $R = 100\,\text{cm}$ (bottom right). These phase-contrast neutron images were obtained with a fixed pinhole to object needle distance $L = 650\,\text{cm}$, a fixed pinhole diameter $d_{pin} = 0.5\,\text{mm}$, and a fixed 180-minute exposure time.

The best resolved image is the bottom-right image with syringe to detector distance $R = 100\,\text{cm}$, which was due to the limited pinhole to detector distance of $L + R = 750\,\text{cm}$. Even the smallest syringe channel is resolved in which the phase-contrast effect and minor magnification $M = 1.15$ overcome the radiographic blurring of a needle 1 m from the detector. Phase-contrast radiography and tomography are thus exciting new prospects in thermal neutron imaging and applications such as nondestructive testing and analysis.

5. EXPERIMENTS IN FAST NEUTRON RADIOGRAPHY

Radiography has become an important part of medicine since the advent of X-rays in the early 1900s, and has expanded from X-rays to gammas, thermal neutrons, fast neutrons (FNs), and energetic ion beams. X-ray and gamma radiography is an important part of medical technology. An important advance occurred in 1964 with the Anger camera, which was designed for imaging gammas from radioisotopes and the 0.511-MeV gammas from positron-electron annihilation from positron sources (see Anger, 1964). The Anger scintillation camera obtained images of the distribution of gamma-ray– and positron-emitting isotopes *in vivo* in which no scanning is used because the camera's field of view is sufficient.

To obtain a gamma image of an object (or neutron image) a collimator first projects a gamma (neutron) image of the object onto a scintillator, which can be a plastic scintillator plate. However, in the original Anger camera (Anger, 1964), the scintillator was a single sodium iodide crystal with an 11.5-inch diameter and 0.55-inch thickness, which was coupled by an optical light guide to a close-packed hexagonal array of 19 multiplier phototubes (Figure 137). The phototubes viewed overlapping areas on the scintillator, and the light emission from each point on the scintillator was divided among the 19 tubes. The combination of scintillator, light guide, and phototubes formed the gamma imaging detector, which can apply to FNs as well. The phototubes were connected to an analog computer that identified the x- and y-coordinates and brightness of emitted scintillation light from the scintillator.

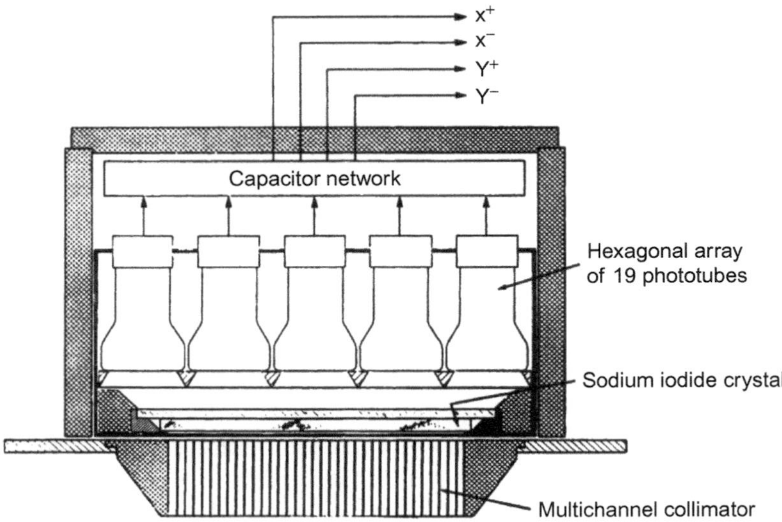

FIGURE 137 Anger scintillation camera image detector with multichannel collimator (Anger, 1964).

For large gamma- or neutron-emitting objects, collimators with large numbers of parallel holes provide the best combination of sensitivity and resolution. Parallel-channel collimators provide a one-to-one size relationship between the object and the scintillator image. Also, the object size is independent of the distance between the object and the collimator. The sharpest image resolution is obtained in the object parts lying closest to the collimator. The multichannel collimators of the Anger camera can be used in many combinations of collimator hole diameter, length, and septal thickness depending on the object, gamma or neutron energy, and imaging requirements. The Anger camera collimators are most efficient when used with low-energy gamma and FNs because the septa can be thinner and more holes can be packed into a given area (Figure 138).

The resolution-distance R in Figure 138 is derived from geometric considerations, where

$$R = d \cdot \frac{a + b + c}{a}. \tag{215}$$

Here distance a is the height of the collimator septa, b is the distance between the object and collimator, c is the distance between the collimator exit and the center of the sodium iodide scintillator, d is the empty space between each pair of adjoining septa through which the gammas or fast

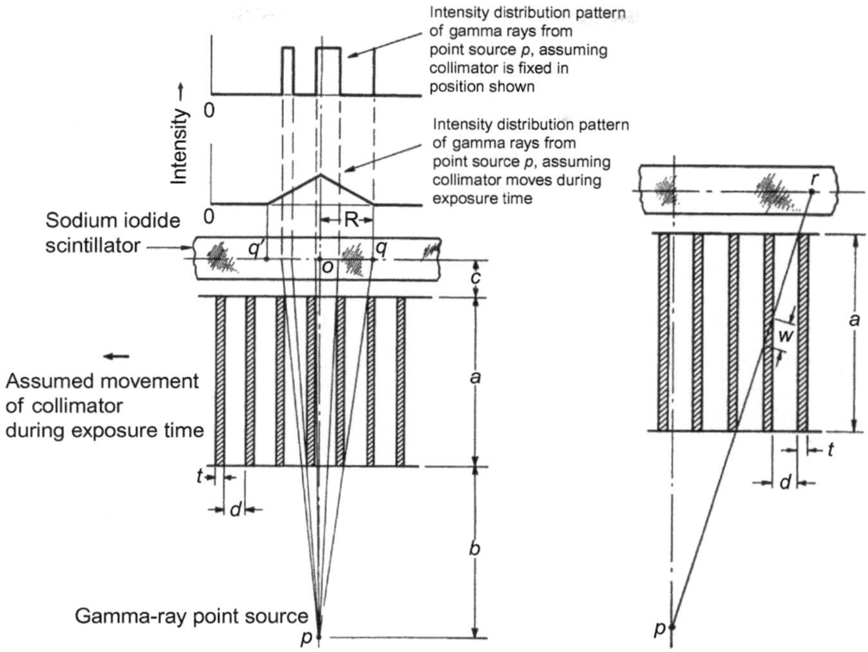

FIGURE 138 (Left) A section view of a parallel-channel collimator in the Anger camera, which shows the paths of the gammas and irradiated areas of the scintillator. (Right) The path of minimum attenuation for gamma rays penetrating collimator septa in the Anger camera (Anger, 1964).

or thermal neutrons can pass, and t is the wall thickness of the individual septa. Small distances b and c yield the sharpest image, which occurs when the resolution distance R is small.

The geometric efficiency g of the collimator is defined as the number of gamma rays (or FN rays) that pass through the channel divided by the number of gamma rays (or FN rays) emitted from the object (source of gamma or neutron rays). Efficiency g does not include the gamma rays that are scattered or pass through the septa of the collimator, and the efficiency g is independent of the distance b between the object and collimator (and experimentally verified), where

$$g = \left[\frac{Kd^2}{a\left(d+t\right)} \right]^2. \tag{216}$$

The dimensionless constant K is determined by the shape of the collimator holes formed by septa boundaries in which for square holes and collimator arrays $K = 0.282$, and for circular holes and collimator arrays $K = 0.238$.

Finally, referring to the diagrams in Figure 138, the shortest travel distance w through a collimator septa along an unwanted gamma or neutron ray path is

$$w = \frac{at}{2d + t}, \tag{217}$$

or

$$t = \frac{2dw}{a - w}. \tag{218}$$

As a demonstration of the overall camera, Anger (1964) placed 5 point sources of 80-keV gamma-emitting barium-133 radioisotope at a distance of 1.0 inch (2.54 cm) in front of the "A" collimator. The sources were placed on the axis of certain holes (as shown in the bottom diagram of Figure 139). The enlarged central portion of the image (top of Figure 139) shows the attained resolution. The gamma rays, which traveled through adjacent holes, produced adjacent white areas in the image. However, the gamma rays that traveled through holes separated by a blank hole in the collimator (no gamma source on blank hole axis) produced clearly separated spots.

Anger logic has also been used with a position-sensitive photomultiplier tube (PSPMT) and attached scintillator plate, which can perform gamma imaging (see Siegel *et al.*, 1996). Anger logic with a PSPMT can be applied to FN imaging, as shown later in the section. These Anger logic–based gammas detectors with PSPMTs can be converted to FN imaging detectors using a plastic scintillator sheet, fiber array, or a capillary array filled with liquid scintillator in place of the gamma scintillator.

The first PSPMT, the Hamamatsu R2486 (Hamamatsu Photonics, K.K., Hamamatsu City, Japan), was developed in 1985 and represented a technological advance for gamma-ray imaging (see Pania *et al.*, 2003). The later-generation Hamamatsu H8500 Flat Panel PMT is a position sensitive PMT and has an 8×8 array of anode pixels, each 5.8 mm $\times$ 5.8 mm square with a 6.08-mm pitch (spacing) between pixel centers, which are contained within a 49-mm $\times$ 49-mm imaging area contained within a 52-mm $\times$ 55-mm $\times$ 27.4-mm enclosure (see Kyushima *et al.*, 2000 and Hamamatsu, 2011).

Anger logic with a position-sensitive PSPMT was applied to gamma imaging by Popov and colleagues (see Popov and Majewski (2006), Popov *et al.*, 2006b,a, 2001; Popov, 2004). The patent by Popov presented a solution for the readout electronics of a PSPMT couple to the light-emitting pixels of a scintillator. Popov's patent shows a resistor matrix network that provides decoupling of all the 256 anode outputs of the 16×16 PSPMT array into a set of 16 analog outputs for the pixel x-coordinates and 16 analog outputs for the pixel y-coordinates, and includes an anode

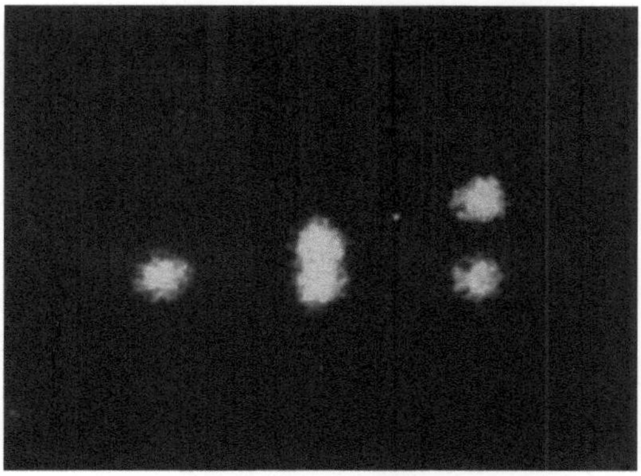

(a)

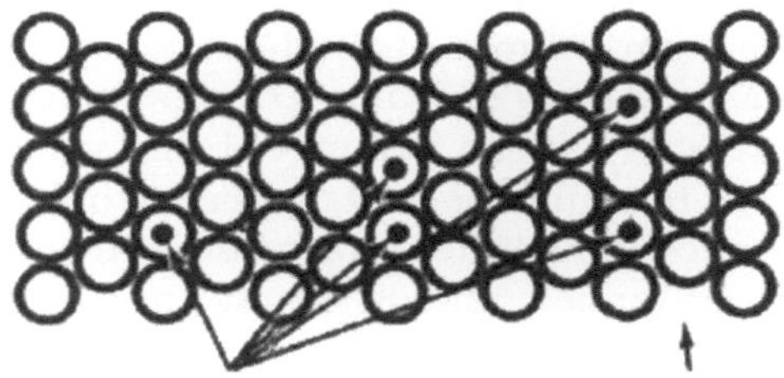

(b)

FIGURE 139 (a) An enlarged portion of an image of five point sources of 80-keV gamma-emitting, Barium-133 radioisotope, which was resolved by multichannel collimator that was set 1 inch from the collimator. (b) The sources are aligned along the axes of the holes (Anger, 1964).

gain nonuniformity correction circuit. Although the output pads have a 2.5-mm center to center spacing, the center of gravity interpolation determines the x- and y-positions and provides a resolution that is a fraction of a millimeter.

The main components of the PSPMT are input window, photocathode, focusing electrodes, dynodes, and an anode. By the photoelectric effect the photocathode converts incoming light into electrons, which are directed by the focusing electrodes toward the dynode stages. A column of dynode electrodes amplify the electrons via secondary electron emission, which provides $10^3 - 10^8$ electron gain depending on the number of dynodes and the interdynode voltage. The amplified electron current terminates on the anode.

The output currents from the 64 anodes of the 8×8 PSPMT array are input into a resistor matrix, which is connected directly to an array of low-impedance amplifiers. Output signals from each of the 64 resistors are sent along x- and y-signal paths that terminate at an end of each of the 8 x columns and 8 y rows. The 8 x signals and 8 y signals are then further processed to determine the location/coordinate information in which the center of gravity interpolation determines the x to y separation. The resulting submillimeter resolution is significantly better than the 2.5-mm spacing of the 8×8 PSPMT array of anodes. The matrix output readout system in the Popov (2004) patent can be applied to larger arrays, such as a 16×16, 32×32, or even 64×64, which requires 128 amplifiers and 64 x and 64 y outputs.

Popov *et al.* (2001) created a larger area gamma detector by using a 4 PSPMT (Hamamatsu R7600 C8) set into a 2×2 array. Each of the 4 Hamamatsu R7600 C8 PSPMTs was a 4×4 array of pixels. Each PSPMT had a 22 mm × 22-mm active area, which was divided among the 16 square pixels with a 4 mm × 4-mm area and 4.86-mm pitch. The 4 PSPMTs were coupled to a 27×27 array of 3 mm × 3 mm × 3-mm cubic crystals of CsI(Tl) to form a detector with an 81 mm × 81-mm scintillator area composed of the 729 CsI(Tl) scintillator crystals. The top of Figure 140 shows a diagram of a conventional charge division readout circuit using a resistive chain and the equation for determining the x-coordinate based on the end output voltages A and B,

$$X = \frac{A - B}{A + B}. \tag{219}$$

The simplified circuit diagram of the analog signal converter for the one coordinate (x- or y-) of the PSPMT output is shown at the bottom left of Figure 140. Figure 140 (bottom middle) shows the block diagram of the connection of a single 4×4 anode Hamamatsu PSPMT to the analog converter and the four-input data acquisition module. The bottom right of Figure 140 shows the image portion produced by 511-keV irradiation of the detector by a Na-122 radioisotope source (positron annihilation).

Popov *et al.* (2011) developed and tested a new type detector for use in the FN imaging radiography applications using the Adelphi Technology DD109 and DD108 neutron generators (http://www.adelphitech.com/products/dd109.html), which produce 2.5-MeV FNs. Their device used a neutron-sensitive plastic scintillator attached to a PSPMT. The PSPMT was operated in an event-by-event readout mode, which allowed acquisition of energy, timing, and pulse shape information for all detected radiation events. The information was used to aid in separation of FN interactions in the scintillator from the background events, which were caused by the electronics noise and other types of background radiation. The selection of pure FN events in the final image can improve image contrast and

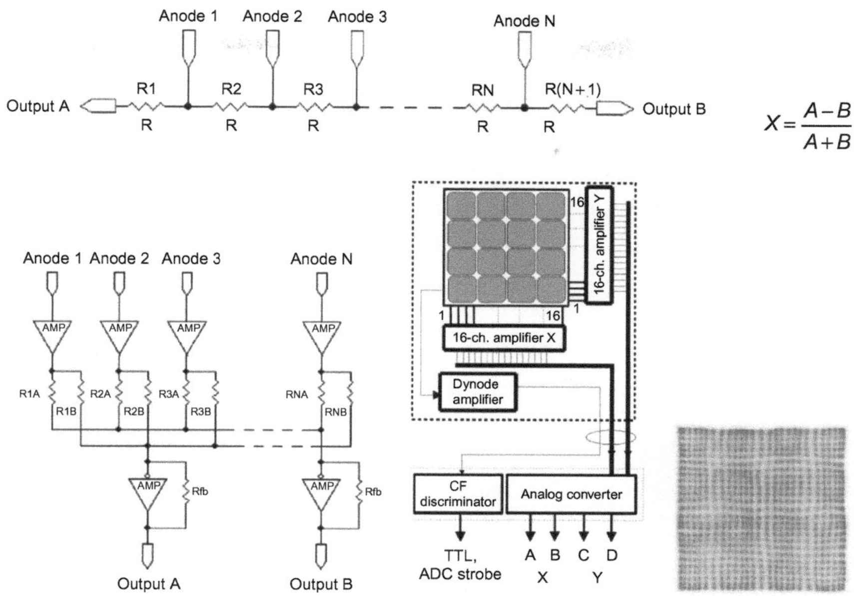

FIGURE 140 (Top) A conventional charge division readout circuit using a resistive chain and the equation for determining the *x*-coordinate based on the end output voltages A and B. (Bottom left) A simplified diagram of the analog signal converter for the one coordinate (*x* or *y*) of the PSPMT output. (Bottom middle) The diagram shows the connection of a single 4 × 4 anode Hamamatsu PSPMT to the analog converter and the four-input data acquisition module. (Bottom right) A small portion of detector, which was illuminated with 511-keV gammas from a Na-22 positron-emitting radioisotope (Popov *et al.*, 2001).

resolution compared with FN imaging detectors, which operate in an integration mode, thereby acquiring dark noise and background radiation without recourse to selection of input radiation.

Popov *et al.* (2011) tested the performance of their FN imaging detector at Adelphi Technology using the Adelphi DD108 neutron generator, which was operated in the continuous mode with 10^8-n/s yield of 2.5-MeV DD neutrons. Test phantoms were made of metal, plastic, wood, and other materials in a variety of shapes and were used to measure the detector resolution, efficiency, and uniformity. In place of the gamma- and X-ray, Anger logic imaging camera of Popov *et al.* (2011), they replaced the gamma scintillator by a 5-mm thick plastic scintillator sheet (Bicron BC-408 from Saint-Gobain Crystals of Hiram, Ohio, USA. http://www.detectors.saint-gobain.com/Plastic-Scintillator.aspx) to allow the same camera to perform FN imaging.

The top of Figure 141 shows photos of the Adelphi Technology DD108 neutron generator and the FN detector and phantoms fabricated by Popov

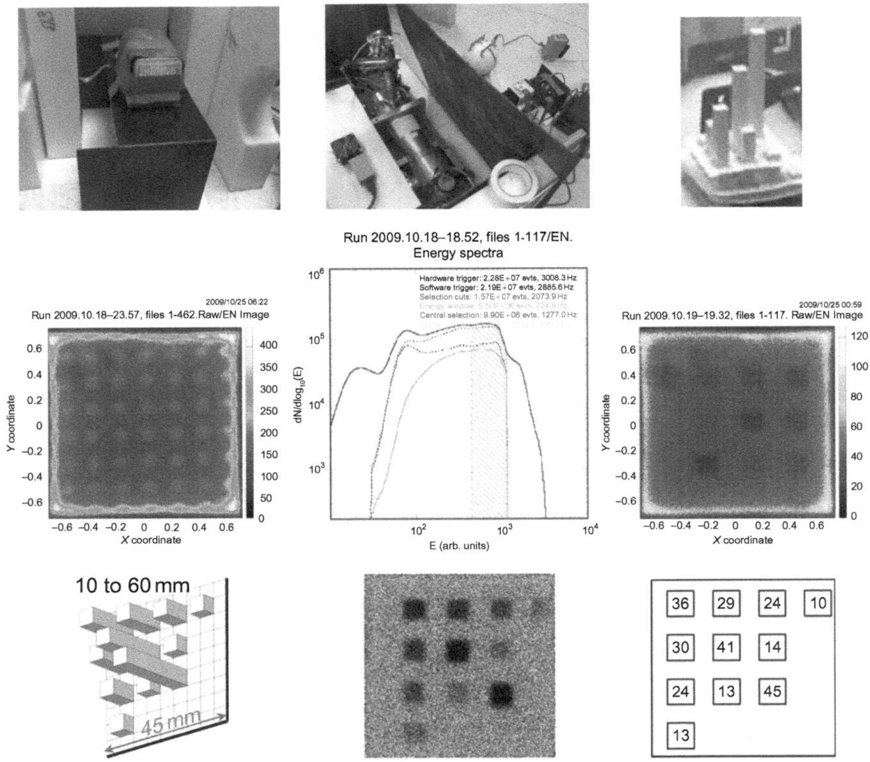

FIGURE 141 (Top) Photos are of Adelphi DD108 generator (10^8 n/s yield) with fast neutron detector and phantoms. (Middle left) Fast neutron image of 3-mm holes with 3-mm walls in 2-cm thick polycarbonate plate. (Middle) The vertical band in the PSPMT voltage spectrum was selected to produce images. (Middle right) False-color neutron image of "NY towers." (Bottom left) NY towers diagram with 5 mm × 5-mm square cross section and heights 10–60 mm. (Middle) The raw fast neutron image of NY towers phantom. (Right) The measured percent attenuation fraction in NY towers image (Popov *et al.*, 2011). (See Color Insert.)

et al. (2011) . The bottom and middle right are the raw and false-color FN radiographic images of the "NY towers" phantom (test object). As shown in the diagram (bottom left), the NY towers phantom was composed of 11 styrene plastic rectangular bars with a square ($4.8 \times 4.8\text{-mm}^2$) cross section and with lengths of 10, 20, 30, 40, 50, and 60 mm. Based on line profiles through the edges of the absorption contrast images, their measured resolution was estimated to be less than 0.5 mm in the transverse x and y directions.

The middle left of Figure 141 is a false-color FN image of 3-mm diameter holes with equidistant spacing and 3-mm walls in 2-cm thick polycarbonate plate obtained by Popov *et al.* (2011). The holes are clearly defined in the FN image as the hydrogen-rich, 2-cm thick polycarbonate attenuates a large fraction of the incident 2.5-MeV neutrons by elastic scatter; thereby producing moderated lower-energy neutrons as well as thermal neutrons. The plastic scintillator produces light via the induced electronic transitions, which are caused by ions and electrons created by X-ray and gamma photo-absorption, Compton scatter, and pair production, as well as energetic recoil protons that are produced by the FNs. However, other than the gamma emission from hydrogen nucleus capture of thermal neutrons in the plastic scintillator, the collision of the low-energy (0.025-eV) thermal neutrons produces no signal in the plastic scintillator.

The middle of Figure 141 shows the voltage spectrum produced in the output of the Popov *et al.* (2011) detector. The FN and background gamma and X-ray radiation, as well as scattered neutrons, produced output voltages in their detector's PSPMT. These voltages are proportional to the energies of the FN, gamma, X-ray, and scattered neutrons, which are incident on the plastic scintillator and viewed by the PSPMT via the Anger collimator. The Anger logic produces an output voltage spectrum that corresponds roughly to the energy spectrum of the incident radiation. The incident radiation spectrum is composed of the sample-transmitted DD neutrons and the noise from X-rays, gammas, as well as scattered FNs, which did not pass through the sample but rather are scattered by the surroundings into the detector.

The advantage of the Popov *et al.* (2011) detector is that the output image can be obtained by selecting the range of voltage output from the PSPMT. The selected voltage range would then be primarily produced by the FNs with a minimum noise content of gamma, X-ray, and non–sample-scattered neutrons. That is, most of the noise from the FN image can be filtered by selecting a particular bandwidth of the voltage output spectrum of the PSPMT, which has been experimentally determined to correspond primarily to nonscattered FNs that have made a straight-line flight from the Adelphi Technology DD generator target source to the detector's plastic scintillator. The light vertical band in the voltage spectrum of Figure 141 (middle) was the range of voltage outputs from the PSPMT that were used to form the images. In addition, the Popov *et al.* (2011) detector has a very fast timing resolution, which allows the detector to be combined with TOF to produce energy-resolved FN imaging.

Eberhardt *et al.* (2005) and Sowerby and Tickner (2007a,b) at the Australian Commonwealth Scientific and Industrial Research Organization (CSIRO) developed a new FN and gamma radiography technique for a full-scale prototype scanner for Brisbane International Airport. The goal of the scanner developed by the CSIRO group was to rapidly scan cargo containers in search of contraband, explosives, or dangerous substances. The bottom of Figure 142 shows a diagram of the prototype air cargo

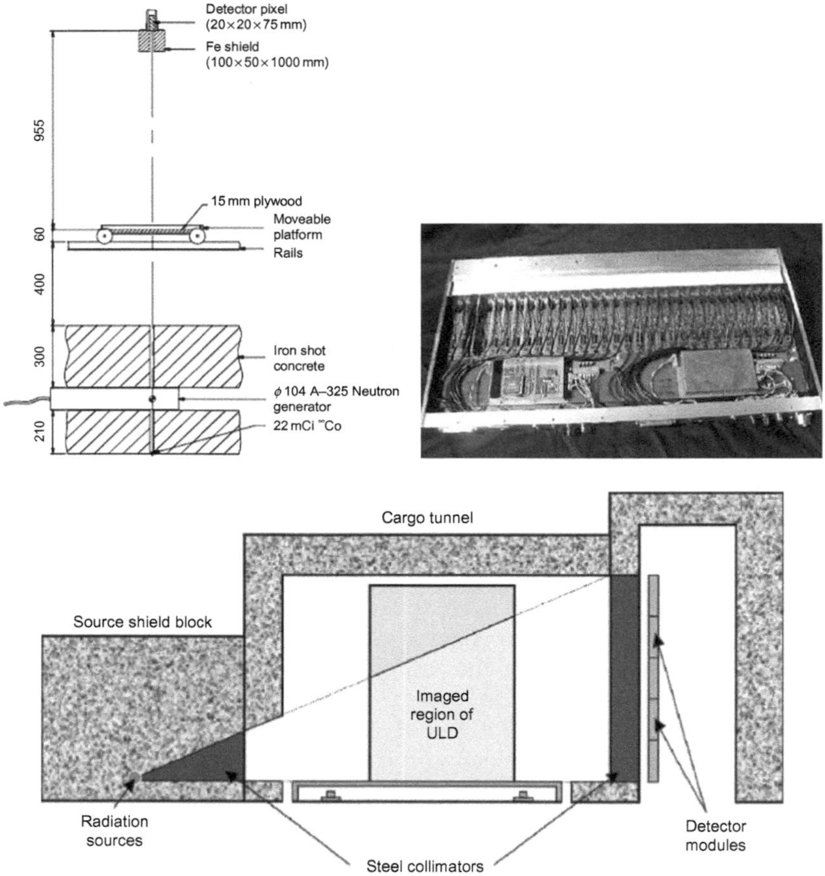

FIGURE 142 (Top left) Cross-sectional view of test facility to conduct fast neutron and gamma scans of objects placed on Movable platform; the length dimensions are in mm units. (Top right) Photo of 32-detector array composed of 2 cm × 2-cm area × 7.5-cm plastic orange scintillators viewed by photodiodes. (Bottom) Diagram of the CSIRO airport scanner used to inspect air cargo containers (ULDs) for the Brisbane Airport (Eberhardt *et al.*, 2005).

scanner, which the CSIRO group fabricated and used for experimental testing and prototype development.

The air cargo is usually packaged into lightweight aluminum containers called unit load devices (ULDs). The 1.6-m to 2.4-m width (beam path length) of the ULD and the average density of its 1.6- to 6.8-ton cargo are two important design constraints, which require the use of high-energy penetrating gammas and FNs rather than lower-energy X-rays and thermal neutrons. The ULD content is normally limited by mass rather than volume, which results in a homogeneous distribution of the load.

The cobalt gamma and DT FN generator, via a fan-shaped collimator, create a fan beam, which passes through the cargo container and onto detector modules. The modules are comprised of arrays of plastic scintillator detectors for neutron imaging and arrays of CsI(Tl) detectors for gamma imaging. The CSIRO group exploited the ratio of the FN cross section divided by gamma cross section to enable imaging and identification of the various materials in an object interrogated by FNs and gammas. Their scanning and imaging detection system was composed of a 22-mCi Co-60 source of 1.17-MeV and 1.33-MeV gammas (5.3-year half-life), and a 0.82-GBq source of 14-MeV DT neutrons provided by a Thermo MF Physics A-235 (Thermo Scientific) neutron generator. The detectors were plastic scintillator arrays for the detection of both gammas and FNs. Alternatively, the CSIRO group used an array of CsI(Tl) detectors for the detection of gammas.

In the prototype scanner the radiation source was approximately 2.5 m from the midpoint of the cargo and 4.5 m from the detector array, and the detector array height was matched to the 1.8-m height of the tunnel through which the scanned cargo containers are moved. In the CSIRO prototype scanner, the target object was moved through the tunnel on rails at a speed of about 0.25 m/s, which thus required 2 hours to scan a 2-m long ULD (cargo container) as a result of the relatively weak 22-mCi ^{60}Co gamma and 0.82-GBq 14-MeV FN sources. Unwanted radiation scattered from the target surroundings and from portions of the target not in the direct line of the beam cause a decrease in contrast and resolution of the gamma and FN transmission images. Hence, steel collimators were used to minimize radiation scatter and accurately define the gamma and FN interrogation beams.

In comparison, the planned commercial air cargo scanner for Brisbane Airport was to use a DT generator with $10^9 - 10^{10}$-n/s yield and a 185-GBq ^{60}Co gamma source for 1- to 2-minute scan times of an entire ULD. The Brisbane Airport scanner would require separate gamma and FN scanning in which 250 CsI(Tl) gamma detectors are arrayed in modules of 16, and the 700 plastic scintillator FN detectors are arrayed in modules of 32 with temperature control to reduce electronic noise. This design would take advantage of the smaller-sized CsI(Tl) gamma detectors to provide better spatial resolution and the gamma radiation induced in the plastic scintillator neutron detectors would be reduced.

Eberhardt *et al.* (2005) calculate the contents of the container (or an individual stowaway) would receive a dose of approximately 5 μS (microsieverts), which is low considering the natural background dose rate at sea level is 0.25 μS/hr and at 10,000-m passenger aircraft altitude the dose rate is 7 to 8 μS/hr, which exceeds the interrogation dose rate after only 1 hour of flight. The residual induced radioactivity after neutron/gamma interrogation is calculated to be over 1000 times lower than the natural activity in a typical cargo.

The dose of naturally occurring potassium-40 (1.25 billion-year half-life) in one banana is 0.1 μS/hr or one banana equivalent dose (BED). The CSIRO neutron/gamma interrogation to an individual stowaway (or smuggled chimpanzee) in a ULD would be equivalent to eating 50 bananas (50 BEDs) during a 13-hour flight from Los Angles to Brisbane at 10,000-m altitude, which delivers a 80- to 100- μS/hr dose to the stowaway (at least 16 to 20 times more dose than the CSIRO interrogation). Homeostasis in the body balances the elimination of ^{40}K from the body with the uptake of ^{40}K to maintain a fairly constant level of potassium in the body.

The detector design was to have the individual scintillators grouped into 16 detector line arrays and 32 detector linear arrays, which could be stacked to form a 2D imaging array. The top-right photo in Figure 142 shows the 32-detector array used by Eberhardt *et al.* (2005) in the prototype scanner. The arrays were composed of individual 2-cm × 2-cm area by 7.5-cm thick orange plastic scintillators (e.g., ELJEN EJ-264; Eljen Technology, Sweetwater, TX), whose light output was better matched in wavelength to the silicon photodiodes than the blue or ultraviolet output of Bicron BC-408 (ELJEN EJ-200) plastic scintillators. However, in the laboratory prototype scanner a single column of 80 plastic scintillator detectors was used for the simultaneous detection of both gammas and FNs.

The silicon photodiodes were used because they were inexpensive, compact, and require only 50 V compared with the higher (kilovolt range) photomultipliers. With photodiodes imaging the plastic scintillators and CsI(Tl) detectors a large area detection array could be fabricated at less cost and complexity. Anger logic was applied to produce a FN image from the 32 plastic scintillator pixels. In Figure 142 (top left) is a diagram showing the cross-sectional view of the small laboratory facility used by Eberhardt *et al.* (2005) to conduct preliminary fat neutron and gamma scan of objects, which were placed on the movable platform. The length dimensions are in millimeter units.

The geometry of a narrow beam was assumed for CSIRO detection system in which the scattered gammas and neutrons from the surroundings do not enter the detector. Only FNs and gammas, which are emitted from the respective DT generator and ^{60}Co sources and pass through the sample with or without scatter, but travel directly to the detector array, are considered. A neutron ray and a gamma ray pass in a line along the z-direction through object of thickness $t_z\left(x,y\right)$, which is dependent on transverse coordinates (x,y). Along that line of X-ray and gamma flight, the object has mass density $\rho\left(x,y,z\right)$ and neutron linear attenuation $\mu_n(x,y,z)$ with units μ_{n} (cm^{-1}).

Given the cross section $\sigma_n(x,y,z)$ for neutron attenuation (with units in cm^2), or in a salute to our agrarian past with units of barns (10^{-24} cm^2), the neutron linear attenuation coefficient $\mu_n(x, y, z)$ is

$$\mu_n(x, y, z) = \rho(x, y, z)\, \sigma_n(x, y, z). \tag{220}$$

Similarly, the gamma linear attenuation coefficient $\mu_g(x, y, z)$ is expressed in terms of the cross section σ_g (cm^2) for gamma attenuation, where

$$\mu_g(x, y, z) = \rho(x, y, z)\, \sigma_g(x, y, z) \tag{221}$$

Without the object, a bundle of FN rays along a flight line arrives at the single, rectangular plastic scintillator unit (pixel) of the neutron scintillator array with intensity $I_n^0(x, y)$. Similarly, without an object a bundle of gamma rays along a flight line arrives at the single, rectangular plastic scintillator or CsI(Tl) unit (pixel) of the gamma scintillator array with intensity $I_g^0(x, y)$.

If the object is present, then the bundle of neutron rays passes through object thickness x and encounters the variable object mass density $\rho(x, y, z)$ and variable neutron linear attenuation $\mu_n(x, y, z)$, which results in a measured neutron intensity at the neutron detector pixel $I_n(x, y)$ given by

$$\begin{aligned} \frac{I_n(x, y, z)}{I_n^0(x, y, z)} &= \exp\left[-\int_0^{t_z(x,y)} \mu_n(x, y, z)dz\right] \\ &= \exp\left[-\int_0^{t_z(x,y)} \rho(x, y, z)\, \sigma_n(x, y, z)dz\right]. \end{aligned} \tag{222}$$

The neutron ratio R_n is defined as the natural log of the above neutron transmission ratio, where

$$R_n \equiv \log\left[\frac{I_n(x, y, z)}{I_n^0(x, y, z)}\right] = -\int_0^{t_z(x,y)} \mu_n(x, y, z)dz = -\int_0^{t_z(x,y)} \rho(x, y, z)\, \sigma_n(x, y, z)dz. \tag{223}$$

Similarly, the bundle of gamma rays passes through the same object thickness x and encounters the same variable object mass density $\rho(x, y, z)$, but with variable gamma linear attenuation $\mu_g(x, y, z)$, which results in a measured gamma intensity at the gamma detector pixel $I_g(x,y)$ given by

$$\frac{I_g(x,y,z)}{I_g^0(x,y,z)} = \exp\left[-\int_0^{t_z(x,y)} \mu_g(x,y,z)dz\right] = \exp\left[-\int_0^{t_z(x,y)} \rho(x,y,z)\,\sigma_g(x,y,z)dz\right]. \tag{224}$$

The gamma ratio R_g is defined as the natural log of the above gamma transmission ratio, where

$$R_g \equiv \log\left[\frac{I_g(x,y,z)}{I_g^0(x,y,z)}\right] = -\int_0^{t_z(x,y)} \mu_g(x,y,z)dz = -\int_0^{t_z(x,y)} \rho(x,y,z)\,\sigma_g(x,y,z)dz. \tag{225}$$

The neutron ratio R_n divided by the gamma ratio R_g is the important ratio R, which is used to characterize the component materials in a fast neutron/gamma interrogated object, where

$$R \equiv \frac{R_n}{R_g} = \frac{\int_0^{t_z(x,y)} \mu_n(x,y,z)dz}{\int_0^{t_z(x,y)} \mu_g(x,y,z)dz} = \frac{\int_0^{t_z(x,y)} \rho(x,y,z)\,\sigma_n(x,y,z)dz}{\int_0^{t_z(x,y)} \rho(x,y,z)\,\sigma_g(x,y,z)dz}. \tag{226}$$

Assume the interrogated object is composed of one material that has uniform mass density $\rho(x,y,z) = \rho$ and uniform neutron and gamma attenuation cross sections with $\sigma_n(x,y,z) = \sigma_n$ and $\sigma_g(x,y,z) = \sigma_g$, then the ratio R is simplified to the ratio of the material neutron attenuation cross section σ_n divided by the gamma attenuation cross section σ_g, where

$$R \equiv \frac{R_n}{R_g} = \frac{\sigma_n}{\sigma_g} = \frac{\ln\left(I_n/I_n^0\right)}{\ln\left(I_g/I_g^0\right)}. \tag{227}$$

Single elements and isotopes can be identified by their R ratio as shown in the plot at the top of Figure 143.

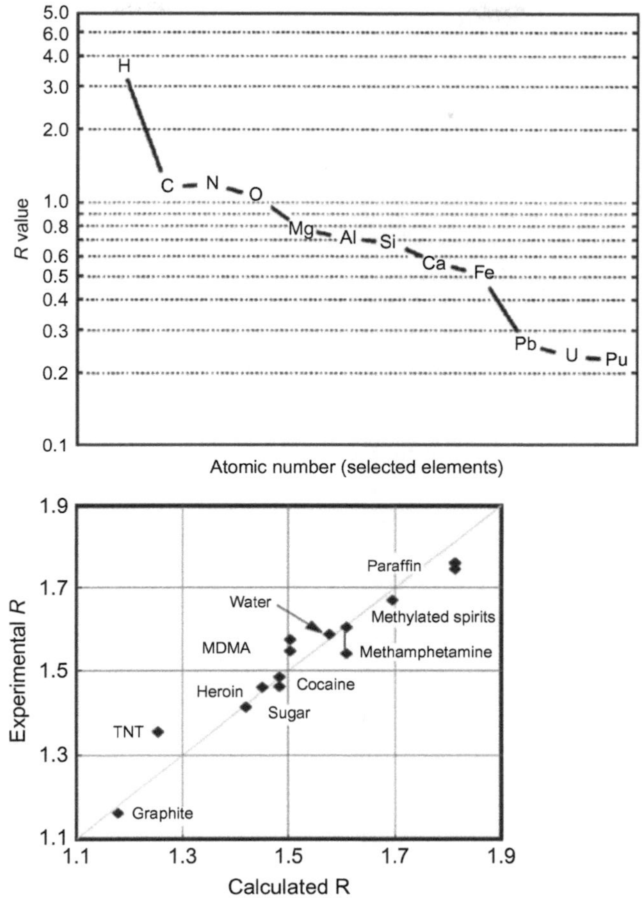

FIGURE 143 (Top) Plot of the ratio of *R* values as a function of atomic number. (Bottom) Plot of the of ratio *R* values for common materials as well as drugs and explosives (Eberhardt *et al.*, 2005).

In reality a neutron/gamma interrogated target is composed of one or more materials with varying target thickness $t_z(x, y)$ and mass density $\rho(x, y, z)$. The variation of the element composition of the target materials causes variation of the element-dependent gamma attenuation cross section $\sigma_g(x, y, z)$. In addition, the variation of the isotope composition of the target material causes variation of the atomic weight–dependent neutron attenuation cross section $\sigma_n(x, y, z)$. The measured ratio R_{exp} is thus determined by the average value of the neutron linear attenuation $<\mu_n(x, y, z)>$ and the average value of the gamma $<\mu_g(x, y, z)>$ linear attenuation along the thickness $t_z(x, y)$ of the neutron/gamma beam ray path through the target, where

$$R_{\mathrm{exp}} = \frac{<\mu_n(x,y,z)>}{<\mu_g(x,y,z)>} = \frac{<\rho(x,y,z)\sigma_n(x,y,z)>}{<\rho(x,y,z)\sigma_g(x,y,z)>}. \tag{228}$$

However, the target measured ratio R_{exp} can be determined via the measured neutron and gamma-ray transmissions yield without the knowledge of neutron/gamma interrogated target mass lying with the radiation beam path. The ratio R can provide a powerful discriminator for different material classifications if the neutron and gamma-ray beam energies are chosen carefully. The high-Z metals have the lowest R values, and hydrogen has the highest R value. Lower-energy neutrons provide better discrimination between elements; however, the lower-energy neurons also have reduced penetration in an interrogated cargo container. The bottom of Figure 143 shows the plotted ratio R values for some common materials, drugs, and explosives. The plot shows the R values of TNT is well separated from the other materials and drugs.

Spatial filtering was done with both the gamma image, which provided density information and the R ratio image created from the FN and gamma images, as well as application of a sharpening filter to enhance the edges and boundaries in the gamma image. A smoothing filter also was applied to the R ratio image to reduce noise, which competed against the relatively low neutron count rates. The use of straightforward linear sharpening and smoothing filters increased the image noise and blurred the edges, which thus required the use of nonlinear filtering algorithms.

The gamma detector–acquired target density and the ratio of the fast detector to gamma detector images, which yield the R images, were combined by mapping R to the pixel hue (color). In the images, regions of low R values of inorganic materials were colored purple or blue, the intermediate R values were green, and the high R values for the organic materials were colored yellow or orange. This color scheme of the CSIRO group was designed to match the color scheme used in dual-energy X-ray scanners for luggage and parcels. The gamma image intensity was used to determine the color brightness in which as the density increased the color darkened to its saturated value, which was set by the R value. A pixel was colored white when there was no intervening material. Very high densities were darkened toward black. The black (for high density) and white (for no density) allowed poorly determined R values, which resulted from insufficient or excessive intervening material and caused the poor R value to be colored black or white. As a result, a poor R value did not produce a noisy image.

With the prototype scanner Eberhardt *et al.* (2005) obtained scanned images of a motorbike (left) and a ULD (right) (Figure 144). The top-left image was a gamma scan of the motorbike and the bottom-left image

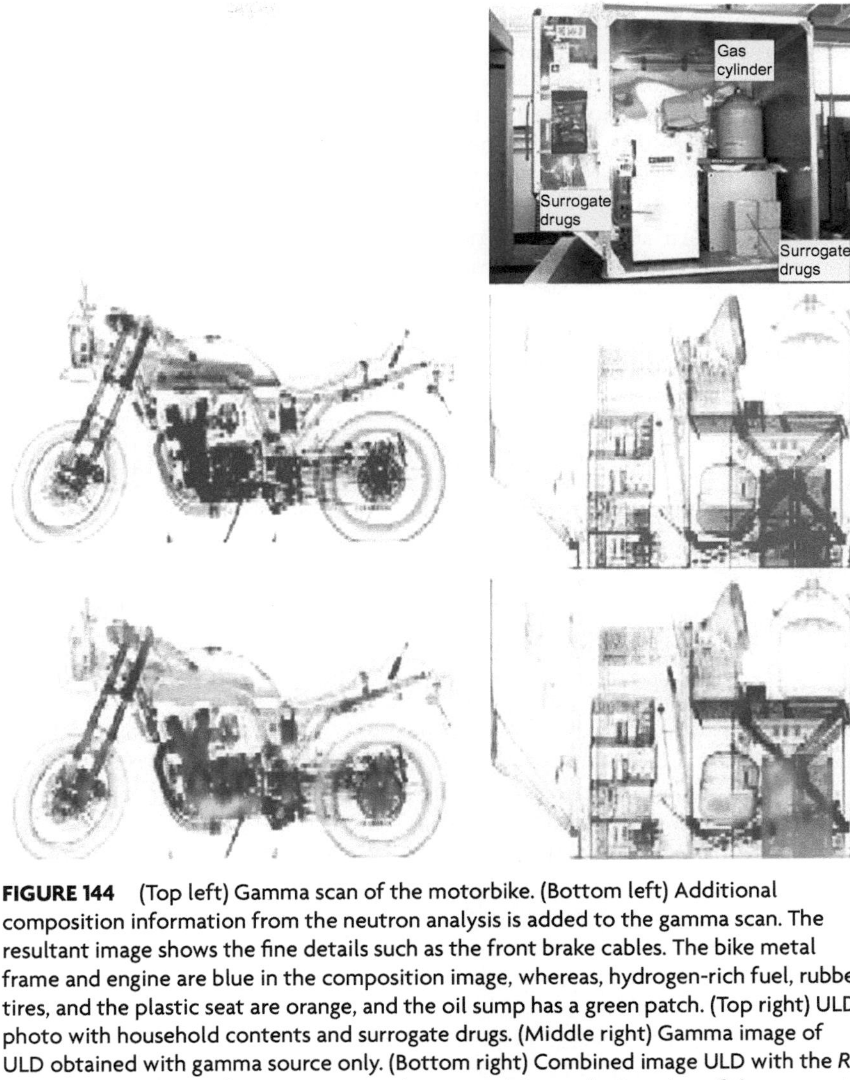

FIGURE 144 (Top left) Gamma scan of the motorbike. (Bottom left) Additional composition information from the neutron analysis is added to the gamma scan. The resultant image shows the fine details such as the front brake cables. The bike metal frame and engine are blue in the composition image, whereas, hydrogen-rich fuel, rubber tires, and the plastic seat are orange, and the oil sump has a green patch. (Top right) ULD photo with household contents and surrogate drugs. (Middle right) Gamma image of ULD obtained with gamma source only. (Bottom right) Combined image ULD with the *R* value image added to the gamma image detects and shows the location of the surrogate drugs (Eberhardt *et al.*, 2005). (See Color Insert.)

was the gamma scan with the added composition information from the neutron analysis. The resultant combined gamma density image (with shape information) with the neutron image, which provided the composition information or *R* value image, provided an excellent shape, density, and composition image of the motorbike. The detector array pixels were 20 mm in size, but application of the Anger collimation, Anger logic and centroiding algorithms produced motorbike and ULD images with considerably better resolution.

The top right of Figure 144 is a photo of the scanned ULD, which contains assorted household items including surrogate drugs (concealed in one of the computers, behind the refrigerator, and in one of the concrete blocks). The middle-right image is a gamma density image (gamma interrogation only), and the bottom-right image is the gamma image with the additional information from the neutron image via the ratio R image (division of neutron image by gamma image with filtering and smoothing). The projection of the contents of the ULD onto the detector added some distortion to the gamma and combined images. However, the contents of the ULD, including the surrogate drugs, are clearly identified.

Kobayasi (1990) first applied the CCD camera to thermal neutron radiography and in subsequent years, the continuing advances in the CCD camera and cooling technology have made thermal as well as FN imaging with the CCD cameras affordable, practical, and effective. As stated by Lewis *et al.* (2003), there are two phases in the installation of a CCD camera system, namely (1) the hardware assembly of the CCD camera, mirror, lens, scintillation screen, and a light-tight enclosure, and (2) testing of the CCD camera system performance. The testing of the CCD-based radiation imaging system (gamma, thermal, or FNs) includes (1) measurement of image resolution and contrast of Rose phantom objects and determination of the required exposure times, (2) measurement of the efficiency of the radiation-scintillation screen interaction, and (3) measurement of resolution, contrast, and exposure times for images of the actual objects of interest (e.g., actual aircraft flight control surfaces in the case of the thermal neutron imaging of Lewis *et al.*, 2003).

McFarland *et al.* (1991) designed, fabricated, and tested a prototype position-sensitive neutron detector for 2D and 3D tomographic imaging of materials and complex assemblies. The detector system consisted of a LiF-ZnS scintillator screen that was optically coupled to a cooled CCD. Their imaging system of consisted of the MIT Research Reactor as the thermal neutron source (nominal flux $\sim 2.4 \times 10^8\,\mathrm{n/cm^2/s}$, a rotating sample holder, and the 2D position-sensitive CCD detector. They calculated a single-screen resolution limit of $100\,\mu\mathrm{m}$, which was limited by the neutron-beam divergence and the scintillator screen resolution. The contrast resolution is the ability to distinguish differences in intensity, and they calculated the contrast resolution in a $(200\,\mu\mathrm{m})^3$ cubic volume element to be 1% of the cross section in small samples. They also obtained 2D and 3D tomograms of phantoms and several objects, including a control valve and an oil well dolomite core sample.

The Medipix2 ASIC is a high spatial, high-contrast resolving CMOS pixel readout chip working in single photon counting mode invented at CERN (2011). The Medipix2 was originally designed to be combined with different semiconductor transducers, which convert X-rays or gammas directly into detectable electric signals. However, the Medipix2 imaging detector was successfully adapted for thermal neutron imaging by Uher *et al.* (2008) via the addition of a ^{6}LiF neutron converter. For testing with thermal neutrons at the Institut Laue-Langevin (ILL) neutron radiography and tomography station in Grenoble, they measured a detector $SNR \cong 8$ with an estimated spatial resolution of 100 μm.

Next, the Medipix2 detector was modified for FN imaging by Uher *et al.* (2008). Their Medipix2 device consisted of a silicon pixel detector (256 × 256 square pixels, 55 μm × 55 μm each), which is bump-bonded to a readout chip. FNs were detected via proton recoil from a 1-mm thick polyethylene layer set on the detector surface. Uher *et al.* (2008) measured the performance of the detector using AmBe neutron sources as well as neutron beam testing with the neutron beams of the Sparrow reactor at the Czech Technical University in Prague and the high-flux reactor at ILL.

Radiography using FNs from accelerators has been known as an excellent method to penetrate high-density thick objects. FN radiography was demonstrated by Hall *et al.* (2001, 2006) and Hall (2001) at Lawrence Livermore National Laboratory (LLNL) using 30 cm × 30 cm × 4-cm thick plastic scintillator plates that were viewed by a CCD camera. Hall *et al.* (2001, 2006) obtained multiple view and "mock" tomographic FN images with a variety of test objects (Figure 145).

Plastic scintillator fiber arrays were used to image DT fusion reactions in imploded pellets (see by Ress *et al.*, 1995). FN imaging with hollow capillary arrays, which were filled with hydrogen-based liquid scintillator and alternatively filled with deuterium-based liquid scintillator, was performed by Annis *et al.* (2000) and Disdier *et al.* (2003, 2004). The motive for a capillary array filled with liquid scintillator is that the inside diameter of the capillary arrays can be made much smaller than solid scintillator fibers, and yet the capillary array walls are sufficiently thin to produce an 85% hollow array for the liquid scintillator. That is, the solid fiber array is limited to 250-μm diameter fibers (pixels); whereas the capillary arrays can be 50-μm inside diameter (25-fold smaller pixel area).

Figure 146 shows a diagram of the experimental setup, FN images, and plotted results from the FN imaging detection system of Disdier *et al.* (2003, 2004). The hollow capillary arrays used by Disdier *et al.* (2003, 2004) were fabricated by Walter Siegmund of Sigma Fiber Optics (Pomfret Center, Conn., USA) and the liquid scintillants were provided by Hurlbut (2011).

To reduce the stopping distance of the recoil ion from a neutron scattering event, the hydrogen-based liquid scintillator is replaced with either a deuterium- or a lithium-based scintillator (see Hurlbut, 2011). For the deuterium-based scintillator, the resulting recoil ion can now be a deuteron, which has a shorter stopping range than a proton. Disdier *et al.*

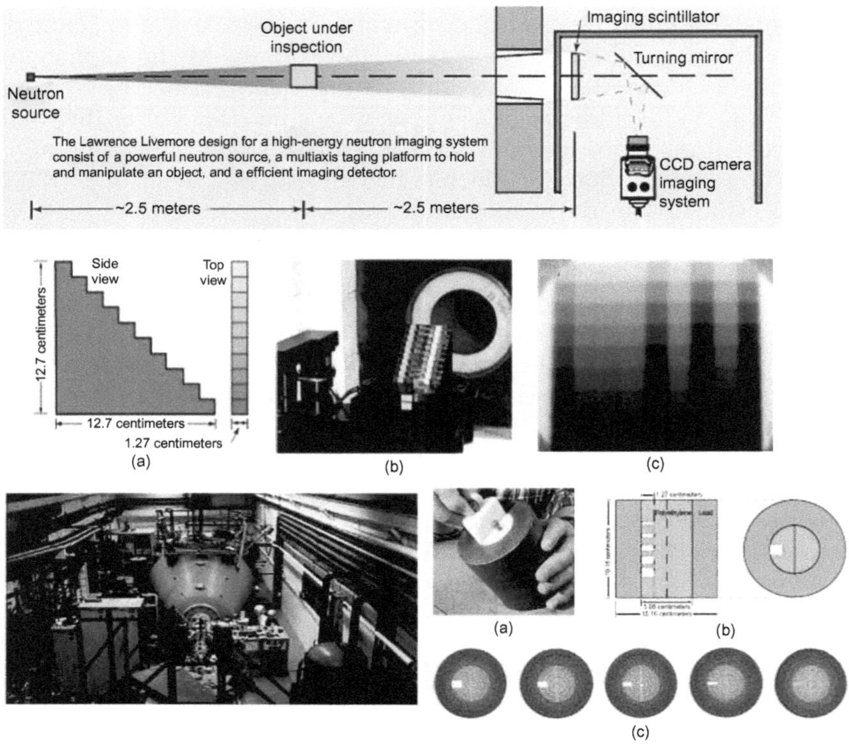

FIGURE 145 The radiographic fast neutron imaging by Hall *et al.* at Lawrence Livermore National Laboratory using the Ohio State University Accelerator Laboratory to perform a set of nine step wedges fabricated from lead, Lucite, mock high explosive, aluminum, beryllium, graphite, brass, polyethylene, and stainless steel. Each step wedge had 10 steps ranging in thickness from 1.27 cm to 12.7 cm. The nine wedges were grouped together and radiographed as a single unit (looking up the steps from thick to thin) in a series of two 1-hour exposures. The radiographs clearly differentiated the different material and step thicknesses. The bottom right shows fast neutron tomographic images of several cylindrical test objects composed of nested shells of high-Z and low-Z materials, with machined features, which was taken in a series of sixty-four 10-minute exposures at angles evenly distributed over 180° (Hall *et al.*, 2001). (See Color Insert.)

(2004) were able to reduce the emitted light point spread function (PSF) of 14-MeV D-T neutrons from 750 μm (FWHM) to μm (FWHM) with a deuterium-based liquid scintillator.

A sufficiently compact, high-yield fast DD and DT neutron source, which can be also moderated to act as a high-yield thermal neutron source, introduces new possibilities for radiographic applications in the laboratory and field deployment. (See Nargolwalla and Przybylowicz,

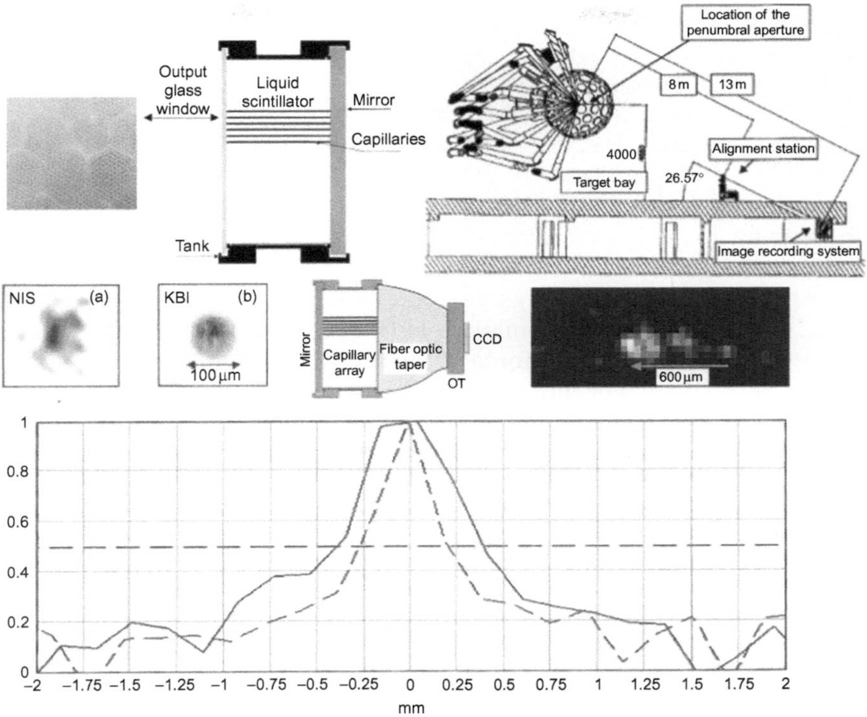

FIGURE 146 (Top) The fast neutron imaging detection system for imaging fusion pellet implosion, which uses a 10 cm × 10-cm area array of liquid scintillator–filled capillaries 50 mm long. The plot compares an 85-μm diameter bore, 75-μm thick wall, capillaries filled with hydrogen-based (solid line), and deuterium-based (dashed line) liquid scintillators. Fast neutron–produced recoil deuterons travel less distance than recoil protons in the liquid scintillator, decreasing the light production distance and thus improving resolution. The streak camera photo shows the 600-μm path of recoil proton produced by a 14-MeV neutron in a single capillary (Disdier *et al.*, 2003 and 2004).

1973; Barletta, 2001; Chichester and Simpson, 2004; Reijonen *et al.*, 2007, 2005; and Qing *et al.*, 2009). The requirement of large-area, robust, and affordable FN imaging with good resolution and contrast favors the use of plastic scintillator sheets over the scintillator fiber arrays and liquid scintillator capillary arrays.

The plastic scintillator fiber and liquid scintillator arrays are limited in area by cost and complexity. However, large-area, fast-response plastic scintillators are simple and inexpensive. A large plastic scintillator can be viewed indirectly via a 45° mirror by a CCD camera with image intensifier, directly by a radiation-hardened CCD camera with image intensifier, or directly by an array of PSPMTs using Anger logic are simpler and less costly. Furthermore, the use of PSPMTs allows one to select the energy range of the PSPMT output, which contains mainly FN signal, thereby rejecting much of the gamma noise, which is normally recorded by the CCD camera in the course of imaging FNs.

The FN generator produces 2.5-MeV neutrons with a yield in the range of $\sim 10^8 - 10^{10}$ n/s from the D-D fusion reaction at a small target in the diode of an RF plasma-based neutron generator. The neutrons pass through the target to a large-area plastic scintillator, which is composed of aromatic (ring) organic molecules (e.g., xylene) in which are dissolved the light-producing, aromatic, organic solutes (about 1% to 2% by weight such as 2,2-p-phenilene-bis(5-pheniloxazole [POPOP]), which includes possibly a wavelength-shifting aromatic organic solute (see Knoll, 1989, and Tsoulfanidis, 1983). For more on liquid and solid plastic scintillators, see Hurlbut (2011).

In the plastic scintillator sheet recoil protons are produced by FN collision with the hydrogen atoms in the scintillator organic polymers. The energetic recoil protons acquire on average 1.25-MeV recoil energy from their elastic collision with the approximate, equal-mass, fast DD neutron of energy 2.5 MeV. Within the stopping distance of the 1.25-MeV recoil proton are created a trial of ions and electrons. The cascade of ionization and electrons leads to a multitude of excited molecular state electrons, which then decay to ground state emitting UV or visible light. The UV light is wavelength-shifted by a wavelength-shifting organic, fluor ring compound dissolved in the solid state (or liquid) scintillator solvent.

For relatively small acceleration energies (80 to120 keV) of deuterium ions into a deuterium or tritium target, the DD and DT reactions produce monoenergetic neutrons with energy of 2.45 MeV and 14 MeV. Titanium is a very good retainer or getter of hydrogen, deuterium, and tritium. The titanium target in a DD, DT, or TT neutron generator can be loaded by a beam of accelerated deuterium and/or tritium ions, or the target can be a titanium hydride compound (TiD_2 or TiT_2). DD and DT neutrons also can be produced by high-energy accelerators, which produce monoenergetic DD neutrons with energy ranging from 1.64 to 7.75 MeV, and monoenergetic DT neutrons with energy ranging from 11.74 to 20.5 MeV.

The TT neutron generator produces a very wideband or white neutron spectrum, which, unlike DT or DD reactions, is a three-body reaction that produces two neutrons and one He-4 atom. The white TT neutron spectrum can be calculated non-relativistically, and the TT neutron generator yield is approximately twice the yield of a DD neutron generator because the TT reaction emits two neutrons in the TT fusion reaction (see Lou, 2003). The top of Figure 147 plots the neutron spectrum produced by the DD, DT, and TT reactions, and the bottom plots the neutron production cross section for DD, DT, and DT reactions. Both plots are from the University of California at Berkeley Ph.D. thesis of Lou (2003).

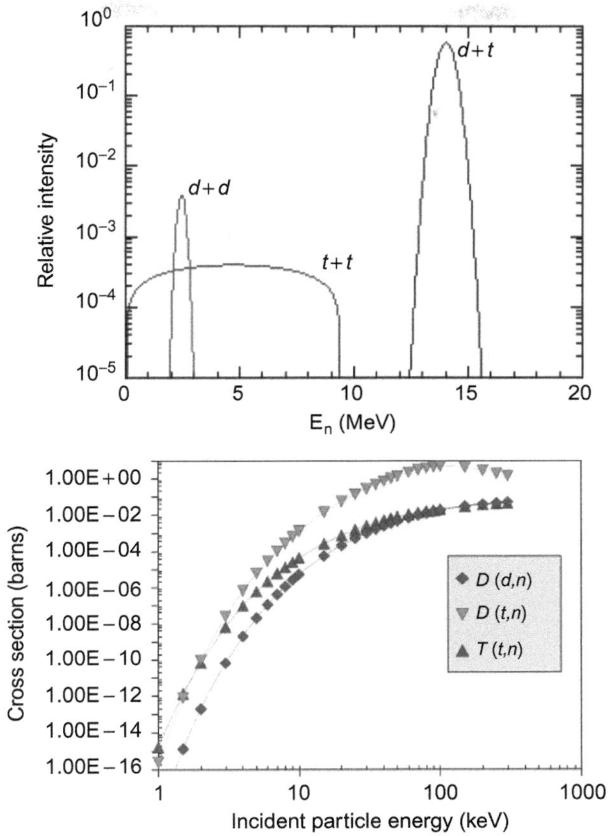

FIGURE 147 (Top) Neutron spectra due to DD, DT, and TT reactions. (Bottom) Neutron production cross section for DD, DT, and DT reactions (Lou, 2003).

The Adelphi Technology fast and thermal neutron imaging system consists of a neutron generator and imaging detector Cremer *et al.* (2005, 2012). The 2.5-MeV FNs are produced from the deuterium-deuterium (DD) fusion reaction in the diode of an RF plasma-based neutron generator. The generator housing was ~91.44 cm long, and the generator's high-density polyethylene (HDPE) shielding weighed about 90 kg. To reduce the background X-ray and gamma radiation, a 6.35-mm thick sheet of lead was placed over the source port of the DD generator. The lead strongly attenuates the bremsstrahlung from accelerated electrons in the diode region and attenuates the X-rays emitted from electrons that fill K-shell vacancies in the copper and titanium target created by the DD reaction's 3-MeV protons.

The top of Figure 148 is an illustration of the DD neutron generator the bottom right is a photo of the generator. The deuteron ions are accelerated by an applied DC voltage and strike a titanium-coated hollow copper target with deuterium atoms embedded in the titanium from the impact of prior D ions. Water flows through channels in the hollow copper target to provide cooling from the 1 kW of deposited energy of the 120-kV, 9.0-mA, D ion beam on the 4.5-cm^2 target area, which produces a total DD neutron yield of 10^9 n/s. Each leg of the vertically oriented V titanium-coated copper target intercepts the 4.0-mm diameter cylindrical D beam to produce a half-ellipse-shaped source with a horizontal (x), major axis of 5.6 cm and vertical (y) minor axis of 0.4 cm. For Monte Carlo simulation, the DD source is represented by a rectangle in the z 0 plane with horizontal dimension x 2.4 cm and vertical dimension y 0.4 cm as illustrated at the bottom left of Figure 148.

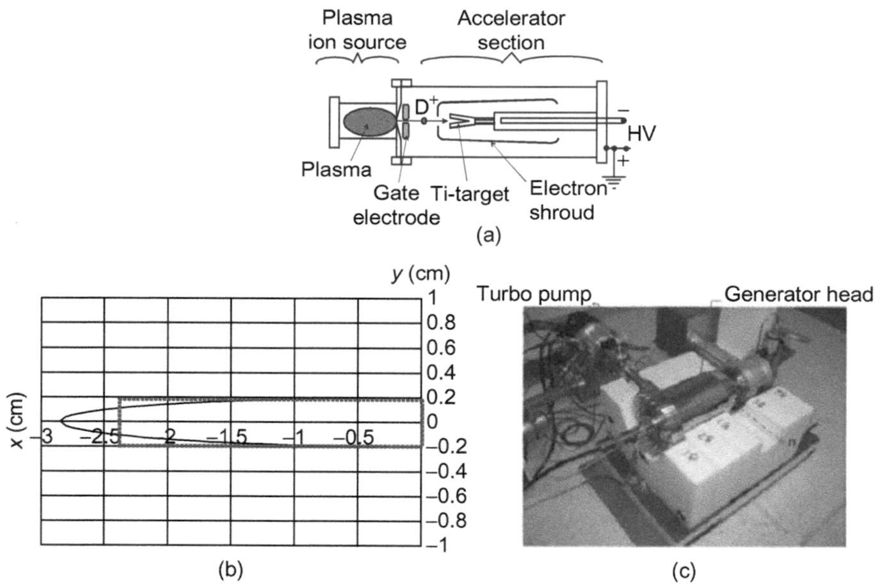

FIGURE 148 (Top) The diagram shows the principal components of the Adelphi Technology DD109 fast neutron generator with $10^9 - 10^{10}$ n/s yield. High voltage (HV) is applied to accelerate D ions from the plasma source to titanium (Ti) target that is embedded with prior-bombarded D atoms. (Bottom left) DD neutron emission area is half-ellipse shape with 5.7-cm major and 0.4-cm minor axes. On each leg of the vertically oriented titanium-coated V target is an ellipsoid neutron emission area. For Monte Carlo simulations, the DD source is represented by the rectangle in the z 0 plane with horizontal y 2.4 cm and vertical x 0.4 cm sides (see the dotted line). (Bottom right) Photo of the DD109 generator (Cremer *et al.*, 2012, 2009). (See Color Insert.)

The emission of neutron flux and energy is slightly anisotropic for the DD and DT reactions relative to the direction of the D^+ ion beam that strikes the titanum-coated DD target (see Nargolwalla and Przybylowicz, 1973). Forward-emitted neutrons have the largest flux with maximum emitted energy of 2.9 MeV. Back-emitted neutrons have a midrange flux at the minimum emitted energy of 2.1 MeV. Neutrons emitted to the side toward the detector have the minimum flux with emitted average energy of 2.5 MeV. A brightness of 8.3×10^6 n/sr/mA was calculated compared with the measured brightness of 6.6×10^6 n/sr/mA.

Thermal neutrons, X-rays, or low-energy gammas have an attenuation cross section dominated by absorption, which allows the use of relatively thin imaging detectors. FN attenuation is dominated by elastic scatter, which requires the use of thicker scintillation detectors, which are hydrogen rich, and increases the probability of interaction. Each elastic collision of an FN with the proton of a hydrogen atom on average transfers 50% of the incident neutron energy to the proton. The average 50% decrease of neutron energy per elastic collision increases the neutron cross section for subsequent scatter and absorption, and significantly increases the probability of a second scatter or absorption interaction in sufficiently thick plastic scintillators. The scatter of FNs from hydrogen produces recoil protons in organic, hydrogen-rich scintillators, which are in the form of plastic scintillation sheets. The recoil protons create a cascade of scatter events in the scintillator, which leads to excited electron states, which transitions to ground with the emission of light.

Figure 149 shows the imaging system schematically. The camera box allows the Stanford Photonics CCD camera (Stanford Photonics, Inc., Palo Alto, Calif., USA; http://www.stanfordphotonics.com) and its electronics to be placed outside the direct neutron beam exposure. This prevents direct neutron beam radiation damage of the camera CCD sensor and electronics. At the base of the light-tight camera box is a removable, 11.25 cm × 11.25-cm area, aluminum cover that is 3 mm thick. A thick knife-edge plate or object can be placed on the outer cover surface and the plastic scintillator on the inner surface. Numerous FN imaging trials were performed with various thicknesses of BC408 plastic scintillators that ranged from 1.0 mm to 25 mm. As a result, the 5.0-mm thickness scintillator appeared to provide the best imaging resolution and contrast for 5- to 60-minute exposure times with 10^8 to 10^9 n/s DD neutron source yields.

The emitted scintillator light is intercepted by a 45° mirror (with reflectivity >95%). The mirror reflects the light upward out of the neutron path to a large 5-cm diameter lens with f-stop $f^{\#} = 1.0$. The lens focuses the light onto the dual-microchannel plate (MCP). The dual-MCP light amplifier and CCD chip are thermoelectrically cooled to –20°C. The dual-MCP amplifier produces high photon gains up to 10^6 at a very low dark count level of 1 or 2 electron counts per frame at 30 frames per second.

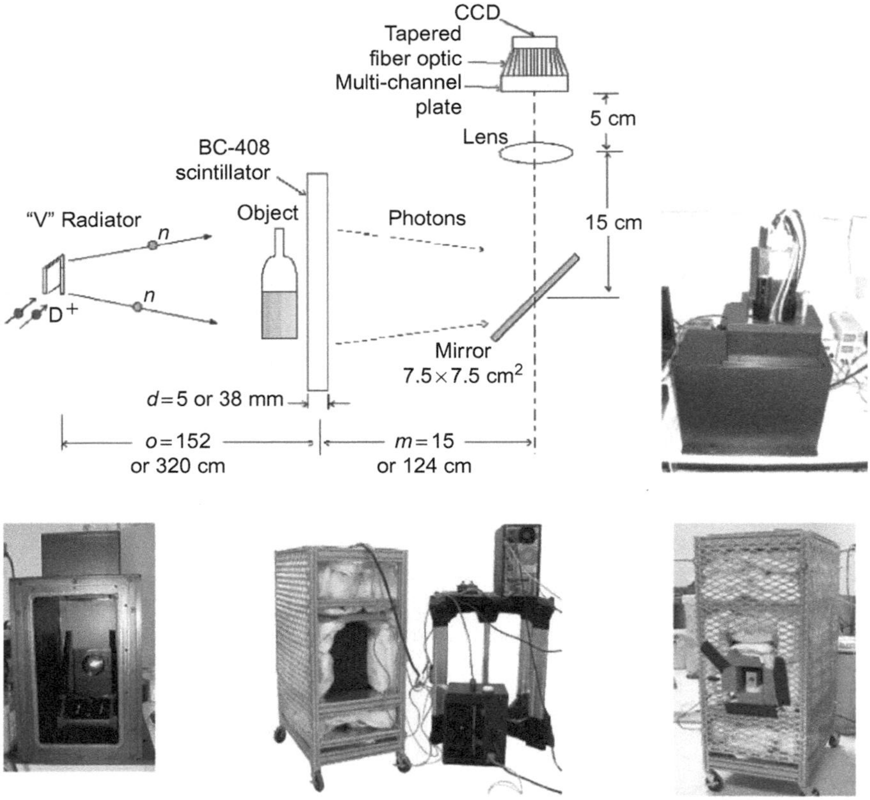

FIGURE 149 The Adelphi Technology fast/thermal neutron contact radiographic imaging system composed of a fast 2.5-MeV neutron source (or moderated with graphite for thermal neutron source) and an imaging detector. The chiller is on the floor and the computer control is on the table. Small objects were placed on the outer surface of 3-mm thick aluminum lid of the CCD camera box. On the inside surface of the aluminum lid of the camera box was placed a plastic scintillator for imaging fast neutrons or gadolinium-based phosphor for imaging thermal neutrons. The camera assembly was surrounded by thermal neutron shielding composed of lithium carbonate powder in zip-top plastic bags shielding to eliminate neutron damage to the camera (Cremer *et al.*, 2012, 2009).

In our experiments, the dual-MCP amplifier was set to an overall photon amplification gain of $G_{mcp} = 5.0 \times 10^5$.

The Stanford Photonics CCD camera uses a Sony XX285 scientific-grade image sensor, which has a full-frame pixel count of 1380K × 1024K in which the $n_p = 1.41 \times 10^6$ pixels are 6.47 μm square. The addition of a 1.6:1 reducing fiber-optic taper between the smaller CCD sensor and the larger dual-MCP image intensifier output creates an effective square pixel with 10.35-μm sides at the input image plane. The dual-MCP image intensifier has a 50 line-pairs/mm resolution, which corresponds to a 10-μm effective pixel size. An image is created by summing multiple frames. Each frame is 66-ms exposure time (15 frames per second).

The short frame times reduce the dark count accumulation to 1 or 2 dark counts per 66-ms frame, and the Piper software (Stanford Photonics, Inc.) for camera control, image acquisition, and image signal processing further reduces noise by filtering out cosmic ray noise and spot noise from each of the incoming 66-ms frames. The dark count of the dual-MCP (multichannel plate) with 1.4×10^6 pixels is then 2.1×10^{-5} electrons per pixel per second. The frames can be stacked to form images composed of 5×10^4 frames or more, which enabled image times from 66 ms to 8 hours or more. The image times in our experiments ranged from 5 to 50 minutes.

The dual-MCP CCD night-vision camera used in our neutron imaging experiment, which stacks short–time-exposure frames, is designed to be limited by the photon noise (shot noise) rather than the dark noise and camera electronic noise. The photon noise is the natural variation of the photon flux emitted from the scintillator. That is, the dark count levels of the CCD pixels can be ignored compared with the Poisson uncertainty of the photon emission from the scintillator, as well as the Poisson uncertainties of the input neutron signal fluence. The dark noise and camera electronic noise levels were tested by operation of the imaging system for the 5-minute to 1-hour exposure times with the neutron generator turned off. The no-generator testing showed each frame and the stacked frames to be dark. Speckle and noise were readily apparent only when the generator was emitting FNs.

Imaging large objects required a large plastic scintillator placed outside the camera box; (Figure 150). A large 63.5 cm × 99-cm area × 38-mm-thick BC408 scintillator was used in which the CCD camera was able to image only a 25 cm × 38-cm area of the large scintillator. The external large plastic scintillator required fabrication of a light-tight tunnel between the open CCD camera box and the large-area scintillator. The scintillator itself was light-tight wrapped with aluminum-coated Mylar on all sides except the interior, light-emission side facing the 45° mirror of the camera box. The assorted objects were placed at a distance of 124 cm from the 45° camera mirror and 3.2 m from the DD generator target. The neutron exposure time of the assorted objects was 5.5 minutes.

The photos in Figure 150 are 5.5-minute FN contact radiographic images (top and middle) and show a variety of objects that were shot together, including the four-step Rose phantoms of lead, silicon, polyethylene, and delrin. The Rose phantom was obtained in four steps with thickness 6.35 mm, 12.7 mm, 25.4 mm, and 50.8 mm. The phantoms also included parallel polycarbonate plates with 4.8-mm vertical thickness, 3-cm horizontal thickness with 4.8-mm air gaps, a 3-cm thick polycarbonate plate

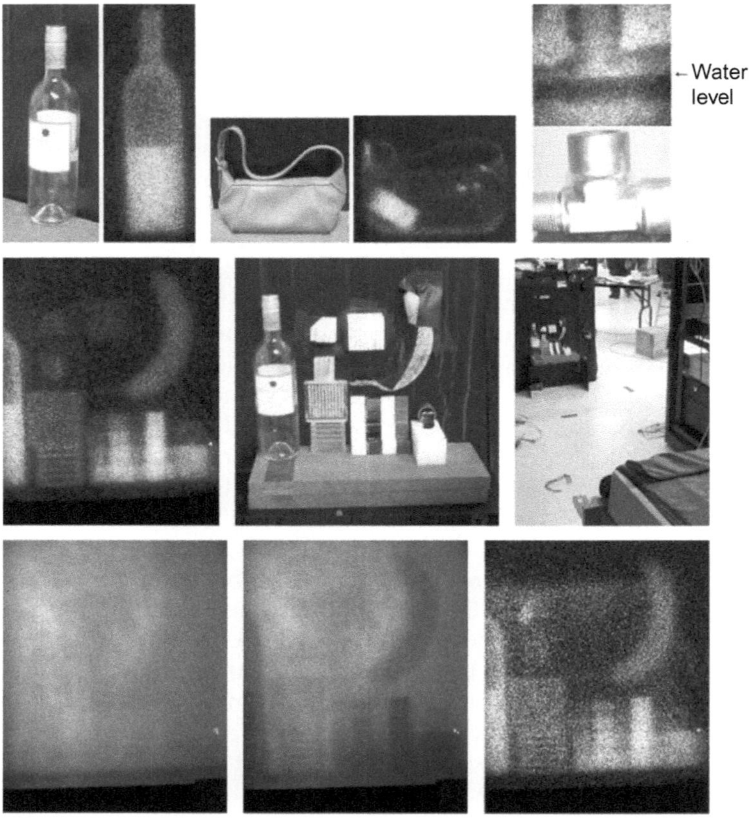

FIGURE 150 Photos and false-color fast neutron contact radiographic images of objects, such as the half-filled wine bottle, which were acquired in a 5.5-minute exposure time with objects placed 3.2 m from the Adelphi Technology DD-109 fast neutron generator target with 10^9 n/s yield emission of 2.5-MeV neutrons. (Bottom left) Fast neutron contact radiographic image of background only. (Bottom middle) Fast neutron contact radiographic image of assorted objects. (Bottom right) Background-subtracted radiographic image produced by pixel-by-pixel subtraction of the background-only image from assorted objects image using CCD camera software (or ImageJ) (Cremer *et al.*, 2012, 2009). (See Color Insert.)

with 3-mm diameter holes, and rods 3 cm long with 3-mm diameter made of polycarbonate, brass, and wood. Other assorted phantoms included the half-filled wine bottle, banana, and a beryllium biconcave neutron lens with a piece of securing black tape. The wine bottle (half-filled with water) object from Geyser Peak Winery was provided by collaborator D. Williams, who recommended this Sonoma County, California wine, and the pink purse, containing a cell phone, was donated by his wife.

One exception in the images was the brass pipe half-filled with water shown at the top right of Figure 150. The brass pipe, half-filled with water, was obtained in a 30-minute exposure with the pipe set at a 30-cm distance from the Adelphi Technology DD-108 generator target. The pipe was imaged using the small 10 cm × 10-cm area, 5-mm thick BC408 plastic scintillator, which was on the inner surface of the 3-mm thick aluminum lid of the neutron camera box.

The contact radiographic FN images can sometimes be enhanced by first obtaining an image of the background without the sample object. Next the sample object is placed adjacent to the plastic scintillator and imaged with an equal exposure time and the background-only image. Finally, the CCD camera software can be used to perform a pixel-by-pixel intensity subtraction of the background-only image from the image with the sample object.

The bottom left of Figure 150 is the FN radiographic image of the background only (no object) produced by the Stanford Photonics CCD camera summation of 5000 stacked frames (15 frames per second frame rate for 0.066-second frames) or 5.5-minute exposure. The background-only radiographic image used in the background subtraction was produced by averaging pixel by pixel six background-only 5.5-minute radiographic exposures. The FN radiographic image of Figure 150 (bottom middle) was a 5.5-minutes exposure (5000 stacked frames) with the assorted objects placed adjacent to the 25 cm × 38 cm active area of the 38 mm thick BC-408 plastic BC-408 scintillator.

The radiographic image of Figure 150 (bottom right) was the background-subtracted radiographic image, which resulted from pixel-by-pixel subtraction of the background-only image from the image of the assorted objects. The background subtraction was performed with the CCD camera software. Background subtraction and other image processing can be performed with ImageJ, which is provided by the U.S. National Institutes of Health (NIH) as freeware (website http://rsbweb.nih.gov/ij/). The overall image resolution of 2.5 mm was determined by a knife-edge measurement, which was compared with a Monte Carlo simulation. In both the thick knife-edge and imaging experiments the resolution was dominated by neutron scatter rather than the neutron divergence angles.

In some radiographic samples the attenuation cross section $\sigma(E)$ varies strongly with energy E of the interrogating X-ray, gamma, or neutron. For FN radiography this allows the possibility of fast neutron resonance radiography (NRR). The fast DD and DT neutron generator sources emit neutrons whose energy E varies as a function of the emission angle θ of the neutron from the generator target. This range of neutron energies emitted over the range of neutron emission angles include the nominal, average 2.5-MeV DD and 14-MeV DT neutron energy. The angle variation of the emitted DD or DT neutron energy can be applied to NRR as will be shown.

Preston *et al.* (1954) presented empirical formulas and measured data for the differential DD neutron cross section $d\sigma/d\Omega$ as a function of polar angle θ and deuteron kinetic energy W for deuteron kinetic energies ranging from 150 to 450 keV. Booth *et al.* (1956) presented empirical formulas and measured data for the differential DD neutron cross section $d\sigma/d\Omega$ as a function of polar angle θ, which is measured relative to the direction of the deuteron ion beam that impacts the deuterated target, and deuteron kinetic energy W for deuteron kinetic energies ranging from 40 to 90 keV. Nargolwalla and Przybylowicz (1973) tabulated DD and DT neutron energy E, yield Y, and the differential DD neutron cross section $d\sigma/d\Omega$ as a function of polar angle θ and deuteron projectile energy W. For example, Table 2.3 in their book tabulates the DD neutron energy E as a function of scattered polar angle θ and deuteron projectile energy W.

The DD reaction in the lab frame produces a modest anisotropic, but azimuthally symmetric neutron emission. The maximum neutron flux is emitted in the 0° forward direction, the minimum flux emitted in the 90° side direction, and almost-average flux emitted in the 180° backward direction. The energy of the emitted DD neutrons is azimuthally symmetric but varies moderately with the polar angle θ of emission relative to the direction of the D^+ ion beam impacting the neutron generator target. For deuterium D^+ ion impact energies on the deuterium target of 120 keV, the forward-emitted (0°) DD neutrons had maximum energy 2.9 MeV, the back-emitted neutrons (180°) had minimum energy 2.1 MeV, and the side-emitted (90°) neutrons had average energy 2.5 MeV.

In a few cases, one can obtain images dominated by one of the elements of a sample via obtaining FN images in two separate directions, provided there is relatively low gamma emission from the DD generator target and surrounding structure. Consider two FN images taken of a single sample of thickness d at angles $\theta_{\min}$ and $\theta_{\max}$, which correspond to emitted DD neutrons energy $E_{\min}$ and $E_{\max}$ in which the interrogated sample material has linear attenuation coefficients are $\mu_{\min}(E_{\min})$ and $\mu_{\max}(E_{\max})$. The relative intensity R_{21} of the neutron image of the sample taken at angle $\theta_{\max}$ relative to angle $\theta_{\min}$ is

$$R_{21} = \exp[-d(\mu_{\max} - \mu_{\min})]. \tag{229}$$

That is, R_{21} is the ratio of the transmitted intensity through the sample at angular positions 2 and 1. If the major contribution to the difference between $\mu_{\max}(E_{\max})$ and $\mu_{\min}(E_{\min})$ is from a single isotope A, then by image subtraction one obtains a crude density map image of isotope A.

Chen and Lanza (2002) devised an elemental imaging method with applications in contraband detection and mineral analysis, which is based on NRR (see also Chen's 2001 Ph.D. dissertation; patent by Chen and

Lanza, 2004; and papers by Chen and Lanza, 2002; Lanza, 2002; Raas *et al.*, 2005; Lanza, 2009; and Dangendorf *et al.*, 2002). The NRR technique provides a 2D elemental mapping of hydrogen, carbon, nitrogen, oxygen, and the sum of other elements in an object, which is interrogated by FNs and imaged by a plastic scintillator with a CCD camera. The object is imaged by FNs at different neutron energies, which are chosen to cover the resonance cross section features of one or more elements.

In preliminary experiments Chen (2001) and Chen and Lanza (2002, 2004) produced NRR images of various simulants using a variable energy beam. They used the University of Massachusetts at Lowell Van de Graaff accelerator to accelerate protons into a lithium target and deuterium ions into a deuterium target, which emitted FNs whose energy varied with the neutron emission angle θ. The neutron emission angle θ was measured relative to the direction of the incident ion beam that impacted the target. They accelerated 2- to 6-MeV protons with currents of 5–10 μA into a lithium-7 target, which produced $5 \times 10^7\,\mathrm{n/sr/s}$ at $\theta = 0°$. The proton beam with lithium target reaction is

$$^{7}Li + p \rightarrow {}^{7}Be + n - 1.646\ MeV, \tag{230}$$

and

$$^{7}Li + p \rightarrow {}^{7}Be^{*} + n - 2.076\ MeV. \tag{231}$$

They also used a DD fusion FN source via acceleration of deuterium ions into a deuterated target. The deuterium beam with deuterium target is

$$D + D \rightarrow {}^{3}He + n + 3.27\ MeV. \tag{232}$$

By using the angular variation in neutron energy in both of these neutron-producing reactions, Chen and Lanza (2002, 2004) scanned through various neutron energies by rotating the detector with the attached object around the FN source (Figure 151). The NRR imaging results (Figure 152) were performed using the 2- to 6-MeV, 5–10 μA proton beam bombardment of lithium to produce the FNs. Single distinct peaks of maximum neutron cross section (resonant peaks) with maximum FN attenuation are found in the range of neutron energies from 1.5 to 2.5 MeV (top left of Figure 151). The resonant peaks are $\sim$10 to 30 keV, which requires a thin lithium target to maintain an emitted neutron beam of almost mono-energetic energy $E(\theta)$ neutrons at each neutron emission angle θ from the target.

The object and attached detector intercept a cone angle of $\Delta\theta = 10°$ in which the cone axis is centered about neutron emission angle θ from

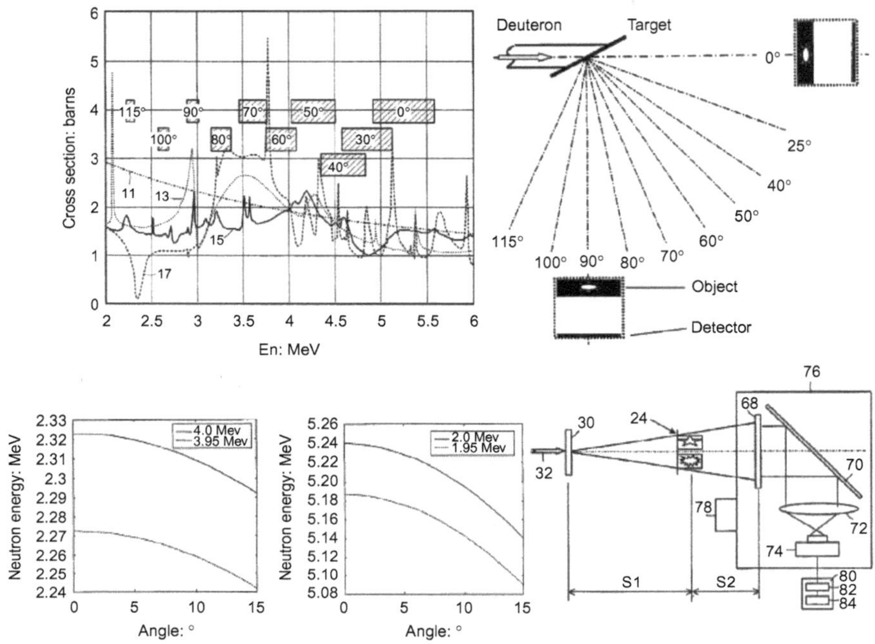

FIGURE 151 (Top left) Plot of the neutron cross section (barns) for 2 to 6 MeV for hydrogen (curve 11), carbon (curve 13), nitrogen (curve 15), and oxygen (curve 17); the neutron emission angle for a particular neutron energy is given in the boxes. (Top right) The schematic shows an object attached to detector was rotated about the target of the neutron generator to obtain radiographs at different neutron energies. (Bottom left) Plot of neutron energy variation versus emission angle for 4.0-MeV and 3.95-MeV protons with lithium target. (Bottom middle) Neutron energy versus emission angle for 2.0-MeV and 1.95-MeV deuterons with deuterium target. (Bottom right) Schematic of D ion beam-32 hitting deuterated target-30 with sample-24 set at source to object distance s_1 30 cm and object to detector distance s_2 30 cm. Object-transmitted neutrons impact scintillator (68) of the detector with shielding (78), and scintillator light is reflected by the 45 mirror (70) to the focusing lens (72) and imaged onto the CCD camera (74) with the computer (80) (Chen and Lanza, 2002, 2004).

the generator target. Thick lithium targets would produce broad range of neutron energies $\Delta E(\Delta\theta$ 10) in the 10 cone angle. However, for NRR imaging the neutrons should be almost mono-energetic with approximate energy $E(\theta)$ within the $\Delta\theta$ 10 cone angle. That is, the variation of neutron energy ΔE within cone angle $\Delta\theta$ 10 must be less than the energy difference ΔE_{on-off} between (1) the neutron energy for the peak absorption cross section (and sample attenuation) and (2) the off-resonance neutron energy in which the absorption cross section and sample attenuation is significantly smaller.

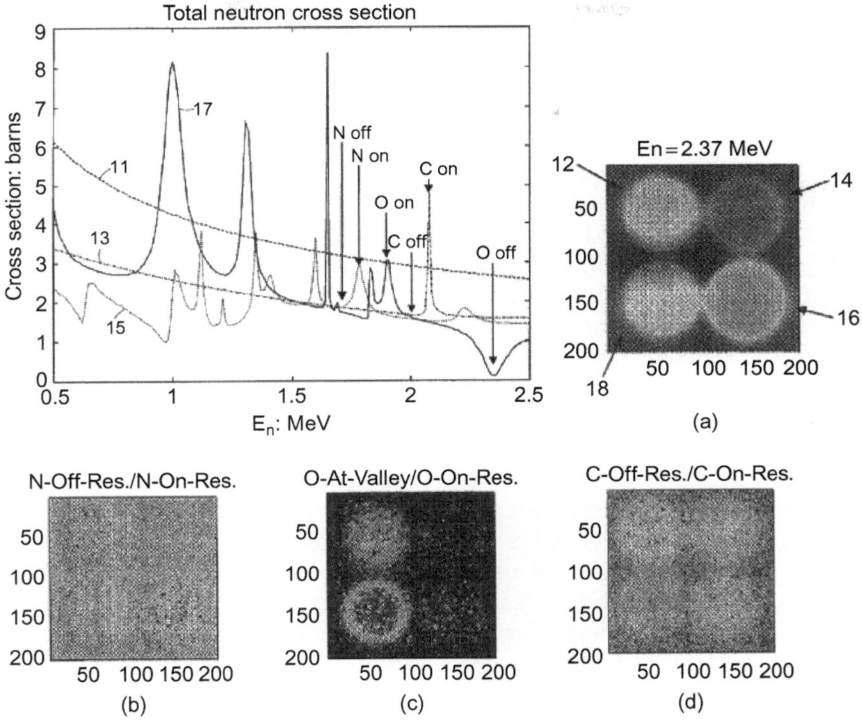

FIGURE 152 (Top left) Plotted are neutron cross section vs. neutron energy in hydrogen (11), carbon (13), nitrogen (15), and oxygen (17). Shown are maximum cross section at on-resonance neutron energy and minimum cross section at off-resonance neutron energy. (top right – 2a) Shown are 2.37 MeV radiographic images of sample comprised of drug stimulant (12), graphite powder (14), explosive stimulant (18), and melamine (16). (bottom) Radiographic subtraction images with on-resonant-energy radiograph subtracted from less attenuated, off-resonant-energy radiograph for nitrogen (left – 2b), for oxygen (middle – 2c), and for carbon (right – 2d) (Chen and Lanza, 2002, 2004).

The top left of Figure 151 plots the neutron cross sections from Chen and Lanza (2002, 2004) for hydrogen (curve 11), carbon (curve 13), nitrogen (curve 15), and oxygen (curve 17) over a range of neutron energies from 2 to 6 MeV. The boxes above the peak cross sections for certain neutron energies show the corresponding angle of emission for the neutrons of that energy. In the Chen and Lanza (2002, 2004) experiments the detector and attached object can be rotated about the neutron generator target to vary the incident neutron energy striking the sample as shown in the top-right schematic in Figure 151.

The bottom-left plot in Figure 151 (from Chen and Lanza, 2002) is the variation of the neutron energy as a function of the neutron emission angle for 4.0-MeV and 3.95-MeV proton bombardment of the lithium target. The bottom-middle plot is a variation of the neutron energy as a function of the neutron emission angle for 2.0-MeV and 1.95-MeV deuterium ion bombardment of a deuterium target. The bottom-right schematic shows the Chen and Lanza (2002, 2004) FN imaging detector and setup.

Four test compositions were imaged with FNs at different DD neutron energies by rotating the four samples to different angles about the generator target (see Chen and Lanza, 2002, 2004). The four samples, contained in small cylindrical containers (35-mm photographic film containers), were a drug stimulant, an explosive stimulant, graphite powder, and melamine. The DD generator target source to sample distance was $s_1 = 30\,\text{cm}$ and sample to detector distance was $s_2 = 30\,\text{cm}$ in which the samples were within a 10° beam cone.

The top-left graph in Figure 152 is the neutron cross section from Chen and Lanza (2002, 2004) in hydrogen (curve 11), carbon (curve 13), nitrogen (curve 15), and oxygen (curve 17) for energy range 0.5 to 2.5 MeV. The plots show the three resonant peaks (cross section and attenuation peak) carbon, nitrogen, and oxygen. The corresponding three neutron energies at the resonant peaks were used to produce three on-resonance radiographic images of the four samples at the respective resonant energies for carbon, nitrogen, and oxygen. Also shown are the three off-resonance neutron energies at minimum cross section and attenuation, which were used to produce the corresponding off-resonant radiographic images for carbon, nitrogen, and oxygen.

The top-right image (labeled Figure 152a) in from Chen and Lanza (2002, 2004) is a 2.37-MeV radiographic image of the four samples each occupying a quadrant of the image. The drug stimulant (12) is in top-left quadrant of the image, the graphite powder (14) is in the top-right quadrant, the explosive stimulant (18) is in the bottom-left quadrant, and melamine (16) is in the bottom0right quadrant. Three radiographic images of the four-simulant sample were made at the on-resonance neutron energies for the peak cross sections of carbon, nitrogen, and oxygen, and three radiographic images were made at the less-attenuating, off-resonance energies for carbon, nitrogen, and oxygen (top left of Figure 152).

Next, for each isotope of interest (carbon, nitrogen, and oxygen), the more-attenuated radiographic image at the isotope on-resonant energy was subtracted from the radiographic image at the isotope off-resonant energy. The three radiographic subtraction images from Chen and Lanza (2002, 2004) are shown at the bottom of Figure for nitrogen (left, labeled Figure 152b), oxygen (middle, labeled Figure 152c), and carbon (right, labeled Figure 152d). The subtraction images for nitrogen, oxygen, and carbon thus map the respective carbon, oxygen, and nitrogen density in the drug stimulant (top-left quadrant image), graphite (top-right quadrant image), explosive (bottom-left quadrant image), and melamine (bottom-right quadrant image).

The nitrogen-rich drug, explosive, and melamine stimulants appear in the nitrogen image (bottom left, labeled Figure 152b) in Figure 152, whereas the carbon-only graphite does not appear. The oxygen-rich drugs and explosive appear in the oxygen image (bottom middle, labeled Figure 152c), but melamine and graphite, which do not contain oxygen, do not appear in the oxygen image. Finally, the carbon image (bottom right, labeled Figure 152d) clearly shows the carbon-based graphite and the carbon-rich drugs, explosives, and melamine.

The radiographic transmission image provided a 2D mapping of the sum of elemental contents, which was weighted by the attenuation coefficients. The set of transmission measurements taken at the different polar angles θ_m with corresponding neutron energy E_m are formed into a set of linear equations. The mth angle θ_m is measured relative to the direction of the deuterium ion beam that strikes the generator target. If M radiographs of the object are taken at different energies E_m, and the object is composed of N isotopes, then the values of MN neutron linear attenuation coefficients $a_{m,n}$ are required. That is, the value of the linear attenuation coefficient $a_{m,n}$ is needed for each nth isotope of the $n = 1, 2, 3, \ldots, N$ isotopes, which are evaluated at each mth energy E_m of the $m = 1, 2, 3, \ldots, M$ neutron energies E_m.

Each mth energy radiograph is obtained by imaging the object at the mth angle of neutron emission with emitted neutron energy E_m. The fraction x_n of the nth isotope atoms in the imaged object, and the measured linear attenuation b_m of the object at the mth energy, are related by the set of M equations from Chen and Lanza (2002, 2004), where

$$\sum_{n=1}^{N} a_{m,n} x_n = b_m \quad \text{for} \quad m = 1, 2, 3, \ldots, M. \tag{233}$$

In the $M \times N$ matrix form the relation is expressed as

$$\begin{bmatrix} a_{1,1} & a_{1,2} & \cdots & a_{1,n} \\ a_{2,1} & a_{2,2} & \cdots & a_{2,n} \\ \cdots & \cdots & \cdots & \cdots \\ a_{m,1} & a_{m,2} & \cdots & a_{m,n} \end{bmatrix} \begin{bmatrix} x_1 \\ x_2 \\ \cdots \\ x_n \end{bmatrix} = \begin{bmatrix} b_1 \\ b_2 \\ \cdots \\ b_m \end{bmatrix}. \tag{234}$$

The simultaneous equations are solved for the fraction x_n of atoms of each nth isotope of N isotopes in the imaged object. This yields a 2D map of the individual isotope contents of the FN interrogated object.

6. A BRIEF HISTORY OF NEUTRON OPTICS BEFORE COMPOUND REFRACTIVE NEUTRON LENSES

This section presents a brief and incomplete history of neutron optics. For more details on the early history of neutrons and neutron optics, see the books by Hughes (1954), Bacon (1987), and Amaldi *et al.* (1998), and papers by Goldberger and Seitz (1947) and Klein and Werner (1983). Neutron optics began shortly after the discovery of the neutron in 1932. The previous research and development of X-ray optics, starting in 1895, served as a precursor guide for the later development of neutron optics after 1932.

Soon after the discovery of the neutron by Chadwick (1932a,b) the wavelike properties of neutrons were demonstrated by measured neutron cross sections for short-range interactions with nuclei. These measured cross sections were unusually large compared with the expected geometrical cross section based on the particle picture (see Amaldi, 1977). Elsasser (1936) first proposed neutron diffraction by crystals should occur based on the quantum mechanical picture of neutrons. The wavelike properties of neutrons and their diffraction from crystals were confirmed in experiments performed by Von Halban and Preiswerk (1936) and Mitchell and Powers (1936). They showed the thermal neutrons had de Broglie wavelike properties with wavelengths in the same range as X-rays (see Klein and Werner, 1983).

Fermi (1936) had developed a semi-empirical theory for modeling neutron scatter interaction with a nucleus. A number of experiments also were performed before 1947 that strongly indicated the scatter of neutrons was different in polycrystalline mediums than in noncrystalline mediums (see Beyer and Whittaker, 1940, and Anderson *et al.*, 1962). Atomic and nuclear collision theory was well developed in 1933 by Mott and Massey (1933), and wave mechanics was well established in the 1930s and 1940s by Heitler (1944) and Schiff (1949).

Wick (1937) asserted the neutron scatter in a crystalline medium was completely different from neutron scatter in gases and amorphous materials due to the interference effect. Prior to 1937 Teller had originally suggested thermal neutron scatter by polyatomic systems could be modeled by wave mechanics and the Breit–Wigner formalism (Breit, 1933), on which basis Schwinger and Teller (1937a,b) published quantitative predictions. These quantitative predictions were experimentally verified by Halpern *et al.* (1937a) and Alvarez and Pitzer (1940). (Also see Feshbach *et al.* (1947) and Blatt and Weisskopf, 1952).

Following Wick (1937) were numerous papers detailing the scatter of neutrons by crystals. More carefully than Wick's work, the influence of temperature on neutron scatter cross section was examined by Pomeranschuk (1938). Neutron scatter by paramagnetic media was investigated by Van Vleck (1939), and Halpern *et al.* (1941) treated elastic, Laue–Bragg neutron scatter from a simple crystal lattice (see also Goldberger and Seitz, 1947). The emission of thermal quanta or phonons due to inelastic neutron scatter with a crystal lattice and the governing selection rules was qualitatively treated by Seeger and Teller (1942). And in 1944, Weinstock produced a more complete theory of inelastic neutron scatter and used his theoretical results to calculate the effect of temperature on the neutron scatter cross section in a polycrystalline medium. Zinn (1947) and Fermi and Marshall (1947) performed neutron scattering experiments with single crystals at Argonne National Laboratory, and Borst *et al.* (1946) and Wollan and Shull (1948) performed similar work at Clinton Laboratories of the Manhattan Project. Their scattering experiments demonstrated a substantial component of the neutron scatter intensity was due to coherent scatter that followed the Laue–Bragg equations.

Van Hove (1954) derived the general intensity of neutrons scattered by a group of atoms; he used Fermi's approach in which the actual neutron-nucleus interaction can be modeled by effective potential (pseudo-potential) that is weaker than the actual interaction potential. This pseudo-potential is sufficiently weak to allow application of the Born approximation. The Born approximation is a perturbation method applied to a scattering group of atoms in which the scattered wave amplitudes from a group of atoms are assumed to be small compared with the amplitude of the wave incident upon the group of atoms. That is, each scattering atom is assumed to encounter only the incident neutron wave, and the scattered waves from all the other scattering atoms, which are incident on the atom, are ignored (see Pynn, 1990).

The various coherent and incoherent effects of multiple scattering of neutrons are reviewed by Lax (1951, 1952) and Foldy (1945). Direct measurements of neutron coherent cross sections were done by Halpern *et al.* (1937a), Sutton *et al.* (1947), McReynolds and Weiss (1951), and Placzek (1952). Much of the early work on the development of neutron powder diffraction was done by Wollan and Shull (1948) at Oak Ridge National Laboratories (ORNL) in the 1940s. Charge independence of nuclear scattering forces and the effective range of nuclear scatter were presented in the theory paper by Bethe (1949). Early inelastic neutron scatter experiments were performed at Chalk River Laboratories in Canada and the Atomic Energy Research Establishment (AERE) in Harwell, England, in which Brockhouse and Hurst (1952) observed neutron energy changes in inelastic scatter of neutrons in cadmium. The Debye–Waller factor for lead was experimentally measured and compared with the theory by Shull and Wollan (1951).

Alvarez and Bloch (1940) provided the first measurement of the neutron magnetic dipole moment. Van Vleck (1939) investigated the scattering of neutrons from paramagnetic materials, and Halpern and Johnson (1939) considered the Laue–Bragg scatter from simple lattices. The interaction of the spin magnetic moment of an atom and a neutron was developed in detail by Halpern and Holstein (1941), Halpern *et al.* (1941), and Hamermesh and Johnson (1942). This work included magnetic neutron scatter from isolated atoms and from ferromagnetic and paramagnetic crystals for both polarized and unpolarized neutron beams.

Neel (1948) generalized his theory of anti-ferromagnetic materials. The zero net magnetic moment of MnO, which was predicted by Neel, was experimentally verified in one of the first applications of neutron diffraction to magnetic materials by Shull and Smart (1949). Also, Shull *et al.* (1951) used magnetic neutron powder diffraction to determine (1) the variation of intensity from magnetite with direction of magnetization and (2) the structure of magnetite Fe_3O_4, which agreed with the anti-ferromagnetic structure proposed by Neel. The quantitative difference of the magnetic form factor with the similar atomic form factor for iron was investigated by Steinberger and Wick (1949). The two indexes of refraction and resulting two angles of reflection from iron mirrors were measured in experiments by Hughes and Burgy (1951), which led to the technique for producing polarized neutron beams.

The storage of ultracold neutrons was predicted by Zeldovich (1959) and experimentally verified by Lushchikov *et al.* (1969) and Steyerl (1969). Thermal, cold, very cold, and ultracold neutron experiments revealed the wave behavior of neutrons—namely, diffraction, refraction, and reflection, as well as elastic and inelastic coherent and incoherent scatter—was similar to that of classical optics. Neutron refraction by gravity and magnetic fields was investigated experimentally and theoretically by Frank (1987, 1989) with ultracold neutrons. A magnetic neutron lens and magnetic neutron prism were fabricated and experiments were performed by Just *et al.* (1973). Stern–Gerlach gradient magnetic field deflection of neutrons was performed by Jones and Williams (1980). Polarized neutrons and magnetic neutron optics for polarized neutrons are discussed thoroughly by Williams (1988).

In early neutron experiments the source of neutrons was nuclear reactions such as radioactive polonium alpha source mixed with powdered beryllium. The alphas from polonium transmuted the beryllium-7 to carbon-12 and a fast neutron, which is moderated by a hydrogen-rich moderator. Early thermal neutron imaging was readily accomplished by a sheet of film placed in tight contact against a foil converter. The thermal neutron capture in the foil of gadolinium or samarium produced gammas, which then exposed the film. Following the first controlled fission reactor by Fermi and his colleagues under the abandoned west stands of the original Alonzo Stagg Field at the University of Chicago stadium (called Chicago Pile-1 or CP-1), the development of fission reactors ("pile neutron sources") during and after World War II took place (see

Hughes, 1953). Pulsed reactors were developed in the Soviet Union (see Shabalin, 1979).

Reactor- and accelerator-based neutron sources for researchers began after World War II in the United States, Soviet Union, and Europe. The Brookhaven Graphite Research Reactor (BGRR) was the first peacetime reactor constructed in the United States after World War II and operated from 1950-1968. With the experience gained with the BGRR, the High Flux Beam Reactor (HFBR) attained a self-sustaining chain reaction on Halloween (October 31, 1965) and operated at 30–60 MW. In the HFBR, the maximum neutron flux was reached outside the reactor core. As a result, the neutrons were readily available for experiments outside the core, rather than inside the core, as with most reactors.

Neutron beams were delivered to experimenters through beam ports with thermal neutron beam fluxes up to $5 \times 10^{14}\,n/cm^2/s$. The neutrons were made to exit the reactor core on a tangent, rather than radiating outward as was typical in previous reactors. This improvement in HFBR yielded slower neutrons, which are desirable for most research, and was an innovation incorporated in later reactors. In 1996 leaked tritium was found in the groundwater south of the HFBR reactor, which led to the shutdown of the Halloween reactor in 1996 (see "Brookhaven History: High Flux Beam Reactor"; http://www.bnl.gov/bnlweb/history/HFBR_main.asp.)

In 1957 the 120-MW National Research Universal Reactor was commissioned at Chalk River in Canada with a thermal neutron flux $5 \times 10^{14}\,n/cm^2/s$. Ten years later the 85-MW High Flux Reactor commissioned in 1967 at ORNL, which produced a thermal neutron beam source with flux $10^{15}\,n/cm^2/s$ and also had a cold source. And in 1967 a 10-MW reactor was commissioned at the National Institute of Standards and Technology (NIST), which included a beam port for cold neutrons. The NIST reactor was upgraded to 20 MW in 1985 with thermal neutron flux $1.5 \times 10^{14}\,n/cm^2/s$ in nine radial beams. A D_2O cold source was added in 1987, a guide hall of neutron beam lines was added in 1990, a liquid hydrogen LH_2 cold source was added in 1995, and an advanced second liquid hydrogen cold source was added in 2002.

In Europe at the Laue–Langevin Institute (ILL) in Grenoble, a neutron user facility was commissioned in 1972 with a 58-MW high flux reactor that produced a thermal neutron beam flux $10^{15}\,n/cm^2/s$. The ILL high flux reactor is a large facility with many cold and thermal neutron guide tubes. The development of neutron guide tubes in the reactor- and accelerator-based neutron facilities allowed the transport of neutrons over large distances from the reactor- or accelerator-based neutron source with minimal loss of neutron intensity. These neutron guide tubes greatly facilitated the evolution of reactor- and accelerator based neutron sources into user facilities in which the available space allows a large variety of experiments to be conducted simultaneously.

In 1973 the Berlin Neutron Scattering Center began operation with a 10-MW reactor that provided thermal neutron fluxes $2 \times 10^{14}\,n/cm^2/s$ and included a cold source. In 1980 the 14-MW ORPHEE reactor began operation at the Laboratoire Léon Brillouin at the CEA (French Atomic Energy Commision) Saclay research facility near Paris. The ORPHEE neutron source has a thermal neutron flux of $2 \times 10^{14}\,n/cm^2/s$ and includes a cold source.

In 1990 the Japan Atomic Energy Agency commissioned the 20 MW JRR-3M with $3 \times 10^{14}\,n/cm^2/s$ thermal flux and included a cold source. Seven years later (in 1997) the South Korea 24-MW HANARO reactor with $2 \times 10^{14}\,n/cm^2/s$ thermal flux began operation. And in 2004 the 20-MW FRM-II reactor of the Technische Universitat München began operation with a cold source and a thermal flux $8 \times 10^{14}\,n/cm^2/s$. This was followed by the 20-MW OPAL reactor of the Australian Nuclear Science and Technology Organisation (ANSTO) with a cold source and a thermal flux $3 \times 10^{14}\,n/cm^2/s$ (see Anderson *et al.*, 2009).

Subsequent development of accelerator-based neutron sources in the 1970s greatly expanded neutron physics research and enabled neutron optics research and development. The first neutron user facility in the United States was the Intense Pulsed Neutron Source (IPNS) at Argonne National Laboratory, where Carpenter (1973, 1977) and fellow Argonne researchers built the IPNS predecessors, the first pulsed spallation source for neutron scattering, called ZING-P and ZING-P′ in the mid-1970s.

In a spallation neutron source, a high-energy (GeV) pulsed proton beam from a proton linac hits a depleted uranium target, which emits a pulse of fast neutrons, which are moderated by a surrounding polyethylene moderator to produce a pulsed thermal neutron source. At present, the typical moderators used to convert the fast spallation neutrons to thermal neutrons are liquid methane, liquid hydrogen, liquid deuterium, and water. The IPNS began operation as a user facility with 12 beam lines in 1981 with a 30-Hz repetition rate, 7-kW proton beam of 450 MeV, and 15 μA. Unfortunately and sadly, IPNS unexpectedly closed in 2008 due to lack of funds, the result of a significant reduction in the budget of the Basic Energy Sciences by the financially strapped U.S government.

In 1977 at Los Alamos National Laboratory, a pulsed spallation neutron source with a 20-Hz, 56-kW, 800-MeV, 70-μA linac proton beam and a tungsten spallation target was commissioned to produce fast and moderated neutrons for time-of-flight experiments. And beginning in 1985, with the completion of the Proton Storage Ring (PSR) that compresses proton pulses from 750 μs to a quarter of a microsecond, the Los Alamos Neutron Scattering Center (LANSCE), now known as the Lujan Center, was established with seven instruments.

In 1980 a pulsed neutron faciltiy (ISIS) at Rutherford Appleton Laboratory was commissioned, which uses a 50-Hz linac proton beam with 400-ns bursts of 2.5×10^{13} p/s to impact a depleted uranium target to produce a fast neutron yield 4×10^{16} n/s. The fast neutrons are thermalized by four moderators set close to a target and embedded in a beryllium reflector. Together they moderate neutrons to epithermal, thermal, and cold neutrons and produced pulse lengths in the range of 1–100 μs (see Fender *et al.*, 1980).

In Japan the High Energy Accelerator Research Organization KENS spallation source, which used a 20-Hz, 4.5-kW, 500-MeV, 9-μA linac proton beam and tungsten spallation target, was commissioned in 1980 and closed in 2005. KENS was followed by the 2008 operation of the Japan Atomic Energy Agency JSNS, which uses a 25-Hz, 1-MW, 3000-MeV, 333-μA linac proton beam with mercury spallation target.

The SINQ spallation source at the Paul Scherrer Institute in Switzerland opened in 1996 with a continuous 1-MW, 590-MeV, 1500-μA cyclotron proton beam that uses a zircaloy spallation target. And 10 years later in 2006, the most intense source of pulse neutron beams, the Spallation Neutron Source at ORNL, was completed (see the "ORNL: Neutron Sciences" website http://www.sns.gov/). The SNS spallation source uses a mercury spallation target and a 60-Hz, 1.4-MW, 1000-MeV, 1400-μA linac proton beam (see Anderson *et al.*, 2009).

The development of chopper and crystal monochromators and cold moderators, neutron wave guides, polarizing Soller guides, helium-3 detectors, and other such neutron instruments and components enabled the use of neutrons as a basic research instrument (see Williams, 1988). The triple-axis spectrometry method was first developed by Bertram Brockhouse at the National Research Experimental (NRX) reactor at the Chalk River Laboratories in Canada. In 1994 the Nobel Prize was jointly awarded to Bertram N. Brockhouse of Chalk River and then McMaster University for the development of neutron spectroscopy and to Clifford G. Shull of ORNL and then MIT for the neutron diffraction technique. The triple-axis spectrometer was further developed for neutron polarization analysis by Moon *et al.* (1969).

The polarized neutron diffraction method of applying polarized neutron scatter from single crystals to determine the distribution of magnetism on an atomic level was developed by Nathans and Paoletti (1959) and Nathans (1959). Polarized neutron diffraction experiments were performed on iron by Shull and Yamada (1962), on cobalt by Moon (1964), and on nickel by Mook (1966).

Refraction of neutrons by prisms, such as the bi-prism interferometer developed by Maier-Leibnitz (1962). The experimental determination of neutron refractive index scatter lengths of materials fabricated into refractive prisms was performed by Shull *et al.* (1967) and Schneider and Shull (1971). Koester (1977) presented a compilation of neutron scatter lengths and compared the various methods of scatter length measurement. At present, the experimental measurement neutron cross sections as a function of imposed conditions (e.g., temperature, pressure, pH) are best measured rather than theoretically predicted, as shown in Freund's (1983) paper on measured and theoretical neutron scatter cross section from single crystals and polycrystals. The properties of polarizing monochromator crystals are summarized by Freund and Forsyth (1979).

Polarizing cold neutron beams by mirror reflection has been a very useful tool ever since Hughes and Burgy (1951) first produced a 100% polarized neutron beam with a cobalt mirror. The reflection properties of neutrons were also exploited in the development of neutron guides, which transport neutrons by multiple reflections with only small intensity attenuation in the same manner as a visible light guide or fiber optic, as first proposed by Christ and Springer (1962) and Maier-Leibnitz and Springer (1962).

The development of neutron guide tubes in the 1960s then allowed the development of small-angle neutron scattering beam lines. In the 1970s multilayer mirrors composed of thin metallic films were developed, which allowed large increases in the reflectivity and range of reflection angles as a function of wavelength (see Schoenborn *et al.*, 1974; Mezei, 1976; Goldberger and Seitz, 1947 and Gukasov *et al.*, 1979). The first polarizing neutron Soller guide, which consisted of a large number of mirrors stacked to form a series of parallel and curved polarizing and reflectively conducting channels, was fabricated Hayter *et al.* (1978). An improved polarizing Soller device was developed by Rekveldt and Kraan (1983).

The refractive properties of neutrons were exploited by Gähler *et al.* (1980, 1982). They used a prism monochromator and biconcave lens in a neutron-optical experiment to measure the neutron electric charge, which was found to have an upper limit charge of $10^{-20}e$ compared with the elementary charge e of an electron or proton. Neutron polarizers, which use the Stern–Gerlach effect with applied magnetic field gradients, can be used to directly measure the polarization of cold neutron beams (see Sherwood *et al.*, 1954, and Jones and Williams, 1980).

Hughes and Burgy (1951) first considered the non-adiabatic spin flipping of neutrons with two direct current coils, and spin-flipping of 5-Å neutrons was accomplished by Drabkin *et al.* 1969. Jones and Williams (1980) further developed the neutron spin-flipper. First proposed by Turchin (1967), the neutron supermirror was developed by Mezei (1976) and Mezei and Dagleish (1977) in the 1970s. In 1972 the technique of spin echo with polarized neutron beams was developed by Mezei (1972). Polarized neutron optics and applications are treated in detail in *Polarized Neutrons* by Williams (1988).

By 1954 neutron optics—that is, neutron diffraction, reflection, refraction, and scatter—was well established as evidenced by two books by Hughes (1953, 1954). These two books were followed by many other book publications between 1954 and 1994. See the Background Reading below for a chronological (albeit incomplete) list of books on neutron optics.

BACKGROUND READING

Books on Neutron Diffraction, Reflection, Refraction, and Scatter (1953–1992)

Hughes, D.J. (1953). *Pile Neutron Research.* Addison-Wesley, Cambridge, MA.
Hughes, D.J. (1954). *Neutron Optics.* Interscience Publishers, New York.
Turchin, V.F. (1965). *Slow Neutrons.* Israel Program for Scientific Translations, Jerusalem.
Egelstaff, P.A. (1965). *Thermal Neutron Scattering.* Academic Press, New York.
Boutin, H., & Yip, S. (1968). *Molecular Spectroscopy with Neutrons.* MIT Press, Cambridge, MA.
Gurevich, I.I., & Tarasov, L.V. (1968). *Low Energy Neutron Physics.* North-Holland, Amsterdam.
Izyumov, Y.A., & Ozerov, R.P. (1970). *Magnetic Neutron Diffraction.* Plenum Press, New York.
Marshall, W., & Lovesey, S.W. (1971). *Theory of Thermal Neutron Scattering.* Clarendon, Oxford, UK.
Willis, B.T.M. (ed.). (1970). *Thermal Neutron Diffraction.* Oxford University Press.
Willis, B.T.M. (ed.). (1973). Chemical Applications of Thermal Neutron Scattering. Oxford University Press, Oxford, UK.
Bacon, G.E. (1975). *Neutron Diffraction,* ed. 3. Clarendon, Oxford, UK.
Koester, L., & Steyerl, A. (1977). *Neutron Physics,* vol 80. Springer-Verlag, Berlin.
Dachs, H. (ed.). (1978). *Neutron Diffraction.* Springer-Verlag, Berlin.
Squires, G.L. (1978). *Introduction to the Theory of Thermal Neutron Scattering.* Dover Publications, Mineola, NY.
Marshall, W., Bonse, U., & Rauch, H. (ed.). (1979). *Neutron Interferometry.* Oxford University Press, Oxford, UK.
Windsor, C.G. (1981). *Pulsed Neutron Scattering.* Taylor and Francis, New York.
Lovesey, S.W. (1986). *Theory of Neutron Scattering from Condensed Matter.* Oxford University Press, Oxford, UK.
Sköld, K., & Price, D. (1986). *Neutron Scattering,* vol. 23A (Methods in Experimental Physics). Academic Press, Orlando, FL.
Lovesey, S.W. (1986). *The Theory of Neutron Scattering from Condensed Matter,* vol. II (International Series of Monographs on Physics). Oxford University Press, New York.
Domanus, J.C., & Markgraf, J.F.W. (eds.) (1987). *Collimators for Thermal Neutron Radiography: An Overview.* Springer-Verlag, Berlin.
Williams, W.G. (1988). *Polarized Neutrons.* Clarendon, Oxford, UK.
Sears, V.F. (1989). *Neutron Optics.* Oxford University Press, New York.
Balcar, E., & Lovesey, S.W. (1989). *Theory of Magnetic Neutron and Photon Scattering* (Oxford Series on Neutron Scattering in Condensed Matter). Oxford University Press, New York.

Collins, M.F. (1989). *Magnetic Critical Scattering* (Oxford Series on Neutron Scattering in Condensed Matter). Oxford University Press, New York.
Aksenov, V.L., Plakida, N.M., Stamenkovic, S., & Pontekorvo, D.B. (1990). *Neutron Scattering by Ferroelectrics.* World Scientific, Singapore.
Golub, R., Richardson, D., & Lamoreaux, S.K. (1991). *Ultra-Cold Neutrons.* Taylor and Francis, New York.
Ignatovich, V.K. (1990). *The Physics of Ultracold Neutrons.* Clarendon, Oxford, UK.
Domanus, J.C. (ed.). (1992). *Practical Neutron Radiography.* Springer-Verlag, Berlin

Books on Neutron, X-Ray, and Gamma Sources, Optics, and Detectors after 1992

More recent publications between 1992 and 2011 on primarily neutron, and secondarily, X-ray and gamma sources, optics and detectors, and related topics are included in the following incomplete list.

Carpenter, J.M., Cline, D.B., & Lanza, R. (eds.). (1993). *Neutrons, X Rays, and Gamma Rays: Imaging Detectors, Material Characterization Techniques, and Applications* (Proceedings of SPIE, Society of Photo Optical Engineering). Bellingham, WA.
Winick, H. (ed.). (1994). *Synchrotron Radiation Sources: A Primer* (World Scientific Series on Synchrotron Radiation Techniques). World Scientific, Singapore.
Egelstaff, P.A. (1994). *An Introduction to the Liquid State* (Oxford Series on Neutron Scattering in Condensed Matter). Oxford University Press, New York.
Glyde, H.R. (1995). *Excitations in Liquid and Solid Helium* (Oxford Series on Neutron Scattering in Condensed Matter). Oxford University Press, New York.
Letokhov, S. (1995). *Atom Optics with Laser Light* (Laser Science and Technology), Harwood Academic Publishers, Chur, Switzerland.
Hoover, R.B., & Doty, F.P. (eds.). (1996). *Hard X-Ray/Gamma-Ray and Neutron Optics, Sensors, and Applications* SPIE-International Society for Optical Engineering, Bellingham, WA.
Neumann, D.A., Russell, T.P., & Wuensch, B.J. (eds.). *Neutron Scattering in Materials Science,* vol. 376 (MRS Proceedings). Materials Research Society.
Higgins, J.S., & Benoît, H.C. (1997). *Polymers and Neutron Scattering* (Oxford Series on Neutron Scattering in Condensed Matter). Oxford University Press, New York.
Berman, P.R. (ed.) (1997). *Atom Interferometry.* Academic Press, Chestnut Hill, MA.
Hoff, J. (ed.) (1998). *Capture Gamma-Ray Spectroscopy* (AIP Conference Proceedings 238). American Institute of Physics. New York.
Macdonald, C.A., Goldberg, K.A., Maldonado, J.R., & Marker, A.J. (eds.). (1999). *EUV, X-Ray, and Neutron Optics and Sources* (Proceedings of SPIE, the International Society for Optical Engineering, vol. 3767). Society of Photo Optical Engineering, Bellingham, WA.
James, R.B., & Schirato, R.C. (eds.). (1999). *Hard X-Ray, Gamma-Ray, and Neutron Detector Physics* (Proceedings of SPIE). Society of Photo Optical Engineering, Bellingham, WA.
Roe, R.J. (2000). *Methods of X-ray and Neutron Scattering in Polymer Science* (Topics in Polymer Science). Oxford University Press, New York.
Wender, S. (ed.). (2000). *Capture Gamma-Ray Spectroscopy and Related Topics* (AIP Conference Proceedings). American Institute of Physics.
Rauch, H., & Werner, S.A. (2000). *Neutron Interferometry: Lessons in Experimental Quantum Mechanics* (Oxford Series on Neutron Scattering in Condensed Matter, 12). Oxford University Press, New York.
Hempelmann, R. (2000). *Quasielastic Neutron Scattering and Solid State Diffusion* (Oxford Series on Neutron Scattering in Condensed Matter). Oxford University Press, New York.
Nield, V.M., & Keen, D.A. (2001). *Diffuse Neutron Scattering from Crystalline Materials* (Oxford Series on Neutron Scattering in Condensed Matter, 14). Oxford University Press, New York.

Fujine, S., Kanda, K., Kobayahi, H., & Kanda, K. (2001). *Sixth World Conference on Neutron Radiography* (6). Taylor and Francis, London.

Wood, J.L., & Anderson, I.S. (eds.). (2001). *Neutron Optics* (Proceedings of SPIE). Society of Photo Optical, Bellingham, WA.

Born, M., & Wolf, E. (2002). *Principles of Optics*, ed 7. Cambridge University Press Cambridge, UK.

Kumakhov, M.A. (ed.). (2002). *International Conference on X-ray and Neutron Capillary Optics* (Proceedings of SPIE), Society of Photo Optical Engineering, Bellingham, WA.

Zemb, Th., & Lindner, P. (eds.) (2002). *Neutron, X-rays and Light. Scattering Methods Applied to Soft Condensed Matter* (North-Holland Delta Series). North Holland, Amsterdam.

Shirane, G., Shapiro, S.M., & Tranquada, J.M. (2002). *Neutron Scattering with a Triple-Axis Spectrometer: Basic Techniques.* Cambridge University Press, Cambridge, UK.

Wiedemann, H. (2003). *Synchrotron Radiation* (Advanced Texts in Physics). Springer-Verlag, Berlin.

Fitzpatrick, M.E., & Lodini, A. (eds.). (2003). *Analysis of Residual Stress by Diffraction using Neutron and Synchrotron Radiation.* Taylor and Francis, London.

Dianoux, A.J., & Lander, G. (2003). *Neutron Data Booklet*, ed. 2 (Institut Laue-Langevin). Old City Publishing, Philadelphia.

Mezei, F., Pappas, C., & Gutberlet, T. (eds.) (2003). *Neutron Spin Echo Spectroscopy: Basics, Trends and Applications.* Springer-Veralg, Berlin.

Clarke, J.A. (2004). *The Science and Technology of Undulators and Wigglers* (Oxford Series on Synchrotron Radiation). Oxford University Press, New York.

Del Rio, M.S. (ed.). (2004). *Advances in Computational Methods for X-ray and Neutron Optics* (Proceedings of SPIE). Society of Photo Optical Engineering, Bellingham, WA.

Molnar, G.L. (ed.). (2004). *Handbook of Prompt Gamma Activation Analysis: with Neutron Beams.* Kluwer Academic Publishers, Dordrecht, Netherlands.

Mitchell, P.C.H., Parker, S.F., & Ramirez-Cuesta, A.J. (2005). *Vibrational Spectroscopy with Neutrons: With Applications in Chemistry.* World Scientific, Singapore.

Hippert, F., Geissler, E., Hodeau, J.L., Lelièvre-Berna, E., & J.R. Regnard J.R. (eds.). (2005). *Neutron and X-ray Spectroscopy.* Springer-Verlag, Berlin.

Kumakhov, M.A. (ed.). (2005). *X-ray and Neutron Capillary Optics II* (Proceedings of SPIE). Society of Photo Optical Engineering, Bellingham, WA.

Chatterji, T. (2006). *Neutron Scattering from Magnetic Materials.* Elsevier Science, Amsterdam.

Paganin, D. (2006). *Coherent X-Ray Optics* (Oxford Series on Synchrotron Radiation). Oxford University Press, New York.

Woehr, A., & Aprahamian, A. (eds.). (2006). *Capture Gamma-Ray Spectroscopy and Related Topics: 12th International Symposium* (AIP Conference Proceedings/High Energy Physics). American Institute of Physics.

Erko, A., Idir, M., Krist, T., & Michette, A.G. (2008). *Modern Developments in X-Ray and Neutron Optics.* Springer-Verlag, Berlin.

kram, A., Purwanto, A., Sutiarso, S., Zulfia, A., Hendrana, S., & Nurachman, Z. (eds.). (2008). *Neutron and X-Ray Scattering 2007: The International Conference* (AIP Conference Proceedings/Mathematical and Statistical Phsyics). American Institute of Physics.

Reimers, W., Pyzalla, A.R., Schreyer, A.K. & Clemens, H. (eds.). (2008). *Neutrons and Synchrotron Radiation in Engineering Materials Science: From Fundamentals to Material and Component Characterization.* Wiley-VCH, Weinheim.

Andersson, R.A. (2008). *Spin-Echo Small-Angle Neutron Scattering Measurements.* IOS Press, Lansdale, PA.

Burger, A., Franks, L.A., & James, R.B. (eds.). (2008). *Hard X-ray, Gamma-ray, and Neutron Detector Physics X* (Proceedings of SPIE). Society of Photo Optical Engineering, Bellingham, WA.

Kisi, E.H., & Howard, C.J. (2008). *Applications of Neutron Powder Diffraction* (Oxford Series on Neutron Scattering in Condensed Matter). Oxford University Press, New York.

Willis, B.T.M., & Carlile, C.J. (2009). *Experimental Neutron Scattering.* Oxford University Press, New York.
Anderson, I.S., McGreevy, R.L., & Bilheux, H.Z. (eds.). (2009). *Neutron Imaging and Applications.* Springer, New York.
Furrer, A., Mesot, J., & Strässle, T. (2009). *Neutron Scattering in Condensed Matter Physics.* World Scientific, Singapore.
Duke, P. (2009). *Synchrotron Radiation: Production and Properties* (Oxford Series on Synchrotron Radiation). Oxford University Press, New York.
Kuzmany, H. (2009). *Solid-State Spectroscopy: An Introduction,* ed. 2. Springer-Verlag, Berlin.
Eckold, G., Schober, H., & Nagler, S.E. (2009). *Studying Kinetics with Neutrons: Prospects for Time-Resolved Neutron Scattering* (Springer Series in Solid-State Sciences). Springer-Verlag, Berlin.
Utsuro, M., & Ignatovich, V.K. (2010). *Handbook of Neutron Optics.* Wiley-VCH, Weinheim.
Bass, M., DeCusatis, C., Enoch, J., & Lakshminarayanan, V. (eds.). *Handbook of Optics,* ed 3, vol. V: *Atmospheric Optics, Modulators, Fiber Optics, X-Ray and Neutron Optics.* McGraw-Hill Professional, New York.
Fitter, J., Gutberlet, T., & Katsaras, J. (eds.). *Neutron Scattering in Biology: Techniques and Applications* (Biological and Medical Physics, Biomedical Engineering). Springer-Verlag, Berlin.
Daillant, J., & Gibaud, A. (eds.). (2010). *X-ray and Neutron Reflectivity: Principles and Applications* (Lecture Notes in Physics). Springer-Verlag, Berlin.
Furrer, A. (2010). *Neutron Scattering in Layered Copper-Oxide Superconductors* (Physics and Chemistry of Materials with Low-Dimensional Structures). Kluwer Academic Publishers, Amsterdam.
Richter, D., Monkenbusch, M., Arbe, A., & Colmenero, J. (2010). *Neutron Spin Echo in Polymer Systems* (Advances in Polymer Science). Springer-Verlag, Berlin.
Liang, L., Rinaldi, R., & Schober, H. (eds.) (2010). *Neutron Applications in Earth, Energy and Environmental Sciences* (Neutron Scattering Applications and Techniques). Springer-Verlag, Berlin.
Sivia, D.S. (2011). *Elementary Scattering Theory for X-ray and Neutron Users.* Oxford University Press, Oxford, UK.
Hurd, A.J., & Sinha, S. (eds.) (2011). *Neutrons and Nanoscience* (Neutron Scattering Applications and Techniques). Springer-Verlag, Berlin.
Imae, T., Kanaya, T., Furusaka, M., & Torikai, N. (eds.). *Neutrons in Soft Matter.* John Wiley and Sons, New York.

7. BRIEF HISTORY AND OVERVIEW OF DIFFRACTIVE AND REFLECTIVE X-RAY OPTICS AND MICROSCOPY

The development of X-ray optics began in December 22, 1895, at the University of Würzburg with the first X-ray radiograph of the hand and ring of Wilhelm Conrad Röntgen's wife, Bertha (15 minute exposure) Figure 153. Roentgen handed his X-ray manuscript with radiographic images on December, 28 1895, to the secretary of the Würzburg Physical Medical Society with the title "Eine neue Art von Strahlen" or "A New Kind of Ray." Due to the Christmas holidays he was scheduled to read the paper before the Würzburg Physical and Medical Society later on January 23, 1896. However, Röntgen received the off-prints on New Year's Day,

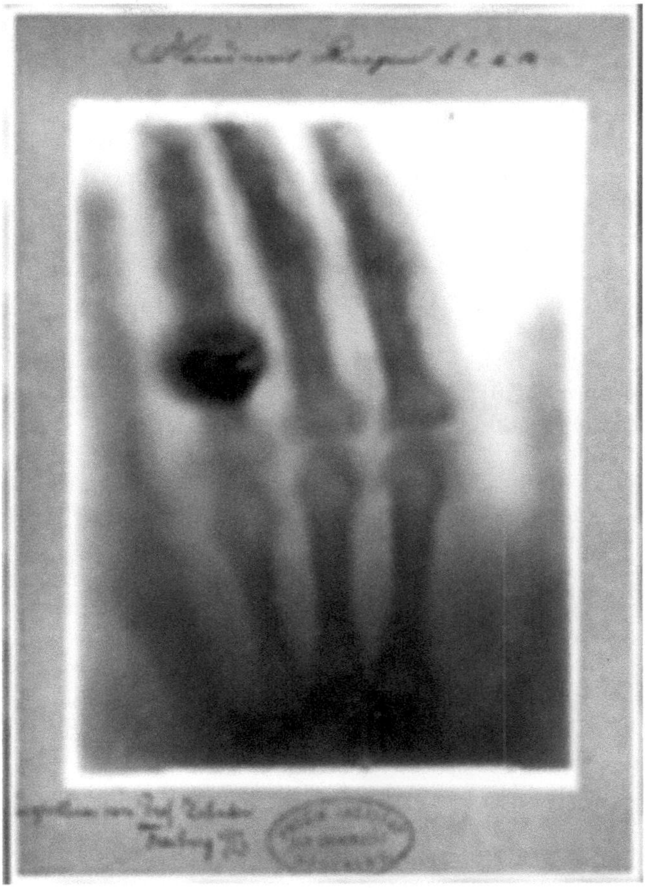

FIGURE 153 Wilhelm C. Roentgen of Würzburg University took this radiograph of his wife Bertha's hand on December 22, 1895, with a 15-minute exposure. The bones of her hand and two rings are clearly shown. This image dates the beginning of radiography. (Image from General Electric Company's Medcyclopaedia website, http://www.medcyclopaedia.com/, W.C Röntgen and the discovery of X-rays).

and he sent several of these copies with a few illustrative pictures to his scientific colleagues.

One colleague and old friend Professor Franz Exner loaned his copy to another colleague, Professor Ernst Lecher of Prague, whose father was editor of a Vienna newspaper *Die Presse,* which published an article on Roentgen's work on Sunday, January 5, 1896, with Roentgen's name misspelled as "Routgen." The next day via London cable the news of "Routgen's" discovery of X-rays was worldwide.

Röntgen's manuscript was then translated by Arthur Stanton into English and published in *Nature* on January 23, 1896 (volume 53, p. 274) on the same day Röntgen finally gave his scheduled oral presentation to the Würzburg Physical and Medical Society (see General Electric, 2011). Before Roentgen's discovery, there had been unexplained observations of exposed photographic plates, which were a result of the ongoing and previous development of electric discharge tubes. This served as another example of a world full of chance occurrences with relatively few prepared observers. The public and the science/medical community became instantly intrigued with these mysterious rays beyond the ultraviolet.

At A. Sommerfeld's institute at the Ludwig Maximilian University of Munich, Walter Friedrich and Paul Knipping acted on a proposed X-ray experiment with crystals by Laue, which produced X-ray diffraction patterns of crystals (Fredrich *et al.*, 1912; Laue, 1912). Their discovery inspired William Henry Bragg and son William Lawrence Bragg (Bragg & Bragg, 1913) to pursue X-ray crystallography in which they developed the X-ray analyis of crystal. A talk with Laue, combined with his later published results, then inspired P.P. Ewald with his nearly aquired doctorate to pursue X-ray diffraction (as Sommerfeld's assistant) in whose laboratory he developed the dynamic theory of X-ray diffraction in three publications (see Ewald, 1916a,b, 1917). Charles Darwin published two key papers on the diffraction of X-rays from perfect crystals (Darwin, 1914a,b, 1922). Debye (1913) and Waller (1923) developed the Debye–Waller factor, which described the attenuation of diffracted X-rays caused by thermal motion in crystals.

In 1912, Ewald was sent by Sommerfeld to mathematician David Hilbert to act as his physics tutor for one year, with the goal of Hilbert developing a standard of mathematical physics. In 1924, Richard Courant took Hilbert's lecture notes and expanded them to produce a two-volume book in which volume 1, "Methoden der mathematischen Physik I," was published in 1924. This book provided the mathematics needed for the development of quantum mechanics.

The early works of Laue, W.H. Bragg and W.L. Bragg, Ewald, Debye, and Darwin gave birth to the field of X-ray crystallography, see Ewald (1962). The de Broglie hypothesis for matter waves, the Schrödinger equation, and quantum mechanics set the stage for the concept of neutron waves and neutron optics (see papers by de Broglie (1923) and Schrödinger (1926); and books by (Born (1969); Dirac (1930); and Compton and Allison (1935)). The diffraction of X-rays covered in books by Buerger (1942), Zachariasen (1945), Guinier (1963), and Guinier and Fournet (1955) prepared the path for the development of neutron diffraction and neutron small-angle scatter.

This section presents a brief, incomplete history and survey of diffractive and reflective X-ray optics and microscopy. In about 1896 work began in the development of water-cooled anodes for X-ray tubes by C. Heinrich and F. Muller for their company, C.H.F. Müller Röntgenwerk. In 1913 William Henry Bragg and his son (William Lawrence) fabricated the first X-ray spectrometer. Between 1924 and 1944, Phillips took over the C.H.F. Müller Röntgenwerk company, consolidated X-ray tube manufacturing in Eindhoven and Hamburg, and patented the safe Metalix tube. North American Philips, in collaboration with the U.S. Naval Research Laboratories, developed the world's first commercial X-ray diffractometer, which was branded Norelco, followed soon by the widely used Philips PW1050 powder diffractometer.

The development of X-ray sources has driven the development of X-ray optics. The emergence of the synchrotron X-ray sources in the 1960s allowed the development of, first, large-scale X-ray optical systems and the later development of microspot X-ray tubes, X-ray area detectors; charge-coupled device (CCD) camera-based imaging in the 1980s allowed the development of tabletop X-ray optical systems. Later chapters cover novel X-ray sources, such as Cherenkov, channeling, transition, and parametric X-ray sources, which were developed in the latter half of the twentieth century and are emerging sources of directed, narrowband, narrow-angle, intense X-rays suitable for X-ray optics. In addition to the synchrotron, cyclic electron/positron accelerator source of X-rays, are linear and cyclic accelerator–based plasma X-ray sources for X-ray optics. The development of high-powered lasers gave rise to laser-produced plasma X-ray sources (Colombant and Tonon, 1973). The laser-produced plasma source is a potential source of X-rays for X-ray imaging systems, and in inertial confinement fusion with lasers, X-ray imaging is used as a diagnostic.

Johann (1931) demonstrated a focusing curved monochromator, and Johansson (1933) proposed an advanced design of a curved crystal with the crystal surface exactly on the focusing circle. The disadvantage of bent crystals is the difficulty in deforming a single-crystal material, which attains good crystalline quality at the Bragg peak and has a large angle of acceptance. However, bent crystals found widespread use in crystallography. Kirkpatrick and Baez (1948) invented an X-ray reflection system with two cylindrical mirrors of equal radius of curvature, which produces a real image point of an object point and an image of an extended object with astigmatism. Kirkpatrick and Baez (father of singer Joan Baez) used their X-ray mirror to create the first X-ray microscope at Stanford in 1948, which had a resolution between the best optical microscope and an electron microscope. Kirkpatick's student, Hussein El-Sum, published a dissertation that would become the first book on holography (see Stanford University News Service 12/28/1992).

Wolter (1952a,b) designed an aplanatic system of grazing incidence mirrors, which satisfied the Abbe sine condition in which both spherical aberration and coma were eliminated, and led to Wolter X-ray telescopes.

X-ray mirrors and X-ray microscopes emerged in the 1950s (see Dyson, 1952; Thathachari, 1953 and Cosslett and Nixon, 1960). The lobster-eye X-ray telescope uses a set of X-ray grazing incidence mirrors to focus an extended X-ray-emitting object onto an image plane in a similar manner to a lobster eye with visible light (see Angel, 1979; and Land and Nilsson, 2002).

Materials are absorbing in the soft X-ray region and Prins (1930) extended Darwin's treatment to handle X-ray absorption in crystals, which was applicable to an infinite number of periodic layers presented by a crystal. To minimize absorption, the synthetic multilayer X-ray mirrors with a finite number of layers can be designed to have nonperiodic structures, which were treated by Parratt (1954). In the 1970s and early 1980s multilayer mirrors for neutrons and X-rays emerged. Spiller *et al.* (1980) fabricated multilayer mirrors for soft X-rays, and Saxena and Shoenborn (1977) developed a multilayer neutron monochromator. Figure 154 shows a schematic of a multilayer X-ray mirror and a plot from Underwood (2009) of the reflectivity of 8048-keV X-rays from a multilayer X-ray mirror as a function of the grazing incidence angle θ_i, which is measured relative to the mirror surface. The X-ray multilayer is comosed of alternating layers of tungsten and boron carbide with a period $d = 4.5\,\text{nm}$ in which the ratio of the high electron density (i.e., high-Z layer) thickness to the lower electron density layer is $\Gamma = 0.4$.

The use of synthetic multilayer mirrors as Bragg diffractors for X-rays and extreme ultraviolet were treated in papers by Spiller *et al.* (1980), Underwood and Barbee (1981a,b), Rosenbluth and Lee (1982), Lee (1981), Marmoret and Andre (1983), Videl and Vincent (1984), and Barbee (1986). In Russia, Kumakhov (1990) developed the use of capillary optics (Kumakhov lenses) for X-rays and neutrons.

Diffraction grating was invented in 1785 by David Rittenhouse and then independently by Joseph von Fraunhofer in the 1800s. The development of circular gratings began in the latter half of the 1800s (see Rayleigh *et al.*, 1888; Rowland, 1883; Soret, 1875 and Wood, 1898). The development of controlled, blazed (triangle-shaped) gratings by R.W. Wood in 1910 allowed the maximum intensity of the diffraction envelope of the grating interference pattern to be directed from the zero ($m = 0$) diffraction order of an unblazed grating to a desired higher diffraction order ($m > 0$ or $m < 0$) in the blazed grating. The later development of laminar gratings with rectangular grooves allowed both the bottom groove surface and the top planar surface of the grating to contribute to the diffracted intensity. The groove depth h, spacing a, and incidence angle on a laminar reflection grating can be adjusted to cause destructive interfere of the zero-order light or X-rays and thereby provide more diffraction intensity in the higher diffraction orders (see Hellwege, 1937; and Franks *et al.*, 1975). As stated by Michette (1986), corresponding to the laminar and blazed X-ray

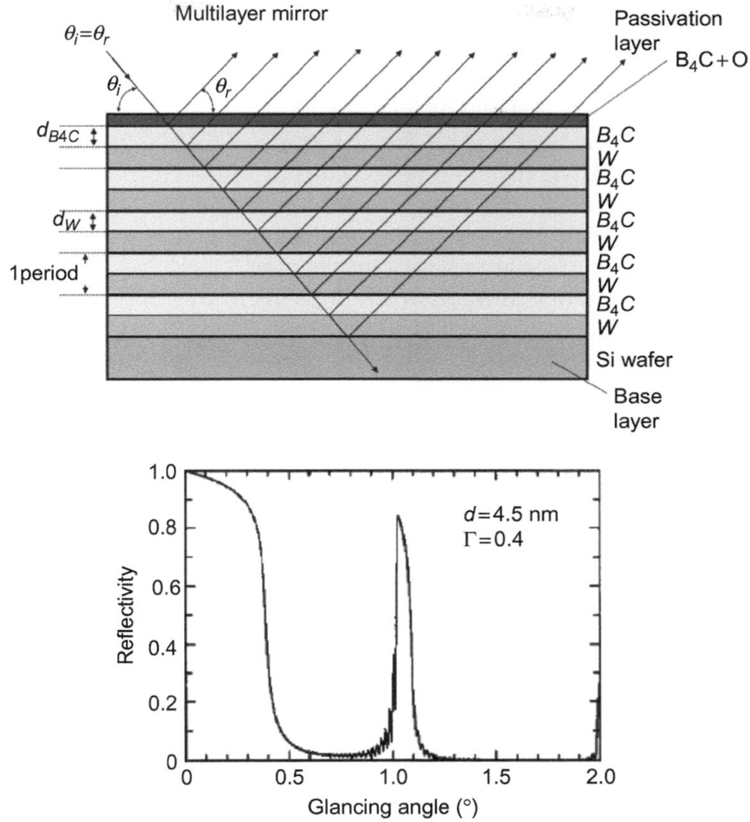

FIGURE 154 (Top) Schematic of a multilayer mirror composed of $N = 5$ periods consisting of bilayers of tungsten with thickness d_w and boron carbide with thickness d_{B_4C}. (Bottom) This plot is from the Center for X-ray Optics and Advanced Light Source *X-ray Data Booklet* (Underwood, 2009) and is the reflectivity of 8048-keV X-rays versus grazing incidence angle θ_i, measured relative to the mirror surface. The multilayer is composed of repeating units of tungsten and boron carbide layers with $d = 4.5$ nm and ratio $\Gamma = 0.4$ of high-Z layer thickness divided by low-Z layer thickness.

reflection gratings are X-ray transmission gratings with a rectangular profile (see Arakawa and Caldwell, 1980) and X-ray transmission gratings with a blaze-like optimized profile (see Tatchyn *et al.*, 1982).

Planar, spherical, toroidal, and ellipsoidal surfaces using the classical Rowland ruling began in 1882 when H. Rowland fabricated gratings for light that combined diffraction and focusing on a spherical concave surface that were far superior to previous gratings (see Rowland, 1883; and Beutler, 1945). A Rowland circle is a circle tangent to the center of the concave spherical Roland grating in which the radius of the circle r is equal to the one-half radius of curvature R of the spherical grating mirror surface, where $r = R/2$. A source point of white light on the Rowland circle emits rays of all wavelengths onto all points of the concave spherical grating, and each wavelength is focused to its own focal point on the Rowland circle. Rowland was among the group of first faculty members in 1875 (including M. Michelson 5 years later) of Johns Hopkins University, which was founded in 1876 with an endowment by philanthropist Johns Hopkins to establish a research-oriented American university based on the existing German research-based universities.

In the 1900s, X-ray diffraction gratings were developed by William Henry Bragg and his son William Lawrence Bragg, who in addition to developing X-ray diffraction, were also pioneers in developing X-ray emission (fluorescence). The Braggs measured the X-ray wavelength spectra of many elements using both high-energy electrons and diffraction gratings, which they fabricated by diamond ruling on glass. In X-ray gratings the X-ray reflectivity is low for all materials, but incidence at less than the critical angle of light, measured relative to the tangent plane to the point of incidence on the surface, provides complete X-ray reflection. X-ray gratings require grazing incidence of the X-rays at an angle, which is less than the X-ray critical angle of reflection, which is typically less than a degree or a few degrees (depending on the X-ray energy) and is measured relative to the smooth reflection surface with wavelength surface quality.

The Bragg X-ray microscope and optical computation, developed by William Lawrence Bragg in 1939, used the following approach: Light and X-rays are both electromagnetic waves, but with different wavelengths in which coherent light is shone on an X-ray diffraction pattern to create an image of the original scatter object. W. Lawrence Bragg used the diffraction pattern of 010 planes of the mineral crystal $CaMg(SiO_3)_2$, which provides intensity without phase (shape) information. Fortunately, for this particular mineral crystal only small phase shifts are needed to reconstruct the spatial image of the lattice atoms. That is, the diffraction pattern needed very little phase information to reconstruct the image of the calcium, magnesium, silicon, and oxygen lattice. Bragg simply drilled holes in a brass plate that matched the diffraction pattern of the 010 planes. The holes were illuminated with coherent light (no lasers existed, so a pinhole source in which the pinhole is smaller than the wavelength of light is required) and the resulting image reproduced an approximate image of the mineral lattice.

More updated developments in diffraction gratings are given by Hutley (1982) and Stroke (1967). X-ray gratings were developed by Stanley *et al.* (1968), Franks *et al.* (1975), and Arakawa and Caldwell (1980). A combination of reflection grating and multilayer mirror is presented in the 1990 U.S. patent by T.W. Barbee (Figure 155).

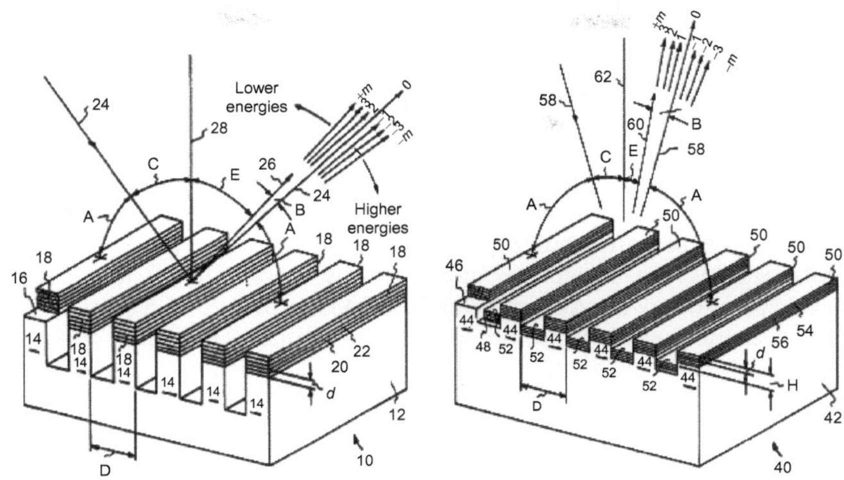

FIGURE 155 Diagram of a combined diffraction grating and multilayer mirror from a patent by T. W. Barbee ("Multilayer Diffraction Grating," U.S. Patent No.4,915,463, April 10, 1990), which references his 1986 paper. Ordinary laminar or phase gratings on both the lands (top surfaces) and groove surfaces contribute to diffracted intensities. By selecting the groove depth h and the incidence angle, the diffraction contributions from the top land and bottom groove surfaces interfere destructively for the $m = 0$ zero order. This allows more of the incident X-rays to be diffracted to he higher orders, $m = \pm1, \pm2, \pm3, \ldots$ (see Michette, 1986). The above invention makes use of the top multilayer surface and groove multilayer surface to direct the incident X-rays into preferred diffraction orders at larger angles of incidence and Bragg diffraction angles, measured relative to the surface.

Holographic gratings are formed by an interference fringe field of two laser beams. A polished substrate with photoresist is exposed to the standing wave interference pattern produced by the two lasers. The exposed photoresist yields a pattern of straight lines with a sinusoidal cross section. Holographic gratings cause less scattered light than ruled gratings except when the ratio of the period to wavelength is near 1, in which case the holographic grating has the same efficiency as the ruled grating. Holographic master gratings are replicated by a process identical to that used for ruled gratings (see Namioka *et al.*, 1973) and holographic light gratings are readily commercially available (see Palmer and Loewen, 2005).

Focusing and imaging X-rays can be accomplished by (1) total reflection (X-ray mirrors or capillaries), (2) refraction with lenses, and (3) diffraction with Bragg–Fresnel and Soret lenses or Fresnel zone plate (FZP) lenses. The development of diffractive X-rays optics led to the first practical X-ray microscopes in the 1970–2000 period. The latter half of the 1900s saw the development of X-ray zone plates. The development of high-brilliance X-ray sources, coupled with advances in manufacturing technologies over the past 50 years, has led to significant improvements in submicrometer probes for spectroscopy, diffraction, and imaging applications. In 1960, working at the Harvard Smithsonian Astrophysical Observatory in Cambridge, Massachusetts, Baez (1961) developed an FZP optic for an X-ray telescope that could focus wavelengths in the soft X-ray region. Baez reported the following in his paper. The zone plate consisted of a set of thin circular gold bands made self-supporting by radial struts, leaving the transparent zones empty. Experimental tests of the $D = 0.26$-cm diameter zone plate at 6700-, 43580, and 2537-Å wavelength λ gave a theoretical minimum angular resolution, Rayleigh criterion, $\sin\theta_{\rm min} = 1.22\lambda/D$. The zone plate contained 19 opaque zones, and the outermost, narrow zone was $\sim 20\,\mu$m across. The zone plate formed an image 40 times faster with sixfold better resolution than an optimum pinhole, and the zone plate produced images that compared favorably with those made with a lens of similar focal length and aperture. A regular lens produced an image about 20 times faster than the zone plate at 4358 Å; however at shorter wavelengths such as 1000 Å, which a regular lens cannot image, the zone plate is faster. The angular resolution of the zone plate at 2537 Å was close to the theoretical value of 1.2×10^{-4} rad and held over a field of at least 1.75×10^{-2} rad, which is twofold larger than the angle subtended by the sun's disk at the earth. Baez then concluded a zone plate telescope, operating in the soft X-ray or extreme ultraviolet region, far above the earth's atmosphere in an orbiting satellite, was possible.

The phase Fresnel lens was proposed by Miyamoto (1961). The phase Fresnel lens is designed to satisfy the Fresnel condition for the phase shift $\phi(u,v)$ of the diffraction of incident waves of wavelength λ, which propagate in the axial z-direction by the N Fresnel zones of the lens. For each kth zone and the lens transverse Cartesian coordinates (x,y) on the Fresnel phase zone plate, the Fresnel condition

$$(k-1)\,\lambda < \phi(u,v) < k\lambda \quad k = 1,2,\ldots,N. \tag{235}$$

The phase FZP is designed to shift the wave surface by the amount $\phi\,(x,y) - (k-1)\,\lambda$ in each kth Fresnel zone. The amount of wave front deformation in each zone is smaller than λ, but the phase FZP is quite equivalent to a lens that deforms the wave surface by $\phi(x,y)$ because of the Fresnel condition.

During the 1970s there was much interest by the U.S. government in the development of X-ray lenses as witnessed by the support of the U.S. Defense Advanced Research Projects Agency (DARPA), the U.S. Army, the U.S. National Science Foundation, and the U.S. Department of Energy, which funded research at various institutions including MIT

Lincoln Laboratory and the Lawrence Livermore National Laboratory to develop X-ray optics and sources. Beginning his X-ray lens work at the MIT Research Laboratory of Electronics, N. Ceglio at the Lawrence Livermore National Laboratory published a paper proposing a blazed Fresnel phase plate (BFPP), which is a Fresnel zone structure with a spatially distributed phase shift within each Fresnel zone (Ceglio and Smith, 1977). The spatial distribution of the phase shifts in the BFPP is designed to concentrate all the unabsorbed energy into a single focal spot. The top of Figure 156 is a diagram of the amplitude-modulating FZP and the bottom is a diagram of a phase-modulating zone plate. The top of Figure 157 is a diagram of a BFPP with the curved lens surfaces; the bottom diagram shows more detail.

The amplitude modulation FZP performs amplitude modulation of the incident X-ray plane wave; the amplitude modulation has the form of a square wave periodic with the squared radial distance r^2 from the zone plane center in which 10% of the incident energy is focused into a first-order focal spot. A better way to remove the zero-order transmitted wave is with a phase-contrast FZP. Rather than absorbing and transmitting incident X-rays in the alternating opaque and transparent zones, the phase modulation zone plate changes the phase of the incident wave by a fixed specified amount. Ideally, the phase modulation zone plate removes the zero-order wave from the focal spot; thereby allowing only the diffracted waves to converge on the first-order focal spot. The FZP is a form of diffraction grating so there are focal spots for each diffraction order, and assuming negligible absorption, the phase modulation FZP can focus approximately 40% of the incident wave energy into a first-order focal spot.

The blazed FZP transforms the incident X-ray wave into a converging spherical wave, which has an amplitude modulation periodic in r^2 because of finite absorption of the X-rays. As a result of the amplitude modulation, the blazed FZP diffracts the incident X-ray plane wave energy into other focal spots in addition to the first-order focal spot. However, if the absorption is negligible, then almost all the unabsorbed energy is diffracted into the first-order focal spot. The blazed FZP focuses X-rays diffractively in which the resolution increases as the radial width of the outer zone decreases. In contrast, a conventional Fresnel lens is a circular symmetric, prismatic lens, composed of a series of concentric circular symmetric prisms of equal annular (radial) width, that focuses light or X-rays by refraction. If the annular width of the annular prisms is sufficiently decreased, then blurring occurs because of diffraction effects. The blazed FZP is equivalent to a phase Fresnel lens, which was proposed Miyamoto as described above.

The development of vacuum ultraviolet and soft X-ray optics was well under way in the late 1960s and 1970s (see Franks, 1971, and Kirz, 1974).

What is a blazed fresnel phase plate?

Illustration in stages: (1) Fresnel zone plate

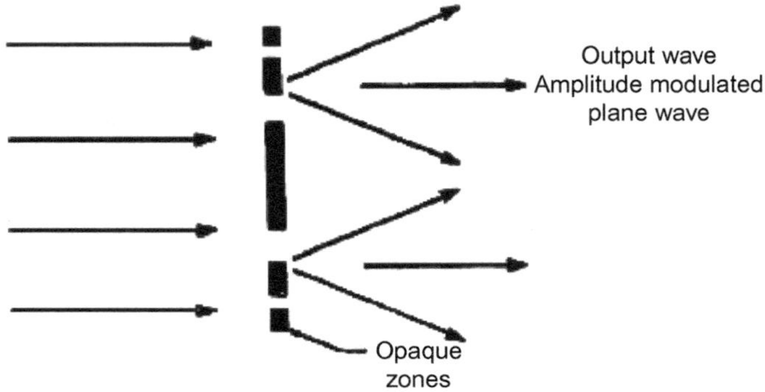

Fractional energy distribution

Device	1st order real focus	Absorbed	Undiffracted	All other foci
FZP	0.1	0.5	0.25	0.15

(a)

What is a blazed fresnel phase plate?

Illustration in stages: (2) Fresnel phase plate

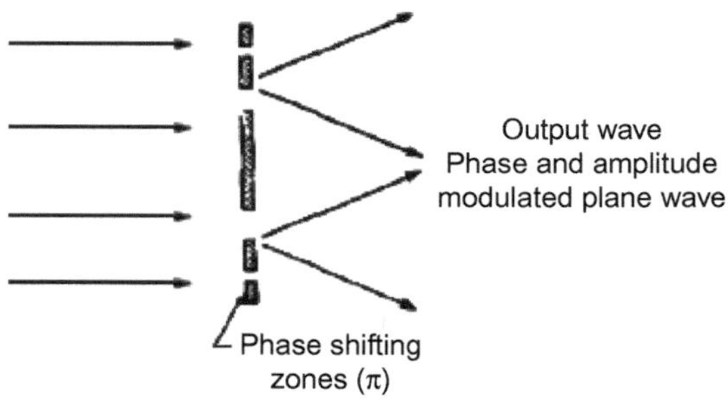

Fractional energy distribution*

Device	1st order real focus	Absorbed	Undiffracted	All other foci
FZP	$\left(\frac{1+\Gamma}{\pi}\right)^2$	$\left(\frac{1-\Gamma^2}{2}\right)$	$\left(\frac{1+\Gamma}{2}\right)^2$	$\frac{\pi^2-4}{4\pi^2}(1+\Gamma)^2$

*$\Gamma = \exp(-\pi\eta)$ is the fractional amplitude transmission through a phase shifting zone of the FPP.

(b)

FIGURE 156 (Top) An amplitude modulation Fresnel zone plate. (Bottom) A phase modulation Fresnel zone plate (Ceglio and Smith, 1977).

What is a blazed fresnel phase plate?

Illustration in stages: (3)

Blazed fresnel phase plate

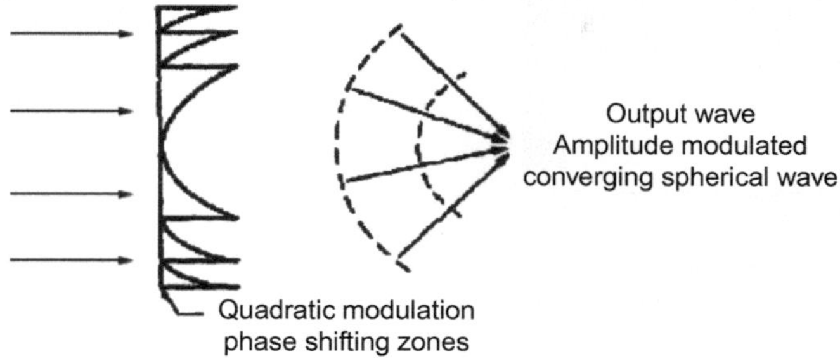

Fractional energy distribution*

Device	1st order real focus	Absorbed	Undiffracted	All other foci
BFPP	$\left(\frac{1-\Gamma^2}{2\pi\eta}\right)^2$	$1-\frac{(1-\Gamma^4)}{4\pi\eta}$	$\frac{\eta^2}{1+\eta^2}\left(\frac{1-\Gamma^2}{2\pi\eta}\right)^2$	$\frac{1-\Gamma^2}{2\pi\eta}\left(\frac{1+\Gamma^2}{2}-\frac{1+2\eta^2}{1+\eta^2}-\frac{1-\Gamma^2}{2\pi\eta}\right)$

$^*\eta \equiv \kappa/\delta;\ \Gamma = \exp(-\pi\eta)$

Blazed fresnel phase plate (description)

A BFPP is a fresnel zone structure with a spatially distributed phase shift within each fresnel zone. The spatial distribution of the phase shifts is chosen to concentrate all unabsorbed energy into a single focal spot.

Converging lens

$$t(r)\cdot\delta(r) = \frac{\lambda(r^2 - r_m^2)}{2r_1^2}$$

$m = 0, 2, 4, \ldots$

$r_m \le r \le r_{m+2}$

Where

$t(r)$ = material thickness

δ = Real part of refractive index ($n = 1-\delta+ik$)

r_1 = Radius of first fresnel zone

λ = X-ray wavelength

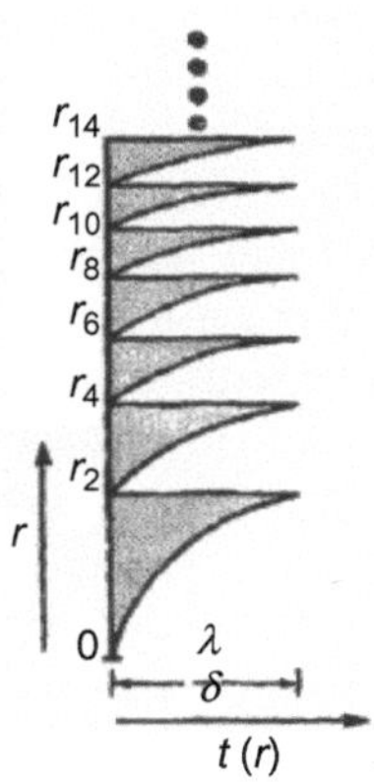

FIGURE 157 The general (top) and more detailed (bottom) diagrams of the blaze Fresnel phase plate proposed in 1977 by Ceglio at the Lawrence Livermore National Laboratory (Ceglio and Smith, 1977).

The practical use of soft X-ray imaging via diffractive optics with an FZP imaging a sample onto a soft X-ray CCD sensor occurred in 1974 at the Georg-August University in Göttingen (see Schmahl and Rudolph, 1969; and Niemann *et al.*, 1974, 1976). Ceglio *et al.* (1983) fabricated the first high-resolution phase lens for use at X-ray wavelengths. The lens fabricated and tested by Ceglio *et al.* (1983) was a 5500-Å, 100-zone Fresnel phase plate made of silver, with a minimum zone width of 3200 Å, that was designed for the aluminum K_α line at approximately 1.5 keV with anticipated first-order 25% diffraction efficiency.

The development of X-ray grating flourished in the 1980s in which fabrication techniques for X-ray zone plates were developed, tested, and optimized. Harada and Kita (1980) and Hettrick (1984) developed variably ruled X-ray gratings. Electron beam and X-ray lithography techniques for zone plates were developed by Aritome *et al.* (1984); Shaver *et al.* (1979); Smith *et al.* (1984); Vladimirsky *et al.* (1984), and Simpson and Michette (1984).

Although the FZP lenses yielded the best spatial resolution, their applications were traditionally limited to rather soft X-rays. Also, the FZP efficiency is much less than 100% because a significant fraction of incident beam is transmitted or nondiffracted into the zero order. At multi-keV energies, the efficiencies are about 10% to 20%, which severely reduces the focal spot intensity. The FZP efficiency can be increased to 100% focusing efficiency by optimization of the structure profile of the zones. Tatchyn (1984) theoretically showed 100% focusing efficiency with the FZP designed as a nonabsorbing lens with a parabolic zone profile (see also Michette, 1986).

The optimized zone plate profile derived by Tatchyn (1984) and covered by Michette, (1986) is a Fresnel lens with outer portions of each zone removed (shown in Figure 158 for the $m = 1$ first diffraction order focusing with focal length f). For each ith zone, the higher diffraction orders $m = 2, 3, 4, \ldots$ focus closer to the zone plate than the $m = 1$ first diffraction order with focal length $f_m = f/m$; the $m = -1, -2, -3, \ldots$ orders are divergent and do not focus; and the $m = 0$ zero order is transmitted nondiffracted (i.e., the $m = 0$ order for each ith zone focuses at infinity). The resolution σ of the optimized zone plate with N zones $i_{\max} = N$ for the $m = 1$ diffraction order is approximately equal to outer zone width $\Delta r = r_N - r_{N-1}$, where $\sigma = 1.22\Delta r/m$. Fabrication of the optimized zone plate with the Fresnel lens–like profile is presently very difficult. However, stepped (blazed) zone plates can be fabricated with two or more steps per zone and thus approximate the quadratic profile of the optimized zone plate.

In addition to their high efficiency, the FZPs with parabolic lens shapes offer the advantages of low background signal and effective reduction of unwanted diffraction orders by introducing selection rules (see Yun

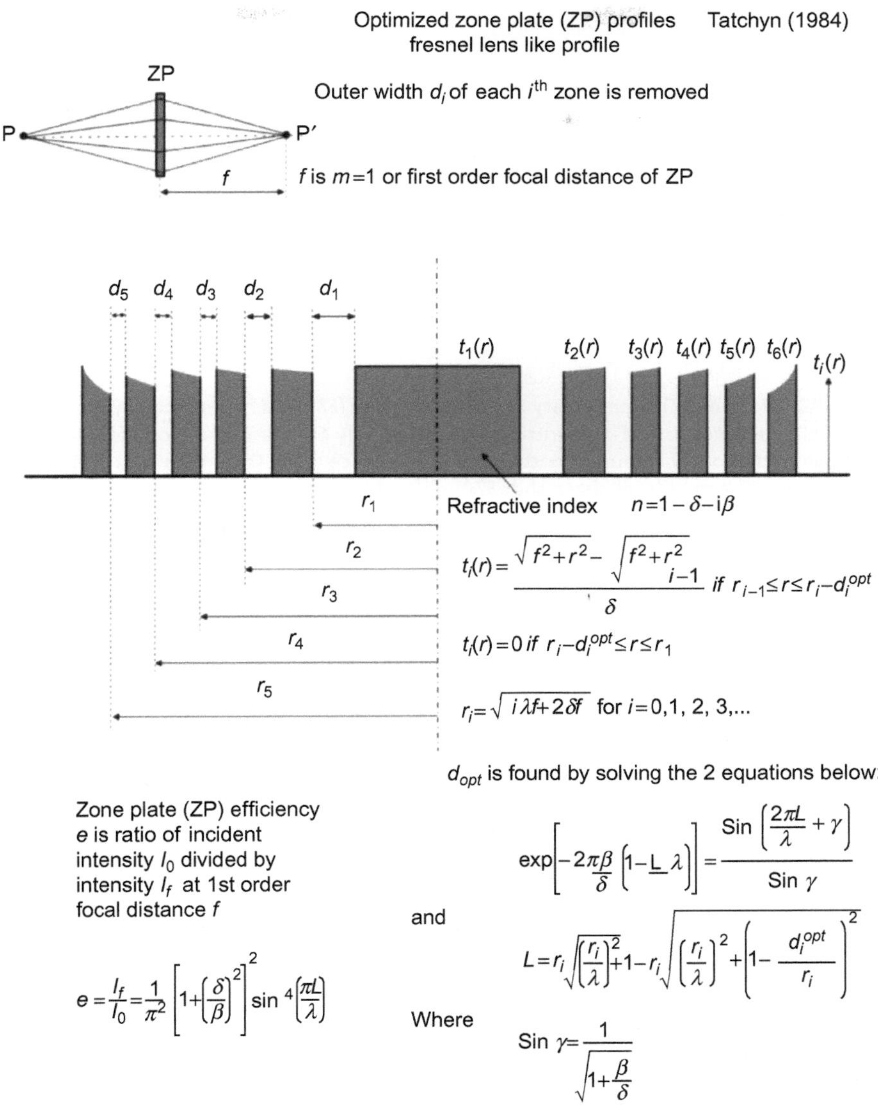

FIGURE 158 Diagram depicting the optimized zone plate profile derived by Tatchyn (1984) and covered in Michette's (1986) book.

et al., 1999). Parabolic profiles are extremely difficult to produce with existing lithography techniques, and hence a multistep profile zone, which approximated the approximated to the parabolic profile, was fabricated in the form of a quaternary, electroplated nickel, blazed zone-plate by

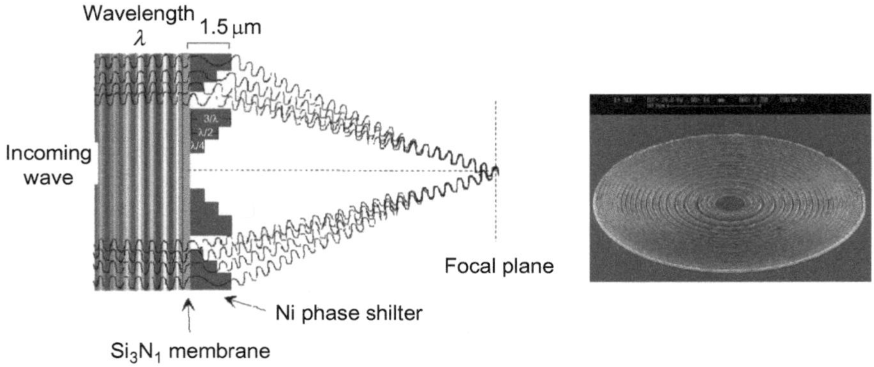

FIGURE 159 (Left) The quaternary Fresnel zone plate (FZP) is a high-efficiency blazed zone plate and consists of concentric rings that amplify and/or phase-modulate the incident radiation. This causes constructive interference along the optical axis at the focal point. The multilevel phase-shift FZPs diffract with phase modulation in which the optical path of the incident radiation increases by $\lambda$and the phase by 2π over one period. The 1.5 μm thickness of refractive nickel was electroplated into three step layers of wavelength thickness $\lambda/4$, λ, and $3\lambda/4$ corresponding to 8 keV. The nickel step layers were electroplated on a thin bottom Si_3N_4 membrane support layer. The three step layers acted as a phase shifter to correct for the phase difference, which occurs between the different X-ray optical path lengths. (Right) The multilevel FZP fabrication requires the deposition and alignment of levels accurate to a fraction of the desired spatial resolution. The 8-keV multilevel FZP with 4 levels and 150-μm diameter and 500-nm outermost zone width was designed for 1-m focal length; the scale bar: 50 μm (Di Fabrizio *et al.*, 1999).

Di Fabrizio *et al.* (1999) in collaboration with CNR-IESS (Rome) and TASC-INFM (Trieste). The multistep zone plate was tested over an energy range of 2 to 8 keV (Figure 159). The FZP efficiency measurements were performed at 5.5, 6.0, 7.0, and 8.0 keV and the efficiencies were 43%, 45%, 57%, and 45%, respectively. The measured optical gain (the ratio between the photon density delivered by the FZP at the focus and the photon density impinging on the FZP) for the nickel FZP is about 2,500 at 7 keV.

Figure 160 shows two diagrams from U.S. patents. The left diagram depicts a recent patent of Yun and Wang (2008) for combined Fresnel lens, FZP, and phase plate for X-ray phase-contrast microscopy. The right diagram depicts an older patent (Schmal and Rudolph, 1996) for a phase-contrast X-ray microscope.

The year 1988 saw the development of the scanning soft X-ray microscope in which the sample is scanned past a focused spot and the image is electronically produced one pixel at a time at SUNY Stony Brook and the Brookhaven National Laboratory (see Rarback *et al.*, 1988; and Jacobsen

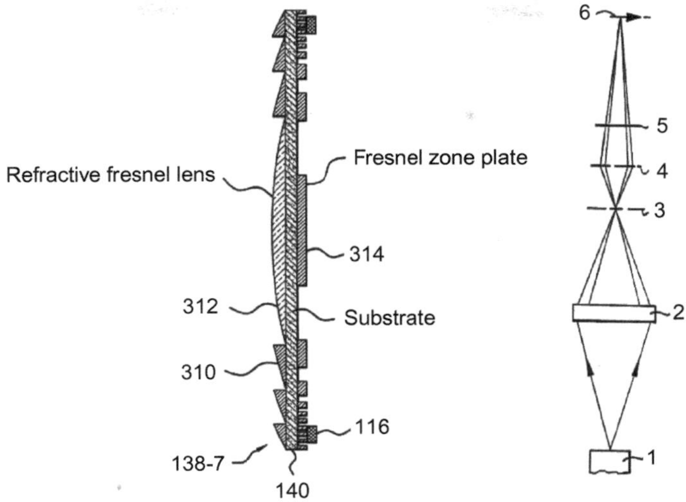

FIGURE 160 (Left) U.S. patent diagram (Yun and Wang, 2008) shows a combination Fresnel lens, Fresnel lens, and a phase plate (item 116), which is used for phase-contrast X-ray microscopy. The patent was assigned to Xradia (in Pleasanton, California). (Right) U.S. patent diagram of Schmal and Rudolph (1996) shows a phase-contrast X-ray microscope. A plasma X-ray source emits X-rays intercepted by an annular condenser zone plate lens or mirror (2) focusing X-rays onto the sample (3). The nonscattered and scattered X-rays are intercepted by the objective zone plate lens (4), and the nonscattered rays pass through the annulus of the phase plate (5) and undergo a 90° or 270° phase shift, while the X-rays scattered and phase-shifted by the sample pass inside the annulus and receive no phase shift by the phase plate. The micro zone plate objective lens has only a 20- to 50-μm diameter and is located ~0.5 to 1.0 mm above the sample. The X-rays scattered and phase-shifted by the sample and the sample-transmitted X-rays then converge, combine, and interfere on the detector (6).

et al., 1991). The 1990s saw the further development of X-ray microscopy (see Jacobsen *et al.*, 1992; Williams *et al.*, 1995; and Jochum and Meyer-Ilse, 1995). Since 2000 progress has continued in X-ray imaging (see Miao *et al.*, 2004; Shapiro *et al.*, 2005; Chao *et al.*, 2005; Jefimovs *et al.*, 2007; and Thibault *et al.*, 2008). Figure 161 shows an example of an X-ray bright-field (BF) microscope.

In dark-field (DF) X-ray microscopy only the X-rays scattered from the specimen are intercepted and imaged by the detector (Figure 162). In the case of soft X-rays, the scattering is predominantly coherent elastic scatter (Rayleigh scatter). The scattering strength of a particle, which is surrounded by an embedding medium, depends on the magnitude of the particle complex scatter factor compared with that of the embedding medium. The scatter factor magnitude increases with increasing atomic

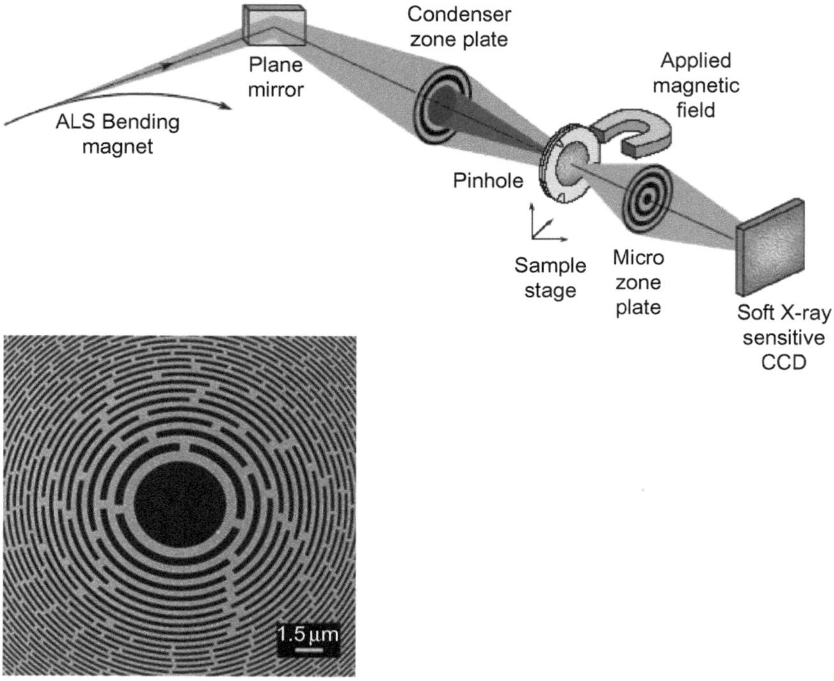

FIGURE 161 (Top) Diagram of the XM-1 bright-field soft X-ray microscope at the Advance Light Source at Lawrence Berkeley National Laboratory. A magnetic field environment can be provided for the sample (from "X-ray Imaging of Nanoscale Structures and Processes"; http://www.cxro.lbl.gov/XM1). (Bottom) The micro–zone plate lens fabricated at the Center of X-ray Optics, Lawrence Berkeley National Laboratory (from "Diffractive Lenses for a New Generation of X-Ray Beamlines"; http://www.cxro.lbl.gov/zoneplate).

electron density; the large changes in atomic electron density, which occur at sharp features and edges, strongly scatter X-rays, resulting in those imaged features and edges having large contrast compared with the surroundings with small spatial gradients in atomic electron density. For example, the higher atomic electron density of gold causes a gold sphere to scatter X-rays more strongly than an identical sphere of carbon. In comparison, BF X-ray images are based on photoelectric absorption and the image is a result of transmitted X-rays. For biological samples the image is largely based on the different absorption of X-rays by water and embedded carbon-based biological structures. Figure 162 is a schematic of a DF X-ray microscope.

Originally introduced with visible light, Zernike phase contrast is a well-established technique in full-field X-ray microscopes for visualization of weakly absorbing samples. The Zernike phase-contrast light microscope technique for imaging weakly absorbing samples was applied

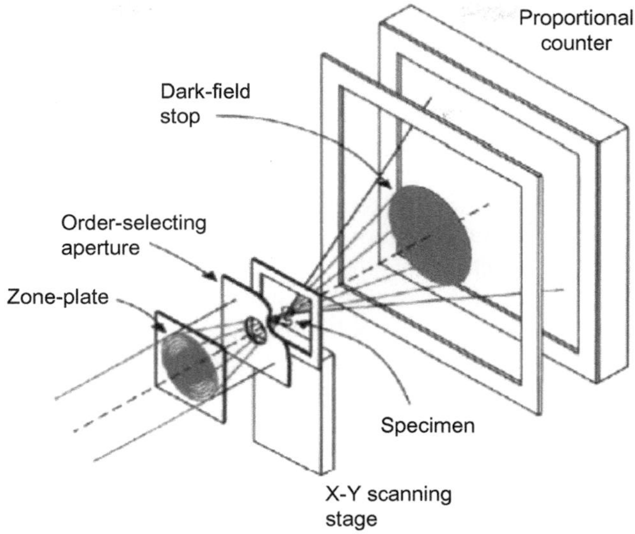

FIGURE 162 The dark-field microscope schematic from Chapman *et al.* (1996).

to the scanning X-ray microscope in 2010 in a collaboration between Stony Brook University, Argonne National Laboratory, and Xradia, Inc. (see Holzner *et al.*, 2010).

Submicron optical resolution can be attained by X-ray imaging using the differential interference contrast (DIC) microscope approach with X-rays (see Murphy, 2001). The DIC microscopes produce contrast within the specimen by using the phase differences between a sample feature light ray and a reference ray. DIC converts gradients in the sample feature optical path length into amplitude differences that can be visualized as amplitude contrast in the resulting image. The sample optical path difference between two points on the optical path of a ray is the product of the refractive index difference (between the sample and surrounding medium) and the ray propagation distance between the two points. Images produced in DIC microscopy have a distinctive shadow-cast appearance, which often renders samples in a pseudo–three-dimensional (3D) relief, and is not an indicator of actual topographical structure.

The DIC light microscope was devised by Smith (1954). He modified a polarized light microscope with the addition of two Wollaston prisms. The first Wollaston prism was at the front focal plane of the condenser lens and the second Wollaston prism at the objective lens rear focal plane. A Wollaston prism is composed of two wedges (prisms) The wedges are a birefringent, uniaxial crytalline material (e.g., quartz or calcite). The wedges are cemented together at their hypotenuse with a film of Canadian basalm. The optical axis is the axis of symmetry in a crystalline material. A birefringent material has ordinary refractive index n_0 perpendicular to the optical axis and extraodinary refractive index n_e parallel to the optical axis.

Incident light with a linear polarization perpendicular to the optical axis is termed the *ordinary ray,* which refracts by Snell's law. Light with a linear polarization component parallel to the optical axis is termed the *extraoordinary ray,* which does not refract by Snell's law. The extrordinary rays refract at an angle that is determined by the angle of incidence and the difference Δn between the extraordinary and ordinary refractive indexes, $\Delta_n = n_e - n_0$. The electric fields of the extraordinary and ordinary rays oscillate in perpendicular directions in which the refractive index differs. Consider an axial-directed, 45° polarized light wave that enters the birefringent material. Inside the birefringent crystal the transverse electric field components of the light travel at different speeds, thereby splitting the incident wave into extraordinary and ordinary waves (see Hecht and Zajac, 1974).

The optical axis of the input (first) wedge of the Wolaston prism is parallel to the light input surface, and the optical axis of the output (second) wedge of the Wollaston prism is parallel to the light output surface. However, the optical axis of the input and output wedges are perpendicular to each other, and the optical axis of each wedge is perpendicular to the axial direction of the incident light. The disadvantage of using a Wollaston prism as a beam combiner is that the Wollaston prism must be located in the back (rear) focal plane located downstream of the objective lens, where rays from infinity focus. The placement of the Wollaston prism in the back focal plane of the objective lens usually places the Wollaston prism inside the body of the short focal length objective lens combination.

In a Nomarski-modified Wollaston prism, the output wedge is cut so its optical axis is parallel to the output light surface, but the optical axis of the output wedge is now at an oblique angle, rather than a right angle, with respect to the optical axis of the input wedge. The Nomarski-modified Wollaston prism can merge the two incident divergent light rays with perpendicular transverse electric field polarizations into two overlapping rays of the same polarization, which is of important use in a microscope. The Nomarski-modified Wollaston prism could be placed downstream of the objective lens back focal plane, and from the back focal plane, the two divergent rays with orthogonal polarizations could be merged into overlapping rays with orthogonal polarization but phase-shifted Figure 163.

The second Nomarski-modified prism causes the reference and sample feature rays to acquire the same polarization but still retain the relative phase shift due to the sample feature. This allows interference and the formation of an amplitude-contrast image of the phase shift induced by the sample feature. Microscopes with Nomarski-type DIC optics were first produced commercially by Zeiss in 1965 (see Nomarski, 1955; and Allen *et al.*, 1969).

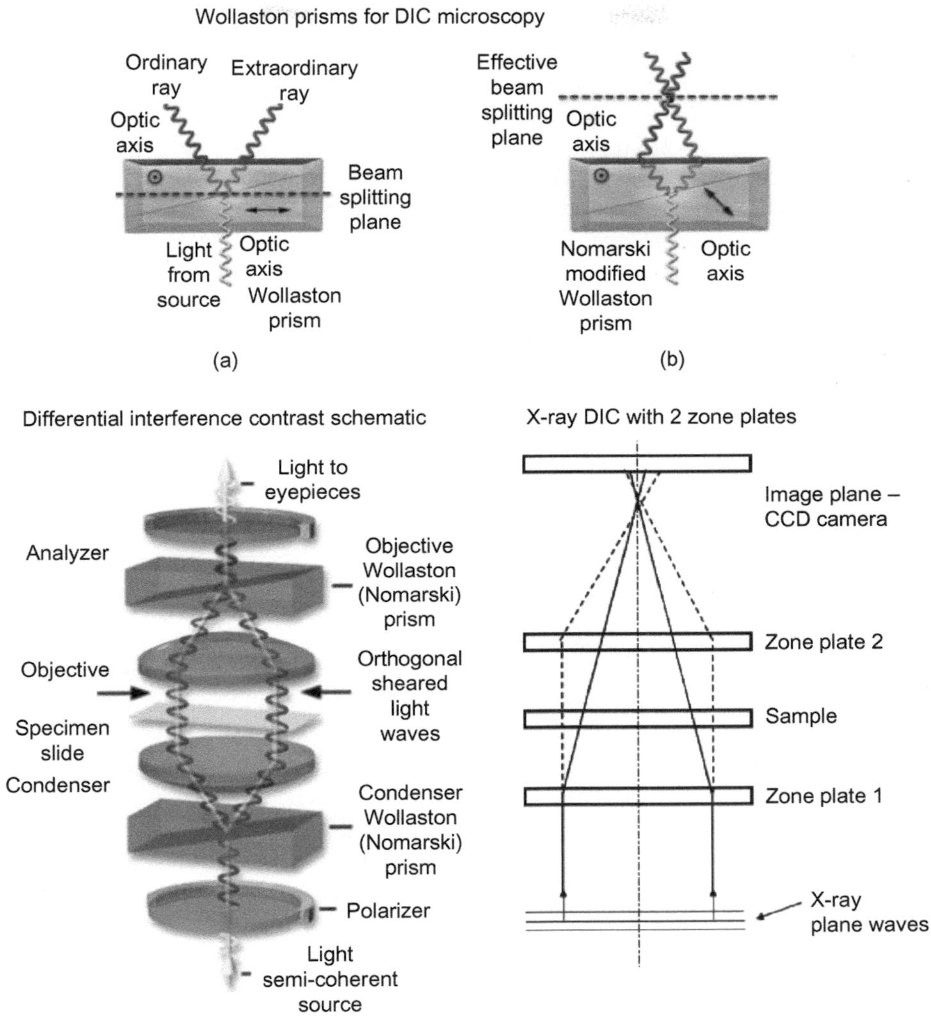

FIGURE 163 (Top left) Wollaston prism and (top right) Nomarski-modified Wollaston prism. (Bottom left) Diagram of the Nomarski or differential interference contrast (DIC) microscope. Both schematics are from Abramowitz and Davidson (2011) from the Olympus Microscopy Resource Center website. (Bottom right) X-ray DIC with two zone plates with shearing occurring between the focused rays of the two zone plates; schematic from Kaulich *et al.* (2002). (See Color Insert.)

We now examine the DIC light microscope in more detail. The DIC microscope produces polarized light from an unpolarized light source and then manipulates the polarized light. For unpolarized light the electric field oscillates in the plane transverse to the direction of light propagation and can oscillate in any direction in that transverse plane. However, in the case of polarized light, the electric field oscillates along a single direction in a plane transverse to the direction of light propagation. For example, the light wave electric field oscillates along the x-axis with light wave propagation in the axial z-direction. An optical component, which creates polarized light from unpolarized light, is termed a *polarizer*, and when the same polarizer is used to examine polarized or partly polarized light, the polarizer is termed an *analyzer.* Two polarizers with polarizations at right angles (orthogonal) are termed *crossed polarizers.* In the DIC microscope the polarizer and downstream analyzer are crossed polarizers, which prevent the unpolarized light from the light source from reaching the CCD imaging detector.

The DIC microscope (see Figure 163) works as follows. A polarizer oriented at 45° to incident light source outputs 45° polarized light. The light wave electric field now has equal electric field components in the transverse x- and y-directions relative to the z-direction of the incident light wave propagation. The light wave with its equal x- and y-transverse electric field components can now be spilt into an x-polarized light wave and a y-polarized light wave with equal electric field amplitudes by an anisotropic crystal-based optical element, such as a Wollaston or Nomarski-modified prism. A single ray of the 45° polarized light is split by the lower (first) Nomarski-modified Wollaston prism into two orthogonally polarized rays—one polarized ray passing through the sample feature (sample feature ray) and the other orthogonally polarized ray passing around the sample feature and interacting with adjacent background (reference ray). This lower Nomarski-modified Wollaston (DIC compensator) compensates for optical path differences between all possible off-axis rays relative to on-axis rays. The phase differences of all light rays emitted from the full width of the condenser aperture are thus compensated, which allows the entire condenser aperture to be used for sample illumination and an overall brighter image.

The z-directed sample and reference rays are focused by the condenser lens onto two adjacent points on the sample, which are separated by a very small distance of ~0.2 μm. The sample is thus illuminated at two closely spaced points by two coherent light beamlets. The reference ray light beamlet is polarized in the transverse x-direction passes through the adjacent background, and the sample feature beamlet is polarized in the transverse y-direction and passes through the sample feature. The sample feature ray undergoes some absorption, scatter, and refraction by the sample feature, which causes the sample feature ray to acquire a phase shift relative to the reference ray. The reference and sample feature rays each contain a BF image of different, but closely spaced, adjoining points of the sample. The x-directed, transverse electric field polarization of the reference ray and the y-directed, transverse electric field polarizations of the sample feature ray prevent interference between these two images after emergence from the sample. The sample feature ray and reference ray propagate to the objective lens and are focused for incidence on the second Nomarski-modified prism.

The second Nomarski-modified prism merges the sample feature and reference rays into overlapping rays and causes the reference ray and the phase-shifted, sample feature ray to acquire the same 135° transverse electric field polarization. The two rays now have the same polarization, which allows constructive and destructive interference between the two rays. This interference produces amplitude contrast in the summed amplitudes of the two rays due to their relative phase shift induced by the sample feature. The overlapped and interfering rays then pass through the analyzer, which is simply a polarizer oriented at 135°. The 135° analyzer and 45° polarizer are crossed polarizers, which are used to prevent unpolarized source light from passing through the microscope onto the image plane. The brightening (constructive interference) and darkening (destructive interference) of the image feature and its immediate surroundings occurs in the image plane as the two 135° polarized sample feature and reference wave fronts converge and interfere at the CCD detector. The interference pattern at the CCD detector is an amplitude-contrast image of the phase shifts induced by the optical density gradients of the sample.

The light DIC microscope as a guiding inspiration and progress in lithography and nanofabrication made possible the application of DIC to X-ray microscopy (see Di Fabrizio *et al.*, 1999). An X-ray doublet lens, which consisted of two specially developed zone plates for BF imaging (absorption contrast) were put to use in an X-ray DIC experiment. The zone plate doublet (or twin zone plate, TZP) in the DIC experiment produces focal spots closely spaced within the TZP spatial resolution of 160 nm (see Wilhein *et al.*, 2001; and Kaulich *et al.*, 2002). Optical path differences are imposed on the X-rays passing through the sample. These X-rays then pass through the TZP and are recorded by a CCD camera for full-field imaging. Alternatively, the images are recorded by an aperture photodiode in the scanning transmission mode. For both the full-field imaging and scanning X-ray microscope methods a large 20-fold increase in image contrast for a low-absorbing specimen is attained with the X-ray DIC in a manner similar to the Nomarski DIC method for visible light microscopy (see Figure 163).

Differential phase-contrast (DPC) imaging has been developed for both neutrons and X-rays (see U.S. patent by David *et al.*, 2010). Figure 164 shows a neutron shearing interferometer developed by Pfeiffer *et al.* (2006a), which produces DPC neutron images similar to Schlieren imaging (see Born and Wolf, 2002) or Shack–Hartmann wave front sensor (see Hartmann, 1904, and Pfeiffer *et al.*, 2006a). Figure 165 shown the

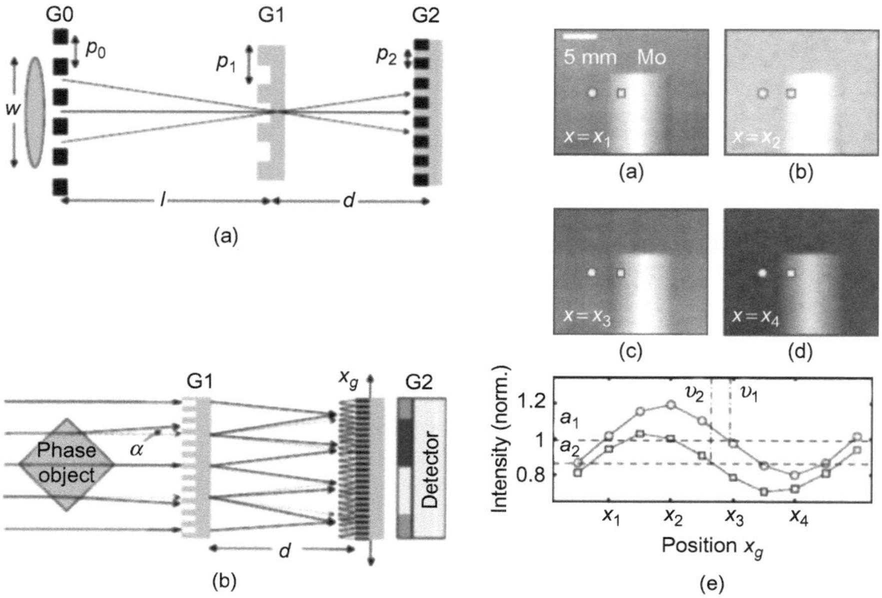

FIGURE 164 Differential phase-contrast (DPC) imaging with neutrons. (Left) Schematic for a neutron shearing interferometer, which produces DPC neutron images similar to Schlieren imaging. The source grating G0 in diagram (a) creates an array of individually coherent but mutually incoherent sources. The phase grating (G1) acts as a beam splitter and divides the incoming beam into the two first diffraction orders in which the neutron wavelength λ is much smaller than the grating pitch, and the two beams then overlap because their angle is sufficiently small. The phase grating G1 in diagram (b) forms a periodic interference pattern in the plane of the analyzer grating G2. A phase object in the incident beam will cause a slight refraction, which results in changes in the locally transmitted intensity through the analyzer grating G2 onto the CCD detector. (Right) The raw neutron image data of a molybdenum metal rod taken at different positions $x_g = x_1, \ldots, x_4$ of the analyzer grating G2. The plot shows the intensity oscillation as a function of position x_g of the translating grating (Pfeiffer *et al.*, 2006a).

X-ray shearing interferometer developed by Pfeiffer *et al.* (2006a), which produces X-ray DPC images (see Pfeiffer *et al.*, 2006b).

By 2011 advancements allowed the development of X-ray optics, microscopes, and telescopes that use the following the X-ray optical elements:

1. Kirkpatrick–Baez microscopes consisting of two grazing incidence mirrors at right angles to form the X-ray mirror optic for focusing an on-axis object point to a conjugate image point source

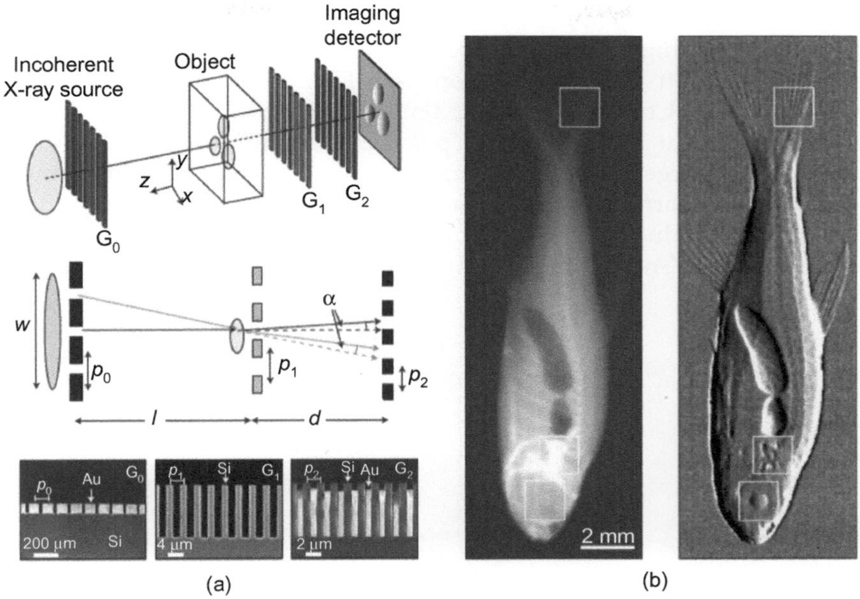

FIGURE 165 Differential phase-contrast (DPC) imaging with X-rays. (Left) An X-ray shearing interferometer, which produces DPC X-ray images in the same manner as the neutron shearing interferometer in Figure 164. Shown below the schematic are scanning electron micrographs of cross sections through the silicon gratings fabricated from silicon wafers. The gratings are made using standard photolithography techniques followed by electroplating that fills the grooves with gold (G0 and G2). (Right) Comparison of the (a) conventional transmission X-ray image on the left of a small fish obtained with a standard X-ray tube operated at 40 kV/25 mA, and (b) the image (Pfeiffer *et al.*, 2006b).

2. Cylindrical grazing incidence Wolter X-ray mirrors composed of two quadratic reflecting surface segments for reflective imaging of extended objects for Wolter microscopes and telescopes
3. Single Kumakhov mono-capillary (X-ray light pipe) to image a single object point
4. Kumakhov poly-capillary array (bundle of X-ray light pipes tapered at one or both end) to focus an object spot to an image point
5. Fresnel X-ray zone plates for focusing and imaging
6. Compound refractive lenses for focusing and imaging
7. Fresnel X-ray lenses
8. Multilayer X-ray mirrors
9. Lobster-eye grazing incidence mirror or Laue diffraction array
10. Coded aperture arrays.

Newly developed multilayer-based mirrors and optical elements allow the imaging of high-resolution structures and ultrafast dynamics of samples with the soft X-ray free electron laser (FLASH) at DESY in Hamburg, Germany (see Chapman *et al.*, 2006). In principle, an ultrashort and extremely bright coherent X-ray pulse can record a single diffraction pattern of a large macromolecule, a virus, or a cell before the sample explodes and turns into a plasma Figure 166. The FLASH free-electron laser (FEL) produces intense, ultrashort soft X-ray pulses, which allows time-resolved X-ray imaging and holography of nanostructures with a

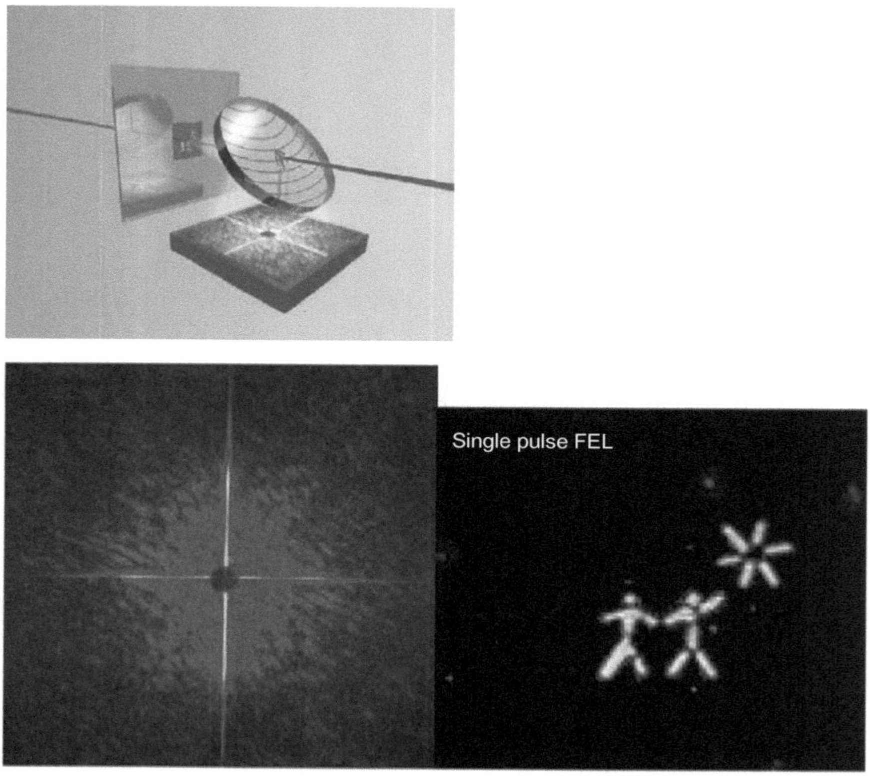

FIGURE 166 (Top) Diagram of the no-lens coherent diffractive imaging, which uses a multilayer-coated mirror that reflects the diffraction pattern onto a CCD. The direct FEL beam passes through the sample and through an exit hole while the multilayer mirror reflects the sample diffraction pattern onto the CCD detector. (Bottom) The X-ray diffraction image at the lower left is a coherent diffraction pattern recorded by a single, $\lambda = 32$-nm FEL pulse from a microfabricated test object. The image is reconstructed from the diffraction pattern by phase retrieval in which the image of resolution 62 nm has 7.5 -μm lateral extent (Bajt and Chapman, 2007).

temporal resolution approaching 10 fs, and enables studies of laser-matter interactions and the dynamics of correlated systems. The ultrashort pulses allow structural data to be obtained before the onset of radiation damage, which greatly increases the radiation dose that can be used to record images of biological samples, thereby improving the image resolution.

Bajt and Chapman (2007) reported the first experimental demonstration of this principle using the FLASH soft-X-ray FEL at DESY. An intense 25-fs 4×10^{13} watt/cm^2 pulse, containing at 10^{12} at 32-nm wavelength, produced a coherent diffraction pattern from a nanostructured nonperiodic object, before destroying it at 60,000°K. A novel X-ray camera with single-photon detection sensitivity filtered out parasitic scattering and plasma radiation (see Figure 166). The reconstructed image was obtained directly from the coherent pattern by phase retrieval through oversampling in which the sample showed no measurable damage, and the sample was reconstructed at diffraction-limited resolution. A 3D dataset can be obtained by assembling a set of images, which are produced with copies of reproducible, identical samples that are each flash-imaged by a single FEL pulse.

The lobster-eye structure and principle of imaging light has been applied to X-ray telescopes and microscopes and applied to neutron focusing mirrors (Figure 167). Concentric mirrors, microchannel plates, or multilayer mirrors can be configured as a lobster eye to produce wide-angle images over a broad energy range from hard X-rays to long-wave infrared. In place of mirrors the lobster-eye focusing of X-rays can be accomplished by Laue diffraction with crystals or synthetic multilayer mirrors (see Gorenstein, 2010), and the lobster eye with multilayer neutron mirrors can be used to focus neutrons (see Johnson and Daymond, 2000).

Kang *et al.* (2006) fabricated a modified linear zone plate, termed a *multilayer Laue lens* (MLL), for nanometer-scale focusing of hard X-rays. They sectioned a multilayer and provided X-ray illumination in Laue diffraction geometry. Coupled wave theory calculations by Kang *et al.* (2006) indicate focusing to 5-nm diameters or smaller with high efficiency may be possible. They fabricated partial MLL structures, which had outermost zone widths as small as 10 nm, and these MLL structures were tested with a 19.5-keV synchrotron X-ray source. The smallest attained focal diameters with these MLL structures were 30 nm and maximum measured efficiencies were 44%.

The scanning electron microscope operates in the 0.06- to 3.0-nm range and produces images with 2-nm best resolution from the surface of imaged samples with a high depth of field. By comparison, visible light microscopy operates from 400 to 800 nn and produces images with 200-nn best resolution from the surface and inside the sample with a low depth of field. The X-ray microscope operates from 0.03 to 50 nn and produces

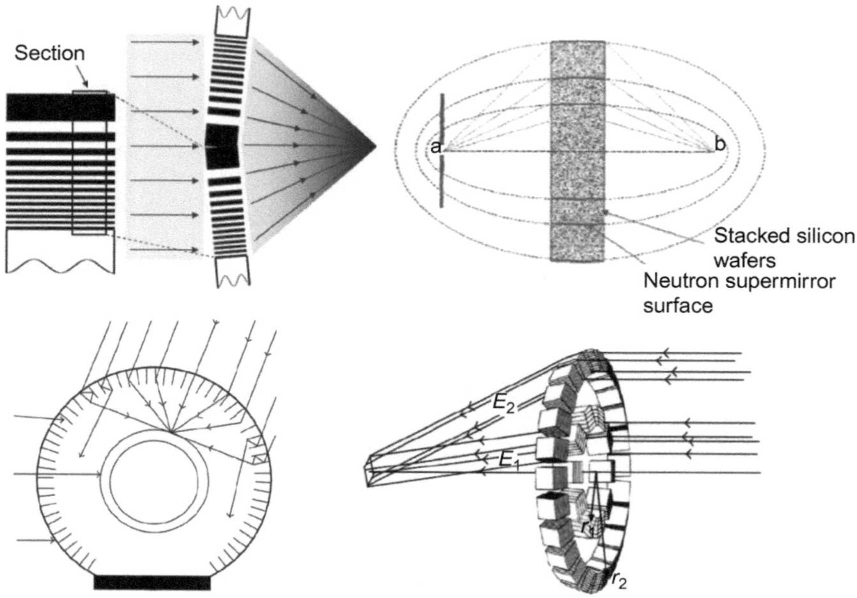

FIGURE 167 (Top left) A lobster eye is accomplished by Laue diffraction from multilayer surfaces rather than grazing incidence reflection from concentric, tilted mirrors (from Kang *et al.*, 2006). (Top right) Diagram of a cylindrical neutron lens based on multiple concentric cylindrical neutron supermirrors supported and separated by silicon wafers (from Johnson and Daymond, 2000). (Bottom) X-ray focusing with a lobster eye with reflecting mirrors in which the maximum grazing angle is much smaller than indicated in the drawing (bottom left) and X-ray focusing with Laue diffracting crystals or multilayers (bottom right). Both bottom figures are from Gorenstein (2010).

images with 20-nn best resolution of the inside of the sample with a medium to high depth of field. (See "Overview over many kinds of X-ray optics!" at the X-ray Optics website, http://www.X-ray-optics.de.) The rotation of the object and the recording of the X-ray amplitude-contrast or X-ray phase-contrast image at each rotation angle for a set of object rotation angles between 0 and 180° is the basis of X-ray tomography, which allows a 3D rendering of the object. Tomography can be performed with thermal and fast neutrons as well. Amplitude-contrast images taken at a set of object rotation angles between 0 and 180° can be done with thermal refractive imaging and radiographic imaging, as well as fast neutron radiographic imaging, to produce 3D images of the neutron interrogated object.

We now discuss the scanning near-field optical microscope (SNOM), which was developed for visible light and then extended to infrared light from a FEL (see Margaritondo, 2002). The SNOM overcomes the Rayleigh diffraction limit of the classical microscope by using a detector and a source that are both smaller than the wavelength λ of the electromagnetic wave. The lateral resolution of a stethoscope is ~1.0 cm because the heart and the stethoscope with its disk-shaped sound antenna and sound guide tube are much smaller than the $\lambda = 300$-m wavelength. The diffraction limit of light is overcome by a light source and detector that are smaller than the light wavelength. The imaging of evanescent infrared FEL light waves produced at the end of a single tapered tip of a Kumakhov X-ray, hollow fiber optic are reflected back into the fiber-optic tip by a tiny region of a light-reflecting sample just adjacent to the tip.

At the tip of a tapered hollow fiber an evanescent wave is emitted that is subject to complex refractive index n of the tip in the direction of lateral (y-directed) expansion of the wave, where $n_r + in_i$. The Heisenberg uncertainty principle limits the resolution in the lateral y-direction at the fiber tip to $\Delta y \geq h/\Delta p_y$ in which the maximum $\Delta p_{y\,\max} = p_y$, and hence $\Delta y \geq h/p_y$. The de Broglie relation relates the total wave momentum p to its wavelength λ by $p = h/\lambda$ in which $p = \sqrt{p_x^2 + p_y^2}$ and the maximum possible value of the transverse y-directed momentum component p_y is the total momentum p, where $p_{y\,\max} = p$ and

$$\Delta y \geq \frac{h}{\Delta p_y} = \frac{h}{p_y} = \frac{h}{p} = \frac{h}{\sqrt{p_x^2 + p_y^2}} = \frac{\lambda}{n_r + in_i}. \tag{236}$$

In the far-field case, the scattered waves from a sample travel over a long image distance r_i from the sample to the detector compared with the light wavelength λ, where $r_i \gg \lambda$. In air or vacuum at the distant detector, the imaginary part of the refractive index is zero $n_i = 0$ and the real part is unity $n_r = 1$. Given the maximum, transverse momentum uncertainty of the wave as equal to the total wave momentum, where $\Delta p_{y\,\max} = p_y = p$, this requires the x-component of the wave momentum to be zero $p_x = 0$ so that

$$\Delta y \geq \frac{h}{\Delta p_y} = \frac{h}{p_y} = \frac{h}{p} = \frac{\lambda}{n_r} = \lambda. \tag{237}$$

This is the normal diffraction limit for far-field imaging. However, in the near-field case with an evanescent wave interacting with an sample with a complex refractive index with a nonzero imaginary component, where $n_i \neq 0$ and $\mathrm{Im}[p_x] \neq 0$, one then obtains an imaginary lateral uncertainty Δy, which is now an invalid limitation on resolution, where

$$\Delta y \geq \frac{h}{\Delta p_y} = \frac{h}{p_y} = \frac{h}{p} = \frac{h}{\sqrt{-p_x^2 + p_y^2}} = \frac{\lambda\,(n_r - in_i)}{n_r^2 + n_i^2} \quad \text{Not valid.} \tag{238}$$

Hence, the near-field resolution of an evanescent wave emitted from a tiny source and interacting with a very nearby adjacent sample is not limited. As a result, a tiny detector could be placed next to a tiny source, each of which are smaller than the wavelength λ of the electromagnetic wave emitted and detected, and thus obtain a higher resolution than the Rayleigh limit. At present the SNOM is limited to visible and infrared light but some day may be extended to X-rays.

The following list contains an incomplete chronological list of papers and books on X-ray optics and X-ray sources, which include reflective, diffractive, and refractive X-ray optics:

Compton, A. H., & Allison, S. K. (1935). *X-rays in Theory and Experiment.* Van Nostrand, Princeton, NJ.

Buerger, M. J. (1942). *X-ray Crystallography.* Wiley, New York.

Zachariasen, W. H. (1945). *Theory of X-ray Diffraction in Crystals,* Wiley, New York.

Guinier, A., & Fournet, G. (1955). *Small-Angle Scattering of X-rays.* Wiley, New York.

Buerger, M. J. (1958). X-ray Crystallography: An Introduction to the Investigation of Crystals by Their Diffraction of Monochromatic X Radiation. Chapman and Hall, London.

Cosslett, V. E., & Nixon, W. C. (1960). *X-ray Microscopy.* Cambridge University Press, Cambridge, UK.

Sptizer, L. (1962). *Physics of Fully Ionized Gases.* Interscience, New York (1962)

Buerger, M. J. (1963). *Elementary Crystallography: An Introduction to the Fundamental Geometrical Features of Crystals.* Wiley.

James, R. W. (1965). *Optical Principles of the Diffraction of X-rays* (Crystalline State, Vol. 2), Cornell University Press, Ithaca, NY.

Ter-Mikaelian, M. L. (1972). *High-Energy Electromagnetic Processes in Condensed Media.* Wiley-Interscience, New York.

Hora, H., & Schwarz, H. (1974). *Laser Interaction and Related Plasma Phenomena,* Vol. 3a. Kluwer Academic, Norwell, MA.

Born, M., & Wolf, E. (1975). *Principles of Optics,* ed 5. Pergamon Press, Elmsford, NY.

Agarwal, B. K. (1979). *X-ray Spectroscopy.* Springer Series in Optical Sciences, Vol 15. Springer, Berlin.

Hora, H. (1981). *Physics of Laser Driven Plasmas.* John Wiley & Sons.

Brodie, I., & Murray, J. J. (1982). *The Physics of Microfabrication.* Plenum Press, NY.

Schmahl, G., & Rudolph, D., eds. (1984). *X-ray Microscopy.* Springer Series in Optical Sciences, vol. 43. Springer, Berlin.

Sáenz, A. W., & Überall, H. eds. (1985). *Coherent Radiation Sources.* Springer, Berlin.

Michette, A. G. (1986). *Optical Systems for Soft X-rays.* Plenum Press, New York.

Warren, B. E. (1990). *X-ray Diffraction.* Dover Publications, New York.

Bragg, L., Phillips, D. C., & Lipson, H. (1992). *The Development of X-Ray Analysis.* Dover Publications.

Spiller, E. (1994). *Soft X-ray Optics.* SPIE Optical Engineering Press, Bellingham, WA.

Winick, H. (ed.). (1994). *Synchrotron Radiation Sources: A Primer* (World Scientific Series on Synchrotron Radiation Techniques). World Scientific, Singapore.

Guinier, A. (1994). *X-Ray Diffraction: In Crystals, Imperfect Crystals, and Amorphous Bodies,* Dover Publications.

Rullhusen, P., Artru, X., & Dhez, P. (eds.). (1997). *Novel Radiation Sources Using Relativistic Electrons, from Infrared to X-Rays: From Infrared to X-Rays* (Series on Synchrotron Radiation Techniques and Applications). World Scientific, Singapore.

Buhrke, V. E., Jenkins, R., & Smith, D. K. (eds.). (1997). *A Practical Guide for the Preparation of Specimens for X-Ray Fluorescence and X-Ray Diffraction Analysis*. Wiley-VCH, New York.
Turcu, I. C. E., & Dance, J. B. (1998). *X-Rays From Laser Plasmas: Generation and Applications*. John Wiley and Sons Ltd, West Sussex, England.
Attwood, D. (1999). *Soft X-rays and Extreme Ultraviolet Radiation, Principles and Applications*. Cambridge University Press.
Jenkins, R. (1999). *X-Ray Fluorescence Spectrometry* (Chemical Analysis: A Series of Monographs on Analytical Chemistry and Its Applications). Wiley-Interscience, New York.
Hora, H. (2000). *Laser Plasma Physics: Forces and the Nonlinearity Principle*. SPIE, Bellingham, WA.
Janssens, K. H. A., Adams, F. C. V., & Rindby, A. (2000). *Microscopic X-Ray Fluorescence Analysis*. John Wiley and Sons, West Sussex, England.
Margaritondo, G. (2002). *Elements of Synchrotron Light for Biology, Chemistry, and Medical Research*. Oxford University Press.
Zemb, Th., & Lindner, P. (eds.). (2002). *Neutron, X-rays and Light. Scattering Methods Applied to Soft Condensed Matter* (North-Holland Delta Series). Elsevier, Amsterdam.
Van Grieken, R., & Markowicz, A. (eds.). (2002). *Handbook of X-Ray Spectrometry*, ed 2. Marcel Dekker, New York.
Watts, J. F., & Wolstenholme, J. (2003). *An Introduction to Surface Analysis by XPS and AES*. John Wiley and Sons, New York.
Ladd, M. F. C., & Palmer, R. A. (2003). *Structure Determination by X-Ray Crystallography*. Kluwer Academic/Plenum, New York.
Wiedemann, H. (2003). *Synchrotron Radiation* (Advanced Texts in Physics). Springer, Berlin.
Tsuji, K., Injuk, J., & Van Grieken, R. (eds.). (2004). *X-Ray Spectrometry: Recent Technological Advances*. John Wiley and Sons, West Sussex, England.
Clarke, J. A. (2004). *The Science and Technology of Undulators and Wigglers* (Oxford Series on Synchrotron Radiation). Oxford University Press, New York.
Birkholz, M. (2006). *Thin Film Analysis by X-Ray Scattering*. Wiley-VCH, Weinheim, Germany.
Paganin, D. (2006). *Coherent X-Ray Optics* (Oxford Series on Synchrotron Radiation). Oxford University Press, New York.
Hippert, F., Geissler, E., Hodeau, J. L., Lelièvre-Berna, E., Regnard, J.-R. (eds.). (2006). *Neutron and X-ray Spectroscopy*. Springer, Dordrecht.
Zschornack, G. H. (2007). *Handbook of X-Ray Data*. Springer, Berlin.
Nickles, P. V., & Janulewicz, K. A. (eds.). (2007). *X-Ray Lasers 2006: Proceedings of the 10th International Conference*, August 20–25, Berlin, Germany (Springer Proceedings in Physics). Springer, Dordrecht.
Freund, L. B., & Suresh, S. (2009). *Thin Film Materials: Stress, Defect Formation and Surface Evolution* (Cambridge Pocket Clinicians). Cambridge University Press, Cambridge, England.
Duke, P. (2009). *Synchrotron Radiation: Production and Properties* (Oxford Series on Synchrotron Radiation). Oxford University Press, New York.
Beyer, H. F., Kluge, J.-J., & Shevelko, V. P. (2010). *X-Ray Radiation of Highly Charged Ions* (Springer Series on Atomic, Optical, and Plasma Physics). Springer, Heidelberg.
Bunker, G. (2010). *Introduction to XAFS: A Practical Guide to X-ray Absorption Fine Structure Spectroscopy*. Cambridge University Press.
Waseda, Y., Matsubara, E., & Shinoda, K. (2011). *X-Ray Diffraction Crystallography: Introduction, Examples and Solved Problems*. Springer, Heidelberg.
Guo, J., (ed.) (2010). *X-Rays in Nanoscience: Spectroscopy, Spectromicroscopy, and Scattering Techniques*. Wiley-VCH, Weinheim, Germany.
Daillant, J., & Gibaud, A. (eds.). (2010). *X-ray and Neutron Reflectivity: Principles and Applications* (Lecture Notes in Physics). Springer, Berlin.

Als-Nielsen, J., & McMorrow, D. (2011). *Elements of Modern X-ray Physics*, ed. 2. John Wiley and Sons.
Kaastra, J., & Paerels, F. (eds.). (2011). *High-Resolution X-Ray Spectroscopy: Past, Present and Future*. Springer, Berlin.
Sivia, D.S. (2011). *Elementary Scattering Theory for X-ray and Neutron Users*. Oxford University Press, Oxford, UK.

REFERENCES

Section 1: Background and Theory

Arthur, J. R., & Tatchyn, R. O. (2000). Radiation properties of the Linac Coherent Light Source: Challenges for x-ray optics. *Proceedings of SPIE, 4143*, 1–8.
Bass, M., DeCusatis, C., Enoch, J., Lakshminarayanan, V., Li, G., MacDonald, C., Mahajan, V., & Van Stryland E. (Eds.). (2010). *Handbook of Optics*, 3rd ed., Vol. V: *Atmospheric Optics, Modulators, Fiber Optics, X-Ray and Neutron Optics*. McGraw-Hill, New York.
Beguiristain, H. R., Cremer, J. T., & Piestrup, M. A. (2004). Fabrication of unit lenses for compound refractive lenses. U.S. Patent No. 6,674,583 B2, January 6, 2004.
Beynon, T. D., & Pink, A. G. (1980). Neutron holography using Fresnel zone plate encoding. *Nature, 283*, 749–751.
Born, M., & Wolf, E. (2002). *Principles of Optics*. (7th ed). Cambridge University Press, Cambridge, UK.
Choi, S. M., Barker, J. G., Glinka, C. J., Cheng, Y. T., & Gammel, P. T. (2000). Focusing cold neutrons with multiple biconcave lenses for small-angle neutron scattering. *Journal of Applied Crystallography, 33*, 793–796.
Daillant, J., & Gibaud, A. (Eds.). (2009). *X-ray and Neutron Reflectivity: Principles and Applications (Lecture Notes in Physics)*. Springer, Berlin.
Eskildsen, M. R., Gammel, P. L., Isaacs, E. D., Detlefs, C., Mortensen K., & Bishop, D. J. (1998). Compound refractive optics for imaging and focusing of low-energy neutrons. *Nature, 391*, 563–566.
Freeman, M. H., & Hull, C. C. (2003). *Optics*. Butterworth-Heinemann, London.
Gabor, D. (1949). Microscopy by reconstructed wave-fronts. *Proceedings of the Royal Society of London Series A-Mathematical Physical and Engineering Sciences, 197*(1051), 454–487.
Hecht, E., & Zajac, A. (1974). *Optics*. Addison-Wesley, New York.
Henke, B. L., Gullikson E. M., & Davis, J. C. (1993). X-ray interactions: photoabsorption, scattering, transmission, and reflection at E = 50-30000 eV, Z = 1-92. *Atomic Data and Nuclear Data Table, 54*(2), 181–342.
Kearney, P. D., Klein, A. G., Opat, G. I., Gähler, R. (1980). Imaging and focusing of neutrons by a zone plate. *Nature, 287*, 313–314.
Kim, K.-J. (October 2009). Synchrotron radiation (Section 2, Fig. 2.9). In *X-Ray Data Booklet*. Lawrence Berkeley National Laboratory, Center for X-ray Optics and Advanced Light Source, LBNL/Pub-490 Rev. 3.
Köhler, A. (1893). Ein neues Beleuchtungsverfahren für mikrophotographische Zwecke. *Zeitschrift für wissenschaftliche Mikroskopie und für Mikroskopische Technik 10*(4), 433–440. http://www.archive.org/stream/zeitschriftfrw10stut.
Lauter-Pasyuk, V. (2007). Neutron grazing incidence techniques for nano-science. *Collection SFN 7*, 221–240. EDP Sciences, Les Ulis, Cedex A, France.
Lawrence Berkeley National Laboratory, Center for X-ray Optics and Advanced Light Source. (2009). Section 4.2: Specular reflectivities for grazing-incidence mirrors. In E. M. Gullikson (Ed.), *X-Ray Data Booklet*. Lawrence Berkeley National Laboratory, Center for X-ray Optics and Advanced Light Source, LBNL/Pub-490 Rev. 3 (October 2009).

Lehmann, E. H. (2000). Neutron imaging. In *Proceedings of the 18th summer school on neutron scattering, August 2000.* World Scientific Publishing, Singapore. http://neutra.web.psi.ch/publication.

Leith, E. N., & Upatnieks, J. J. (1964). Wavefront reconstruction with diffused illumination and three-dimensional objects. *Journal of the Optical Society of America, 54,* 1295–1301.

Lodha, G. S., Yamashita, K., Kunieda, H., Tawara, Y., Yu, J., Namba, Y., & Bennett, J. M. (1998). Effect of surface roughness and subsurface damage on grazing-incidence x-ray scattering and specular reflectance. *Applied Optics, 37*(22), 5239–5252.

Mertz, L., & Young, N. (1961). Fresnel transformations of images. In K. J. Habell (Ed.), *Proceedings of the international conference on optical instruments and techniques* (p. 305). Chapman & Hall, London.

Nieman, B., Rudolph, D., Schmahl, G. (1974). Soft X-ray imaging zone plates with large zone numbers for microscopic and spectroscopic applications. *Optics Communications, 12,* 160.

Pantell, R. H., Feinstein, J., Beguiristain, H. R., Piestrup, M. A., Gary, C. K., & Cremer, J. T. (2001). Refractive lenses for coherent x-ray sources. *Applied Optics, 40*(28), 5100–5105.

Piestrup, M. A., Pantell, R. H., & Beguiristain, H. R. (2004). Methods of imaging, focusing, and conditioning neutrons, U.S. Patent No. 6,765,197 B2, July 20, 2004.

Piestrup, M. A., Pantell, R. H., Cremer, J. T., & Beguiristain, H. R. (2001). Compound refractive lens for x-rays. U.S. Patent No. 6,269,145 B1, July 31, 2001.

Rarback, H., & Shu, D., Feng, S. C., Ade, H., Kirz, J., McNulty, I., ... Iskander, N. (1988). Scanning X-ray microscope with 75-nm resolution. *Review of Scientific Instruments, 59*(1), 52–59.

Savage, D. E., Kleiner, J., Schimke, N., Phang, Y.-H., Jankowski, T., Jacobs, J., Kariotis, R., & Lagally, M. G. (1991). Determination of roughness correlations in multilayer films for x-ray mirrors. *Journal of Applied Physics, 69*(3), 1411–1424.

Schmahl, G., & Rudolph, D. (1969). High power zone plates as image forming systems for soft X-rays (in German). *Optik, 29,* 577.

Schulte-Schrepping, H. (2000). Photon beamlines at TESLA x-ray-FEL undulators. *Proceedings of SPIE, 4143,* 9.

Skold, K., & Price, D. L. (Eds.). (1986). *Neutron Scattering.* Academic Press, New York.

Snigerev, A., Kohn, V., Snigereva, I., & Lengeler, B. (1996). A compound refractive lens for focusing high-energy x-rays. *Nature,* 384, 49–51.

Squires, G. L. (1996). *Introduction to the Theory of Thermal Neutron Scattering.* Dover Publications, New York.

Windt, D. L., Waskiewicz, W. K., & Griffith, J. E. (1994). Surface finish requirements for soft x-ray mirrors. *Applied Optics, 33*(10), 2025–2031.

Section 2: Experiments with Neutron Optics

Beguiristain, H. R., Anderson, I. S., Dewhurst, C. W., Piestrup, M. A., Cremer, J. T., & Pantell, R. H. (2002). A simple neutron microscope using a compound refractive lens. *Applied Physics Letters, 81*(22), 4290–4292.

Chen, H., Downing, R. G., Mildner, D., Gibson, W. M., Kumakhov, M. A., Ponomarev, I. Y., & Gubarev, M. V. (1992). Guiding and focusing neutron beams using capillary optics. *Nature, 357,* 391–393.

Choi, S. M., Barker, J. G., Glinka, C. J., Cheng, Y. T., & Gammel, P. L. (2000). Focusing cold neutrons with multiple biconcave lenses for small-angle neutron scattering. *Journal of Applied Crystallography, 33,* 793–795.

Cremer, J. T., Park, H., Piestrup, M. A., Gary, C. K., Pantell, R. H., Flocchini, R. G., Egbert, H. P., Kloh, M. D., & Walker, R. B. (2007a). Simple microscope using a compound refractive lens and a wide-bandwidth thermal neutron beam. *Applied Physics Letters, 90,* 141113.

Cremer, J. T., Park, H., Piestrup, M. A., Gary, C. K., Pantell, R. H., Flocchini, R. G., Egbert, H. P., Kloh, M. D., & Walker, R. B. (2007b). Imaging with thermal neutrons using a compound refractive lens. Unpublished results.
Cremer, J. T., Piestrup, M. A., Gary, C. K., Pantell, R. H., & Glinka, C. J. (2004). Biological imaging with a neutron microscope. *Applied Physics Letters, 85*(3), 494–496.
Cremer, J. T., Piestrup, M. A., Gary, C. K., Pantell, R. H., Glinka, C. J., & Barker, J. G. (2005). Imaging hydrogenous materials with a neutron microscope. *Applied Physics Letters, 87*(16), 161913.
Cremer, J. T., Williams, D. L., Fuller, M. J., Gary, C. K., Piestrup, M. A., Pantell, R. H., ... Egbert, H. P. (2010). Periodic magnetic field as a polarized and focusing thermal neutron spectrometer and monochromator. *Review of Scientific Instruments, 81*(1), 013902.
Eskildsen, M. R., Gammel, P. L., Isaacs, E. D., Detlefs, C., Mortensen K., & Bishop, D. J. (1998). Compound refractive optics for imaging and focusing of low-energy neutrons. *Nature, 391*, 563–566.
Forgan, E. M., & Cubitt, R. (1998). Cancellation of gravity for slow neutrons in small angle scattering experiments. *Neutron News, 9*, 25–31.
FujiFilm (2012). FujiFilm Medical Systems USA, Inc., Stamford CT.
Hirota, K., Shinohara, T., Ikeda, K., Mishima, K., Adachi, T., Morishima, T., ... Sasao, H. (2005). Development of a neutron detector based on a position-sensitive photomultiplier. *Physics of Chemical Chemical Physics, 7*, 1836–1838.
Just, W., Schneider, C. S., Ciszewski, R., & Shull, C. G. (1973). Refraction of thermal neutrons by shaped magnetic fields. *Physical Review B: Condensed Matter, 7*(9), 4142–4145.
Kumakhov, M. A. & Sharov, V. A. (1992). A neutron lens. *Nature, 357*, 390–391.
Lal, J., Lang, E., Carpenter, J., Geltenbort, P., Cremer, T., Shimizu, H., Hirota, K., & Mishima, K. (2007). Unpublished results of material and magnetic CRL imaging experiment at Pf2 ILL beam line in August 2007.
Littrell, K. C., Te Velthuis, S. G. E., Fletcher, G. P., Park, S., Kirby, B. J., & Fitzsimmons, M. R. (2007). A magnetic compound refractive lens for focusing and polarizing cold neutron beams. *Review of Scientific Instruments, 78*, 035101.
Majkrzak, C. F., & Wood, J. L. (Eds). (1992). *Neutron optical devices and applications*, 22–24 July 1992. Society of Photo-optical Instrumentation Engineers.
Mikula, P., Kulde, J., Horalik, L., Chalupa, B., & Lukas, P. (1986). The spatial condensation of the neutron beam by an asymmetric diffraction in thermal-neutron monochromatization. *Journal of Applied Crystallography, 19*, 324–330.
Pantell, R. H., Feinstein., J., Beguiristain, H. R., Piestrup, M. A., Gary, C. K., & Cremer, J. T. (2003). Characteristics of the thick, compound refractive lens. *Applied Optics, 42*(4), 719–723.
Park, H., Cremer, J. T., Piestrup, M. A., Gary, C. K., Hjelm, R. P., Sellyey, W. C. L. J., & Pantell, R. H. (2006). Measured operational neutron energies of compound refractive lenses. *Nuclear Instruments and Methods in Physics Research Section B: Beam Interactions with Materials and Atoms, 251*, 507–511.
Sears, V. F. (1989). *Neutron Optics*. Oxford Univiversity Press, Oxford, UK.
Shimizu, H. M. (2002). Development of neutron optical devices. *Applied Physics, A 74*, S326–S328.
Shimizu, H. (2004). Neutron optics. In *Conference on Neutron Optics*. Tokyo, Japan, January 18, 2004 [slide presentation].
Shimizu, H. M., Adachi, T., Furusaka, M., Kiyanagi, Y., Oku, T., Sasao, H., Shinohara, T., & Suzuki, J. (2004). Neutron optics and a superconducting magnetic lens for small angle neutron scattering with focusing geometry. *Nuclear Instruments and Methods in Physics Research Section A: Accelerators, Spectrometers, Detectors and Associated Equipment, 529*(1–3), 5–9.
Shimizu, H., Suzuki, J., & Oku, T. (2010). Neutron polarization apparatus. U.S. Patent application publication No. 2010/0032554 A1, February 11, 2010.

Shinohara, T., Hirota, K., Adachi, T., Ikeda, K., Shimizu, H. M., Suzuki, J., & Oku, T. (2006). Thermal neutron refraction by material prism. *Physica B, 385*, 1232–1235.

Squires, G. L. (1996). *Introduction to the Theory of Thermal Neutron Scattering*. Dover Publications, New York.

Williams, W. G. (1988). *Polarized Neutrons* (Oxford Studies in Neutron Scattering and Condensed Matter, No. 1). Clarendon Press.

Section 3: Experiments with X-ray Optics

Anderson, R. H. (1979). Close-up imaging of documents and displays with lens arrays. *Applied Optics, 18*(4), 477–484.

Arms, D. A., Dufresne, E. M., Clarke, R., Dierker, S. B., Pereira, N. R., Foster, D. (2002). Refractive optics using lithium metal. *Review of Scientific Instruments, 73*(3), 1492–1494.

Artemiev, A., Snigirev, A., Kohn, V., Snigireva, I., Artemiev, N., Grigoriev, M., ... Kvardakov, V. (2006). X-ray parabolic lenses made from glassy carbon by means of laser. *Review of Scientific Instruments, 77*(6), 063113.

Beguiristain, H. R., Cremer, J. T., Piestrup, M. A., Gary, C. K., & Pantell, R. H. (2002). X-ray focusing using compound lenses made of beryllium. *Optics Letters, 27*(9), 778–780.

Beguiristain, H. R., Piestrup, M. A., Pantell, R. H., Gary, C. K., Cremer, J. T., Tatchyn, R. (2000a). Development of compound refractive lenses for X-rays. In P. Pianetta, J. Arthur, & S. Brennan (Eds.), *Synchrotron radiation instrumentation: SRI99: Eleventh US national conference. Proceedings of the 11th Synch. Rad. Conference* (Vol. 521, pp 258–266). AIP Press, New York.

Beguiristain, H. R., Cremer, J. T., Piestrup, M. A., Pantell, R. H., Gary, C. K., & Feinstein, J. (2000b). Compound X-ray refractive lenses made of polyimide. In C. A. MacDonald & A. M. Khounsary (Eds.), *Advances in laboratory-based X-Ray sources and optics. Proceedings of SPIE* (Vol. 4144, pp. 155–164). SPIE, Bellingham, WA.

Berger, M. J., Hubbell, J. H., Seltzer, S. M., Chang, J., Coursey, J. S., Sukumar, R., Zucker, D. S., & Olsen, K. (1998). XCOM: Photon Cross Sections Database. NIST Standard Reference Database 8 (XGAM). National Institute of Standards and Technology; http://www.nist.gov/pml/data/xcom/index.cfm.

Bohic, S., Simionovici, A., Snigirev, A., Ortega, R., Devès, G., Heymann, D., & Schroer, C. G. (2001). Synchrotron hard x-ray microprobe: Fluorescence imaging of single cells. *Applied Physics Letters, 78*(22), 3544–3546.

Bonse, U., Johnson, H., Nichols, M., Nuβhardt, R., Krasnicki, S., & Kinney, J. (1986). High resolution tomography with chemical specificity. *Nuclear Instruments and Methods in Physics Research Section A: Accelerators, Spectrometers, Detectors and Associated Equipment, 246*(1–3), 644–648.

Bonse, U., Nuβhardt, R., Busch, F., Pahl, R., Johnson, Q. C., Kinney, J. H., Saroyan, R. A., & Nichols, M. C. (1989). Optimization of CCD-based energy-modulated x-ray microtomography. *Review of Scientific Instruments, 60*(7), 2478–2481.

Bragg, L., & Nye, J. F. (1947). A dynamic model of crystal structure. *Proceedings of the Royal Society of London, 190*, 474–481.

Cadbury-Schweppes (2005). 7 UP: the making of a legend. In *Cadbury Schweppes: America's beverages*. Copyright 2005–2007. http://web.archive.org/web/20080430022917/http://www.brandspeoplelove.com/csab/?TabId=148.

Cederström, B. (2002). *A Multi-Prism Lens for Hard X-rays*. Ph.D. thesis, Kungl Tekniska Hogskolan (KTH), Stockholm, Sweden.

Cederström, B. (2003). Refractive X-ray arrangement. U.S. Patent No. 6,668,040 B2, December 23, 2003, assigned to Mamea Imaging AB, Stockholm, Sweden.

Cederström, B., Cahn, R. N., Danielsson, M., Lundqvist, M., & Nygren, D. R. (2000). Focusing hard X-rays with old LPs. *Nature, 404*, 951.

Ceglio, N. M., Hayryluk, A. M., & Shattenburg. M. (1983). X-ray phase lens design and fabrication. *Journal of Vacuum Science & Technology B, Journal of Vacuum Science & Technology B, 1*, 1285–1288.

Cremer, J. T., Piestrup, M. A., Beguiristain, H. R., Gary, C. K., & Pantell, R. H. (2003). Large aperture compound lenses made of lithium. *Review of Scientific Instruments, 74*(4), 2262–2266.

Cremer, J. T., Piestrup, M. A., Beguiristain, H. R., Gary, C. K., Pantell, R. H., & Tatchyn, R. (1999). Cylindrical compound refractive x-ray lenses using plastic substrates. *Review of Scientific Intruments, 70*(9), 3545–3548.

Cremer, J. T., Piestrup, M. A., Gary, C. K., & Pantell, R. H. (2004). Large-aperture compound refractive lenses. In R. O. Tatchyn, Z. Chang, J.-C. Kieffer, & J. B. Hastings (Eds.), *Fourth-Generation X-ray sources and ultrafast X-ray detectors. Proceedings of SPIE* (Vol. 5194, pp. 62–89). SPIE, Bellingham, WA.

Csonka, P. L. (1976). Suggested method for coherent x-ray production by combined x-ray and low-energy photon pumping. *Physical Review A, 13*, 405–410.

Dudchik, Y. I., & Kolchevsky, N. N. (1999). A microcapillary lens for X-rays. *Nuclear Instruments and Methods in Physics Research Section A: Accelerators, Spectrometers, Detectors and Associated Equipment, 421*(1–2), 361–364.

Dudchik, Y. I., & Kolchevsky, N. N. (2001). Fabrication technique and ray tracing calculations for microcapillary x-ray lens. In A. K. Freund, T. Ishikawa, A. M. Khounsary, D. C. Mancini, A. G. Michette, & S. Oestreich (Eds.), *Advances in X-ray optics. Proceedings of SPIE* (Vol. 4145, pp. 235–242). SPIE, Bellingham, WA.

Dudchik, Y. I., Kolchevsky, N. N., Komarov, F. F., Kohmura, Y., Awaji, M., Suzuki, Y., & Ishikawa, T. (2000). Glass capillary x-ray lens: Fabrication technique and ray tracing calculations. *Nuclear Instruments and Methods in Physics Research Section A: Accelerators, Spectrometers, Detectors and Associated Equipment, 454*(2–3), 512–519.

Dudchik, Y. I., Kolchevsky, N. N., Komarov, F. F., Piestrup, M. A., Cremer, J. T., Gary, C. K., & Pantell, R. H. (2004a). Short-focal-length compound refractive lenses for X-rays. In R. O. Tatchyn, Z. Chang, J.-C. Kieffer, & J. B. Hastings (Eds.), *Fourth-Generation X-ray sources and ultrafast X-ray detectors. Proceedings of SPIE* (Vol. 5194, pp.56–61). SPIE, Bellingham, WA.

Dudchik, Y. I., Kolchevsky, N. N., Komarov, F. F., Piestrup, M. A., Cremer, J. T., Gary, C. K., Park, H., & Khounsary, A. M. (2004b). Microspot x-ray focusing using a short focal-length compound refractive lenses. *Review of Scientific Instruments, 75*(11), 4651–4655.

Dudchik, Y. I., Komarov, F. F., Piestrup, M. A., Gary, C. K., Park, H., Cremer, J. T. (2007). Using of a microcapillary refractive X-ray lens for focusing and imaging. *Spectrochimica Acta Part B, 62*, 598–602.

Dufresne, E. M., Arms, D. A., Clarke, R., Pereira, N. R., Dierker, S. B., & Foster, D. (2001). Lithium metal for X-ray refractive optics. *Applied Physics Letters, 79*(25), 4085–4087.

Elleaume, P. (1998). Two-plane focusing of 30 keV undulator radiation. *Journal of Synchrotron Radiation, 5* (Part 1), 1–5.

Eskildsen, M. R., Gammel, P. L., Isaacs, E. D., Detlefs, C., Mortensen K., & Bishop, D. J. (1998). Compound refractive optics for imaging and focusing of low-energy neutrons. *Nature, 391*, 563–566.

Feynman, R. P., Leighton, R. B., & Sands, M. (1964). *The Feynman Lectures on Physics.* (Vol. 2, pp. 30–10). Addison-Wesley, Reading, MA.

Gary, C. K., Doggett, D. E., Park, H., Harris, J. L., Pikuz, S. A., Mitchell, M. D., Chandler, K. M., Shelkovenko, T. A., & Hammer, D. A. (2005). A refractive imaging system for the characterization of x-ray sources. In *Laser-Generated, synchrotron, and other laboratory X-ray and EUV sources, optics, and applications II. Proceedings of SPIE* (Vol. 5918, p. 591812). SPIE, Bellingham, WA.

Gary, C. K., Park, H., Lombardo, L. W., Piestrup, M. A., Cremer, J. T., Pantell, R. H., & Dudchik, Y. I. (2007a). High resolution x-ray microscope. *Applied Physics Letters, 90*, 181111.

Gary, C. K., Park, H., Pantell, R. H., Dudchik, Y. I., Lombardo, L. W., & Cremer, J. T. (2007b). Radiography and tomography system using refractive lenses. In F. P. Doty, H. B. Barber, & H. Roehrig (Eds.), *Penetrating radiation systems and applications VIII. Proceedings of SPIE* (Vol. 6707, p. 67070R). SPIE, Bellingham, WA.

Gary, C. K., Pikuz, S. A., Mitchell, M. D., Chandler, K. M., Shelkovenko, T. A., Hammer, D. A., & Dudchik, Y. I. (2004). X-ray imaging of an X-pinch plasma with a bubble compound refractive lens. *Review of Scientific Instruments, 75*(10), 3950–3952.

Hogan, J. P., Gonsalves, R. A., & Krieger, A. S. (1991). Fluorescent computer tomography: A model for correction of x-ray absorption. *IEEE Transactions on Nuclear Science, 38*(6), 1721–1727.

Kirkpatrick, P. (1949). X-ray images by refractive focusing. *Journal of the Optical Society of America, 39*(9), 796.

Kirkpatrick, K., & Baez, A. V. (1948). Formation of optical images by x-rays. *Journal of the Optical Society of America, 38*(9), 766–773.

Lengeler, B., Schroer, C. G., Benner, B., Günzler, T. F., Kuhlmann, M., Tummler, J., Simionovici, A. S., Drakopoulos, M., Snigirev, A., & Snigireva, I. (2001). Parabolic refractive x-ray lenses: a breakthrough in x-ray optics. *Nuclear Instruments and Methods in Physics Research Section A: Accelerators, Spectrometers, Detectors and Associated Equipment, 467–468* (Part 2), 944–950.

Lengeler, B., Schroer, C., Tümmler, J., Benner, B., Richwin, M., Snigirev, A., Snigireva, I., & Drakopoulos, M. (1999a). Imaging by parabolic refractive lenses in the hard x-ray range. *Journal of Synchrotron Radiation, 6*(Part 6), 1153–1167.

Lengeler, B., Schroer, C. G., Richwin, M., Tümmeler, J., Drakopolulos, M., Snigirev, A., & Snigireva, I. (1999b). A microscope for hard x rays based on parabolic compound refractive lenses. *Applied Physics Letters, 74*, 3924–3926.

Lengeler, B., Tümmeler, J., Snigirev, A., Snigireva, I., & Raven, C. (1998). Transmission and gain of singly and doubly focusing refractive x-ray lenses. *Journal of Applied Physics, 84*, 5855–5861.

Michette, A. (1991). No-X-ray lens. *Nature, 353*, 510.

Mitchell, P. B., & Hadzi-Pavlovic, D. (2000). Lithium treatment for bipolar disorder. *Bulletin of the World Health Organization, 78*(4), 515–517.

Pantell, R. H., Feinstein, J., Beguiristain, H. R., Piestrup, M. A., Gary, C. K., & Cremer, J. T. (2001). Refractive lenses for coherent x-ray sources. *Applied Optics, 40*(28), 5100–5105.

Park, H., Piestrup, M. A., Gary, C. K., Fuller, M. K., Cremer, J. T., Pantell, R. H., & Dudchik, Y. I. (2007). X-ray dark-field imaging using a bubble compound refractive lens. Adelphi Technology, Inc. Redwood City, CA, USA. Unpublished.

Pereira, N. R., Arms, D. A., Clarke, R., Dierker, S. B., Dufresne, E., & Foster, D. (2001). Li metal for x-ray refractive optics. In A. M. Khounsary & C. A. MacDonald (Eds.), *Advances in laboratory-based X-ray sources and optics II. Proceedings of SPIE* (Vol. 4502, pp. 173–183). SPIE, Bellingham, WA.

Pereira, N. R., Dufresne, E. M., Clarke, R., & Arms, D. A. (2004). Parabolic lithium refractive optics for x rays. *Review of Scientific Instruments, 75*(1), 37–41.

Piestrup, M. A. (2003). X-ray and neutron imaging. U.S. Patent No. 2003/0081724 A1, May 1, 2003.

Piestrup, M. A. (2004). Large area x-ray and neutron imaging using three-dimensional arrays of microlenses. *Review of Scientific Instruments, 75*(11), 4769–4774.

Piestrup, M. A., Beguiristain, H. R., Gary, C. K., Cremer, J. T., Pantell, R. H., Tatchyn, R. (2001). Compound refractive lenses for novel x-ray sources. *Nuclear Instruments and Methods in Physics Research Section B: Beam Interactions with Materials and Atoms, 173*(1–2), 170–177.

Piestrup, M. A., Cremer, J. T., Beguiristain, H. R., Gary, C. K., & Pantell, R. H. (2000). Two-dimensional x-ray focusing from compound lenses made of plastic. *Review of Scientific Instruments, 71*(12), 4375–4379.

Piestrup, M. A., Gary, C. K., Park, H., Harris, J. L., Cremer, J. T., Pantell, .H., Dudchik, Y. I., Kolchevsky, N. N., & Komarov, F. F. (2005). Microscope using an x-ray tube and a bubble compound refractive lens. *Applied Physics Letters, 86*, 131104.
Piestrup, M. A., Pantell, R. H., Cremer, J. T., & Beguiristain, H. R. (2001). Compound refractive lens for x-rays. U.S. Patent No. 6,269,145 B1, July 31, 2001.
Posey, R., Brown, D., Broadwater, K., Swint, T., Valentine, J., & Ebert, P. J. (1998). Imaging soft X-rays with opaque spheres. *Journal of X-ray Science and Technology, 8*, 105–116.
Ribbing, C., and Cederström, B., and Lundqvist, M. (2003). *Journal of Micromechanics and Microengineering, 13*, 714–720.
Rozsnyai, B., Watanabe, H., & Csonka, P. L. (1985). Numerical simulation of a soft-x-ray Li laser pumped with synchrotron radiation. *Physical Review A 32*(1), 357–366.
Rust, G.-F., & Weigelt, J. (1998). X-ray fluorescent computer tomography with synchrotron radiation. *IEEE Transactions on Nuclear Science, 45*(1), 75–88.
Said A. H., & Shastri S. D. (2010). Silicon saw-tooth refractive lens for high-energy x-rays made using a diamond saw. *Journal of Synchrotron Radiation, 17*, 425–427.
Schroer, C. G., Benner, B., Günzler, T. F., Kuhlmann, M., Zimprich, C., Lengeler, B., Snigireva, I. (2002a). High resolution imaging and lithography with hard X-rays using parabolic compound refractive lenses. *Review of Scientific Instruments, 73*(3), 1640–1642.
Schroer, C. G., Kuhlmann, M., Günzler, T. F., Lengeler, B., Richwin, M., Griesebock, B., Lützenkirchen-Hecht, D., Frahm, R., Ziegler, E., Mashayekhi, A., Haeffner, D. R., Grunwaldt, J.-D., & Baiker, A. (2003a). Mapping the chemical states of an element inside a sample using tomographic x-ray absorption spectroscopy. *Applied Physics Letters, 82*(19), 3360–3362.
Schroer, C. G., Kuhlmann, M., Hunger, U. T., Günzler, T. F., Kurapova, O., Feste, S., Frehse, F., Lengeler, B., Drakopoulos, M., Somogyi, A., Simionovici, A. S., Snigirev, A., Snigireva, I., Schug, C., & Schröder, W. H. (2003b). Nanofocusing parabolic refractive x-ray lenses. *Applied Physics Letters, 82*(9), 1485–1487.
Schroer, C. G., Kuhlmann, M., Lengeler, B., Günzler, T. F., Kurapova, O., Benner, B., Rau, C., Simionovici, A. S., Snigirev, A., & Snigireva, I. (2002b). Beryllium parabolic refractive x-ray lenses. In Derrick C. Mancini (Ed.), *Design and microfabrication of novel X-ray optics. Proceedings of SPIE* (Vol. 4783, pp.10–18, 150–163). SPIE, Bellingham, WA.
Schroer, C. G., Lengeler, B., Benner, B., Günzler, F., Tümmler, J., Drakopoulos, M., Weitkamp, T., Snigirev, A., & Snigireva, I. (2000a). A microscope for hard x-rays based on parabolic compound refractive lenses. In W. Meyer-Ilse, T. Warwick, & D. Attwood (Eds.), *X-ray microscopy: Proceedings of the sixth international conference. AIP conference proceedings,* (Vol. 507, pp. 340–345).
Schroer, C. G., Tümmler, J., Günzler, F., Lengeler, B., Schröder, W. H., Kuhn, A. J., Simionovici, A. S., Snigirev, A., & Snigireva, I. (2000b). Fluorescence microtomography: external mapping of elements inside biological samples. In F. P. Doty, H. B. Barber, H. Roehrig, & E. J. Morton (Eds.), *Penetrating radiation systems and applications II. Proceedings of SPIE* (Vol. 4142, pp. 287–296). SPIE, Bellingham, WA.
Schroer, C. G., Lengeler, B., Benner, B., Tümmler, J., Günzler, F., Drakopoulos, M., Simionovici, A. S., Snigirev, A., & Snigireva, I. (2000c). Compound refractive lenses for x-ray microanalysis. In W. Meyer-Ilse, T. Warwick, & D. Attwood (Eds.), *X-Ray Microscopy: Proceedings of the Sixth International Conference. AIP Conference Proceedings,* American Institute of Physics, New York. (Vol 507, pp. 694–699).
Schroer, C. G., Lengeler, B., Benner, B., Kuhlmann, M., Günzler, T. F., Tümmler, J., Rau, C., Weitkamp, T., Snigirev, A., & Snigireva, I. (2001). Parabolic compound refractive lenses for hard X-rays. In A. Freund, T. Ishikawa, A. M. Khounsary, D. C. Mancini, A. G. Michette, & S. Oestreich, (Eds.), *Advances in X-ray optics. Proceedings of SPIE* (Vol. 4145, pp. 274–284). SPIE, Bellingham, WA.

Schroer, C. G., Meyer, J., Kuhlmann, M., Benner, B., Günzler, T. F., Lengeler, B., Rau, C., Weitkamp, T., Snigirev, A., & Snigireva, I. (2002b). Nanotomography based on hard x-ray microscopy with refractive lenses. *Applied Physics Letters, 81*(8), 1527–1529.

Shastri S. D., Almer J, Ribbing C, & Cederström B. (2007). High-energy x-ray optics with silicon saw-tooth refractive lenses. *Journal of Synchrotron Radiation, 14*, 204–211.

Shastri, S. D., Mashayekhi, A., Cremer, J. T., & Piestrup, M. A. (2003). X-ray optics for 50–100 keV undulator radiation using crystals and refractive lenses. In T. Ishikawa, A. T. Macrander, & J. L. Wood (Eds.), Crystals, multilayers, and other synchrotron optics. *Proceedings of SPIE* (Vol. 5195, pp. 63–75). SPIE, Bellingham, WA.

Simionovici, A., Chukalina, M., Drakopoulos, M., Snigireva, I., Schroer, C. G., Lengeler, B., Janssens, K., & Adams, F. (2000). 2D imaging by X-ray fluorescence microtomography. In W. Meyer-Ilse, T. Warwick, & D. Attwood (Eds.), *X-ray microscopy: Proceedings of the sixth international conference. AIP conference proceedings*, American Institute of Physics, New York. (Vol. 507, pp. 539–544).

Simionovici, A., Chukalina, M., Drakopoulos, M., Snigireva, I., Snigirev, A., Schroer, C., Lengeler, B., Janssens, K., & Adams, F. (1999). X-ray fluorescence microtomography: Experiment and reconstruction. In U. Bonse (Ed.), *Developments in X-ray tomography II. Proceedings of SPIE* (Vol. 3772, pp. 328–337). SPIE, Bellingham, WA.

Smither, R. K., Khounsary, A. M., & Xu, S. (1997). Potential of a beryllium x-ray lens. In A. T. Macrander & A. M. Khounsary (Eds.). *High heat flux and synchrotron radiation beamline. Proceedings of SPIE* (Vol. 3151, pp. 150–163). SPIE, Bellingham, WA.

Snigirev, A., Filseth, B., Elleaume, P., Klocke, Th., Kohn, V., Lengeler, B., Snigireva, I., Souvorov, A., & Tümmeler, J. (1997b). Refractive lenses for high-energy x-ray focusing. In A. T. Macrander & A. M. Khounsary (Eds.), *High heat flux and synchrotron radiation beamline. Proceedings of SPIE* (Vol. 3151, pp. 164-170). SPIE, Bellingham, WA.

Snigirev, A., Kohn, V., Snigireva, I., & Lengeler, B. (1996). A compound refractive lens for focusing high-energy X-rays. *Nature, 384*, 49–51.

Snigerev, A., Snigereva, I., Bösecke, P., Lequien, S., & Schelokov, I. (1997a). High energy x-ray phase contrast microscopy using a circular Bragg-Fresnel lens. *Optics Communications, 135*(4–6), 378–384.

Snigerev, A., Snigereva, I., Kohn, V., Kuznetsov, S., & Schelokov, I. (1995). On the possibilities of x-ay phase contrast microimaging by coherent high-energy synchrotron radiation *Review of Scientific Instruments, 66*(12), 5486–5492.

Snigirev, A., Snigireva, I., Kohn, V., Yunkin, V., Kuznetsov, S., Grigoriev, M. B., Roth, T., Vaughan, G., & Detlefs, C. (2009). X-ray nanointerferometer based on Si refractive bilenses. *Physics Review Letters, 103*(064801), 1–4.

Snigerev, A., Filseth, B., Elleaume, P., Klocke, TRh., Kohn, V., Lengeler, B., Snigerva, I., Souvorov, A., & Tümmler, J. (1997b). Refractive lenses for high energy X-ray focusing: SPIE Vol 3151.

Suehiro, S., Miyaji, H., & Hayashi, H. (1991). Refractive lens for X-ray focus. *Nature, 352*, 385–386.

Tatchyn, R., Arthur, J., Boyce, R., Cremer, T., Fasso, A., Montgomery, J., . . . Freund, A. K. (1997). X-ray optics design studies for the 1.5 to 15-A Linac Coherent Light Source (LCLS) at Stanford Linear Accelerator Laboratory. In A. K. Freund & M. R. Howells (Eds.), *Coherent electron-beam X-ray sources: Techniques and applications. Proceedings of SPIE* (Vol. 3154, pp. 174–222). SPIE, Bellingham, WA.

Tomie, T. (1994). X-ray Lens Japan Patent No. 6-045288, February 18, 1994.

Tomie, T. (1995). Roentgenstrahllinsen. German Patent DE 195 05 433 C2, 1995.

Tomie, T. (1997). X-ray lens. U.S. Patent No. 5,594,773, January 14, 1997; and U.S. Patent No. 5,684,852, November 4, 1997.

Tomie, T. (2010). The birth of the X-ray refractive lens. *Spectrochimica Acta Part B, 65*(3), 192–198.

Tümmler, J. (2000). *Development of Compound Refractive Lenses for Hard X-rays: A Novel Instrument in Hard X-ray Analysis*. Ph.D. dissertation, RWTH, Aachen.
Yang, B. X. (1993). Fresnel and refractive lenses for x-rays. *Nuclear Instruments and Methods in Physics Research Section A: Accelerators, Spectrometers, Detectors and Associated Equipment, 328*(3), 578–587.
Yuasa, T., Akiba, M., Takeda, T., Kazama, M., Hoshino, A., Watanabe, Y., Hyodo, K., Dilmanian, F. A., Akatsuka, T., & Itai, Y. (1997). Reconstruction method for fluorescent x-ray computed tomography by least-squares method using singular value decomposition. *IEEE Transactions on Nuclear Science, 44*(1), 54–62.

Section 4: Thermal Neutron Radiography, Coded Aperture, and Phase-Contrast Imaging

Accorsi, R., & Lanza, R. C. (2001). Near-field artifact reduction in planar coded aperture imaging. *Applied Optics, 40*(26), 4697–4705.
Allman, B. E., McMahon, P. J., Nugent, K. A., Paganin, D., Jacobson, D. L., Arif, M., & Werner, S. A. (2000a). Imaging: Phase radiography with neutrons. *Nature, 408*, 158–159.
Allman, B. E., McMahon, P. J., Tiller, J. B., Nugent, K. A., Paganin, D., Barty, A., McNulty, I., Frigo, S. P., Wang, S., & Retsch, C. C. (2000b). Non-interferometric quantitative phase imaging with soft x-rays. *Journal of the Optical Society of America, A17*(10), 1732–1743.
Barty, A., Nugent, K. A., Paganin, D., & Roberts, A. (1998). Quantitative optical phase microscopy. *Optics Letters, 23*(11), 817–819.
Berger, H. (1964). Characteristics of a thermal neutron beam for neutron radiography. *International Journal of Applied Radiation and Isotopes, 21*, 407–414.
Berger, H. (1969). Some experiments in fast neutron radiography. *Materials Evaluation, 27*, 245–253.
Berger, H. (1970). Image detection methods for 14.5 MeV neutrons: Techniques and applications. *International Journal of Applied Radiation and Isotopes, 21*, 59–70.
Berger, H. (1971). Neutron Radiography. *Annual Review of Nuclear Science, 21*, 335–364.
Cao, L. (2007). Development of a high spatial resolution neutron imaging system and performance evaluation. Ph.D. dissertation, University of Texas at Austin.
Caroli, E., Stephen, J. B., Di Cocco, G., Natalucci, L., & Spizzichino, A. (1987). Coded aperture imaging in x- and gamma-ray astronomy. *Space Science Reviews, 45*, 349–403.
Casall, F., Chirco, P., & Zanarini, M. (1995). Advanced imaging techniques: A new deal for neutron physics. *Rivista Del Nuovo Cimento, 18*(10), 1–69.
Cremer, J. T., Park, H., Piestrup, M. A., Pantell, R. H., Gary, C. K., Flocchini, R. G., Egbert, H. P., Kloh, M. D., & Walker, R. B. (2007). Thermal Neutron Radiography. Unpublished results.
Criscuolo, E. L., & Polansky, D. (1961). Fast neutron radiography. In *Proceedings of the Missiles and Rockets Symposium*. U.S. Naval Ammunition Depot, Concord, CA, pp. 112–115.
Dicke, R. H. (1968). Scatter-hole cameras for X-rays and gamma rays. *Astrophysics Journal, 153*, L101–L106.
Domanus, J. C. (1985). *Research and Standardization Activities of the EUROATOM Neutron Radiography Working Group, RISØ-M-2524*. Risø National Laboratory, Roskilde, Denmark.
Donnelley, M., Siu, K. K. W., Jamison, R. A., & Parsons, D. W. (2011). Synchrotron phase-contrast X-ray imaging reveals fluid dosing dynamics for gene transfer into mouse airways. *Gene Therapy* (advance online publication, June 9, 2011; doi:10.1038/gt.2011.80).
Downing, R. G., Feller, W. B., & White, P. L. (2000). High-spatial-resolution imaging of low-energy neutrons with microchannel plates. *Transactions of the American Nuclear Society, 82*, 83–84.
Fezzaa, K., & Wang, Y. (2008). Ultrafast X-ray phase-contrast imaging of the initial coalescence phase of two water droplets. *Physical Review Letters, 100*, 104501.
Gureyev, T. E., Nesterets, Y. I., Stevenson, A. W., Miller, P. R., Pogany, A., & Wilkins, S. W. (2008). Some simple rules for contrast, signal-to-noise and resolution in in-line x-ray phase-contrast imaging. *Optics Express, 16*(5), 3223–3241.

Gureyev, T. E., Roberts, A., & Nugent, K. A. (1995). Phase retrieval with the transport-of-intensity equation: Matrix solution with use of Zernike polynomials and partially coherent fields, the transport-of-intensity equation, and phase uniqueness. *Journal of the Optical Society of America, A12*(9), 1932–1946.

Gureyev, T. E., & Wilkins, S. W. (1998a). On x-ray phase imaging with a point source. *Journal of the Optical Society of America, A15*(3), 579–585.

Gureyev, T. E., & Wilkins, S. W. (1998b). On X-ray phase retrieval from polychromatic images. *Optics Communication, 147*(4–6), 229–232.

Hawkesworth, M. R., & Walker, J. (1969). Review: Radiography with neutrons. *Journal of Materials Science, 4*, 817–835.

Kapadia, A. J., Sharma, A. C., Tourassi, G. D., Bender, J. E., Howell, C. R., Crowell, A. S., Kiser, M. R., Harrawood, B. P., Pedroni, R. S., & Floyd C. E. Jr. (2008). Neutron stimulated emission computed tomography for diagnosis of breast cancer. *IEEE Transactions on Nuclear Science, 55*(1), 501–509.

Kardjilov, N. (2003). Further developments and applications of radiography and tomography with thermal and cold neutrons. Ph.D. dissertation, Technical University of Munich.

Kardjilov, N., Lee, S. W., Lehmann, E., Lim, I. C., Sim, C. M., & Vontobel, P. (2005). Improving the image constrast and resolution in the phase-contrast neutron radiography. *Nuclear Instruments and Methods in Physics Research Section A: Accelerators, Spectrometers, Detectors and Associated Equipment, 542*, 100–105.

Kobayashi, H., Tomura, K., Harasawa, S., & Hattori, M. (1990). Neutron radiography using cooled CCD-camera. In *Neutron Radiography 3: Proceedings of the 3rd World Conference on Neutron Radiography* (S. Fujine, ed.). Kluwer Academic, Dordrecht, pp. 421–428.

Lamarsh, J.R. (1983). *Introduction to Nuclear Engineering*, ed. 2. Addison-Wesley, New York, Chapter 9, p. 433.

Lehmann, E. H. (2008). Neutron imaging: A non-destructive tool for materials testing. Report of a coordinated research project 2003–2006, IAEA-TECDOC-1604. International Atomic Energy Agency; http://www-pub.iaea.org/MTCD/publications/PDF/te_1604_web.pdf.

Nugent, K. A., Gureyev, T. E., Cookson, D., Paganin, D., & Barnea, Z. (1996). Quantitative phase imaging using hard X-rays. *Physical Review Letters, 77*(14), 2961–2964.

Paganin, D., & Nugent, K. A. (1998). Noninterferometric phase imaging with partially coherent light. *Physical Review Letters, 80*(12), 2586–2589.

Parks, P. B., Brown, M., & Hammer, D. (1969). Problems of fast neutron radiography. *Biomedical Sciences Instrumentation, 6*, 118–126.

Pleinert, H., Lehmann, E., & Körner, S. (1997). Design of a new CCD-camera neutron radiography detector. *Nuclear Instruments and Methods in Physics Research Section A: Accelerators, Spectrometers, Detectors and Associated Equipment, 399*(2–3), 382–390.

Polansky, D., & Criscuolo, E. L. (1964). Radiographic aspects of fast neutron detection. *Proceedings of the 24th National Conference of the American Society for Nondestructive Testing*, Evanston, IL.

Richardson, A. E. (1977). Improved images in 14.5 MeV neutron radiography. *Materials Evaluation, 35*, 52–58.

Roth, H., & Lehman, D. K. (2006). Choosing the right X-ray tool for the job. *Evaluation Engineering*; http://www.evaluationengineering.com/search/article.php?aid=5375.

Skinner, G. K. (1984). Imaging with coded-aperture masks. *Nuclear Instruments and Methods in Physics Research, 221*(1), 33–40.

Teague, M. R. (1983). Deterministic phase retrieval: A Green's function solution. *Journal of the Optical Society of America, 73*(11), 1434–1441.

Tochilin, E. (1965). Photographic detection of fast neutrons: Application to neutron radiography. *Physics in Medicine and Biology, 10*(4), 477–490.

Tremsin, A. S., McPhate, J. B., Vallerga, J. V., Siegmund, O. H. W., Feller, W. B., Lehmann, E., & Dawson, M. (2011). Improved efficiency of high resolution thermal and cold neutron imaging. *Nuclear Instruments and Methods in Physical Research* A *628*(1), 415–418.

von der Hardt, P., & Rottger H. (eds.). (1981). *Neutron Radiography Handbook.* D. Reidel Publishing, Dordrecht.

Weon, B. M., Je, J. H., Hwu, Y., & Margaritondo, G. (2006). Phase contrast X-ray imaging. *International Journal of Nanotechnology, 3*(2–3), 280–297.

Wilkins, S. W., Gureyev, T. E., Gao, D., Pogany A., & Stevenson, A. W. (1996). Phase-contrast imaging using polychromatic hard X-rays. *Nature, 384*, 335–338.

Wood, D. E. (1967). Fast neutron radiography with a neutron generator. *Transactions of the American Nuclear Society, 10*, 423–444.

Ziock, K. P., Collins, J. W., Fabris, L., Gallagher, S., Horn, B. K. P., Lanza, R. C., & Madden, N. W. (2006). Source-search sensitivity of a large-area, coded-aperture, gamma-ray imager. *IEEE Transactions on Nuclear Science, 53*(3), 1614–1621.

Zoofan, B., Kim, J.-Y., Rokhlin, S. I., & Frankel, G. S. (2006). Phase-contrast x-ray imaging for nondestructive evaluation of materials. *Journal of Applied Physics, 100*, 014502.

Section 5: Fast Neutron Radiography

Anger, H. O. (1964). Scintillation camera with multichannel collimators. *Journal of Nuclear Medicine, 5*(7), 515–531.

Annis, P., Bay, A., Benussi, L., Bruski, N., Buontempo, S., Currat, C., D'Ambrosio, N., van Dantzig, R., Dupraz, J., Ereditato, et al. (2000). High resolution tracking using large capillary bundles filled with liquid scintillator. *Nuclear Instruments and Methods in Physics Research Section A: Accelerators, Spectrometers, Detectors and Associated Equipment, 449*(1–2), 60–80.

Barletta, W. (2001). Compact neutron sources for pure and applied science. Lawrence Berkeley National Laboratory. In *International Seminar on Nuclear War and Planetary Emergencies, 26th Session* (R. Ragaini, ed.) (The Science and Culture Series, Nuclear Strategy and Peace Technology). World Scientific E books, pp. 236–241.

Booth, V. D. L., Preston, G., & Shaw, P. F. D. (1956). The cross section and angular distributions of the D-D reactions between 40 and 90 keV. *Proceedings of the Physical Society. Section A, 69*(3), 265–270.

CERN (European Organization for Nuclear Research). (2011). Knowledge transfer group; http://medipix.web.cern.ch/medipix/.

Chen, G. (2001). Fast neutron resonance radiography for elemental imaging: Theory and applications. Ph.D. thesis, Massachusetts Institute of Technology, Department of Nuclear Engineering.

Chen, G., & Lanza, R. C. (2002). Fast neutron resonance radiography for elemental imaging: Theory and applications. *IEEE Transactions on Nuclear Science, 49*(4), 1919–1924.

Chen, G., & Lanza, R. C. (2004). Fast neutron resonance radiography for elemental mapping. U.S. Patent No. 6,693,281 B2, February 17, 2004.

Chichester, D. L., & Simpson, J. D. (2004). Compact accelerator neutron generators. *Industrial Physicist, 9*(6), 22–25.

Cremer, J., Piestrup, M., & Wu, X. (2005). Fast and thermal neutron radiography. In *Penetrating Radiation Systems and Applications VII* (F. P. Doty, H. B., & Barber, H. Roehrig, eds.). *Proceedings of SPIE*, Vol. 5923. SPIE, Bellingham, WA, pp. 59230A1-19.

Cremer, J. T, Williams, D. L., Gary, C. K., Piestrup, M. A., Faber, D. R., Fuller, M. J., Vainionpaa, J. H., Apodaca, M., Pantell, R. H., & Feinstein, J. (2012). Fast neutron imaging of hydrogenous materials using a DD fusion generator, *Nuclear Instruments and Methods in Physics Research Section A*, 675, 51–55.

Cremer, J. T., Williams, D. L., Gary, C. K., Piestrup, M. A., Faber, D. R., Fuller, M. J., Vainionpaa, J. H., Apodaca, M., Pantell, R. H., & Feinstein, J. (2009). Efficient high-resolution fast-neutron detector. Unpublished results of Phase II SBIR administered by J. Figeroa, U.S. National Science Foundation, Contract No. IIP-0724503 (2008).

Dangendorf, V., Laczko, G., Kersten, C., Jagutzki, O., & Spillmann, U. (2002). Fast neutron resonance radiography in a pulsed neutron beam. In *7th World Conference On Neutron Radiography,* Rome, September 2002; http://arxiv.org/ftp/nucl-ex/papers/0301/0301001.pdf.

Disdier, L., Lerche, R. A., Bourgade, J. L., & Glebo, V. Y. (2004). Capillary detector with deuterated scintillator for inertial confinement fusion neutron images. *Review of Scientific Instruments, 75*(6), 2134–2139.

Disdier, L., Rouyer, A., Fedotoff, A., Bourgade, J. L., Marshall, F. J., Glebov, V. Y., & Stoeckl, C. (2003). Neutron imaging of ICF target plasmas. *Review of Scientific Instruments, 74*(3), 1832–1836.

Eberhardt, J. E., Rainey, S., Stevens, R. J., Sowerby, B. D., & Tickner, J. R. (2005). Fast neutron radiography scanner for the detection of contraband in air cargo containers. *Applied Radiation and Isotopes, 63*(2), 179–188.

Hall, J. M. (2001). Uncovering hidden defects with neutrons. *Science and Technology Review,* Lawrence Livermore National Laboratory; http://www.llnl.gov/str/May01/Hall.html.

Hall, J., Dietrich, F., Logan, C., & Rusnak, B. (2001). Recent results in the development of fast neutron imaging techniques. Lawrence Livermore National Laboratory, LLNL Report No. UCRL-MI-140345; https://e-reports-ext.llnl.gov/pdf/240834.pdf.

Hall, J. M., Rusnak, B., Shen, S., & Fitsos, P. J. (2006). High-energy neutron imaging development at LLNL. Lawrence Livermore National Laboratory, LLNL Report No. UCRL-TR-217993; https://e-reports-ext.llnl.gov/pdf/328752.pdf.

Hamamatsu Photonics (2011). Electron Tube Division. Flat panel type multianode photomultiplier tube assembly. H8500 series, Bridgewater, NJ; http://sales.hamamatsu.com/assets/pdf/parts_H/H8500C_H8500D_TPMH1308E01.pdf.

Hurlbut, C. (2011). ELJEN Technology, Sweetwater, TX; http://www.eljentechnology.com.

Knoll, G. F. (1989). *Radiation Detection and Measurement.* John Wiley & Sons, New York, pp. 219–227.

Kobayashi, H., Tomura, K., Harasawa, S., & Hattori M. (1990). Neutron radiography using cooled CCD camera. In *Neutron Radiography (3) Proceedings of the Third World Conference on Neutron Radiography* (S. Fujine, K. Kanda, G. Matsumoto, & J. Barton, eds.) Kluwer Academic, Dordrecht, Netherland, pp. 421–428.

Kyushima, H., Shimoi, H., Atsumi, A., Ito, M., Oba, K., & Yoshizawa, Y. (2000). The development of flat panel PMT. In *IEEE Nuclear Science Symposium Conference Record,* Vol. 1, pp. 7/3–7/7 (IEEE Xplore Digital Library).

Lanza, R. C. (2002). Neutron resonance radiography for security applications. In *Penetrating Radiation Systems and Applications IV* (H. B. Barber, H. Roehrig, F. P. Doty, L. J. Porter, E. J. Morton, eds.). Proceeding of SPIE, Vol 4786. SPIE, Bellingham, WA, pp. 50–51.

Lanza, R. C. (2009). Homeland security and contraband detection. In *Neutron Imaging and Applications,* Chap. 18 (I. S. Anderson, R. L. McGreevy, & H. Z. Bilheux, eds.). Springer Science and Business Media, New York.

Lewis, W. J., Bennett, L. G. I., Chalovick T. R., & Francescone, O. (2003). Neutron radiography of aircraft composite flight control surfaces. *e-Journal of Nondestructive Testing, 8*(2); http://www.ndt.net/article/ecndt02/1/1.htm.

Lou, T.-P. (2003). "Compact D-D/D-T neutron generators and their applications." Ph.D. thesis, Nuclear Engineering Department, University of California–Berkeley.

McFarland, E. W., Lanza, R. C., & Poulos, G. W. (1991). Multi-dimensional neutron computed tomography using cooled charge-coupled devicies. *IEEE Transactions on Nuclear Science, 38*(2), 612–622.

Nargolwalla S. S., & Przybylowicz, E. P. (1973). *Activation Analysis with Neutron Generators.* John Wiley & Sons, New York.

Pania, P., Cinti, M. N., Pellegrini, R., Trotta, C., Trotta, G., Montani, L., Ridolfi, S., Garibaldi, F., Scafe, R., Belcari, N., et al. (2003). Evaluation of flat panel PMT for gamma ray imaging.

Nuclear Instruments and Methods in Physics Research Section A: Accelerators, Spectrometers, Detectors and Associated Equipment, 504(1–3), 262–268.

Popov, V. (2004). Matrix output device readout system. U.S. Patent No. 6,747,263 B1, assigned to Southwest University Research Assn., Inc., June 8, 2004.

Popov, V., & Majewski, S. (2006c). A compact high performance readout electronics solution for H9500 Hamamatsu 256 multianode photomultiplier tube for application in gamma cameras. In *Nuclear Science Symposium Conference Record,* Vol. 5, pp. 2981–2985 (IEEE Xplore Digital Library).

Popov, V., Degtiarenko, P., & Musatov, I. (2011). New detector for use in fast neutron radiography. *Journal of Instrumentation, 6,* C01029.

Popov, V., Degtiarenko, P., Musatov, I., & Williams, M. (2006b). Data readout and processing toolkit for small-size gamma cameras. In *Nuclear Science Symposium Conference Record,* Vol. 5, pp. 2929–2932 (IEEE Xplore Digital Library).

Popov, V., Majewski, S., & Welch, B.L. (2006a). A novel readout concept for multianode photomultiplier tubes with pad matrix anode layout. *Nuclear Instruments and Methods in Physics Research Section A: Accelerators, Spectrometers, Detectors and Associated Equipment, 567*(1), 319–322.

Popov, V., Majewski, S., Weisenberger A.G., & Wojcik, R. (2001) Analog readout system with charge division type output. In *IEEE Nuclear Science Symposium Conference Record,* Vol. 4, pp. 1937–1940 (IEEE Xplore Digital Library).

Preston, G., Shaw, P. F. D., & Young, S. A. (1954). The cross-sections and angular aistributions of the D—D Reactions between 150 and 450 keV. *Proceedings of the Royal Society of London. Series A, Mathematical and Physical Sciences, 226*(1165), 206–216.

Qing, J., Ying, W., Regis, M., & Kwan, J. W. (2009). Initial testing of a compact portable microwave-driven neutron generator. *IEEE Transactions on Nuclear Science, 56*(3), 1312–1315.

Raas, W. L., Blackburn, B. W., Boyd, E., Hall, J., Kohse, G., Lanza, R.C., Rusnak, B., & Watterson, J. I. W. (2005). Neutron resonance radiography for explosives detection: technical challenges. In *Nuclear Science Symposium Conference Record,* Vol. 1, 2005, pp. 129–133 (IEEE Xplore Digital Library).

Reijonen, J., Andresen, N., Gicquel, F., Gough, R., King, M., Kalvas, T., Leung, K.-N., Lou, T.-P., Vainionpaa, H., Antolak, A., et al. (2007). Development of advanced neutron/gamma generators for imaging and active interrogation applications. In *Optics and Photonics in Global Homeland Security III* (T. T. Saito, D. Lehrfeld, & M. J. DeWeert, eds.). *Proceedings of SPIE,* Vol. 6540. SPIE, Bellingham, WA, p. 65401P.

Reijonen, J., Gicquel, F., Hahto, S. K., King. M., Lou, T.-P., & Leung, K.-N. (2005). D-D neutron generator development at LBNL. *Applied Radiation and Isotopes, 63*(5-6), 757–763.

Ress, D., Lerche, R. A., Ellis, R. J., Heaton, G. W., & Lehr, D. E. (1995). High-sensitivity scintillating-fiber imaging detector for high-energy neutrons. *Review of Scientific Instruments, 68*(10), 4943–4948.

Siegel, S., Silverman, R. W., Shao, Y., & Cherry, S. R. (1996). Simple charge division readouts for imaging scintillator arrays using a multi-channel PMT. *IEEE Transactions on Nuclear Science, 43*(3), 1634–1641.

Sowerby, B. D., & Tickner. J. R. (2007a). Recent advances in fast neutron radiography for cargo inspection. *Nuclear Instruments and Methods in Physics Research, Res. 580*(1), 799–802.

Sowerby, B. D., & Tickner, J. R. (2007b). Radiographic equipment. U.S. Patent No. 7,313,212 B2, December 25, 2007, assigned to Commonwealth Scientific and Industrial Research Organization (CSIRO), Campbell, Australia.

Tsoulfanidis, N. (1983). *Measurement and Detection of Radiation.* McGraw-Hill, New York, pp. 201–207, 314–316, 451.

Uher, J., Jakůbek, J., Celine, L., Claude, L., Pospíšil, S., Škoda, R., & Vykydal, Z. (2008). Detection of fast neutrons with the Medipix-2 pixel detector. *Nuclear Instruments and Methods in Physics Research Section A: Accelerators, Spectrometers, Detectors and Associated Equipment, 591*(1), 71–74.

Section 6: A Brief History of Neutron Optics

Alvarez, L. W., & Bloch, F. (1940). A quantitative determination of the neutron moment in absolute nuclear magnetons. *Physical Review, 57*(2), 111–122.

Alvarez, L. W., & Pitzer, K. S. (1940). Scattering of 20° neutrons in ortho- and parahydrogen. *Physical Review, 58*(11), 1003–1004.

Amaldi, E. (1977). Personal notes on neutron work in Rome in the 1930s. In *History of Twentieth Century Physics* (C Weiner, ed.). Academic Press, New York, pp. 294–339.

Amaldi, E., Battimelli, G., & Paoloni, G. (1998). *20th Century Physics: Essays and Recollections: A Selection of Historical Writings* (Edoardo Amaldi Foundation Series, Vol. 3). World Science Publishing Company, London.

Anderson, H. L., Fermi, E., & Wood, J. A. (1962). CP-718, CP-781 (Manhattan District Reports), and by Rainwater and Havens, CP-1962 (Manhattan District Reports, USA).

Anderson, I. S., McGreevy, R. L., & Bilheux, H. Z. (eds.). (2009). *Neutron Imaging and Applications* (see table 2.1 on p. 17). Springer, New York.

Bacon, G. E. (ed.). (1987). *Fifty Years of Neutron Diffraction. The Advent of Neutron Scattering.* Hilger with the assistance of the International Union of Crystallography, Bristol.

Bethe, H. A. (1949). Theory of the effective range in nuclear scattering. *Physical Review, 76*(1), 38–50.

Beyer, H. G., & Whittaker, W. D. (1940). Interference phenomena in the scattering of slow neutrons. *Physical Review, 57*(11), 976–981.

Blatt, J. M., & Weisskopf, V. F. (1952). *Theoretical Nuclear Physics*. John Wiley & Sons, New York.

Borst, L. B., Ulrich, A. J., Osborne, C. L., & Hasbrouck, B. (1946). Neutron diffraction and nuclear resonance structure. *Physical Review, 70*(7–8), 557–557.

Breit, G. (1933). Quantum theory of dispersion (continued). Parts VI and VII. *Reviews of Modern Physics, 5*(2), 91–140.

Brockhouse, B. N., & Hurst, D. G. (1952). Energy distribution of slow neutrons scattered from solids. *Physical Review, 88*(3), 542–547.

Carpenter, J. M. (1973). High intensity, pulsed thermal neutron source. U.S. Patent No. 3,778,627, December 11, 1973.

Carpenter, J. M. (1977). Pulsed spallation neutron sources for slow neutron scattering. *Nuclear Instruments and Methods, 145*(1), 91–113.

Chadwick, J. (1932a). Possible existence of a neutron. *Nature, 129*(3252), 312.

Chadwick, J. (1932b). The existence of a neutron. *Proceedings of the Royal Society of London Series A-Mathematical Physical and Engineering Sciences, 136*(830), 692–708.

Christ, J., & Springer, T. (1962). Über die entwicklung neutronenleiters am FRM reactor. *Nukleonik, 4*, 23.

Drabkin, G. M., Zabidarov, E. I., Kasman, Ya. A., & Okorokov, A. I. (1969). Investigation of a phase transition in nickel with polarized neutrons. *Soviet Journal of Experimental and Theoretical Physics, 29*, 261.

Elsasser, W. M. (1936). *Comptes Rendus de l'Académie des Sciences (Paris), 202*, 1029.

Fender, B. E. F., Hobbis, L. C. W., & Manning, G. (1980). The U.K. Spallation Neutron Source. *Philosophical Transactions of the Royal Society of London Series B-Biological Sciences, B290*(1043), 657–672.

Fermi, E. (1936). Sul moto dei neutroni nelle sostanze idrogenate. *Ricerca Scientifica, 7*(Part 2), 13–52.

Fermi, E., & Marshall, L. (1947). Interference phenomena of slow neutrons. *Physical Review, 71*(10), 666–667.

Feshbach, H., Peaslee, D. C., & Weisskopf, V. F. (1947). On the scattering and absorption of particles by atomic nuclei. *Physical Review, 71*(3), 145–158.

Foldy, L. L. (1945). The multiple scattering of waves. I. General theory of isotropic scattering by randomly distributed scatterers. *Physical Review, 67*(3–4), 107–119.

Frank, A. I. (1987). Optics of ultracold neutrons and the neutron microscope problem. (*Usp. Fiziol. Nauk 151*) *Soviet Physics Uspekhi, 30*(2), 110–132.

Frank, A. I. (1989). Ultracold-neutron microscopy. *Atomnaya Energiya, 66*(2), 93–99.

Freund, A., & Forsyth, J. B. (1979). Materials problems in neutron devices. In *Treatise on Materials Science and Technology, Vol. 15: Neutron Scattering* (G. Kostorz, ed.). Academic Press, New York.

Freund, A. K. (1983). Cross-sections of materials used as neutron monochromators and filters. *Nuclear Instruments and Methods in Physics Research, 213*(2–3), 495–501.

Gähler, R., Kalus, J., & Mampe, W. (1980). An optical instrument for the search of a neutron charge. *Journal of Physics E: Scientific Instruments, 13*(5), 546–548.

Gähler, R., Kalus, J., & Mampe, W. (1982). Experimental limit for the charge of the free neutron. *Physical Review, D: Particles and Fields, 25*(11), 2887–2894.

Goldberger, M. L., & Seitz, F. (1947). Theory of the refraction and the diffraction of neutrons by crystals. *Physical Review, 71*(5), 294–310.

Gukasov, A. G., Deriglazov, V. V., Kezerashvili, V. Y., Krutov, G. A., Kudryashov, V. A., Peskov, B. G., Syromyatnikov, V. G., Trunov, V. A., Kharchenkov, V. P., & Shchebetov, A. F. (1979). Multilayer neutron monochromator-polarizer. *Soviet Journal of Experimental and Theoretical Physics, 50*, 862–864.

Halpern, J., Estermann, I., & Stern, O. (1937a). The scattering of slow neutrons by liquid ortho- and parahydrogen. *Physical Review, 52*(2), 142.

Halpern, J., Estermann, I., Simpson, O. C., & Stern, O. (1937b). The scattering of slow neutrons by liquid ortho- and parahydrogen. *Physical Review, 52*(2), 142.

Halpern, O., & Holstein, T. (1941). On the passage of neutrons through ferromagnets. *Physical Review, 59*(12), 960–981.

Halpern, O., & Johnson, M. H. (1939). On the magnetic scattering of neutrons. *Physical Review, 55*(10), 898–923.

Halpern, O., Hamermesh, M., & Johnson, M. H. (1941). The passage of neutrons through crystals and polycrystals. *Physical Review, 59*(12), 981–996.

Hamermesh, M., & Johnson, M. H. (1942). Magnetic scattering of neutrons. *Physical Review, 61*(1–2), 17–18.

Hayter, J. B., Penfold, J., & Williams, W. G. (1978). Compact polarising Soller guides for cold neutrons. *Journal of Physics E: Scientific Instruments, 11*(5), 454–458.

Heitler, W. (1944). *Elementary Wave Mechanics*. Oxford University Press, Oxford, UK.

Hughes, D. J. (1954). *Neutron Optics*. Interscience Publishers, London.

Hughes, D. J. (1953). *Pile Neutron Research*. Addison-Wesley, New York.

Hughes, D. J., & Burgy, M. T. (1951). Reflection of neutrons from magnetized mirrors. *Physical Review, 81*(4), 498–506.

Jones, T. J. L., & Williams, W. G. (1980). A Stern-Gerlach polarimeter for cold neutrons. *Journal of Physics E: Scientific Instruments, 13*(2), 227–232.

Just, W., Schneider, C. S., Ciszewski, R., & Shull, C. G. (1973). Refraction of thermal neutrons by shaped magnetic fields. *Physical Review B: Condensed Matter, 7*(9), 4142–4145.

Klein, A. G., & Werner, S. A. (1983). Neutron optics. *Reports on Progress in Physics, 46*(3), 259–335.

Koester, L. (1977). Neutron physics. *Springer Tracts in Modern Physics, 80*, 1–56.

Lax, M. (1951). Multiple scattering of waves. *Reviews of Modern Physics, 23*(4), 287–310.

Lax, M. (1952). Multiple scattering of waves. II. The effective field in dense systems. *Physical Review, 85*(4), 621–629.

Lushchikov, V. I., Pokotilovskii, Yu. N., Strelkov, A. V., & Shapiro, F. L. (1969). Observation of ultracold neutrons. *Soviet Physics JETP Letters, 9*(1), 23.

Maier-Leibnitz, H. (1962). Neutronenrefraktometers zur Absolutbestimmung kohärenter Streuquerschnitte. *Zeitschrift für Angewandte Physik, 14*, 738.

Maier-Leibnitz, H., & Springer, T. (1962). Ein Interferometer für langsame Neutronen. *Zeitschrift für Physik, 167*(4), 386–402.

McReynolds, A. W., & Weiss, R. J. (1951). Coherent neutron scattering cross section of V^{51}. *Physical Review, 83*(1), 171–172.

Mezei, F. (1972). Neutron spin echo: A new concept in polarized thermal neutron techniques. *Zeitschrift fur Physik, 255*(2), 146–160.

Mezei, F. (1976). Novel polarized neutron devices: Supermirror and spin component amplifier. *Communications on Physics, 1*, 81–85.

Mezei, F., & Dagleish, P. A. (1977). Corrigendum and first experimental evidence on neutron supermirror. *Communications on Physics, 2*, 41–43.

Mitchell, D. P., & Powers, P. N. (1936). Bragg reflection of slow neutrons. *Physical Review, 50*(5), 486–487.

Mook, H. A. (1966). Magnetic moment distribution of nickel metal. *Physical Review, 148*(2), 495–501.

Moon, R. M. (1964). Distribution of magnetic moment in hexagonal cobalt. *Physical Review, 136*(1A), A195–A202.

Moon, R. M., Riste, T., & Koehler, W. C. (1969). Polarization analysis of thermal-neutron scattering. *Physical Review, 181*(2), 920–931.

Mott, N. F., & Massey, H. S. W. (1965). *The Theory of Atomic Collisions*. Oxford University Press, Amen House, London E.C.A.

Nathans, R. (1959). The use of polarized neutrons in determining the magnetic scattering by iron and nickel. *Journal of Physics and Chemistry of Solids, 10*(2–3), 138–146.

Nathans, R., & Paoletti, A. (1959). Magnetic form factor of cobalt. *Physical Review Letters, 2*(6), 254–256.

Neel, L. (1948). Propriétés magnétiques des ferrites. Ferrimagnétisme et antiferromagnétisme. *Annals of Physics, 3*, 137–198.

Placzek, G. (1952). The scattering of neutrons by systems of heavy nuclei. *Physical Review, 86*(3), 377–388.

Pomeranschuk, L. (1938). On scattering of slow neutrons in crystalline lattice. *Physics Zeitschrift Sowjetunion, 13*, 65.

Pynn, R. (1990). Neutron scattering: A primer. Los Alamos Neutron Science Center; http://www.fas.org/sgp/othergov/doe/lanl/pubs/00326651.pdf.

Rekveldt, M. Th., & Kraan, W. H. (1983). Curved neutron mirror system as an improved polarizer. *Physica B+C, 120*(1–3), 81–85.

Schiff, L. I. (1949). *Quantum Mechanics*. McGraw-Hill, New York.

Schneider, C. S., & Shull, C. G. (1971). Forward magnetic scattering amplitude of iron for thermal neutrons. *Physical Review, B3*, 830–835.

Schoenborn, B. P., Caspar, D. L. D., & Kammerer, O. F. (1974). A novel neutron monochromator. *Journal of Applied Crystallography, 7*, 508–510.

Schwinger, J., & Teller, E. (1937a). The scattering of neutrons by ortho- and parahydrogen. *Physical Review, 51*(9), 775.

Schwinger, J., & Teller, E. (1937b). The scattering of neutrons by ortho- and parahydrogen. *Physical Review, 52*(4), 286–295.

Seeger, R. J., & Teller, E. (1942). On the inelastic scattering of neutrons by crystal lattices. *Physical Review, 62*(1–2), 37–40.

Shabalin, E. P. (1979). *Fast Pulsed and Burst Reactors*. Pergamon Press, Oxford.

Sherwood, J. E., Stephenson, T. E., & Bernstein, S. (1954). Stern-Gerlach experiment on polarized neutrons. *Physical Review, 96*(6), 1546–1548.

Shull, C. G., & Smart, J. S. (1949). Detection of antiferromagnetism by neutron diffraction. *Physical Review, 76*(8), 1256–1257.

Shull, C. G., & Wollan, E. O. (1951). Coherent scattering amplitudes as determined by neutron diffraction. *Physical Review, 81*(4), 527–535.

Shull, C. G., Billmann, K. W., & Wedgwood, F. A. (1967). Experimental limit for the neutron charge. *Physical Review, 153*(5), 1415–1422.

Shull, C. G., Wollan, E. O., & Koehler, W. J. (1951). Neutron scattering and polarization by ferromagnetic materials. *Physical Review, 84*(5), 912–921.

Shull, C. G., & Yamada, Y. (1962). *Journal of the Physical Society of Japan, 17*(Suppl BIII), p. 1.
Steinberger, J., & Wick, G. C. (1949). On the polarization of slow neutrons. *Physical Review, 76*(7), 994–995.
Steyerl, A. (1969). Measurements of total cross sections for very slow neutrons with velocities from 100 m/sec to 5 m/sec. *Physical Letters B, 29*(1), 33–35.
Sutton, R. B., Hall, T., Anderson, E. E., Bridge, H. S., DeWire, J. W., Lavatelli, L. S., Long, E. A., Synder, T., & Williams, R. W. (1947). Scattering of slow neutrons by ortho- and parahydrogen. *Physical Review, 72*(12), 1147–1156.
Turchin, V. F. (1967). Diffraction of slow neutrons by stratified systems. *Soviet Atomic Energy* [translated from *Atom-naya Energiya, 22*(2), 119].
Van Hove, L. (1954). Correlations in space and time and born approximation scattering in systems of interacting particles. *Physical Review, 95*(1), 249–262.
Van Vleck, J. H. (1939). On the theory of the forward scattering of neutrons by paramagnetic media. *Physical Review, 55*(10), 924–930.
Von Halban, H., & Preiswerk P. (1936). Experimental evidence of neutron diffraction. *Comptes Rendus de l Academie des Sciences (Paris), 203*, 73.
Weinstock, R. (1944). Inelastic scattering of slow neutrons. *Physical Review, 65*(1–2), 1–20.
Wick, G. C. (1937). Über die Streuung von Neutronen an Kristallgittern. *Physikalische Zeitschrift*. Band 28, p. 403 & p. 689
Williams, W. G. (1988). *Polarized Neutrons*. Clarendon, Oxford.
Wollan, E. O., & Shull, C. G. (1948). The diffraction of neutrons by crystalline powders. *Physical Review, 73*(8), 830–841.
Zel'dovich, Ya. B. (1959). Storage of cold neutrons. *Soviet Physics. Journal of Experimental and Theoretical Physics Letters (JETP)* Vol. 36, p. 1952–1953.
Zinn, W. H. (1947). Diffraction of neutrons by a single crystal. *Physical Review, 71*(11), 752–757.

Section 7: Diffractive and Reflective Optics

Abramowitz, M., & Davidson, M. W. (2011). Brief overview of DIC microscopy. Olympus America Microscopy Resource Center; http://www.olympusmicro.com/primer/techniques/dic/dichome.html.
Allen, R. D., David, G. B., & Nomarski, G. (1969). The Zeiss-Nomarski differential interference equipment for transmitted-light microscopy. *Zeitschrift für wissenschaftliche Mikroskopie und mikroskopische Technik, 69*, 193–221.
Angel, J. R. P. (1979). Lobster eyes as x-ray telescopes. *Astrophysical Journal, 233*, 364–373.
Arakawa, E. T., & Caldwell, P. J. (1980). Holographically produced transmission gratings for soft x-rays. *Nuclear Instruments and Methods, 172*(1–2), 292–296.
Aritome, H., Aoki, H., & Namba, S. (1984). Focusing characteristics of x-ray zone plates fabricated by electron beam lithography and reactive ion etching. *Japanese Journal of Applied Physics Part 2-Letters, 23*, L406–L408.
Baez, A. V. (1961). Fresnel zone plate for optical image formation using extreme ultraviolet and soft x radiation. *Journal of the Optical Society of America, 51*(4), 405–412.
Bajt, S., & Chapman, H. N. (2007). Multilayers are enabling new science with x-ray free electron lasers. Lawrence Livermore National Laboratory Report, UCRL-JRNL-232981, July 23, 2007; http://e-reports-ext.llnl.gov/pdf/350072.pdf.
Barbee, T. W. Jr. (1986). Multilayers for x-ray optics. *Optical Engineering, 25*, 893–915.
Barbee, T. W. Jr. (1990). Multilayer diffraction grating. U.S. Patent No. 4,915,463, April 10, 1990.
Beutler, H. G. (1945). The theory of the concave grating. *Journal of the Optical Society of America, 35*(5), 311–350.
Born, M. (1969). *Atomic Physics*. Dover Publications, New York.

Born, M., & Wolf, E. (2002). *Principles of Optics*, edition 7. Cambridge University Press Cambridge, UK.
Bragg, W. H., & Bragg, W. L. (1913). The reflection of x-rays by crystals. *Proceedings of the Royal Society of London Series A*, *88*(605), 428–438.
Buerger, M. J. (1942). *X-ray Crystallography*. John Wiley & Sons, New York.
Ceglio, N. M., & Smith, H. I. (1977). An efficient lensing element for x-rays. Lawrence Livermore National Laboratory. UCRL No. 79966. Presented at the VIII International Conference on X-rays Optics and Microanalysis, Boston, MA, August 18–24, 1977; http://www.osti.gov/bridge/servlets/purl/5228445-aafXGW/5228445.pdf.
Ceglio, N. M., Hayryluk, A. M., & Schattenburg, M. (1983). X-ray phase lens design and fabrication. *Journal of Vacuum Science & Technology B*, *1*(4), 1285–1288.
Chao, W. L., Harteneck, B. D., Liddle, J. A., Anderson, E. H., & Attwood, D. T. (2005). Soft X-ray microscopy at a spatial resolution better than 15nm. *Nature*, *435*(7046), 1210–1213.
Chapman, H. N., Barty, A., Bogan, M. J., Boutet, S., Frank, M., Hau-Riege, S. P., Marchesini, S., Woods, B. W., Bajt, S., Benner, W. H., et al. (2006). Femtosecond diffractive imaging with a soft-X-ray free-electron laser. *Nature Physics*, *2*(12), 839–843.
Chapman, H. N., Fu, J., Jacobsen, C., & Williams, S. (1996). Dark-field x-ray microscopy of immunogold-labeled cells. *Journal of the Microscopy Society of America*, *2*(2), 53–62.
Colombant, D., & Tonon, G. F. (1973). X-ray emission from laser-produced plasmas. *Journal of Applied Physics*, *44*(8), 3524–3527.
Compton A. K., & Allsion, S. K. (1935). *X-rays in Theory and Experiment*. Van Nostrand, New York.
Cosslett, V. E., & Nixon, W. C. (1960). *X-ray Microscopy*. Cambridge University Press, London.
Darwin, C. G. (1914a). The theory of X-ray reflexion. Part I. *Philosophical Magazine*, *27*(158), 315–333.
Darwin, C. G. (1914b). The theory of X-ray reflexion. Part II. *Philosophical Magazine*, *27*(160), 675–690.
Darwin, C. G. (1922). The reflexion of x-rays from imperfect crystals. *Philosophical Magazine*, *43*(257), 800–829.
David, C., Donath, T., Hempel, E., Hoheisel, M., Pfeiffer, F., & Popescu, S. (2010). X-ray system for x-ray phase contrast and/or x-ray dark field imaging. U.S. Patent No. US 2010/0091936 A1, April 15, 2010.
de Broglie, L. (1923). Ondes et quanta. *Comptes Rendus des Sciences de lAcademie des Sciences*, *117*, 507–510.
Debye, P. (1913). Interferenz von Röntgenstrahlen und wärmebewegung. *Annalen der Physik*, *348*(1), 49–92.
Di Fabrizio, E., Romanato, F., Cabrini, S., Gentili, M., Kaulich, B., Susini, J., & Barrett, R. (1999). High-efficiency multilevel zone plates for keV X-rays. *Nature*, *401*(6756), 895–898.
Dirac, P. A. M. (1930). *The Principles of Quantum Mechanics*. Clarendon Press, Oxford.
Dyson, J. (1952). Optics of the focusing x-ray microscope. *Proceedings of the Physical Society of London Section B*, *65*(8), 580–589.
Ewald, P. P. (1916a). Zur begründung der kristalloptik. Teil I. *Annalen der Physik*, *354*(1), 1–38.
Ewald, P. P. (1916b). Zur begründung der kristalloptik. Teil II. *Annalen der Physik*, *354*(2), 117–143.
Ewald, P. P. (1917). Zur begründung der kristalloptik. Teil III. *Annalen der Physik*, *359*(23), 519–556.
Ewald, P. P. (ed.). (1962). *50 Years of X-Ray Diffraction*. Reprinted for the IUCr XVIII Congress, Glasgow, Scotland: http://www.iucr.org/publ/50yearsofxraydiffraction.
Franks, A., & Lindsey, K., & Bennett, J. M., & Speer, R. J., & Turner, D., & Hunt, D. J. (1975). The theory, manufacture, structure, and performance of N.P.L. x-ray gratings. *Proceedings of the Royal Society of London Series A*, *277*(1271), 503–543.
Franks, A. I. (1971). X-ray optics. *Science Progress (Oxford)*, *64*, 371–422.

Fredrich, W., Knipping, P., & van Laue, M. (1912). Interferenz-erscheinungen bei Röntgenstrahlen. *Proceedings* of the *Bavarian Academy of Science*, p. 303–322; reprinted in *Annalen der Physik, 346*(10), 971–988.

General Electric. W.C Roentgen and the discovery of X-rays; http://www.medcyclopaedia.com/library/radiology/chapter01.aspx.

Gorenstein, P. (2010). Focusing x-ray optics for astronomy. *X-Ray Optics and Instrumentation*, 109740.

Guinier, A. (1963). *X-ray Diffraction*. W.H. Freeman and Company, San Francisco.

Guinier, A., & Fournet, G. (1955). *Small-Angle Scattering of X-rays*. John Wiley and Sons, New York.

Harada, T., & Kita, T. (1980). Mechanically ruled aberration-corrected concave gratings. *Applied Optics, 19*(23), 3987–3993.

Hartmann, J. (1904). Bemerkungen über den Bau und die Justirung von Spektrographen. *Zeitschrift fur Instrumentenkunde, 29*, 47–58.

Hecht, E., & Zajac, A. (1974). *Optics*, Chap. 8. Addison-Wesley, Reading, MA.

Hellwege, K. H. (1937). Über rasterförmiger reflexionsglitter. *Zeitschrift für Physik A, 106*(9–10), 588–596.

Hettrick, M. C. (1984). Aberration of varied line-space grazing incidence gratings. *Applied Optics, 23*(18), 3221–3235.

Holzner, C., Feser, M., Vogt, S., Hornberger, B., Baines, S. B., & Jacobsen, C. (2010). Zernike phase contrast in scanning microscopy with x-rays. *Nature Physics, 6*(11), 883–887.

Hutley, M. C. (1982). *Diffraction Gratings*. Academic Press, New York.

Jacobsen, C., Kirz, J., & Williams, S. (1992). Resolution in soft x-ray microscopes. *Ultramicroscopy, 47*(1–3), 55–79.

Jacobsen, C., Williams, S., Anderson, E., Browne, M., Buckley, C. J., Kern, D., Kirz, J., Rivers, M., & Zhang, X. (1991). Diffraction-limited imaging in a scanning transmission x-ray microscopy. *Optics Communications, 86*(3–4), 351–364.

Jefimovs, K., Bunk, O., Pfeiffer, F., Grolimund, D., van der Veen, J. F., & David, C. (2007). Fabrication of Fresnel zone plates for hard x-rays. *Microelectronic Engineering, 84*(5–8), 1467–1470.

Jochum, L., & Meyer-Ilse, W. (1995). Partially coherent image formation with x-ray microscopes. *Applied Optics, 34*(22), 4944–4950.

Johann, H. H. (1931). Die Erzeugung lichtstarker Röntgenspektren mit Hilfe von Konkavkristallen. *Zeitschrift für Physik A, 69*(3–4), 185–206.

Johansson, T. (1933). Über ein neuartiges, genau fokussierendes Röntgenspektrometer Erste Mitteilung. *Zeitschrift für Physik A, 71*(7–8), 507–528.

Johnson, M. W., & Daymond, M. R. (2000). The neutron silicon lens: A new lens design for thermal neutrons. *Physica B, 283*(4), 308–313.

Kang, H. C., Maser, J., Stephenson, G. B., Liu, C., Conley, R., Macrander, A. T., & Vogt, S. (2006). Nanometer linear focusing of hard x rays by a multilayer Laue lens. *Physical Review Letters, 96*(12), 127401.

Kaulich, B., Wilhein, T., di Fabrizio, E., Romanato, F., Altissimo, M., Cabrini, S., Fayard, B., & Susini, J. (2002). Differential interference contrast x-ray microscopy with twin zone plates. *Journal of the Optical Society of America A: Optics, Image Science, and Vision, 19*(4), 797–806.

Kirkpatrick, P., & Baez, A. V. (1948). Formation of optical images by x-rays. *Journal of the Optical Society of America, 38*(9), 766–773.

Kirz, J. (1974). Phase zone plates for x-rays and extreme UV. *Journal of the Optical Society of America, 64*(3), 301–309.

Kumakhov, M. A. (1990). Channeling of photons and new x-ray optics. *Nuclear Instruments and Methods in Physics Research Section B: Beam Interactions with Materials and Atoms, 48*(1–4), 283–286.

Land, M. F., & Nilsson, D. E. (2002). *Animal Eyes*. Oxford University Press, Oxford, UK.

Lee, P. (1981). X-ray diffraction in multilayers. *Optics Communications, 37*(3), 159–164.
Margaritondo, G. (2002). *Elements of Synchrotron Light.* Oxford University Press, New York, pp. 240–246.
Marmoret, R., & Andre, J. M. (1983). Bragg reflectivity of layered synthetic microstructures in the x-ray anomalous scattering region, *Applied Optics, 22*(1), 17–19.
Miao, J., Chapman, H. N., Kirz, J., Sayre, D., & Hodgson, K. O. (2004). Taking x-ray diffraction to the limit: Macromolecular structures from femtosecond x-ray pulses and diffraction microscopy of cells with synchrotron radiation. *Annual Review of Biophysics and Biomolecular Structure, 33*, 157–176.
Michette, A. G. (1986). *Optical Systems for Soft X-rays.* Chapter 7: The design and manufacture of X-ray Diffraction Gratings. Plenum Press, New York, pp. 154–156.
Miyamoto, K. (1961). The phase Fresnel lens. *Journal of the Optical Society of America, 51*(1), 17–20.
Murphy, D. B. (2001). *Fundamentals of Light Microscopy and Electronic Imaging.* John Wiley & Sons, New York.
Namioka, T., Noda, H., & Seya, M. (1973). Possibility of using holographic gratings in vacuum monochromators. *Science of Light (Tokyo), 22*, 77–99.
Niemann, B., Rudolph, D., & Schmahl, G. (1974). Soft x-ray imaging zone plates with large zone numbers for microscopic and spectroscopic applications. *Optics Communications, 12*(2), 160–163.
Niemann, B., Rudolph, D., & Schmahl, G. (1976). X-ray microscopy with synchrotron radiation. *Applied Optics, 15*(8), 1883–1884.
Nomarski, G. (1955). Microinterféromètre différentiel à ondes polarisées. *Journal de Physique et Le Radium, Paris, 16*, 9S–11S.
Palmer, C., & Loewen, E. (2005). Holographic gratings. In *Diffraction Grating Handbook*, Chap. 4. Newport Corporation; http://gratings.newport.com/information/handbook.
Parratt, L. G. (1954). Surface studies of solids by total reflection of x-rays. *Physical Review, 95*(2), 359–369.
Pfeiffer, F., Grünzweig, C., Bunk, O., Frei, G., Lehmann, E., & David, C. (2006a). Neutron phase imaging and tomography. *Physical Review Letters, 96*(21), 215505.
Pfeiffer, F., Weitkamp, T., Bunk, O., & David, C. (2006b). Phase retrieval and differential phase-contrast imaging with low-brilliance x-ray sources. *Nature Physics, 2*, 258–261.
Prins, J. A. (1930). Die Reflexion von Röntgenstrahlen an absorbierenden idealen kristallen. *Zeitschrift für Physik A, 63*(7–8), 477–493.
Rarback, H., Shu, D., Feng, S. C., Ade, H., Kirz, J., McNulty, I., Kern, D. P., Chang, T. H. P., Vladimirsky, Y., Iskander, N., et al. (1988). Scanning X-ray microscope with 75-nm resolution. *Review of Scientific Instruments, 59*(1), 52–59.
Rayleigh, Lord (1888). Wave theory. In *Encyclopedia Britannica*, ed 9, Vol. 24, pp. 429–451.
Rosenbluth, A. E., & Lee, P. (1982). Bragg condition in absorbing x-ray multilayers. *Applied Physics Letters, 40*(6), 466–468.
Rowland, H. A. (1883). On concave gratings for optical purposes. *Philososophical Magazine Series 5, 16*(99), 197–210.
Rowland, H. A. (1893). Gratings in theory and practice. *Philososophical Magazine Series 5, 35*(216), 397–419.
Saxena, A. H., & Shoenborn, B. P. (1977). Multilayer neutron monochromators. *Acta Crystallographica Section A, 33*(Part 5), 805–813.
Schmahl, G., & Rudolph, D. (1969). Lichtstarke Zonenplatten als abbildende Systeme für weiche. *Röntgenstrahlen Optik, 29*, 577.
Schmal, G., & Rudolph, D. (1996). Phase contrast x-ray microscope. U.S. Patent No. 5,550,887, August 27, 1996, assigned to Carl-Zeiss Stiftung of Germany.
Schrödinger, E. (1926). Quantisierung als eigenwertproblem. *Physical Review, 28*(6), 1049–1070.

Shapiro, D., Thibault, P., Beetz, T., Elser, V., Howells, M., Jacobsen, C., Kirz, J., Lima, E., Miao, H., Neiman, A. M., et al. (2005). Biological imaging by soft x-ray diffraction microscopy. *Proceedings of the National Academy of Science USA*, *102*(43), 15343–15346.
Shaver, D. C., Flanders, D. C., Ceglio, N. M., & Smith, H. I. (1979). X-ray zone plates fabricated using electron-beam and x-ray lithography. *Journal of Vacuum Science & Technology*, *16*(6), 1626–1630.
Simpson, M. J., & Michette, A. G. (1984). Imaging properties of modified Fresnel zone plates. *Optica Acta*, *31*(4), 403–413.
Smith, F. H. (1954). Two half-shade devices for optical polarizing instruments. *Nature*, *173*(4399), 362–363.
Smith, H. I., Anderson, E. H., Hawryluk, A. M., & Schattenburg, M. L. (1984). Planar techniques for fabricating x-ray diffraction gratings and zone plates. In *X-ray Microscopy* (G. Schmahl, & D. Rudolph, eds.) (Springer Series in Optical Sciences, Vol. 43). Springer, Berlin, pp. 51–62.
Soret, J. L. (1875). Concerning diffraction by circular gratings. *Annual Review of Physical Chemistry*, *156*, 99–106.
Spiller, E., Segmüller, A., Rife, J., & Haelbich, R. P. (1980). Controlled fabrication of multilayer soft x-ray mirrors. *Applied Physics Letters*, *37*(11), 1048–1050.
Stanford News Service. (1992). Press release: Paul Kirkpatrick, inventor of the X-ray microscope, dead at 98, December 28, 1991; http://www.stanford.edu/dept/news/pr/92/921228Arc2007.html.
Stanley, V. W., Franks, A., & Lindsey, K. (1968). A simple ruling engine for x-ray gratings. *Journal of Physicis E: Scientific Instruments*, *1*, 643–645.
Stroke, G. W. (1967). Diffraction gratings. In *Handbuch der Physik XXIX* (S. Flugge, ed.). Springer, Berlin, pp. 426–754.
Tatchyn, R. O. (1984). Optimum Zone Plate Theory and Design. In "*X-ray Microscopy*." (G. Schmahl, D. Rudolph, eds.), *Springer Series in Optical Sciences*, Vol. 43, Berlin, pp. 40–50.
Tatchyn, R., Csonka, P. L., & Lindau, I. (1982). optimization of planar metallic nonrefracting transmission-grating profiles for mth-order intensity maximization in the soft-x-ray range. *Journal of the Optical Society of America*, *72*(12), 1630–1638.
Thathachari, Y. T. (1953). Theory of image formation in combinations of x-ray focusing mirrors. *Proceedings of the Indian Academy of Sciences: Mathematical Sciences*, *37*, 41–62.
Thibault, P., Dierolf, M., Menzel, A., Bunk, O., David, C., & Pfeiffer, F. (2008). High-resolution scanning x-ray diffraction microscopy. *Science*, *321*(5887), 379–382.
Underwood, J. H. (2009). Optics and Detectors (Section 4). In *X-Ray Data Booklet*. Lawrence Berkeley National Laboratory, Center for X-ray Optics and Advanced Light Source; http://xdb.lbl.gov/Section4/index.html.
Underwood, J. H., & Barbee, T. W. Jr. (1981a). Synthetic multilayers as Bragg diffractors for x-rays and extreme ultraviolet: Calculation of performance. In *Low Energy X-Rays Diagnostics* (D. T. Underwood, & B. L. Henke, eds.). *AIP Conference Proceedings*, Vol. 75. American Institute of Physics, New York, pp. 170–178.
Underwood, J. H., & Barbee, T. W. Jr. (1981b). Layered synthetic microstructures as Bragg diffractors for x-rays and extreme ultraviolet: Theory and predicted performance. *Applied Optics*, *20*(17), 3027–3034.
Van Laue, M. (1912). Eine quantitative prüfung der theorie für die interferenzerscheinungen bei Röntgenstrahlen. *Proceedings of Bavarian Academy of Sciences*, p. 363–373, reprinted in *Naturwissenschaften* (1952), *39*(16), 368–372.
Videl, B., & Vincent, P. (1984). Metallic multilayers for x-rays using classical thin-film theory. *Applied Optics*, *23*(11), 1794–1801.
Vladimirsky, Y., Källne, E., & Spiller, E. (1984). Fabrication of free standing x-ray transmission gratings and zone plates. In *X-ray Lithography and Applications of Soft X-rays to Technology* (A. D. Wilson, ed.). *Proceedings of SPIE*, Vol. 448. Bellingham, WA, pp. 25–37.

Waller, I. (1923). Zur frage der einwirkung der wärmebewegung auf die interferenz von Röntgenstrahlen. *Zeitschrift für Physik A, 17*(1), 398–408.

Wilhein, T., Kaulich, B., di Fabrizio, E., Romanato, F., Cabrini, S., & Susini, J. (2001). Differential interference contrast x-ray microscopy with submicron resolution. *Applied Physics Letters, 78*(14), 2082–2084.

Williams, S., Jacobsen, C., Kirz, J., Maser, J., Wirick, S., Zhang, X., Ade, H., & Rivers, M. (1995). Instrumentation developments in scanning soft x-ray microscopy. *Review of Scientific Instruments, 66*(2), 1271–1275.

Wolter, H. (1952a). Spiegelsysteme streifenden einfalls als abbildende optiken für Röntgenstrahlen. *Annalen der Physik, 445*(1–2), 94–114.

Wolter, H. (1952b). Verallgemeinerte Schwarzschildsche Spiegelsysteme streifender Reflexion als Optiken für Röntgenstrahlen. *Annalen der Physik, 445*(4–5), 286–295.

Wood, R. W. (1898). Phase-reversed zone plates and diffraction telescopes. *Philosophical Magazine Series 5, 45*(277), 511–522.

Yun, W., & Wang, Y. (2008). Phase contrast microscope and short wavelength radiation imaging method. U.S. Patent No. 7,414,787 B2, August 19, 2008.

Yun, W., Lai, B., Krasnoperova, A. A., di Fabrizio, E., Cai, Z., Cerrina, F., Chen, Z., Gentili, M., & Gluskin, E. (1999). *Review of Scientific Instruments, 70*(9), 3537–3541.

Zachariasen, W. H. (1945). *Theory of X-ray Diffraction in Crystals.* John Wiley & Sons, New York.

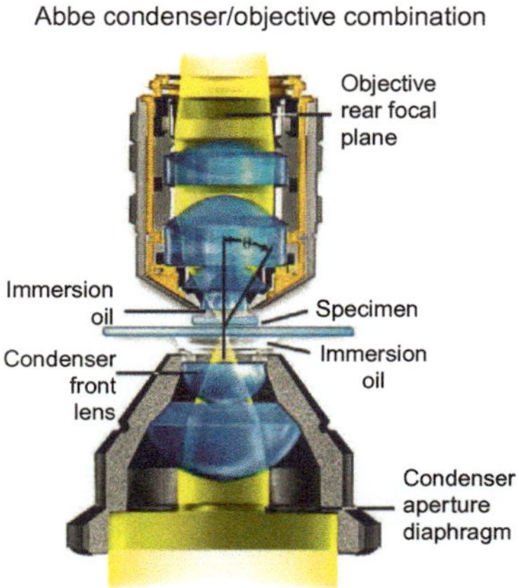

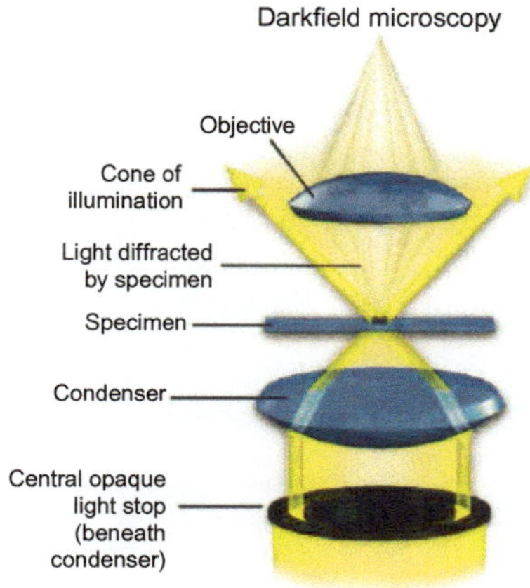

PLATE 1 (Figure 6 on page 20 of this volume)

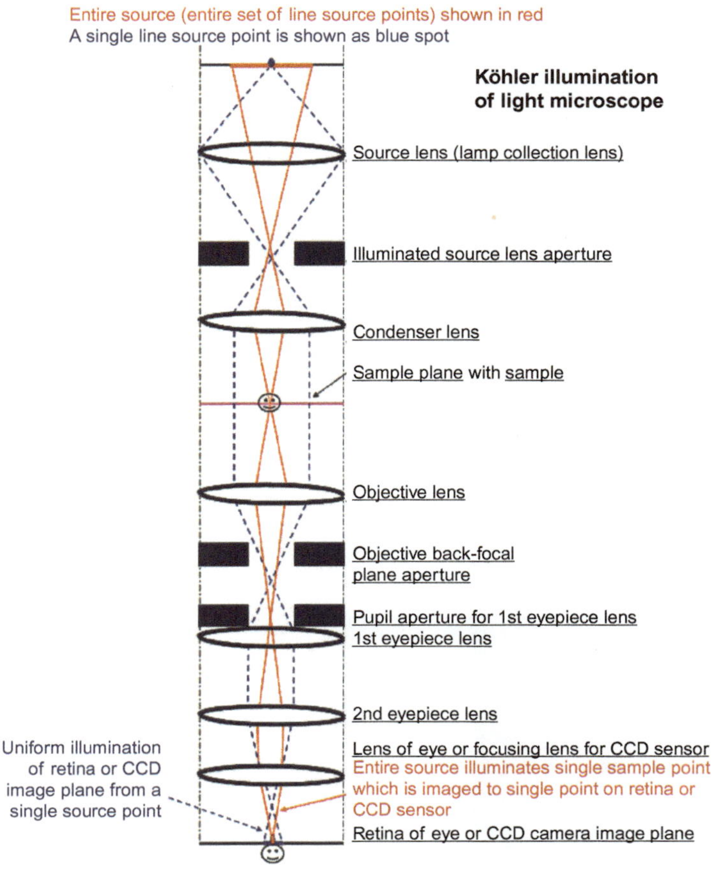

PLATE 2 (Figure 7 on page 21 of this volume)

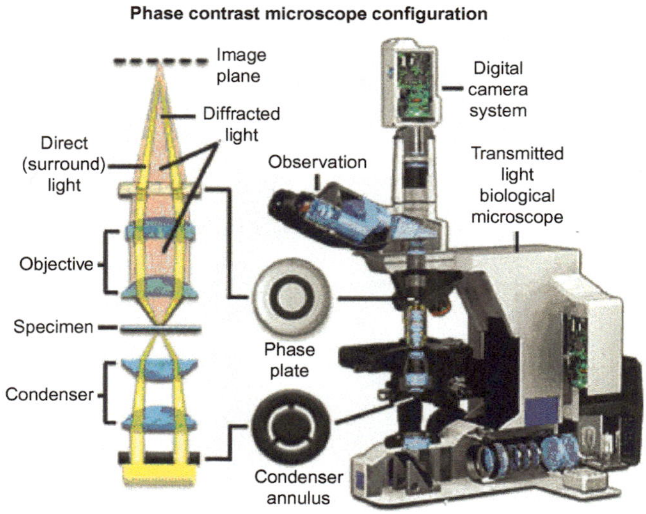

PLATE 3 (Figure 8 on page 23 of this volume)

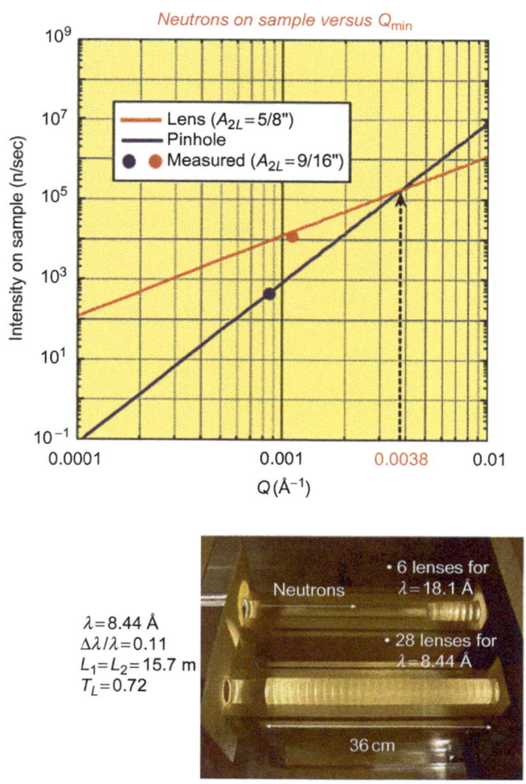

PLATE 4 (Figure 22 on page 51 of this volume)

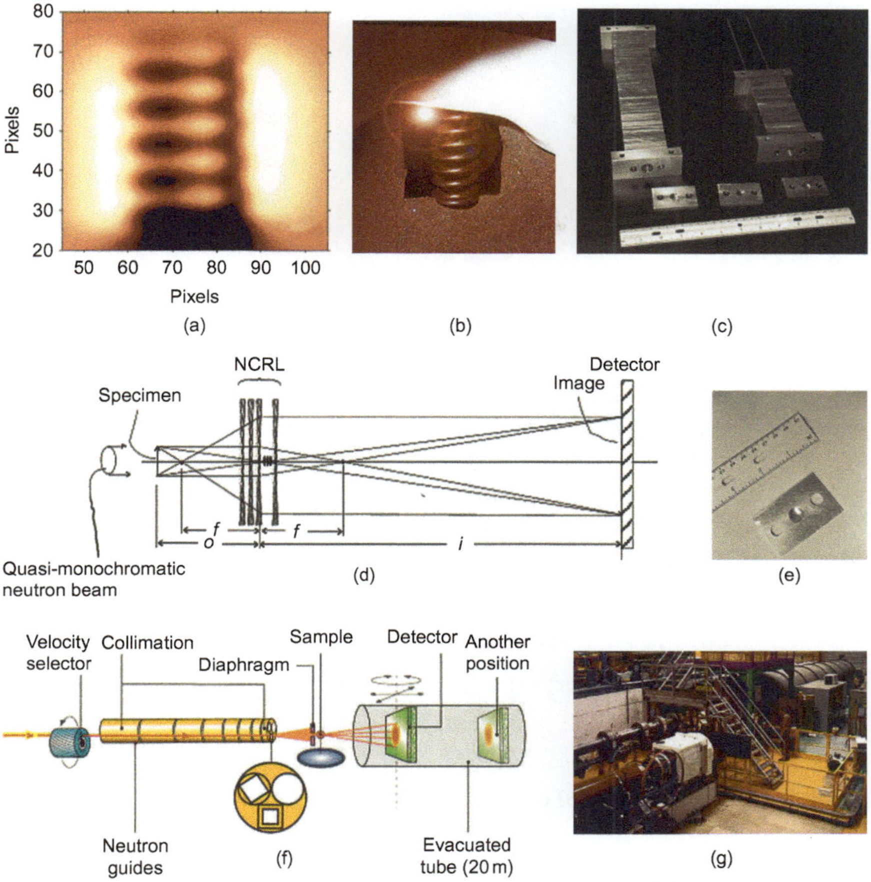

PLATE 5 (Figure 23 on page 52 of this volume)

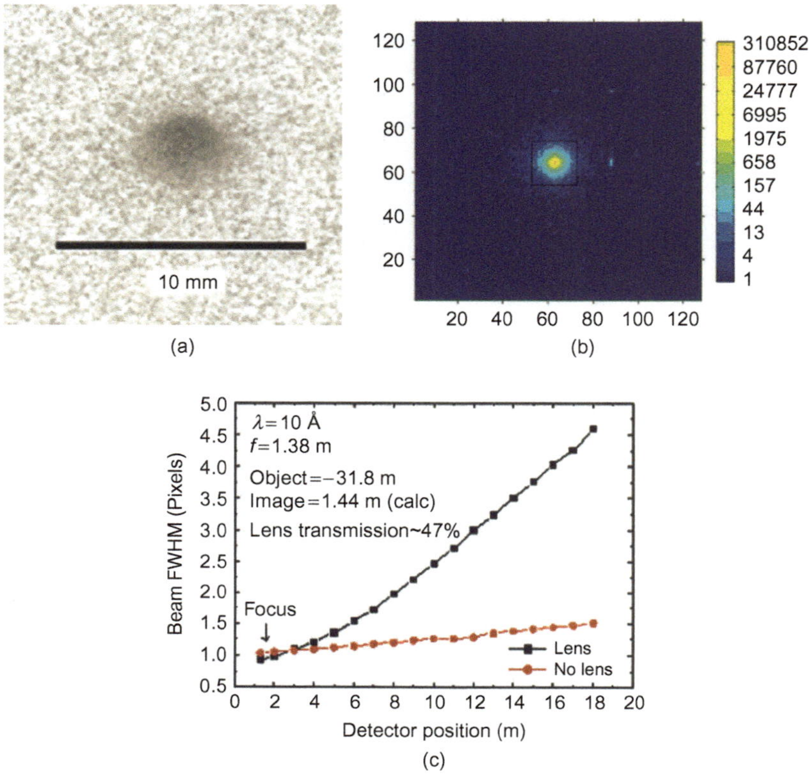

PLATE 6 (Figure 24 on page 53 of this volume)

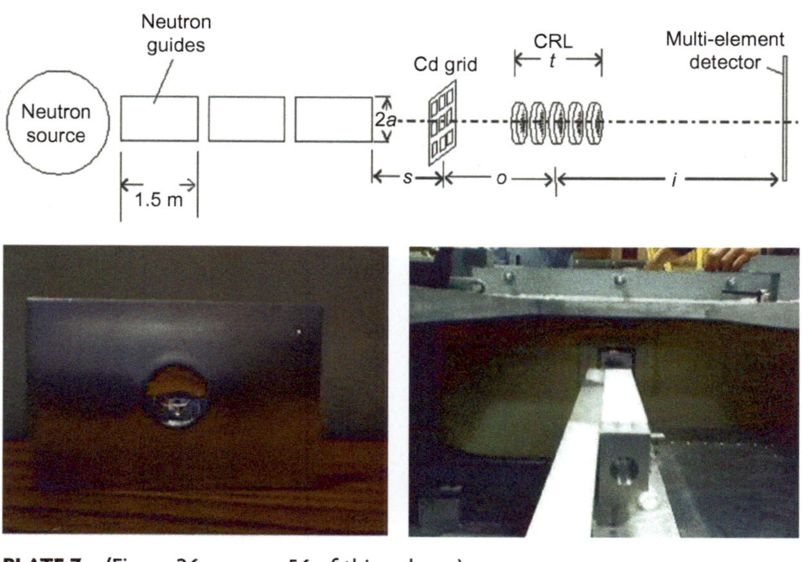

PLATE 7 (Figure 26 on page 56 of this volume)

Measured and Calculated values of the FOV

Source-to-object Distance, s	Calculated FOV	Measured FOV
0.35 m	39.3 mm	41.4 mm
3.0 m	20.8 mm	20.5 mm
6.0 m	15.1 mm	11 - 16 mm

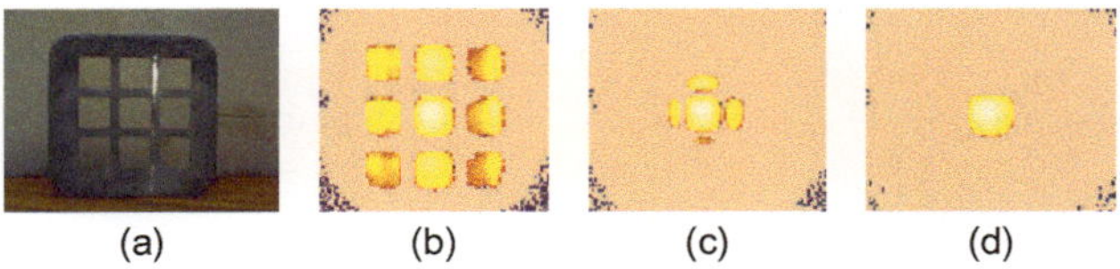

PLATE 8 (Figure 27 on page 57 of this volume)

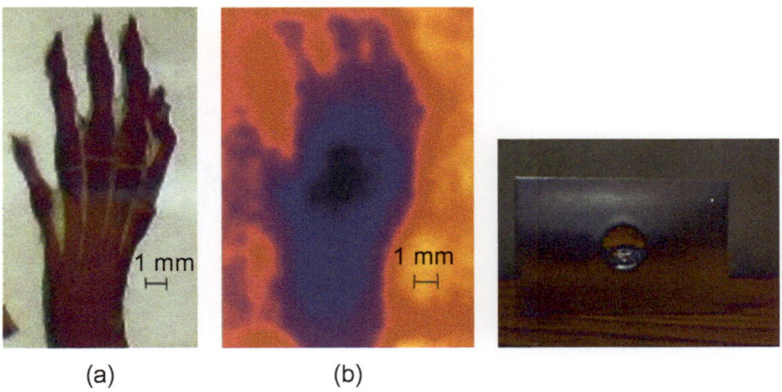

PLATE 9 (Figure 28 on page 58 of this volume)

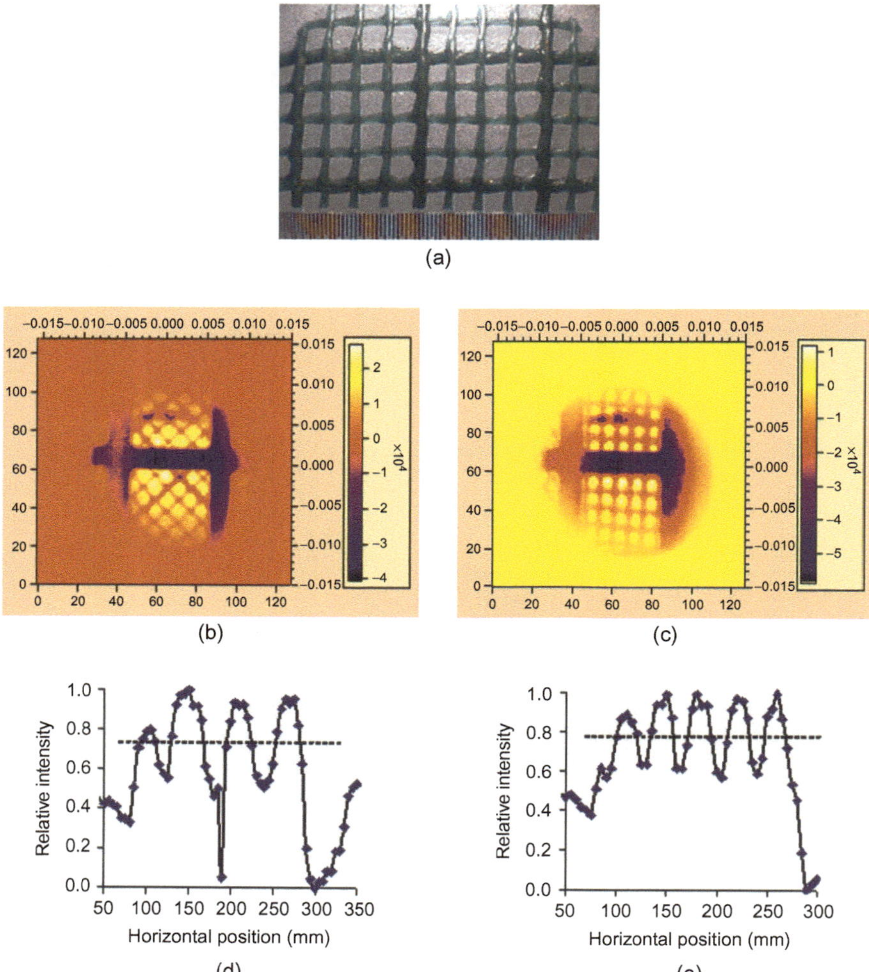

PLATE 10 (Figure 31 on page 62 of this volume)

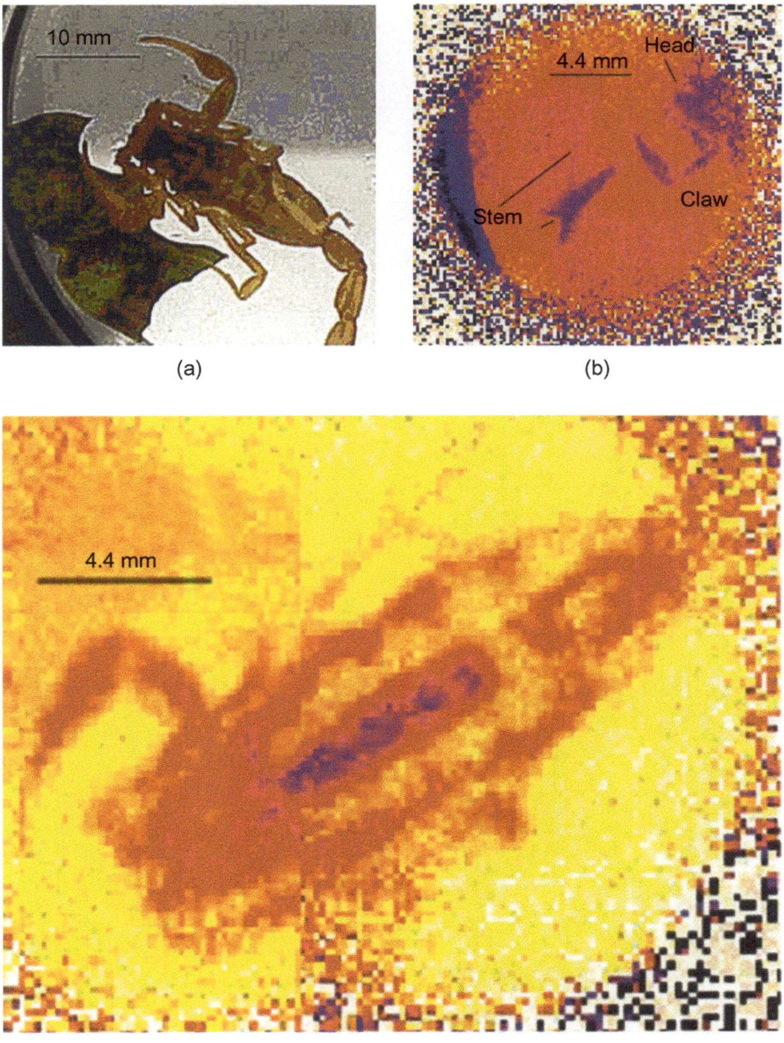

PLATE 11 (Figure 32 on page 63 of this volume)

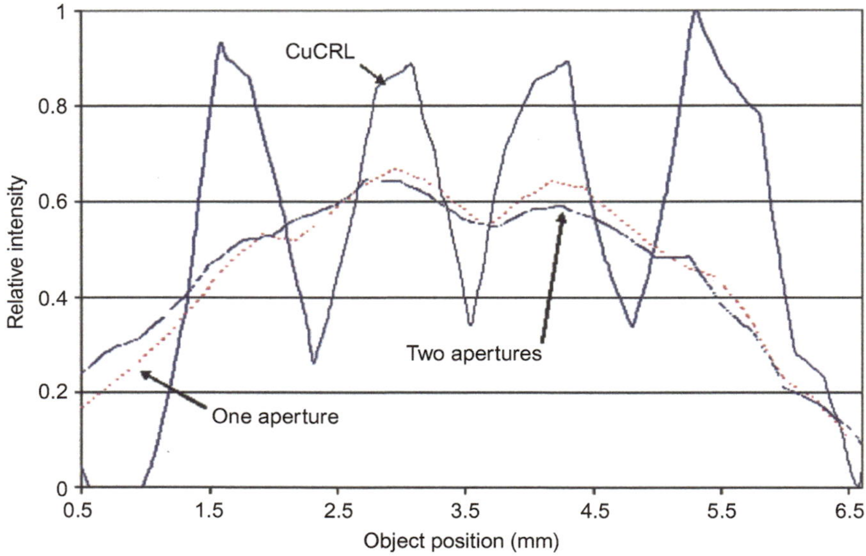

PLATE 12 (Figure 36 on page 69 of this volume)

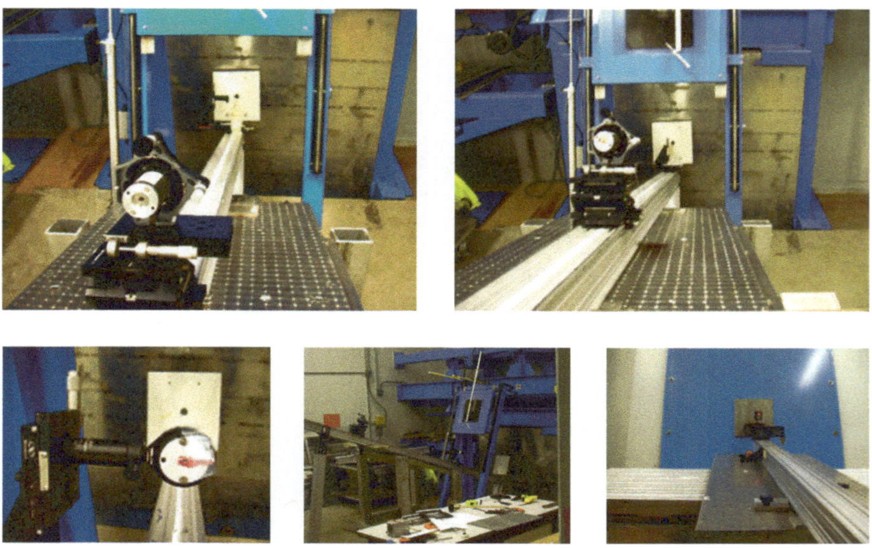

PLATE 13 (Figure 40 on page 74 of this volume)

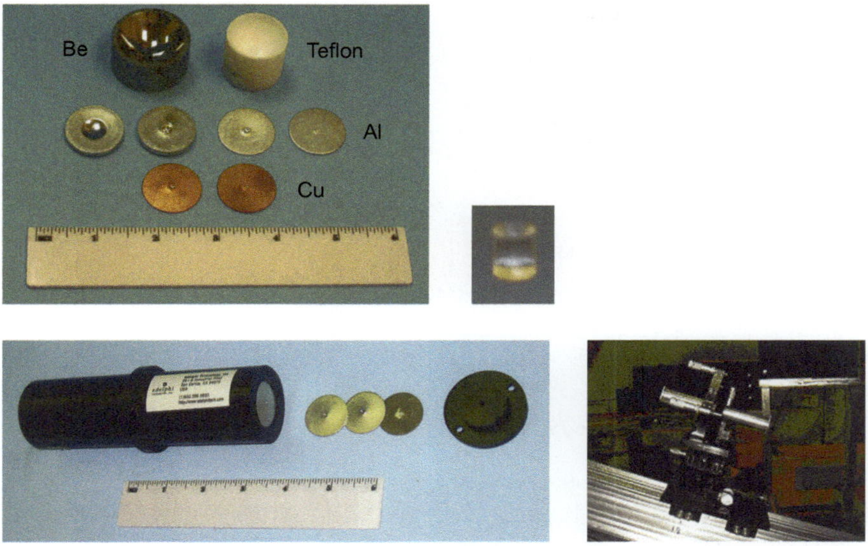

PLATE 14 (Figure 41 on page 75 of this volume)

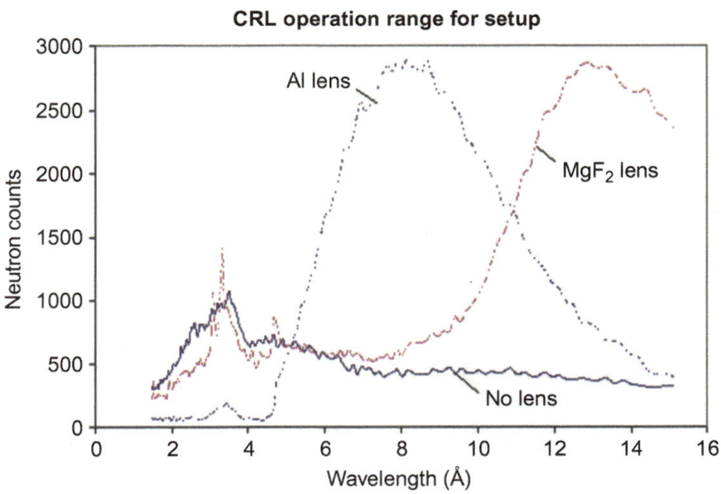

PLATE 15 (Figure 43 on page 78 of this volume)

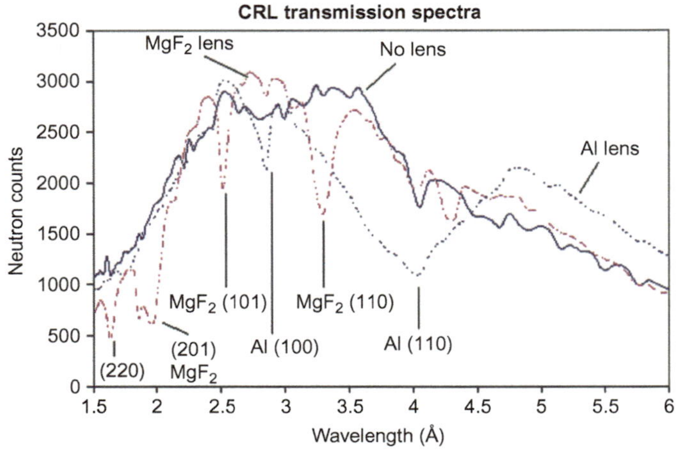

PLATE 16 (Figure 44 on page 79 of this volume)

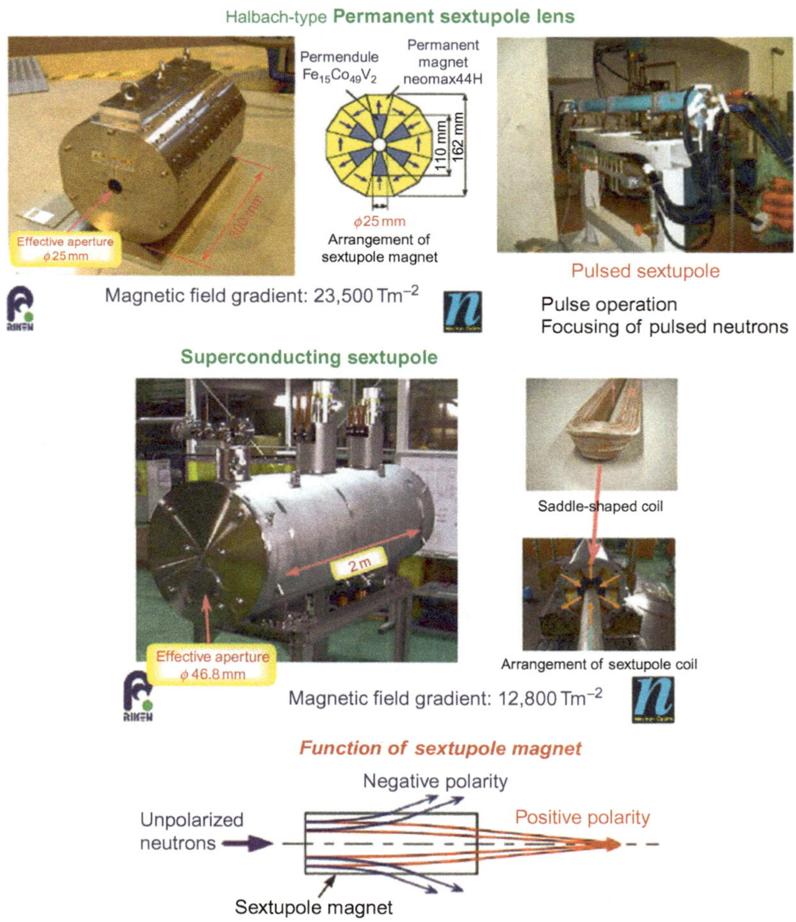

PLATE 17 (Figure 49 on page 85 of this volume)

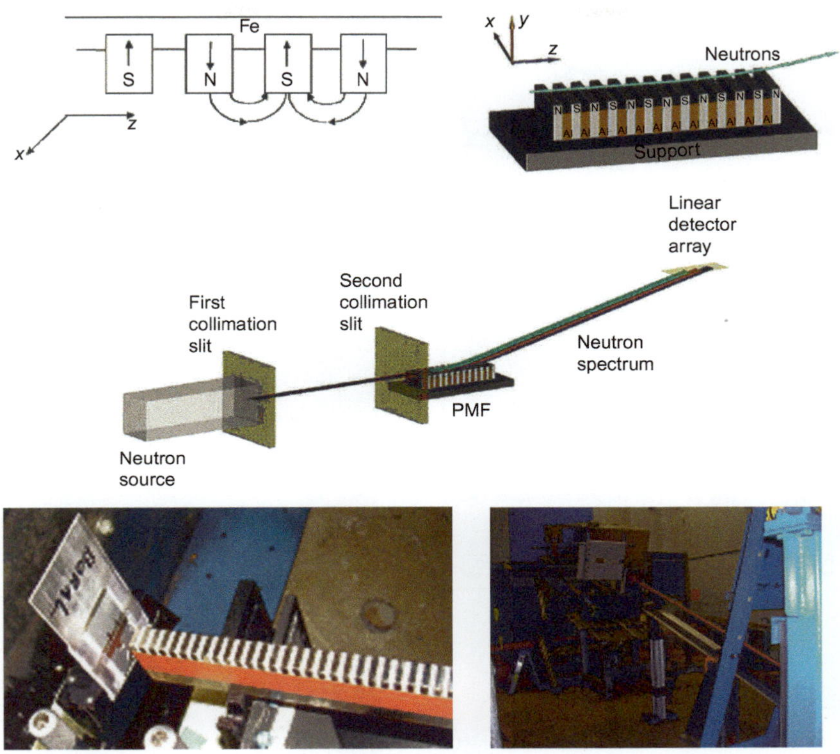

PLATE 18 (Figure 52 on page 88 of this volume)

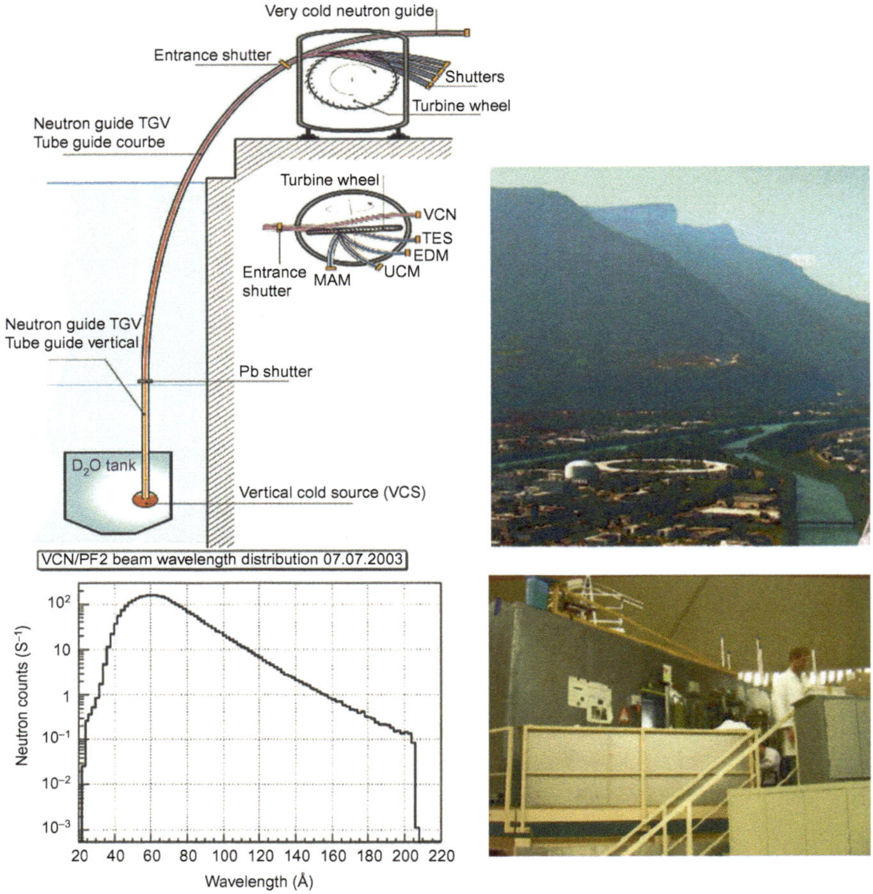

PLATE 19 (Figure 55 on page 93 of this volume)

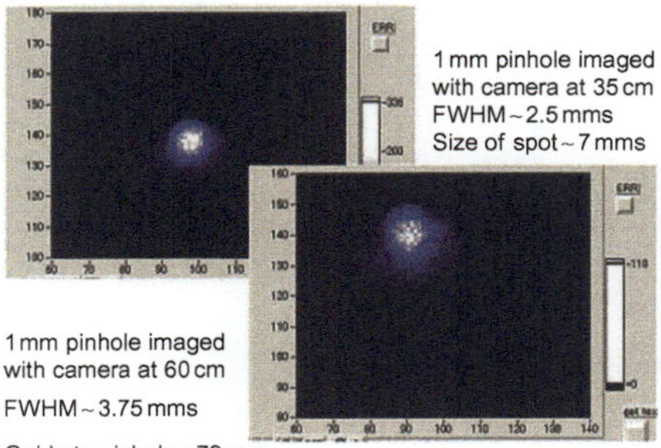

1 mm pinhole imaged
with camera at 35 cm
FWHM ~ 2.5 mms
Size of spot ~ 7 mms

1 mm pinhole imaged
with camera at 60 cm
FWHM ~ 3.75 mms

Guide to pinhole = 79 cms

Ted Cremer, Adelphi Tech.
Jack Carpenter, E. Lang, J. Lal (ANL)
Hirohiko Shimizu KEK, Japan
Katsuya Hirota, Kenji Mishima RIKEN

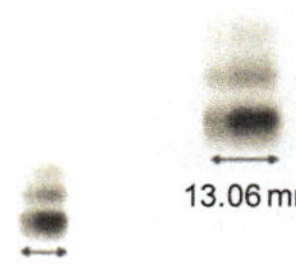

13.06 mm

8.7 mm

1 mm pinhole
imaged at 39 cms

1 mm pinhole
imaged at 67 cms

IMAGE PLATE (1 pixel = 200 μm)

PLATE 20 (Figure 56 on page 95 of this volume)

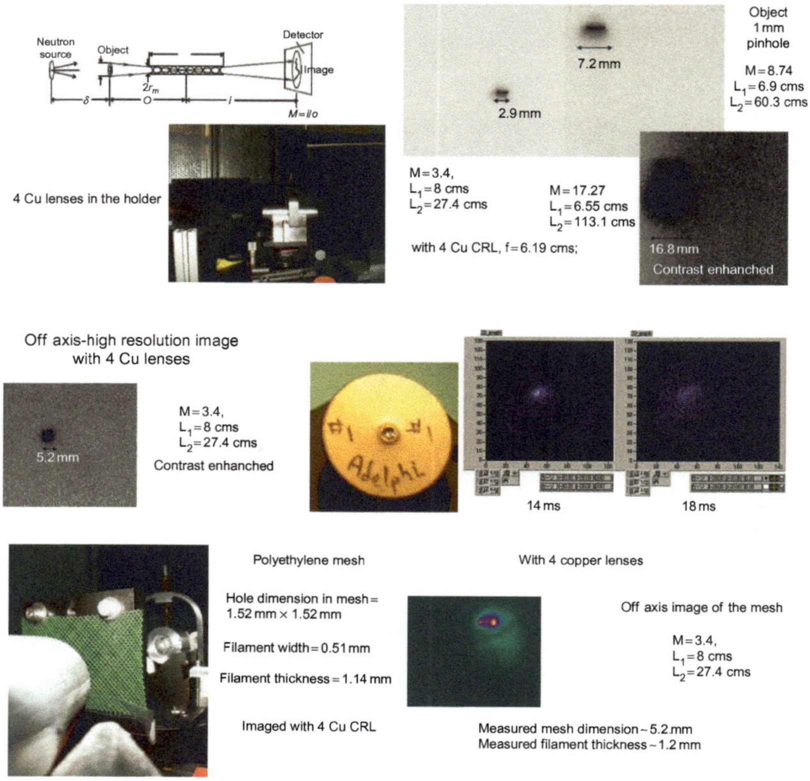

PLATE 21 (Figure 57 on page 96 of this volume)

TABLE I. Lens parameters.

Lens No.	Lens material	R (mm)	N	Min. lens thickness d (μm)	Photon energy (keV)	Calc. f (cm)	Calc. image distance (cm)	Meas. image dist (cm)	Phys. apert. (μm)	Calc. abs. apert. (μm)	Meas. abs. apert (μm)	Peak transmission (%)	Gain calc.	Gain exp.
M5	Mylar	1.6	193	10	8	99	105	62	645	287	250	12	2.5	1.5
M5	Mylar	1.6	193	10	9	125	134	82	645	343	296	22	3.8	-
M5	Mylar	1.6	193	10	10	153	169	92	645	405	330	33	4.9	2.0
K5	Kapton	1.98	400	15.8	11	97	103	72	370	375	330	11	2.0	2.0
K5	Kapton	1.98	400	15.8	12	100	107	85	370	381	360	19	2.7	2.5
K7	Kapton	1.98	400	12.5	10	60	62	65	286	266	303	11	2.7	1.6
K7	Kapton	1.98	400	12.5	12	81	85	103	286	345	364	27	6.4	3.8
K6	Kapton	1.98	600	15.8	13	77	81	70	370	348	350	21	2.3	2.2
K6	Kapton	1.98	600	15.8	14	87	92	75	370	383	—	19	3.2	—
K11[a]	Kapton	1.35	330	12.5	9	48	49	47	350	230	262	8	2.9	1.7
K11[a]	Kapton	1.35	330	12.5	10	59	61	62	350	270	291	16	5.4	3.8
K11[a]	Kapton	1.35	330	12.5	11	69	72	72	350	310	310	24	8.0	5.0

[a]Parabolic lens.

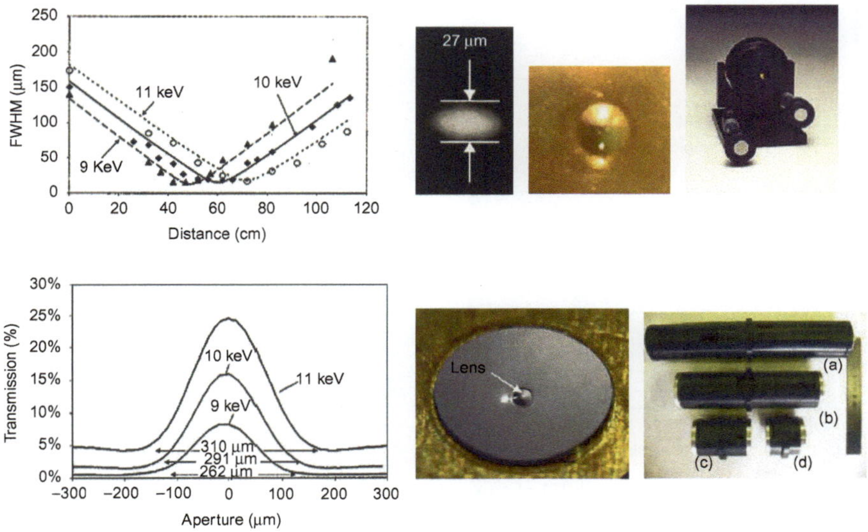

PLATE 22 (Figure 88 on page 150 of this volume)

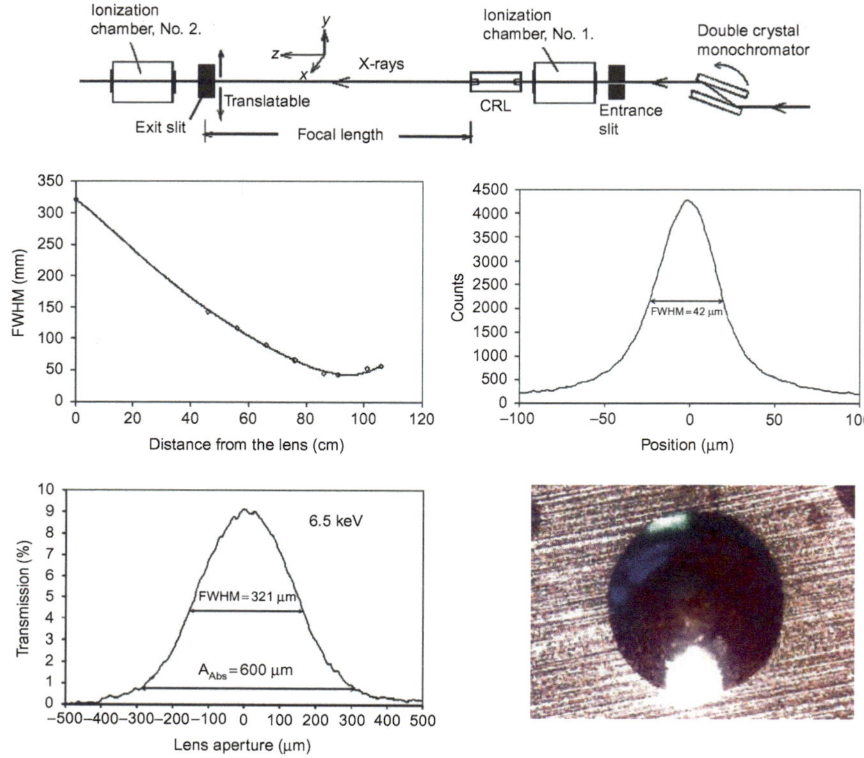

PLATE 23 (Figure 89 on page 151 of this volume)

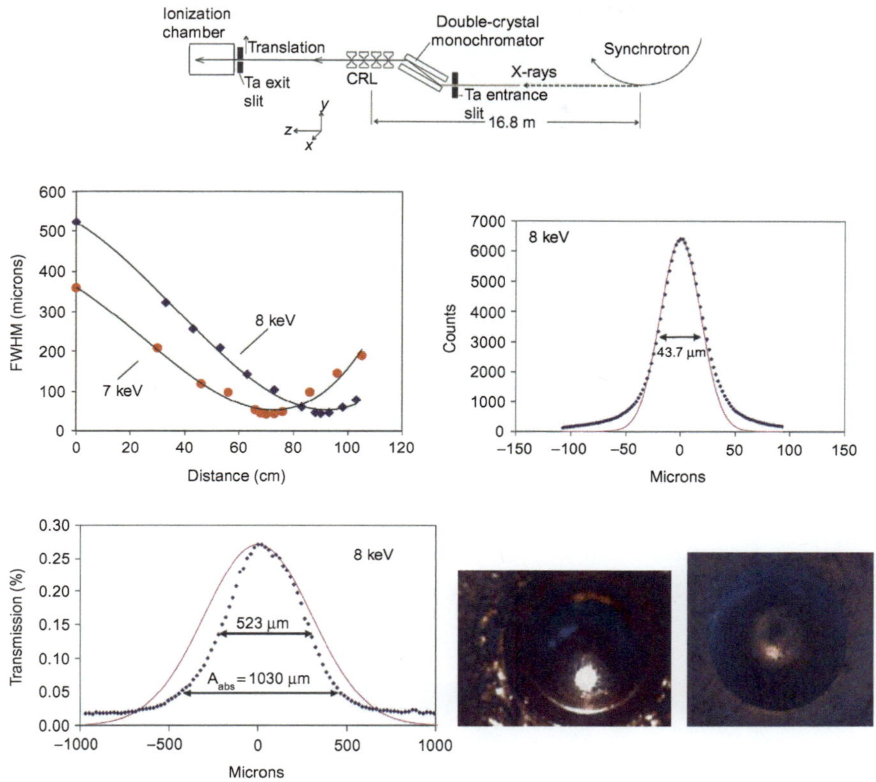

PLATE 24 (Figure 90 on page 152 of this volume)

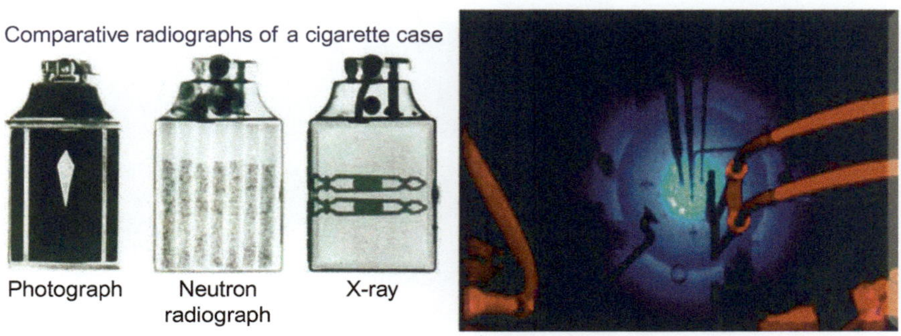

PLATE 25 (Figure 115 on page 201 of this volume)

PLATE 26 (Figure 124 on page 217 of this volume)

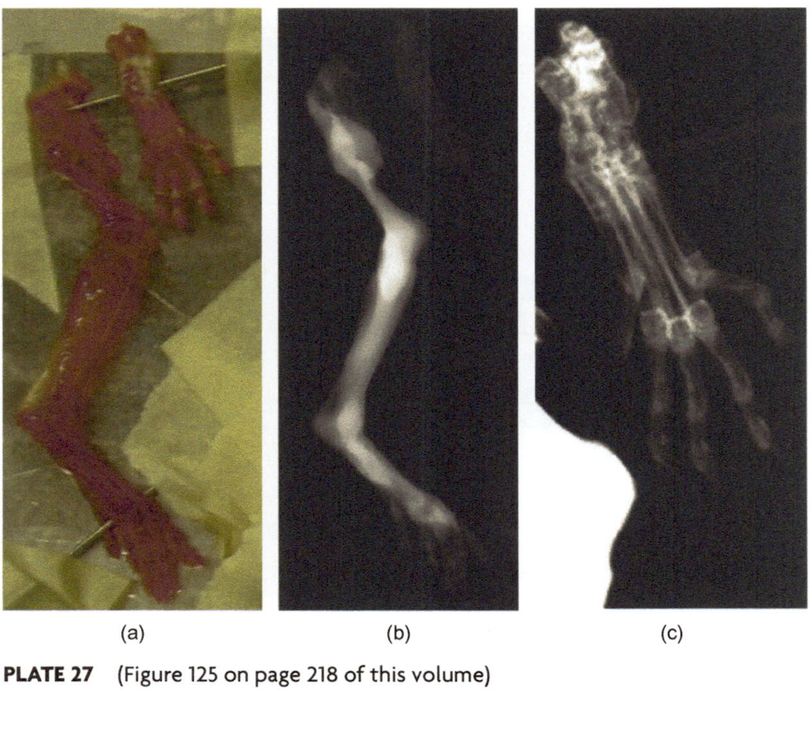

(a) (b) (c)

PLATE 27 (Figure 125 on page 218 of this volume)

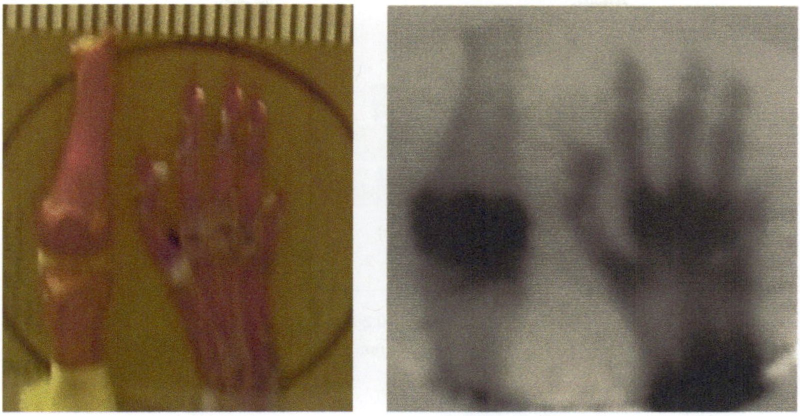

PLATE 28 (Figure 126 on page 219 of this volume)

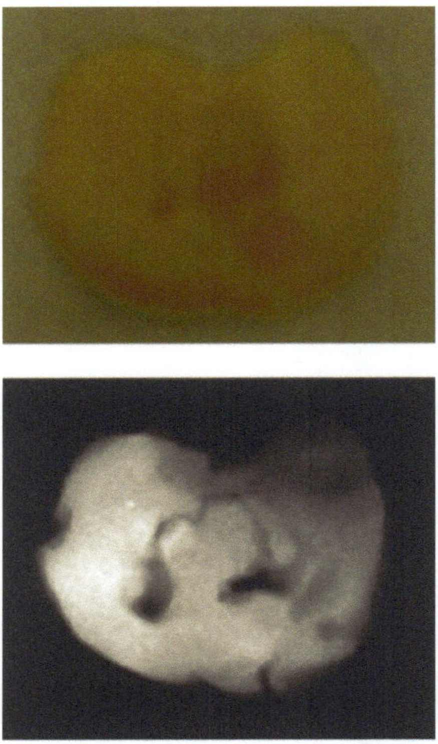

PLATE 29 (Figure 128 on page 221 of this volume)

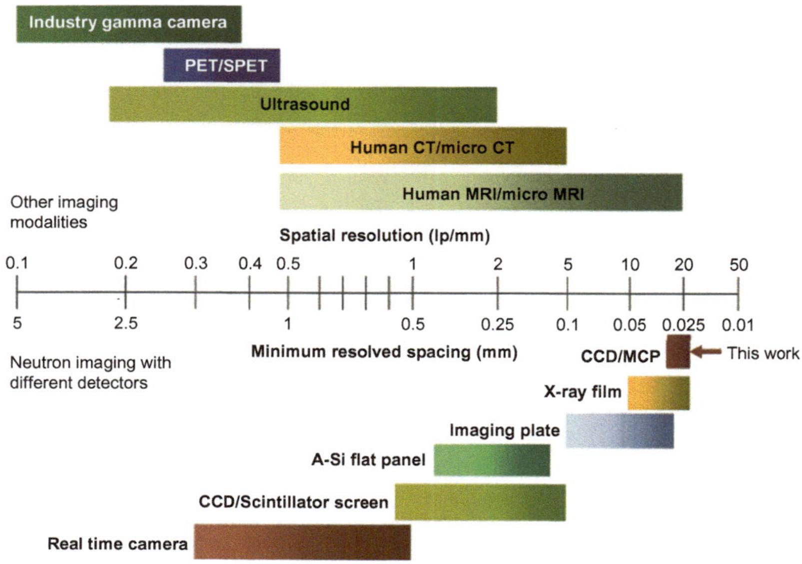

PLATE 30 (Figure 129 on page 222 of this volume)

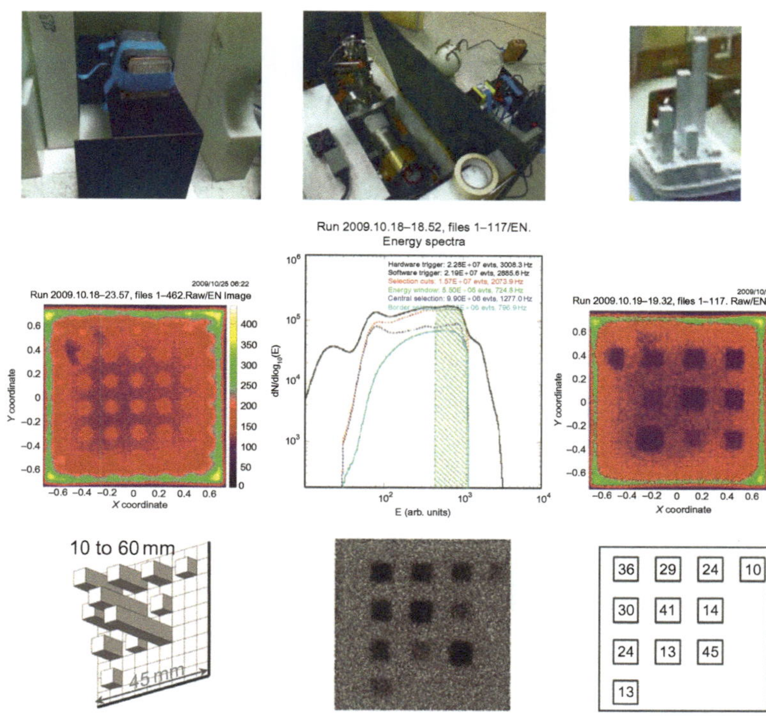

PLATE 31 (Figure 141 on page 244 of this volume)

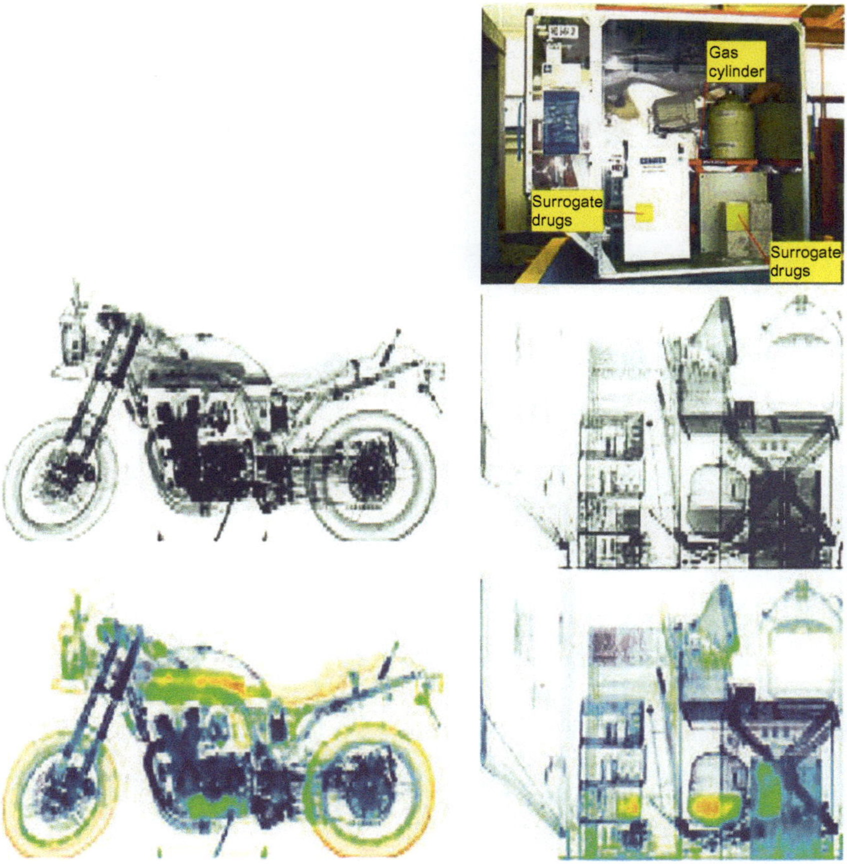

PLATE 32 (Figure 144 on page 253 of this volume)

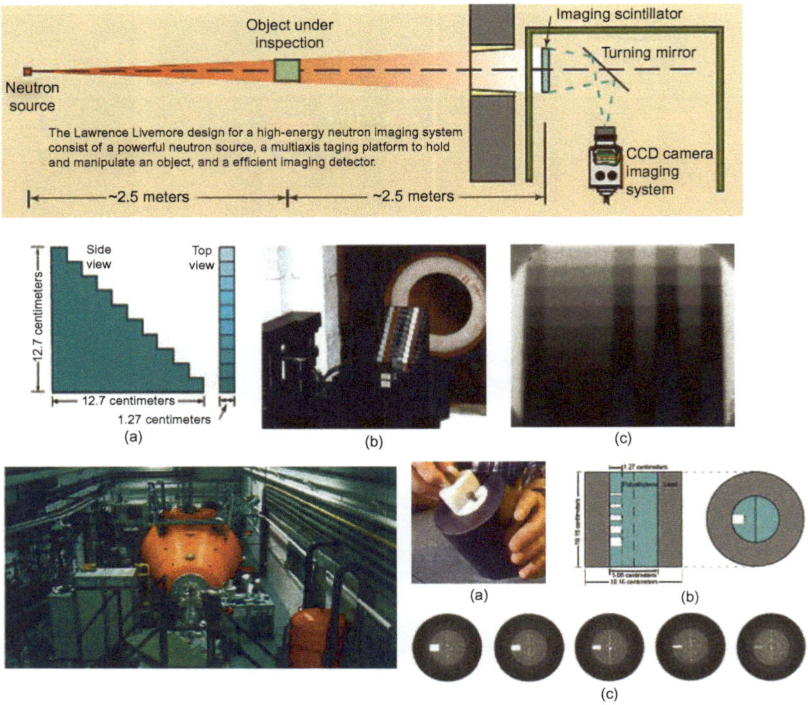

PLATE 33 (Figure 145 on page 256 of this volume)

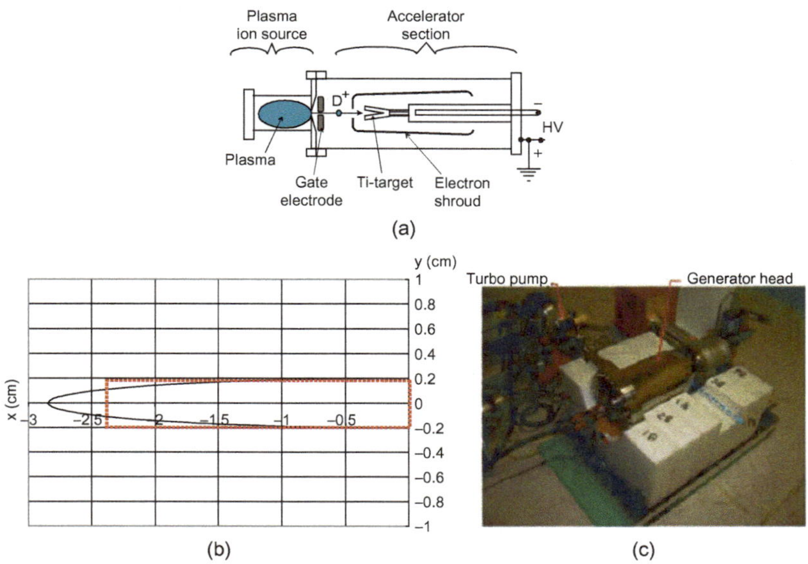

PLATE 34 (Figure 148 on page 260 of this volume)

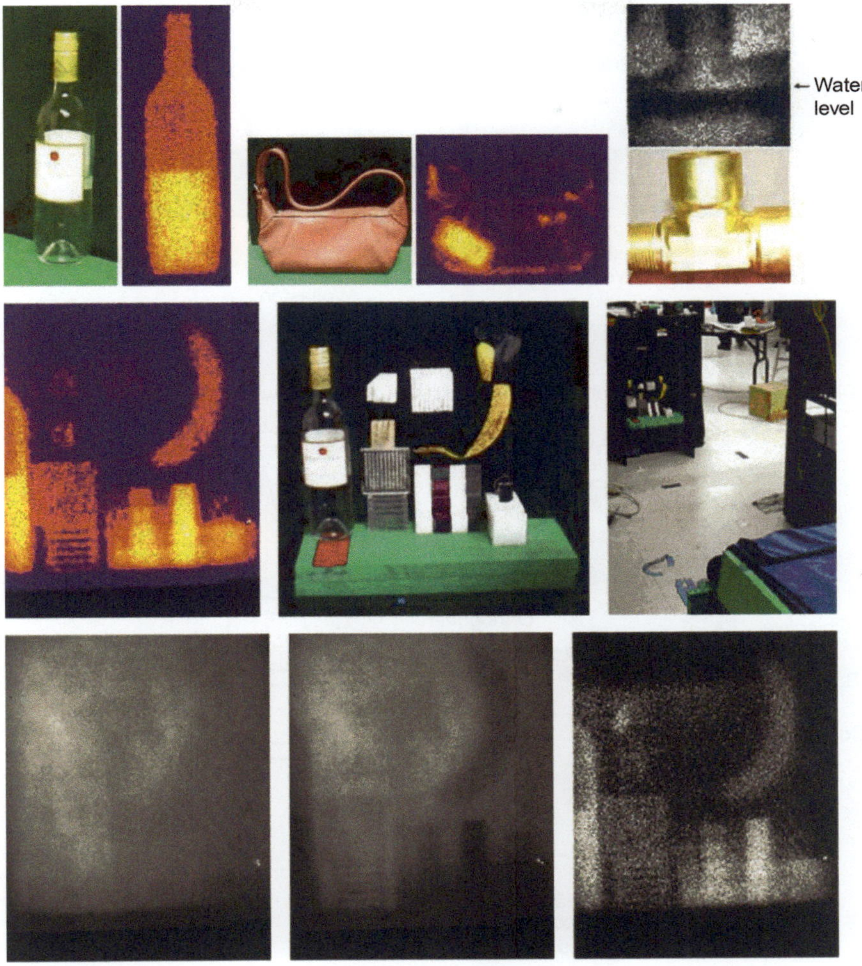

PLATE 35 (Figure 150 on page 264 of this volume)

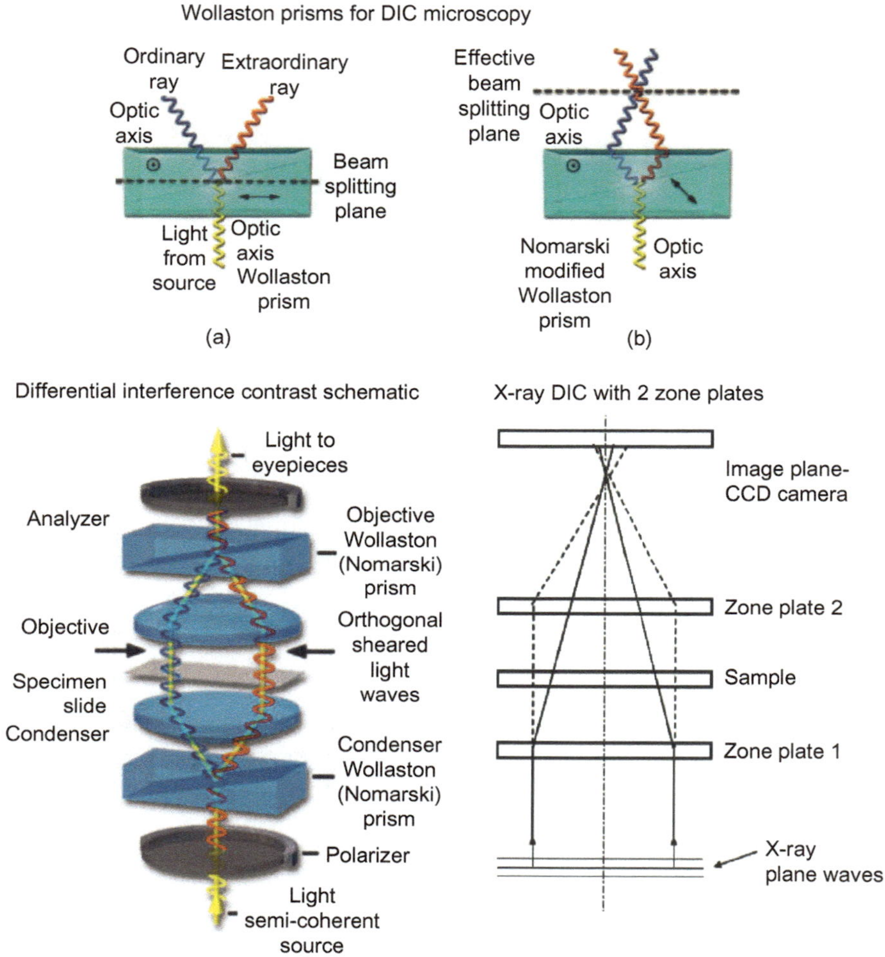

PLATE 36 (Figure 163 on page 301 of this volume)

Chapter 2

Compound Refractive Lenses and Prisms

Jay Theodore Cremer, Jr.

Contents

1. The Compound Refractive Lens 336
2. Numerical Aperture and Thin Lenses 338
3. Biconcave Parabolic Lens: Path Length, Focal Length, and Absorption Aperture Radius 339
4. Biconcave Spherical Lens: Path Length, Focal Length, and Absorption Aperture Radius 346
5. Parabolic Aperture Radius in Spherical Biconcave CRLs 349
6. Matrix Representation of X-ray and Neutron Optics with Paraxial Approximation 350
7. Thin Lens Matrix and Ray Angle Deviation in a Thin Lens 353
8. Average CRL Transmission for X-rays or Neutrons 358
9. CRL Intensity Gain Including Attenuation 361
10. CRL Transverse and Axial Magnification 368
11. CRL Depth of Field and Depth of Focus 371
12. Modulation Transfer Function for CRL Resolution Determination 373
13. Calculation of CRL Modulation Transfer Function by Line Profile Measurements of Knife-Edge Images 377
14. CRL Field of View 380
15. Thick Lens CRLs 382
16. CRL Surface Roughness and Lens Alignment 390
17. Compound Refractive Prisms: X-ray and Neutron Deflection by a Single Prism 398
18. Deflection of X-rays and Neutrons in a Compound Refractive Prism 402
19. Calculation of X-ray or Neutron Absorption Aperture Depth and Transmission in a Compound Refractive Prism 404
20. Use of a Compound Refractive Prism to Offset Neutron Gravity Droop in Small-Angle Scattering 406

Advances in Imaging and Electron Physics, Volume 172, ISSN 1076-5670, DOI: 10.1016/B978-0-12-394422-1.00002-5.
Copyright © 2012 Elsevier Inc. All rights reserved.

21. Curvature and Radius of Curvature of Lens Surface and Wave Surface 409
22. Measurement of Spherical and Parabolic Lens Curvature 411
23. Wave Vergence and Wave Vergence Change by Lens Surface 413
24. Refractive Index, Snell's Law, Huygens' Principle, and Fermat's Principle of Least Time 414
25. The Paraxial Approximation and the Fundamental Paraxial Equation 417
26. The Gaussian Lens Equation for Thin Lenses 418
27. Thick Lens Power and Focal Length, and Wave Vergence Change in Thick Lenses 420
28. Wave Vergence Change and Power and Focal Length of a Separated Lens Doublet 424
References 426

1. THE COMPOUND REFRACTIVE LENS

The compound refractive lens (CRL) is a stack of N lenses whose total focal length f is the focal length of a single unit lens f_1 divided by the number of lenses N, where

$$f = \frac{f_1}{N}. \quad (1)$$

The material index of refraction n of the CRL determines the X-ray or neutron refractive and attenuation properties via the decrement δ and the absorption index β, where

$$n = 1 - \delta + i\beta. \quad (2)$$

The absorption index β is expressed in terms of the linear attenuation coefficient μ (cm^{-1}), where

$$\beta = \frac{\mu\lambda}{4\pi}. \quad (3)$$

The refractive index n is then

$$n = 1 - \delta + i\frac{\mu\lambda}{4\pi}. \quad (4)$$

The linear attenuation coefficient μ (cm^{-1}) contains the attenuation properties—namely, absorption, incoherent scatter, and inelastic coherent scatter, such that for thermal neutron absorption,

$$\mu \propto \lambda. \quad (5)$$

The decrement δ contains the refractive properties and can be positive or negative, but is usually positive, and is denoted by

$$\delta = \frac{\lambda^2 b\rho}{2\pi}. \tag{6}$$

For X-rays and neutrons the decrement is very small, where

$$\delta \ll 1. \tag{7}$$

The decrement δ depends on wavelength λ, atom density ρ, and bound coherent scattering length b. The bound scattering length is usually positive for neutrons and X-rays.

As ordinary optical lenses, CRLs can be used for imaging by placing the lens between an illuminated object and a detector. The lens or lensmaker's equation applies to the CRL composed of thin lenses, with distance r_0 between the object and the CRL center, distance r_i between the CRL center and the detector, denoted as follows:

$$\frac{1}{f} = \frac{1}{r_0} + \frac{1}{r_i}. \tag{8}$$

Newton's lens equation is a useful formula that can be applied to thin lens CRLs and is derived as follows (Figure 1). A diagonal ray from an object point at radial distance x from the axis passes through the first focal point of the thin lens situated at downstream distance s_0 from the object plane and upstream distance f_0 from the center plane of the lens. The ray intercepts the center plane at radial distance y. The focal ray is refracted by the CRL center plane to become a parallel ray, which is parallel to the CRL

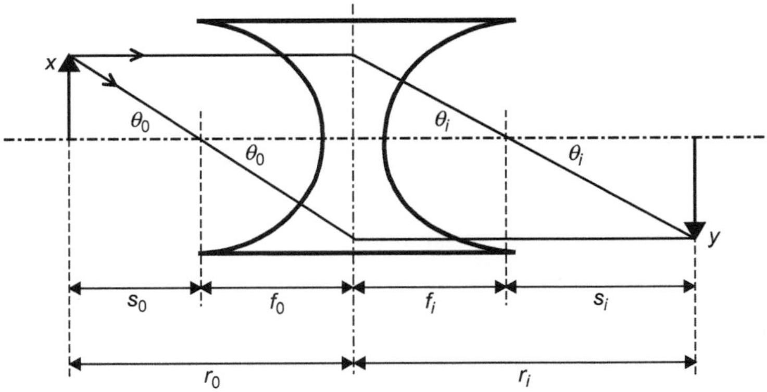

FIGURE 1 A diagram used to derive Newton's lens equation.

center axis, and terminates at a point on the image plane at radial distance y. With the center z-axis, the focal ray creates two similar triangles on the input side of the lens so that

$$\theta_0 = \frac{x}{s_0} = -\frac{y}{f_0}. \tag{9}$$

Alternatively, from the same object point at radial distance x, a ray parallel to the center axis travels to a point on the CRL center plane, also at radial distance x. At the CRL center plane the parallel ray is refracted and passes through the second focal point of the lens at downstream distance f_i and upstream of the image plane by axial distance s_i. The refracted ray then terminates at the same image point as the focal ray in the previous case at the radial distance. This refracted ray forms two similar triangles on the output side of the CRL, where

$$\theta_i = \frac{y}{s_i} = -\frac{x}{f_i}. \tag{10}$$

Eliminating the common ratio x/y in the two similar triangle expressions yields Newton's lens equation:

$$s_0 s_i = f_0 f_i. \tag{11}$$

For biconcave lenses one assumes in general the upstream (input) focal length f_0 and downstream (output) focal length f_i are equal, where

$$f_0 = f_i = f. \tag{12}$$

2. NUMERICAL APERTURE AND THIN LENSES

Most useful X-ray and neutron lenses are plano-concave or biconcave because the index of refraction for most materials for neutrons and X-rays is less than 1 so that focusing is accomplished by concave lenses. In contrast, lenses for visible light in which the decrement is greater than 1 for most materials are plano-convex or bi-convex. The relatively strong attenuation of X-rays and lesser attenuation of neutrons by refractive lenses favor concave lenses in that the central transmission is larger through the center, thinner part of the lens. In the case of titanium neutron lenses in which the index of refraction is greater than 1 and requires a convex lens, the transmission of neutrons through the central part of the titanium lens is less than its periphery. Consequently, a titanium zone plate is better than a titanium lens.

The CRL consists of N biconcave lenses and acts as a cylinder of length l and effective aperture diameter $2r_e$. The CRL has a limited acceptance angle θ_a given by

$$\theta_a = \frac{2r_e}{l} \tag{13}$$

The CRL numerical aperture NA is its aperture radius r_e divided by its focal length f:

$$NA = \frac{r_e}{f}. \tag{14}$$

If the focal length f is much less than the CRL length l, then the thin lens approximation applies to the CRL, where

$$f \ll l \quad \text{thin lens.} \tag{15}$$

A X-ray or neutron ray, passing through a thin lens CRL, changes direction but not its radial distance r from the CRL center axis, provided each individual lens has a radius of curvature R much larger than the CRL radially dependent thickness $t(r)$, where

$$t(r) \ll R \quad 0 < r < r_e. \tag{16}$$

The thickness $t(r)$ of each biconcave lens increases with r from its minimum center thickness d, where

$$t(0) = d. \tag{17}$$

3. BICONCAVE PARABOLIC LENS: PATH LENGTH, FOCAL LENGTH, AND ABSORPTION APERTURE RADIUS

The thickness $t(r_a)$ of a single biconcave lens in an N-lens CRL at the absorption aperture radius r_a is given by

$$t(r_a) = d + \frac{2}{N\mu}. \tag{18}$$

With respect to the midplane of a biconcave lens, the left (−) and right (+) lens surfaces are defined by the parabolic lens half-thickness $\tau(r)$:

$$\tau(r) = \pm\frac{t(r)}{2}. \tag{19}$$

The slope of the line tangent to a lens surface at radial distance r from the lens center axis is the differential change of lens radius per lens thickness change $dr/d\tau$. The tangent line is perpendicular to the normal line at a point on the lens surface at radial distance r. As a result, the slope of the normal line at a point on the lens surface is $-d\tau/dr$.

Figures 2 and 3 demonstrate optical lenses and light imaging systems were well under way in the second half of the nineteenth century with corrections for monochromatic and chromatic aberrations, wide-field imaging, and so on. In fact, the treatise on electricity and magnetism published in 1873 by the first director of Cavendish Laboratory of Cambridge University, James Clerk Maxwell, is a remarkable two-volume book, which then (as now) is a very good introduction to electricity and magnetism (Maxwell, 1954).

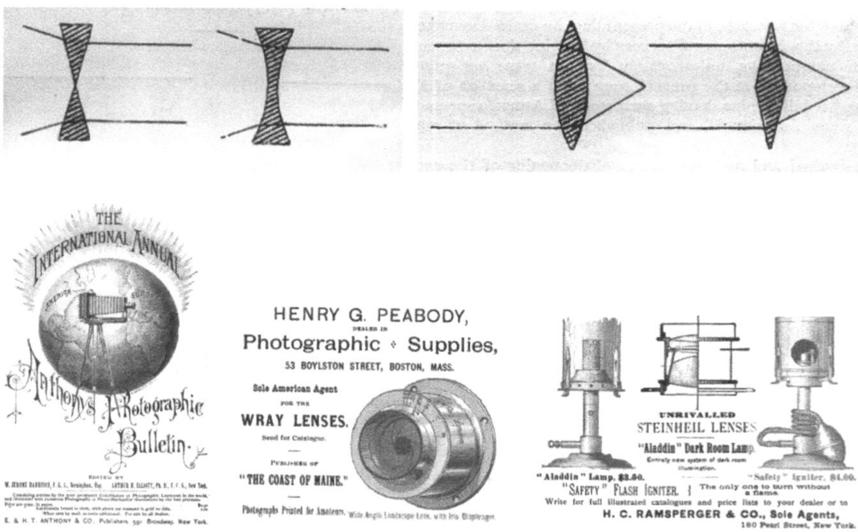

FIGURE 2 The refractive index for visible light is greater than 1, whereas the index of refraction for X-rays and neutrons is less than 1. These two figures from E. J. Wall's 1889 photography book illustrate the use of a bottom prism and top inverted prism to represent a biconcave lens, and a top prism and inverted bottom prism to represent a biconvex lens (Wall, 1889). The bottom figures indicate lenses, irises, and sources. Photography was the precursor, followed by imaging by X-rays, radio waves, electrons, radar, protons, ions, synthetic aperture radar, lidar, infrared, terahertz, gamma, neutron, and, as the King of Siam in the musical. "The King and I" states, "and etc., etc., etc.... you fill in the details." The brilliant idea may be difficult to come by, but producing the filling in the details—the physical realization of the idea, and then a useful application or even commercially viable product—takes tenacious, hard work lasting anywhere from years to decades or longer (e.g., commercial nuclear fusion energy source).

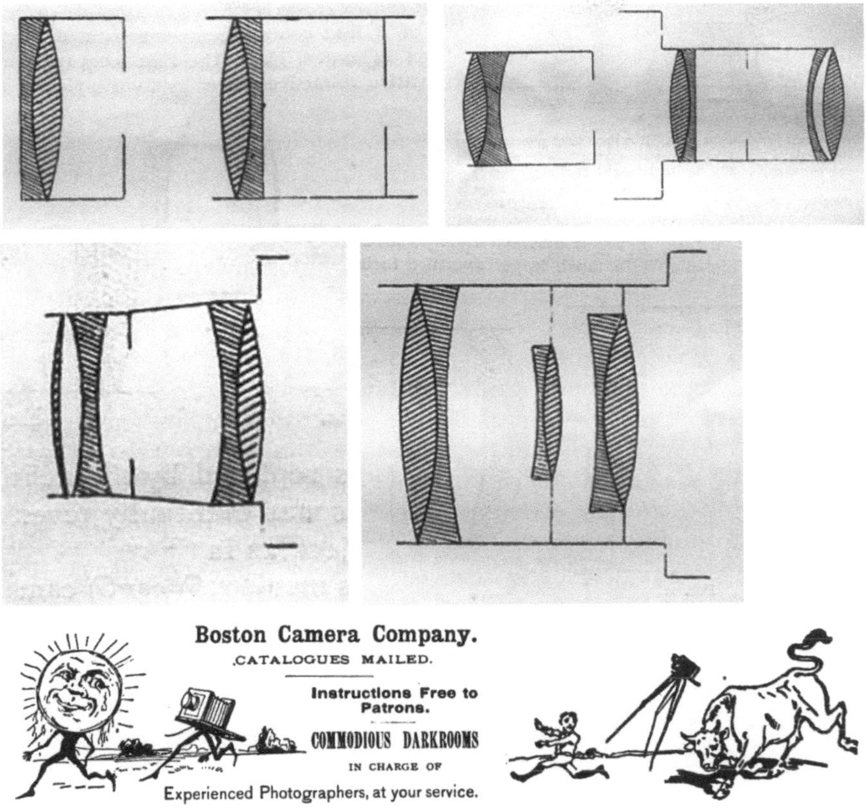

FIGURE 3 The above figures from E. J. Wall's 1889 photography book illustrate the use of combinations of lenses to produce an achromatic lens (Wall, 1889). The bottom two figures are advertisements at the end of Wall's book, which illustrate the search for brighter sources then, as now, goes on and then, as now, be aware of health and safety issues regarding imaging.

The remarkable physical intuition of Maxwell was able to treat charges and currents in a manner fitting reality and the mathematics despite the fact that the electron had not yet been discovered by J. J. Thomson. The third director of Cavendish Labs (Lord Rayleigh was the second director of Cavendish Labs, which was established through the £6,300 generosity of the Chancellor of Cambridge University, William Cavendish, the seventh Duke of Devonshire, in 1874). Thomson also handled the third revision of Maxwell's treatise. Ernest Rutherford succeeded Thomson in 1919 followed by Lawrence Bragg in 1938, then Nevuk Mott in 1954, and then Brian Pippard in 1971 (http://www.phy.cam.ac.uk/history/). The double-helix structure and recognized functioning ("It has not escaped our notice...") of DNA was determined by Watson and Crick (1953) at Cavendish Labs using X-ray crystallography, which ushered in the widespread of X-ray crystallography applied to the molecules and macromolecules of biology.

Figure 2 shows the representation of a biconvex lens as a prism and inverted prism joined at their bases and a biconcave lens as a prism and inverted prism joined at their apexes. Most materials have a refractive index less than 1 ($n < 1$) for both X-rays and neutrons and thus require a biconcave lens for focusing. Among notable exceptions of a refractive index larger than $1 (n > 1)$ for neutrons is titanium, which requires the use of titanium biconvex lenses to focus neutrons.

Figure 4 provides a diagram of a biconcave spherical or parabolic lens, which shows the refraction of an X-ray or neutron ray, incident parallel to the center optical or z-axis, as the ray passes through both surfaces of the lens. The ray incidence at radial distance r and the ray maintain an approximate constant radial distance r as it passes through the thin lens. However, the drawing greatly exaggerates the change in radial position r of the ray through the lens to allow depicting the tiny angle changes $\Delta\theta$

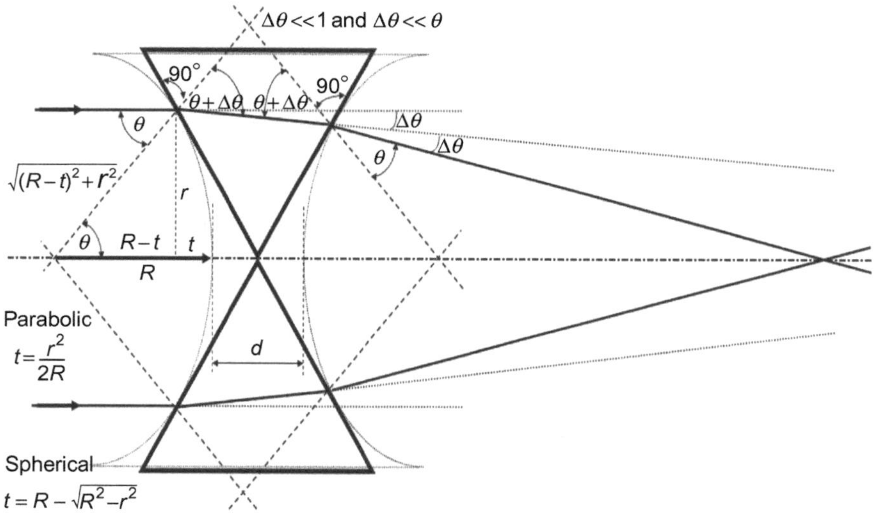

FIGURE 4 The diagram shows the refraction of an X-ray or neutron ray at the two surfaces of a symmetric biconcave spherical or parabolic lens, which is inscribed by a bottom isosceles prism and inverted top isosceles prism whose faces are tangent to the surface of the lens at the point of the ray incidence at radial distance r. The ray maintains an approximately constant radial distance r as it passes through the thin lens. The drawing greatly exaggerates the change in ray radial position r as it passes through the lens. This allows depicting the tiny angle changes $\Delta\theta$ of the ray at each interface.

of the ray at each lens surface. The tiny angle change is due to the very small decrement $\delta \ll 1$ of the lens refractive index $n = 1 - \delta$, which yields ry little refraction, unless N lenses are combined to form a CRL.

The X-ray or neutron ray is refracted at a lens surface by a very small angle $\Delta\theta$ toward the lens center axis, from incident angle θ_1 to refracted angle θ_2 with respect to the biconcave lens surface normal at radius r, where

$$\theta_2 = \theta_1 + \Delta\theta. \quad \Delta\theta \ll 1 \tag{20}$$

On the left side of the biconcave lens, Snell's law gives

$$\sin\theta_1 = (1 - \delta)\sin(\theta_1 + \Delta\theta). \tag{21}$$

With $\Delta\theta \ll 1$, the small angle approximation gives for the left lens surface is

$$\sin\theta_1 = (1 - \delta)(\sin\theta_1 + \Delta\theta\cos\theta_1) \cong \sin\theta_1 + \Delta\theta\cos\theta_1 - \delta\sin\theta_1. \tag{22}$$

Solving for $\Delta\theta$ on the left lens surface gives

$$\Delta\theta = \delta\tan\theta_1. \tag{23}$$

On the right side of the biconcave lens, Snell's law gives

$$(1 - \delta)\sin(\theta_1 + \Delta\theta) = \sin\theta_1. \tag{24}$$

With $\Delta\theta \ll 1$, the small angle approximation gives for the right lens surface is

$$(1 - \delta)(\sin\theta_1 + \Delta\theta\cos\theta_1) \cong \sin\theta_1 + \Delta\theta\cos\theta_1 - \delta\sin\theta_1 = \sin\theta_1. \tag{25}$$

Solving for $\Delta\theta$ on the left lens surface gives

$$\Delta\theta = \delta\tan\theta_1. \tag{26}$$

The left and right surfaces of the biconcave lens each cause ray angular deviation $\Delta\theta$ toward the center lens axis, where

$$\Delta\theta = \delta\tan\theta_1, \tag{27}$$

where

$$1 - \delta \cong 1. \tag{28}$$

Consider an incident parallel ray at radial distance r that is refracted and intersects the axis at focal distance f_1, where

$$f_1 = \frac{r}{2\Delta\theta}. \tag{29}$$

For an N-lens CRL the focal distance f_{crl} is then

$$f_{crl} = \frac{r}{2N\Delta\theta}. \tag{30}$$

The tangent of incident angle θ_1 at an interface is the negative differential rate of lens thickness change $d\tau$ with change in radial distance dr from the lens center axis:

$$\tan\theta_1 = -\frac{d\tau}{dr}. \tag{31}$$

Hence, the angular change $\Delta\theta$ of the X-ray or neutron wave vector $\mathbf{k}$ by refraction at each lens surface interface with the surrounding air or vacuum becomes

$$\Delta\theta = -\delta\frac{d\tau}{dr}. \tag{32}$$

Consider a single thin lens, which is placed between an object and image plane, which are respectively at object distance r_0 and image distance r_i from the center single principal plane. In particular, we examine a pair of object O and image I points, which are on the lens center axis at object distance r_0 and image distance r_i. Between the object point O and image point I an on-axis ray travels path length distance D_z along the center z-axis of the lens, where

$$D_z = r_0 + r_i. \tag{33}$$

However, for an off-axis ray path from object point O to image point I a path length D_{off}, which is the sum of (1) the divergent diagonal ray from the on-axis object point O to an off-axis point L lying on the lens center plane at radial distance r, and (2) the diagonal convergent ray from the off-axis lens point L to the on-axis image point I, is given by

$$D_{off} = \sqrt{r_0^2 + r^2} + \sqrt{r_i^2 + r^2}. \tag{34}$$

One assumes radial distance r of the off-axis lens point L is much less than the object r_0 or image r_i distances so that the off-axis ray path length D_{off}

can undergo the binomial approximation, where

$$D_{off} = r_0\sqrt{1 + \frac{r^2}{r_0^2}} + r_i\sqrt{1 + \frac{r^2}{r_i^2}} = r_0 + r_i + \frac{r^2}{2r_0} + \frac{r^2}{2r_i}. \tag{35}$$

The path length difference D is then

$$D = D_{off} - D_z = \frac{r^2}{2}\left(\frac{1}{r_0} + \frac{1}{r_i}\right). \tag{36}$$

The neutron phase velocity is slower for air/vacuum than a lens, whose thickness $t(r)$ varies to provide equal time of flight for all ray paths in which the path length difference D becomes

$$D = \frac{r^2}{2}\left(\frac{1}{r_0} + \frac{1}{r_i}\right) = \frac{r^2}{2f} = t\delta. \tag{37}$$

We now consider parabolic biconcave lenses, which are characterized by the on-axis radius of curvature R, where $R/2$ is the parabola vertex to directrix distance. The biconcave parabolic lens has total thickness t_p given by

$$t_p = d + \frac{r^2}{R}. \tag{38}$$

The thickness $t_c(r)$ of the curved portion of the parabolic lens is the center minimum thickness d subtracted from the total parabolic lens thickness t_p, where

$$t_c = t_p - d = \frac{r^2}{R}. \tag{39}$$

Substitution into Eq. 37 for $t(r)$, expressed by Eq. 39, in the equation for the path length difference D yields the familiar lensmaker's equation:

$$D = \frac{r^2}{2}\left(\frac{1}{r_0} + \frac{1}{r_i}\right) = \frac{r^2}{2f} = \frac{r^2}{R}\delta. \tag{40}$$

This yields the focal length f_p of a single parabolic biconcave lens for light, X-rays, or neutrons and the focal length f_{par} for an N-lens parabolic biconcave lens CRL, where

$$f_p = \frac{R}{2\delta} \tag{41}$$

and

$$f_{par} = \frac{R}{2N\delta}. \tag{42}$$

The thickness $t_p(r_a)$ of a single biconcave parabolic lens in an N-lens CRL at the absorption aperture radius r_a is given by

$$t_p(r_a) = d + \frac{2}{N\mu} = d + \frac{r_a^2}{R}. \tag{43}$$

The absorption aperture radius r_{ap} for the parabolic biconcave N-lens CRL is the radial distance from the center CRL axis in which the light, X-ray, or neutron intensity transmission decreases by e^{-2} compared with transmission along the center axis, where

$$r_{ap} = \sqrt{\frac{2R}{\mu N}} = 2\sqrt{f_{par}\frac{\delta}{\mu}}. \tag{44}$$

The angular change $\Delta\theta$ of the X-ray or neutron wave vector **k** by refraction at the input interface surface of the parabolic biconcave lens increases relative to the normal pointing out of the lens into the air of vacuum, where

$$\Delta\theta = -\delta\frac{d\tau}{dr} = \frac{\delta}{2}\frac{dt_p}{dr} = \frac{r\delta}{R}. \tag{45}$$

At the output interface surface of the parabolic biconcave lens, angular change $\Delta\theta$ of the wave vector **k** decreases relative to the normal pointing out of the lens into the air of vacuum is

$$\Delta\theta = -\delta\frac{d\tau}{dr} = -\frac{\delta}{2}\frac{dt}{dr} = -\frac{r\delta}{R} \tag{46}$$

At each surface the wave vector **k** is refracted toward the center axis of the lens, and substitution into Eq. 30 of the magnitude $\Delta\theta$ of Eq. 46 confirms the parabolic lens CRL focal length:

$$f_{crl} = \frac{r}{2N\Delta\theta} = \frac{R}{2N\delta}. \tag{47}$$

4. BICONCAVE SPHERICAL LENS: PATH LENGTH, FOCAL LENGTH, AND ABSORPTION APERTURE RADIUS

We now consider biconcave spherical lenses, which are characterized by the radius of curvature R. Figure 5 compares the parabolic biconcave lens

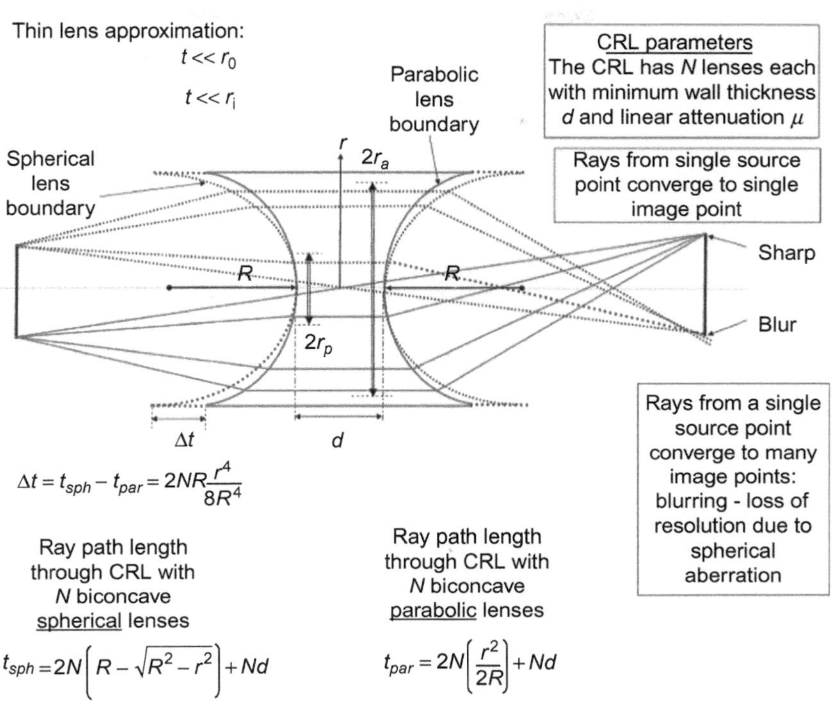

FIGURE 5 In the thin lens approximation, a ray passing through a spherical or parabolic biconcave lens is assumed to maintain a constant radial distance r from the lens center in which the ray changes direction upon exit from the lens. The center radial portion $r < r_p$ of the spherical biconcave lens closely approximates a parabolic biconcave lens in which r_p is the parabolic aperture radius. The rays outside the parabolic aperture radius $r > r_p$ are more strongly refracted by the larger spherical curvature than the rays inside the parabolic aperture radius $r \le r_p$ with the smaller parabolic curvature. Hence, the outer rays focus closer to the lens than the inner paraxial rays.

with a spherical biconcave lens. The biconcave spherical lens has total thickness $t_s(r)$ given by

$$t_s(r) = d + 2R - 2\sqrt{R^2 - r^2}. \tag{48}$$

The thickness $t_c(r)$ of the curved portion of the spherical lens is the center minimum thickness d subtracted from the total spherical lens thickness t_s, where

$$t_c = t_s - d = 2\left(R - \sqrt{R^2 - r^2}\right). \tag{49}$$

The angular change $\Delta\theta$ of the X-ray or neutron wave vector **k** by refraction at the input interface surface of the spherical biconcave lens increases relative to the normal pointing out of the lens into the air of vacuum, where

$$\Delta\theta = -\delta\frac{d\tau}{dr} = \frac{\delta}{2}\frac{dt_s}{dr} = \frac{r\delta}{\sqrt{R^2 - r^2}}. \tag{50}$$

At the output interface surface of the parabolic biconcave lens, angular change $\Delta\theta$ of the wave vector **k** decreases relative to the normal pointing out of the lens into the air of vacuum is

$$\Delta\theta = -\delta\frac{d\tau}{dr} = -\frac{\delta}{2}\frac{dt_s}{dr} = -\frac{r\delta}{\sqrt{R^2 - r^2}}. \tag{51}$$

At each surface the X-ray, neutron, or light wave vector **k** is refracted toward the center axis of the lens, and substitution into Eq. 47 with the magnitude $\Delta\theta$ of Eq. 51 yields the biconcave spherical lens CRL focal length f_{sp}, where

$$f_{sp} = \frac{r}{2N\Delta\theta} = \frac{\sqrt{R^2 - r^2}}{2N\delta}. \tag{52}$$

The absorption aperture radius r_{as} for the biconcave spherical N-lens CRL is obtained from the expression for the thickness $t_s(r_{as})$ of a single biconcave spherical lens in an N-lens CRL, where

$$t_s(r_{as}) = d + \frac{2}{N\mu} = d + 2R - 2\sqrt{R^2 - r_{as}^2}. \tag{53}$$

Solving for absorption aperture radius r_{as} for the biconcave spherical N-lens CRL gives

$$r_{as} = \sqrt{\frac{2R}{N\mu} - \left(\frac{1}{N\mu}\right)^2}. \tag{54}$$

In terms of the biconcave parabolic N-lens absorption aperture radius r_{ap}, the spherical lens CRL absorption aperture radius r_{as} is

$$r_{as} = r_{ap}\sqrt{1 - \left(\frac{r_a}{2R}\right)^2}. \tag{55}$$

5. PARABOLIC APERTURE RADIUS IN SPHERICAL BICONCAVE CRLS

The small radius inner portion of the spherical biconcave lens is approximately parabolic in shape, but beyond the inner portion of the spherical lens, the spherical lens shape is no longer parabolic. As a result, X-rays or neutrons incident in the middle to outer radius of the lens undergo spherical aberration. X-rays and neutrons incident upon the inner radial portion of the spherical biconcave lens do not undergo spherical aberration. The inner radius of the biconcave spherical lens, which is effectively a parabolic biconcave lens, is termed the *parabolic aperture radius* r_p (see Figure 5).

Spherical aberration occurs outside the center parabolic portion of a spherical lens and causes loss of image resolution, but not intensity. At radial distance r from the lens center axis, the additional lens thickness Δt encountered by neutrons propagating in the spherical lens, compared with a parabolic lens, is

$$\Delta t = t_s - t_p = \left(d + 2R - 2\sqrt{R^2 - r^2}\right) - \left(d + \frac{r^2}{R}\right). \tag{56}$$

Since the radius of curvature of a single biconcave spherical lens is much greater than the effective aperture radius r_e of the lens, where $R \gg r_e$ with $r \leq r_e$, and thus $R \gg r$, then the binomial expansion of the square root as a sum of the linear and quadratic terms in $(r/R)^2$ gives

$$\Delta t = \frac{r^4}{4R^3}. \tag{57}$$

The additional thickness Δt causes phase shift $\Delta\phi$, where

$$\Delta\phi = 2\pi \cdot \frac{\Delta t}{\lambda} \cdot \delta \leq \pi \quad \text{amplitude summation.} \tag{58}$$

Summation of ray amplitudes occurs at a single focal point, if the phase shift angles $\Delta\phi$ between the ray paths from an on-axis object point O to an on-axis conjugate image point I are less than π, where substitution into Eq. 58 of Δt from Eq. 57 gives

$$\Delta\phi = 2\pi \cdot \frac{r^4}{4R^3\lambda} \cdot \delta \leq \pi \quad \text{amplitude summation.} \tag{59}$$

Solving for the radial distance r from the lens center axis gives the parabolic lens radius or maximum radius r_p, which contains ray paths from object point O to image point I, where

$$r_p = \left(\frac{2R^3\lambda}{\delta}\right)^{1/4}. \tag{60}$$

For a biconcave spherical N-lens CRL, the parabolic aperture radius r_p with path length difference $N\Delta t$ is then

$$r_p = \left(\frac{2R^3\lambda}{N\delta}\right)^{1/4}. \tag{61}$$

Ray paths from on-axis object point O, which fall outside the parabolic aperture radius $r > r_p$, are focused in a locus of on-axis image points upstream and downstream of on-axis image point I. The outer rays beyond the spherical lens CRL parabolic aperture radius $r > r_p$ have focal length f_{sp} that decreases with radial distance r from the lens center axis, where one recalls

$$f_{sp} = \frac{\sqrt{R^2 - r^2}}{2N\delta}. \tag{62}$$

For design purposes, the actual or physical aperture radius r_m of the spherical biconcave lens does not need to be much larger than the absorption aperture radius r_a; usually the absorption aperture radius is significantly larger than the parabolic aperture radius r_p, where $r_m > r_a > r_p$. An X-ray or neutron CRL composed of biconcave parabolic lenses eliminates the spherical aberration, so the physical radius of the lens is equal to the parabolic radius, which is usually designed to be only slightly larger than the absorption aperture radius. Hence, for a parabolic biconcave lens, the physical aperture is equal to the parabolic aperture radius, which is greater than or equal to the absorption radius, where $r_m = r_p > r_a$. The effective aperture radius r_e of an X-ray or neutron CRL without spherical aberration is the minimum of the physical aperture radius r_m, absorption aperture radius r_a, and parabolic aperture radius r_p, where

$$r_e = Min\left(r_m, r_a, r_p\right). \tag{63}$$

6. MATRIX REPRESENTATION OF X-RAY AND NEUTRON OPTICS WITH PARAXIAL APPROXIMATION

The trajectory of X-ray or neutron rays through a lens, drift space, or a mirror is readily described by matrix representation. Azimuthal symmetry is assumed in the optical system about the center z-axis for a simple matrix representation, and variation is only along the axial z-direction or transverse radial direction. The trajectory of an X-ray or neutron ray is described by its transverse or radial distance r and angle θ_z, which are both measured from the positive z-axis. Above the z-axis the radial distance r is positive, where $r > 0$. Below the z-axis the radial distance r is negative, where $r < 0$. The positive z-axis originates at the center of the source plane, passes through the center of the object plane and along the center lens (CRL) axis, and through the center point of the image plane.

To enable a simplified matrix representation of ray trajectories we consider only planar, spherical, or parabolic surfaces of refraction or reflection and uniform media of refractive, reflective, or drift space materials. Snell's law is applied to refractive surfaces and the law of reflection to reflective surfaces. Both laws are described by angles measured relative to the surface normals at medium interfaces. The normal at a point on an interface, which occurs between the entrance and exit media (labeled, respectively medium 1 and medium 2), points toward entrance medium 1 of the incident ray.

Consider first the drift space matrix. The drift space is assumed empty or filled with medium with uniform refractive index n. Matrix representation has the incident X-ray or neutron ray enter at point L on the entrance (left side) of the drift space of axial thickness Z. Point L is at radial distance radius r_1 and at angle θ_{z1} relative to the center z-axis. The ray exits the drift space at point R at radius r_2 and at angle θ_{z2}. In the drift space, only the radial distance of the ray changes and the angle relative to the z-axis remains unchanged, where

$$r_1 \neq r_2 \quad \text{and} \quad \theta_{z1} = \theta_{z2}. \tag{64}$$

The drift space matrix M_d for drift space of thickness Z is then

$$\begin{bmatrix} r_2 \\ \theta_{z2} \end{bmatrix} = M_d \begin{bmatrix} r_1 \\ \theta_{z1} \end{bmatrix}, \tag{65}$$

where

$$M_d = \begin{bmatrix} 1 & Z \\ 0 & 1 \end{bmatrix}. \tag{66}$$

The angles relative to the surface normals need to be expressed in terms of angles measured relative to the center z-axis to facilitate matrix representation. Consider the incidence of a ray on a refractive surface at radius r_1 and angle θ_{z1}. Refraction deflects the incident ray to refracted ray at radius r_2 and angle θ_{z2} at the surface interface between two media of different refractive index. At the refractive surface only the ray angle changes; the ray radius is unchanged, where

$$r_1 = r_2 \quad \theta_{z1} \neq \theta_{z2}. \tag{67}$$

The ray from the entrance (left) medium 1 of refractive index n_1 is refracted at the interface surface share with exit (right) medium 2 with refractive index n_2. The matrix M_s for refraction at a spherical interface between two media with incident refractive index n_1 and exit refractive index n_2 is

$$\begin{bmatrix} r_2 \\ \theta_{z2} \end{bmatrix} = M_s \begin{bmatrix} r_1 \\ \theta_{z1} \end{bmatrix}. \tag{68}$$

The matrix M_{thk} for refraction by a thick, biconcave spherical lens requires the matrices for refraction at the left M_{ls} and right M_{rs} biconcave spherical surfaces and a drift space matrix M_d, where

$$\begin{bmatrix} r_2 \\ \theta_{z2} \end{bmatrix} = M_{thk} \begin{bmatrix} r_1 \\ \theta_{z1} \end{bmatrix} \tag{69}$$

and

$$M_{thk} = M_{rs} M_d M_{ls}. \tag{70}$$

If the lens is a thin lens, then the drift space distance Z is zero and the drift space matrix becomes the unity matrix, where

$$M_d \Rightarrow \begin{bmatrix} 1 & 0 \\ 0 & 1 \end{bmatrix} \quad Z = 0. \tag{71}$$

The matrix M_{thin} for refraction by a thin, biconcave spherical lens then requires only the two matrices for refraction at the left M_{ls} and right M_{rs} biconcave spherical surfaces, where

$$\begin{bmatrix} r_2 \\ \theta_{z2} \end{bmatrix} = M_{thin} \begin{bmatrix} r_1 \\ \theta_{z1} \end{bmatrix} \tag{72}$$

and

$$M_{thin} = M_{rs} M_{ls}. \tag{73}$$

The ray paths through an optical system can be determined by the matrix methods in which each optical surface, optical element, or drift space is characterized by a matrix. Multiplication of the 2×2 matrices of the sequence of optical elements, surfaces, and drift spaces can thus yields a 2×2 matrix characterizing the ray path from a point $P(r_s, \theta_{zs})$ in the source plane to image point I (r_i, θ_{zi}) in the image plane. The ray trajectory through optical system comprised of Q optics is then calculated via

$$\begin{bmatrix} r_i \\ \theta_{zi} \end{bmatrix} = M_1 M_2 M_3 \ldots M_{Q-1} M_Q \begin{bmatrix} r_s \\ \theta_{zs} \end{bmatrix}. \tag{74}$$

For example, the ray trajectory through CRL composed of N identical thin lenses would be given by

$$\begin{bmatrix} r_i \\ \theta_{zi} \end{bmatrix} = M_{thin}^{N} \begin{bmatrix} r_s \\ \theta_{zs} \end{bmatrix}. \tag{75}$$

7. THIN LENS MATRIX AND RAY ANGLE DEVIATION IN A THIN LENS

The matrices for the left M_{ls} and right M_{rs} spherical surfaces of a thin, biconcave spherical lens and their multiplication to produce the thin lens matrix M_{thin} are derived as follows. Consider an X-ray or neutron ray emitted from source point S that is incident on the entrance (left) surface of a spherical biconcave lens at point L, situated at radial distance r from the center z-axis. The entrance angle θ_{1l} of the ray is measured relative to the entrance (left) surface normal in which the normal at point L is directed outward toward the source (object) plane. The angle θ_{nzl} of the outward-directed normal at point L on the left lens surface is measured relative to the positive z-direction of the lens center axis. For the assumed thin lens approximation, the ray trajectory from the left lens entrance surface at point L to right lens exit surface at point R is assumed to maintain a constant radius r, measured relative to the lens center axis.

The point L on the left surface of the lens lies on a portion of a spherical surface of radius R, which is the distance of the point L from the center point of the sphere that lies on the z-axis. The transverse of radial distance of the point L from the z-axis is r so that the angle θ_{nzl} of the outward-directed unit, normal vector at point L is then

$$\tan\theta_{nzl} = \frac{r}{\sqrt{R^2 - r^2}}. \tag{76}$$

Since the radius or magnitude of the radius of curvature R of the left biconcave, spherical surface is much larger than the off-axis transverse distance r to point L, then

$$\tan\theta_{nzl} = \frac{r}{R} \quad |r| \ll R. \tag{77}$$

Hence,

$$\theta_{nzl} \cong \frac{r}{R} \quad |r| \ll R. \tag{78}$$

The ray exit angle θ_{1r} is measured relative to the normal on the lens exit (right) surface at point R in which the normal is directed inward toward the lens interior. The angle θ_{nzr} of the inward-directed normal at point R on the right lens surface is measured relative to the positive z-direction of the lens center axis. The point R on the right surface of the lens lies on a portion of a spherical surface of radius R, which is the distance of the point R from the center point of the sphere that lies on the z-axis. The transverse or radial distance of the point R from the z-axis is r so that the angle θ_{nzr} of the inward-directed unit, normal vector at point R is then

$$\tan\theta_{nzr} = \frac{r}{\sqrt{R^2 - r^2}}. \tag{79}$$

Since the radius or magnitude of the radius of curvature R of the left biconcave spherical surface is much larger than the off-axis transverse distance r to point L, then

$$\tan\theta_{nzr} = \frac{r}{R} \quad |r| \ll R. \tag{80}$$

Hence,

$$\theta_{nzr} \cong \frac{r}{R} \quad |r| \ll R. \tag{81}$$

The incident X-ray or neutron ray at point L on the left entrance surface of the lens is also specified by entrance angle θ_{1lz} that is measured relative to the positive z-directed, center axis (optical axis) of the lens. At the left lens surface, ray incidence angle θ_{1lz} (relative to positive z-axis) is equal to the incidence angle θ_{1l} (relative to the normal) subtracted by the normal angle ϕ_{nzl} (relative to positive z-axis), where

$$\theta_{1lz} = \theta_{1l} - \phi_{nzl} = \theta_{1l} - \frac{r}{R} \quad \text{left}. \tag{82}$$

At the left lens surface, the ray incidence angle θ_{1l} (relative to the normal) is the sum of the ray incidence θ_{1lz}(relative to positive z-axis) and the normal angle ϕ_{nzl} (relative to the positive z-axis). Transposing the previous equation gives

$$\theta_{1l} = \theta_{1lz} + \frac{r}{R} \quad \text{left}. \tag{83}$$

The incident X-ray or neutron ray at point R on the right exit surface of the lens is also specified by an incidence angle θ_{1rz} that is measured relative to the positive z-directed, center axis (optical axis) of the lens. At the right lens surface, the ray incidence angle θ_{1rz} (relative to positive z-axis) is equal to the incidence angle θ_{1r}(relative to the normal) subtracted by the normal angle θ_{nzr} (relative to positive z-axis), where

$$\theta_{1rz} = \theta_{1r} - \phi_{nzr} = \theta_{1r} - \frac{r}{R} \quad \text{right.} \tag{84}$$

Transposing, at the right lens surface, the ray incidence angle θ_{1r} (relative to the normal) is the sum of the ray incidence θ_{1rz} (relative to positive z-axis) and the normal angle ϕ_{nzr} (relative to positive z-axis), where

$$\theta_{1r} = \theta_{1rz} + \frac{r}{R} \quad \text{right.} \tag{85}$$

One now applies a paraxial or small-angle form of Snell's law to the left and right surfaces of the lens. Snell's law is

$$n_1 \sin\theta_1 = n_2 \sin\theta_2. \tag{86}$$

The small-angle or paraxial form of Snell's law is

$$n_1\theta_1 = n_2\theta_2. \tag{87}$$

For higher accuracy the first two terms (linear and cubic) of the Taylor series expansion of the sine function can be used, where

$$\sin\theta = \theta - \frac{\theta^3}{3!}. \tag{88}$$

Snell's small-angle expression with the two-term sine expansion becomes

$$n_1\left(\theta_1 - \frac{\theta_1^3}{6}\right) = n_2\left(\theta_2 - \frac{\theta_2^3}{6}\right). \tag{89}$$

Transposing yields a cubic equation, whose solution gives refracted angle θ_2 in terms of refracted angle θ_1, where

$$\theta_2 - \frac{\theta_2^3}{6} - \frac{n_1}{n_2}\left(\theta_1 - \frac{\theta_1^3}{6}\right) = 0 \tag{90}$$

For the very small angles of X-ray and neutron CRLs commonly encountered the cubic term is ignored and the linear form of Snell's law is used in the paraxial approximation.

We now apply the paraxial from of Snell's law to the left (entrance) and right (exit) surfaces of the biconcave spherical lens. On the left lens surface, application of the paraxial (linear) form of Snell's law relates incident angle θ_{1l} of the ray in air medium 1 to refracted angle θ_{2l} in lens medium 2, where application of the binomial approximation gives

$$\theta_{2l} = \frac{1}{1-\delta}\theta_{1l} = (1+\delta)\,\theta_{1l} \quad \delta \ll 1 \quad \text{left.} \tag{91}$$

On the right lens surface, the incident ray in the lens medium 2 with refractive index n_2 is refracted into air medium 3, which is identical to medium 1 with refractive index n_1, so that

$$\theta_{2r} = (1-\delta)\,\theta_{1r} \quad \text{right.} \tag{92}$$

The refracted ray angle θ_{2lz} relative to the z-axis is equal to the refracted ray angle θ_{2l} at the left lens surface relative to the surface normal subtracted by the angle of the normal relative to the z-axis, where

$$\theta_{2lz} = \theta_{2l} - \phi_{nzl} = \theta_{2l} - \frac{r}{R} \quad \text{left.} \tag{93}$$

At the right lens surface,

$$\theta_{2rz} = \theta_{2r} - \phi_{nzr} = \theta_{2r} - \frac{r}{R} \quad \text{right.} \tag{94}$$

At the left lens surface, substitution into Eq. 93 for θ_{2l} from the Snell's law paraxial expression of Eq. 91 gives

$$\theta_{2lz} = (1+\delta)\,\theta_{1l} - \frac{r}{R} \quad \text{left.} \tag{95}$$

At the right lens surface, substitution into Eq. 94 for θ_{2r} from the Snell's law paraxial expression of Eq. 92 gives

$$\theta_{2rz} = (1-\delta)\,\theta_{1r} - \frac{r}{R} \quad \text{right.} \tag{96}$$

Finally, at the left lens surface, substitution for θ_{1l} of Eq. 83 into Eq. 95 gives

$$\theta_{2lz} = (1+\delta)\left(\theta_{1lz} + \frac{r}{R}\right) - \frac{r}{R} \quad \text{left.} \tag{97}$$

And, at the right lens surface, substitution for θ_{1r} of Eq. 85 into Eq. 96 gives

$$\theta_{2rz} = (1-\delta)\left(\theta_{1rz} + \frac{r}{R}\right) - \frac{r}{R} \quad \text{right.} \tag{98}$$

The matrices for the left M_{ls} and right M_{rs} spherical surfaces of a thin biconcave, spherical lens are

$$\begin{bmatrix} r_2 \\ \theta_{z2} \end{bmatrix} = M_{ls} \begin{bmatrix} r_1 \\ \theta_{z1} \end{bmatrix} \quad \text{and} \quad M_{ls} = \begin{bmatrix} 1 & 0 \\ \delta/R & 1+\delta \end{bmatrix} \cong \begin{bmatrix} 1 & 0 \\ \delta/R & 1 \end{bmatrix} \quad \text{left} \tag{99}$$

and

$$\begin{bmatrix} r_2 \\ \theta_{z2} \end{bmatrix} = M_{rs} \begin{bmatrix} r_1 \\ \theta_{z1} \end{bmatrix} \quad \text{and} \quad M_{rs} = \begin{bmatrix} 1 & 0 \\ -\delta/R & 1-\delta \end{bmatrix} \cong \begin{bmatrix} 1 & 0 \\ -\delta/R & 1 \end{bmatrix} \quad \text{right.} \tag{100}$$

The multiplication of the matrices for the left and right spherical surfaces yields the matrix M_{thin} for thin, biconcave, spherical lens, where

$$\begin{bmatrix} r_2 \\ \theta_{z2} \end{bmatrix} = M_{thin} \begin{bmatrix} r_1 \\ \theta_{z1} \end{bmatrix} \quad \text{lens.} \tag{101}$$

Let us assume the left and right surfaces have different radius of curvatures in which the left lens surface has radius of curvature R_l and the right lens surface has radius of curvature R_r so that

$$M_{thin} = M_{ls}M_{rs} = \begin{bmatrix} 1 & 0 \\ -\delta/R_l & 1 \end{bmatrix} \begin{bmatrix} 1 & 0 \\ \delta/R_r & 1 \end{bmatrix} \quad \text{lens.} \tag{102}$$

Carrying out the multiplication yields the matrix M_{thin} for a thin, biconcave spherical lens as well as a thin, biconcave parabolic lens, where

$$M_{thin} = \begin{bmatrix} 1 & 0 \\ \delta\left(\frac{1}{R_r} - \frac{1}{R_l}\right) & 1 \end{bmatrix} \quad \text{lens.} \tag{103}$$

If the center point CP of the concave lens spherical or parabolic surface is on the left (or source) side of the lens surface, the radius of curvature is positive, where

$$R_l > 0. \tag{104}$$

If the center point CP of the concave lens spherical or parabolic surface is on the right (or image) side of the lens surface, the radius of curvature is negative, where

$$R_r < 0. \tag{105}$$

Assume the magnitude of the radius of curvature R is equal on both sides of the lens:

$$R_l = -R_r = R. \tag{106}$$

The matrix M_{thin} for thin, biconcave, spherical lens as well as a thin, biconcave parabolic lens becomes

$$M_{thin} = \begin{bmatrix} 1 & 0 \\ -2\frac{\delta}{R} & 1 \end{bmatrix}. \tag{107}$$

The focal length f of the single lens is

$$f = \frac{R}{2\delta}. \tag{108}$$

The matrix M_{thin} for a thin, biconcave, spherical or parabolic lens is then

$$M_{thin} = \begin{bmatrix} 1 & 0 \\ -1/f & 1 \end{bmatrix}. \tag{109}$$

8. AVERAGE CRL TRANSMISSION FOR X-RAYS OR NEUTRONS

The biconcave nature of the individual spherical or parabolic lenses of a CRL for X-rays or neutrons causes the absorption to increase with radial distance r from the center lens axis z-axis at

$$r = 0 \quad \text{or} \quad (l_x, l_y) = (0, 0). \tag{110}$$

For lens transverse coordinates (l_x, l_y), the radial distance r from the lens center axis is

$$r = \sqrt{l_x^2 + l_y^2} \tag{111}$$

The transmission $T_p(l_x, l_y)$ varies with distance from the center axis with in x and y for a parabolic lens CRL as

$$T_p(l_x, l_y) = \exp[-\mu N d_p]. \tag{112}$$

Here $d_p(l_x, l_y)$ is the path length through a single parabolic lens, with center thickness d and on-axis radius of curvature R, at distance r from the center axis, where

$$d_p = d + \frac{l_x^2 + l_y^2}{2R}. \tag{113}$$

By comparison, the path length $d_s(l_x, l_y)$ at radius r for a single spherical lens with center thickness d and on-axis radius of curvature R is

$$d_s = d + \left[R - \sqrt{R^2 - \left(l_x^2 + l_y^2\right)}\right], \tag{114}$$

and the transmission $T_s(l_x, l_y)$ through the biconcave spherical lens CRL is

$$T_s(l_x, l_y) = \exp[-\mu N d_s]. \tag{115}$$

The binomial series approximation is applied to the spherical lens CRL transmission $T_s(r)$ and yields the parabolic lens CRL transmission, $T_p(r)$, where

$$R \gg \sqrt{l_x^2 + l_y^2}. \tag{116}$$

For both spherical and parabolic N-lens CRLs with center thickness (minimum) d, the on-axis ($r = 0$) transmission is the maximum, and

$$T_p(0) = T_s(0) = \exp(-N\mu d). \tag{117}$$

For a parabolic lens CRL the transmission at the lens absorption aperture radius, where amplitude $T_p(r_a)$ falls off by $1/e$ of the center axis amplitude, is

$$T_p(0) = \exp(-N\mu d). \tag{118}$$

The intensity (squared amplitude) falls of as $1/e^2$ of that along the center axis, where

$$r = r_a = \frac{2R}{\mu N} \tag{119}$$

The intensity (squared amplitude) transmission through the parabolic biconcave lens CRL at radius $r_a\sqrt{2}$ is

$$\left[T_p\left(r_a\sqrt{2}\right)\right]^2 = 0.018. \tag{120}$$

The amplitude transmission through the CRL for radial distances r, which exceeds$\sqrt{2}r_a$, can be approximated as zero, where

$$T_p(r) = 0 \quad r > r_a\sqrt{2}. \tag{121}$$

For example, at lens radius $r = 2r_a$ the intensity transmission drops to

$$\left|T_p(2r_a)\right|^2 = 3.3 \times 10^{-4}. \tag{122}$$

Integration over transverse x- and y-direction coordinates l_x and l_y from $-2r_a$ to $2r_a$ gives over 99.97% of the N-lens CRL contribution to the wave amplitude at the image plane. Integration of l_x and l_y from $-\infty$ to $-2r_a$ and $2r_a$ to $+\infty$ yields the remaining 0.03% of the amplitude contribution. Hence, the integral of l_x and l_y from $-\infty$ to $+\infty$ is a close approximation to the integral of l_x and l_y from $-2r_a$ to $2r_a$.
Integration of the transmission over the radius r of the parabolic lens CRL yields its average transmission $< T_p >$ of the ray amplitude:

$$< T_p > = \frac{1}{R}\int_0^R T(r)dr = \frac{1}{R}\int_0^R \exp\left[-\mu N\left(d - \frac{r^2}{2R}\right)\right]dr. \tag{123}$$

As discussed above, most of the integrand contribution to the integral occurs for values of r from 0 to $\sqrt{2}r_a$. The absorption aperture radius r_a is normally much smaller than the parabolic lens aperture radius r_p, which closely approximates the parabolic lens on-axis radius of curvature R, where $r_a \ll R$. The upper integration limit R can be replaced by ∞, and the effective parabolic lens aperture radius term $1/R$ can be replaced by $1/\sqrt{2}r_a$, so that

$$< T_p > = \frac{1}{r_a\sqrt{2}}\int_0^\infty \exp\left[-\mu N\left(d - \frac{r^2}{2R}\right)\right]dr = \frac{\sqrt{\pi}}{4}\exp(-N\mu d). \tag{124}$$

That is, the average transmission $< T_p >$ through the parabolic lens CRL is approximately 44.3% of its on-axis transmission,

$$< T_p > = 0.443 \cdot T_p(0) = 0.443\exp\left(-N\mu d\right). \tag{125}$$

At the same radial distance r, path length $d_s(r)$ through a spherical lens CRL radius is found via binomial series expansion to be larger than the parabolic lens CRL path length $t_p(r)$,

$$t_s(r) > t_p(r). \tag{126}$$

Thus, the average transmission $< T_s >$ of the spherical lens CRL will be less than that of the parabolic lens CRL $< T_p >$; that is,

$$< T_s > << T_p >. \tag{127}$$

The average transmission through a spherical lens CRL $< T_s >$ can be calculated numerically. However, a rough estimate of the average transmission $< T_s >$ through a parabolic lens or spherical lens CRL can be obtained by using

$$< T_s > = 0.443 \cdot \exp(-N\mu d). \tag{128}$$

9. CRL INTENSITY GAIN INCLUDING ATTENUATION

One application of CRLs is to increase the image intensity of a distant X-ray or neutron source. The CRL intensity gain G is defined as the ratio of the on-axis intensity at the CRL image plane divided by the on-axis intensity at the same plane without the CRL. The CRL two-dimensional (2D) intensity gain G is derived as follows. Consider an elliptical source with horizontal half-width w_s and vertical half-height h_s and source area

$$A_s = \pi w_s h_s. \tag{129}$$

That is, the source or object half-height measured from the center optical axis is h_s, and the source or object half-width measured from the optical z-axis is w_s; see the top diagram of Figure 6.

The source emits X-rays or neutrons through an aperture of radius r_e in the lens plane, which is located at object distance r_0 from the source. The aperture without a lens produces an image with corresponding horizontal half-width w_{i0} and vertical half-height h_{i0}, which are measured from the center optical z-axis and situated at image distance r_i. The aperture is in a wall that is opaque to X-rays and neutrons, and a plane wave of X-rays or neutrons from the source is assumed incident on the aperture. The plane wave assumption requires the object distance r_0 to be much larger than the effective aperture radius r_e, where

$$r_0 \gg r_e. \tag{130}$$

Without a CRL in the aperture, the elliptical image area A_{i0} is

$$A_{i0} = \pi w_{i0} h_{i0}. \tag{131}$$

The image half-width w_{i0} and half-height h_{i0} are found by similar triangles with common width angle ϕ and height angle θ (see the bottom

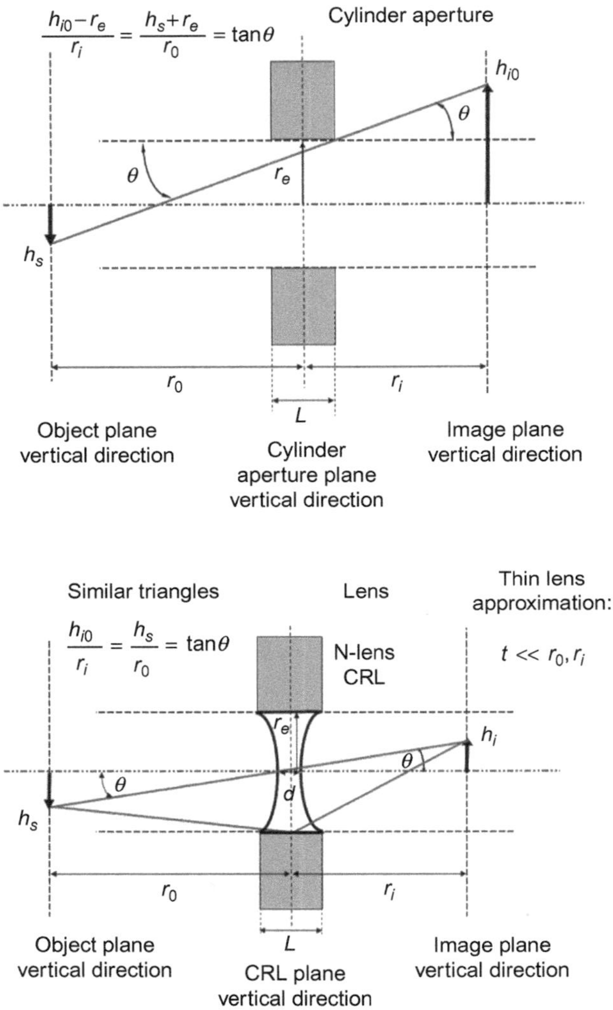

FIGURE 6 (Top) Calculation of intensity gain of cylinder aperture of length L and effective aperture radius r_e for X-rays or neutrons. (Bottom) Calculation of intensity gain of CRL with length L and effective aperture radius r_e equal to the cylinder length L and effective aperture radius r_e.

diagram in Figure 6). With only an aperture of radius r_e for imaging, the pair of similar triangles in the xz-plane with the object half-width w_s and image half-width w_{i0}, measured from the center axis, gives

$$\frac{w_{i0} - r_e}{r_i} = \frac{w_s + r_e}{r_0} = \tan\phi. \tag{132}$$

And with only the aperture the pair of similar triangles in the yz-plane with the object half-height h_s and image half-height h_{i0} is

$$\frac{h_{i0} - r_e}{r_i} = \frac{h_s + r_e}{r_0} = \tan\theta. \tag{133}$$

Solving for image width w_{i0} and height h_{i0}, one obtains

$$w_{i0} = \frac{r_i}{r_0}(r_e + w_s) + r_e \tag{134}$$

and

$$h_{i0} = \frac{r_i}{r_0}\left(r_e + h_s\right) + r_e. \tag{135}$$

Substitution for w_{i0} and h_{i0} of Eqs. 134 and 135 in the elliptical image area expression A_{i0} of Eq. 131 for a cylinder of the same length L and aperture radius r_e as the CRL length and effective aperture radius r_e is then

$$A_{i0} = \pi w_{i0} h_{i0} = \pi\left[\frac{r_i}{r_0}(r_e + w_s) + r_e\right]\left[\frac{r_i}{r_0}\left(r_e + h_s\right) + r_e\right]. \tag{136}$$

The cylinder of length L and effective aperture radius r_e is now replaced with a CRL with equal length L and equal effective aperture radius r_e. The CRL consists of N biconcave spherical or parabolic lenses of center thickness d (cm) and the lens material has linear attenuation coefficient μ (cm^{-1}). The CRL physical aperture radius r_m is assumed to be less than the absorption aperture radius r_a of the N-lens biconcave CRL. Recall, the transmitted intensity at the CRL absorption aperture radius r_a is e^{-2} of the transmitted intensity along the CRL center axis, where

$$T_0 = \exp\left(-N\mu d\right). \tag{137}$$

The gain of a 2D lens must also include the absorptive losses of the X-rays passing through the lens. The approximate average transmission $< T >$ through a parabolic lens or spherical lens CRL is 44% of its center transmission, or

$$< T > = 0.44 T_0 = 0.44 \exp\left(-N\mu d\right). \tag{138}$$

Hence, in this case the effective aperture radius r_e of the CRL is approximately equal to the physical aperture radius r_m, where $r_e = r_m$. The hollow cylinder aperture is replaced with a CRL with an equal length L and equal effective aperture radius of radius r_e (see the top of Figure 6). The object half-width w_s and image half-width w_{i0} are measured from the center axis, and similar triangles give

$$\frac{w_{i0}}{r_i} = \frac{w_s}{r_0} = \tan\phi. \tag{139}$$

With the CRL in place the pair of similar triangles in the yz-plane formed from the object half-height h_s and image half-height h_{i0} gives

$$\frac{h_{i0}}{r_i} = \frac{h_s}{r_0} = \tan\theta. \tag{140}$$

The elliptical source spot is imaged by the CRL to an elliptical image spot with transverse magnification M given by

$$M = \frac{r_i}{r_0} \tag{141}$$

The elliptical image spot area A_i with the CRL in place is

$$A_i = \pi w_i h_i \tag{142}$$

The image half-width w_i and half-height h_i is that of the corresponding source of half-width w_s and half-height h_s multiplied by the transverse magnification M of the CRL, where

$$w_i = Mw_s \quad \text{and} \quad h_i = Mh_s. \tag{143}$$

Substitution for w_i and h_i of Eq. 143 in the elliptical image area expression A_i of Eq. 142 for the CRL of length L and effective aperture radius r_e is then

$$A_i = \pi w_i h_i = \pi M^2 w_s h_s. \tag{144}$$

The 2D CRL intensity gain G is then the ratio of the area A_{i0} of the image without a CRL, divided by the area A_i of the image produced by the CRL, where

$$G = \frac{A_{i0}}{A_i} = \frac{0.44\exp(-N\mu d)}{M^2 w_s h_s}\left[\frac{r_i}{r_0}(r_e + w_s) + r_e\right]\cdot\left[\frac{r_i}{r_0}(r_e + h_s) + r_e\right]. \tag{145}$$

With $M = r_i/r_0$ the 2D CRL gain G becomes

$$G = \frac{0.44 \exp(-N\mu d)}{M^2 w_s h_s} \left[M \left(r_e + w_s\right) + r_e\right] \cdot \left[M \left(r_e + h_s\right) + r_e\right]. \tag{146}$$

Assume the source half dimensions (w_s, h_s) to be much smaller than the CRL effective aperture radius r_e, where

$$r_e \gg w_s, h_s. \tag{147}$$

Then, at image distance r_i from the illuminated circular CRL aperture, a circular image spot of the elliptical source spot is produced, where the image half-width w_i and image half-height h_i of the CRL are approximated by

$$w_i = h_i \cong r_e \left(1 + \frac{r_i}{r_0}\right). \tag{148}$$

Hence, if the CRL effective aperture radius aperture r_e is much larger than the source dimensions (w_s, h_s), then the CRL intensity gain becomes

$$G = 0.44 \frac{r_e^2 \exp(-N\mu d)}{w_s h_s} \left(\frac{M+1}{M}\right)^2 \quad r_e \gg w_s, h_s. \tag{149}$$

Multiplication of the numerator and denominator of gain G by π yields the circular aperture area of the CRL in the numerator A_{crl} and the elliptical source area A_s in the denominator, where

$$A_{crl} = \pi \cdot r_e^2 \quad \text{and} \quad A_s = w_s h_s. \tag{150}$$

For a CRL effective aperture area A_{crl} much larger than the source area A_s, the CRL 2D intensity gain G becomes

$$G = 0.44 \frac{A_{crl}}{A_s} \left(1 + \frac{1}{M}\right)^2 \exp(-N\mu d) \quad A_{crl} \gg A_s. \tag{151}$$

An example of a source whose area is usually much smaller than the CRL effective aperture area is a synchrotron radiation X-ray source. A synchrotron X-ray source usually provides a very long object distance r_0 in relation to the image distance r_i, so that the CRL image distance r_i approaches its focal length f, where

$$f \cong r_i \ll r_0. \tag{152}$$

The CRL can image not only in this application, but the CRL can also focus source X-rays to high peak intensities at a sample, such as a crystal in X-ray crystallography. This applies also to neutrons in which the neutron source is at relatively large distance from the CRL and the CRL is used to focus neutrons onto the sample to provide more flux, such as in neutron crystallography. For X-rays or neutrons the 2D intensity gain of the CRL is

$$G = 0.11 \cdot \frac{A_{crl}}{A_s} \cdot \left(1 + \frac{r_0}{f}\right)^2 \cdot \exp(-N\mu d) \quad f \cong r_i \ll r_0 \quad \text{and} \quad A_{crl} \gg A_s. \tag{153}$$

In a spherical N-lens CRL, the 2D gain increases with increasing CRL effective aperture radius r_e as long as the absorption radius r_a is larger than the CRL physical aperture radius r_m:

$$r_e = r_m < r_a. \tag{154}$$

However, when the CRL effective aperture radius r_e exceeds the CRL parabolic aperture radius r_p, where $r_e > r_p$, then the resolution begins to decrease despite the increasing intensity.

Next we consider the case in which the source area A_s is much larger than the CRL aperture area A_{crl}, as in the case of a thermal neutron source and neutron lens:

$$r_e \ll w_s, h_s \quad \text{and} \quad A_s \gg A_{crl}. \tag{155}$$

We start with the CRL intensity gain equation:

$$G = \frac{0.44 \exp(-N\mu d)}{M^2 w_s h_s} \left[M\left(r_e + w_s\right) + r_e\right] \cdot \left[M\left(r_e + h_s\right) + r_e\right]. \tag{156}$$

If the source area is larger than the CRL effective aperture area, then the CRL 2D intensity gain G for neutrons or X-rays in the CRL focal plane becomes

$$G = \frac{A_{i0}}{A_i} < T >= 0.44 \cdot \left[1 + \frac{r_e}{M w_s}\right] \cdot \left[1 + \frac{r_e}{M h_s}\right] \cdot \exp(-N\mu d). \tag{157}$$

Now suppose the source half-width w_s and source half-height h_s each greatly exceed the CRL effective aperture radius r_e divided by the CRL magnification M, where

$$M w_s \gg r_e \quad \text{and} \quad M h_s \gg r_e. \tag{158}$$

Then, when the CRL aperture area is much smaller than the source area, and for sufficiently large CRL magnification M, the CRL 2D-dimensional intensity gain G for X-rays or neutrons is less than unity, where

$$G = 0.44 \exp(-N\mu d) < 1. \tag{159}$$

Finally, consider again the case of the source area A_s, which is much larger than the CRL aperture area A_{crl}, where

$$r_e \ll w_s, h_s \quad \text{and} \quad A_s \gg A_{crl}. \tag{160}$$

The above assumptions are applied to the CRL intensity gain equation,

$$G = \frac{0.44 \exp(-N\mu d)}{M^2 w_s h_s} \left[M\left(r_e + w_s\right) + r_e\right] \cdot \left[M\left(r_e + h_s\right) + r_e\right]. \tag{161}$$

However, in this case, one also assumes the image distance r_i is much smaller than the object distance r_0 in which the CRL focal length f is much smaller than the object distance r_0. As a result, the CRL magnification M is much less than 1, where

$$r_0 \gg r_i \cong f \quad \text{and} \quad M \ll 1. \tag{162}$$

The very small magnification $M \ll 1$ yields

$$Mw_s \ll r_e \quad \text{and} \quad Mh_s \ll r_e. \tag{163}$$

Hence, if the X-ray or neutron source area A_s is much larger than the CRL effective aperture area A_{crl}, then the CRL 2D intensity gain G becomes

$$G = 0.44 \cdot \frac{A_{crl}}{A_s} \cdot \left(\frac{r_0}{f}\right)^2 \cdot \exp(-N\mu d) \quad f \cong r_i \ll r_0 \quad \text{and} \quad A_{crl} \ll A_s. \tag{164}$$

The above CRL intensity gain G is almost identical to the CRL gain when the CRL effective aperture area A_{crl} is much larger than the source area A_s, where $A_{crl} \gg A_s$, and again with the assumption of the much larger object than image distance, where $f \cong r_i \ll r_0$, and thus $M \ll 1$. Hence, for a CRL aperture area A_{crl} much smaller or much larger than the source area A_s, and for an object distance r_0 much greater than the image distance r_i, where the magnification is much less than 1 with $M \ll 1$, one obtains approximately identical CRL intensity gain G expressions.

10. CRL TRANSVERSE AND AXIAL MAGNIFICATION

The CRL has both a transverse magnification and an axial magnification. In a thin lens, the chief ray is the ray that propagates from the object point in a straight line through the center point of the lens and onward without bending by the lens. The chief ray forms two similar triangles with the center z-axis of the lens such that

$$\frac{y}{r_i} = \frac{x}{r_0}. \tag{165}$$

The transverse magnification M_T is defined as

$$M_T = \frac{y}{x}; \tag{166}$$

hence

$$M_T = \frac{r_i}{r_0}. \tag{167}$$

In contrast, the axial magnification M_A is defined as the ratio of change the image distance divided by the corresponding change in the object distance for a fixed focal length f corresponding to wavelength λ or

$$M_A = \frac{\Delta r_i}{\Delta r_0} \tag{168}$$

For paraxial rays in which the angles are sufficiently small such that the sines of the angles are equal to the angles themselves and application of Snell's law yields the Gauss's formula for a thin lens,

$$\frac{1}{f} = \frac{1}{r_0} + \frac{1}{r_i}. \tag{169}$$

The Gauss thin lens is equivalent to adding the vergence of the incident wave to the sum of the powers of the two lens surfaces to obtain the vergence of the exit wave. Recall the vergence is the wave curvature (reciprocal of radius of curvature), multiplied by the refractive index in which the wave propagates. The power of a surface is its curvature multiplied by the change in the refractive index upon crossing the surface, and the focal length is the reciprocal of the sum of the powers of the two lens surfaces. From the lensmaker's formula one obtains a useful expression relating the lens image distance r_i to the lens focal length f and transverse magnification M_T, which is also denoted by M, where

$$r_i = (1 + M_T)f. \tag{170}$$

The differential form of the lensmaker's equation for a fixed focal length f is

$$\frac{\Delta r_i}{r_i^2} = -\frac{\Delta r_0}{r_0^2}. \tag{171}$$

From this the axial magnification M_A is found to be related to transverse magnification M by

$$M_A = -M^2. \tag{172}$$

The transverse magnification M_T is equal to the ratio of the image area A_i divided by the object area A_0. Consider the image height y_1 or y_2 divided by object height x_1 or x_2 as shown in Figure 6, where

$$M_{T1} = \frac{y_1}{x_1} = \frac{y_1}{x} \quad \text{and} \quad M_{T2} = \frac{y_2}{x_2} = \frac{y_2}{x}. \tag{173}$$

The chief ray from object point x_1 through the lens center point to a conjugate, inverted image point y_1, together with the center optical z-axis, forms a pair of similar triangles with common apex angle θ_1 that yields transverse magnification M_{T1}. Similarly, the chief ray from object point x_2 through the lens center point to a conjugate, inverted image point y_2, together with the center optical z-axis, forms a pair of similar triangles with common apex angle θ_2 that yields transverse magnification M_{T2}. The Newton lens equation is obtained from this pair of similar triangles.

Magnification in the axial direction is the ratio of the distance $i_2 - i_1$ between two axial image points i_1 and i_2, divided by the difference $o_2 - o_1$ between their corresponding object points o_1 and o_2 (Figure 7). The axial magnification M_A reverses the image axially and is negative, and is defined by

$$M_A \equiv \frac{i_2 - i_1}{o_2 - o_1}. \tag{174}$$

By similar triangles shown in Figure 6 one finds

$$\frac{i_1}{f} = -\frac{f}{o_1} \tag{175}$$

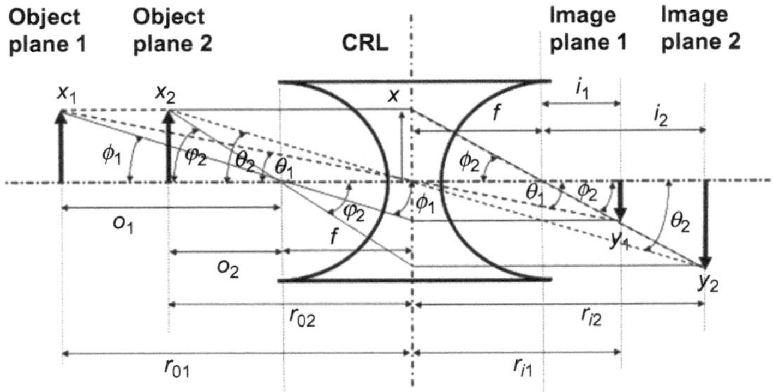

Axial magnification definition

$$M_A \equiv \frac{i_2-i_1}{o_2-o_1}$$

Substitution for i_2 and i_1 gives

$$M_A = -\frac{f^2}{o_1 o_2}$$

Substitution for o_2 and o_1 gives

$$M_A = -M_{T1}\cdot M_{T2}$$

$$\text{if } o_1 \cong o_2$$

$$M_A = -M_T^2$$

Axial magnification derivation by similar triangles

$$\tan\phi_1 = \frac{x_1}{o_1} = \frac{x}{o_1} = -\frac{y_1}{f}$$

$$\tan\phi_2 = \frac{x}{f} = -\frac{y_1}{i_1} = -\frac{y_2}{i_2}$$

$$\tan\varphi_2 = \frac{x_2}{o_2} = \frac{x}{o_2} = -\frac{y_2}{f}$$

$$M_{T1} = -\frac{y_1}{x} = \frac{i_1}{f} = -\frac{f}{o_1} \longrightarrow i_1 = -\frac{f^2}{o_1}$$

$$M_{T2} = -\frac{y_2}{x} = \frac{i_2}{f} = -\frac{f}{o_2} \longrightarrow i_2 = -\frac{f^2}{o_2}$$

Transverse magnification

$$M_{T1} = -\frac{y_1}{x_1} = -\frac{y_1}{x} = \frac{i_1}{o_1} = \tan\theta_1$$

$$M_{T2} = -\frac{y_2}{x_2} = -\frac{y_2}{x} = \frac{i_2}{o_2} = \tan\theta_2$$

FIGURE 7 The diagram shows the definition and calculation of the axial magnification M_A of a lens or compound refractive lens (CRL) and the relationship between the lens or CRL axial magnification M_A and transverse magnification M_T, where $M_A = -M_T^2$, if $o_1 \cong o_2$.

and

$$\frac{i_2}{f} = -\frac{f}{o_2}. \tag{176}$$

With substitution into Eq. 174 for i_1 and i_2 of Eqs. 175 and 176, the axial magnification M_A becomes

$$M_A \equiv \frac{i_2 - i_1}{o_2 - o_1} = \frac{f^2}{o_2 - o_1}\left(\frac{1}{o_2} - \frac{1}{o_1}\right) = -\frac{f^2}{o_1 o_2}. \tag{177}$$

One can then relate the axial magnification M_A to the transverse magnification M_T by substitution into Eq. 177 for $M_{T1} = -f/o_1$ and $M_{T2} = -f/o_2$. With the approximation of almost equal object distances, $o_1 \cong o_2$, one then obtains

$$M_A = -M_{T1}M_{T2} \cong -M_T^2. \tag{178}$$

11. CRL DEPTH OF FIELD AND DEPTH OF FOCUS

Given a CRL of physical aperture diameter $D = 2r_m$ and incident X-rays or neutrons of individual wavelength λ, the depth of focus can be determined via the total image resolution σ_i. This total resolution σ_i includes contributions from the Rayleigh diffraction limit, the detector-limited pixel resolution, spherical aberration, and chromatic aberration. For large-aperture CRLs with off-axis rays, there are additional contributions to the total image resolution σ_i—namely, coma, astigmatism, curvature of field, and distortion.

The CRL depth of focus Δf, about focal length f, is the axial extent that the detector can be dithered about the image plane, corresponding to a fixed object plane, such that the blur diameter of image points is within twice the total image resolution σ_i for wavelength λ. Two triangles share the same vertex angle, with the vertex at the focal point f, in which the opposite two sides of the two triangles are the total resolution diameter σ_i and the lens aperture diameter $D = 2r_m$, such that

$$\frac{D}{f + \Delta f} = \frac{\sigma_i}{\Delta f}. \tag{179}$$

Assume

$$\Delta f \ll f. \tag{180}$$

And with

$$r_i = M_T r_o = (1 + M_T)f, \tag{181}$$

The focal length f is then

$$f = \frac{2\pi R}{N\rho b\lambda^2}. \tag{182}$$

The CRL depth of focus Δf is

$$\Delta f = \frac{7.66 M r_o R}{D^2 N\rho b\lambda}. \tag{183}$$

Depending on the depth of focus Δf, there is a corresponding range of object distances for an object point (depth of field [DOF]), in which the conjugate focal points fall within $f + \Delta f$, and the corresponding conjugate image point is in focus and falls within $r_i + \Delta r_i$, where

$$r_i + \Delta r_i = (M+1)\left(f + \Delta f\right). \tag{184}$$

The DOF is related to the depth of focus Δf via the CRL axial magnification:

$$M_A = -M_T^2. \tag{185}$$

Expressing the DOF in terms of the transverse magnification M_T is

$$DOF = \frac{\Delta f}{M_A} = \frac{0.61\lambda}{M_T^2 NA^2} \tag{186}$$

Figure 8 is a diagram showing the DOF and depth of focus Δ of a CRL in which two axial object points A and B produce conjugate axial image

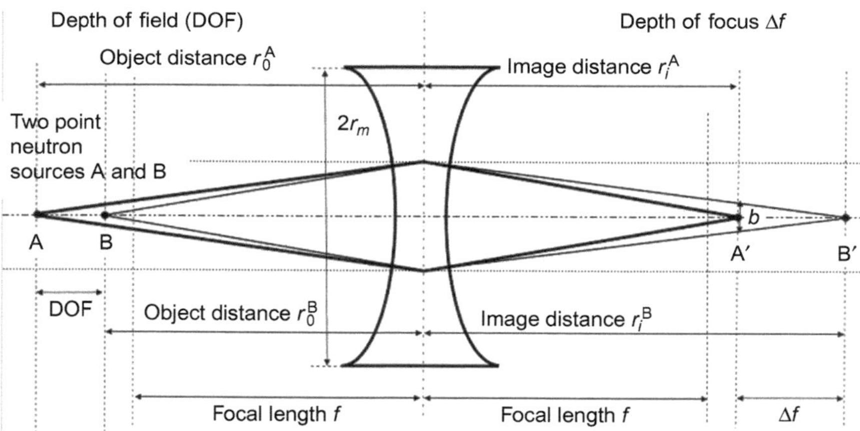

FIGURE 8 The depth of field (DOF) and depth of focus Δf of a lens or CRL. The blur diameter b is set to the lens image resolution diameter, which is the sum of resolution-limiting contributions from Rayleigh diffraction, spherical and chromatic aberration, and the detector pixel size. The axial separation between two object points A and B produces corresponding focal points A′ and B′ separated by the axial distance Δf. The depth of focus Δf is set to produce the lens detector–limited resolution spot diameter b at focal plane A′. The focal plane A′ then corresponds to the object point A, which then determines the DOF or maximum axial separation between object points A and B.

points A′ and B′. The depth of focus Δf for a single object point is the axial range of the focal planes, which geometrically form a single focal point to a locus of focal points whose radius is equal to the minimum image resolution diameter b of the lens. The minimum imaging resolution spot diameter is the sum of contributions from the lens Rayleigh diffraction, spherical and chromatic aberration, and the detector pixel size. The DOF corresponding to the set of focal planes is the axial distance between two object points A and B, which produce image points A′ and B′ at different focal distances. The resolution spot diameter b limits the DOF or axial separation of object points.

12. MODULATION TRANSFER FUNCTION FOR CRL RESOLUTION DETERMINATION

Spatial resolution is the smallest distance that two objects can be separated and still be distinguished. *Contrast* is the relative signal difference between and object and its background. *Radiographic contrast* is the relative difference between the neutron flux emerging from adjacent portions of a specimen, whereas *image contrast* is the relative difference in image brightness between the corresponding specimen portions imaged on the detector by the CRL.

From the Kirchhoff equation, which accounts for both monochromatic and chromatic aberrations, the CRL point spread function (PSF) and the modulation transfer function (MTF) can be calculated. The MTF is a quantitative and standard measure of image quality and resolution and describes the ability of a lens or system to transfer object contrast to the image. The *resolution* of a microscope generally means the capability to distinguish two point objects of about equal intensity. In the Rayleigh criterion the PSF center maximum of a second point is placed at the zero intensity position of the PSF of the first point. The Rayleigh resolution is the reciprocal of the spatial frequency s or period $1/s$ corresponding to $MTF = 0.09$. However, the period $1/s$ at $MTF = 0.05$ is a common resolution limit.

Consider a sinusoidal variation of the image intensity between the minimum intensity I_{min} and the maximum intensity I_{max}. The intensity variation ΔI is defined by

$$\Delta I = I_{\text{max}} - I_{\text{min}}. \tag{187}$$

The average intensity I_0 is defined by

$$I_0 = \frac{I_{\text{max}} + I_{\text{min}}}{2}. \tag{188}$$

Contrast C is defined by for maximum I_{max} and minimum I_{min} transmitted intensities, where

$$C = \frac{I_{max} - I_{min}}{I_{max} + I_{min}} = 2\frac{\Delta I}{I_0} \tag{189}$$

The modulation M is defined as the ratio of the intensity variation ΔI divided by the mean intensity I_0, where

$$M = \frac{\Delta I}{I_0}. \tag{190}$$

One performance test for the CRL is how much modulation can be transferred from the object to the image. The modulation transfer function $MTF(f)$ is defined by the ratio of the image intensity modulation $M_{image}(f)$ divided by the object intensity modulation $M_{object}(f)$, which are dependent on the spatial frequency f, where (Meyer-Arendt, 1984)

$$MTF(f) = \frac{M_{image}(f)}{M_{object}(f)}. \tag{191}$$

The spatial frequency f with units of mm^{-1} or line pairs per millimeter is the reciprocal of the spatial period in units of millimeters, for example. Using Eqs. 189 and 190 to express the image and object intensity modulations $M_{image}(f)$ and $M_{object}(f)$ in the $MTF(f)$ expression of Eq. 191, then yields the $MTF(f)$ as the ratio of the image contrast $C_{image}(f)$ divided by the object contrast $C_{object}(f)$, where

$$MTF(f) = \frac{C_{image}(f)}{C_{object}(f)}. \tag{192}$$

The top of Figure 9 shows the profile along the x-direction of the intensity modulation imparted to an incident X-ray or neutron wave, which passes through the object and variable undergoes amplitude attenuation as a function of position x. The bottom diagram is the phase shift between the relative intensity profiles of the object-emitted X-rays or neutrons and the detector-imaged X-rays or neutrons as a function of position along the x-direction.

For spatial frequencies greater than zero $f > 0$, the $MTF(f)$ describes the range of detail of the object, which is expressed as the contrast variation transferred from the detector input to the detector output. At zero spatial frequency the MTF is unity:

$$MTF(0) = 1. \tag{193}$$

As the spatial frequency increases from zero, the $MTF(f)$ decreases from 1 to zero:

$$MTF(f) < 1. \quad f > 0 \tag{194}$$

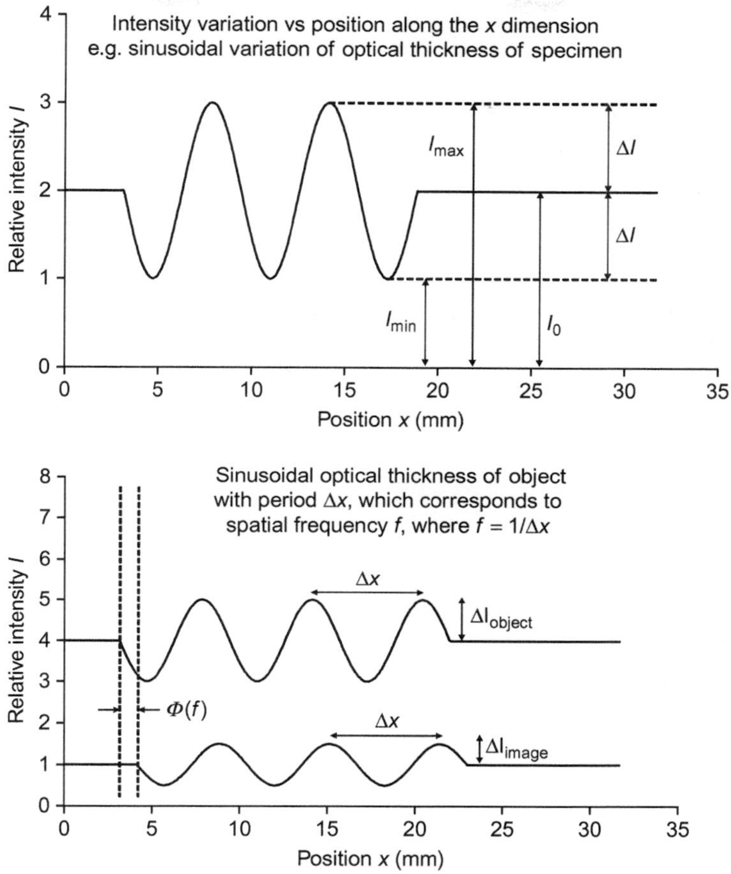

FIGURE 9 (Top) Amplitude modulation of the transmitted intensity of an object specimen as a function of position *x* (mm), which corresponds to the spatial frequency *f* (1/mm) of the reciprocal space. (Bottom) The phase shift $\phi(f)$ of the image intensity modulation compared with the object intensity modulation as a function of position *x* (mm).

The spatial frequency f characterizes contrast variation or the detail of the object, and f is the reciprocal of a spatial period representing an object detail of size $1/f$.

The $MTF(f_x)$, as a function of the spatial frequency f_x along the x-direction, is the Fourier transform of the x-directed line spread function (LSF) $L_x(x)$, where $f_x = 1/x$. And the $MTF(f_y)$ as a function of the spatial frequency f_y along the y-direction is the Fourier transform of the y-directed LSF $L_y(y)$, where $f_y = 1/y$. The LSFs can be measured using line sources of X-rays or neutrons.

A second method of experimentally measuring the MTF is by imaging a set of periodic slit arrays with different spatial periods $1/f$. Each slit array is a set of equally spaced slits in an X-ray- or neutron-absorbing material in which the slit widths are equal to the slat widths. The images of the slit arrays varying from small to large spatial periods $1/f$ allows one to determine the imaging system resolution and *MTF*.

A third method of experimentally measuring the MTF is via a knife edge. That is, half the image plane is covered with a completely absorbing X-ray or neutron material with a precise edge and the other half of the image plane is uncovered. An X-ray-absorbing thin tungsten plate or thin thermal neutron absorbing gadolinium plate with a precise edge (knife edge) along the y-axis that covers the $x \leq 0$ half of the image plane. This allows measurement of the LSF $L_x(x)$ along the perpendicular x-direction. Similarly, a knife edge along the x-axis of an X-ray- or neutron-absorbing plate covers the $y \leq 0$ half of the image plane.

In an actual knife-edge test, the change in the image intensity ΔI from the covered to uncovered region depends on (1) the energy-dependent transmission fraction T of the X-rays through a thin absorbing sheet of tungsten or lead or (2) the transmission fraction T of neutrons through a thin absorbing sheet of gadolinium or cadmium, where

$$\Delta I = 1 - T. \tag{195}$$

The measurement with a y-directed knife edge or a y-directed line source produces a line intensity profile in the x-direction $I_x(x)$. The LSF $L_x(x)$ in the x-direction is the slope or x-derivative of $I_x(x)$, where

$$L_x(x) = \frac{dI_x(x)}{dx}, \tag{196}$$

and the measurement with an x-directed knife edge or an x-directed line source produces a line intensity profile in the y-direction $I_y(y)$. The LSF $L_y(y)$ in the y-direction is the slope or y-derivative of $I_y(y)$, where

$$L_y(y) = \frac{dI_y(y)}{dy}. \tag{197}$$

The $MTF(f_x)$ along the x-direction is then the Fourier transform of the x-directed LSF $L_x(x)$, and the $MTF(f_y)$ along the y-direction is then the Fourier transform of the y-directed LSF $L_y(y)$, where

$$MTF(f_x) = \int_{-\infty}^{\infty} L_x(x) \exp(-i2\pi \cdot f_x \cdot x)\, dx \tag{198}$$

and

$$MTF\left(f_y\right) = \int_{-\infty}^{\infty} L_y\left(y\right) \exp(-i2\pi \cdot f_y \cdot y)\, dy. \tag{199}$$

Aberrations of the CRL may cause some image pixels to become dislocated as in coma and distortion. The extent of the dislocations is measured by the phase shift in the x- and y-directions $\varphi(f_x)$ and $\varphi(f_y)$, which are functions, respectively, of the x and y spatial frequencies f_x and f_y. Hence, $\varphi(f_x)$ and $\varphi(f_y)$ are the x-directed and y-directed phase factors of the object modulation, which is imaged by the CRL into an image modulation.

Together the magnitudes of the x-directed and y-directed MTFs, $MTF(f_x)$ and $MTF(f_y)$, and the associated phase-shift factors $\varphi(f_x)$ and $\varphi(f_y)$, lead to the x-directed and y-directed complex optical transfer functions $OTF(f_x)$ and $OTF(f_y)$ given by

$$OTF(f_x) = MTF(f_x) \exp[i\varphi(f_x)] \tag{200}$$

and

$$OTF(f_y) = MTF(f_y) \exp[i\varphi(f_y)]. \tag{201}$$

The optical transfer function $OTF(f)$ is the ratio of the Fourier transform of the image intensity distribution divided by the Fourier transform of the object intensity distribution as discussed by Meyer-Arendt (1984).

13. CALCULATION OF CRL MODULATION TRANSFER FUNCTION BY LINE PROFILE MEASUREMENTS OF KNIFE-EDGE IMAGES

The MTF can be calculated from easily measuring the knife-edge response $K(x)$ in the image plane produced by a knife edge in the object plane that separates a region of complete neutron or X-ray transmission from that of total absorption. The knife-edge X-ray or neutron transmission $K(x)$ for an ideal knife edge would be a step function $u(x)$, where

$$K\left(x\right) = u\left(x\right). \tag{202}$$

But in physical reality the knife edge is a more gradual transition from complete transmission to absorption and can be described by a hyperbolic tangent function, where m adjusts the boundary sharpness of the knife edge, where

$$K(x) = \frac{\tanh(mx) + 1}{2}. \tag{203}$$

The LSF $H(x)$ is simply the derivative of the knife-edge response, where

$$H = \frac{dK}{dx}. \tag{204}$$

For an ideal knife edge response, $H(x)$ is a delta function, where

$$H(x) = \delta(x) \tag{205}$$

For the less-sharp boundary described by the hyperbolic tangent, the LSF is Gaussian-like so that

$$H(x) = \frac{m \cdot \sec h^2(mx)}{2}. \tag{206}$$

Thus, the MTF is the Fourier transform of the derivative in the direction of intensity variation of the measured knife-edge function.

A convenient analytic expression of the MTF in terms of the measured knife-edge response is derived as follows. The measured knife-edge response is

$$K(x) \propto \int_{-\infty}^{x} du \int_{-\infty}^{\infty} dv PSF(u,v). \tag{207}$$

Here $PSF(u,v)$ is the PSF or image that is produced by a single point in the object plane—that is, the PSF is a Green's function. The line spread function $LSF(u)$ is

$$LSF(u) \propto \int_{-\infty}^{\infty} PSF(u,v)\, dv. \tag{208}$$

Hence, the knife-edge response is related to the LSF by

$$K(x) \propto \int_{-\infty}^{x} LSF(u)\, du \tag{209}$$

Thus,

$$\frac{dK(x)}{dx} \propto LSF(x) \tag{210}$$

The MTF is the normalized Fourier transform of the LSF, where

$$MTF(f) = \frac{\int_{-\infty}^{\infty} LSF(u)\cos\left(\frac{2\pi u}{f}\right)du}{\int_{-\infty}^{\infty} LSF(u)\,du} \tag{211}$$

Substitution for the LSF give

$$MTF(f) = \frac{\int_{-\infty}^{\infty} \frac{dK}{du}\cos\left(\frac{2\pi u}{f}\right)du}{\int_{-\infty}^{\infty} \frac{dK}{du}du}. \tag{212}$$

The MTF is the Fourier transform of the derivative of the knife-edge function. Since dK/du approaches zero near the edge, the infinite limits in the above MTF equation can be replaced with finite limits of integration $[a, b]$, and then integration by parts yields

$$MTF(f) = \frac{K(b)\cos\left(\frac{2\pi b}{f}\right) - K(a)\cos\left(\frac{2\pi a}{f}\right) + \frac{2\pi}{f}\int_a^b K(u)\sin\left(\frac{2\pi u}{f}\right)du}{K(b) - K(a)}. \tag{213}$$

The knife-edge measurement may be normalized so that

$$K(b) = 1 \quad \text{and} \quad K(a) = 0. \tag{214}$$

Hence,

$$MTF(f) = \cos\left(\frac{2\pi b}{f}\right) + \frac{2\pi}{f}\int_a^b K(u)\sin\left(\frac{2\pi u}{f}\right)du. \tag{215}$$

This equation may be readily applied to the measured knife-edge data.

14. CRL FIELD OF VIEW

We now consider the FOV of a CRL for X-rays or neutrons. A CRL of aperture radius r_m and length L has a FOV F_{crl} determined by the CRL's angular acceptance of an X-ray- or neutron-illuminated object. One assumes a CRL thin lens with a single principal plane at the CRL center plane in which the object distance r_0 is assumed to be much larger than the focal length f_{crl}, where

$$r_0 \gg f_{crl}. \tag{216}$$

The CRL-limiting effective aperture radius r_e is given by the minimum of the CRL absorption aperture radius r_a and physical aperture radius r_m, where the CRL effective radius r_e is

$$r_e = MIN(r_m, r_a). \tag{217}$$

The CRL acts like a cylinder with radius r_e and length l, which restricts its angular acceptance of the neutrons coming from the object and passing through it, where

$$r_a \leq r_m. \tag{218}$$

This limits the FOV of the object that can be seen by the CRL. If the X-rays or neutrons, which pass through the CRL, are not greatly refracted, then the aspect ratio $2r_a/l$ of the cylinder, determines what range of angles can be accepted.

As shown in Figure 10, the CRL acts as a long cylinder of effective aperture radius r_e and length L. The CRL FOV, FOV_{crl}, is the product of its

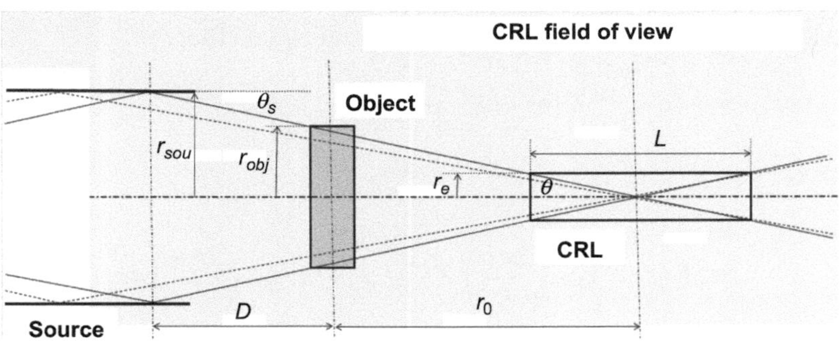

FIGURE 10 The diagram depicts the compound refractive lens (CRL) field of view (FOV) with an object illuminated by a neutron or X-ray source. The source is assumed to be a wave guide source of neutrons or X-rays.

acceptance angle θ_{crl} and the object distance r_0, so that

$$FOV_{crl} = \frac{4r_e r_0}{L}, \tag{219}$$

where

$$\theta_{crl} = \frac{4r_e}{L}. \tag{220}$$

The following conditions are required for the CRL to image and object within its FOV. Rays from the source must pass through every object point within the CRL FOV, and each object point must send a ray to each point within the CRL effective aperture diameter. These rays are refracted by the CRL to a single conjugate image point. The image-forming neutrons or X-rays must pass straight from the source to the object and through the CRL effective aperture to the image plane.

An upper limit of the FOV is established by the last point on the object from which a neutron can be traced from the source to the CRL. This FOV limit can be caused by either the size of the source or its divergence. The source to object distance is D, and the maximum source divergence angle is θ_s. That is, a source perimeter ray exits the source at source radius r_{sou} and passes through the object at the object perimeter radius r_{obj}. The perimeter ray then strikes the CRL at its effective aperture radius r_e. This perimeter ray defines the maximum CRL acceptance angle θ. For the CRL to view the entire source, the maximum source divergence angle θ_s must be greater than or equal to the maximum CRL acceptance angle θ, where

$$\theta = \tan^{-1}\left[\frac{r_{sou} - r_e}{D + r_0}\right] \leq \theta_s. \tag{221}$$

Since the CRL acceptance angles are usually very small, one can apply the small angle approximation, where

$$\theta \cong \frac{r_{sou} - r_e}{D + r_0} \leq \theta_s. \tag{222}$$

The source divergence angle θ_s of the X-ray or neutrons exiting the wave guide source is the critical angle of X-ray or neutron reflection θ_c in the guide walls, where

$$\theta_s = \theta_c. \tag{223}$$

The guide walls are composed of surface reflective material with refractive index $n = 1 - \delta$ and decrement δ, where

$$\theta_c = \sqrt{2\delta}. \tag{224}$$

For 5-Å neutrons and a nickel mirror, the grazing critical angle for total reflection θ_s is 8.5 mrad, measured relative to the mirror surface.

The perimeter X-ray or neutron wave vector **k** passes through the object at radius r_{obj}. The FOV of the source and source-illuminated object FOV_{so} is thus limited to the radius r_{obj} in the object plane, where

$$FOV_{so} = r_{obj} = r_e + r_0 \tan\theta. \tag{225}$$

The FOV of the source and object FOV_{so} by the CRL is dependent on the source to CRL distance $D + r_0$ and the CRL to object distance r_0, where

$$FOV_{so} \leq 2r_e + 2r_0\theta \cong 2\left(r_e + r_0\frac{r_{sou} - r_e}{D + r_0}\right). \tag{226}$$

The resultant CRL FOV of the object FOV is the minimum of the CRL FOV of the object FOV_{sou} and the intrinsic CRL FOV, FOV_{crl}, where

$$FOV = MIN\left(FOV_{crl}, FOV_{sou}\right) \cong MIN\left[4r_e\frac{r_0}{L}, 2\left(r_e + r_0\frac{r_{sou} - r_e}{D + r_0}\right)\right]. \tag{227}$$

15. THICK LENS CRLS

So far we have assumed thin lens CRLs in which the ray changes direction but maintains approximately the same radial distance form the center z-axis of the CRL during the passage of the X-ray or neutron ray through the CRL. However, in a thick lens CRL a ray changes not only direction but also radial distance from the lens center axis. This creates the need for two principal planes separated by the lens at midplane. In a thick lens the center principal plane is split into primary and secondary principal planes. The primary principal plane is shifted to the left away from the lens center plane toward the source (object), and the secondary principal plane is shifted to the right away from the lens center plane toward the image. In a thin lens the straight line of the chief ray extends from an object point through the lens midpoint to the image point.

In a thick lens the chief ray is no longer a straight line from the object to image point. Instead, the chief ray extends in a straight line from the object point to the lens first principal point at the intersection of the lens center axis and the lens first principal plane. Within the thick lens the chief ray travels parallel to the center lens axis from a point on the first principal plane to a point on the second principal plane. From the second principal plane point the chief ray is refracted and extends in a straight line to the image point. That is, the thick lens CRL is geometrically modeled as a thick lens with two principal planes in which the refraction takes place at the secondary (downstream) principal plane facing the image plane.

Figure 11a is a diagram of a thick CRL depicted relative to the CRL FOV and CRL image plane. Figure 11b shows a thick lens CRL diagram

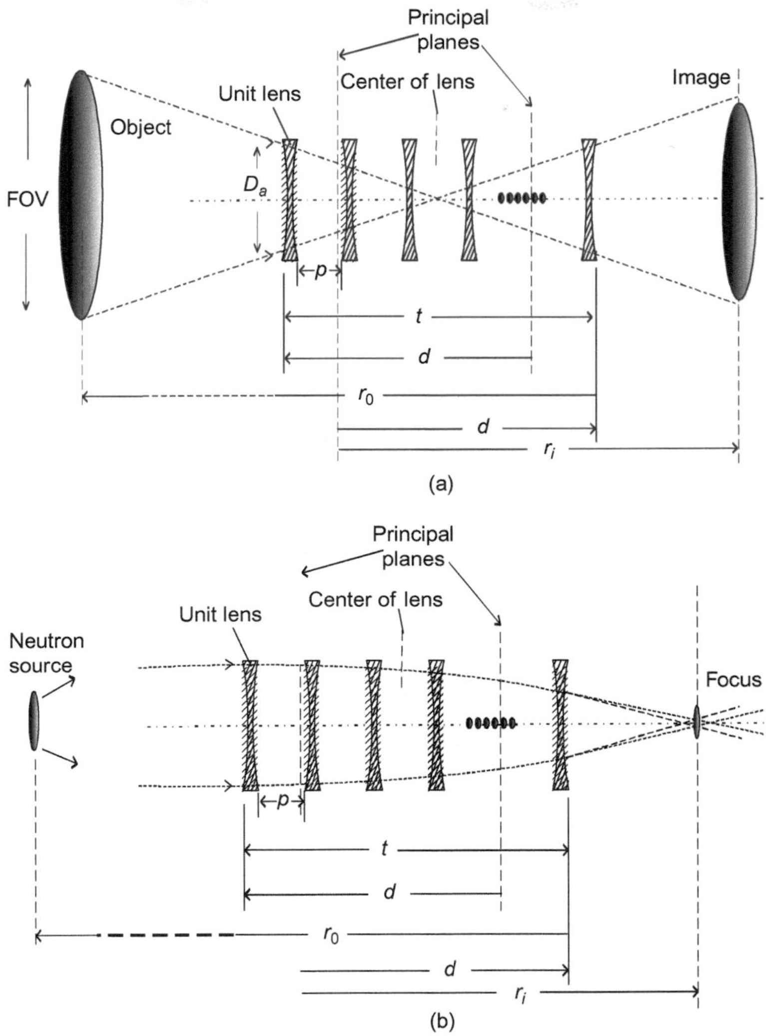

FIGURE 11 (a) A thick lens CRL for neutrons or X-rays with object plane and image plane. When the image distance r_i is measured from the left (primary) principal plane and the object distance r_0 is measured from the right (secondary) principal plane, then the lensmaker's formula f^{-1} r_0^{-1} r_i^{-1} holds for thick lens focal length f, which is measured from the left (primary) principal plane. (b) A thick lens CRL for neutrons or X-rays with source plane and focal plane. The ray trajectory is curved in a thick lens, in contrast with the straight-line trajectory in a thin lens.

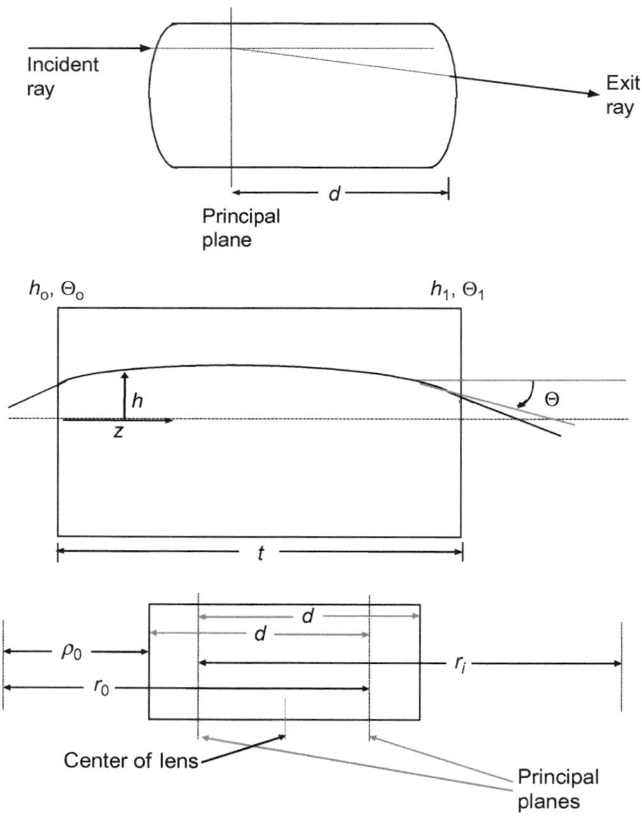

FIGURE 12 An X-ray or neutron ray has radial coordinate h, measured from the center optical axis, and direction angle coordinate θ with the incident ray radial and angle coordinates (h_0, θ_0) at CRL entrance and ray coordinates (h_1, θ_1) at CRL exit. Principal plane locations are for a symmetrical thick lens; and with focal length f and image r_i and source r_0 distances measured from principal planes, the matrix representation for a thick lens becomes the matrix for a thin lens.

in which the CRL is depicted relative to a source plane and the CRL focal plane. The three schematics in Figure 12 show the positions of the thick lens CRL primary (left or input) principal plane and secondary (right or output) principal planes, the ray trajectory, and trajectory parameters, which are used in the derivation of the thick lens CRL focal length and positions of the principal planes.

Principal planes are defined for the thick optical lens in which the object distance r_0 is measured from primary principal plane, and the thick lens focal length f and the image distance r_i are measured from the secondary principal plane. The primary principle plane, which measures object distance r_0, is at distance d from the back (exit) surface of the lens (image side). The principal plane, which measures image distance r_i and measures the thick lens focal length f, is at the same distance d from the front (entrance) surface of the lens (object side). These object r_0 and image r_i distances and focal length f then obey the lensmaker's equation. In a thin lens the two planes coincide at the lens center plane (see Hecht and Zajac, 1974).

We now mathematically model the refraction of X-ray or neutrons in a thick lens CRL in which the X-ray or neutron ray changes direction continuously as the ray passes through the entire length t of the CRL. The individual lenses of the CRL are closely stacked together so the focusing of neutrons or X-rays can be considered continuous. For a lens segment dz in length, the focal length f is expressed in terms of the thin lens focal length f_0 and the CRL length t by

$$f = \frac{f_0 t}{dz}. \tag{228}$$

Recall, the thin lens focal length for a CRL composed of N biconcave lenses with radius of curvature R with lens material decrement δ is

$$f_0 = \frac{R}{2N\delta}. \tag{229}$$

Because the change in direction of the ray passing through a lens segment is given by the ratio of the ray height h to the focal length of the segment f, we obtain

$$d\theta = \frac{h}{f_0 t} dz \tag{230}$$

and

$$dh = -\theta dz. \tag{231}$$

The solution of the coupled differential equations below yields the radial distance $h(z)$ and direction $\theta(z)$ as a function of axial distance z from the front surface of the CRL, where

$$\frac{d^2\theta}{dz^2} = -w\theta \tag{232}$$

and

$$\frac{d^2h}{dz^2} = -wh, \tag{233}$$

where

$$w = \frac{1}{f_0 t}. \tag{234}$$

The initial slope of the ray, incident on the front surface of the CRL at radial position h_0, is determined by the radial position s of a source point, and the axial distance ρ_0 of the source or object point from the CRL front (input) surface. Hence, the radial position $h(z)$ and direction $\theta(z)$ of the neutron ray inside the CRL from $z = 0$ to $z = t$ are

$$h(z) = h_0 \left[\cos \frac{z}{\sqrt{f_0 t}} + \frac{\sqrt{f_0 t}}{\rho_0} \sin \frac{z}{\sqrt{f_0 t}} \right] - s \left[\frac{\sqrt{f_0 t}}{\rho_0} \sin \frac{z}{\sqrt{f_0 t}} \right] \tag{235}$$

and

$$\theta(z) = \frac{dh}{dz} = h_0 \left[-\frac{1}{\sqrt{f_0 t}} \sin \frac{z}{\sqrt{f_0 t}} + \frac{1}{\rho_0} \cos \frac{z}{\sqrt{f_0 t}} \right] - s \left[\frac{1}{\rho_0} \cos \frac{z}{\sqrt{f_0 t}} \right]. \tag{236}$$

Here ρ_0 is the distance between the source and the front surface of the lens and s is the radial position of the ray at the source, h_0 is the radius at which the ray enters the lens, and $h(z)$ is the radius as a function of longitudinal position. The radial position $h(z)$ and angle $\theta(z)$ of a ray at the CRL entrance are denoted by radial distance h_0 and angle θ_0 and at the CRL exit by radial distance h_1 and angle θ_1. The ray radial position and angle at the CRL entrance and CRL exit are related by the matrix equation

$$\begin{bmatrix} h_1 \\ \theta_1 \end{bmatrix} = M \cdot \begin{bmatrix} h_0 \\ \theta_0 \end{bmatrix}, \tag{237}$$

where

$$M = \begin{bmatrix} \cos \sqrt{\frac{t}{f_0}} & -\sqrt{f_0 t} \sin \sqrt{\frac{t}{f_0}} \\ \frac{1}{\sqrt{f_0 t}} \sin \sqrt{\frac{t}{f_0}} & \cos \sqrt{\frac{t}{f_0}} \end{bmatrix}. \tag{238}$$

A parallel neutron ray $\theta_0 = \theta(0) = 0$ emitted from a source point with radial position s at the CRL entrance will be bent through the CRL to the focal point by angle θ_1, where

$$s = h_0 = h(0) \tag{239}$$

and

$$\theta_1 = \frac{h_0}{f}. \tag{240}$$

One equates θ_1 to $\theta(z = t)$. This yields the solution for the thick lens focal length f that is expressed by the thin lens focal length f_0 and the length t of the thick lens CRL. This solution is conveniently expressed as the ratio f/f_0, where

$$\frac{f}{f_0} = \frac{\sqrt{\frac{t}{f_0}}}{\sin\sqrt{\frac{t}{f_0}}}. \tag{241}$$

The primary principal plane is distance d from the thick lens exit surface plane and the secondary principal plane is displaced by the same distance d from the thick lens entrance surface plane. The drift space matrix for a ray over axial drift distance d is

$$\begin{bmatrix} h_1 \\ \theta_1 \end{bmatrix} = D \cdot \begin{bmatrix} h_0 \\ \theta_0 \end{bmatrix} = \begin{bmatrix} 1 & d \\ 0 & 1 \end{bmatrix} \cdot \begin{bmatrix} h_0 \\ \theta_0 \end{bmatrix}. \tag{242}$$

The matrix expression for a thin lens with the same focal length f as the thick lens is

$$\begin{bmatrix} h_1 \\ \theta_1 \end{bmatrix} = \begin{bmatrix} 1 & 0 \\ \frac{1}{f} & 1 \end{bmatrix} \cdot \begin{bmatrix} h_0 \\ \theta_0 \end{bmatrix}. \tag{243}$$

The thin lens matrix can be equated to the thick lens matrix M, which is pre-multiplied and post-multiplied by the drift space matrix D for the drift distance d of the primary and secondary principal planes from the respective exit and entrance surfaces of the thick lens CRL, where

$$\begin{bmatrix} 1 & d \\ 0 & 1 \end{bmatrix} M \begin{bmatrix} 1 & d \\ 0 & 1 \end{bmatrix} = \begin{bmatrix} 1 & 0 \\ \frac{1}{f} & 1 \end{bmatrix}. \tag{244}$$

Solving the two equations for two unknowns d and f in the above matrix expression yields the prior solved thick lens focal length expression f/f_0 and the distance d between the primary principal plane and CRL exit, as well as the same distance d between the secondary principal plane and the CRL entrance plane, where (Pantell *et al.*, 2003)

$$d = f\left[1 - \cos\sqrt{\frac{t}{f_0}}\right]. \tag{245}$$

The primary principal plane is shifted from the CRL center plane toward the source (object) and the secondary principal plane shifted from the CRL center plane toward the image. For $f < d$ the specified range of CRL thickness t is

$$\frac{3\pi}{2} > \sqrt{\frac{t}{f_0}} - 2m\pi > \frac{\pi}{2}. \tag{246}$$

Here m is a positive integer. Imposing the condition that the image plane be outside the lens; that is, $r_i \geq d$ the maximum value for r_0/f, is

$$\left(\frac{r_o}{f}\right)_{\max} = 1 - \sec\sqrt{\frac{t}{f_0}}. \tag{247}$$

If

$$\sqrt{\frac{t}{f_0}} < \frac{\pi}{2}, \tag{248}$$

then the CRL thickness t is outside the above specified range and so object distance r_0 is unlimited. The fact that r_0/f may have a maximum value limits the gain of the CRL. If

$$r_0 < \frac{(2r_e)^2}{\lambda}, \tag{249}$$

then the one dimensional CRL gain G is

$$G = \frac{2r_eT}{\sigma}\frac{r_0}{f}. \tag{250}$$

Here T is the CRL transmission, σ_s is the source diameter, and $2r_e$ is the CRL effective aperture diameter, which is the minimum of the CRL absorption aperture diameter $2r_a$ and the CRL physical aperture diameter $2r_m$. Therefore, to have a gain greater than unity it is necessary that $2r_eT/\sigma$ be greater than f/r_0.

Assuming negligible thickness of the CRL lens element along its axis, the absorption radius for the CRL is defined as the radial position on the lens where the power transmitted through the lens is reduced by the factor e^{-2}. For a thin lens the absorption aperture diameter D_{a0}, which is twice the absorption radius, has the following value:

$$D_{a0} = 4\sqrt{\frac{f_0\delta}{\mu}}. \tag{251}$$

Here μ is the linear attenuation coefficient of X-rays or neutrons for the lens material. In a thick CRL, if a ray passed through the lens at a fixed distance from the lens axis, then the thin lens absorption aperture diameter D_{a0} would still be the correct expression for the absorption aperture. However, the ray does not remain at a fixed distance from the axis, so that the thick lens absorption aperture D_a becomes

$$\frac{D_a}{D_{a0}} = \frac{h_0}{\sqrt{\int_0^t \frac{dz}{t} h^2(z)}}. \tag{252}$$

Here h_0 is the radius at which the ray enters the lens and $h(z)$ is the radius as a function of longitudinal position.

Consider the case where the source or object distance ρ_0 is far from the front entrance of the CRL, such that

$$\frac{\sqrt{f_0 t}}{\rho_0} \ll 1. \tag{253}$$

As the thickness of the CRL increases, the absorption aperture diameter increases because the ray is focused toward the lens axis where the loss is less. The curve is terminated at

$$\frac{t}{f_0} = \left(\frac{\pi}{2}\right)^2. \tag{254}$$

Termination occurs because the source distance r_0 is restricted to larger values of t/f_0 so that the image plane lies outside the CRL. In the above case for t/f_0, the correction factor for the aperture is 1.4, which means that the one-dimensional gain for the lens is increased by the same factor; the resolution of the CRL, which is inversely proportional to the thick lens absorption aperture diameter D_a, is improved by 40%, and the 2D gain, which is proportional to D_a^2, increases by a factor of two. In addition, at this value for t/f_0 the chromatic aberration for the CRL, expressed by $\Delta f/f$, is half the value for the thin lens.

16. CRL SURFACE ROUGHNESS AND LENS ALIGNMENT

Surface roughness of a lens causes a deviation of the lens thickness from the thickness of an ideally smooth lens (Figure 13b). The deviation in lens thickness dT, and hence a neutron ray path length through a lens due to surface roughness, causes a deviation in its phase shift $d\varphi$. The additional phase shift $\Delta\phi$, introduced by the overall lens surface roughness, results

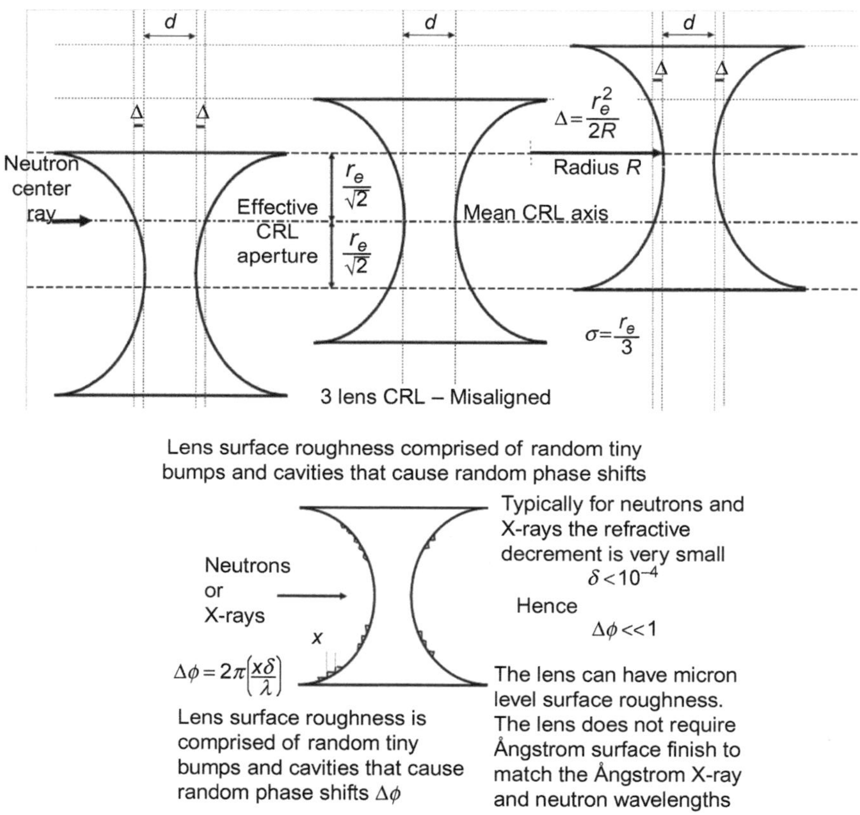

FIGURE 13 (Top) Misalignment of biconcave lenses of a CRL leads to a loss of X-ray or neutron intensity and decreases the CRL effective aperture radius r_e. (Bottom) The effect of lens surface roughness, which causes additional, usually random phase shifts, on the wave fronts of the refracted X-ray or neutron waves. The bumps' thickness or cavity depth x in a lens with decrement δ cause phase shift $\Delta\phi = 2\pi\ (x\delta/\lambda)$ in the X-rays or neutrons of wavelength λ (typically $\delta < 10^{-4}$) and hence X-ray and neutron lenses can have relatively rough surfaces.

in the standard deviation σ of the difference of the lens thickness from ideally smooth surface lens. A perfectly aligned N-lens CRL has $2N$ surfaces, whose standard deviations σ add in quadrature.

The overall phase shift $\Delta\phi$ through the CRL, caused by an individual lens surface roughness standard deviation σ, is

$$\Delta\phi = \frac{2\pi\,\delta\sigma\sqrt{2N}}{\lambda}. \tag{255}$$

Lens surface roughness effects on resolution are small if the resultant additional phase shift $\Delta\phi$ is less than $\pi/2$, or

$$\frac{4\delta\sigma\sqrt{2N}}{\lambda} < 1. \tag{256}$$

Since the refractive decrement δ of lens materials is typically very small; that is, $\delta < 10^{-4}$, the X-ray and neutron lenses can have surface roughness on the order of 100 μm without significant degradation of the image. That is, the Ångstrom range of X-ray and neutron wavelengths does not require Ångstrom-level surface smoothness.

The required maximum lens surface roughness is proportional to the square root of the number of lenses, N. Consider an aluminum neutron CRL ($N = 100$) at 10 Å ($\delta = 4 \times 10^{-5}$). If the surface error is entirely random, then a root mean square error of $\sigma_s \leq 6\,\mu\text{m}$ can be tolerated. This lens surface roughness is quite large and greater than that of even visible optics lenses. This is counterintuitive given that we are using lenses of optical or even infrared quality to focus neutrons whose wavelengths are anywhere from 1000 to 10,000 times smaller.

The critical angle of reflection of neutrons is $\theta_c = \sqrt{2\delta}$, measured relative to the surface tangent plane. Complete neutron reflection occurs from lens and prism surfaces at incident angles less than θ_c. Reflection becomes significant when θ_c becomes sufficiently large at the very long neutron wavelengths (very cold and ultracold), where typically $\delta > 10^{-3}$ for $\lambda > 40$ Å.

The effect of lens misalignment on the CRL performance due to unwanted additional phase shifting of the X-ray or neutron wave is minimal, if the phase shift is less than $\pi/2$ for the entire N-lens CRL. The increase in thickness Δd of a parabolic biconcave lens with an on-axis radius of curvature R, as a function of radial distance r from the CRL center axis, is

$$\Delta d = 2\left(\frac{r^2}{2R}\right) = \frac{r^2}{R}. \tag{257}$$

This result is also valid for a spherical lens of radius R for radial distances less than the parabolic aperture radius ($r < r_p$). Lens misalignment leads to an additional absorption factor on-axis given by

$$\Delta T = \exp(-N\mu \Delta d). \tag{258}$$

This leads to an overall average absorption T_{avg} over the absorption aperture diameter given by

$$T_{avg} \cong 0.43 \exp(-N\mu (d + \Delta d)). \tag{259}$$

Let us consider in more detail the phase shift and change of amplitude due to misalignment of parabolic biconcave lenses in an X-ray or neutron CRL. The squared amplitude magnitude in intensity and the change in phase leads to a change in amplitude, which leads to a change in the image intensity when lenses are misaligned in a CRL. The thickness t_i of the ith parabolic biconcave lens with minimum center or base thickness b and on-axis radius of curvature R is expressed in terms of the lens displacement by displacement d_i in a transverse direction from the mean, center z-axis of the N-lens CRL. At transverse distance h from the mean center axis, the thickness t_i of the ith parabolic biconcave lens is

$$t_i = b + \frac{(h - d_i)^2}{R}. \tag{260}$$

That is, for transverse distance h of the neutron ray or X-ray from the mean center z-axis of the CRL, the axial thickness of the displaced ith lens is t_i.

The displacement of the CRL lenses has transverse x- and y-component directions, which are calculated separately. That is, the transverse lens displacement distances d_i for $i = 1, 2, \ldots, N$ along the positive/negative transverse x-direction are calculated first Then, transverse lens displacement distances d_i for $i = 1, 2, \ldots, N$ along the positive/negative transverse y-direction are separately calculated. If the N lenses of the CRL are perfectly aligned, where the lens transverse displacement the mean center CRL z-axis is zero for each lens; that is, $d_i = 0$ for $i = 1, 2, \ldots, N$, then the CRL focal length f is

$$f = \frac{R}{2N\delta}, \tag{261}$$

where

$$t_i = b + \frac{(h - d_i)^2}{2f\,N\delta}. \tag{262}$$

In each transverse direction the displacement distance d_i of the ith lens center from the mean center z-axis of the CRL can be positive $d_i > 0$, negative $d_i < 0$, or zero $d_i = 0$. In each transverse direction, the distance h of the propagating X-ray or neutron ray is measured from the mean center z-axis of the N-lens CRL. And in each transverse direction, the average displacement distance $\bar{d}$ of the randomly offset lenses about the mean center z-axis is zero, where

$$\sum_{i=1}^{N} d_i = 0/ \tag{263}$$

Propagating through the ith lens thickness t_i, the wave amplitude ψ of the X-rays or neutrons is proportional to the $\exp(j\phi_i)$ in which the phase shift ϕ_i in the ith biconcave parabolic lens is

$$j\phi_i = \left(j\frac{2\pi}{\lambda}\delta - \frac{\mu}{2}\right) \cdot t_i. \tag{264}$$

The lens material refractive index n is given by

$$n = 1 - \delta - j\beta = 1 - \delta - j\frac{\mu\lambda}{4\pi}. \tag{265}$$

Substitution into Eq. 264 for lens thickness t_i, of Eq. 262, which occurs at transverse distance h from the mean center z-axis of the CRL, then gives the -(or y-) component contribution to the phase shift $j\phi_i$ that is due to the transverse x- (or y-) displacement d_i of the lenses, where

$$j\phi_i = \left(j\frac{2\pi}{\lambda}\delta - \frac{\mu}{2}\right)\left(b + \frac{(h - d_i)^2}{2f\,N\delta}\right), \tag{266}$$

where

$$\psi_i \propto \exp(j\phi_i). \tag{267}$$

Summing over the N parabolic biconcave lenses of the CRL, the contribution to the wave amplitude Ψ resulting from the lens misalignment in the transverse x- (or y-) directions is then

$$\Psi \propto \exp\left(j\sum_{i=1}^{N}\phi_i\right) = \exp\left[\left(j\frac{2\pi}{\lambda}\delta - \frac{\mu}{2}\right)\sum_{i=1}^{N}\left(b + \frac{(h - d_i)^2}{2f\,N\delta}\right)\right]. \tag{268}$$

Expansion of the argument in the exponential gives

$$\Psi \propto \exp\left[\left(j\frac{2\pi}{\lambda}\delta - \frac{\mu}{2}\right)\left(Nb + \frac{h^2}{2f\delta} - \frac{h}{fN\delta}\sum_{i=1}^{N} d_i + \frac{\sigma^2}{2f\delta}\right)\right]. \tag{269}$$

Here σ^2 is the variance of the distribution of transverse displacement distances d_i of the lens from the mean center z-axis of the CRL in the transverse x- or y-direction, where

$$\sigma^2 = \frac{1}{N}\sum_{i=1}^{N} d_i^2. \tag{270}$$

In the case of a spherical, biconcave lens CRL, a power series expansion can be performed on the thickness t_i expression for the ith spherical biconcave lens with minimum center thickness b in which the lens is displaced by transverse distance d_i from the mean center z-axis of the N-lens CRL. At transverse distance h from the mean center z-axis of the CRL, the axial thickness of the spherical biconcave lens t_i is then

$$t_i = b + \frac{(h - d_i)^2}{R} + \frac{(h - d_i)^4}{4R^3}. \tag{271}$$

The phase shift ϕ_i through a single spherical biconcave lens is

$$j\phi_i = \left(j\frac{2\pi}{\lambda}\delta - \frac{\mu}{2}\right)\cdot\left(b + \frac{(h - d_i)^2}{R} + \frac{(h - d_i)^4}{4R^3}\right). \tag{272}$$

One expands the above terms. The transverse displacement of the lenses are assumed random with a zero mean transverse distance $\bar{d}$ of the lenses from the mean center z-axis of the CRL, where

$$\bar{d} = \frac{1}{N}\sum_{i=1}^{N} d_i = 0. \tag{273}$$

The second σ^2, third σ^3, and fourth σ^4 moments about the mean center z-axis are

$$\sigma^2 = \frac{1}{N}\sum_{i=1}^{N} d_i^2 \quad \sigma^3 = \frac{1}{N}\sum_{i=1}^{N} d_i^3 \quad \text{and} \quad \sigma^4 = \frac{1}{N}\sum_{i=1}^{N} d_i^4. \tag{274}$$

One sums over the N spherical biconcave lenses of the CRL to obtain the x- (or y-) contribution to the wave amplitude Ψ resulting from the random lens displacement in the transverse x- (or y-) direction and its associated the -x (or y-) phase shift $j\phi$, where

$$\Psi \propto \exp(j\phi)\,, \tag{275}$$

where

$$j\phi = \left(j\frac{2\pi}{\lambda}\delta - \frac{\mu}{2}\right) \times \left(\frac{h^4}{8f\delta R^2} + \frac{h^2}{2f\delta}\left(1+\frac{3\sigma^2}{2R^2}\right) - \frac{h\sigma^3}{2f\delta R^2} + \left(Nb + \frac{\sigma^2}{2f\delta} + \frac{\sigma^4}{8f\delta R^2}\right)\right). \tag{276}$$

The coefficient of the h^4 term is the spherical aberration $S.A.$ for a perfectly aligned CRL with $d_i = 0$ for $i = 1, 2, \ldots, N$, where

$$S.A. = \frac{1}{8f\delta R^2}. \tag{277}$$

For the misaligned CRL, the coefficient of the h^2 term defines the misaligned focal length f' in terms of the focal length $f$of a perfectly aligned CRL, where

$$f' = f\left(1+\frac{3\sigma^2}{2R^2}\right)^{-1}. \tag{278}$$

If the variance σ^2 the spherical lens displacement is not the same in the two transverse x- and y- directions from the mean axial center line of the CRL, then the focal lengths in the xz and yz planes will be different, thereby giving rise to spherical lens astigmatism. The coefficient of the h term is the transverse displacement Δy of the image from the mean center CRL axis, where given image distance r_i,

$$\Delta y = \frac{r_i h\sigma^3}{2f\delta R^2}. \tag{279}$$

If the third-moment σ^3 is not the same in both the x- and y- transverse directions, then the CRL image will have distortion due to the different amounts of image translations in the transverse x- and y- directions. The last term in the exponential argument is a constant phase-shift offset in amplitude Ψ.

We now consider lens misalignment together with lens surface roughness for the case of a parabolic lens CRL, or spherical lens CRL with paraxial approximation in which the lenses are misaligned from the center mean axis with a Gaussian standard deviation σ and zero mean displacement $\bar{d} = 0$. That is, we assume CRL is effectively composed of N biconcave parabolic lenses. The surface roughness of each lens can be modeled by adding thickness Δt_i to the ith lens thickness t_i at distance h in the transverse x- (or y-) direction from the CRL mean center z-axis. The deviation of the lens thickness due to the surface roughness is assumed to be periodic with identical amplitude r and period p for each lens, where

$$\Delta t_i = r \cos\left[\frac{2\pi}{p}\left(h - d_i\right)\right]. \tag{280}$$

In addition to surface roughness, each ith lens has a transverse displacement d_i about the mean center z-axis, which is calculated separately for the transverse x- and y- directions. The x- (or y-) component of the phase shift $j\phi$ and wave amplitude Ψ due to the displacement of the N rough surface, biconcave parabolic lenses of the CRL in the x- (or y-) transverse direction are given by

$$j\phi = \sum_{i=1}^{N}\left(j\frac{2\pi}{\lambda}\delta\right)\left\{b + \frac{(h - d_i)^2}{2f N\delta} + r\cos\left[\frac{2\pi}{p}\left(h - d_i\right)\right]\right\}, \tag{281}$$

where

$$\Psi \propto \exp\left(j\phi_i\right). \tag{282}$$

Next we calculate the average phase shift due to the lens surface roughness only, which is determined by the last term in the $j\phi$ expression. The last term is the x- (or y-) phase contribution due to lens surface roughness with lens displacement in the transverse x- (or y-) directions about the mean center z-axis. One assumes the lens transverse displacement in the transverse x- and y-directions are Gaussian distributed with zero mean with respective variances σ_x^2 and σ_y^2. The Gaussian probability density ρ_x and ρ_y for the displacement of the parabolic biconcave lenses in the transverse x- and y-directions are

$$\rho_x = \frac{1}{\sqrt{2\pi\sigma_x^2}}\exp\left(-\frac{d_{x,i}^2}{2\sigma_x^2}\right) \quad \text{and} \quad \rho_y = \frac{1}{\sqrt{2\pi\sigma_y^2}}\exp\left(-\frac{d_{y,i}^2}{2\sigma_y^2}\right). \tag{283}$$

The cosine summation term is evaluated via the cosine angle difference identity. The simplifying assumption is made that the net displacement of net displacement of the lenses in the transverse x- (or y-) displacement $d_{x,y}$ in the CRL multiplied by the N lenses, where

$$\sum_{i=1}^{N}\cos\left[\frac{2\pi}{p}\left(h-d_i\right)_{x,y}\right]=N\left[\cos\left(\frac{2\pi}{p}h\right)\cos\left(\frac{2\pi}{p}d_{x,y}\right)+\sin\left(\frac{2\pi}{p}h\right)\sin\left(\frac{2\pi}{p}d_{x,y}\right)\right]. \qquad (284)$$

The mean phase-shift angle $\bar{\phi}_x$ and $\bar{\phi}_y$, which are due only to surface roughness, are calculated for the Gaussian-distributed, mean-displacement of a biconcave parabolic lens in the N-lens CRL in the respective transverse x- and y-directions, where

$$\bar{\phi}_x=\frac{2\pi}{\lambda}\frac{rN\delta}{\sqrt{2\pi\sigma_x^2}}\cos\left(\frac{2\pi}{p}h\right)\int_{-\infty}^{\infty}\cos\left(\frac{2\pi}{p}d_x\right)\exp\left(-\frac{d_x^2}{2\sigma_x^2}\right)d(d_x). \qquad (285)$$

$$\bar{\phi}_y=\frac{2\pi}{\lambda}\frac{rN\delta}{\sqrt{2\pi\sigma_y^2}}\cos\left(\frac{2\pi}{p}h\right)\int_{-\infty}^{\infty}\cos\left(\frac{2\pi}{p}d_y\right)\exp\left(-\frac{d_y^2}{2\sigma_y^2}\right)d(d_y). \qquad (286)$$

The integrals with the sine terms are zero

$$\int_{-\infty}^{\infty}\sin(bx)\exp\left(-a^2x^2\right)dx=0. \qquad (287)$$

To evaluate the cosine integral the tabulated integral rule for $ab\neq 0$ is applied, where

$$\int_{-\infty}^{\infty}\cos(bx)\exp\left(-a^2x^2\right)dx=2\int_{0}^{\infty}\cos(bx)\exp\left(-a^2x^2\right)dx=2\frac{\sqrt{\pi}}{2a}\exp\left(-\frac{b^2}{4a^2}\right), \qquad (288)$$

where

$$a^2=\frac{1}{2\sigma_{x,y}^2}\quad b=\frac{2\pi}{p} \qquad (289)$$

(see mathematic tables such as Selby and Girling (1965)). Integration yields mean phase-shift angles $\bar{\phi}_x$ and $\bar{\phi}_y$ (abbreviated $\bar{\phi}_{x,y}$) that are due only to the surface roughness of the Gaussian random displacement of the biconcave parabolic lenses in the transverse x- and y-directions about the zero-mean, center z-axis of the N-lens CRL, where

$$\bar{\phi}_{x,y} = \frac{2\pi}{\lambda}\delta N r \cos\left(\frac{2\pi}{p}h\right)\exp\left[-2\left(\frac{\pi\sigma_{x,y}}{p}\right)^2\right]. \tag{290}$$

The above result is only approximate because of the assumed possible transverse displacements of up to $\pm\infty$ in the transverse x- and y-directions by the choice of the $\pm\infty$ integration limits. Depending on the parameters, the above result is usually a good approximation for finite integration limits of at least twice the standard deviation in the x- and y directions, $\pm 2\sigma_{x,y}$. That is, one can neglect the lower and upper integration contributions from $-\infty$ to $-2\sigma_{x,y}$ and from $2\sigma_{x,y}$ to ∞ (for more details see Pantell *et al.*, 2001).

17. COMPOUND REFRACTIVE PRISMS: X-RAY AND NEUTRON DEFLECTION BY A SINGLE PRISM

A material or magnetic compound refractive prism (CRP) of length l_1 is composed of N material or magnetic prisms with $2N$ prism-air interfaces (Figure 14). Each material or magnetic prism has an apex angle α. The material prisms are created by fabricating a saw-tooth pattern in a rectangular bar of material or by stacking material prisms in one or more rows. In the magnetic prism, the prism-shaped magnetic field is produced by a pair triangular magnetic pole pieces, which are separated by a gap that is small compared with the pole surface dimensions.

A single prism of the CRP is an isosceles triangle with apex half-angle $\alpha/2$ (i.e., full vertex angle α) and height h with base half-width

$$w = h\tan\left(\frac{\alpha}{2}\right). \tag{291}$$

The base of the prism lies on the z-axis. The incident X-ray or neutron beam propagates in the positive z-direction parallel to the base of the prism or at a small angle ε relative to the z-axis and prism base. The X-ray or neutron ray is refracted at each material or magnetic prism surface by angle $\Delta\theta$ from angle θ_1 to θ_2 with respect to the prism surface normal, where

$$\theta_2 = \theta_1 + \Delta\theta \tag{292}$$

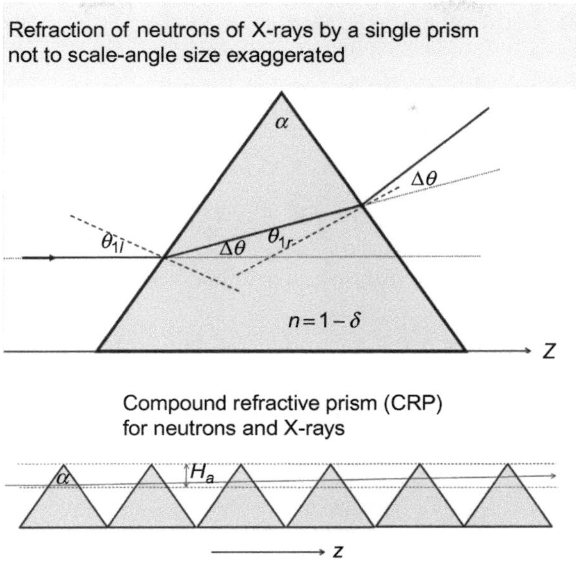

FIGURE 14 (Top) The refraction of a neutron ray through a single prism. (Bottom) A compound refractive prism (CRP) composed of a row of *N* single prisms. The critical depth H_a is indicated in which the transmitted intensity at the critical depth is $1/e^2$ of the transmitted intensity by a ray that passes through the vertex points of the *N* prisms.

and

$$\Delta\theta \ll 1. \tag{293}$$

At the left prism surface, the ray incidence angle is θ_{1l} (relative to the normal), and Snell's law is applied at the entrance or left face of the prism, which gives

$$\sin\theta_{1l} = (1-\delta)\sin(\theta_{1l} + \Delta\theta_l). \tag{294}$$

The incident ray on the left prism surface is bent or refracted by angle $\Delta\theta_l$. At the right prism surface, the ray incidence angle is θ_{1r} (relative to the normal), and Snell's law is applied at the exit or right face of the prism, which gives

$$(1-\delta)\sin\theta_{1r} = \sin(\theta_{1r} + \Delta\theta_r). \tag{295}$$

The incident ray on the right prism surface is refracted by angle $\Delta\theta_r$. The sine angle-sum identity is used:

$$\sin\theta_2 = \sin(\theta_1 + \Delta\theta) = \sin\theta_1\cos\Delta\theta + \cos\theta_1\sin\Delta\theta. \tag{296}$$

The refraction condition at the left and right surfaces of the prism becomes

$$\sin\theta_{1l} = (1-\delta)\left(\sin\theta_{1l}\cos\Delta\theta_l + \cos\theta_{1l}\sin\Delta\theta_l\right) \quad \text{left} \tag{297}$$

and

$$(1-\delta)\sin\theta_{1r} = \sin\theta_{1r}\cos\Delta\theta_r + \cos\theta_{1r}\sin\Delta\theta_r \quad \text{right.} \tag{298}$$

The small angle $\Delta\theta$ approximation is invoked, where

$$\cos\Delta\theta \cong 1 \quad \text{and} \quad \sin\Delta\theta \cong \Delta\theta. \tag{299}$$

The product term with both δ and $\Delta\theta_l$ is much smaller than the other terms and is thus neglected by setting to zero, where

$$\delta\cos\theta_{1l}\Delta\theta_l \cong 0. \tag{300}$$

The deviation angles of the ray on the left $\Delta\theta_l$ and right $\Delta\theta_r$ sides of the prism is then

$$\Delta\theta_l = \delta\tan\theta_{1l} \quad \text{left} \tag{301}$$

and

$$\Delta\theta_r = -\delta\tan\theta_{1r} \quad \text{right.} \tag{302}$$

Ray angles are measured relative to the direction of the positive z-axis. A ray angle measured counterclockwise from a surface normal or line of ray incidence is positive and measured clockwise the angle is negative. Suppose the neutrons propagate parallel to the z-axis and the base of the prism. The angle sign convention (simple trigonometry) can now determine the angle of incidence of the neutron beam at the left prism face θ_{1l}, which is measured relative to the surface normal pointing diagonally upward and back toward the source plane, and within the prism material the angle of incidence θ_{1r} of the neutron beam at the right prism face, which is measured relative to the surface normal pointing diagonally upward and back into the prism material, where

$$\theta_{1l} = -\theta_{1r} = \frac{\alpha}{2}. \tag{303}$$

The incident angles θ_{1l} and θ_{1r} relative to their surface normals, which are now expressed in terms of the prism apex half-angle $\alpha/2$, are inserted into the expressions for the X-ray or neutron ray deviation angles $\Delta\theta_l$ and $\Delta\theta_r$ for the left and right prism surfaces, where

$$\Delta\theta_l = \delta \tan\frac{\alpha}{2} \quad \text{left} \tag{304}$$

and

$$\Delta\theta_r = \delta \tan\frac{\alpha}{2} \quad \text{right.} \tag{305}$$

The change of the X-ray or neutron ray angle $\Delta\theta$ in the ray passage across both sides of the prism is then

$$\Delta\theta = \Delta\theta_l + \Delta\theta_r = 2\delta \tan\frac{\alpha}{2} \quad \text{prism.} \tag{306}$$

The X-ray or neutron ray in its passage through the prism is deflected counterclockwise relative to the line of incidence, which is vertically upward, away from the prism base if the decrement is positive (for most isotopes, $\delta > 0$). If the decrement is negative ($\delta < 0$, e.g., titanium for neutrons), then the ray deflection is downward toward the prism base. Note that a biconcave lens is crudely described by a bottom upright prism whose top vertex touches the bottom vertex of the inverted top prism in which the two prism vertices touch at the center of the "triangular" biconcave lens.

Now suppose the incident X-ray or neutron beam travels in the general positive z-direction but at an angle ε relative to the positive z-axis on which sits the prism base. The incident angles θ_{1l} and θ_{1r} relative to their surface normals, now become

$$\theta_{1l} = -\theta_{1r} = \frac{\alpha}{2} + \varepsilon. \tag{307}$$

Substitution of Eq. 307 into Eqs. 304 and 305 gives the deviation angles of the ray on the left $\Delta\theta_l$ and right $\Delta\theta_r$ sides of the prism is then

$$\Delta\theta_l = \delta \tan\left(\frac{\alpha}{2} + \varepsilon\right) \quad \text{left} \tag{308}$$

and

$$\Delta\theta_r = \delta \tan\left(\frac{\alpha}{2} + \varepsilon\right) \quad \text{right.} \tag{309}$$

The change of the X-ray or neutron ray angle $\Delta\theta$ in the ray passage across both sides of the prism is then with the added mathematical inconvenience of angle ε:

$$\Delta\theta = \Delta\theta_l + \Delta\theta_r = 2\delta \tan\left(\frac{\alpha}{2} + \varepsilon\right) \quad \text{prism.} \tag{310}$$

From trigonometry,

$$\tan(z_1 + z_2) = \frac{\tan z_1 + \tan z_2}{1 - \tan z_1 \tan z_2} \cong \tan z_1 + z_2 \quad z_2 \ll 1, \tag{311}$$

where

$$z_1 = \frac{\alpha}{2} \quad \text{and} \quad z_2 = \varepsilon \ll 1. \tag{312}$$

The change of the X-ray or neutron ray angle $\Delta\theta$ in the ray passage across the prism becomes

$$\Delta\theta = \Delta\theta_l + \Delta\theta_r = 2\delta\left(\tan\frac{\alpha}{2} + \varepsilon\right) \quad \text{prism}. \tag{313}$$

18. DEFLECTION OF X-RAYS AND NEUTRONS IN A COMPOUND REFRACTIVE PRISM

Due to the very small deflection angle $\Delta\theta$ by each prism (because $\delta \ll 1$), the X-ray or neutron ray is nearly parallel to the prism bases in its passage through an N-prism CRP, so the total CRP deflection angle is approximated by

$$\theta_p = 2N\Delta\theta. \tag{314}$$

Each prism of the CRP has with vertex angle α. Each prism of the CRP deflects an incident ray by angle

$$2\delta \tan\left(\frac{\alpha}{2}\right). \tag{315}$$

For an N-prism CRP of total length l_1, the deflection angle $\phi(z)$, as a function of propagation distance z in the CRP, is

$$\phi(z) = 2N\left(\frac{z}{l_1}\right)\Delta\theta. \tag{316}$$

The deflection angle axial gradient $d\phi/dz$ is then

$$\frac{d\phi}{dz} = \frac{2N\delta}{l_1}\tan\left(\frac{\alpha}{2}\right). \tag{317}$$

For X-rays and neutrons δ is much less than 1 and the critical angle of reflection ϕ_c of X-rays or neutrons is thus also very small, where

$$\theta_c = \sqrt{\delta} \ll 1 \quad \delta \ll 1. \tag{318}$$

The critical angle of reflection ϕ_C is measured relative to the tangent plane of the surface. Reflection is significant for very small, grazing incidence angles. This not the case with a prism in which the X-rays or neutrons are incident at large angles ϕ relative to the tangent plane of the prism surface, where

$$\theta_C \ll \phi. \tag{319}$$

Thus, the prism completely transmits the X-rays or neutrons from the geometric optics view.

X-ray or neutron attenuation occurs in passage through the prism material. One must choose a prism material with minimum X-ray or neutron losses due to absorption and inelastic scatter, but with maximum refractivity due to coherent elastic scatter. However, the neutron attenuation coefficient in optical materials is much smaller than that of X-rays so that neutrons are more readily transmitted than X-rays. X-rays are strongly attenuated in trajectories through a CRP compared with neutrons, even through the CRP apexes where the least amount of prism material is encountered. As a result, CRPs are usually used with neutrons and not X-rays.

The X-ray or neutron refraction depends on decrement δ, which increases quadratically with X-ray or neutron wavelength λ, where

$$\delta \propto \lambda^2. \tag{320}$$

The X-ray or neutron deflection angle of the CRP is dependent on X-ray or neutron wavelength via

$$\Delta\theta \propto \lambda^2. \tag{321}$$

Hence, a prism can be used as a monochromator for X-rays and neutrons (as with visible light). A series of N prisms forms a compound refractive prism with $\theta = \alpha/2$, and yields total deflection angle,

$$\Delta\theta_{CRP} = 2N\Delta\theta. \tag{322}$$

Substitution into Eq. 322 for $\Delta\theta$ via Eqs. 304 and 305 gives the total upward neutron deflection angle produced by the N prisms of the CRP, where

$$\Delta\theta_{CRP} = 2N\delta \tan\left(\frac{\alpha}{2}\right) \tag{323}$$

In this simple case, one assumes the incident direction of the neutrons is parallel to the CRP base.

For example, at neutron wavelength $\lambda = 1.8$ Å the decrement δ for aluminum is $\delta = 1.07 \times 10^{-6}$, and an aluminum CRP with $N = 140$ prisms each with apex angle $\alpha = 120°$ deflects a beam of neutrons (incident parallel to the CRP base) by total deviation angle $\Delta\theta_{CRP}$ of 0.53 mrad. This corresponds to a neutron vertical deflection per one meter of neutron propagation downstream of the CRP exit of

$$D = 1000\Delta\theta_{CRP}. \tag{324}$$

With deflection angle 0.26 mrad the vertical deflection is $D = 0.53\,\text{mm/m}$. By comparison at 1.8 Å for X-rays aluminum has almost the same decrement $\delta = 1.14 \times 10^{-5}$.

19. CALCULATION OF X-RAY OR NEUTRON ABSORPTION APERTURE DEPTH AND TRANSMISSION IN A COMPOUND REFRACTIVE PRISM

The CRP absorption aperture depth H_a is the effective aperture width of the CRP (see Figure 14). The critical depth H_a is defined as the distance between the apex of any one of its prisms and a plane parallel to the prism base in which the transmitted X-ray or neutrons intensity has decreased by a factor e^{-2} compared with the intensity passing through the apex. The length of neutron flight d through the prism at distance (depth) h from the prism apex is

$$d = 2h \tan\left(\frac{\alpha}{2}\right). \tag{325}$$

For a single prism the transmitted X-ray or neutron intensity transmitted $I(h)$ at depth h compared with the transmitted intensity at the prism apex, which is also the incident neutron intensity $I(0)$, is

$$I(h) = I(0)\exp(-\mu d) \tag{326}$$

or

$$I(h) = I(0)\exp\left[-2\mu h \tan\left(\frac{\alpha}{2}\right)\right]. \tag{327}$$

For the N-prism CRP, the transmitted intensity of X-rays or neutrons at depth h below the prism apexes is

$$I(h) = I(0)\,T(h). \tag{328}$$

With transmission

$$T(h) = \exp\left[-2N\mu h \tan\left(\frac{\alpha}{2}\right)\right]. \tag{329}$$

The absorption aperture depth H_a is

$$I(h = H_a) = I(0)\exp(-2). \tag{330}$$

The absorption aperture depth H_a is

$$H_a = \frac{1}{N\mu}\tan\left(\frac{\alpha}{2}\right). \tag{331}$$

For example, at neutron wavelength $\lambda = 1.8$ Å the linear attenuation for aluminum is $\mu = 0.104\,\text{cm}^{-1}$, and an aluminum CRP with $N = 140$ prisms, each with apex angle $\alpha = 120°$, has an absorption aperture of $H_a = 0.4\,\text{mm}$. An N-prism CRP whose grooves are cut with a physical aperture depth H has a length

$$L = 2Nh\tan\left(\frac{\alpha}{2}\right), \tag{332}$$

For example, with a physical aperture depth $H = 1.0\,\text{mm}$ the CRP has length $L = 48.5\,\text{cm}$. By comparison at 1.8 Å for X-rays aluminum has a much larger linear attenuation $\mu = 202\,\text{cm}^{-1}$.

We now consider the average X-ray or neutron transmission through a CRP with

$$A = 2N\mu\tan\left(\frac{\alpha}{2}\right) \tag{333}$$

and

$$T(h) = \exp(-Ah). \tag{334}$$

The average transmission $< T >$ of X-rays or neutrons through the N-prism CRP for aperture depth $h = H$ is given by

$$< T > = \frac{1}{H}\int_0^H T(h)dh. \tag{335}$$

Integration yields

$$< T > = \frac{1 - T(H)}{AH} \tag{336}$$

or

$$< T > = \frac{1 - \exp(-AH)}{AH}. \tag{337}$$

For a CRP aperture from prism apex to absorption aperture depth H_a in the above example, the average neutron transmission fraction is 0.2.

20. USE OF A COMPOUND REFRACTIVE PRISM TO OFFSET NEUTRON GRAVITY DROOP IN SMALL-ANGLE SCATTERING

Neutrons have a mass m and hence are pulled downward by gravity, especially cold neutrons with wavelength $\lambda > 5\text{Å}'$, which travel down long distances on small-angle scattering (SANS) beam lines. In SANS the neutrons are scattered by a specimen material such that the initial neutron wave number vector k_i is changed to k_f, where

$$Q = k_f - k_i. \tag{338}$$

The neutron initial energy E_i is changed to final energy E_f, where

$$\Delta E = E_f - E_i. \tag{339}$$

The neutron momentum p is related to its wave number k by $p = \hbar k$ via Planck's constant $\hbar = h/2\pi$, and the neutron energy E is related to the neutron frequency f via $E = \hbar w$ with $w = 2\pi f$. The scattering process measures the change of neutron momentum $\hbar Q$ in which elastic scattering has $\Delta E = 0$ and inelastic scattering has $\Delta E \neq 0$. The change in neutron energy and momentum measured by SANS allows determination of the positions and motions of atoms in condensed matter.

Gravity droop of cold neutrons in the long SANS beam lines limits the minimum scattered-induced neutron momentum change $\hbar Q$, which can be measured at the SANS detector. That is, gravity droop limits the minimum resolution of the SANS. However, the CRP with its base at bottom and apex at top vertically deflects a neutron beam of wavelength λ upward by angular deflection in the direction opposite the gravity droop, as follows:

$$\Delta\theta_{CRP} \propto \lambda^2 \tag{340}$$

An inserted CRP can be designed to offset the gravitational droop of cold neutrons independent of their wavelength, such that $q_{\min}$ is minimized, thus improving the resolution of the SANS for cold neutrons.

We calculate the prism parameters needed to offset gravitational droop in SANS as follows. A neutron beam is collimated between a source aperture and a specimen aperture separated by distance L_1. The N-prism CRP is placed adjacent to the sample. From the specimen aperture to the SANS detector is distance L_2. Neutrons from the source pass distance L_1 and pass through the CRP into the adjacent sample and then travel distance L_2 to the detector. The downward velocity acquired by the neutrons prior to the CRP due to gravity is compensated by the upward velocity given to the neutrons by the CRP.

The vertical drop y_g of the neutrons due to gravitational acceleration $g = 9.8\ \mathrm{m/sec}^2$ in its propagation over distance L_2 from the specimen aperture to detector is

$$y_g = \frac{gm^2}{2h^2}\,(L_1 + L_2)\left(L_2\lambda^2\right). \tag{341}$$

Here h is Planck's constant and m is the neutron mass. These expressions can be simplified via evaluation of constants, where

$$G_0 = \frac{gm^2}{2h^2} = 3.11 \times 10^7 \times \mathrm{cm}^{-3} = 3.11 \times 10^{-9}\,\mathrm{cm}^{-1}A^{-2}. \tag{342}$$

The prism with apex full angle α upwardly deflects the neutrons over distance L_2 from the specimen aperture to the detector by vertical upward distance y_p, where

$$y_p = L_2\Delta\theta_{CRP}. \tag{343}$$

The downward drop y_g due to gravity is set equal to the upward deflection y_p due to the CRP, where

$$y_g = y_p. \tag{344}$$

Substitution of Eqs. 341 and 343 into Eq. 344 gives

$$G_0\,(L_1 + L_2)\left(L_2\lambda^2\right) = L_2\Delta\theta_{CRP}. \tag{345}$$

The required upward neutron deflection angle $\Delta\theta_{CRP}$ required of the CRP to offset the gravity droop over the SANS source to detector distance $L_1 + L_2$ is

$$\Delta\theta_{CRP} = G_0\,(L_1 + L_2)\,\lambda^2. \tag{346}$$

The neutron deflection angle produced $\Delta\theta_{CRP}$ by a CRP is given by

$$\Delta\theta_{CRP} = 2N\delta \tan\left(\frac{\alpha}{2}\right). \tag{347}$$

Equating the expression for CRP neutron ray deflection angle $\Delta\theta_{CRP}$ with the required neutron ray deflection angle $\Delta\theta_{CRP}$ required to offset gravity of the source to detector distance then gives

$$2N\delta \tan\left(\frac{\alpha}{2}\right) = G_0 (L_1 + L_2) \lambda^2. \tag{348}$$

The decrement δ depends on wavelength λ, atom densityρ, bound coherent scattering length b, and wavelength λ, where

$$\delta = \frac{\lambda^2 b\rho}{2\pi}. \tag{349}$$

One requires only the scatter length b_i (cm) and atom density ρ_i (#/cm^3) for each ith isotope of the Q isotopes that compose the prism material to obtain the scatter length density s_l (#/cm^2) for the prism material, where

$$s_l = b\rho = \sum_{i=1}^{Q} b_i \rho_i \tag{350}$$

and

$$\delta = \frac{s_l \lambda^2}{2\pi}. \tag{351}$$

Substitution of δ of Eq. 351 into Eq. 348 for the equated pair of expressions for $\Delta\theta_{CRP}$ results in removal by division of the wavelength dependence, where

$$\tan\left(\frac{\alpha}{2}\right) = \frac{G_0 \pi}{s_l N} (L_1 + L_2). \tag{352}$$

For individual prism height h and base width w,

$$\tan\left(\frac{\alpha}{2}\right) = \frac{w}{2h}. \tag{353}$$

The CRP design equation becomes

$$\frac{w}{2h} = 9.77 \times 10^{-9} \frac{L_1 + L_2}{s_l N}. \tag{354}$$

The CRP material the neutron scatter length density s_l is in units of $Å^{-2}$, and each prism of the N-prism CRP has base width w (cm) and height h (cm). The source to CRP/sample distance L_1 (cm) and CRP/sample to detector distance L_2 (cm) are adjusted with w (cm) and h (cm) to eliminate gravity droop of the neutrons at the SANS detector.

For example, a CRP composed of magnesium fluoride MgF_2 prisms has scatter length density s_l of $5.13 \times 10^{-6} A^{-2}$ so that

$$\frac{w}{2h} = 1.90 \times 10^{-3} \frac{L_1 + L_2}{N}. \tag{355}$$

If there are 10 prisms ($N = 10$) in the CRP and the source to SANS detector distance $L_1 + L_2$ is 3×10^3 cm, then the prism base divided by height is approximately unit, where $w/h = 1.14$ so that the full prism apex angle $\alpha = \pi/2$. The isosceles MgF_2 prism with a 90° full apex angle is thus used to allow a neutron to maintain a constant height in its trajectory from the source through the prism and adjacent sample to the SANS detector. Any deviation in the height of the neutron is then attributed to small-angle scatter by the sample.

21. CURVATURE AND RADIUS OF CURVATURE OF LENS SURFACE AND WAVE SURFACE

The refraction of photons and neutrons can be considered from the viewpoint of the surfaces of the X-ray or neutron wave and the surface interfaces between material media or material medium and vacuum. The ray is the wave vector perpendicular to a small patch of wave surface. X-ray or neutron waves can be described by ray paths or by the curvature and change of curvature of the wave surface as the wave propagates through regions of different media whose interface boundaries present refractive surfaces with curvature. Geometric optics can be described by the ray trajectories, which are the trajectories of the normal vectors (i.e., wave vectors) of the electromagnetic wave surfaces or by the curvature and change of curvature of the wave surfaces.

One can describe by geometric optics the wave front shape or wave curvature as the wave propagates through a series of media and optical elements. The X-ray or neutron wave surface is described by a tangent plane. At each point on the wave surface is a tangent plane and from that point is a normal or perpendicular line to the tangent plane. This perpendicular line or ray points in the direction of propagation of the infinitesimal wave surface patch centered about the point on the wave surface. The following sections describe the wave propagation in terms of the curvature of the wave and the interfaces and optical elements encountered. Optometry and many optical devices are designed and described in terms of powers, radius of curvature, curvature, and vergence. X-ray and neutron optical elements, and X-ray and neutron waves can be described similarly.

A *plane surface* is described as a surface where the perpendicular lines or normals at all points are parallel to each other. A *spherical surface* is one in which all the normals pass through a single point. A spherical surface has a center of curvature O and a radius of R, defined as the distance from the center of the circle O to a point P1 on the surface. The radial line from the center of the circle O passing through point P1 on the spherical surface is perpendicular (or normal) to a plane tangent to the surface at point P1. The arc length p along the surface between two points P1 and P2 is the radius R multiplied by the angle $\Delta\theta$ between the radial lines originating at O and passing through P1 and P2; that is,

$$\Delta\theta = \frac{p}{R}. \tag{356}$$

The *curvature* of the surface C is defined as the angle through which the surface turns in a unit length of arc $p = 1$ from surface point P1 to point P2. That is, curvature C is the value of angle $\Delta\theta$ when the arc length p is equal to 1, such that

$$\mathrm{C} = \frac{1}{R}. \tag{357}$$

The radius of curvature ρ is the reciprocal of the curvature. The radius of curvature ρ of one half of a biconcave spherical lens is thus the radius R of the spherical lens surface; that is,

$$\rho = R. \tag{358}$$

A parabola can also be defined as the locus of points in a plane that are equidistant from a given point (the focus) and a given line (the directrix). A parabolic surface of one half of a biconcave parabolic lens can be described by a parabolic curve, which is rotated azimuthally about the center axis of the lens (z-axis). The left half or right half parabolic lens thickness along z, t_z is a function of the radial position l on the lens. The radius l lies in the lens transverse midplane, which is perpendicular to the z-axis. This midplane separates the parabolic lens into its left half (facing the $-z$ axis) and right half (facing the $+z$ axis). One half of the parabolic lens surface, which is described by its axial half-thickness t_z as a function of the lens radial position l, is given by

$$t_z = \frac{d}{2} + \frac{l^2}{2R}. \tag{359}$$

From elementary calculus the equation of curvature C for a 2D curve is

$$C = \frac{d^2 t_z}{dl^2} \cdot \left[1 + \left(\frac{dt_z}{dl}\right)^2\right]^{-3/2}. \tag{360}$$

For the parabola the curvature C and its reciprocal, the radius of curvature ρ, are given by

$$C = \frac{1}{R}\left[1 + \left(\frac{l}{R}\right)^2\right] \tag{361}$$

and

$$\rho = \frac{1}{C} = R\left[1 + \left(\frac{l}{R}\right)^2\right]^{3/2}. \tag{362}$$

The on-axis $(l = 0)$ curvature of a parabola $C_{par}\,(l = 0)$ is

$$C_{par}\,(l = 0) = \frac{1}{R}. \tag{363}$$

The parabola on-axis radius of curvature $\rho_{par}\,(l = 0)$ is

$$\rho_{par}\,(l = 0) = R. \tag{364}$$

22. MEASUREMENT OF SPHERICAL AND PARABOLIC LENS CURVATURE

To measure the radius of curvature of a spherical or parabolic lens, one need not know the location of the origin of the radius of curvature, provided one can measure a chord length and corresponding sag of the lens surface. In the case of a parabolic lens, a small area of a parabolic lens surface can be approximated as a spherical surface. Consequently, one need measure only a chord and its sag (see Freeman and Hull, 2003). The chord is the line segment between the two endpoints of an arc on the spherical concave surface. Bisecting and perpendicular to the chord is the sag, which is the line segment from the midpoint of the chord C1 to the midpoint of the arc A1. That is, the sag is a radial segment of length s, which is much less than the radius of curvature R of the spherical surface (Figure 15).

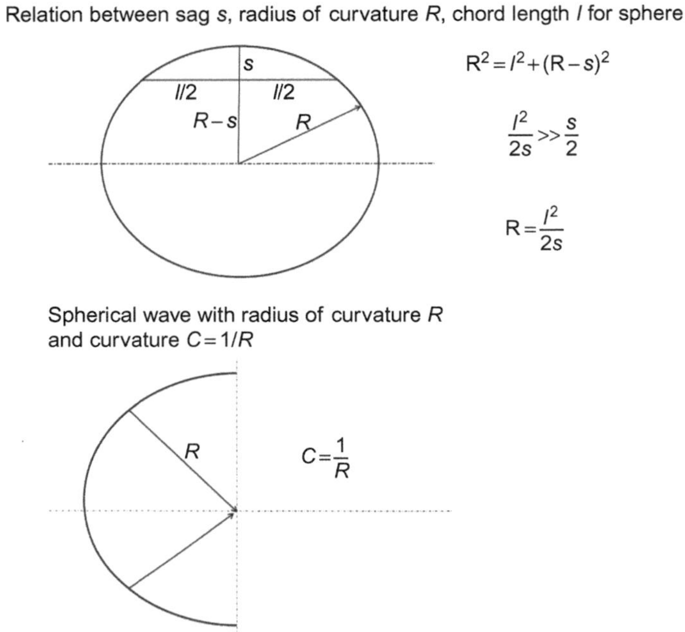

FIGURE 15 (Top) The relation among sag s, chord length l, and radius of curvature R for a sphere of radius R. (Bottom) A convergent spherical wave of radius of curvature R and curvature $C = 1/R$.

The chord has length l, the sag has length s, and the radius of the concave spherical surface from center O to A1 has length R; these are related by the Pythagorean theorem, where

$$R^2 = l^2 + (R - s)^2 . \tag{365}$$

If

$$\frac{l^2}{2s} \gg \frac{s}{2}, \tag{366}$$

then

$$R = \frac{l^2}{2s}. \tag{367}$$

The corresponding measured curvature $C_{sph}(l)$ for the spherical surface is then

$$C_{sph}(l) = \frac{2s}{l^2}. \tag{368}$$

One need only to contact the concave or convex lens at its surface with three pins: a pin at each end of the chord (and arc) and a pin at the center point of the arc. The pins are spring-loaded and connected to dial gauges. Placement of this spherometer against the lens surface presses the outer two pins against the lens surface, causing them to retract, while the inner pin protrudes into the gap between the chord center and arc center, until it to touches the lens surface. The difference in the reading of the protruding center pin with respect to the outer two retracted pins yields the sag length s, and the distance between the two outer pins (set by the spherometer instrument) determines the chord length l.

23. WAVE VERGENCE AND WAVE VERGENCE CHANGE BY LENS SURFACE

Lenses change the curvature of a wave front and change the direction of the rays, which are perpendicular to the wave front. An incident wave front has a curvature C_0 when it encounters the surface of an X-ray or neutron optic. For example, a planar incident wave with zero curvature $C_0 = 0$ encounters a spherical lens with radius of curvature R and curvature,

$$C = \frac{1}{R}. \tag{369}$$

The curvature of the wave is changed in its passage through the surface of the optic. X-ray and neutron wave fronts and lenses each can be described by radius of curvature R and curvature C and the changes of R and C during propagation through an optical system. (For radius of curvature, curvature, vergence, and refractive power, see optics texts by Pedrotti and Pedrotti (1993) and Freeman and Hull (2003)).

Consider a spherical wave that is emitted from a point O. In time τ the spherical wave expands to radius R_0 in vacuum or air ($n_0 = 1$):

$$R_0 = \frac{v_{p0}\tau}{n_0}. \tag{370}$$

In the same time interval τ the spherical wave expands to radius R, where

$$R = \frac{v_{p0}\tau}{n}. \tag{371}$$

That is, in time interval τ

$$\tau = n_0 R_0 = nR. \tag{372}$$

Substitution into Eq. 372 of Eq. 4 for n , neglecting attenuation, yields

$$R = (1 + \delta) R_0. \tag{373}$$

If $\delta > 0$, then the spherical wave that expands in the material has a larger radius of curvature R, but smaller curvature $C = 1/R$, than the spherical wave that expands in vacuum or air.

The wave travels through a medium of refractive index n. The product of the medium refractive index n and the reciprocal of the radius of curvature R of the wave is called the vergence V, where

$$V = \frac{n}{R}. \tag{374}$$

In air or vacuum ($n = 1$) the vergence V is simply the reciprocal of the radius of curvature of the wave, where $V = 1/R$.

The vergence V can also be defined as refractive index n multiplied by the curvature C:

$$V = nC. \tag{375}$$

Recall that the curvature C of the wave or lens surface is the reciprocal of the radius of curvature R of the wave or lens surface, where

$$C = \frac{1}{R}. \tag{376}$$

Curved surfaces and lenses cause a change in the vergence of a wave front, and their effect on a neutron wave or X-ray wave can be calculated easily by using the concept of vergence. In visible light optics the unit of vergence is the diopter. The radius of curvature R of the wave fronts, curved surfaces, and lenses are measured in meters. The ability or power of lenses or optical surfaces to change the vergence of an incident neutron or X-ray wave front can thus be measured in diopters, as in visible light optics.

24. REFRACTIVE INDEX, SNELL'S LAW, HUYGENS' PRINCIPLE, AND FERMAT'S PRINCIPLE OF LEAST TIME

X-ray optics and neutron optics define the refractive index n as the velocity of the neutron (X-ray) wave in the material v divided by its velocity in vacuum v_0, where

$$n = \frac{v}{v_0} \leq 1 \quad \text{X-ray/neutron.} \tag{377}$$

The vacuum velocity is the speed of light for X-rays ($v_0 = c$) and for vacuum group velocity v_g of the neutron wave is equal to the neutron particle velocity v_0, where

$$v_g = v_0 = \frac{h}{m\lambda} \quad \text{neutron.} \tag{378}$$

Note that the neutron phase velocity v_p is half the neutron group velocity v_g, so that the neutron phase velocity v_p is one half the neutron particle velocity v_0, where

$$v_p = \frac{v_g}{2} = \frac{v_0}{2} = \frac{h}{2m\lambda} \quad \text{neutron.} \tag{379}$$

Historically, the relative refractive index of a material is defined as the velocity of the light wave in vacuum c divided by the velocity of the light wave in a material v so that

$$n = \frac{c}{v} \geq 1 \quad \text{light.} \tag{380}$$

The refractive index of materials for visible light is thus greater than 1 because the velocity light in materials v is less than the speed of light in vacuum c.

Snell's law of refraction can be derived by examining the incidence of a plane wave traveling from a medium 1 with refractive index n_1 through a planar boundary into a medium 2 with refractive index n_2. For a given distance d along the planar surface, the angle of incidence θ_1 between the wave front of the plane wave and the planar surface yields the distance d_1 of the lagging edge of the wave front from the surface, when the leading edge of the wave front initially contacts the surface, where

$$d_1 = d \sin \theta_1. \tag{381}$$

This is analogous to the calculation of the far-field diffraction pattern produced by a plane wave passing through a slit.

By Huygens' principle each point on the wave front surface is a source of spherical waves, and the constructive and destructive interference of the spherical wavelets from each point on the wave front produce the subsequent wave front. The time τ_1 required for the lagging wave, moving at phase velocity v_0/n_1 in medium 1 to reach the surface of medium 2 is

$$\tau_1 = \frac{d_1 n_1}{v_0}. \tag{382}$$

In the meantime, the point of the leading edge of the wave front has become a source or center of a spherical wave that has expanded at phase velocity v_0/n_2 to a radius of d_2 in the same time τ_2, where

$$\tau_2 = \frac{d_2 n_2}{v_0}. \tag{383}$$

The distance d_2 of the leading edge of the expanded wave front from the surface is

$$d_2 = d \sin\theta_2. \tag{384}$$

The refracted wave front has changed direction so that its angle with respect to the surface of medium 2 is θ_2. Equal times of propagation yield unequal distances of propagation because of the differing phase velocities of medium 1 and 2, so that

$$\tau = \tau_1 = \tau_2 \tag{385}$$

and

$$\tau = \frac{d_1 n_1}{v_0} = \frac{d_2 n_2}{v_0}. \tag{386}$$

Substitution for d_1 and d_2 of Eqs. 381 and 384 into Eq. 386, yields Snell's law of refraction:

$$n_1 \sin\theta_1 = n_2 \sin\theta_2. \tag{387}$$

The laws of refraction and reflection are a result of Fermat's principle of least time, which states the actual path traveled by a wave in going from one point to another is that which requires the least time. The time τ_o taken to travel distance z at vacuum phase velocity v_o is

$$\tau_o = \frac{z}{v_0}. \tag{388}$$

The time τ to travel the same distance z in a medium of refractive index n is

$$\tau = \frac{nz}{v_0} \tag{389}$$

Here $\tau < \tau_o$ for $n < 1$, when for X-rays and neutrons $\delta > 0$. The distance nz is termed the *optical path length* or optical thickness. The ray paths defined by Fermat's principle, and therefore by the laws of refraction and reflection, do not depend on the direction of the wave from point A to point B. This is the principle of *reversibility of optical path* and can be used to simplify the calculation of a ray path from an object point through a lens to an image point.

25. THE PARAXIAL APPROXIMATION AND THE FUNDAMENTAL PARAXIAL EQUATION

The fundamental paraxial equation can be applied to single refracting surfaces, lenses, and mirrors and is derived as follows. Snell's law of refraction is

$$n_1 \sin\theta_1 = n_2 \sin\theta_2. \tag{390}$$

This equation can be simplified by the small-angle paraxial approximation in which

$$\sin\theta_1 \cong \theta_1 \quad \text{and} \quad \sin\theta_2 \cong \theta_2. \tag{391}$$

Hence,

$$n_1\theta_1 = n_2\theta_2. \tag{392}$$

The resulting paraxial or Gaussian optics includes the approximate relation of the wave front radius curvature R to a small wave front arc and its corresponding chord length l and sag length s; that is,

$$R = \frac{l^2}{2s}. \tag{393}$$

The paraxial approximation assumes all object points are near the center (optical) axis, the lens apertures are small, and that point objects produce point images.

The paraxial approximation leads us to the derivation of the fundamental paraxial equation, which calculates the change in the vergence of a wave front by a spherical surface or mirror or a spherical lens. This requires the definition of surface power P. The surface power P is a constant for a given surface with a particular radius of curvature R, where

$$P = \frac{n_2 - n_1}{R}. \tag{394}$$

The surface power P is the reciprocal of the focal length of the surface. The ray incident ray crosses the surface from medium 1 with refractive index n_1 into medium 2 with refractive index n_2. This surface causes the same total change in vergence of an incident wave front, regardless of the initial wave front radius of curvature R_1 and vergence V_1, where

$$V_1 = \frac{n_1}{R_1}. \tag{395}$$

Surface power P is measured in diopters. The surface power P is positive for a wave surface that converges toward a downstream image point, and the surface power P is negative for a wave surface that diverges away from an upstream source point. The fundamental paraxial equation relates the change in the vergence of the initial wave front V_1 to the vergence of the refracted (or reflected wave front) V_2 by the vergence of the surface or surface power P, and is given by

$$V_2 = \frac{n_2}{R_2} = V_1 + P \quad \text{fundamental paraxial equation.} \tag{396}$$

The vergence of the surface or surface power P, which is the reciprocal of the focal length of the surface, is

$$P = \frac{n_2 - n_1}{R}. \tag{397}$$

26. THE GAUSSIAN LENS EQUATION FOR THIN LENSES

The basic lensmaker's equation of Gaussian lens equation is derived for a thin lens using lens surface power and the X-ray or neutron wave vergence. The convention for the sign or polarity of distances, angles, and directions in an optical system is defined first. Clockwise angles measured relative to the positive z-direction are negative, and counterclockwise angles are positive. Distances along the axis are measured from the lens, so that those in the same direction of the incident wave have positive values, and the distances in the direction opposite the incident wave have negative values. Distances perpendicular to the axis are measured from the axis so that those distances above the axis are positive, and the distances below the axis are negative. The spherical surface intersects the z-axis at its vertex V and has center point CP also on the z-axis. The on-axis radius of curvature R extends from spherical surface center point CP to vertex V on the z-axis.

The radius of curvature R_L of the left, concave spherical surface (exposed to the source plane) of the biconcave spherical lens is negative because R_L points from the vertex V toward the CP in the direction of the source plane, which is opposite to the incident direction of the X-ray or neutron beam, where

$$R_L < 0. \tag{398}$$

The radius of curvature R_R of the right, concave spherical surface (exposed to the image plane) of the biconcave spherical lens is positive, because the direction of the radius of curvature from right surface vertex V to surface center point CP is in the same direction as the incident beam:

$$R_R > 0. \tag{399}$$

A point has zero radius of curvature ($R = 0$), and as the radius of curvature of the sphere increases to infinity ($R \to \infty$) the spherical surface becomes a plane.

The radius of curvature R_3 of a convergent refracted wave that exits the spherical lens crosses the z-axis at distance R_3 to the right of the lens mid-plane, such that R_3 can be considered the lens image distance r_i. A divergent incident wave front with radius of curvature R_1 entering the left side of the lens originates from a point on the z-axis at distance R_1 to the left of the lens midplane, such that R_1 can be considered the lens image distance r_o. In terms of vergences and powers, the fundamental paraxial equation gives

$$V_i = V_0 + P_L + P_R. \tag{400}$$

The incident wave propagating in air or vacuum from the source point arrives at the lens at object distance r_0. The object distance is measured from the lens toward the source in the opposite direction of beam propagation and thus object distance r_0 measured from the lens is negative so that vergence of the incident wave V_0 at the entrance surface of the lens is

$$V_0 = -\frac{1}{r_0}. \tag{401}$$

The refracted wave propagating in air or vacuum with from the lens toward the image plane at distance r_i has vergence V_i in which the image distance measured from the lens to image plane is in the direction of the incident beam and is thus positive, where

$$V_i = \frac{1}{r_i}. \tag{402}$$

The lens has refractive index n_{lens} and left P_L and right P_R lens surface powers, where

$$P_L = -\frac{n_{lens} - n_0}{|R_L|} \quad R_L < 0 \tag{403}$$

$$P_R = \frac{n_0 - n_{lens}}{|R_R|} \quad R_R > 0. \tag{404}$$

The power P_{thin} of the thin lens is the sum of the left P_L and right P_R surface powers, where

$$P_{thin} = P_L + P_R = (n_0 - n_{lens})\left(\frac{1}{|R_R|} + \frac{1}{|R_L|}\right). \tag{405}$$

Since

$$n_0 = 1 \quad \text{and} \quad n_{lens} = 1 - \delta, \tag{406}$$

The power P_{thin} of the thin lens becomes

$$P_{thin} = P_L + P_R = \delta \cdot \left(\frac{1}{|R_R|} + \frac{1}{|R_L|}\right). \tag{407}$$

Substitution of Eqs. 401, 402, and 407 into Eq. 400 gives

$$\frac{1}{r_i} + \frac{1}{r_0} = \delta \cdot \left(\frac{1}{|R_R|} + \frac{1}{|R_L|}\right). \tag{408}$$

The thin lens focal length f_{thin} is the reciprocal of the lens power P_{thin} and thus one obtains the Gaussian lens formula or lensmaker's equation:

$$\frac{1}{r_i} + \frac{1}{r_0} = \frac{1}{f_{thin}}, \tag{409}$$

where

$$P_{thin} = \frac{1}{f_{thin}}. \tag{410}$$

27. THICK LENS POWER AND FOCAL LENGTH, AND WAVE VERGENCE CHANGE IN THICK LENSES

The change of the vergence of an X-ray or neutron wave as the wave passes through a thick lens and the power and focal length of the thick lens are now calculated. A biconcave spherical lens with material index of refraction n_{lens} and lens thickness D is centered on the z-axis or optical axis. The lens consists of a left surface with radius of curvature R_L and a right surface with radius of curvature R_R. The radius of curvature of each lens surface is relative to a center point, which is situated on the z-axis. From this center point the radius of curvature, for example, the left surface radius of curvature R_L, generates the spherical left surface by sweeping the radius through a range of polar and azimuth angles.

The medium to the left of the lens has refractive index n_1 and the medium to the right has refractive index n_0. An incident wave front with radius of curvature R_1, which originates from the left medium, crosses the left lens surface and is refracted to radius of curvature R_2 by the power of the left surface P_L, where

$$P_L = \frac{n_{lens} - n_0}{R_L} = -\frac{\delta}{R_L}. \tag{411}$$

The expressions for V_1, V_2, and P_L then yield

$$V_2 = V_1 + P_L. \tag{412}$$

If the refracted wave is convergent, it would converge to a point on or near (paraxial) the center z-axis at approximate distance R_2 to the right of the point of intersection of the left lens surface and the center z-axis, provided the point lies within a sufficiently thick lens. Otherwise, the wave front encounters the second surface and is refracted to a new radius of curvature and now converges to a new point on the z-axis. When the refracted wave front crosses the lens, the wave front radius of curvature R_2 is reduced by the thickness of the lens D, so that the radius of curvature of the wave front incident on the right surface of the thick lens R_{2D} is

$$R_{2D} = R_2 - D. \tag{413}$$

The wave attains vergence V_{2D} after it crosses the lens to the right thick lens surface is then

$$V_{2D} = \frac{n_{lens}}{R_2 - D}. \tag{414}$$

With

$$V_2 = \frac{n_{lens}}{R_2} \tag{415}$$

one obtains

$$V_{2D} = \frac{V_2}{1 - \dfrac{DV_2}{n_{lens}}}. \tag{416}$$

One substitutes into Eq. 416 the vergence V_2 of the refracted wave on the left thick lens surface via the paraxial equation:

$$V_2 = V_1 + P_L. \tag{417}$$

This yields the vergence of the wave refracted on the right surface of the thick lens:

$$V_{2D} = \frac{n_{lens}}{R_2 - D} = \frac{V_1 + P_L}{1 - (V_1 + P_L)\dfrac{D}{n_{lens}}}. \tag{418}$$

The right surface power P_R refracts the wave front from radius of curvature R_{2D} to radius of curvature R_3 in the medium to the right of the lens, where

$$P_R = \frac{n_0 - n_{lens}}{R_R} = \frac{\delta}{R_R}. \tag{419}$$

The vergence V_3 of the wave upon exit from the thick lens is

$$V_3 = \frac{n_0}{R_3}. \tag{420}$$

From the expressions for V_3 and P_R one obtains via the paraxial equation

$$V_3 = V_{2D} + P_R. \tag{421}$$

Substitution for V_{2D} of Eq. 418 into Eq. 421, and rearrangement yields the vergence V_3 of the wave in the medium to the right of the thick lens of thickness D, where

$$V_3 = \frac{V_1\left(1 - \dfrac{P_R D}{n_{lens}}\right) + P_L + P_R - \dfrac{P_R P_L D}{n_{lens}}}{1 - \dfrac{D(V_1 + P_L)}{n_{lens}}} \quad \text{thick lens} \quad D \neq 0. \tag{422}$$

Since refractive decrement δ in materials for X-rays and neutrons is much less than 1, then the refractive index of the lens can be approximated as 1, where

$$n_{lens} = 1 - \delta \cong 1 \quad \delta \ll 1. \tag{423}$$

The vergence V_3 becomes

$$V_3 = \frac{V_1(1 - P_R D) + P_L + P_R - P_R P_L D}{1 - D(V_1 + P_L)} \quad \text{thick lens} \quad D \neq 0. \tag{424}$$

The effective power of the thick lens can be calculated without regard to the incident wave by dropping the incident wave vergence V_1 from the above expression; that is, $V_1 = 0$, but this corresponds to an incident plane wave. Hence, the power of a thick lens P_{thk} with maximum thickness D and refractive index n_{lens} is

$$P_{thk} = V_3\,(V_1 = 0) \tag{425}$$

or

$$P_{thk} = \frac{P_L + P_R - DP_R P_L}{1 - DP_L} \quad \text{thick lens} \quad D \neq 0. \tag{426}$$

The focal length of the thick lens f_{thk} is the reciprocal of the thick lens power P_{thk}, where

$$f_{thk} = \frac{1}{P_{thk}} = \frac{1 - DP_L}{P_L + P_R - DP_R P_L} \quad \text{thick lens} \quad D \neq 0. \tag{427}$$

Substitution into Eq. 427 for left P_L and right P_R surface powers of Eqs. 411 and 419, yields the power of the thick lens in terms of the radius of curvature of the left R_L and R_R right surfaces and the decrement δ of the lens material

$$f_{thk} = \frac{1}{P_{thk}} = \frac{1 + \delta \dfrac{D}{R_L}}{\delta\left(\dfrac{1}{R_R} - \dfrac{1}{R_L}\right) + \delta^2\left(\dfrac{D}{R_R R_L}\right)} \quad \text{thick lens} \quad D \neq 0. \tag{428}$$

With f_L and f_R as the focal lengths of the left and right surfaces of the lens, the thick lens focal length becomes

$$f_{thk} = \frac{1 - \dfrac{D}{f_L}}{\dfrac{1}{f_L} + \dfrac{1}{f_R} - \dfrac{D}{f_L f_R}} \quad \text{thick lens} \quad D \neq 0. \tag{429}$$

If the thickness D of the thick lens zero ($D = 0$), then the thick lens becomes a thin lens. In this case, the wave exits the thin lens with vergence

V_3, which is the sum of the powers of the left P_L and right P_R lens surfaces given by

$$V_3 = P_L + P_R \quad \text{thin lens} \quad D = 0. \tag{430}$$

The thin lens power P_{thin} is the sum of the lens left surface power P_L and the lens right surface power P_R, where

$$P_{thin} = P_L + P_R \quad \text{thin lens} \quad D = 0. \tag{431}$$

The focal length of the thin lens f_{thin} is the reciprocal of the lens power P_{thin}, where

$$f_{thin} = \frac{1}{P_{thin}} = \frac{1}{P_L + P_R} \quad \text{thin lens} \quad D = 0. \tag{432}$$

Expressed in terms of the focal lengths left f_L and right f_R lens surfaces, the thin lens focal length becomes

$$f_{thin} = \left(\frac{1}{f_L} + \frac{1}{f_R}\right)^{-1} \quad \text{thin lens} \quad D = 0. \tag{433}$$

28. WAVE VERGENCE CHANGE AND POWER AND FOCAL LENGTH OF A SEPARATED LENS DOUBLET

The power and the focal length of a pair of separated lenses are derived for X-ray and neutron waves. Consider two lenses (each lens thin or thick) separated by distance S and surrounded by drift space medium with refractive index n_0, such as air or vacuum with $n_0 = 1$. In the doublet lens system, the incident X-ray or neutron wave front with radius of curvature R_1 (and vergence V_1) enters the first lens with power P_1. The wave front exits the lens into the drift space with vergence V_2. Exit wave vergence V_2 is given by paraxial equation as the sum of the first lens power P_1 and the vergence of the entrance wave vergence V_1, where

$$V_2 = V_1 + P_1, \tag{434}$$

where

$$V_1 = \frac{n_1}{R_1}. \tag{435}$$

The exit wave radius R_2 of curvature is related to its vergence V_2 by the refractive index n_0 of medium in which the wave propagates, where

$$R_2 = \frac{n_0}{V_2}. \tag{436}$$

The wave front propagates distance S through drift space to a second lens, which results in a change in its radius of curvature to

$$R_{2S} = R_2 \pm S. \tag{437}$$

Here the spacing S between the lenses is added if the wave front is divergent from some point on or off the z-axis, and lens spacing S is subtracted if the wave front is convergent to some point on or off the z-axis. At the second lens the incident wave front has vergence V_{2S}:

$$V_{2S} = \frac{n_0}{R_{2S}}. \tag{438}$$

Substitution into Eq. 438 for R_{2S} of Eq. 437, and using Eqs. 434 and 436, gives

$$V_{2S} = \frac{n_0}{\dfrac{n_0}{V_1 + P_1} - S} = \frac{V_1 + P_1}{1 - \dfrac{S(V_1 + P_1)}{n_0}}. \tag{439}$$

The wave front is then refracted by second lens with power of the P_2 and exits with radius of curvature R_3 or vergence V_3, which is obtained by the paraxial equation, where

$$V_3 = \frac{n_0}{R_3} = V_{2S} + P_2. \tag{440}$$

Substitution into Eq. 440 for V_{2S} of Eq. 439 gives the vergence V_3 of the wave exiting the second lens in which the medium surrounding the two lenses is air or vacuum with refractive index $n_0 = 1$ and

$$V_3 = \frac{V_1 + P_1}{1 - \dfrac{S(V_1 + P_1)}{n_0}} + P_2 \quad \text{lens doublet} \quad n_0 = 1. \tag{441}$$

To obtain the resultant power P_3 of the two lenses separated by drift space S, the incident wave is assumed to be a plane wave ($R_1 = \infty$).

With $V_1 = 1/R_1$, the vergence of the incident plane wave is zero ($V_1 = 0$), where

$$P_3 = V_3. \tag{442}$$

The first lens power P_1, second lens power with P_2, the lens spacing S, and air refractive index $n_0 = 1$ together yield the power of the lens doublet:

$$P_3 = \frac{P_1}{1 - SP_1} + P_2 \quad \text{lens doublet.} \tag{443}$$

The focal length f_3 of the lens doublet is the reciprocal of the power of the lens doublet P_3

$$f_3 = \frac{1}{P_3} = \left(\frac{P_1}{1 - SP_1} + P_2\right)^{-1} \quad \text{lens doublet.} \tag{444}$$

With substitution for the first f_1 and second f_2 lens focal lengths, which are the reciprocal of the respective powers of the thin lenses as given in Eq. 410, the focal length expression for the lens doublet becomes

$$f_3 = \left(\frac{1}{f_1 - S} + \frac{1}{f_2}\right)^{-1} \quad \text{lens doublet.} \tag{445}$$

If the spacing between the two lenses is zero $S = 0$, the power of the lens doublet is then the sum of the first P_1 and second P_2 lens powers, where

$$P_3 = P_1 + P_2 \quad \text{lens doublet} \quad S = 0. \tag{446}$$

The focal length f_3 of the lens doublet with zero spacing $S = 0$ is then

$$f_3 = \left(\frac{1}{f_1} + \frac{1}{f_2}\right)^{-1} \quad \text{lens doublet} \quad S = 0. \tag{447}$$

REFERENCES

Freeman, M. H., & Hull, C. C. (2003). *Optics*. Butterworth-Heinemann, London, UK.

Hecht, E., & Zajac, A. (1974). *Optics*. Addison-Wesley Publishing, New York.

Maxwell, J. C. (1954). *A Treatise on Electricity and Magnetism* (Vol. 1 and 2). Dover Publications, New York.

Meyer-Arendt, J. R. (1984). *Introduction to Classical and Modern Optics*. Prentice Hall, New York.

Pantell, R. H., Feinstein, J., Beguiristain, H. R., Piestrup, M. A., Gary, C. K., & Cremer, J. T. (2001). The effect of unit lens alignment and surface roughness on X-ray compound lens performance. *Review of Scientific Instruments, 72*(1), 48–52.

Pantell, R. H., Feinstein, J., Beguiristain, H. R., Piestrup, M. A., Gary, C. K., & Cremer, J. T. (2003). Characteristics of the thick, compound refractive lens. *Applied Optics, 42*(4), 719–723.

Pedrotti, L., & Pedrotti, L. S. (1993). *Introduction to Optics.* Prentice Hall, Englewood Cliffs, NJ.

Selby, S. M., & Girling, B. (1965). *Standard Mathematical Tables* (3rd ed., p. 346). CRC Press, Boca Raton, FL.

Wall, E. J. (1889). *A Dictionary of Photography for the Amateur and Professional Photographer* (Anthony's Photo Series No. 25). E. & H. T. Anthony & Co., New York.

Watson, J. D., & Crick, F. H. (1953). A structure for deoxyribose nucleic acids. *Nature, 171,* 737–738.

Chapter 3

Geometric Neutron and X-ray Optics – Aberrations

Jay Theodore Cremer, Jr.

Contents

1.	Chromatic Aberration in Spherical and Parabolic Biconcave Lenses	430
2.	Spherical Aberration in Spherical Biconcave Lenses and Total Object Resolution	434
3.	CRL Image and Object Resolution: Spherical and Chromatic Aberration	438
4.	CRL Aberrations: Astigmatism, Curvature of Field, Coma, and Distortion	444
5.	Parabolic and Spherical Waves Converge to Different Focal Points by CRL	450
6.	Monochromatic and Chromatic Aberrations Occur in Parabolic and Spherical Lenses	453
7.	Aberrations from an Imperfect Spherical Lens Surface	456
8.	Derivation of the Five Seidel Monochromatic Aberrations of a Spherical Biconcave Lens	461
9.	Relation of the Optical Path Length Difference to its Associated Aberration	470
10.	Spherical Aberration	473
11.	Coma	475
12.	Coma: The Abbe Sine Condition and the Coddington Shape Factor	477
13.	Curvature of Field and the Petzval Condition	479
14.	Astigmatism	481
15.	Combined Astigmatism and Curvature of Field	483
16.	Astigmatism: Sturm's Interval and Remedy by Upstream Aperture Stop	485
17.	Distortion	487
18.	Images without and with Distortion	490
19.	Prevention of Distortion and the Petzval Condition	492
20.	Distortion in Compound Refractive Lens Imaging with Synchrotron X-ray Sources	494
	References	495

Advances in Imaging and Electron Physics, Volume 172, ISSN 1076-5670, DOI: 10.1016/B978-0-12-394422-1.00003-7.
Copyright © 2012 Elsevier Inc. All rights reserved.

1. CHROMATIC ABERRATION IN SPHERICAL AND PARABOLIC BICONCAVE LENSES

Chromatic aberration arises from the variation of the focal length with the X-ray or neutron wavelength. For example, a parabolic compound refractive lens (CRL) has focal length

$$f = \frac{R}{2N\delta}. \tag{1}$$

The decrement $\delta \propto \lambda^2$ is proportional to the square of the neutron or X-ray wavelength. The chromatic aberration for a CRL is derived as follows. An object point in the object plane, at transverse distance x from the center optical axis, is at a fixed object distance r_o upstream from a lens point on the lens center plane, and the corresponding image point in the image plane is at transverse distance y from the optical axis and at image distance r_i downstream from the lens center plane. A ray bundle from the fixed object propagates to the lens point. The ray bundle is composed of rays of varying wavelength that are refracted to various focal points on the optical axis. These rays then intersect the chief ray to create corresponding image points in a range of image planes, whose image distance r_i, varies with wavelength λ. For all wavelengths, the single chief ray passes through the center z-axis at the lens center transverse plane. If there is a pupil in front of the lens, the single chief ray passes through the center z-axis at the entrance pupil plane for all wavelengths.

For a thin lens the chief ray is the ray that propagates from an object point at radial distance x and object distance r_0 in a straight line through a point on the center axis at the lens midplane and then terminates at an image point at image distance r_i and radial distance y. The two similar triangles produced by the thin lens center axis and the lens chief ray, which passes from an object point through the lens center point to an image point, yields

$$\frac{y}{x} = \frac{r_i}{r_0}. \tag{2}$$

Given the thin lens equation,

$$\frac{1}{f} = \frac{1}{r_0} + \frac{1}{r_i}, \tag{3}$$

one obtains

$$y = \frac{x r_i}{f} - x. \tag{4}$$

For a fixed object point, where object radius x and object distance r_0 are constant, the variation of the image radius y of Eq. 4 with respect to the variation of the focal length f is

$$\Delta y = -\frac{\Delta f}{f}\frac{xr_i}{f}. \tag{5}$$

From the lens equation the image distance r_i is related to the lens focal length f, for constant transverse magnification M, by

$$r_i = (M+1)f. \tag{6}$$

Substitution of r_i of Eq. 6 into the expression for the image blur diameter Δy of Eq. 5 yields the blur diameter:

$$\Delta y = -x \cdot (M+1) \cdot \frac{\Delta f}{f} \tag{7}$$

The maximum transverse distance x from the center optical axis, which is transmitted by the lens to the image plane, is assumed to be the effective aperture radius of the lens r_e. Here the effective lens aperture radius r_e is the minimum of the physical aperture radius r_m and absorption aperture radius r_a of the lens, where

$$r_e = MIN\,(r_m, r_a). \tag{8}$$

A bundle of rays of bandwidth Δf passing through the lens at its perimeter at radius r_e experiences the maximum chromatic aberration, and thus this bundle of rays produces the maximum image blur width Δy_c, where

$$\Delta y_c = -r_e \cdot (M+1) \cdot \frac{\Delta f}{f}. \tag{9}$$

Suppose plane waves arrive at the parabolic biconcave lens CRL with focal length f or

$$f = \frac{R}{2N\delta} = \frac{2\pi R}{N\rho b\lambda^2}, \tag{10}$$

or alternatively, plane waves arrive at the spherical biconcave lens CRL with focal length f or

$$f_{sp} = \frac{\sqrt{R^2 - r^2}}{2N\delta} = \frac{R}{2N\delta}\sqrt{1 - \left(\frac{r}{R}\right)^2}. \tag{11}$$

Variation of focal length Δf with wavelength variation $\Delta\lambda$ yields

$$\frac{\Delta f}{\Delta\lambda} = -\frac{2f}{\lambda}. \tag{12}$$

This applies to spherical lens and parabolic lens CRLs in that the variation of $\Delta f/f$ is due to variation of only the wavelength-dependent decrement δ. For effective lens aperture diameter $D = 2r_e$ and wavelength bandwidth $B = \Delta\lambda/\lambda$, the chromatic image blur width Δy_c becomes

$$\Delta y_c = D\,(M+1)\,B. \tag{13}$$

The variation in object point radial distance from the lens center axis Δx_m produces a corresponding change in the radial distance of its conjugate image that is related by the transverse magnification $M = \Delta y_c/\Delta x_c$, so that the chromatic object resolution Δx_c is

$$\Delta x_c = \frac{\Delta y_c}{M} = \frac{D\,(M+1)\,B}{M}. \tag{14}$$

The parameter σ in the expressions for the spherical and chromatic aberration is defined in terms of the CRL effective aperture radius r_e and transverse magnification M, where

$$\sigma = \frac{r_e(1+M)}{M}. \tag{15}$$

For chromatic aberration, the object resolution $\sigma_c = \Delta x_c$ is then

$$\sigma_c = 2\sigma B = 2\sigma\frac{\Delta\lambda}{\lambda} = 2\frac{r_e(1+M)}{M}\frac{\Delta\lambda}{\lambda}. \tag{16}$$

Chromatic aberrations are caused by the increase of CRL focal lengths with increasing wavelengths. As a result, the object resolution σ_c due to chromatic aberration decreases with increasing effective aperture radius r_e of the CRL and increasing bandwidth B, where

$$B = \frac{\Delta\lambda}{\lambda}. \tag{17}$$

Correction for transverse chromatic aberration can be accomplished by using two CRLs of the same material and with identical or different focal lengths. This requires the proper separation distance d between the two CRLs. The separation distance between the two CRLs needed to correct the transverse chromatic aberration is derived as follows. The refractive power of the CRL doublet P_T is obtained from refractive powers P_1 and P_2 of the two CRLs separated by distance d, where

$$P_T = P_1 + P_2 - P_1 P_2 d. \tag{18}$$

The above equation is derived in a later section. The power of a lens P is the reciprocal of its focal length f, where

$$P = \frac{1}{f} \tag{19}$$

In terms of focal lengths f_1 and f_2 of the two CRLs and their separation distance d, the focal length of the double CRL f_T is then

$$P_T = \frac{1}{f_T} = \frac{1}{f_1} + \frac{1}{f_2} - \frac{d}{f_1 f_2}. \tag{20}$$

Assume the two biconcave spherical or parabolic lens CRLs are composed of the same material with common refractive index $n = 1 - \delta$. The power of the first CRL P_1 is the reciprocal of the CRL focal length f_1. With N_1 biconcave spherical or parabolic lenses, each with radius of curvature R_1, the power of the first CRL is

$$P_1 = \frac{1}{f_1} = \frac{2N_1\delta}{R_1}. \tag{21}$$

The power of the second CRL P_2 is the reciprocal of its focal length f_2 and is calculated from the N_2 biconcave spherical or parabolic lenses, each with radius of curvature R_2:

$$P_2 = \frac{1}{f_2} = \frac{2N_2\delta}{R_2}. \tag{22}$$

Setting to zero the derivative of the total power with respect to the refractive index gives the minimum variation of the power P_T of the CRL doublet relative to a variation of refractive index. That is, each X-ray or neutron ray has a different refractive index in the material common to both CRLs. The power of the doublet CRL is different for each X-ray or neutron wavelength. With the doublet CRL one desires to minimize the variation of the power of the focal length of the doublet CRL as a function of the X-ray or neutron wavelength. This is accomplished by determining the CRL separation distance d, which is determined by the minimization condition:

$$\frac{dP_T}{dn} = \frac{dP_T}{d\delta} = 0. \tag{23}$$

Substitution of Eq. 21 for P_1 and Eq. 22 for P_2 into Eq. 18 for P_T, and then applying Eq. 23 gives

$$0 = \frac{2N_1}{R_1} + \frac{2N_2}{R_2} - 2\delta\left(\frac{2N_1}{R_1}\right)\left(\frac{2N_2}{R_2}\right)d. \tag{24}$$

Multiplication by δ on both sides of the equation gives

$$0 = \frac{1}{f_1} + \frac{1}{f_2} - \frac{2d}{f_1 f_2}. \tag{25}$$

Solving for the spacing d between the two lenses yields the condition for the minimum variation of the doublet CRL focal length for incident X-ray or neutron wavelength variation,

$$d = \frac{f_1 + f_2}{2}. \tag{26}$$

Refracted X-ray or neutron rays from the first CRL are intercepted by the second CRL. The second CRL intercepts the diverging rays of different wavelengths and bends them to the extent that they are separated, but parallel, thereby reducing the lateral chromatic aberration at the image plane (especially for long image distances). However, the longitudinal chromatic aberration remains in which each neutron wavelength has a separate image plane. Chromatic aberration is an issue with neutron optics used with a reactor source in which the source is not wavelength filtered. Neutron beam choppers can reduce the bandwidth of the neutron beam to about 10%, and crystal and multilayer neutron monochromators can attain bandwidths of 1% to 5%. In the case of X-rays, the X-ray crystal and multilayer monochromators can produce bandwidths well below 0.1%.

2. SPHERICAL ABERRATION IN SPHERICAL BICONCAVE LENSES AND TOTAL OBJECT RESOLUTION

We now consider the spherical aberration from a biconcave spherical N-lens CRL. A parallel ray from an object point at radius x, which is equal to the CRL effective aperture radius r_e, is refracted by CRL, where $x = r_e$.

The convergent diagonal ray originates from a point on the CRL effective aperture radius r_e and crosses the center CRL axis at focal distance f_{sp}. The ray terminates at image radius y, thereby forming two similar triangles between the lens and image plane, where

$$-\frac{r_e}{f_{sp}} = \frac{y}{r_i - f_{sp}}. \tag{27}$$

Solving for the image radius y gives

$$y = -r_e\left(\frac{r_i}{f_{sp}} - 1\right). \tag{28}$$

The variation of the image radius Δy, which is due to the variation of the image distance Δr_i for fixed object radius $x = r_e$, is given by

$$\Delta y = -r_e\frac{\Delta r_i}{f_{sp}}. \tag{29}$$

The lensmaker's formula gives

$$r_i = (M+1)f_{sp}. \tag{30}$$

This allows the image distance Δr_i variation to be expressed in terms of the focal length variation Δf_{sp}. The resulting image radius variation Δy_{sp} due to spherical aberration is

$$\Delta y_{sp} = r_e \cdot (M+1) \cdot \left(\frac{-\Delta f_{sp}}{f_{sp}}\right) \tag{31}$$

The variation of the biconcave spherical N-lens CRL focal length Δf_{sp} with the radial distance variation Δr of an incident ray from the center CRL axis is

$$\Delta f_{sp} = \Delta\left(\frac{\sqrt{R^2 - r^2}}{2N\delta}\right) = -\frac{r\Delta r}{2N\delta R\sqrt{1-(r/R)^2}}. \tag{32}$$

The ratio $\Delta f_{sp}/f_{sp}$ is then

$$\frac{\Delta f_{sp}}{f_{sp}} = -\frac{r\Delta r}{R^2 - r^2} = -\frac{r\Delta r}{R^2\left(1-(r/R)^2\right)}. \tag{33}$$

To obtain the image blur radius Δy_{sp} for the spherical aberration of the biconcave, spherical N-lens CRL, one sets the radius r of ray incidence on the lens to the maximum radius that corresponds to no spherical aberration—that is, the parabolic aperture radius, $r = r_p$. Next one sets

$$\Delta r = r_e - r_p. \tag{34}$$

Here Δr is the radial portion of the lens between its parabolic aperture radius r_p and its effective aperture radius r_e (the minimum of the lens physical aperture r_m and the absorption aperture radius r_a). With $r = r_p$, and substitution of Eq. 34 into Eq. 33, and substitution of the resulting expression into Eq. 31 gives the image blur radius Δy_{sp} from spherical aberration as

$$\Delta y_{sp} = r_e \cdot (M+1) \cdot r_p \cdot \frac{r_e - r_p}{R^2 - r_p^2} \tag{35}$$

The corresponding spherical aberration object resolution σ_{sp} is obtained from the image resolution Δy_{sp} by dividing by the CRL magnification M, where

$$\sigma_{sp} = \frac{\Delta y_{sp}}{M} \tag{36}$$

Substitution for image resolution Δy_{sp} of Eq. 35 into Eq. 36 yields the object resolution σ_{sp} contribution from spherical aberration,

$$\sigma_{sp} = r_e \cdot \frac{M+1}{M} \cdot r_p \cdot \frac{r_e - r_p}{R^2 - r_p^2} \tag{37}$$

Spherical lens CRLs have slightly better resolution than parabolic lens CRLs for wide-bandwidth neutron sources, such as thermal Maxwell–Boltzmann-distributed neutron sources. Also, the geometric optics–derived object resolution σ_{tot} yields larger values for modulation transfer function–measured and calculated object resolutions. Kirchoff analysis shows that chromatic aberration is ameliorated by spherical aberration and absorption, resulting in better resolution than expected by σ_{tot} because the interactions among absorption, spherical, and chromatic aberration are not included in the geometric optics resolution result σ_{tot}. With increasing radial distance r from the CRL center axis, the CRL thickness $t_s(r)$ increases and the transmission fraction $T(r)$ decreases for longer-wavelength incident rays because $\mu(\lambda) \propto \lambda$, where

$$T(r) \propto \exp\left(-t_s(r) N \mu(\lambda)\right). \tag{38}$$

The longer-wavelength neutrons focus closer to the CRL than those of shorter wavelengths, given the CRL focal length f_{sp} is inversely proportional to the squared wavelength λ^2, where

$$f_{sp} \propto \frac{1}{\lambda^2}. \tag{39}$$

However, with increasing radial distance r from the CRL center axis, the focal lengths f_{sp} of the shorter-wavelength rays are reduced by spherical aberration, where

$$f_{sp} \propto \sqrt{R^2 - r^2} = R\sqrt{1 - (r/R)^2}. \tag{40}$$

The resulting CRL focal length deviation about the CRL mean focal length is decreased, which yields a finer CRL resolution.

A CRL used with a pulsed neutron source and a synchronized time-of-flight (TOF) imaging detector can improve resolution by causing the chromatic aberration to be dependent on the frame acquisition time of the detector, rather than the CRL. In a neutron pulse the neutrons arrive with steadily decreasing velocities and increasing wavelengths during the frame acquisition time. A short acquisition time Δt can result in a sufficiently small bandwidth, where

$$\frac{\Delta\lambda}{\lambda} = 0.025 \quad \text{and} \quad \Delta\lambda \propto \Delta t. \tag{41}$$

For example, the beryllium spherical lens CRL resolution could increase by 23-fold to $\sigma_{tot} = 47\,\mu m$. Spherical aberration is removed by a parabolic lens CRL.

Consider now the total object resolution σ_{tot} for a CRL. Spherical aberration causes peripheral rays to focus closer to the spherical lens CRL than the central rays, whereas chromatic aberration leads to increasing focal lengths with increasing neutron energies, both leading to aberrations that cause blurring and dominate the CRL resolution. Derived with geometric optics and neglecting interactions among absorption, spherical, and chromatic aberrations, the CRL object resolution σ_{tot} is

$$\sigma_{tot} = \sqrt{\sigma_{df}^2 + \sigma_c^2 + \sigma_{sp}^2 + \sigma_{od}^2}. \tag{42}$$

The total CRL object resolution σ_{tot} also includes contributions from diffraction-limited Rayleigh resolution σ_{df} and the detector pixel widths

σ_{od}. The diffraction-limited Rayleigh object resolution given object distance r_0 and effective lens aperture radius r_e is

$$\sigma_{df} = 0.61\frac{r_0\lambda}{r_e}. \quad (43)$$

The detector-limited resolution σ_{od} for a pixel width σ_{id} full width half maximum (FWHM) given lens transverse magnification M is

$$\sigma_{od} = \frac{\sigma_{id}}{M}. \quad (44)$$

Derived in the previous section, the object resolution σ_c contribution from chromatic aberration is given by

$$\sigma_c = 2\frac{r_e(1+M)}{M}\frac{\Delta\lambda}{\lambda} \quad (45)$$

3. CRL IMAGE AND OBJECT RESOLUTION: SPHERICAL AND CHROMATIC ABERRATION

Derived with geometric optics and neglecting interactions among absorption, spherical, and chromatic aberrations, the spherical lens CRL object resolution for X-rays, neutrons, or light σ_{tot} (in μm) is the quadrature sum of resolution contributions from the diffraction-limited Rayleigh resolution σ_{df}, the detector limited resolution σ_{od}, spherical aberration σ_{sp}, and chromatic aberration σ_{ch}, where

$$\sigma_{tot} = \sqrt{\sigma_{df}^2 + \sigma_{ch}^2 + \sigma_{sp}^2 + \sigma_{od}^2}. \quad (46)$$

The diffraction-limited Rayleigh object resolution σ_{df} is

$$\sigma_{df} = 0.61\frac{r_0}{r_e}\lambda. \quad (47)$$

The detector-limited object resolution σ_{od} for a pixel FWHM width σ_{id} is

$$\sigma_{od} = \frac{\sigma_{id}}{M}. \quad (48)$$

The top of Figure 1 shows a diagram of the diffraction-limited or Rayleigh image resolution. Division of the diffraction-limited image resolution $1.22\,f\lambda/r_m$ by the CRL magnification M yields the diffraction-limited object resolution σ_{df}. The bottom diagram of Figure 1 shows

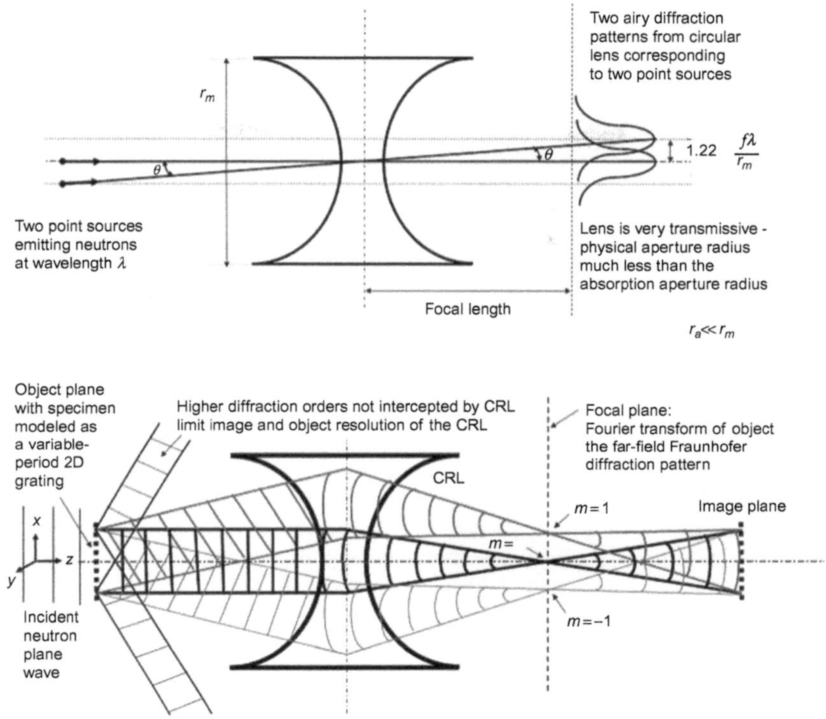

FIGURE 1 (Top) Diffraction-limited or Rayleigh image resolution. Division of the diffraction-limited image resolution $1.22\,f\lambda/r_m$ by the CRL magnification M yields the diffraction-limited object resolution σ_{df}. (Bottom) Abbe's theory of imaging of an object by a lens in which the object is modeled as a variable-period diffraction grating. The detail of an image of the object depends on the number of diffraction orders of emitted X-rays or neutrons from the object that are intercepted by the lens and focused to the image plane.

Abbe's theory of imaging of an object by a lens in which the object is modeled as a variable-period diffraction grating (Hecht and Zajac, 1974). The ability of the lens to produce a detailed image of the object depends on the number of diffraction orders that are intercepted by the lens and focused to the image plane. Spherical aberration is depicted in Figure 2.

The object resolution contributions from spherical aberration σ_{sp} and from chromatic aberration σ_{ch} for X-rays, neutrons, or light are derived as follows. A parallel ray from the object radius x is refracted by a spherical lens. The convergent diagonal ray from the lens radius x crosses the center lens axis at focal distance f_{sp} and terminates at image radius y, thereby forming two similar triangles between the lens and image plane. The apex

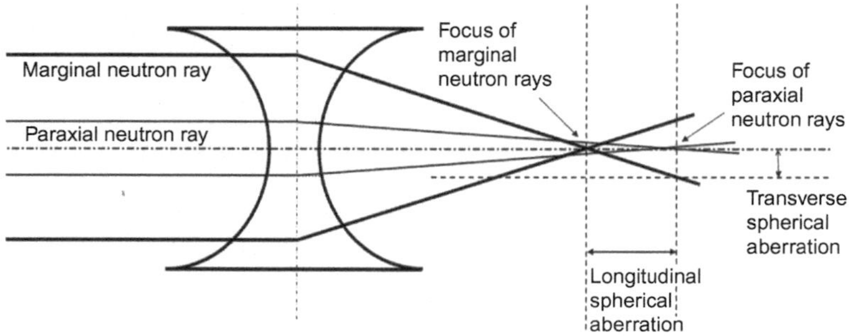

FIGURE 2 This figure shows spherical aberration, which has an axial (longitudinal) and transverse (lateral) component. The paraxial rays are within a close radial distance of the lens center axis, and the paraxial rays pass through the center, parabola-shaped portion of a spherical lens. The paraxial rays are focused according to the parabolic lens focal length $f = R/2N\delta$. The outer rays or marginal rays lie outside the parabolic aperture radius in a spherical lens and thus focus closer to the spherical lens than the inner paraxial rays. Divergent paraxial and marginal rays emitted from an object point on the center optical axis are refracted by a parabolic lens to a single focal point on the optical axis. For a spherical lens, the paraxial and marginal rays from a single object point on the axis produce a locus of focal points—that is, a blurred spot.

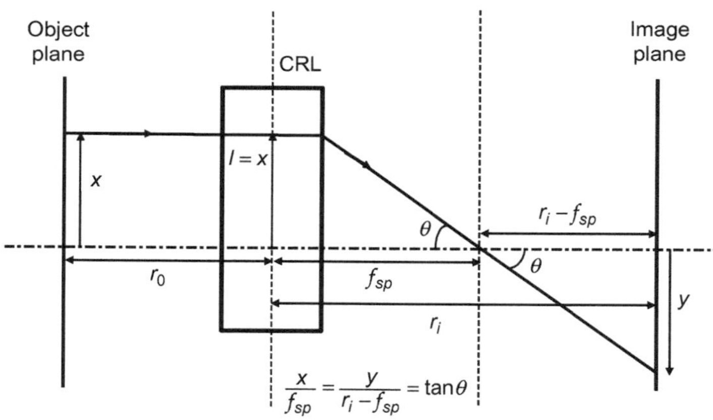

FIGURE 3 This diagram guides the calculation of the resolution contribution from the spherical aberration of a spherical lens CRL and the chromatic aberration contribution from a spherical or parabolic lens CRL.

of the two similar triangles meet at the focal point f_{sp} and share the same common apex angle θ (shown in Figure 3), where

$$\frac{x}{f_{sp}} = \frac{y}{r_i - f_{sp}} = \tan\theta. \tag{49}$$

For image radius y and fixed object radius x, the similar triangle relation gives

$$y = \frac{r_i x}{f_{sp}} - x = x\left(\frac{r_i}{f_{sp}} - 1\right) \tag{50}$$

Assume the fixed radius x of the object point is equal to the effective aperture radius of the lens r_e, where $x = r_e$. The uncertainty or image resolution Δy caused by spherical aberration is

$$\Delta y = r_i r_e \Delta\left(\frac{1}{f_{sp}}\right) = -r_i r_e \frac{\Delta f_{sp}}{f_{sp}^2} \tag{51}$$

From the lensmaker's formula one obtains

$$r_i = (M+1) f_{sp}. \tag{52}$$

Then image distance r_i of Eq. 52 is substituted into the expression for image resolution Δy of Eq. 51. The resulting expression for image resolution Δy applies to both spherical aberration and chromatic aberration, where

$$\Delta y = r_i r_e \Delta\left(\frac{1}{f_{sp}}\right) = -r_e \cdot (M+1) \cdot \frac{\Delta f_{sp}}{f_{sp}} \tag{53}$$

With this single equation one sets the wavelength constant for calculated of spherical aberration and the focal length varies as a function of radial distance r from the lens center axis. Alternatively, one sets the radial distance r constant and equal to the lens effective aperture radius, where $r = r_e$, and then one varies the wavelength of the incoming X-rays or neutrons.

In the case of spherical aberration, the wavelength λ is fixed, and in the lens principal plane the radial distance r of the incident ray from the lens center axis is varied, where

$$\Delta f_{sp} = \Delta\left(\frac{\sqrt{R^2 - r^2}}{2N\delta}\right) = -\frac{1}{2N\delta}\frac{r}{\sqrt{R^2 - r^2}} = -\frac{f_{sp} r \Delta r}{R^2 - r^2}. \tag{54}$$

The variation in radial distance Δr occurs between the parabolic aperture radius r_p and the effective aperture radius r_e of the biconcave spherical lens CRL, where

$$\Delta r = r_e - r_p. \tag{55}$$

One evaluates Δy at the parabolic aperture radius $r = r_p$ to obtain the image resolution $(\Delta y)_{r=r_p}$ expression:

$$(\Delta y)_{r=r_p} = r_e \cdot (M+1) \cdot r_p \cdot \frac{r_e - r_p}{R^2 - r_p^2} \tag{56}$$

The object resolution σ_{sp} for spherical aberration is obtained from the image resolution $(\Delta y)_{r_p}$ for spherical aberration by dividing by the lens magnification M, where

$$\sigma_{sp} = \frac{(\Delta y)_{r_p}}{M} = r_e r_p \frac{M+1}{M} \cdot \frac{r_e - r_p}{R^2 - r_p^2}. \tag{57}$$

To obtain the maximum chromatic aberration σ_{ch} contribution to the object resolution, the radial distance r of the ray from the lens center axis as the ray passes through the lens is fixed at the maximum radial distance—namely, the effective aperture radius, where $r = r_e$. One then varies the wavelength λ in the expression for the spherical lens CRL focal length f_{sp} of the spherical lens CRL to obtain the corresponding variation Δf_{sp}, where

$$\Delta f_{sp} = \frac{\sqrt{R^2 - r_e^2}}{2N} \Delta \left(\frac{1}{\delta} \right) = -\frac{\sqrt{R^2 - r_e^2}}{2N} \frac{\Delta \delta}{\delta^2}. \tag{58}$$

Recall the expression for the refractive index decrement δ, where

$$\Delta \delta = \Delta \left(\frac{\lambda^2 b \rho}{2\pi} \right) = 2\delta \frac{\Delta \lambda}{\lambda}. \tag{59}$$

Substitution for the decrement variation $\Delta\delta$ of Eq. 59 into the focal length variation Δf_{sp} of Eq. 58 yields

$$\Delta f_{sp} = -2 f_{sp} \frac{\Delta \lambda}{\lambda}. \tag{60}$$

We recall image resolution Δy of Eq. 53 is due to variation in the spherical lens focal length Δf_{sp} and substitute the expression for Δf_{sp} in Eq. 60 that applies to chromatic aberration, where

$$\Delta y = -r_e \cdot (M+1) \cdot \frac{\Delta f_{sp}}{f_{sp}} = 2 r_e \cdot (M+1) \cdot \frac{\Delta \lambda}{\lambda} \tag{61}$$

The object resolution σ_{ch} for spherical aberration is obtained from the image resolution Δy for chromatic aberration by dividing by the lens magnification M, where

$$\sigma_{ch} = 2r_e \cdot \frac{M+1}{M} \cdot \frac{\Delta\lambda}{\lambda} \tag{62}$$

The top diagram in Figure 4 shows chromatic aberration and its two component aberrations in the axial and transverse directions relative to the

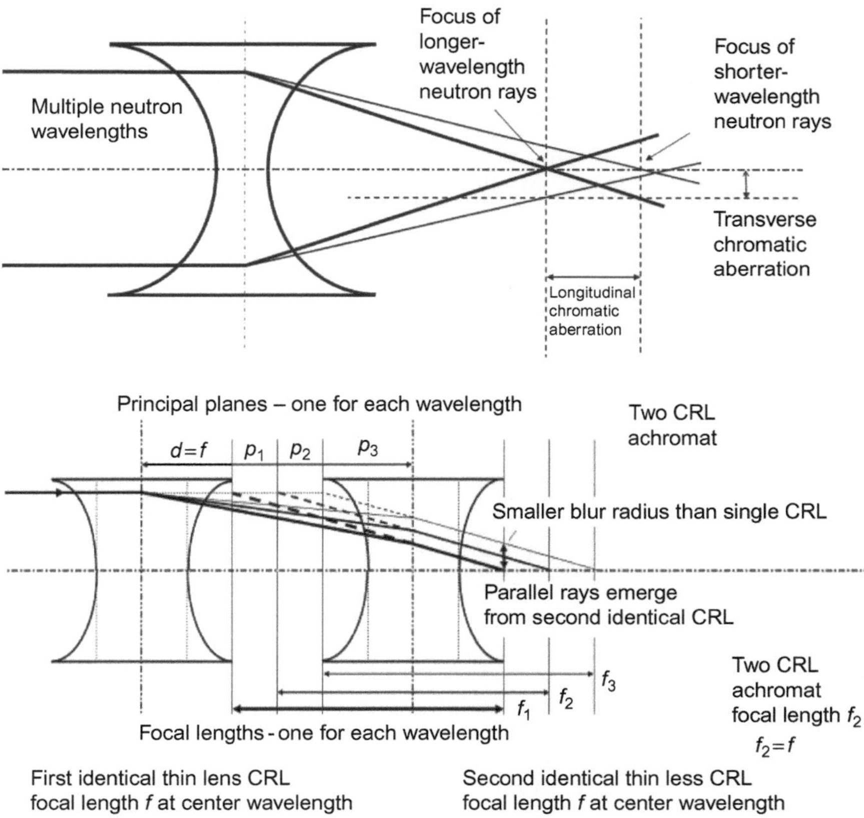

FIGURE 4 (Top) The diagram shows chromatic aberration, which has an axial (longitudinal), and transverse (lateral) component. (Bottom) Two-CRL achromat–lateral chromatic aberration is reduced by placement of two identical CRLs separated by one focal length but the longitudinal chromatic aberration remains. The chromatic-induced divergence of the multiple wavelength rays from the first CRL is refracted by the second CRL to parallel rays that produce a smaller blur radius in a set image plane (Jenkins and White, 1957).

center optical z-axis of the lens or CRL. The bottom diagram in Figure 4 shows a means to reduce chromatic aberration by using two identical lenses, each with focal length f, in which the center principal plane's two thin lenses are separated by a distance d equal to focal length f, where $d = f$. This simple two-lens or two-CRL achromat reduces chromatic aberration by the second CRL refracting the divergent multiple wavelength rays from the first CRL into parallel rays. These parallel rays form a smaller blur radius than the divergent rays of a single CRL at an image plane set for the center wavelength.

4. CRL ABERRATIONS: ASTIGMATISM, CURVATURE OF FIELD, COMA, AND DISTORTION

Monochromatic aberrations become significant for larger aperture lenses and off-axis object points. In Figure 5 incident plane waves from a distant object are refracted by the biconcave lens into curved wave fronts that are shaped by the curved lens surfaces and the different path lengths of the rays through the lens material. Since the refractive index $n = 1 - \delta$ of most materials is less than 1 for X-rays and neutrons, then the X-rays and neutrons have faster phase velocities in the materials than vacuum and air. As a result, the biconcave lens curvature and the different path lengths experienced by parallel incident rays through the lens together produce a curved wave front of exiting X-rays or neutrons. The various portions of the emerging curved wave front of X-rays or neutrons have different curvatures and convergent focal points due to lens aberrations.

The top of Figure 5 shows the relation between the sag s, the half-chord length w, and the radius of curvature r of a divergent spherical wave front or the spherical surface of lens in one dimension, which is depicted as a circle of radius r, where

$$r^2 = w^2 + (r - s)^2 \tag{63}$$

Solving for the radius of curvature r of the lens surface in one-dimension yields

$$r = \frac{w^2}{2s} + \frac{s}{2}. \tag{64}$$

In the case of X-ray and neutron CRLs, one can assume the chord half-width w, which equals the lens or CRL effective aperture radius r_e,

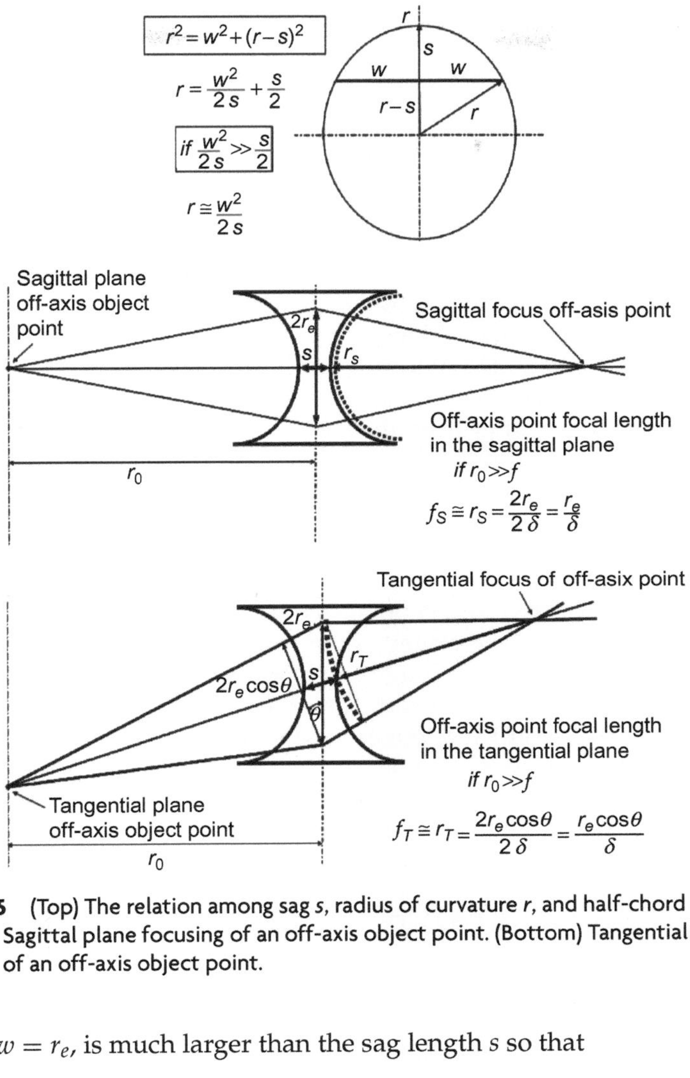

FIGURE 5 (Top) The relation among sag s, radius of curvature r, and half-chord width w. (Middle) Sagittal plane focusing of an off-axis object point. (Bottom) Tangential plane focusing of an off-axis object point.

where $w = r_e$, is much larger than the sag length s so that

$$\frac{w^2}{2s} \gg \frac{s}{2}. \tag{65}$$

The curvature of the lens or CRL imparts a corresponding curvature to a plane wave front of X-rays or neutrons form a distant source. The sag length of the refracted of curved wave front is also assumed much less than the chord half-width of the wave front. Consequently, the radius of curvature r of the lens, CRL, or the refracted X-ray or neutron wave front is then approximated as

$$r \cong \frac{w^2}{2s}. \tag{66}$$

Without lens aberrations all portions of the emergent wave front would have the same curvature and converge to the same focal point. With a parabolic biconcave lens the divergent spherical wave front from an on-axis object is refracted by the lens to a convergent spherical wave front, which converges to a single focal point. However, only the parabolic center portion of a spherical lens, which is less than the parabolic aperture radius r_p, refracts the incident, divergent spherical wave front to a spherically convergent wave front that converges to a single focal point. The emerging wave front from the various outer portions of the spherical lens have differing radius of curvatures and thus different convergent focal points. Spherical aberration becomes more pronounced for the rays, which are incident on the outer radial portions of a large-aperture lens that greatly exceed the lens parabolic aperture radius r_p.

Astigmatism is caused by the different focal points and image points for a single off-axis fixed object point. As shown in the middle and bottom diagrams of Figure 5, the center ray from a single, fixed object point O extends as a straight line through the lens center and to a downstream point I that lies between the shorter focal length of rays in the tangential plane (Figure 5, bottom) and longer focal length rays in the sagittal plane (Figure 5, middle). Hence, a single common sag length s, which lies on the center ray, corresponds to both tangent and sagittal plane wave fronts that emerge from the lens, where (Freeman and Hull, 2003)

$$s = \frac{r_e^2}{2r_s} = \frac{r_e^2 \cos^2\theta}{2r_T}. \tag{67}$$

Incident plane waves from a distant object are refracted by the biconcave lens in which the object distance r_0 is much greater than the lens focal length f, where $r_0 \gg f$. The radius of curvature of the emerging wave fronts is then approximately equal to the focal length, where

$$f_S \cong r_S \quad \text{and} \quad f_T \cong r_T. \tag{68}$$

Hence the focal lengths f_S for the rays in the sagittal plane and the focal lengths f_T of the rays in the tangential plane are different and related via

$$s = \frac{r_e^2}{2f_s} = \frac{r_e^2 \cos^2\theta}{2f_T}, \tag{69}$$

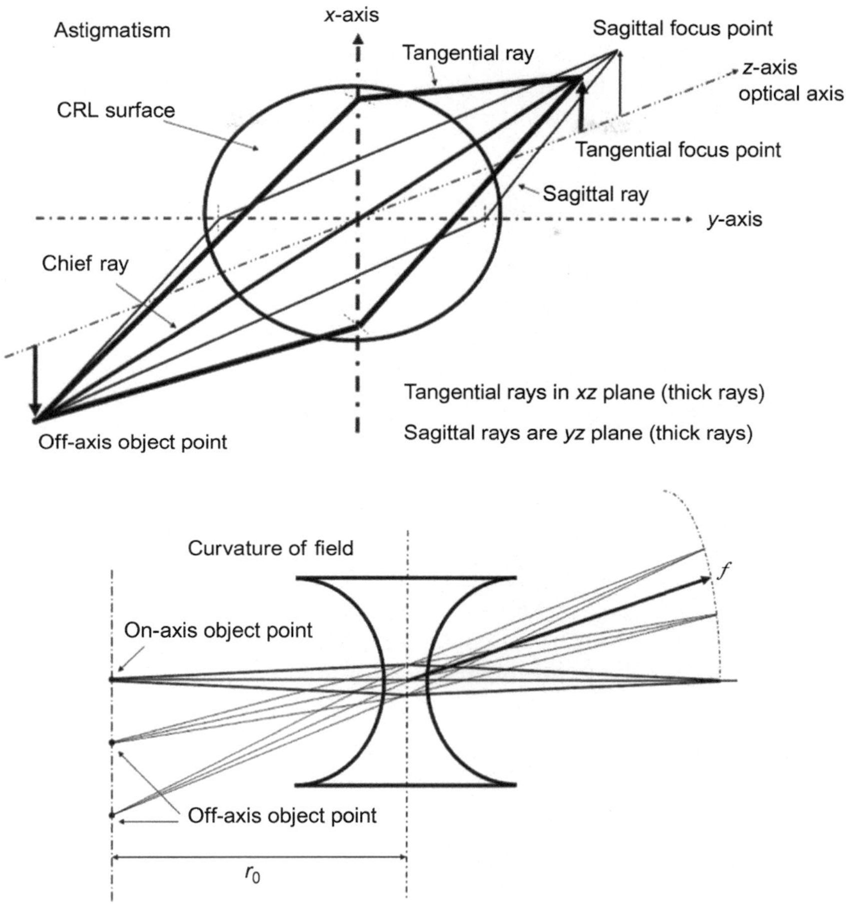

FIGURE 6 (Top) Astigmatism is caused by different image distances and focal lengths in the sagittal and tangential planes for a single off-axis object point. (Bottom) The curvature of field aberration in which on-axis and off-axis narrow bundles of rays, which pass through the center of the biconcave N-lens CRL, have the same paraxial focal length $f = R/2N\delta$, which then requires the focal plane to be curved rather than planar.

where

$$f_T = f_s \cos^2\theta = f_s\left(1 - \sin^2\theta\right). \tag{70}$$

The top diagram in Figure 6 shows the sagittal plane and tangential plane focal points of a single off-axis object point and the separation distance or Sturm interval between the focal points. The detector is usually placed at a point between the sagittal and tangential focal planes to obtain an image of the least confusion. Astigmatism causes the spokes of a bicycle wheel to be focused in a different plane than the rim. The astigmatic distant or Sturm's interval Δf between the focal points the sagittal rays and tangential rays is given by

$$\Delta f = f_s - f_T = -f_s \sin^2\theta \cong f \sin^2\theta = \frac{R}{2N\delta}\sin^2\theta. \tag{71}$$

Here, the sagittal focal length was approximated by the paraxial focal length f of the CRL.

If the lens were free of all optical aberrations, the aberration of curvature of field would remain because each image point corresponds to an object point on a planar object plane. A narrow bundle of rays emerging from each on-axis point or each off-axis point of a distant object plane, and then passing through the center of the lens, has approximately the same image distance r_i and thus focal length f. Assuming the object distance r_0 is large compared with the focal length f, where $r_0 \gg f$, then the image distance r_i of each object point is approximately equal to the single lens paraxial focal length f, where $r_i \cong f$, and for a biconcave N-lens CRL, $f = R/2N\delta$. As the object points move to larger radial distances from the center z-axis of the optical system, the rays become more diagonal and the corresponding image point moves closer to the center lens plane (shown

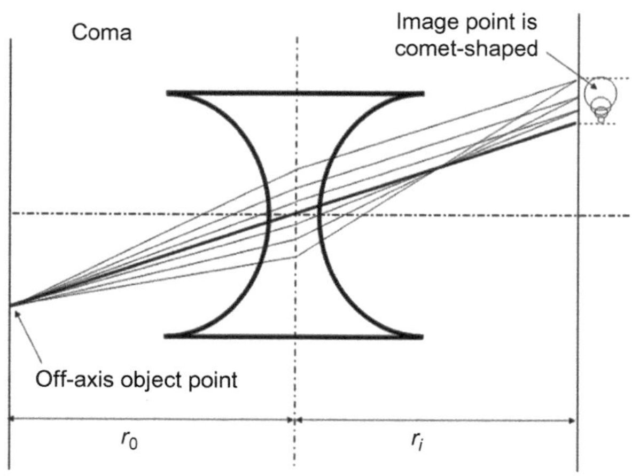

FIGURE 7 Coma aberration occurs because the principal planes of the lens are actually curved surfaces rather than the idealized planar surfaces. Coma can be reduced by positioning an aperture at the proper location. Coma was considered one of the most severe aberrations because of the asymmetrical comet shape of the locus of image points corresponding to a single fixed object point.

at the bottom of Figure 6), which then requires the focal plane to be curved rather than planar.

The aberration of *coma* arises because the lens principal planes are actually curved surfaces rather than planar surfaces. The principal planes are planar only for the center bundle of paraxial rays. For more off-axis positions, a single object point is focused into a locus of image points forming a comet shape (Figure 7). The aberration of *distortion* occurs when an aperture is positioned upstream of the lens or downstream from the lens (Figure 8). An aperture placed between the object plane and the lens leads to barrel distortion in which the transverse magnification M_T decreases with the radius y of an image point I, where

$$y\uparrow \Rightarrow M_T\downarrow \quad \text{barrel distortion.} \tag{72}$$

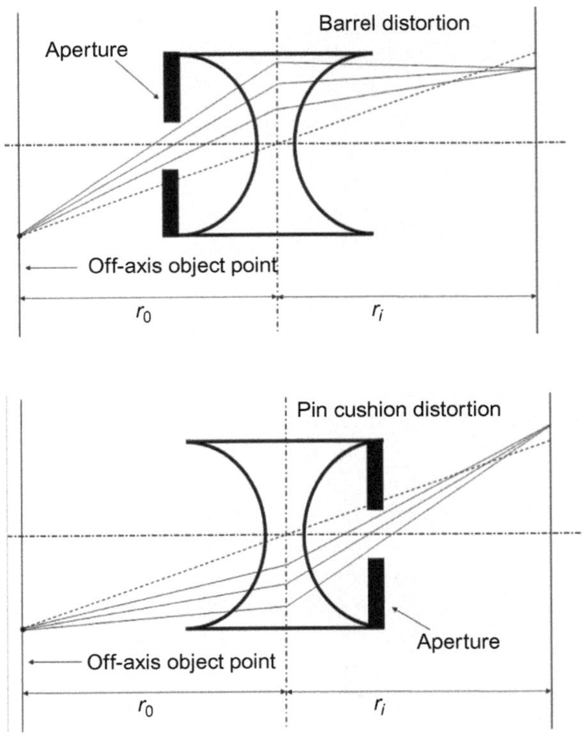

FIGURE 8 (Top) Barrel distortion is caused by placement of an aperture between the object plane and lens. (Bottom) Pincushion distortion is caused by placement of an aperture between the lens and image plane.

An aperture placed between the lens and the image plane leads to pincushion distortion in which the transverse magnification M_T increases with the radius y of an image point I, where

$$y\uparrow \Rightarrow M_T\uparrow \quad \text{pincushion distortion.} \tag{73}$$

5. PARABOLIC AND SPHERICAL WAVES CONVERGE TO DIFFERENT FOCAL POINTS BY CRL

Consider the case of an X-ray or neutron ray emitted from an object point at transverse distance s from the center z-axis. This ray travels axial object distance r_o to a lens point at transverse distance l from the center axis. The lens point lies on the upstream surface of the first spherical lens of a CRL with N spherical biconcave lenses. The CRL refracts the ray, such that the ray changes direction.

As a simplifying approximation the ray is assumed to maintain a constant transverse distance l from the center z-axis of the CRL (optical axis). The ray then is directed by the CRL onto the image point at transverse distance h from the z-axis and at axial image distance r_i from the last lens of the CRL. If the object and image distances are each much greater than the axial extent of the lens, then the object and image distances can be measured from the center point of the CRL. The index of refraction of the lens material is

$$n = 1 - \delta. \tag{74}$$

For air or vacuum that surrounds the lens the index of refraction is $n = 1$.

Non-paraxial rays are off-axis rays with large slopes or distances with respect to the center z-axis of a lens. Monochromatic non-araxial rays, which propagate from an object through a lens to the image plane, cause a blurred and distorted image of the object as a result of monochromatic aberrations. Aberrations can be visualized as perturbations of the wave front from its spherical or planar form by the interaction of the wave front with a material in which the index of refraction varies along the wave front via the shape of the lens interface with the incident medium (air or vacuum).

The normal or perpendicular to the surface of the wave at each point of the wave surface is the line of propagation for that point. The normals of a spherical wave all are directed to the center point of the spherical wave, and each point on the spherical wave converges to the center point. The lines of the normals to a parabolic wave surface do not converge to a single point, and thus each point on the parabolic wave converges to a different point on the center optical axis.

Consider a converging spherical wave with radius of curvature R and a converging parabolic wave of on-axis radius of curvature R. Recall the parabola has radius of curvature ρ as a function of radial distance l from the center z-axis is

$$\rho = R\left[1+\left(\frac{l}{R}\right)^2\right]^{3/2} \cong R+\frac{3l^2}{2R}. \tag{75}$$

Since $R \gg l$, the binomial series approximation was applied.

The radius of curvature along the center z-axis ($l = 0$) is

$$\rho = R. \tag{76}$$

The two converging waves contact only at one point of tangency R, which lies on the center z-axis at axial distance $-R$ to the left of the center point of the spherical wave that is situated at the z-axis origin O, $z = 0$. The spherical wave is converging to its center point O. However, each point on the parabolic wave front is converging to a different point on the z-axis, with the exception of the single on-axis point R on the parabolic wave front that converges to the same point O as the spherical wave front. Each off-axis point on the parabolic wave front P, situated at radial distance l from the axis, converges to an on-axis point Z to the right of the spherical wave convergence point O by an axial distance increment Δz, which is obtained from above equation, so that

$$\Delta z = \frac{3l^2}{2R}. \tag{77}$$

This axial distance increment Δz is termed the *longitudinal aberration.* A point S on the spherical wave front, situated at the same radial distance l, converges to point O in the z-axis and then proceeds to an off-axis point Y in the transverse plane that intercepts point Z at the z-axis. The radial distance ΔY between the on-axis point Z and off-axis point Y is termed the *transverse* or *lateral aberration.*

The radius of curvature R at point S is along the normal or the line perpendicular to the spherical wave front at point S, which is situated at radial distance l from the z-axis. The slope n_s of the normal to the spherical wave front at point S, and the angle θ_n^{sph} of the normal with respect to the z-axis, is given by

$$n_s = \tan\theta_n^{sph} = -\frac{l}{\sqrt{R^2-l^2}} = -\frac{l}{R}-\frac{l^3}{2R^3}. \tag{78}$$

The tangent line to the spherical wave front at point S is perpendicular to the normal line. From analytic geometry the tangent line has slope t_s or angle θ_t^{sph}, where

$$t_s = \tan\theta_t^{sph} = -\frac{1}{n_s}. \tag{79}$$

Similarly, the radius of curvature ρ at point P on the parabolic wave front is along the normal or the line perpendicular to the parabolic wave front at point P, which is situated at the same radial distance l from the z-axis on the parabolic wave front as the point S on the spherical wave front. The slope n_p of the normal to the parabolic wave front at point P is n_p. The angle θ_n^{par} of the normal with respect to the z-axis is

$$n_p = \tan\theta_n^{par} = -\frac{l}{\sqrt{\rho^2 - l^2}} = -\frac{l}{R} + \frac{l^3}{R^3}. \tag{80}$$

The tangent line to the parabolic wave front at point S is perpendicular to the normal line and has slope t_p or angle θ_t^{par}, where

$$t_p = \tan\theta_t^{par} = -\frac{1}{n_p}. \tag{81}$$

For small angles θ_n^{sph} and θ_n^{par}, one can approximate

$$\tan\theta_n^{sph} \cong \theta_n^{sph} \tag{82}$$

and

$$\tan\theta_n^{par} \cong \theta_n^{spar}. \tag{83}$$

The angle difference $\Delta\theta_n$ between the normal angle of the parabolic and spherical waves, with respect to the z-axis, is

$$\Delta\theta_n = \theta_n^{sph} - \theta_n^{par} = -\frac{3l^3}{2R^3}. \tag{84}$$

The radial distance ΔY or transverse aberration is then

$$\Delta Y = \Delta Z \cdot \Delta\theta_n \tag{85}$$

or

$$\Delta Y = -\frac{9l^5}{4R^4}. \tag{86}$$

The angle difference $\Delta\theta_t$ between the tangential angles of the parabolic and spherical wave, with respect to the z-axis, is

$$\Delta\theta_t = \theta_t^{sph} - \theta_t^{par}. \tag{87}$$

Tangential angle difference $\Delta\theta_t$ is also equal to the difference of their normal angles $\Delta\theta_n$ due to the right-angle relation between the tangent and normal lines, where

$$\Delta\theta_t = \Delta\theta_n. \tag{88}$$

In a medium of index of refraction n, the incremental change in the radius of curvature $\Delta\rho$ of the spherical wave front to the parabolic wave front per incremental change Δl in the transverse distance l from the center z-axis is then

$$\Delta\rho = n\Delta\theta_t\Delta l = (1-\delta)\Delta\theta_t\Delta l. \tag{89}$$

Applying Eqs. 84 and 88 to Eq. 89, the rate of curvature change between the spherical and parabolic wave fronts as a function of transverse distance from the z-axis is thus

$$\frac{\Delta p}{\Delta l} = -\frac{3l^3}{2R^3}(1-\delta). \tag{90}$$

6. MONOCHROMATIC AND CHROMATIC ABERRATIONS OCCUR IN PARABOLIC AND SPHERICAL LENSES

To obtain the monochromatic aberrations commonly associated with a lens, one calculates the third-order expansion of the radical expression for the optical path length Γ_T for a ray path that originates at object point (s_x, s_y), which passes through a point on the lens (l_x, l_y) and terminates on its conjugate image point (x, y). The second-order expansion of the square root expressions for the optical path length of a ray path from object to image point yields idealized sharp images in which the lens produces constant magnification for all object points. However, the validity of the second-order expansion is limited to rays traveling close to the center z-axis of the lens from object point to image point.

For object points and ray paths, which are significantly off-axis, accompanied by larger ray angles a third-order expansion of the radical expressions is required. The third-order expansion adds five terms to the optical path length expression for a ray path from object point to image point. These five terms are the third-order monochromatic Seidel aberrations: spherical aberration, coma, astigmatism, curvature of field, and distortion. These monochromatic aberrations cause the lateral and longitudinal magnifications to no longer be a constant dependent on the object distance r_o and image distance r_i. In addition, the focal length varies depending on the ray point of incidence. Chromatic aberration also causes the lens focal length to vary with wavelength. Consequently, the five monochromatic aberrations along with chromatic aberration lead to blurring and distortion of the image.

A spherical or parabolic lens or imperfect lens and the differing path lengths between the object and image point produce a wave front whose radius of curvature varies, causing it to converge to a locus of image points. Each ray path has an optical path length Γ_T, which is the sum of the products of the M path length segments L_k immersed in medium of refractive index n_k, where

$$\Gamma_T = \sum_{k=1}^{M} L_k n_k. \tag{91}$$

The optical path length Γ_T expression uses the Pythagorean theorem, which is applied to the diagonal ray path lengths. This results in square root terms in the imaginary arguments of the exponential terms that express the drift space phase shifts of the ray path. The exponential terms arise from application of Euler's theorem. Exponentials are more computationally convenient for representation of the sinusoidal variation of the waves in space and time, and for the wave attenuation by material media as well, where Euler's theorem is

$$\exp(i\theta) = \cos\theta + i\sin\theta. \tag{92}$$

The thickness of the spherical lens introduces a square root term in imaginary argument of the exponential term that contains the lens optical path length. Binomial series expansion of the spherical lens optical path length yields the optical path length of a parabolic lens.

Independent of whether the lens is parabolic, spherical, or aspherical, the diagonal ray paths in the drift spaces between the lens and object and image planes lead to radical terms in the imaginary arguments of their exponential expressions. Third-order expansion of these radicals compared with the second-order expansion introduces a change in the optical path length $\Delta\Gamma_T$ from which arise the five Seidel aberrations. Corresponding to the optical path length difference $\Delta\Gamma_T$ is the corresponding phase shift $\Delta\Phi$ of the amplitude $E(x, y, z, t)$ of a point on the wave front, with wavelength λ, which travels on a ray path from a single object point through the lens to a single image point, where

$$\Delta\Phi = \frac{2\pi\,\Delta\Gamma_T}{\lambda}. \tag{93}$$

In optics, each of these monochromatic and chromatic aberrations is commonly reduced by using a pair of lenses, which are separated by drift space, with an iris (or stop) placed midway between the lenses. This spaced doublet with a center stop reduces blur and distortion by limiting the off-axis angles and distances of the ray path from its object point through the lens to its image point.

The resulting phase shift $\Delta\,\Phi$ associated with added optical path length arising from the third-order expansion of the ray path for a spherical, aspherical, or parabolic lens CRL causes aberrations of blurring and distortion via unwanted phase shifts at the wave front due to the CRL and the geometry. A spherical wave originates and diverges from a single object point. The spherical wave front propagates through empty drift space to the CRL. At the CRL the spherical wave front is refracted, such that the CRL changes the wave front divergent curvature to a convergent curvature.

However, the imperfect lens does not produce a wave front at its exit plane that has a constant convergent radius of curvature, which allows the entire wave front to converge to a single image point. Instead, the lens and the differing ray path lengths cause the wave front to acquire a variable radius of curvature, such that each part of the wave front converges to a different single image point. The lens is supposed to equalize the path lengths for each point on the wave front, which travels from the single object point through the lens to a single image point. This requires the ideal lens to have an aspheric shape and thickness and/or material variation, which provides via a variable refractive index and path length, an optical path length offset that allows each ray path to acquire the same optical path length from object point to image point. In this manner, a spherical wave emitted at a single time instant from a single object point converges simultaneously to a single image point.

As a result of the imperfect lens and the different path lengths traveled by each point on the wave front, the wave front emitted from a single object point is refracted by the lens to focus onto a set or locus of points, causing blur. Wave fronts emitted from a particular distribution object points, which form a shape or pattern, are refracted by the lens, such that the convergent wave fronts are distorted and converge to a different distribution image points, thereby forming a different shape (i.e., distortion). A perfect lens allows a single spherical wave emitted from an object point to converge simultaneously onto the image point. However, an imperfect lens leads to a wave front that converges to a distribution of image points located not only in the ideal lens image plane, but also in front of and behind the plane—blurring and distortion.

A *ray* is a line that is perpendicular to the wave front at a single point and a ray traces the path of the wave front point from the object point origin through a lens point to an image point terminus. The function of an ideal lens is to ensure that the optical path length of each ray path is equal. In this manner, along each ray path the phase of the wave front point is the same as the other wave front points when they converge at a particular point in the image plane. This allows constructive interference of wave amplitudes from each ray path from an object point, such that the amplitudes sum to produce the maximum amplitude sum at the single image point.

Recall that a coherent source or object is said to emit coherent waves at each instant of time, if each object point emits a wave, which has a fixed, spatial phase-shift relationship with respect to the other object points. A coherent object can produce an interference pattern as a result of the fixed relationship of the phases. For an object that emits coherent waves, the resultant amplitude contribution is the sum of the amplitudes from each of the ray paths between the single object point and the single image point. The squared magnitude of this summed amplitude yields the image point intensity.

In contrast, an incoherent source or object is said to emit incoherent waves at each instant of time if each object point emits a wave that has a random, spatial phase-shift relationship with respect to the other object points. An incoherent source cannot produce an interference pattern unless it is a single point source, or a sufficiently small source, emitting sufficiently long wavelengths that are intercepted at a sufficiently large distance. In the case of an incoherent source, the resultant amplitude contribution (summed over all rays paths) from each object point is first squared to obtain its intensity contribution. The intensity of each object point contribution to a single image point is then summed to produce the image point intensity for an incoherent object.

7. ABERRATIONS FROM AN IMPERFECT SPHERICAL LENS SURFACE

We now examine the transverse and longitudinal aberration imposed on an incident plane wave that is focused by a spherical lens whose surface is not perfectly spherical (see the treatment of aberration in optics texts by

Pedrotti and Pedrotti (1993), Klein and Furtak (1986), and O'Neill (1991). Consider an approximate plane wave created by a very distant point source. This plane wave propagates through air and intercepts a material with a perfect spherical surface radius of curvature R. This perfect spherical surface of material, which is described by refractive decrement δ, converts the plane wave to a spherical wave that converges to single point P at origin $(0,0,0)$. Given point (u,v,w) on a perfect spherical surface with respect to the origin, the initial convergent spherical wave at this spherical surface $S\,(u,v,w)$ is then described by

$$S\,(u,v,w) = u^2 + v^2 + w^2 = R^2. \tag{94}$$

This spherical wave will converge from $S\,(u,v,w)$ to the origin $(0,0,0)$.

If the surface is not a perfect sphere, then the plane wave is converted to a converging imperfect spherical wave with aberrations, and at the surface, the initial imperfect convergent spherical wave is described by a distorted sphere of radius $R+\Delta$. The initial convergent aspherical wave surface $S'\,(u,v,w)$ is then described by

$$S'(u,v,w) = u^2 + v^2 + w^2 = (R+\Delta)^2. \tag{95}$$

Binomial series expansion to the first order in Δ yields

$$S'(u,v,w) = u^2 + v^2 + w^2 = 2R\Delta + R^2. \tag{96}$$

Taking the gradient on both sides of the above equation yields the normal vector to the deviated wave surface $S'(u,v,w)$, where

$$\nabla S' = 2u\hat{\mathbf{u}} + 2v\hat{\mathbf{v}} + 2w\hat{\mathbf{w}} = 2R\frac{\partial\Delta}{\partial u}\hat{\mathbf{u}} + 2R\frac{\partial\Delta}{\partial v}\hat{\mathbf{v}} + 2R\frac{\partial\Delta}{\partial w}\hat{\mathbf{w}}. \tag{97}$$

This distorted spherical wave will converge from $S'(u,v,w)$ to an off-axis image point Q at position (x,y,z) that is relatively close to $(0,0,0)$ compared with R. With

$$w \cong -R \quad u \ll R \quad v \ll R, \tag{98}$$

the approximate distance R from (u,v,w) to (x,y,z) is

$$R \cong \frac{|\nabla S'|}{2}. \tag{99}$$

Similar triangles show the following relations:

$$\frac{x-u}{R} \cong \frac{u - R\dfrac{\partial \Delta}{\partial u}}{|\nabla S'|/2}, \tag{100}$$

$$\frac{y-v}{R} \cong \frac{v - R\dfrac{\partial \Delta}{\partial v}}{|\nabla S'|/2}, \tag{101}$$

and

$$\frac{z-w}{R} \cong \frac{w}{|\nabla S'|/2}. \tag{102}$$

Combining these three relations yields

$$\frac{x-u}{u - R\dfrac{\partial \Delta}{\partial u}} = \frac{y-v}{v - R\dfrac{\partial \Delta}{\partial v}} = \frac{z-w}{w} \cong -\frac{z}{R} - 1. \tag{103}$$

In the above equation the second-order products of z/R with $\partial\Delta/\partial u$ and $\partial\Delta/\partial v$ were dropped, and the transverse x- and y-component transverse distances x and y are solved relative to the center z-axis of the imperfect spherical wave convergence point. The expression for the coordinates (x, y, z) of the ray of the imperfect spherical wave are obtained as shown below:

$$x \cong -u\frac{z}{R} + R\frac{\partial \Delta}{\partial u} \tag{104}$$

$$y \cong -v\frac{z}{R} + R\frac{\partial \Delta}{\partial v}. \tag{105}$$

The ray of the perfect spherical wave converges to point (0,0,0) in the image plane at $z = 0$ of the perfect spherical wave. The ray of the imperfect spherical wave intercepts the image plane of the perfect wave at an off-axis point with coordinates $(x_0, y_0, 0)$. That is, in the image plane ($z = 0$) of the perfect spherical wave, the distorted spherical wave converges to an off-axis point with x- and y-component distances x_0 and y_0 from the center z-axis, where

$$x_0 \cong R\frac{d\Delta}{du} \tag{106}$$

$$y_0 \cong R\frac{d\Delta}{dv}. \tag{107}$$

Distances x_0 and y_0 represent the transverse or lateral aberration of the imperfect spherical wave.

The ray of the imperfect wave crosses the z-axis at distance z from the image plane of the perfect spherical wave ($z = 0$). The ray of the imperfect spherical wave crosses the z-axis at $x = 0$ and $y = 0$. Hence, one sets $x = y = 0$ in the x and y expressions for the ray coordinates (x, y, z) of the imperfect spherical wave. Solving for z one obtains the z-axis intercept distance z of the imperfect spherical wave ray from the perfect spherical wave image plane ($z = 0$), where

$$z = \frac{R^2}{u}\frac{\partial\Delta}{\partial u} = \frac{R^2}{v}\frac{\partial\Delta}{\partial v}. \tag{108}$$

In polar coordinates, lens Cartesian coordinates (u, v) are expressed in terms of the cylindrical coordinates (l, θ) of the spherical surface of the material, where

$$u = l\cos\theta \qquad \text{and} \quad v = l\sin\theta \tag{109}$$

$$l = \sqrt{u^2 + v^2} \quad \text{and} \quad \theta = \tan^{-1}\frac{v}{u}. \tag{110}$$

The plane wave incident at the spherical material surface is refracted to a converging spherical wave. A perfect spherical surface refracts the incident plane wave to produce a perfect converging spherical wave, and an imperfect spherical surface produces an imperfect converging spherical wave. In the $z = 0$ image plane of the perfect spherical wave, the off-axis point coordinates x_0 and y_0 of the ray of the imperfect spherical wave (i.e., the transverse aberration in polar coordinates) are obtained via the chain rule, where

$$x_0 \cong R\frac{d\Delta}{dl}\frac{\partial l}{\partial u} + R\frac{d\Delta}{d\theta}\frac{\partial\theta}{\partial u} \tag{111}$$

$$y_0 \cong R\frac{d\Delta}{dl}\frac{\partial l}{\partial v} + R\frac{d\Delta}{d\theta}\frac{\partial\theta}{\partial v}. \tag{112}$$

Evaluation of the derivatives for x_0 gives

$$\frac{dl}{du} = \frac{d}{du}\sqrt{u^2 + v^2} = \frac{u}{l} = \frac{l\cos\theta}{l} = \cos\theta \tag{113}$$

$$\frac{d\theta}{du} = \frac{d}{du}\tan^{-1}\frac{v}{u} = -\frac{vu^{-2}}{1 + (v/u)^2} = -\frac{v}{u^2 + v^2}$$

$$= -\frac{l\sin\theta}{l^2\left(\cos^2\theta + \sin^2\theta\right)} = -\frac{\sin\theta}{l}. \tag{114}$$

Evaluation of the derivatives for y_0 gives

$$\frac{dl}{dv} = \frac{d}{dv}\sqrt{u^2 + v^2} = \frac{v}{l} = \frac{l \sin\theta}{l} = \sin\theta \tag{115}$$

$$\frac{d\theta}{dv} = \frac{d}{dv}\tan^{-1}\frac{v}{u} = \frac{u^{-1}}{1 + (v/u)^2} = \frac{u}{u^2 + v^2} = -\frac{l\cos\theta}{l^2\left(\cos^2\theta + \sin^2\theta\right)} = \frac{\cos\theta}{l}. \tag{116}$$

Substitution of Eqs. 113–116 into Eqs. 111 and 112 gives the coordinates (x_0, y_0) of the ray of the imperfect spherical wave at the image plane ($z = 0$) of the perfect spherical wave:

$$x_0 \cong R\left(\frac{\partial\Delta}{\partial l}\cos\theta - \frac{\partial\Delta}{\partial\theta}\frac{\sin\theta}{l}\right) \tag{117}$$

$$y_0 \cong R\left(\frac{\partial\Delta}{\partial l}\sin\theta + \frac{\partial\Delta}{\partial\phi}\frac{\cos\theta}{l}\right). \tag{118}$$

If the variation Δ in the surface of the converging imperfect spherical wave is independent of azimuth angle θ in the lens plane, then the ray of the imperfect spherical wave intercepts the image plane ($z = 0$) of the perfect spherical wave at transverse radial distance r from the center z-axis, where

$$r = \sqrt{x_0^2 + y_0^2} \cong R\sqrt{\left(\frac{\partial\Delta}{\partial l}\right)^2 + \frac{1}{l^2}\left(\frac{\partial\Delta}{\partial\theta}\right)^2}. \tag{119}$$

The distance z of the z-axis intercept of the ray of the imperfect spherical wave from the image plane ($z = 0$) of the perfect spherical wave is the ratio of radial distance r divided by the tangent of angle ψ, where

$$z = \frac{r}{\tan\psi}. \tag{120}$$

The tangent of angle ψ is the slope of the ray at the imperfect spherical material surface, which is the ratio of the radial distance of the ray at the imperfect spherical material surface divided by the radius of curvature $R + \Delta$ of the imperfect spherical material surface, which is approximated by the radius of curvature R of the perfect spherical material surface due to the small radius of curvature deviation Δ, where

$$\tan\psi = \frac{l}{R} \quad R \gg \Delta. \tag{121}$$

In the interval distance z between the z-axis crossing point of the imperfect spherical ray and the perfect spherical wave image plane ($z = 0$), the rays of the imperfect spherical wave converge to a locus of on-axis and off-axis points, rather than to a single on-axis point. This interval distance z is the axial or longitudinal transverse aberration, where

$$z = \frac{R}{l}r = \frac{R^2}{l}\sqrt{\left(\frac{\partial\Delta}{\partial l}\right)^2 + \frac{1}{l^2}\left(\frac{\partial\Delta}{\partial\theta}\right)^2}. \tag{122}$$

The above relations are modified and used in later sections on five third-order or Seidel lens aberrations. A spherical wave produced by a single object point at an arbitrary off-axis position, which even as a plane wave for the case of a distant object point, is refracted by a spherical, parabolic, or aspheric biconcave lens to a locus or set of image points (i.e., spot) rather than a single conjugate image point.

8. DERIVATION OF THE FIVE SEIDEL MONOCHROMATIC ABERRATIONS OF A SPHERICAL BICONCAVE LENS

The five monochromatic Seidel aberrations for a spherical biconcave lens CRL are derived as follows. The thin lens approximation is applied to the CRL in which the neutron ray is assumed to change direction, but not radial position, within the CRL. In a thin lens, a ray is refracted or changes directed at the principal plane of the lens, which is at its midplane. At the CRL, the rays are refracted are incident at transverse lens coordinates (l_x, l_y). While propagating within the CRL, the ray is assumed to maintain a constant radial position (l_x, l_y). At the center plane of the CRL, the ray is refracted and changes direction, but not transverse position, within the CRL. Outside the CRL the ray is assumed to change transverse position but retains its refracted direction.

For a thin lens, the object distance r_0 is the axial distance of the object point from the axial center of the CRL (midplane), and the image distance r_i is the axial distance of the image point from the CRL midplane. The thickness of the CRL $L_{lens}(l_x, l_y)$ is assumed to be much less than the object r_o and image r_i distances, where

$$L_{lens}(l_x, l_y) \ll r_0 + r_i. \tag{123}$$

Thus the object distance r_0 can be approximated as the axial distance from the object point to the input plane or left bounding plane of the CRL, and the image distance r_i can be approximated as the axial distance from the output plane or right bounding plane of the CRL to the image point.

The length of flight L_0 in the drift space from an illuminated object point at transverse coordinates s_x, s_y to a transverse coordinates l_x, l_y at the surface of the first lens of the CRL, with both coordinate pairs with respect to the center z-axis, is given by

$$L_0 = \sqrt{r_0^2 + \left(l_x - s_x\right)^2 + \left(l_y - s_y - o_g\right)^2} \tag{124}$$

For simplicity and application to X-rays, as well as neutrons, one neglects gravity,

$$o_g = i_g = 0. \tag{125}$$

One also assumes the object distance is much greater than the transverse distances in x and y, where

$$r_0 \gg l_x - s_x \tag{126}$$

$$r_0 \gg l_y - s_y. \tag{127}$$

In this case, expansion of the radical to the third order yields

$$L_0 \approx r_0 \left[1 + \frac{1}{2r_0^2}\left(l_x - s_x\right)^2 + \frac{1}{2r_0^2}\left(l_y - s_y\right)^2 - \frac{1}{8r_0^4}\left[\left(l_x - s_x\right)^2 + \left(l_y - s_y\right)^2\right]^2 \right]. \tag{128}$$

If a neutron travels at phase velocity v in vacuum and velocity v/n in a material with refractive index n, then the TOF of the neutron is identical for path length L in a vacuum and path length nL in the material. The optical path length through the drift space between the object point and a point on the surface of the first lens of the CRL Γ_0 is

$$\Gamma_0 = L_0 n = L_0. \tag{129}$$

Here $n = 1$ for air or a vacuum.

The case for lens aberrations involves lens distortion, which arises from upstream and downstream apertures, which are either extrinsic to the CRL and/or intrinsic to the CRL via its thickness. In this case, the transverse x and y distances of the object point (s_x, s_y) are related to its corresponding image point (x, y) by the respective proportionality factors, m_x and m_y, where

$$s_x = m_x x \tag{130}$$

$$s_y = m_y y. \tag{131}$$

With distortion the chief diagonal ray passes from an off-axis object point O above the optical axis through an off-axis lens point L to an off axis image point I below the optical axis. Similar triangles for the x and y component transverse distances give

$$\frac{s_x - l_x}{r_0} = -\frac{x - l_x}{r_i} \tag{132}$$

$$\frac{s_y - l_y}{r_0} = -\frac{y - l_y}{r_i} \tag{133}$$

Solving for s_x and s_y gives

$$s_x = -\frac{r_0}{r_i}x + l_x\left(1 + \frac{r_0}{r_i}\right) \tag{134}$$

$$s_y = -\frac{r_0}{r_i}x + l_y\left(1 + \frac{r_0}{r_i}\right). \tag{135}$$

Comparison of the equations for s_x and s_y gives the expressions for the proportionality factors, m_x and m_y:

$$m_x = -\frac{r_0}{r_i} + \frac{l_x}{x}\left(1 + \frac{r_0}{r_i}\right) \tag{136}$$

$$m_y = -\frac{r_0}{r_i} + \frac{l_y}{y}\left(1 + \frac{r_0}{r_i}\right). \tag{137}$$

If there is no distortion such that the chief ray passes through the center of the lens, then $(l_x, l_y) = (0, 0)$ and

$$m_x = m_y = -\frac{r_0}{r_i}. \tag{138}$$

Substitution for s_x and s_y of Eqs. 130 and 131 into Eq. 128, and substitution of the resulting expression into Eq. 129, gives the optical path length for the drift space between the object and lens

$$\Gamma_0 \approx r_0 + \frac{1}{2r_0}\left([l_x - m_x x]^2 + [l_y - m_y y]^2\right) \\ - \frac{1}{8r_0^3}\left[(l_x - m_x x)^2 + (l_y - m_y y)^2\right]^2. \tag{139}$$

In a single spherical lens with radius of curvature R and minimum wall or center thickness d, a neutron ray incident at transverse x and y distances l_x and l_y from the lens center axis travels an approximate axial distance, or length of flight L_{lens}, through the lens material, denoted by

$$L_{lens} = 2\left(R - \sqrt{R^2 - l_x^2 - l_y^2}\right) + d. \tag{140}$$

The optical path length Γ_{lens} through the lens is related to the lens refractive index n and the neutron path length L_{lens}, where

$$\Gamma_{lens} = L_{lens} n. \tag{141}$$

Between the biconcave lens left plane and its left surface is air, and between the lens right surface and the lens right plane is air. The left and right planes of each biconcave lens are separated by axial distance $d_{\max}$, which is the lens perimeter or maximum thickness. The thickness L_{air} of the air space of each spherical lens is

$$L_{air} = d_{\max} - L_{lens} \tag{142}$$

or

$$L_{air} = d_{\max} - 2\left(R - \sqrt{R^2 - l_x^2 - l_y^2}\right) - d. \tag{143}$$

The optical path length Γ_{CRL} of the ray through the spherical lens CRL is

$$\Gamma_{CRL}^{sph} = N\left(d_{\max} - d\delta\right) - 2N\delta\left(R - \sqrt{R^2 - l_x^2 - l_y^2}\right). \tag{144}$$

The paraxial approximation requires both

$$R \gg l_x \quad \text{and} \quad R \gg l_y. \tag{145}$$

Application of the paraxial approximation gives the third-order expression for spherical lens CRL:

$$\Gamma_{CRL}^{sph} \approx N\left(d_{\max} - d\delta\right) - 2N\delta\left[\frac{l_x^2 + l_y^2}{2R} + \frac{\left(l_x^2 + l_y^2\right)^2}{8R^3}\right]. \tag{146}$$

In the case of a parabolic lens CRL, the exact optical path length is

$$\Gamma_{CRL}^{par} = N\left(d_{\max} - d\delta\right) - \frac{N\delta}{R}\left(l_x^2 + l_y^2\right). \tag{147}$$

The length of flight L_i in the drift space from the point on surface of the last lens of the CRL at transverse coordinates l_x, l_y to a transverse coordinates x, y at the image plane or detector, with both coordinate pairs with respect to the center z-axis, is given by

$$L_i = \sqrt{r_i^2 + (x - l_x)^2 + (y - l_y)^2}. \tag{148}$$

The optical path length through the drift space between the point on the surface of the last lens of the CRL and the image point is Γ_i, where $n = 1$ for air or vacuum, is given by

$$\Gamma_i = L_i n = L_i. \tag{149}$$

Assume the image distance is much greater than the transverse excursions in x and y, where

$$r_i \gg x - l_x \tag{150}$$

$$r_i \gg y - l_y. \tag{151}$$

In this case, expansion to the third order yields

$$\Gamma_i \approx r_i + \frac{1}{2r_i}\left[(x - l_x)^2 + (y - l_y)^2\right] - \frac{1}{8r_i^3}\left[(x - l_x)^2 + (y - l_y)^2\right]^2. \tag{152}$$

The third-order expansion of the radical expression for the optical path length Γ_{3rd} between an object and an image point, separated by a CRL, is thus

$$\Gamma_{3rd} \cong \Gamma_0 + \Gamma_{CRL}^{sph} + \Gamma_i. \tag{153}$$

Substitution into Eq. 153 of (1) the optical path length for the drift space between an object point and a point on the input surface of the CRL first lens from Eq. 139, (2) the third-order expression of Eq. 146 for the optical path length through the spherical lens CRL, which was obtained by the paraxial approximation, and (3) the optical path length of Eq. 152 for the drift space between a point on the output surface of the CRL last lens and an image point, yields

$$\Gamma_{3rd} = r_0 + r_i + N\left(d_{\max} - d\delta\right) + \frac{1}{2r_0}\left(\left[l_x - m_x x\right]^2 + \left[l_y - m_x y\right]^2\right)$$
$$- \frac{1}{8r_0^3}\left[\left(l_x - m_x x\right)^2 + \left(l_y - m_y y\right)^2\right]^2 - 2N\delta\left[\frac{l_x^2 + l_y^2}{2R} + \frac{\left(l_x^2 + l_y^2\right)^2}{8R^3}\right]$$
$$+ \frac{1}{2r_i}\left[\left(x - l_x\right)^2 + \left(y - l_y\right)^2\right] - \frac{1}{8r_i^3}\left[\left(x - l_x\right)^2 + \left(y - l_y\right)^2\right]^2. \quad (154)$$

Next we calculate the second-order expansion of the above radical expression for the optical path length Γ_{2nd} of a ray path from an object point, through the CRL to an image point, for the case of a spherical lens or parabolic lens CRL, as follows:

$$\Gamma_{2nd} \cong r_0 + r_i + N\left(d_{\max} - d\delta\right) - \frac{N\delta}{R}\left(l_x^2 + l_y^2\right) + \frac{1}{2r_i}\left[\left(x - l_x\right)^2 + \left(y - l_y\right)^2\right]$$
$$+ \frac{1}{2r_0}\left[\left[l_x - m_x x\right]^2 + \left[l_y - m_y y\right]^2\right]. \quad (155)$$

The optical path length difference $\Delta\Gamma$ in the third-order and second-order expressions for the optical path length of a ray path originating at an object point and passing through the CRL to an image point yields the five Seidel monochromatic aberrations, where

$$\Delta\Gamma = \Gamma_{3rd} - \Gamma_{2nd}. \quad (156)$$

Substitution into Eq. 156 for the 2nd order expansion of Eq. 155 and the third order expansion of Eq. 154 yields the optical path difference $\Delta\Gamma_{sph}$ for a spherical lens CRL

$$\Delta\Gamma_{sph} = -\frac{1}{8r_0^3}\left[\left(l_x - m_x x\right)^2 + \left(l_y - m_y y\right)^2\right]^2$$
$$- \frac{N\delta}{4R^3}\left(l_x^2 + l_y^2\right)^2 - \frac{1}{8r_i^3}\left[\left(x - l_x\right)^2 + \left(y - l_y\right)^2\right]^2. \quad (157)$$

The optical path difference $\Delta\Gamma_{par}$ for a parabolic lens CRL is

$$\Delta\Gamma_{par} = -\frac{1}{8r_0^3}\left[\left(l_x - m_x x\right)^2 + \left(l_y - m_y y\right)^2\right]^2 - \frac{1}{8r_i^3}\left[\left(x - l_x\right)^2 + \left(y - l_y\right)^2\right]^2. \quad (158)$$

With azimuthal symmetry we convert to cylindrical coordinates. The radial distance s of the object point from the center z-axis is

$$s = \sqrt{s_x^2 + s_y^2}. \tag{159}$$

The radial distance l of the lens point from the center axis is

$$l = \sqrt{l_x^2 + l_y^2}. \tag{160}$$

A straight line ray or chief ray extends from the object point (s_x, s_y) in a straight line through the center of the lens (without aperture stops) to the conjugate image point (x, y). In the case of a lens with aperture stop, the chief ray extends in a straight line from the object point through the center of the aperture stop and through an off-axis lens point to the conjugate image point at (x, y). In either case, the radial distance h of the chief ray at the conjugate image point from the center optical or z-axis axis is

$$h = \sqrt{x^2 + y^2}. \tag{161}$$

The image plane Cartesian coordinates x, y can be expressed by conjugate image point radius h and azimuthal angle ϕ in the image plane, measured from the transverse image plane x-axis, where

$$x = h \cos \phi \tag{162}$$

$$y = h \sin \phi. \tag{163}$$

The Cartesian lens x and y coordinates l_x, l_y in the plane perpendicular to the center z-axis (optical axis) can be expressed lens point radius l and lens plane azimuthal angle θ, measured from the transverse lens plane x-axis, where

$$l_x = l \cos \theta \tag{164}$$

$$l_y = l \sin \theta. \tag{165}$$

Substitution of Eqs. 162–165 into Eq. 157 yields the optical path length difference between the third-order and second-order expansions for a spherical lens CRL:

$$\Delta\Gamma_T = -l^4\left(\frac{N\delta}{4R^3}\right) - \frac{1}{8r_0^3}\left[\left(l\cos\theta - hm_x\cos\varphi\right)^2 + \left(l\sin\theta - hm_y\sin\varphi\right)^2\right]^2$$
$$- \frac{1}{8r_i^3}\left[\left(h\cos\varphi - l\cos\theta\right)^2 + \left(h\sin\varphi - l\sin\theta\right)^2\right]^2. \tag{166}$$

Expansion and collection of terms yields aberrations resulting from the third-order expansion, and their associated path lengths for a spherical lens CRL. The aberration equation is

$$\Delta\Gamma_T = -\frac{l^4}{8}\left[\frac{2N\delta}{R^3} + \frac{1}{r_0^3} + \frac{1}{r_i^3}\right]$$
$$+ \frac{l^3h}{2}\left[\frac{m_x\cos\theta\cos\varphi + m_y\sin\theta\sin\varphi}{r_0^3} + \frac{\cos(\theta-\varphi)}{r_i^3}\right]$$
$$- \frac{l^2h^2}{4}\left[\frac{m_x^2\cos^2\phi + m_y^2\sin^2\phi}{r_0^3} + \frac{1}{r_i^3}\right]$$
$$- \frac{l^2h^2}{2}\left[\frac{\left(m_x\cos\theta\cos\varphi + m_y\sin\theta\sin\varphi\right)^2}{r_0^3} + \frac{\cos^2(\theta-\varphi)}{r_i^3}\right]$$
$$+ \frac{h^3l}{2}\left[\frac{\left(m_x\cos\theta\cos\varphi + m_y\sin\theta\sin\varphi\right)\left(m_x^2\cos^2\phi + m_y^2\sin^2\phi\right)}{r_0^3}\right.$$
$$\left. + \frac{\cos(\theta-\varphi)}{r_i^3}\right] - \frac{h^4}{8}\left[\frac{\left(m_x^2\cos^2\phi + m_y^2\sin^2\phi\right)^2}{r_0^3} + \frac{1}{r_i^3}\right] \tag{167}$$

In the case of no distortion, the chief ray or principal ray is a straight line from the object to the image point through the center lens (principal) point $(l_x, l_y) = (0, 0)$; then the proportionality factors m_x and m_y from Eq. 138 are

$$m_x = m_y = -\frac{r_0}{r_i} \tag{168}$$

Substitution of the above no-distortion proportionality factors of Eq. 168 in the above aberration expression of Eq. 167 cause the fifth term with the

$h^3 l$ coefficient to be zero. The no-distortion case—that is, no stop aperture upstream or downstream from the lens—yields the following expression for the optical path length difference between the third-order and second-order expansions:

$$\Delta\Gamma_T = -\left[\frac{l^4}{8}\left(\frac{2N\delta}{R^3}+\frac{1}{r_0^3}+\frac{1}{r_i^3}\right)+\frac{h^4}{8r_i^3}\left(\frac{r_0}{r_i}+1\right)\right] + \frac{l^3 h\cos(\theta-\varphi)}{2r_i}\left[\frac{1}{r_i^2}+\frac{1}{r_0^2}\right]-\frac{l^2h^2}{4r_i^2}\left[\frac{1}{r_0}+\frac{1}{r_i}\right] - \frac{l^2h^2\cos^2(\theta-\varphi)}{2r_i^2}\left[\frac{1}{r_0}+\frac{1}{r_i}\right]. \quad (169)$$

At azimuthal angle $\phi = 0$ in the image plane, the optical path length difference between the third-order and second-order expansions becomes

$$\Delta\Gamma_T = -\left[\frac{l^4}{8}\left(\frac{2N\delta}{R^3}+\frac{1}{r_0^3}+\frac{1}{r_i^3}\right)+\frac{h^4}{8r_i^3}\left(\frac{r_0}{r_i}+1\right)\right] + \frac{l^3 h\cos\theta}{2r_i}\left[\frac{1}{r_i^2}+\frac{1}{r_0^2}\right] - \frac{l^2h^2}{4r_i^2}\left[\frac{1}{r_0}+\frac{1}{r_i}\right]-\frac{l^2h^2\cos^2\theta}{2r_i^2}\left[\frac{1}{r_0}+\frac{1}{r_i}\right]. \quad \phi = 0 \quad (170)$$

The above aberration equation does not have distortion because the chief ray passes through the center of the lens, and thus the aberrations are composed of only four of the five Seidel aberrations.

Now consider the case of distortion in which an aperture stop is placed upstream or downstream from the biconcave spherical lens CRL. Assume azimuthal symmetry in which the chief ray from object point O at radial distance s passes through the lens at off-axis point L at radial distance l and terminates at image point I at radial distance h, where

$$s = mh. \quad (171)$$

Azimuthal symmetry allows the azimuthal angle ϕ to be dropped by setting it to zero. Using Eqs. 136 and 137 with $x = y = h$, the proportionality factor m is then

$$m = m_x = m_y = -\frac{r_0}{r_i}+\frac{l}{h}\left(1+\frac{r_0}{r_i}\right) \quad \text{and} \quad \phi = 0. \quad (172)$$

In the case of distortion with Eq. 172, the optical path length difference between the third-order and second order expansions of Eq. 167 becomes

$$\Delta\Gamma_T = -\left[\frac{l^4}{8}\left(\frac{2N\delta}{R^3} + \frac{1}{r_0^3} + \frac{1}{r_i^3}\right) + \frac{h^4}{8}\left(\frac{m^4}{r_0^3} + \frac{1}{r_i^3}\right)\right]$$
$$+ \frac{l^3 h\cos\theta}{2}\left[\frac{m}{r_0^3} + \frac{1}{r_i^3}\right] - \frac{l^2h^2}{4}\left[\frac{m^2}{r_0^3} + \frac{1}{r_i^3}\right]$$
$$- \frac{l^2h^2\cos^2\theta}{2r_i^2}\left[\frac{m^2}{r_0^3} + \frac{1}{r_i^3}\right] + \frac{h^3 l\cos\theta}{2}\left[\frac{m^3}{r_0^3} + \frac{1}{r_i^3}\right]. \tag{173}$$

This aberration equation includes distortion and thus has the five Seidel aberrations.

9. RELATION OF THE OPTICAL PATH LENGTH DIFFERENCE TO ITS ASSOCIATED ABERRATION

In the following sections, each of the five Seidel aberration terms in Eq. (173) is examined and evaluated via the following equations. Recall the results of the transverse and longitudinal aberration induced on an incident wave by an imperfect spherical surface. A spherical wave emitted from a source point intercepts a spherical biconcave lens CRL whose surfaces have radius of curvature R. The spherical wave front from the source point is refracted by the CRL to produce a convergent wave that intercepts the image plane at image distance r_i.

In the case of no aberration, the convergent wave from the lens converges to a single image point in the image plane. However, in the case of aberration, the convergent wave converges to a locus of points or spot, rather than a single, focused image point. If the source point is on-axis, then in the case of aberration the conjugate image spot can be off-axis. In the image plane, the off-axis image point I coordinates x_0 and y_0 for the transverse aberrated rays are expressed in polar coordinates, where the transverse aberration in the x and y directions are then

$$x_0 \cong r_i\left(\frac{\partial\Delta}{\partial l}\cos\theta - \frac{\partial\Delta}{\partial\theta}\frac{\sin\theta}{l}\right) \tag{174}$$

$$y_0 \cong r_i\left(\frac{\partial\Delta}{\partial l}\sin\theta + \frac{\partial\Delta}{\partial\theta}\frac{\cos\theta}{l}\right). \tag{175}$$

The chief ray extends in a straight line from the object point O through the center of the lens $l = 0$ to the chief ray conjugate image point CI. If there is an aperture stop, one uses an ideal chief ray that extends in a straight line from object point O through the center of the lens $l = 0$ and to the chief ray image point CI. The ideal chief ray passes through the barrier of the aperture stop placed before or after the lens—hence the term *ideal chief ray*. If the lens has no aperture stops, the real or actual chief ray is also the ideal chief ray and extends in a straight line form object point O through the center of the lens $l = 0$ to chief ray image point CI.

Other diverging rays from the same object point O, which pass though the lens but are not focused to the chief ray image point CI, converge to other image point I in the image plane. The other image point I is located at transverse x and y distances x_0 and y_0 from the chief ray image point CI. The length of the chief ray from its intercept point on the lens center plane L to the conjugate image from the object CI is approximated by the axial image distance r_i between the center lens plane and the image plane.

If there is an aperture stop before or after the lens, then the real chief ray is no longer a straight line but is bent at the lens. In the case of an aperture stop placed before the lens (object side aperture), the real chief ray extends from object point O through the center of the aperture stop to an off-axis point L on the lens ($l \neq 0$), where the chief ray slope changes and then proceeds to image point I, whose position in the image plane is dependent on the aperture placement. In the case of an aperture stop placed after the lens (image side aperture), the real chief ray extends from object point O through an off-axis point L on the lens ($l \neq 0$), where the chief ray changes slopes and proceeds to image point I, whose position in the image plane is dependent on the aperture placement.

Lens aberrations cause the diverging rays emitted from the object to converge to image points I at (x_0, y_0), which is measured from the image point CI of the chief ray. Relative to the center z-axis, the x and y coordinates of chief ray image point CI are h_x and h_y. That is, the chief ray extends from an object point O and intercepts the conjugate image point at radial distance h from the center z-axis. In the image plane, the azimuthal angle ϕ of the radius h of the chief ray CI image point is measured relative to the transverse image plane x-axis, where

$$h_x = h\cos\phi \quad \text{and} \quad h_y = h\sin\phi. \tag{176}$$

In the lens plane, divergent rays from the object point O have radius l, and the ray azimuthal angle θ at lens radius l is measured relative to the lens transverse x-axis. In the image plane, the ray terminates at image point I, which is off-axis from the line of the chief ray image point CI. Off-axis image point I has coordinates (x_0, y_0) and the radius r is measured relative to the x and y position of the chief ray intercept (h_x, h_y) at the image plane point CI, where

$$r = \sqrt{x_0^2 + y_0^2} \cong r_i \sqrt{\left(\frac{\partial \Delta}{\partial l}\right)^2 + \frac{1}{l^2}\left(\frac{\partial \Delta}{\partial \theta}\right)^2}. \tag{177}$$

A divergent ray from the object point O is refracted by the lens to an off-axis image point I at coordinates (x_0, y_0) measured relative to the chief ray image point CI coordinates (h_x, h_y). The image plane is situated at image distance r_i from the center lens plane. The chief ray from the same object point O intercepts the image plane at (h_x, h_y), where $(x_0, y_0) = (0, 0)$.

The angle ψ is measured between the z-axis and the convergent ray from the lens that intercepts the image plane at $(x_0, y_0) \neq (0, 0)$. The tangent of angle ψ is the ratio of the radius l of the object O ray at the center lens plane divided by the image distance r_i from the center lens plane to the image plane, where

$$\tan \psi = \frac{l}{r_i}. \tag{178}$$

The divergent rays from an object point O are focused by the lens onto a locus of image points I in a set of image planes because of lens aberration. The axial extent z of the image planes for the set of image points I that correspond to a single object point O is the longitudinal or axial aberration, where

$$z = \frac{r}{\tan \psi} = \frac{r_i}{l} r = \frac{r_i^2}{l} \sqrt{\left(\frac{\partial \Delta}{\partial l}\right)^2 + \frac{1}{l^2}\left(\frac{\partial \Delta}{\partial \theta}\right)^2}. \tag{179}$$

The refraction of the incident spherical wave from an arbitrary, single object point O by a spherical, parabolic, or aspherical lens produces an imperfect convergent wave that images to a set of points distributed in several image planes, thus creating a three-dimensional locus of image points I or an image cluster. If the lens was somehow perfect, then the lens would refract the incident divergent spherical wave from the single object point O to a convergent perfect spherical wave that would converge to a single image point I. The optical path length difference $\Delta\Gamma$ between the third-order and second-order expressions gives the range of optical path length difference for various ray paths from object point O through the N-lens biconcave lens CRL to the image plane.

Rays originate from a single object point O at object distance r_0 from the CRL center plane. The rays pass through the N-lens biconcave spherical lens CRL to the image plane at distance r_i and produce a locus of image points I in the image plane at coordinates (x_0, y_0). This locus of image points I is centered about the center chief ray whose image point *CI* in the image plane has coordinates $(x_0, y_0) = (0, 0)$. The chief ray coordinates $(x_0, y_0) = (0, 0)$ in the image plane are located at image plane coordinates (h_x, h_y) measured relative to the center z-axis (optical axis). The cluster of image points I rather than a single image point *CI* marked by the chief ray is due to aberration.

Aberration arises because of the range of optical path length differences for various ray paths from object point O through the N-lens biconcave lens CRL to the image plane. These optical path length differences produce constructive interference at a cluster of points rather than at a single point. In the aberration expression of Eq. 173, each of the five types of aberration produces a set of path length differences between the ray paths from single object point O to the cluster of image points I. The ray path difference Δ can be evaluated for each aberration in turn. Then, the range of path length differences Δ for a particular aberration and the equations for the image plane positions x_0, y_0 and cluster radius r yields the shape of the locus of image points. The shape of the locus of image points corresponds to a specific locus of points in the lens center plane (e.g., an annular ring) or, corresponds to a single point or locus of points O in the object plane.

10. SPHERICAL ABERRATION

The first aberration term of the third-order aberration expression of Eq. 173 is for spherical aberration Δ_{sp} in a spherical biconcave lens that includes distortion (see discussion following Eq. 129 in section 8)

$$\Delta_{sp} = -\left[\frac{l^4}{8}\left(\frac{2N\delta}{R^3} + \frac{1}{r_0^3} + \frac{1}{r_i^3}\right) + \frac{h^4}{8}\left(\frac{m^4}{r_0^3} + \frac{1}{r_i^3}\right)\right]. \tag{180}$$

Without a limiting aperture stop before or after the lens, there is no distortion. The first aberration term for spherical aberration Δ_{sp} for the case of no distortion becomes

$$\Delta_{sp} = -\left[\frac{l^4}{8}\left(\frac{2N\delta}{R^3} + \frac{1}{r_0^3} + \frac{1}{r_i^3}\right) + \frac{h^4}{8r_i^3}\left(\frac{r_0}{r_i} + 1\right)\right]. \tag{181}$$

The radial distance of the ray incident on the lens is l, and the radial distance of the refracted ray at the image plane is h. The l^4 and h^4 terms are a result of spherical aberration. However, the l^4 portion can sufficiently depict the lens spherical aberration.

For a spherical, aspherical, or parabolic lens CRL with on-axis radius of curvature R, the path length deviation term Δ_{sp} represents spherical aberration, where

$$\Delta_{sp} = c_l l^4 + c_h h^4. \tag{182}$$

Spherical aberration constants c_l and c_h contain distortion variable m, where

$$c_l = -\frac{1}{8}\left(\frac{2N\delta}{R^3} + \frac{1}{r_0^3} + \frac{1}{r_i^3}\right) \tag{183}$$

$$c_h = -\frac{1}{8}\left(\frac{m^4}{r_0^3} + \frac{1}{r_i^3}\right). \tag{184}$$

In the image plane, the path length deviation Δ_{sp} of a ray depends on the ray's radial distance l at the lens plane. Spherical aberration causes the outer rays from an object point to focus closer to the lens than the inner rays. The chief ray from an object point O intercepts the image plane at the conjugate image point CI. Measured relative to the image plane intercept point CI of the chief ray, the other associated image points I of object point O have image plane coordinates x_0 and y_0. That is, the x- and y-component distances of image points I from chief ray image point CI are expressed in polar coordinates by

$$x_0 \cong r_i\left(\frac{\partial \Delta_{sp}}{\partial l}\cos\theta - \frac{\partial \Delta_c}{\partial \theta}\frac{\sin\theta}{l}\right) = 3c_l r_i l^3 \cos\theta \tag{185}$$

$$y_0 \cong r_i\left(\frac{\partial \Delta_c}{\partial l}\sin\theta + \frac{\partial \Delta_c}{\partial \theta}\frac{\cos\theta}{l}\right) = 3c_l r_i l^3 \sin\theta. \tag{186}$$

In the image plane, the radial distance r of each point image point I of single object point O is measured relative to the chief ray image point CI of object point O in which r is dependent on the radial distance l of the object point O ray at the center lens plane. In the image plane, the blur radius r or transverse aberration caused by spherical aberration is

$$r = \sqrt{x_0^2 + y_0^2} = r_i\frac{\partial \Delta_{sp}}{\partial l} = c_l 3 l^3 r_i. \tag{187}$$

The outer rays from an on-axis object O are focused to a single point on-axis in a set of image planes that are closer to the lens than the inner paraxial rays. The paraxial rays from the same object point O converge to an image point I in the paraxial image plane, designated as the single lens image plane at image distance r_i from the center lens plane. The axial distance z between the image planes of the outer and inner (paraxial) rays, which are produced by a single on-axis object point O, is the longitudinal or axial extent of the spherical aberration, where

$$z = \frac{r_i}{l} r = c_l 3 l^2 r_i^2. \tag{188}$$

Spherical aberration can be remedied by using only the parabolic center portion of the spherical lens via aperture stops on each side of the CRL or by the use of parabolic lenses.

11. COMA

The second aberration term of the third-order aberration expression of Eq. 173 is for coma Δ_c in a spherical biconcave lens, which with distortion caused by an upstream or downstream aperture stop, is (see discussion following Eq. 129 in section 8)

$$\Delta_c = \frac{l^3 h \cos\theta}{2}\left(\frac{m}{r_0^3} + \frac{1}{r_i^3}\right). \tag{189}$$

In the case of no aperture stops and thus no distortion, the coma aberration term Δ_c at image plane azimuthal angle ϕ becomes

$$\Delta_c = \frac{l^3 h \cos(\theta - \phi)}{2 r_i}\left(\frac{1}{r_i^2} + \frac{1}{r_0^2}\right). \tag{190}$$

For a spherical, aspheric, or parabolic lens CRL the path length deviation term Δ_c represents coma, which is an off-axis aberration that is nonsymmetrical about the optical axis (center z-axis) and increases rapidly with lens aperture radius l, where

$$\Delta_c = c_c l^3 h \cos\theta. \tag{191}$$

Coma aberration constant c_c contains distortion variable m, where

$$c_c = \frac{1}{2}\left(\frac{m}{r_0^3} + \frac{1}{r_i^3}\right). \tag{192}$$

Recall that angle θ is the azimuth angle of the radius l of lens point L with coordinates (l_x, l_y) in the lens plane, and in the image plane h is the corresponding radial distance of the coordinates (x, y) of an image point I. In the image plane, the path length deviation Δ_c of a ray depends on the ray's radial distance h at its image point I, and Δ_c is also dependent on the ray's radial distance l and azimuthal angle θ at the lens plane.

Each circular zone of the lens forms a circular image (comatic circle) in which the north and south rays in the lens plane focus at the north point of the image circle, and in which the east and west rays in the lens plane focus at the south point of the image circle. The lens points in the other directions produce corresponding image points that complete the image circle. The image circle radius increases with the radius of the corresponding circle of rays emitted from the lens. A single traversal of an annular ring of lens points corresponds to a double traversal of the corresponding circle of image points in the image plane.

Given a single object point O in the image plane, the x- and y-component distances x_0 and y_0 of the corresponding image points I, which are measured from chief ray image point CI, become for the coma aberration

$$x_0 \cong r_i\left(\frac{\partial \Delta_c}{\partial l}\cos\theta - \frac{\partial \Delta_c}{\partial \theta}\frac{\sin\theta}{l}\right) = c_c r_i l^2 h\left(3\cos^2\theta + \sin^2\theta\right)$$
$$= c_c r_i l^2 h\,(2 + \cos 2\theta) \tag{193}$$

$$y_0 \cong r_i\left(\frac{\partial \Delta_c}{\partial l}\sin\theta + \frac{\partial \Delta_c}{\partial \theta}\frac{\cos\theta}{l}\right) = 2c_c r_i l^2 h\cos\theta\sin\theta = c_c r_i l^2 h\sin 2\theta. \tag{194}$$

With trigonometric identities and transposing one obtains

$$x_0 - 2c_c r_i l^2 h = c_c r_i l^2 h\cos 2\theta \tag{195}$$

$$y_0 = c_c r_i l^2 h\sin 2\theta. \tag{196}$$

Squaring and adding shows that an annular ring of points of radius l in the lens plane produces a circle in the image plane. The center of the circle of points in the image plane is offset along the transverse x-axis from the center z-axis by $2r_i l^2 h$. The radius r of the image point circle increases with increasing lens annular ring radius l^2 and increasing radius h of the conjugate image point that is connected to its object point by the chief ray, where

$$\left(x_0 - 2c_c r_i l^2 h\right) + y_0^2 = \left(c_c r_i l^2 h\right)^2. \tag{197}$$

Both lens radius l and conjugate image point of the chief ray h are measured relative to the z-axis, which passes through the center of the object, lens, and image plane. In the image plane, the radial distance r of each point image point I of single object point O is measured relative to the chief ray image point CI of object point O. And in the case of coma radial distance r is dependent on the radial distance l of the object point O ray at the center lens plane and the radial distance h of the chief ray intercept in the image plane from the center z-axis. In the image plane, the blur radius r or transverse aberration caused by coma is

$$r = c_C r_i l^2 h. \tag{198}$$

The divergent rays from an object point O are focused by the lens onto a locus of image points I in a set of image planes because of coma aberration of the lens. The axial extent z of the image planes is the longitudinal or axial aberration caused by coma, where

$$z = \frac{r_i}{l} r = 2c_C r_i^2 lh \cos\theta. \tag{199}$$

The axial aberration is not azimuthal symmetric due to the azimuthal lens angle dependence of $\cos\theta$. Also, the axial aberration increases with increasing radial distance of rays from the center optical axis.

12. COMA: THE ABBE SINE CONDITION AND THE CODDINGTON SHAPE FACTOR

The Abbe sine condition for the reduction of coma aberration in a lens or CRL is derived as follows. Spherical aberration occurs for object points located on-axis, and coma occurs for object points located off-axis. In the object plane the aperture of diameter D_0 is centered on the center optical axis (z-axis). A plane wave is emitted from the object plane aperture D_0 at angle γ_0 relative to the z-axis. The leading portion (bottom point) of the plane wave front has propagated distance Δr_0 from the object aperture D_0, and the lagging portion of the plane wave at the aperture plane is touching the top point of D_0.

The angle γ_0 of the plane wave is related to aperture diameter D_0 and the path length difference Δr_0 between the leading and lagging portions of the aperture-emitted plane wave in the same manner as the diffraction of a plane wave from a circular aperture, where

$$\sin\gamma_0 = \frac{\Delta r_0}{D_0}. \tag{200}$$

The lower rays of the plane wave that encounter the inner portion of a downstream lens are better focused than the upper rays of the plane wave that encounter the outer portion of the lens. The image is blurred for those rays from the periphery of the lens and less blurred for the rays from the center portion of the lens.

The center portion of the single principal plane of a thin lens or the two principal planes of a thick lens are approximated as planar in basic geometric optics. However, these principal planes in reality are curved surfaces whose curvature becomes significant outside the paraxial radius of the lens. Coma arises because these principal planes are curved surfaces and not planes. The principal planes of a lens are more accurately termed *principal curved surfaces.* This curvature of the principal plane of the thin lens (or the primary and secondary principal planes of a thick lens) results in the comet-shaped aberration.

The point at the head of the comet is formed by rays from an off-axis point that passed through the center portion of the lens. The locus of points at the comet's tail is formed by rays from the same off-axis object point that pass through the outer portion of the lens. That is, rays emitted by the object point that pass through the center portion of the lens are focused to the point at the head of the comet, and the rays that pass through the outer portion of the lens are not well focused and form the locus of points that make up the radial outward-pointed comet tail. The comet's tail is pointed radially outward because the outer portion of the image plane receives the most off-axis rays from the object and lens. However, a single object point close to the optical axis is focused to a single image point near the optical axis, whereas an off-axis peripheral object point is focused to a locus of points shaped like a comet in the image plane.

The lens refracts the incident plane wave toward the image plane at angle γ_i, relative to the optical axis. At the image plane, the refracted wave from the lens forms an image of diameter D_i. The path length difference Δr_i between the leading and lagging portions of a refracted wave front incident on the image plane is related to the image diameter D_i and the wave front angle γ_i (relative to the optical axis) by

$$\sin \gamma_i = \frac{\Delta r_i}{D_i}. \tag{201}$$

To eliminate coma the path length difference Δr_i between the leading and lagging portions of a refracted wave front and the path length difference Δr_0 between the leading and lagging portions of the wave front emerging from the object should be equal, where

$$\Delta r_0 = \Delta r_i. \tag{202}$$

As a result,

$$D_0 \sin \gamma_0 = D_i \sin \gamma_i. \tag{203}$$

The ratio of the lens-produced image diameter D_i divided by the corresponding object diameter D_0 is a constant c_0, where

$$\frac{D_i}{D_0} = c_0. \tag{204}$$

Elimination of coma requires the direction angle γ_0 of the incident wave (ray) and the direction angle γ_i of the refracted wave front (ray) to fulfil the Abbe sine condition, where

$$\frac{\sin \gamma_0}{\sin \gamma_i} = c_0. \tag{205}$$

Coma is corrected by using a combination of lenses symmetric about a center stop or by bending the lens or shape of the lens while maintaining the lens focal length. The shape of the lens can be changed by changing the radius of curvature on its object R_0 and image R_i sides. A measure of changing the lens shape is the Coddington shape factor σ, denoted by

$$\sigma = \frac{R_i + R_0}{R_i - R_0}. \tag{206}$$

Different combinations of the radius of curvature on the lens object R_0 and image R_i sides (i.e., different-shaped lenses) can produce the same focal length.

13. CURVATURE OF FIELD AND THE PETZVAL CONDITION

The third aberration term of the third-order aberration expression of Eq. 173 is for curvature of field Δ_{cf} in a spherical biconcave lens for distortion due to upstream or downstream stops (see discussion following Eq. 129 in section 8)

$$\Delta_{cf} = -\frac{l^2 h^2}{4}\left(\frac{m^2}{r_0^3} + \frac{1}{r_i^3}\right). \tag{207}$$

In the case of no aperture stops, the third aberration term for curvature of field Δ_{cf} for the case of no distortion becomes

$$\Delta_{cf} = -\frac{l^2 h^2}{4 r_i^2}\left(\frac{1}{r_0} + \frac{1}{r_i}\right). \tag{208}$$

For a spherical, aspheric, or parabolic lens CRL the path length deviation term Δ_{cf} represents the aberration of curvature of field, where

$$\Delta_{cf} = c_{cf} l^2 h^2. \tag{209}$$

Curvature of field aberration constant c_{cf} contains distortion variable m, where

$$c_{cf} = -\frac{1}{4}\left(\frac{m^2}{r_0^3} + \frac{1}{r_i^3}\right). \tag{210}$$

In curvature of field, rays that are emitted from a planar object do not arrive at the same time at a planar image plane, but rather arrive simultaneously on a curved image plane. That is, a ray along the z-axis from a center object point through the center of the lens and then to the center image point travels a shorter optical path length than a ray from an off-axis object point through the lens to an off axis image point. Hence, the set of image points, which correspond to a planar set of object points, must form a curved surface, so that all the optical path lengths between each object-image point pair, are equal.

Given a single object point O in the image plane, the x- and y-component distances x_0 and y_0 of the corresponding image points I, which are measured from chief ray image point CI, become for the curvature of field aberration

$$x_0 \cong r_i\left(\frac{\partial \Delta_{cf}}{\partial l}\cos\theta - \frac{\partial \Delta_{cf}}{\partial \theta}\frac{\sin\theta}{l}\right) = 2c_{cf} r_i l h^2 \cos\theta \tag{211}$$

$$y_0 \cong r_i\left(\frac{\partial \Delta_{cf}}{\partial l}\sin\theta + \frac{\partial \Delta_{cf}}{\partial \theta}\frac{\cos\theta}{l}\right) = 2c_{cf} r_i l h^2 \sin\theta. \tag{212}$$

In the image plane, the radial distance r of each point image point I of single object point O is measured relative to the chief ray image point CI of object point O in which r is dependent on the radial distance l of the object point O ray at the center lens plane and the radial distance h of the chief ray intercept in the image plane from the center z-axis. In the image plane, the blur radius r or transverse aberration caused by curvature of field is

$$r = \sqrt{x_0^2 + y_0^2} = 2c_{cf} r_i l h^2. \tag{213}$$

A single object point O emits rays, which converge to image points I in a set of image planes, rather than a single image plane because of the lens curvature of field aberration. The axial extent z of the image planes is the longitudinal or axial aberration, where

$$z = \frac{r_i}{l} r = 2c_{cf} r_i^2 h^2. \tag{214}$$

Curvature of field can be eliminated with two lenses—the first lens with refractive index n_1 and focal length f_1 and the second lens with refractive index n_2 and focal length f_1—if the Petzval condition is met, where

$$n_1 f_1 = -n_2 f_2. \tag{215}$$

An individual lens has an azimuthal symmetric curved image surface in which the object points are focused to their conjugate image points. The use of two lenses of different refractive index and focal lengths produces a flat image plane.

14. ASTIGMATISM

The fourth aberration term of the third-order aberration expression of Eq. 173 is for astigmatism Δ_a in a spherical biconcave lens. When either upstream or downstream stops cause distortion, the astigmatism aberration term Δ_a is (see discussion following Eq. 129 in section 8)

$$\Delta_a = \frac{l^2 h^2 \cos^2 \theta}{2r_i^2} \left(\frac{m^2}{r_0^3} + \frac{1}{r_i^3} \right). \tag{216}$$

When there are no aperture stops and thus no distortion, the astigmatism aberration term Δ_a at image plane azimuthal angle ϕ becomes

$$\Delta_a = \frac{l^2 h^2 \cos^2 (\theta - \phi)}{2r_i^2} \left(\frac{1}{r_0} + \frac{1}{r_i} \right). \tag{217}$$

For a spherical, aspherical, or parabolic lens CRL the path length deviation term Δ_a represents the aberration of astigmatism, where

$$\Delta_a = c_a l^2 h^2 \cos^2 \theta. \tag{218}$$

Astigmatism aberration constant c_a contains distortion variable m, where

$$c_a = \frac{1}{2r_i^2} \left(\frac{m^2}{r_0^3} + \frac{1}{r_i^3} \right). \tag{219}$$

In astigmatism, the rays in the vertical plane (the xz-tangential plane) focus closest to the lens, and if the vertical plane is now rotated about the z-axis, the rays focus farther from the lens. Finally, rays in the horizontal plane (the yz-sagittal plane) focus farthest from the lens. The wave front in the vertical plane, which is refracted by the lens, has a smaller radius of curvature than the refracted wave front in the horizontal plane.

That is, assume a wave front is diagonally incident on a lens input surface in which the incident angle of the wave front's wave vector is measured relative to the lens center z-axis. As the angle of incidence of the wave vector increases, so decreases the projected length of the lens chord diameter upon the portion of the wave front, which contacts and is refracted by the lens. A decrease in the lens chord diameter projected on the wave front portion, which passes through and is refracted by the lens, represents a decrease in the effective lens radius of curvature that refracts the wave front. The lens focal length is proportional to the lens radius of curvature. A decrease of the lens radius of curvature, which acts on the diagonally incident wave front, causes the wave front to be focused to a shorter focal length downstream of the lens.

The projection of the lens chord diameter onto the vertical plane wave front, which passes through the lens, is less than the lens chord diameter projection onto the horizontal plane wave front that also passes through the lens. The vertical wave front section experiences a smaller lens radius of curvature than the horizontal wave front section. As a result, the vertical wave front section focuses closer to the lens than the horizontal wave front section.

Given a single object point O in the image plane, the x- and y-component distances x_0 and y_0 of the corresponding image points I, which are measured from chief ray image point CI, become for the astigmatism aberration

$$x_0 \cong r_i\left(\frac{\partial \Delta_a}{\partial l}\cos\theta - \frac{\partial \Delta_a}{\partial \theta}\frac{\sin\theta}{l}\right) = 2c_a r_i l h^2\left(\cos^3\theta + \cos\theta\sin^2\theta\right)$$
$$= 2c_a r_i l h^2 \cos\theta \tag{220}$$

$$y_0 \cong r_i\left(\frac{\partial \Delta_a}{\partial l}\sin\theta + \frac{\partial \Delta_a}{\partial \theta}\frac{\cos\theta}{l}\right) = 0. \tag{221}$$

In the image plane, the radial distance r of the one or more image points I, which correspond to single object point O, are measured relative to the chief ray image point CI that corresponds to single object point O. Here r is dependent on the radial distance l and azimuth angle θ at the center lens plane for an object point O to image point I ray. Also, r is dependent on the radial distance h of the chief ray intercept from the center z-axis in the image plane. In the image plane, the blur radius r or transverse aberration caused by astigmatism is

$$r = \sqrt{x_0^2 + y_0^2} = 2c_a r_i l h^2 \cos\theta. \tag{222}$$

A single object point O emits rays that converge to image points I in a set of image planes, rather than a single image plane, because of the lens astigmatism aberration. The axial extent z of the image planes is the longitudinal or axial aberration, where

$$z = \frac{r_i}{l} r = 2c_a r_i^2 h^2 \cos\theta. \tag{223}$$

15. COMBINED ASTIGMATISM AND CURVATURE OF FIELD

Astigmatism and curvature of field are usually analyzed together. The summed aberrations Δ_{ac} of astigmatism Δ_a and curvature of field Δ_{cf} are denoted as

$$\Delta_{ac} = \Delta_a + \Delta_{cf} = c_a l^2 h^2 \cos^2\theta + c_{cf} l^2 h^2. \tag{224}$$

In the image plane, the off-axis image point I coordinates x_0 and y_0 for the transverse aberration are expresses in polar coordinates by

$$x_0 \cong r_i \left(\frac{\partial \Delta_{ac}}{\partial l} \cos\theta - \frac{\partial \Delta_{ac}}{\partial \theta} \frac{\sin\theta}{l} \right) = 2\left(c_a + c_{cf}\right) r_i l h^2 \cos\theta \tag{225}$$

$$y_0 \cong r_i \left(\frac{\partial \Delta_{ac}}{\partial l} \sin\theta + \frac{\partial \Delta_{ac}}{\partial \theta} \frac{\cos\theta}{l} \right) = 2c_{cf} r_i l h^2 \sin\theta. \tag{226}$$

In the image plane, the radial distance r of the image points I of single object point O is measured relative to the chief ray image point *CI* for the combined astigmatism and curvature of field. In the image plane, the blur radius r or transverse aberration caused by astigmatism and curvature of field combined is

$$r = \sqrt{x_0^2 + y_0^2} = 2r_i l h^2 \sqrt{c_{cf}^2 + c_a\left(c_a + 2c_{cf}\right)\cos^2\theta}. \tag{227}$$

Alternatively, the expressions for x_0 and y_0 can be combined to obtain the equation for an ellipse, where

$$\frac{x_0^2}{\left(c_{cf} + c_a\right)^2} + \frac{y_0^2}{c_{cf}^2} = \left(2r_i l h^2\right)^2. \tag{228}$$

That is, an annular ring of points in the lens plane is refracted to an ellipse of corresponding image points in the image plane. A single object O point emits rays that converge to a set of image points I in a set of image planes, rather than a single image plane, because of the lens curvature of field aberration. The axial extent z of the image planes is the longitudinal or axial aberration, where

$$z = \frac{r_i}{l} r = \frac{r_i}{l} \sqrt{x_0^2 + y_0^2} = 2r_i^2 h^2 \sqrt{c_{cf}^2 + c_a \left(c_a + 2c_{cf}\right) \cos^2 \theta}. \tag{229}$$

The image plane is at distance r_i from the lens center plane. Refracted rays from the lens in the vertical tangential plane $\theta = 0$ focus to a horizontal line at distance $r_i - z_T$ upstream of the image plane. The axial extent of the axial aberration over distance z_T is given by

$$z_T = \frac{r_i}{l} y_0 = 2 \left(c_a + c_{cf}\right) r_i l h^2. \quad \theta = 0 \tag{230}$$

Refracted rays from the lens in the horizontal sagittal plane $\theta = \pi/2$ focus to a vertical line at distance $r_i + z_S$ downstream from the image plane. The axial extent of the axial aberration over distance z_T is given by

$$z_S = \frac{r_i}{l} r = \frac{r_i}{l} y_0 = 2c_{cf} r_i l h^2 \quad \theta = \pi/2. \tag{231}$$

The axial extent Δz of aberration between the horizontally focused tangential (vertical) plane rays and the vertically focused sagittal (horizontal) plane rays is given by

$$\Delta z = z_T - z_S = 2c_a r_i l h^2. \tag{232}$$

The astigmatism constant c_a thus gives the distance between the tangential and sagittal focal surfaces.

The circle of least confusion lies halfway z_c between the tangential z_T and sagittal z_S focal surfaces, where

$$z_c = \frac{z_T + z_S}{2} = \left(c_a + 2c_{cf}\right) r_i l h^2. \tag{233}$$

At the circle of least confusion the ellipse of image points, which correspond to a circle of lens points, becomes a circle of image points. The pane of the circle of least confusion is an image plane with the least astigmatic and curvature of field aberration.

16. ASTIGMATISM: STURM'S INTERVAL AND REMEDY BY UPSTREAM APERTURE STOP

We now examine the difference in the focal length for incident rays in the vertical plane and incident rays in the horizontal plane that are refracted by a spherical biconcave lens or CRL. Recall that *sag* is the radial distance along the perpendicular bisector line of the chord from the chord bisection point to the spherical surface point. The length of the sag radial segment extends from the chord bisection point to the surface point of the sphere. For wave fronts incident along the z-axis the sag lengths of the emerging wave front from a spherical or parabolic lens are equal. However, for wave fronts in the tilted vertical and the tilted horizontal planes; the sag lengths of the emerging waves differ, which is the cause of astigmatism.

The wave front tilted in the tangential plane intercepts the full diameter D_v of the biconcave lens of radius r_m in which the sag length s_v is the difference in the lens center minimum thickness $t_{\min}$ and the lens maximum perimeter thickness $t_{\max}$, where

$$D_v = 2r_m \quad \text{and} \quad s_v = t_{\max} - t_{\min}. \tag{234}$$

The sag length s_v of the refracted wave emerging from the lens is related to the encountered lens diameter D_v and the acquired wave radius of curvature R_v by

$$s_v = \frac{r_m^2}{2R_v}. \tag{235}$$

The sag length equation is that of a parabola. Recall that the spherical lens is parabola shaped at the center portion of the lens, where the diameter of the lens $2r_m$ is much smaller than the spherical lens radius of curvature R_v. The center portion of spherical lens surface is a parabolic surface and R_v is also the on-axis radius of curvature of the parabolic surface. The spherical biconcave lens with radius of curvature R_v imposes the same radius of curvature R_v onto the incident plane wave to produce a converging refracted spherical wave.

An incident plane wave tilted at angle α in the horizontal or sagittal plane of the spherical biconcave lens also emerges as a spherical wave with sag length s_h, which is equal to sag length s_v of the refracted spherical wave of the incident plane tilted wave front in the vertical plane, where

$$s = s_h = s_v. \tag{236}$$

However, the incident plane wave titled in the horizontal plane encounters the projected lens diameter D_h, which is perpendicular to the propagation direction of the incident plane wave, where

$$D_h = 2r_m \cos\alpha \tag{237}$$

and

$$D_h \geq D_v. \tag{238}$$

Upon refraction the incident plane wave tilted in the horizontal plane of the biconcave lens becomes a spherical wave with acquired radius of curvature R_h. The sag length of the refracted wave tilted in the horizontal plane is related to wave radius of curvature R_h, and the reduced radius of the projected lens diameter is

$$s_h = \frac{(r_m \cos\theta)^2}{2R_h}. \tag{239}$$

The sag length s of the refracted spherical wave emerging from the biconcave spherical lens is identical for the incident plane wave tilted in the horizontal or vertical plane, so that

$$s = \frac{r_m^2}{2R_h} = \frac{r_m^2 \cos^2\alpha}{2R_v}. \tag{240}$$

The radius of curvature of a wave is also the focal length or distance from the lens at which the wave converges to a single point. The focal lengths of the tilted horizontal f_h and tilted vertical wave fronts f_v are, respectively,

$$f_h = R_h \quad \text{and} \quad f_v = R_v. \tag{241}$$

The lens paraxial focal length is

$$f = \frac{R}{2\delta} \cong \frac{r_m^2}{2s}. \tag{242}$$

Horizontal plane radius of curvature R_h and vertical plane radius of curvature R_v are each expressed by s in Eq. 240. Using Eq. 241 yields the astigmatic difference Δf in the focal lengths for the tilted horizontal (sagittal) and tilted vertical (tangential) wave fronts, where

$$\Delta f = f_h - f_v = R_h - R_v = \frac{r_m^2}{2s}\left(1 - \cos^2\alpha\right) = \frac{r_m^2}{2s}\sin^2\alpha. \tag{243}$$

With substitution for f from Eq. 242 the astigmatic difference in the horizontal (sagittal) f_h and vertical (tangential) f_v focal lengths, which is called Sturm's interval, becomes

$$f_v - f_h = f \sin^2 \alpha. \tag{244}$$

The astigmatic aberration of a lens causes a bicycle wheel–shaped object to have its rim in focus at the vertical (tangential) focal length f_v, and its spokes in focus at the horizontal (sagittal) focal length f_h. The elimination of astigmatism requires the sagittal and tangential surfaces to overlap in a single surface termed the *Petzval surface.*

The Petzval surface can be created by altering the lens shape. Tangential and sagittal rays are thus both focused onto the Petzval surface; however, the aberration of curvature of field still remains. That is, the Petzval surface is curved rather than flat, and thus the object points focus onto image points on the curved Petzval surface. Unfortunately, the image plane of the detector is usually flat so that astigmatism is eliminated but curvature of field remains.

However, the curvature of field of a single lens (or CRL) can be eliminated by using an aperture stop, which is placed between the lens and the object. With the aperture stop placed between the lens and object, the chief ray does not pass through the center of the lens. Instead, the chief ray passes from an object point through the center of the stop, through an off-axis lens point, and onto the image plane. The sagittal and tangential surfaces now have opposite curvature in which the sagittal image surface is concave and the tangential image surface is convex. Thus, with the upstream aperture stop, the circle of least confusion is now in a flat image plane. Without the upstream aperture stop the center chief ray passes through the center of lens, and the sagittal and tangential image surfaces now have different curvatures but both curvatures are concave. Thus, without the upstream aperture stop, the circle of least confusion is a concave image surface. In visible light optics an upstream aperture stop is used in simple box cameras to reduce astigmatism.

17. DISTORTION

The fifth aberration term of the third-order aberration expression of Eq. 173 is for distortion Δ_d in a spherical biconcave lens, where

$$\Delta_d = \frac{h^3 l \cos\theta}{2}\left(\frac{m^3}{r_0^3} + \frac{1}{r_i^3}\right). \tag{245}$$

For a spherical, aspheric, or parabolic lens CRL the path length deviation term Δ_d represents the aberration of distortion, where

$$\Delta_d = c_d h^3 l \cos\theta. \tag{246}$$

Distortion aberration constant c_d contains distortion variable m, where

$$c_d = \frac{1}{2}\left(\frac{m^3}{r_0^3} + \frac{1}{r_i^3}\right). \tag{247}$$

In distortion the transverse (lateral) magnification of an object point varies with the distance of the object point s from the center z-axis or lens center axis. If the transverse magnification increases with the object radial distance from the axis, then pin-cushion distortion occurs. If the transverse magnification decreases with the object radial distance from the axis, then barrel distortion occurs. An aperture stop placed between the CRL and image plane can lead to pincushion distortion, whereas an aperture stop placed between the object and CRL can cause barrel distortion.

Given a single object point O in the image plane, the x- and y-component distances x_0 and y_0 of the corresponding image points I, which are measured from ideal chief ray image point CI, for the distortion aberration become

$$x_0 \cong r_i\left(\frac{\partial\Delta_d}{\partial l}\cos\theta - \frac{\partial\Delta_d}{\partial\theta}\frac{\sin\theta}{l}\right) = c_d r_i h^3\left(\cos^2\theta + \sin^2\theta\right) = c_d r_i h^3 \tag{248}$$

$$y_0 \cong r_i\left(\frac{\partial\Delta_d}{\partial l}\sin\theta + \frac{\partial\Delta_d}{\partial\theta}\frac{\cos\theta}{l}\right) = c_d r_i h^3\left(\cos\theta\sin\theta - \sin\theta\cos\theta\right) = 0. \tag{249}$$

In distortion the radial distance r of an aberrated image point I is measured from the image plane intercept point CI of the ideal chief ray. Recall that the ideal chief ray is a straight line from object point O through the center of the lens ($l = 0$) to the image point CI in which the ideal chief ray passes through the off-axis point in the aperture stop material.

In the image plane, the radial distance r for each of the image points I of single object point O is measured relative to the ideal chief ray image point CI of object point O in which r is dependent on the radial distance h of the ideal chief ray intercept in the image plane from the center z-axis. In the image plane, the blur radius r or transverse aberration caused by distortion is

$$r = \sqrt{x_0^2 + y_0^2} = c_d r_i h^3. \tag{250}$$

A single object O point emits rays that converge to a set of image points I in a set of image planes, rather than a single image plane, because of the lens distortion created by the stop aperture. The aperture stop associated with the lens in distortion leads to the significant displacement of the image points, and a dependence of magnification on the radial distance h of the image point from the z-axis. However, the blurring due to the aberration aspect of distortion is not significant because a smaller-diameter bundle of rays from object point O is refracted to a smaller diameter spot in the image plane at the nominal image distance r_i from the center lens plane. Distortion causes the following longitudinal aberration or the axial extent z of the image planes:

$$z = \frac{r_i}{l} r = \frac{c_d r_i^2 h^3}{l}. \tag{251}$$

We now examine the dependence of magnification on the radial distance h of the ideal, straight line chief ray connecting an object point O to chief ray image point CI. The source or object coordinates (s_x, s_y) in cylindrical coordinates are

$$s_x = s\cos\theta \tag{252}$$

$$s_y = s\sin\theta. \tag{253}$$

The lens magnification of the object to the image in the x and y directions are M_x and M_y:

$$M_x = \frac{x_0}{s_x} = \frac{c_d r_i h^3}{s} \tag{254}$$

$$M_y = \frac{y_0}{s_y} = \frac{c_d r_i h^3}{s}. \tag{255}$$

The magnification of object points decreases as their distance from the z-axis increases; this distortion is called *barrel distortion.* As distance s of a source point from the z-axis increases, the distance h of the corresponding image point from the z-axis decreases, so that magnification also decreases, where

$$M_x = M_y = \frac{c_d r_i h^3}{s}. \tag{256}$$

18. IMAGES WITHOUT AND WITH DISTORTION

We now further discuss the physicality of an image without distortion and then an image with distortion. Recall that rays are the vector normals of the spherical wave surface in which spherical waves are emitted from the object points. Each ray is the vector normal for the particular differential area of the wave front surface to which the ray is perpendicularly attached. A ray path shows the trajectory of a differential area of the wave front surface. And a portion of the bundle of isotropic emitted rays from a single object point O is actually a portion of the spherical wave surface area that is made to converge to a single image point I by the differential phase velocity imposed by the lens shape and material.

The simultaneous emission of divergent rays from each object point O that pass through each point L in the lens entrance plane and by the shape and material of the lens are refracted via Snell's law to convergent rays. These rays arrive simultaneously at a single inverted image point I, which causes all the ray paths to have the same TOF. Divergent rays that are emitted at the same time from a single object point O and then collected by the entirety of the lens surface and refracted to a single image point at the same time produce equal TOF for all the ray paths connecting the single object O to its inverted image point I. Hence, all the wave amplitudes from the single object point arrive at the same phase at the inverted image point. That is, all the N wave amplitudes, each of amplitude A, then sum to total maximum amplitude NA and intensity $|NA|^2$ for each image point.

We now determine the magnification $M = x_i/x_0$ for the case of distortion in which x_0 is the transverse distance of object point O and x_i is the transverse image distance of image point I, both measured from the center optical axis, which is also the z-axis. The lens has a center plane that is perpendicular to the z-axis, and the center plane passes through the z-axis at lens center point C. An off-axis lens point R lies in the lens center plane at transverse distance x_R from the lens center point C.

The diagonal object line of length s_0, which passes from object point O to off-axis lens point R, forms angle ϕ_0 with respect to the negative z-direction. The object line angle ϕ_0 is positive for clockwise angles relative to the negative z-direction and negative for counterclockwise angles. The diagonal image line of length s_i, which passes from lens off-axis point R to image plane point I, makes angle ϕ_i with respect to the positive z-direction. Image line angle ϕ_i is positive for counterclockwise angles relative to the positive z-direction and negative for clockwise angles. In distortion the chief ray line from object point O to image plane point I is a broken line with the break (change of slope) at lens off-axis point R, and the object and image lines are unequal, where

$$|\phi_i| \neq |\phi_0| . \tag{257}$$

The axial distance r_0 of the object plane from the lens center plane and the axial distance r_i of the image plane from the lens center plane are both fixed. Above the z-axis the object point, image point, and off-axis lens point R transverse distances are positive:

$$x_i > 0, \quad x_0 > 0 \quad \text{and} \quad x_R > 0. \tag{258}$$

Below the z-axis, these transverse distances are negative:

$$x_i < 0, \quad x_0 < 0 \quad \text{and} \quad x_R < 0. \tag{259}$$

In the object plane the transverse distance x_0 of the object point O from the z-axis is

$$x_0 = r_0 \tan\phi_0 + x_R. \tag{260}$$

In the image plane, the transverse distance x_i of the image point I from the z-axis is

$$x_i = r_i \tan\phi_i + x_R. \tag{261}$$

In distortion the magnification M is

$$M = \frac{x_i}{x_0} = \frac{r_i \tan\phi_i + x_R}{r_0 \tan\phi_0 + x_R}. \tag{262}$$

When there is no distortion, the object line and image line angles are equal, where

$$\phi_i = -\phi_0 \quad \text{and} \quad x_R = 0. \tag{263}$$

The magnification is uniform for all object-image point transverse distances, where

$$M = -\frac{r_i}{r_0}. \tag{264}$$

With a single lens, an object is inverted in the magnified image, as expected, if the chief ray from an object point O crosses the center z-axis at the center point C of the lens to terminate on the inverted conjugated image point I, where $M < 0$ for a single lens. A pinhole camera shows no distortion because the chief ray always passes through the center point of the pinhole. The constant lateral magnification of a pinhole, as well as a lens, requires the chief ray be in a straight line between each object point O, through the center point C of the lens, and on to the conjugate image point I.

Each path from object point O to image point I has a bundle of rays with a central chief ray diverging from the object point O and refracted by the lens to converging rays that meet at the image point I. In a thin or thick lens, the center ray or chief ray must pass in a straight line from the object point O through the lens center point C and emerge from the center lens point C as the same straight line, or emerge from an offset lens point C_{off} as a parallel ray, such that the emerging ray in either case terminates on the conjugate image point I. For a thin lens with no stops, the line sections of the chief ray in the object and image drift space regions lie on the same straight, diagonal line, which crosses the center z-axis at the lens center plane. And for a thick lens with no stops, the chief ray line sections in the object and image drift space regions are parallel but slightly offset, and are joined by a line segment that passes through the center z-axis at the lens center plane. In distortion, arising from stop apertures placed before or after the thin or thick lens, the line sections of the chief ray in the object and image drift space regions have different slopes and intersect at an off-axis point R at the lens center plane.

19. PREVENTION OF DISTORTION AND THE PETZVAL CONDITION

To prevent distortion in a two-lens optical system, a stop or iris must be placed between the two lenses. No distortion in the two-lens system requires

$$\tan\phi_i = \tan\phi_0. \tag{265}$$

Recall that the diagonal object line passes from object point O to off-axis lens point R, forms angle ϕ_0 with respect to the negative z-direction, and the diagonal image line passes from lens off-axis point R to image plane point I and makes angle ϕ_i with respect to the positive z-direction. Also, the no-distortion condition requires the two lenses to meet the Petzval condition—namely, given lens 1 refractive index n_1 and focal length f_1 and lens 2 refractive index n_2 and focal length f_2, then

$$n_1 f_1 = -n_2 f_2. \tag{266}$$

Between the two lenses, the rays are parallel and pass through lens 1 to be converged to the object point or pass through lens 2 to be converged to the image point. This two-lens system is called *orthoscopic;* however, in the orthoscopic setup spherical and astigmatism aberrations remain.

Distortion is created by placing a stop either between the object plane and the lens or between the lens and image plane. If a stop is placed between the object plane and lens, then object angle ϕ_0 increases relative to image angle ϕ_i as the object transverse distance x_0 is increased. Increasing object point transverse distance x_0 increases the transverse distance x_R of off-axis lens point R equally in both the numerator and denominator of magnification M. As a result, the denominator of M increases faster than its numerator, and magnification M decreases with increasing transverse object point distance x_0, which is barrel distortion.

If a stop is placed between the lens and image plane, then the image angle ϕ_i increases relative to object angle ϕ_0 as the object transverse distance x_0 is increased. As before, increasing object point transverse distance x_0 increases the transverse distance x_R of the off-axis lens point R equally in both the numerator and denominator of magnification M. As a result, the numerator of M increases faster than its denominator, and magnification M increases with increasing transverse object point distance x_0, which is pincushion distortion.

Distortion occurs when the chief ray is not allowed to pass through the center of the lens, such that (1) only a portion of the lens surface collects divergent rays emitted from a single object point or (2) only a portion of the lens refracts rays to a single image point I. Equal TOF paths of ray paths from the object point O to image point I constructively sum amplitudes to produce a conjugate image point I for object point O. The set of object points or the set of lens points that contribute amplitudes at an image point differ when a stop is placed either between the lens and the object plane or between the lens and the image plane.

The stops prevent the chief ray of the bundle of emitted rays of an object point from passing through the center of the lens. In distortion, different object points O are then collected by different sets of points L on the lens entrance surface, and different points L on the lens entrance surface are sampled by each image point I. To prevent distortion every point L on the lens entrance surface must be able to receive rays from the same set of object points O, and every lens point L must be able to receive rays from the same set of image points I. Hence, no stops can be placed between the object plane and lens, and no stops can be placed between the lens and image plane.

The *object shape* is defined by the ratios of the x-, y-, and z-component distances of each object point relative to a single fixed reference point at the center of the object plane. The *object contrast* is the intensity of the object points relative to the intensity of the reference point. To image an object without distortion, the ratio of intensity and the x-, y-, and z-component separation distances of each object point must be preserved. In distortion theses ratios are not preserved from the object plane to the image plane.

Distortion can also result from objects that transmit or scatter source X-rays at such small angles that their chief rays are prevented from passing through the lens center point C. Instead, these chief rays pass through the center z-axis downstream from lens center point C. The source rays that are transmitted or scattered by each object point form angle θ with respect to the center z-axis. If an object point O transmits or scatters X-rays within a range of angles θ that is less than the angle ϕ (which is the angle of the straight line that passes unobstructed from object point O though the lens center point C to a image point I), then the image has pincushion distortion. Each bundle of parallel rays with its center chief ray passes through the lens, but these rays do not pass through the z-axis at the lens center point C. Instead by lens refraction, these rays may pass through the lens center axis but at points downstream from the lens center point C.

If there are no rays transmitted or scattered from object point O at an angle θ that equals or exceeds the angle ϕ of the line from object point O through the lens center point C to image plane point I, then this is equivalent to the placement of a downstream stop between the lens and the image plane. This is a condition for pincushion distortion. In order to remove pincushion distortion every lens point L must be able to receive rays from the same set of points I in the image plane. That is, the object must be illuminated by a source of rays in which the object scattered or object transmitted rays have a sufficient angle θ, which allows each object point O to produce a chief ray that passes unobstructed through lens center point C to a point I in the image plane. Removal of distortion implies angle θ of the ray scattered from the object is larger than the angle ϕ of the line from object point O through the lens center point C to image plane point I, where

$$\theta \geq \phi. \tag{267}$$

20. DISTORTION IN COMPOUND REFRACTIVE LENS IMAGING WITH SYNCHROTRON X-RAY SOURCES

As an example, an X-ray CRL that images objects illuminated by extremely parallel X-rays from the tiny and distant synchrotron radiation source can exhibit pincushion distortion. In pincushion distortion, each point L on the lens surface images only a portion of the object rather than the entire object within the lens field of view. The lens field of view consists of the object points O that can be connected by a straight line that passes unobstructed through the lens center point C to a point I in the image plane.

If, however, a thin X-ray scattering plate is placed between the X-ray source and the object, then the scattered rays from the scattering plate strike the object at a sufficient range of angles, which allows each point L on the lens entrance surface to receive a ray from the same set of object points O as the other points L on the lens entrance surface. As a result, with the X-ray scattering plate placed between the source and object, the X-ray image of the object no longer has pincushion distortion. With the X-ray scatter plate the transmitted or scattered ray angles θ from each object point O are now sufficient to allow a chief ray from each object point O to pass unobstructed through the lens center point C to a point I in the image plane.

Another example of distortion can be seen in dark field X-ray microscopy in which the near parallel rays of a distant, tiny synchrotron source illuminates an object, such as a wire mesh, which is imaged by the lens onto an image plane. A small blocking disk, centered on the optical axis, is placed between the lens and image plane. The center disk stops the paraxial X-rays directed from the X-ray source to the image plane from reaching the image plane. As a result, only nonparaxial rays scattered from the object that are not in a direct line to the blocking disk are refracted by the lens and pass outside the radius of the blocking disk onto the image plane. Without an object in place the lens images the distant X-ray source. However, the small blocking disk between the lens and image plane is equivalent to a stop that is set in front of the lens. That is, the blocking disk between the lens and image plane, like the stop placed between the X-ray source and lens, limits the paraxial rays from the source to image plane. Nonparaxial source rays with larger incident angles ϕ_0 can be refracted by the lens outside the disk boundary onto the image plane. Avoiding the blocking disk requires that the increase of the incident ray angle ϕ_0 (effectively an increase in the source to lens or object distance) be more than the resulting increase of the image angle ϕ_i (effectively the lens to detector or image distance), and thus the magnification (image distance divided by object distance) decreases with increasing object angle ϕ_0, which is barrel distortion.

If a wire mesh object is then placed between the source and the lens, and if the rays of the distant source that are transmitted or scattered by the object are sufficiently parallel such that no chief rays pass through the lens center point C, then the image of the mesh object has pincushion distortion. And in the image plane the barrel distortion image of the X-ray source is superimposed on the pincushion distortion image of wire mesh object.

REFERENCES

Freeman, M. H., & Hull, C. C. (2003). *Optics.* Butterworth-Heinemann, London.
Hecht, E., & Zajac, A. (1974). *Optics.* Addison-Wesley Publishing, New York.
Jenkins, F. A., & White, H. E. (1957). *Fundamentals of Optics.* McGraw-Hill, New York.
Klein, M. V., & Furtak, T. E. (1986). *Optics.* John Wiley & Sons, New York.
O'Neill, E. L. (1991). *Introduction to Statistical Optics.* Dover Publications, New York.
Pedrotti, L., & Pedrotti, L. S. (1993). *Introduction to Optics.* Prentice Hall, Englewood Cliffs, NJ.

Chapter 4

X-ray Optics

Jay Theodore Cremer, Jr.

Contents

1. Damped Resonance for Driven Atomic Electron Oscillation and X-ray Emission 498
2. The Complex Atomic Scatter Factor for a Single Element 502
3. The Complex Atomic Scatter Factor for a Mixture or Compound 503
4. Maxwell's Equations Yield the Electromagnetic Vector Wave Equation in Material Media 505
5. The Electromagnetic Vector Wave Equation Contains the Refractive Index 506
6. The Complex Refractive Index and the Complex Atomic Scatter Factor 508
7. The Complex Refractive Index—Decrement, Absorption Index, and Linear Attenuation 510
8. X-ray Complex Index of Refraction—Kramers–Kronig Relation 513
9. X-ray Rayleigh Scatter Total Cross Section and the Complex Atomic Scatter Factor 514
10. Differential Cross Section for the Rayleigh Scatter and Complex Atomic Scatter Factor 517
11. Rayleigh and Thomson X-ray Scatters are Coherent, Elastic Scatter 519
12. Compton Scatter of X-rays is Incoherent and Inelastic 521
13. X-ray Attenuation by Photoelectric Absorption 525
14. X-ray Absorptive Attenuation by Pair Production 527
15. Collision Losses of Charged Particle Passage in Matter 528
16. Bremsstrahlung from Charged Particle Passage in Matter 536
17. Forward-Directed Bremsstrahlung from Relativistic Electrons Undergoing Acceleration 538
18. X-ray Wave Packets—Phase Velocity and Group Velocity, Spatial and Spectral Widths 541
19. X-ray Wave Packets—Dispersion and Group Velocity, and Relation to Real Part of Complex Atomic Scatter Factor 544

Advances in Imaging and Electron Physics, Volume 172, ISSN 1076-5670, DOI: 10.1016/B978-0-12-394422-1.00004-9.
Copyright © 2012 Elsevier Inc. All rights reserved.

20. X-rays in Material Medium—Phase and Group Velocity and Pulse Broadening 548
21. Reflection and Transmission of X-rays at a Planar Interface of Two Media 550
22. Fraction of Transmitted and Reflected Polarized X-rays with an Electric Field Component Parallel to the Planar Surface Interface 554
23. Fraction of Transmitted and Reflected Polarized X-rays with a Magnetic Field Component Parallel to the Planar Surface Interface 556
24. X-ray Brewster Angle Occurs Only for Electric Field Parallel to Plane of Incidence 557
References 559

1. DAMPED RESONANCE FOR DRIVEN ATOMIC ELECTRON OSCILLATION AND X-RAY EMISSION

The electric field **E** emitted from a charged particle of charge e, which undergoes acceleration **a**, is derived from the Lienard-Wiechert potentials. The derivation of the radiated electric field **E** from the accelerated charged particle; that is, an electron is non-trivial and thus is simply stated in Eq. 1. The reader can find the derivation in Attwood (1999), Eyges (1972), Feynman (1964), Jackson (1975), Marion (1965), Landau and Lifshitz (1975), Schwartz (1972), Shadowitz (1975), Sommerfeld (1964), and Stratton (1941). The electric field **E** emitted from the electron of charge e undergoing acceleration **a** is proportional to the electron's perpendicular acceleration component $a_{s\perp}(t_r)$. The electron's perpendicular acceleration component $a_{s\perp}(t_r)$ is the component of the electron acceleration **a**, which occurs at the retarded time t_r, and is perpendicular to the propagation direction of the electric field to the observer, who measures the electric field at a later time t. The time lag $t - t_r$ is due to the finite separation distance between the charged particle and observer and the finite speed of light c. The electric field $\mathbf{E}(\mathbf{r}, t)$ at the present observer position **r** and time t is related to the perpendicular electron acceleration component $a_{s\perp}(t_r)$ evaluated at the earlier or retarded time t_r and electron position when the electron acceleration occurred. For vacuum electric permittivity ε_0 the observer electric field $\mathbf{E}(\mathbf{r}, t)$ due to a distant accelerating electron is

$$\mathbf{E}(\mathbf{r}, t) = \frac{e a_{s\perp}(t_r)}{4\pi \varepsilon_0 c^2 r} \tag{1}$$

Only the component of the charged particle acceleration, which is perpendicular to the direction of the observer, produces an electric field at the observer. Hence, the observer-measured electric field amplitude has an angular dependence relative to the direction of charged particle acceleration.

The acceleration of the atomic electron of charge e by an imposed source electric field can be calculated from the equation of motion of the electron of rest mass m_0 due to the Lorentz force exerted by incident electromagnetic field composed of electric E_i and magnetic B_i field components. The second-order differential equation of this driven electron oscillation is

$$m_0 \frac{d^2 x_s}{dt^2} + m_0 \gamma \frac{dx_s}{dt} + m_0 w_s^2 x_s = -e\,(E_i + v_s \times B_i). \tag{2}$$

The first term of Eq. 2 is the acceleration term for the electron oscillation. The second term is the dissipative force term that inhibits electron oscillation and accounts for energy loss via the damping factor γ, with assumption of small dissipation compared to the driving frequency w of the incident electromagnetic wave, where $\gamma / w \ll 1$.

The third term of Eq. 2 is due to the restoring force of an oscillator with resonant frequency w_s of the bound atomic electrons of the material. Here, x_s is the electron displacement from equilibrium in the direction parallel to the incident electric field, and in a direction perpendicular to the incident magnetic field as well as to the electron to observer direction.

The plane wave magnetic field amplitude $\mathbf{B}$ is much smaller than the electric field amplitude $\mathbf{E}$ by a factor of the speed of light c, where $\mathbf{E} = c\mathbf{B}$. The electric field induced electron velocity $\mathbf{v}_s$ follows the direction of the electric field $\mathbf{E}$, but $\mathbf{v}_s$ is perpendicular to the magnetic field B. Hence, the force $e\mathbf{v}_s \times \mathbf{B}$ acts on the electron in the a direction perpendicular to both the electric and magnetic fields of the incident plane wave. This $e\mathbf{v}_s \times \mathbf{B}$ force of the magnetic field and associated acceleration is much smaller than the electric field force $e\mathbf{E}$ and associated acceleration. Hence, on the right side of Eq. 2, the second term is much smaller than the first term, and thus the second term can be approximated to zero, where

$$\mathbf{v}_s \times \mathbf{B}_i \sim 0. \tag{3}$$

The periodic, incident electric field that E_i imposed on the electron causes electron displacement x_s, velocity dx_s/dt, and acceleration d^2x_s/dt with the same time or reciprocal time (frequency) dependence. The incident electric field E_i at vector position $\mathbf{r}_s$ of the electron is

$$E_i = E_0 \exp[-i\,(wt - \mathbf{k}_1 \cdot \mathbf{r}_s)]. \tag{4}$$

Hence, the electron displacement in the direction of the imposed electric field is

$$x_s = x_{os} \exp[-i\,(wt - \mathbf{k}_1 \cdot \mathbf{r}_s)]. \tag{5}$$

Substitution into Eq. 2 for x_s of Eq. 5 and E_i of Eq. 4 yields the amplitude of the electron position x_{os}:

$$x_{os} = \frac{1}{-w^2 - iw\gamma + w_s^2} \frac{eE_0}{m_0}. \tag{6}$$

The electron position x_s is then

$$x_s = \frac{\exp[-i\,(wt - \mathbf{k}_1 \cdot \mathbf{r}_s)]}{-w^2 - iw\gamma + w_s^2} \frac{eE_0}{m_0}. \tag{7}$$

The electron acceleration a_s is

$$a_s = \frac{w^2 \exp[-i\,(wt - \mathbf{k}_1 \cdot \mathbf{r}_s)]}{w^2 + iw\gamma - w_s^2} \frac{eE_0}{m_0}. \tag{8}$$

The acceleration component $a_{s\perp}$, perpendicular to the direction from electron to observer, is

$$a_{s\perp} = a_s \sin\theta. \tag{9}$$

Here, θ is the angle between the direction of the electron acceleration a_s and the scattered wave direction $\mathbf{k}_2$ toward the observer at position $\mathbf{r}$ from the scattering electron at position $\mathbf{r}_s$. The electron acceleration is in the direction of the applied, incident electric field. The acceleration component $a_{s\perp}(t_r)$, perpendicular to the direction from electron to observer, is evaluated at retarded time

$$t_r = t - \frac{r - r_s}{c}. \tag{10}$$

The electron acceleration by the incident electromagnetic wave occurs before the observer receives the electron-scattered wave. That is, (1) the separation distance $(\mathbf{r} - \mathbf{r}_s)$ between the electron position $\mathbf{r}_s$ at the prior time t_r of wave scatter and the observer position $\mathbf{r}$ at the present time t of detection, and (2) the finite speed of light c, together cause a time lag $(\mathbf{r} - \mathbf{r}_s)/c$ to occur between the wave scatter by the electron and the detection of this scattered wave by the distant observer. Hence, the expression for the electron acceleration component, which is perpendicular to the direction between the electron and observer, must be evaluated at the prior or retarded time t_r given by

$$a_{s\perp}(t_r) = \frac{-w^2}{w^2 - w_s^2 + iw\gamma} \frac{eE_0 \sin\theta \exp[-i\,(wt_r - \mathbf{k}_1 \cdot \mathbf{r}_s)]}{m_0}. \tag{11}$$

Elastic scatter is assumed in which the scattered wave number k_2 is equal to the incident wave number k_1. That is, the direction of the incident wave vector $\mathbf{k}_1$ changes direction by polar angle θ upon scatter to the direction of scattered wave vector $\mathbf{k}_2$, in which the incident and scattered wave vectors have identical magnitude or wave numbers, where

$$k_1 = k_2 = \frac{w}{c} \tag{12}$$

The difference wave vector or the resulting change in direction of the incident wave vector upon scatter from the electron is,

$$\Delta\mathbf{k} = \mathbf{k}_2 - \mathbf{k}_1 \tag{13}$$

The magnitude of the change of wave vector direction is,

$$|\Delta\mathbf{k}| = 2k_1 \sin\theta. \tag{14}$$

The retarded time expression of Eq. 10 is substituted into Eq. 11 for $a_{s\perp}(t_r)$ Next, w/c in the resulting expression is replaced by k_2 using Eq. 12, which then allows substitution for the difference wave vector $\mathbf{k}_2 - \mathbf{k}_1$ of Eq. 13. The expression for the electron acceleration component that is perpendicular to the direction from electron to observer becomes

$$a_{s\perp}(t_r) = \frac{-w^2}{w^2 - w_s^2 + iw\gamma} \frac{eE_0 \sin\theta \exp\left[-iw\left(t - \frac{r}{c}\right)\right] \exp\left(-i\Delta\mathbf{k}\cdot\mathbf{r}_s\right)}{m_0}. \tag{15}$$

Substitution for the electron's perpendicular acceleration component $a_{s\perp}(t_r)$ of Eq. 15 into Eq. 1 for the observer electric field expression $\mathbf{E}[t]$ yields the scattered far-field electric field $E_Z(t)$ from an atom of Z bound electrons to an observer point:

$$E_Z(t) = \frac{e^2 E_0 \sin\theta}{4\pi\varepsilon_0 c^2 m_0} \frac{1}{r} \exp\left[-iw\left(t - \frac{r}{c}\right)\right] \sum_{s=1}^{Z} \frac{w^2 \exp[-i(\Delta\mathbf{k}\cdot\mathbf{r}_s)]}{w^2 - w_s^2 + iw\gamma}. \tag{16}$$

The classical electron radius r_e is contained in Eq. 16 and is defined by

$$r_e = \frac{e^2}{4\pi\varepsilon_0 c^2 m_0} = 2.817 \times 10^{-15}\,\text{m}. \tag{17}$$

The vacuum permittivity ε_0 for electric fields and the electron charge e are expressed in meters/kilograms/seconds (mks) farads and Coulomb units, where

$$\varepsilon_0 = 8.8542 \times 10^{-14}\,\text{F/cm} \quad \text{and} \quad e = 1.6 \times 10^{-19}\text{C}. \tag{18}$$

2. THE COMPLEX ATOMIC SCATTER FACTOR FOR A SINGLE ELEMENT

Suggested by the form of Eq. 16, the complex atomic scattering factor $f(\Delta\mathbf{k}, w)$ is defined as

$$f(\Delta\mathbf{k}, w) = \sum_{s=1}^{Z} \frac{w^2 \exp\left[-i\left(\Delta\mathbf{k} \cdot \mathbf{r}_s\right)\right]}{w^2 - w_s^2 + iw\gamma}. \tag{19}$$

The complex atomic scattering factor $f(\Delta\mathbf{k}, w)$, which is contained in Eq. 16, is the ratio of the scatter of an incident electric field due to a multi-electron atom divided by the electric field scatter due to a single free electron.

To adjust for the discrete resonant energy states of the atomic electrons, one needs to adjust the scattering factor with a weighting of the number of atomic electrons per atomic energy level; namely, the oscillator strength g_s for each sth electron. The sum of the oscillator strengths for the electrons of an atom is equal to the atomic number Z; that is,

$$\sum_{s=1}^{Z} g_s = Z. \tag{20}$$

The complex atomic scattering factor $f(\Delta\mathbf{k}, w)$ becomes

$$f(\Delta\mathbf{k}, w) = \sum_{s=1}^{Z} \frac{g_s w^2 \exp[-i\left(\Delta\mathbf{k} \cdot \mathbf{r}_s\right)]}{w^2 - w_s^2 + iw\gamma}. \tag{21}$$

From Eq. 16, the scattered far-field electric field $E_Z(t)$ from an atom of Z bound electrons to an observer point is then expressed in terms of the complex atomic scattering factor $f(\Delta\mathbf{k}, w)$, where

$$E_Z(t) = \frac{e^2 E_0 \sin\theta}{4\pi\varepsilon_0 c^2 m_0} \frac{1}{r} \exp\left[-iw\left(t - \frac{r}{c}\right)\right] f(\Delta\mathbf{k}, w) \tag{22}$$

The atomic electron position $\mathbf{r}_s$ distance from its nucleus is contained within approximately the Bohr radius a_0. In the long-wavelength λ limit, the X-ray wavelength is much larger than the Bohr radius a_0, where

$$\frac{a_0}{\lambda} \ll 1 \tag{23}$$

In the forward-scattering limit, the angle θ, which is measured relative to the incident wave vector $\mathbf{k}_1$ direction along the positive z-axis, is nearly zero (or very small), so the scattered wave vector $\mathbf{k}_2$ is in the forward direction:

$$\theta \ll 1 \tag{24}$$

In the long-wavelength limit, $a_0/\lambda \ll 1$, and/or forward-scatter limit, $\theta \ll 1$,

$$|\Delta\mathbf{k}\cdot\mathbf{r}_s| \to 0 \tag{25}$$

Finally, in either the long-wavelength or forward-scatter limit, the complex atomic scattering factor becomes

$$f^0(w) = \sum_{s=1}^{Z} \frac{w^2 g_s}{w^2 - w_s^2 + iw\gamma} \quad \frac{a_0}{\lambda} \ll 1 \quad \text{or} \quad \theta \ll 1. \tag{26}$$

That is, $f(\Delta\mathbf{k}, w)$ becomes $f_q^0(w)$ if in the long-wavelength limit and/or the forward-scatter limit. In addition, if, the long-wavelength limit and/or forward-scatter direction, the frequency w of the incident X-ray exceeds the oscillator with resonant frequency w_s, then the atomic-scattering factor $f(\Delta\mathbf{k}, w)$ is equal to the atomic number Z of the atom:

$$f^0(w) = \sum_{s=1}^{Z} g_s = Z \quad w^2 \gg w_s^2 \quad \text{and} \quad \frac{a_0}{\lambda} \ll 1 \quad \text{or} \quad \theta \ll 1. \tag{27}$$

3. THE COMPLEX ATOMIC SCATTER FACTOR FOR A MIXTURE OR COMPOUND

The above-mentioned complex atomic scattering factor $f(\Delta\mathbf{k}, w)$ applies to a scattering material with a single element. The expression is now modified for a material composed of a compound or a mixture of elements. The material has mass density $\rho_m[\text{g/cm}^3]$ and is comoised of Q elements, in which each element has weight fraction w_q, atomic weight A_q, and atom number density $\rho_q[\text{atom/cm}^3]$. Each qth element contributes electric field component $\mathbf{E}_q[t]$ to the total scattered far electric field $\mathbf{E}[t]$ at the observer, where

$$\mathbf{E}[t] = \sum_{q=1}^{Q} \mathbf{E}_q[t] \tag{28}$$

and

$$\mathbf{E}_q[t] = -\frac{r_e E_o \rho_q \sin\theta}{r} \exp\left[-iw\left(t - \frac{r}{c}\right)\right] f_q(\Delta\mathbf{k}, w). \tag{29}$$

One calculates the density $\rho_q(\text{cm}^{-3})$ for the qth element of the material via multiplication of Avogadro's number N_A by the molar density $n_q(\text{mole/cm}^3)$, where

$$\rho_q = n_q N_{av} \quad N_A = 6.02 \times 10^{23} \text{ atoms/mole.} \tag{30}$$

Atomic weight A_q is the gram atomic weight or the number of grams per mole of the qth element of the Q elements comprising the material. The molar density n_q is the mass density $w_q \rho_m$ of qth element divided by its gram atomic weight A_q, where

$$n_q = \frac{w_q \rho_m}{A_q}. \tag{31}$$

The scattering factor $f_q\,(\Delta\mathbf{k}, w)$ for a mixture or compound is a function of the wave frequency w, the various bound electron resonant frequencies w_s, and the variation of the phase $\Delta\mathbf{k} \cdot \mathbf{r}_{s,q}$ of the scattered wave, which arises from the differing electron positions seen by the observer. For each qth element, the scatter factor contribution $f_q\,(\Delta\mathbf{k}, w)$ is the sum of the contributions of the Z electrons, where

$$f_q\,(\Delta\mathbf{k}, w) = \sum_{s=1}^{Z} \frac{g_{s,q} w^2 \exp\left[-i\left(\Delta\mathbf{k} \cdot \mathbf{r}_{s,q}\right)\right]}{w^2 - w_{s,q}^2 + iw\gamma_q}. \tag{32}$$

The total scatter factor for the mixture or compound is

$$f(\Delta\mathbf{k}, w) = \sum_{q=1}^{Q} f_q\,(\Delta\mathbf{k}, w). \tag{33}$$

In the scattering factor expression $w_{s,q}$ is the resonant frequency of the sth electron and γ_q is the electron oscillation-damping constant in the atom of the qth element, which comprises the material that scatters an incident electromagnetic field. And $\mathbf{r}_{s,q}$ is the position of the sth electron, measured relative to its "stationary" nucleus in the qth element.

4. MAXWELL'S EQUATIONS YIELD THE ELECTROMAGNETIC VECTOR WAVE EQUATION IN MATERIAL MEDIA

The electromagnetic wave equation in a material medium is now derived from Maxwell's equations. The wave equation in a material medium contains a material-dependent term, which can be expressed by the complex refractive index n. And the complex refractive index n is expressed by the complex atomic scatter factor $f(\Delta\mathbf{k}, w)$. The atomic electrons are driven by the incident electromagnetic field to oscillate and emit electromagnetic waves, and the complex atomic scatter factor $f(\Delta\mathbf{k}, w)$ is the ratio of the emitted electric field due to a multi-electron atom divided by the electric field scatter due to a single free electron.

Consider an electromagnetic wave passing through a material such as a beryllium lens. The electromagnetic wave equation is derived from Maxwell's equations taking the cross product of the gradient operator ∇ with Faraday's law for the electric field $\mathbf{E}$ and the magnetic induction vector $\mathbf{B}$:

$$\nabla \times \mathbf{E} = -\frac{\partial \mathbf{B}}{\partial t} \tag{34}$$

and

$$\nabla \times (\nabla \times \mathbf{E}) = \nabla \times \left(-\frac{\partial \mathbf{B}}{\partial t}\right). \tag{35}$$

Recall the vector identity for arbitrary vector $\mathbf{A}$

$$\nabla \times \nabla \times \mathbf{A} = \nabla (\nabla \cdot \mathbf{A}) - \nabla^2 \mathbf{A}. \tag{36}$$

Ampere's law is

$$\nabla \times \mathbf{H} = \frac{\partial \mathbf{D}}{\partial t} + \mathbf{J}. \tag{37}$$

The vacuum (or air) constitutive relation between the electric field vector $\mathbf{E}$ and the electric displacement vector $\mathbf{D}$ is

$$\mathbf{D} = \varepsilon_0 \mathbf{E}. \tag{38}$$

The constitutive relation between the magnetic field vector $\mathbf{H}$ and the magnetic density or magnetic induction vector $\mathbf{B}$ is

$$\mathbf{B} = \mu_0 \mathbf{H}, \tag{39}$$

Eq. 39 for magnetic induction vector B is substituted into Eq. 35 (the curl of Faraday's Law), and Eq. 38 for electric displacement vector **D** is substituted into Eq. 37 (Ampere's Law). The double cross product vector identity of Eq. 36 is applied to the left side of Eq. 35 with subsequent substitution of charge density ρ for $\nabla \cdot \mathbf{D}$ using Coulombs's Law $\nabla \cdot \mathbf{D} = \rho$. The resulting right side of Eq. 35 explicitly expresses the time derivative of the right side of Eq. 37 for Ampere's Law. Substitution of the time derivative of the right side of Eq. 37 into the resulting right side of Eq. 35, and subsequent transposing, yields the vector wave equation for the electric field **E**,

$$\left(\frac{\partial^2}{\partial t^2} - c^2\nabla^2\right)\mathbf{E}(\mathbf{r},t) = -\frac{1}{\varepsilon_0}\left[\frac{\partial \mathbf{J}(\mathbf{r},t)}{\partial t} + c^2\nabla\rho(\mathbf{r},t)\right]. \tag{40}$$

The electromagnetic wave phase velocity w/k in vacuum is c, where

$$c = \frac{1}{\sqrt{\mu_0\varepsilon_0}}. \tag{41}$$

For transverse electromagnetic waves, the electric field is perpendicular to the wave number vector; that is, $\mathbf{E} \perp \mathbf{k}$, and, assuming no charge density gradient in the material,

$$\nabla\rho(\mathbf{r},t) = 0 \tag{42}$$

Only the perpendicular component of the current density **J** contributes to the transverse wave, so that the equation for a transverse electromagnetic wave is

$$\left(\frac{\partial^2}{\partial t^2} - c^2\nabla^2\right)\mathbf{E}_\perp(\mathbf{r},t) = -\frac{1}{\varepsilon_0}\frac{\partial \mathbf{J}_\perp(\mathbf{r},t)}{\partial t}, \tag{43}$$

where $\mathbf{E}_\perp$ and $\mathbf{J}_\perp$ are the electric field and current density component vectors perpendicular to the direction of wave propagation. The atomic electrons of the material, which are induced to oscillate by the imposed electric field, are the source of the transverse current density vector $\mathbf{J}_\perp$.

5. THE ELECTROMAGNETIC VECTOR WAVE EQUATION CONTAINS THE REFRACTIVE INDEX

In the previously derived differential equation for an electromagnetic wave in a material medium, the material-dependent term is now expressed via the refractive index n of the material, which is expressed in terms of the complex atomic scatter factor $f(\Delta\mathbf{k}, w)$. Recall that the displacement vector $\mathbf{x}(\mathbf{r}, t)$ of an atomic electron of mass m_0 and charge e from its equilibrium position vector $\mathbf{r}$, relative to the system origin, as a function of time t, by force of an external electric field, is given by

$$\mathbf{x}(\mathbf{r}, t) = \frac{e}{m_0} \frac{1}{\left(w^2 - w_0^2\right) + i\gamma w} \mathbf{E}_\perp(\mathbf{r}, t). \tag{44}$$

The time variation of the transverse electric field thus induces electron velocity $\mathbf{v}(\mathbf{r}, t)$; that is,

$$\mathbf{v}(\mathbf{r}, t) = \frac{e}{m_0} \frac{1}{\left(w^2 - w_0^2\right) + i\gamma w} \frac{\partial \mathbf{E}_\perp(\mathbf{r}, t)}{\partial t}. \tag{45}$$

The transverse current density vector $\mathbf{J}_\perp$ includes contributions of the Z electrons of each single atom of an element (Z, A) with weight fraction w_q in a material with mass density ρ_m. Also, transverse current density vector $\mathbf{J}_\perp$ includes the oscillator strength g_s contribution of each of the sth electron, so that

$$\mathbf{J}_\perp(\mathbf{r}, t) = -e \sum_{q=1}^{Q} \rho_q \sum_{s=1}^{Z} g_{s,q} \mathbf{v}_{s,q}(\mathbf{r}, t). \tag{46}$$

The summed oscillator strengths of the electrons in an atom equals the atomic number; that is,

$$\sum_{s=1}^{Z} g_s = Z. \tag{47}$$

Also, $\mathbf{v}_{s,q}$ and $w_{s,q}$ are the velocity and resonant frequency of the sth electron, and γ_q is the electron oscillation-damping constant in the atom of the qth element in the material. The velocity of the sth electron of the qth element that is situated at position $\mathbf{r}$, relative to the system origin at time t, is

$$\mathbf{v}_{s,q}(\mathbf{r}, t) = \frac{e}{m_0} \frac{1}{\left(w^2 - w_{s,q}^2\right) + i\gamma_q w} \frac{\partial \mathbf{E}_\perp(\mathbf{r}, t)}{\partial t}. \tag{48}$$

Substitution of the $\mathbf{v}_{s,q}$ expression of Eq. 48 into the transverse current density $\mathbf{J}_\perp$ of Eq. 46 yields

$$\mathbf{J}_\perp(\mathbf{r}, t) = -\frac{e^2}{m_0} \sum_{q=1}^{Q} \rho_z \sum_{s=1}^{Z} \frac{g_{s,q}}{\left(w^2 - w_{s,q}^2\right) + i\gamma_q w} \frac{\partial \mathbf{E}_\perp(\mathbf{r}, t)}{\partial t}. \tag{49}$$

Next, substitution of transverse current density $\mathbf{J}_\perp$ of Eq. 49 into the wave equation of Eq. 43 gives

$$\left(\frac{\partial^2}{\partial t^2} - c^2\nabla^2\right)\mathbf{E}_\perp(\mathbf{r},t) = \frac{e^2}{\varepsilon_0 m_0}\sum_{q=1}^{Q}\rho_q\sum_{s=1}^{Z}\frac{g_{s,q}}{\left(w^2 - w_{s,q}^2\right) + i\gamma_q w}\frac{\partial^2\mathbf{E}_\perp(\mathbf{r},t)}{\partial t^2}. \tag{50}$$

Rearrangement yields

$$\left[\left(1 - \frac{e^2}{\varepsilon_0 m_0}\sum_{q=1}^{Q}\rho_q\sum_{s=1}^{Z}\frac{g_{s,q}}{\left(w^2 - w_{s,q}^2\right) + i\gamma_q w}\right)\frac{\partial^2}{\partial t^2} - c^2\nabla^2\right]\mathbf{E}_\perp(\mathbf{r},t) = 0. \tag{51}$$

This expression can be simplified with frequency parameter $n(w)$, which is the index of refraction, where

$$n(w) = \sqrt{1 - \frac{e^2}{\varepsilon_0 m_0}\sum_{q=1}^{Q}\rho_q\sum_{s=1}^{Z}\frac{g_{s,q}}{\left(w^2 - w_{s,q}^2\right) + i\gamma_q w}}. \tag{52}$$

With the index of refraction $n(w)$, the wave equation becomes

$$\left(\frac{\partial^2}{\partial t^2} - \frac{c^2}{n^2(w)}\nabla^2\right)\mathbf{E}_\perp(\mathbf{r},t) = 0. \tag{53}$$

Since the electromagnetic wave phase velocity v_p is modified from c (in vacuum or air) to c/n, the parameter $n(w)$ is indeed the index of refraction, where

$$v_p = \frac{c}{n}. \tag{54}$$

6. THE COMPLEX REFRACTIVE INDEX AND THE COMPLEX ATOMIC SCATTER FACTOR

The transverse electric field $\mathbf{E}_\perp(\mathbf{r},t)$ is the sum of the forward-scattered radiation from the bound electrons of all the material atoms and the incident wave. This sum or the interference of the forward-scattered waves from the material atomic electrons and the incident wave produce a resultant wave that propagates differently than that of a wave in a vacuum. The scattering process has both elastic (lossless) and inelastic (dissipative) components, and hence the resultant index of refraction $n(w)$ is complex. The index of refraction $n(w)$ describes both the modified phase velocity and attenuation of the wave electric (and magnetic) field amplitude in the material. The second term in the radical of the index of refraction $n(w)$ is much less than 1, so that the index of refraction is binomially approximated as

$$n(w) = 1 - \frac{e^2}{2\varepsilon_0 m_0} \sum_{q=1}^{Q} \rho_q \sum_{s=1}^{Z} \frac{g_{s,q}}{\left(w^2 - w_{s,q}^2\right) + i\gamma_q w}. \tag{55}$$

The medium of refractive index $n(w)$ is composed of Q elements. For each sth electron of the Z electrons in each atom of the qth element, g_s is the oscillator strength, $w_{s,q}$ is the resonant frequency, and γ_q is the damping constant for all the electrons in the atom.

The complex atomic scattering factor $f_q(\Delta\mathbf{k}, w)$ defined by

$$f_q\left(\Delta\mathbf{k}, w\right) = \sum_{s=1}^{Z} \frac{w^2 g_{s,q} \exp\left[-i\left(\Delta\mathbf{k}\cdot\mathbf{r}_{s,q}\right)\right]}{\left(w^2 - w_{s,q}^2\right) + iw\gamma_q}. \tag{56}$$

In the long-wavelength limit and/or in the forward-scattering limit,

$$n(w) = 1 - \frac{r_e\lambda^2}{2\pi} \sum_{q=1}^{Q} \rho_q f_q^0, \tag{57}$$

where

$$f_q^0 = \sum_{s=1}^{Z} \frac{w^2 g_{s,q}}{\left(w^2 - w_{s,q}^2\right) + iw\gamma_q}. \tag{58}$$

That is, $f(\Delta\mathbf{k}, w)$ becomes f_q^0 in the long-wavelength limit.

We have thus calculated transverse electric field $\mathbf{E}_\perp(\mathbf{r}, t)$ as a function of the material index of refraction $n(w)$, in which $\mathbf{E}_\perp(\mathbf{r}, t)$ is the summation of the incident wave with material scattered waves in the forward direction ($\theta \ll 1$). In the forward direction $\theta \ll 1$, the scattering factor $f_q^0(w)$ is independent of the positions of the individual electrons $\mathbf{r}_{s,q}$. Hence, in the forward-scattered direction, all electrons of the same resonant frequency produce forward-scattered waves all with the same phase, which leads to constructive interference and complete summation in the forward direction. This is similar to Huygen's principle, in that the secondary waves produced by points on a primary wave front have the same phase only in the forward direction. As a result, constructive interference (and thus complete summation of the wavelets) occur only in the forward direction. In other directions, the secondary waves undergo only partial constructive interference or destructive interference. The net effect is the creation of a secondary wave front in the forward direction, and hence a forward-propagating wave.

The complex atomic scattering factor in the forward direction $f_q^0\,(w)$ can be written in terms of real $f_1^0\,(w)$ and imaginary $f_2^0\,(w)$ parts, so that

$$f_q^0\,(w) = f_{1q}^0\,(w) - if_{2q}^0\,(w)\,. \tag{59}$$

The real part $f_{1q}^0\,(w)$ of the complex atomic scattering factor $f_q^0\,(w)$ is

$$f_{1q}^0(w) = \sum_{s=1}^{Z} \frac{g_{s,q}w^2\left(w^2 - w_{s,q}^2\right)}{\left(w^2 - w_{s,q}^2\right)^2 + w^2\gamma_q^2}. \tag{60}$$

The imaginary part $f_{2q}^0\,(w)$ of the complex atomic scattering factor $f_q^0\,(w)$ is

$$f_{2q}^0\,(w) = \sum_{s=1}^{Z} \frac{g_{s,q}w^3\gamma}{\left(w^2 - w_{s,q}^2\right)^2 + w^2\gamma_q^2}. \tag{61}$$

Substitution of Eqs. 59–61 into Eq. 57 for the index of refraction $n(w)$ yields

$$n(w) = 1 - \frac{r_e\lambda^2}{2\pi}\sum_{q=1}^{Q}\rho_q\left(f_{1q}^0\,(w) - if_{2q}^0\,(w)\right). \tag{62}$$

For more information, see the books by Attwood (1999) and Als-Nielsen and McMorrow (2011), and the reviews by Hubbell *et al.* (1975) and Henke *et al.* (1993).

7. THE COMPLEX REFRACTIVE INDEX—DECREMENT, ABSORPTION INDEX, AND LINEAR ATTENUATION

The dimensionless complex index of refraction $n(w)$ is expressed by a dimensionless, real part $1 - \delta$ and a dimensionless imaginary part β. The decrement $\delta\,(w)$ is responsible for refraction of the wave. The imaginary part or absorption index $\beta(w)$ is responsible for attenuation of the electromagnetic wave by incoherent scatter and absorption. The decrement $\delta(w)$ and absorption index $\beta(w)$ are each dependent on the X-ray frequency w. The frequency-dependent complex refractive index $n(w)$ is, then,

$$n = 1 - \delta + i\beta. \tag{63}$$

Then, comparison of the above $n(w)$ and the expression for $n(w)$ derived in the previous section gives the decrement $\delta$and absorption index β:

$$\delta = \frac{r_e\lambda^2}{2\pi}\sum_{q=1}^{Q}\rho_q f_{1q}^0(w) \tag{64}$$

and

$$\beta = \frac{r_e\lambda^2}{2\pi}\sum_{q=1}^{Q}\rho_q f_{2q}^0(w). \tag{65}$$

The explicit angular frequency w dependence of δ and β is given by

$$\delta(w) = \frac{r_e c^2 2\pi}{w^2}\sum_{q=1}^{Q}\rho_q f_{1q}^0(w) \tag{66}$$

and

$$\beta(w) = \frac{r_e c^2 2\pi}{w^2}\sum_{q=1}^{Q}\rho_q f_{2q}^0(w), \tag{67}$$

where

$$\lambda^2 = \frac{4\pi^2 c^2}{w^2}. \tag{68}$$

In the long-wavelength limit, the decrement δ becomes

$$\delta = \frac{r_e\lambda^2}{2\pi}\sum_{q=1}^{Q}\rho_q Z_q. \tag{69}$$

The dimensionless absorption index β can be expressed in terms of a linear attenuation coefficient $\mu\,(\mathrm{cm}^{-1})$, which can be measured for materials via transmission measurements. To derive β in terms of μ, consider a plane wave $E(\mathbf{r}, t)$ with initial amplitude E_0 that has a complex dispersion relation, where

$$E(\mathbf{r}, t) = E_0 \exp\left[-i\left(wt - \mathbf{k} \cdot \mathbf{r}\right)\right]. \tag{70}$$

The phase velocity v_p of the X-ray wave in this case is complex, where

$$v_p = \frac{w}{k} = \frac{c}{n} = \frac{c}{1 - \delta + i\beta} \tag{71}$$

Solving for the wave number k, we obtain

$$k = \frac{w}{c}\left(1 - \delta + i\beta\right) \tag{72}$$

With substitution from Eqs. 71 and 72, the electric field expression of Eq. 70 becomes

$$E(\mathbf{r}, t) = E_0 \exp\left[-iw \cdot \left(t - \frac{r}{c}\right)\right] \exp\left[-i \cdot \left(\frac{2\pi\delta}{\lambda}\right) \cdot r\right] \exp\left[-\left(\frac{2\pi\beta}{\lambda}\right) \cdot r\right]. \tag{73}$$

The first exponential term in this equation represents the incident wave propagating in the vacuum, the second exponential term represents the phase shift produced in the incident vacuum wave by the material, and the third exponential term represents the attenuation of the wave as it passes through the material. From the second exponential term, we can derive the phase shift $\Delta\phi$, which is produced in a wave with wavelength λ, which passes though a material of thickness Δr with decrement δ; that is,

$$\Delta\phi = \left(\frac{2\pi\delta}{\lambda}\right) \Delta r. \tag{74}$$

From the third exponential term, we can derive the linear attenuation $\mu\,(\mathrm{cm}^{-1})$ of the wave with wavelength λ that passes through the material of thickness Δr in terms of β. In a transmission measurement, the incident intensity $|E_0|^2$ and the transmitted intensity $|E(\mathbf{r}, t)|^2$are measured to obtain the transmission T, where

$$T = \left|\frac{E(\mathbf{r}, t)}{E_0}\right|^2 = \exp\left[-\left(\frac{4\pi\beta}{\lambda}\right) \Delta r\right] = \exp\left[-2\mu\Delta r\right]. \tag{75}$$

From the measured transmission T for a material, one obtains the measured material linear attenuation $\mu\,(\mathrm{cm}^{-1})$, where

$$\mu = \frac{4\pi\beta}{\lambda} \tag{76}$$

The Lawrence Berkeley Laboratory Center for X-ray Optics (CXRO) has online databases for the atomic scattering factors, X-ray properties, index of refraction, and attenuation length for the elements; they can be found at the center's website (http://henke.lbl.gov/optical_constants).

8. X-RAY COMPLEX INDEX OF REFRACTION—KRAMERS–KRONIG RELATION

The decrement delta δ of the real part of the complex index of refraction results from the oscillation of atomic electrons that leads to the refractive effects, and the absorption index β or linear attenuation μ represents the collision losses in the oscillations that lead to the absorptive effects. The complex index of refraction $n(w)$ can be written in terms of decrement δ and linear attenuation μ (cm^{-1}), where

$$n(w) = 1 - \delta + i\frac{\lambda\mu}{4\pi}. \tag{77}$$

Transmission measurements of materials can measure linear attenuation μ and critical reflection angle measurements can measure decrement δ. From μ, one can determine β, and from β, one can determine the imaginary part $f_2^0\,(w)$ of the forward complex scattering factor $f_z^0\,(w)$ of an element (Z, A). The Kramers–Kronig relation allows the calculation of the real part $f_1^0\,(w)$ of the forward complex scattering factor $f_z^0\,(w)$ from its imaginary part $f_2^0\,(w)$. Thus, the real part $f_1^0\,(w)$, which then determines the decrement δ, is calculated from its imaginary, absorptive part β via the Kramers–Kronig relation, where

$$f_1^0\,(w) = Z - \frac{2}{\pi}\mathrm{P_C}\int_0^\infty \frac{u f_2^0\,(u)}{u^2 - w^2}du. \tag{78}$$

The $\mathrm{P_C}$ indicates integration only over the nondivergent parts of the line along u, and avoiding the singular points, $u = \pm w$, such that $\mathrm{P_C}\,(1/x)$ is defined by Cauchy's principal value theorem:

$$\lim_{\varepsilon \to 0} \frac{1}{x \mp i\varepsilon} = \mathrm{P_C}\left(\frac{1}{x}\right) \pm i\pi\delta\,(x). \tag{79}$$

The integrand is converted to the $1/x$ form appropriate for application of Cauchy's principal value theorem via

$$\frac{2u}{u^2 - w^2} = \frac{1}{u - w} + \frac{1}{u + w}. \tag{80}$$

Note that $\delta(x)$ is a Dirac delta function and not the decrement. The principal value $\mathrm{P_C}(1/x)$ is a function that behaves like $1/x$ everywhere except at $x = 0$. The discontinuity at the singularity at $x = 0$ is described by the Dirac delta function $\delta(x)$.

Similarly, if the decrement δ has been measured from the critical angle of reflection of a material, one knows the real part $f_1^0(w)$ of the complex index of refraction, and, from the Kramers–Kronig relation, can also determine imaginary part $f_2^0(w)$, and thus β and μ, via

$$f_2^0(w) = \frac{2w}{\pi} \mathrm{P_C} \int_0^{\infty} \frac{f_1^0(u) - Z}{u^2 - w^2} du. \tag{81}$$

The integrand is converted to the $1/x$ form appropriate for application of Cauchy's principal value theorem via

$$\frac{2w}{u^2 - w^2} = \frac{1}{u - w} - \frac{1}{u + w}. \tag{82}$$

For lithium ($Z = 3$) and beryllium ($Z = 4$), all the atomic electronic electrons have binding energies that are less than X-ray energies, and hence the real scattering factor f_1^0 equals the atomic number Z. To calculate f_2^0, one can start with the estimated f_1^0, where

$$f_1^0 = Z. \tag{83}$$

The linear attenuation coefficient μ is approximately proportional to the element's atomic density ρ_z, the 4.3 power of the element's atomic number, and to the 3.2 power of the wavelength of the X-rays. One can estimate f_2^0 from the measured linear attenuation μ and calculate f_1^0 where

$$\mu \propto \rho_z Z^{4.3} \lambda^{3.2}. \tag{84}$$

This expression is obtained from empirical data for low Z materials near 10 keV.

9. X-RAY RAYLEIGH SCATTER TOTAL CROSS SECTION AND THE COMPLEX ATOMIC SCATTER FACTOR

The total and differential cross sections for X-ray scatter can be determined from materials, which are expressed by the complex atomic scatter factor $f(\Delta \mathbf{k}, w)$ in the long-wavelength limit $f_z^0(w)$. The long-wavelength limit applies to X-ray wavelengths much larger than the atomic diameter. The long-wavelength limit of the complex scatter factor $f_z^0(w)$ is also the forward-scatter limit of the complex atomic scatter factor $f(\Delta\mathbf{k}, w)$.

From the electric field at the observer, which is produced by the scattering of an incident electric field from electrons in a lattice of atoms that lies between the source and the observer, one can calculate the scattering factor of the lattice of atoms for photons. The photon represents an electromagnetic field comprised of an electric field component. The electric field component is then used to calculate the magnetic field component of the electromagnetic wave and its power density. From the incident power density, and forward-scattered power density from a lattice of atoms, one can calculate the forward-scattering factor of the lattice of atoms.

From the calculated scattered electric field $\mathbf{E}[t]$ and incident electric field $\mathbf{E}_i[t]$, one can calculate the differential scattering cross section $d\sigma/dw$ and the total scattering cross section σ for X-rays. The total scatter cross section σ is defined as the average power-radiated P divided by the magnitude of the average incident power per unit area:

$$|\mathbf{S}_i| = \frac{dP_i}{dA_i} \tag{85}$$

That is, incremental incident power dP_i crosses incremental area dA_i. Recall that $\mathbf{S}_i$ is the Poynting vector, which represents the vector flow of incident energy per unit area in the direction perpendicular to the electric $\mathbf{E}_i[t]$ and magnetic fields $\mathbf{H}_i[t]$, or equivalently, in the direction of the electromagnetic wave denoted by the unit's wave number vector, $\hat{\mathbf{k}}_i$:

$$\sigma = \frac{P}{|\mathbf{S}_i|}. \tag{86}$$

The incident power $\mathbf{S}_i$ is given by

$$\mathbf{S}_i = \frac{Re\left(\mathbf{E}_i \times \mathbf{H}_i^*\right)}{2} \tag{87}$$

The incident plane wave magnetic flux density B_i, obtained from the magnetic field intensity H_i via vacuum permeability components, is

$$B_i = \mu_0 H_i \tag{88}$$

In the plane wave, the magnetic field flux density B_i is obtained from the electric field intensity E_i via dividing by the speed of light c in a vacuum:

$$B_i = \frac{E_i}{c} \tag{89}$$

The incident plane wave electric field intensity E_i is given by

$$E_i = E_0 \exp\lfloor -i\,(wt - \mathbf{k}_1 \cdot \mathbf{r}_s)\rfloor. \qquad (90)$$

The magnitude of the incident power $|\mathbf{S}_i|$ is then

$$|\mathbf{S}_i| = \frac{1}{2}\,|E_0|^2\sqrt{\frac{\varepsilon_0}{\mu_0}}. \qquad (91)$$

The conductance G_0 of X-rays in vacuum space, which is the reciprocal of impedance Z_0, is

$$G_0 = \frac{1}{Z_0} = \sqrt{\frac{\varepsilon_0}{\mu_0}}. \qquad (92)$$

Recall that the complex atomic scatter factor $f_z^0\,(w)$ in the long-wavelength limit is

$$f_z^0\,(w) = \sum_{s=1}^{Z}\frac{w^2 g_s}{w^2 - w_s^2 + iw\gamma} \quad \frac{a_0}{\lambda} \ll 1 \qquad (93)$$

For a material comprised of Q elements, the scattered electric field $\mathbf{E}\,[t]$ is given by

$$\mathbf{E}\,[t] = -\frac{r_e E_0 \sin\theta}{r}\left[\sum_{q=1}^{Q} f_q^0\,(w)\right]. \qquad (94)$$

Recall that θ is the angle between the electron acceleration a_s direction and the scattered wave direction $\mathbf{k}_2$. The magnitude of the scattered power per unit area $|\mathbf{S}|$, then, is

$$|\mathbf{S}| = \frac{1}{2}\left|\frac{r_e E_0 \sin\theta}{r}\sum_{q=1}^{Q} f_q^0\,(w)\right|^2\sqrt{\frac{\varepsilon_0}{\mu_0}} \qquad (95)$$

Integration of the scattered power per unit area $S(w)$, over a spherical surface that encloses the radiating medium, then yields the total radiated power P. The incremental solid angle $d\Omega$ of the scatter sphere is subtended by the differential spherical surface area dA, where

$$dA = r^2 \sin\theta d\theta d\phi = r^2 d\Omega \quad \text{and} \quad d\Omega = \sin\theta d\theta d\phi. \qquad (96)$$

The radiated power $P(w)$ is

$$P = \iint |S|\, dA = \frac{r_e^2 E_0^2}{2}\sqrt{\frac{\varepsilon_0}{\mu_0}} \cdot \left|\sum_{q=1}^{Q} f_q^0 (w)\right|^2 \cdot \int_0^{2\pi}\int_0^{\pi} \sin^3 \theta d\theta d\phi. \tag{97}$$

Integration is done over θ from 0 to 2π, and the resulting equation is as follows:

$$\sin^3 \theta = (1 - \cos\theta)\sin\theta. \tag{98}$$

This yields $4/3$, so that the total scattered power P is

$$P = \frac{4\pi r_e^2 E_0^2}{3} \cdot \sqrt{\frac{\varepsilon_0}{\mu_0}} \cdot \left|\sum_{q=1}^{Q} f_q^0 (w)\right|^2. \tag{99}$$

Hence, the total scatter cross section $\sigma(w)$ is

$$\sigma(w) = \frac{P(w)}{|\mathbf{S}_i (w)|} \tag{100}$$

Substitution of Eqs. 91 and 99 into Eq. 100 gives the total cross section:

$$\sigma = \frac{8\pi}{3} \cdot r_e^2 \cdot \left|\sum_{q=1}^{Q} f_q^0 (w)\right|^2. \tag{101}$$

In the long-wavelength limit for X-ray frequencies w above the resonant frequencies w_s, a material composed of a single element ($Q = 1$) has a total cross section of

$$\sigma = \frac{8\pi}{3} \cdot r_e^2 \cdot Z^2 \quad w^2 \gg w_s^2 \quad \text{and} \quad \frac{a_0}{\lambda} \ll 1. \tag{102}$$

10. DIFFERENTIAL CROSS SECTION FOR THE RAYLEIGH SCATTER AND COMPLEX ATOMIC SCATTER FACTOR

Next, the Rayleigh differential cross section $d\sigma/d\Omega$ for coherent, elastic X-ray scatter in materials is determined. One starts with the differential cross section $d\sigma$, equal to the differential scatter power dP divided by the incident power $|\mathbf{S}_i|$, where

$$d\sigma = \frac{dP}{|\mathbf{S}_i|}. \tag{103}$$

Division by the differential solid angle $d\Omega$ and substitution for dP from Eq. 97 with Eq. 96 gives

$$\frac{d\sigma}{d\Omega} = \frac{1}{|\mathbf{S}_i|}\frac{dP}{d\Omega} = \frac{1}{|\mathbf{S}_i|}\frac{d\left(\iint |\mathbf{S}|\, r^2 d\Omega\right)}{d\Omega} = r^2\frac{|\mathbf{S}|}{|\mathbf{S}_i|} \tag{104}$$

Assuming the long-wavelength limit or forward-scatter limit, substitution for $|\mathbf{S}|$ of Eq. 95 and for $|\mathbf{S}_i|$ of Eq. 91 into Eq. 104 yields the differential cross section $d\sigma/d\Omega$ for the scattering material, where

$$\frac{d\sigma}{d\Omega} = \left|\sum_{q=1}^{Q} f_q^0(w)\right|^2 r_e^2 \sin^2\theta \quad \frac{a_0}{\lambda} \ll 1 \quad \text{and} \quad \theta \ll 1 \tag{105}$$

The forward, elastic scatter case for the differential cross section that applies for all wavelengths of X-rays is identical to the differential cross section in the non-forward directions. For forward-scattered X-rays, or X-ray frequencies above the resonant frequencies and scattered in non-forward directions by a material comprised of a single element ($Q = 1$), the differential cross section $d\sigma/d\Omega$ becomes

$$\frac{d\sigma}{d\Omega} = r_e^2 Z^2 \sin^2\theta \quad w^2 \gg w_s^2 \quad \text{and} \quad \frac{a_0}{\lambda} \ll 1 \quad \text{and} \quad \theta \ll 1. \tag{106}$$

We now wish to express the differential cross section in terms of the polar angle Θ between the incident wave vector $\mathbf{k}_1$ and the scattered wave vector $\mathbf{k}_2$. Angle θ between the direction of the incident electric field acceleration of the electron a_s and the scattered wave direction $\mathbf{k}_2$ can be expressed in terms of polar angle Θ and azimuthal angle ϕ by

$$\cos\theta = \cos\phi \sin\Theta \tag{107}$$

with

$$\sin^2\theta = 1 - \cos^2\phi \sin^2\Theta. \tag{108}$$

One obtains

$$\frac{d\sigma}{d\Omega} = r_e^2 Z^2 \left(1 - \cos^2\phi \sin^2\Theta\right). \tag{109}$$

The Rayleigh differential cross section $d\sigma/d\Omega$ is now averaged over the $0 - 2\pi$ azimuth angles ϕ via averaging $\cos^2\phi$ to obtain the average azimuth angle $<\phi>$; and thus obtain the averaged squared cosine of the azimuth angle $<\cos^2\phi>$, where

$$<\cos^2\phi> = \frac{1}{2\pi}\int_0^{2\pi} d\phi \cos^2\phi = \frac{1}{2\pi}\int_0^{2\pi} d\phi \frac{1-\cos 2\phi}{2} = \frac{1}{2}. \tag{110}$$

The azimuth angle averaged differential cross section $< d\sigma/d\Omega >$ is, then, the approximate Rayleigh differential cross section

$$<\frac{d\sigma}{d\Omega}> = r_e^2 Z^2 \left(1 - \frac{\sin^2\Theta}{2}\right) = \frac{r_e^2 Z^2}{2}\left(1+\cos^2\Theta\right). \tag{111}$$

Integration of the averaged differential cross section $<d\sigma/d\Omega>$ over solid angle $d\Omega$ yields the cross section for the scattering of an unpolarized plane electromagnetic wave by Z free electrons, where the approximate total Rayleigh cross section is

$$\begin{aligned} <\sigma> &= \int_0^{2\pi} d\phi \int_0^{\pi} d\Theta < \frac{d\sigma}{d\Omega} > \sin\Theta \\ &= \frac{r_e^2 Z^2}{2}\int_0^{2\pi} d\phi \int_0^{\pi} d\Theta \left(1+\cos^2\Theta\right)\sin\Theta = \frac{8\pi}{3} r_e^2 Z^2. \end{aligned} \tag{112}$$

For the case of a single free electron, with $Z = 1$, the above result is known as the Thomson scattering cross section $<\sigma_T>$, in which

$$<\sigma_T> = \frac{8\pi}{3} r_e^2 = 0.665\,\text{barn} \tag{113}$$

where

$$1\,\text{barn} = 10^{-24}\,\text{cm}^2 \tag{114}$$

11. RAYLEIGH AND THOMSON X-RAY SCATTERS ARE COHERENT, ELASTIC SCATTER

In the limit of zero photon energy (<100 keV), which applies to the low-energy X-ray wavelengths 0.5 to 2.0 Å used typically with compound X-ray lenses, the X-ray scatter follows the classical Thomson radiation scatter, in which the electron vibrates at the same frequency as the incident X-ray, and thus produces emission of secondary electromagnetic radiation of the same frequency. The Thomson cross section serves as the precursor for the calculation of Rayleigh coherent scatter and Compton incoherent scatter. The Rayleigh differential coherent scatter cross section $d\sigma_R/d\Omega$ is the Thomson differential cross section $d\sigma_T/d\Omega$ for scattering angle θ multiplied by the square of the atomic form factor F, which is approximately the atomic number Z of the nucleus, where

$$F = Z. \tag{115}$$

These Rayleigh and Thomson cross sections assume that the incident electromagnetic radiation has a sufficiently long wavelength that the energy of the incident wave is much less than the rest mass of the free electron. This is the case for photon energies up to and including the X-ray region. Although the atomic electrons are bound, they are approximated as free for the Thomson and Rayleigh calculations of cross section for the low-energy incident photons. For higher-energy photons (i.e., gamma, whose energy is comparable to or exceeds the rest mass of the electron), the scattering is called Compton scatter. In Compton scatter, quantum mechanics must be used to determine the differential and total cross sections, which are known as the Klein–Nishina cross sections for a free electron.

A high-energy gamma will eject the atomic electron rather than causing the electron to oscillate, as in the case of Rayleigh and Thomson scatter. In Rayleigh scatter, the sufficiently low energy of the incident photon allows the relatively "free" atomic electron to oscillate. However, the electron is bound to the atom, and the momentum recoil of the oscillating electron is transferred to the atom as a whole for conservation of momentum. And with conservation of energy, the much more massive nucleus gains almost no kinetic energy, and hence almost no energy is transferred from the photon to the electron and to the atom. The Thomson and Rayleigh scatter of photons is elastic, whereas in Compton scatter, the momentum of the photon is transferred to a single electron, which is not shared with the atom. The electron is ejected, and the photon loses significant energy by this Compton inelastic scatter, except in the forward-scatter direction.

X-rays undergo coherent elastic scatter in the form of Rayleigh scatter. The conservation of momentum in the X-ray scatter from the atom as a whole allows the large mass M of the atom to attain an extremely small recoil velocity V_r to match the X-ray momentum p imparted to the atom as a whole in Rayleigh scatter, where

$$V_r = \frac{p}{M}. \tag{116}$$

The recoil energy E_r imparted to the atom as a whole by the X-ray, as we will see, is very small:

$$E_r = \frac{MV_r^2}{2} = \frac{p^2}{2M}. \tag{117}$$

By conservation of energy, the decrease in the X-ray energy due to scatter ΔE_{xr} is equal to the energy gained E_r by the recoiling atom, where

$$\Delta E_{xr} = -E_r \tag{118}$$

The atom has rest energy E_0, given by

$$E_0 = Mc^2. \tag{119}$$

The ratio of the energy loss ΔE_{xr} of the X-ray to recoil energy transferred to the atom divided by the incident X-ray energy E_{xr} is very small:

$$\frac{\Delta E_{xr}}{E_{xr}} = -\frac{E_r}{E_{xr}} = -\frac{p^2}{2ME_{xr}} = \frac{E_{xr}}{2E_0}, \tag{120}$$

where

$$E_{xr} = pc. \tag{121}$$

If the incident X-ray energy E_{xr} is much less than the rest mass energy of the scattering atom E_0, then the X-ray loses approximately no energy (elastic scatter) and has a small scattering angle. The Rayleigh scatter or X-ray coherent, elastic scatter is responsible for the X-ray refraction, reflection, and diffraction.

12. COMPTON SCATTER OF X-RAYS IS INCOHERENT AND INELASTIC

Compton scatter of X-rays is incoherent and inelastic. To derive the Compton scatter of an X-ray from a free stationary electron, we will use the expression derived for the total momentum conservation, which is relativistically correct, where

$$P_t^2 = P_{p0}^2 - 2P_pP_{p0}\cos\theta + P_p^2. \tag{122}$$

The initial projectile X-ray momentum is P_{p0}, and after scatter at angle θ from the stationary, target electron, the scattered X-ray momentum is

P_p. The relativistic expression of the total projectile, photon energy E_p is related to its momentum P_p and rest mass M_{prm} by

$$E_p = \sqrt{P_p^2 c^2 + M_{prm}^2 c^4}. \tag{123}$$

The rest mass of a photon M_{prm} is assumed to be zero, which in our case is the X-ray photon as follows:

$$M_{prm} = M_{p0rm} = 0. \tag{124}$$

Hence, the total relativistic energy of the photon E_p is equal to its kinetic energy K_p and is related to its momentum P_p by

$$E_p = P_p c = K_p. \tag{125}$$

The rest mass energy E_{trm} of the scattered target electron with rest mass M_{trm} is

$$E_{trm} = M_{trm} c^2. \tag{126}$$

And the rest mass M_{t0rm} of the stationary electron is unchanged by the photon scatter:

$$M_{trm} = M_{t0rm}. \tag{127}$$

The scattered electron relativistic kinetic energy K_t is related to its momentum P_t and rest mass M_{trm} by subtracting electron rest mass energy E_{trm} from total energy E_t, where

$$K_t = E_t - E_{trm} = \sqrt{P_t^2 c^2 + M_{trm}^2 c^4} - M_{trm} c^2, \tag{128}$$

Solving for the scattered target electron momentum P_t in terms of the scattered target electron energy K_t gives

$$P_t = \frac{1}{c}\sqrt{K_t\,(K_t + 2E_{trm})}. \tag{129}$$

One now equates the above relativistic momentum P_t expression for the scattered electron with P_t^2 derived from the conservation of momentum, where

$$P_t^2 = \frac{1}{c^2} K_t\,(K_t + 2E_{trm}) = P_{p0}^2 - 2P_p P_{p0}\cos\theta + P_p^2. \tag{130}$$

The X-ray photon incident momentum P_{p0} and energy K_{p0}, as well as the scatter X-ray photon momentum P_p and energy K_p, are related by

$$K_{p0} = cP_{p0} \tag{131}$$

and

$$K_p = cP_p \tag{132}$$

Substitution of the expressions in Eqs. 129, 131, 132 into Eq. 130 gives the relation between the incident K_{p0} and scattered K_p X-ray kinetic energies to that of the scattered electron kinetic energy K_t and its rest mass M_{trm}, where

$$K_t\left(K_t + 2E_{trm}\right) = K_{p0}^2 - 2K_pK_{p0}\cos\theta + K_p^2. \tag{133}$$

By conservation of energy, the incident kinetic energy K_{p0} of the X-ray photon is equal to the sum of the scattered X-ray photon energy K_p and the relativistic kinetic energy K_t of the scattered, target electron, where

$$K_{p0} = K_p + K_t \tag{134}$$

Substitution for K_t from the conservation of energy expression of Eq. 134 into Eq. 133 gives

$$\left(K_{p0} - K_p\right)\left(K_{po} - K_p + 2E_{trm}\right) = K_{p0}^2 - 2K_pK_{p0}\cos\theta + K_p^2. \tag{135}$$

Rearrangement yields the scattered X-ray kinetic energy K_p in terms of its scatter angle θ and initial kinetic energy K_{p0} and the rest mass E_{trm} of the scattered electron, where

$$K_p = \frac{K_{p0}}{1 + \frac{K_{p0}}{E_{trm}}\left(1 - \cos\theta\right)}. \tag{136}$$

The total or kinetic X-ray incident K_{p0} and scattered K_p photon energy are related to their respective wavelengths λ_{p0} and λ_p by

$$K_{p0} = \frac{hc}{\lambda_0} \tag{137}$$

and

$$K_p = \frac{hc}{\lambda}. \tag{138}$$

Then, the scattered X-ray wavelength λ_p, as a function of its incident wavelength λ_{p0} and scatter angle θ and the Compton wavelength λ_c, is

$$\lambda_p = \lambda_{p0} + \lambda_c (1 - \cos\theta). \tag{139}$$

The Compton wavelength, in terms of scattered electron rest mass M_{trm}, is

$$\lambda_c = \frac{hc}{M_{trm}} = 2426\,\text{fm}. \tag{140}$$

To calculate the scattered Compton electron kinetic energy as a function of its scatter angle ϕ and rest mass E_{trm}, as before, one obtains an explicit expression for the total scattered projectile X-ray momentum P_p in terms of its initial total X-ray momentum P_{p0} and the scattered target electron scatter angle ϕ and total momentum P_t:

$$P_p^2 = P_{p0}^2 - 2P_{p0}P_t\cos\phi + P_t^2. \tag{141}$$

Expression of the incident P_{p0} and scattered P_p X-ray momentum in terms of their kinetic energies K_{p0} and scattered K_p gives

$$K_p^2 = K_{p0}^2 - 2cK_{p0}P_t\cos\phi + c^2P_t^2. \tag{142}$$

The conservation of energy is

$$K_p = K_{p0} - K_t. \tag{143}$$

Substitution of the expression for K_p of Eq. 143 into Eq. 142 gives

$$\left(K_{p0} - k_t\right)^2 = K_{p0}^2 - 2cK_{p0}P_t\cos\phi + c^2P_t^2, \tag{144}$$

and substitution of Eq. 130 for the relativistic scattered, target electron momentum P_t in terms of its kinetic energy K_t and rest mass E_{trm} into Eq. 144 gives

$$\left(K_{p0} - K_t\right)^2 = K_{p0}^2 - 2K_{p0}\sqrt{K_t\left(K_t + 2E_{trm}\right)}\cos\phi + K_t\left(K_t + 2E_{trm}\right). \tag{145}$$

Rearrangement yields the explicit expression of the scattered, target electron kinetic energy K_t in terms of its scatter angle ϕ, rest mass E_{trm}, and initial projectile X-ray kinetic energy K_{p0}:

$$K_t = \frac{2E_{trm}}{\left(\dfrac{K_{p0} + E_{trm}}{K_{p0}\cos\phi}\right)^2 - 1}. \tag{146}$$

The neutron and X-ray cross sections are based on different underlying physical scattering and absorption processes. We now briefly examine the X-ray scatter and absorption cross sections. X-rays undergo incoherent scatter from individual electrons. The Thomson cross section $d\sigma_T/d\Omega$ is the low-energy limit of Compton scatter. The Compton cross section σ_c for incoherent X-ray scatter is electron scatter for a free electron, the Klein–Nishina cross section σ_{KN} multiplied by the atomic number Z of the atom to which the electron belongs yields

$$\sigma_c = Z\sigma_{KN}. \tag{147}$$

For unpolarized X-rays below 100 keV, the differential Klein–Nishina cross section per electron is

$$\frac{d\sigma_{KN}}{d\Omega}(\theta) = \frac{r_e^2\left(1+\cos^2\theta\right)}{2\cdot\left[1+\frac{E}{mc^2}(1-\cos\theta)\right]^2}. \tag{148}$$

The X-ray energy E is measured in units of the electron rest mass energy as follows:

$$mc^2. \tag{149}$$

The total Klein–Nishina cross section is approximately

$$\frac{d\sigma_{KN}}{d\Omega}(\theta) = \frac{8\pi r_e^2\left[1+2\frac{E}{mc^2}+1.2\left(\frac{E}{mc^2}\right)^2\right]}{3\cdot\left[1+2\frac{E}{mc^2}\right]^2}. \tag{150}$$

13. X-RAY ATTENUATION BY PHOTOELECTRIC ABSORPTION

For X-rays, photoelectric absorption is the analog of thermal neutron capture. The X-ray interacts with the electron cloud of the entire atom, and as a result, a photoelectron is ejected from the atom, usually the K-shell. The difference between the incident X-ray energy E_0 and the electron binding energy E_b is shared between the ejected electron and the resulting atom recoil energy; however, most of the energy goes to the kinetic energy K of the ejected photoelectron, which is shown as follows. The electron leaves the atom and transfers recoil electron momentum p to the more massive nucleus of mass M, and the nucleus velocity V_r needed to conserve momentum via $MV_r = p$ is very small, where

$$V_r = \frac{p}{M} \ll 1. \tag{151}$$

Thus, the recoil energy carried away by the nucleus E_r is

$$E_r = \frac{1}{2}MV_r^2 = \frac{1}{2}\frac{p^2}{M}. \tag{152}$$

Assume that the incident gamma energy E_0 is much larger than the electron binding energy E_b:

$$E_0 \gg E_b. \tag{153}$$

Thus, the energy available, ΔE, is as follows:

$$\Delta E = E_0 - E_b. \tag{154}$$

Then the ratio of the energy carried away by the recoil nucleus E_r to the energy available ΔE is

$$\frac{E_r}{E_0 - E_b} \cong \frac{E_r}{E_0}. \tag{155}$$

If one also assumes that the incident gamma energy E_0 is much less than the rest mass energy E_a of the atom, where

$$E_0 \ll E_a, \tag{156}$$

then the ratio of energy carried away by the recoil nucleus to the available energy is very small, where

$$\frac{E_r}{E_0} = \frac{p^2}{2M}\frac{1}{pc} = \frac{1}{2}\frac{pc}{Mc^2} = \frac{1}{2}\frac{E_0}{E_a} \ll 1. \tag{157}$$

Hence, in the photoelectric effect, most of the energy is carried away by the photoelectron as kinetic energy K, where

$$K = E_0 - E_b. \tag{158}$$

The vacancy of the ejected photoelectron is filled by an electron from an outer shell, and as a result, an Auger electron or fluorescent X-rays are emitted. The X-ray photoelectric absorption cross section in general decreases with increasing X-ray energy. However, as the X-ray energy exceeds the binding energy of the next atomic energy shell (say, the transition from the M shell to the L shell), the cross section then increases after crossing from M into the next higher energy shell L. As the X-ray energy increases beyond the L-edge energy, the photoelectric absorption cross section resumes monotonic decrease until the X-ray energy exceeds the binding energy threshold of the next shell, the K-shell at the K-edge.

The cross section then increases abruptly at the K-edge, and with further increase in X-ray energy, the photoelectric absorption cross section then decreases monotonically thereon because there are no atomic energy shells to cross. The photoelectric X-ray cross section σ_{ph} is proportional to the fourth power of the atomic number Z of the scattering atom and inversely proportional to the third power of the incident X-ray energy, where

$$\sigma_{ph} \propto \frac{Z^4}{E^3}. \tag{159}$$

14. X-RAY ABSORPTIVE ATTENUATION BY PAIR PRODUCTION

In pair production, the incident photon is completely absorbed in the vicinity of a nucleus, provided that the incident photon energy E_0 exceeds twice the electron mass energy $2m_ec^2$ or 1.02 MeV. The photon of energy E_0 in pair production is converted into an electron and positron with a widely varying distribution of energy between them when

$$E_0 > 2m_ec^2. \tag{160}$$

The positron and electron move at angle θ_{pair} relative to the incident photon direction, but the electron and positron are separated by about 180° azimuthally, where

$$\theta_{pair} = \frac{m_ec^2}{E_0} \tag{161}$$

As the positron propagates through a material matrix, the positron loses kinetic energy by almost entirely electrostatic collision interactions with a trace radiative emission until it has almost stopped, at which point the positron annihilates and creates two photons of m_ec^2 or 511 keV moving in opposite directions. The pair production cross section in the vicinity of a nucleus σ_{pair} is proportional to the squared atomic number Z^2 of the scattering nucleus:

$$\sigma_{pair} \propto Z^2. \tag{162}$$

In the vicinity of the electric field of an electron, the incident photon energy must be four times the rest mass of the electron $4m_ec^2$. Pair production does not occur with X-rays because they are well below 1 MeV. However, thermal neutron capture gammas typically can be several MeV or more, and they can undergo pair production.

These various gamma and higher-energy X-ray cross sections are obtained from J. H. Hubbell and M. J. Berger (National Institute of Standards and Technology [NIST] X-ray cross section tables [XCOM], http://physics.nist.gov), and the Center for X-ray Optics at Lawrence Berkeley National Laboratory has online databases for the atomic scattering factors, X-ray property of elements, index of refraction, and attenuation length for the elements (http://henke.lbl.gov/optical_constants). The thermal neutron cross sections for elastic and inelastic scatter and absorption are obtained from the NIST Neutron Scattering Lengths and Cross Sections (http://www.ncnr.nist.gov/resources/n-lengths). Upon thermal neutron capture or absorption by an isotope nucleus, prompt gammas whose energies are characteristic of that isotope are emitted. The cross sections for prompt gamma emission following thermal neutron capture can be found at the website of the International Atomic Energy Agency (IAEA) under the heading "Prompt Gamma Ray Neutron Activation Analysis (PGAA) Database Files" (http://www-nds.iaea.org/pgaa/). The Compton scatter produces Compton electrons and pair production produces positrons and electrons. Some thermal neutron capture reactions produce energetic alphas or protons. The NIST stopping tables ESTAR, PSTAR, and ASTAR (http://physics.nist.gov) tabulate the electron, proton, and alpha stopping distances. Also tabulated are the total energy dissipation per unit propagation distance, as well as the energy dissipation by collisions with the atomic electron clouds and nuclei or dissipation by electromagnetic radiation emission (bremsstrahlung).

15. COLLISION LOSSES OF CHARGED PARTICLE PASSAGE IN MATTER

We now examine the attenuation of charged particles via collisions as a function of propagation distance into a material medium. The collision of a charged particle in its passage through matter leads to energy loss per propagation distance due to collisions with atoms in the materials. The radiation-stopping power of a material for a charged particle is the average particle energy loss per propagation distance in the material. The charged particle also emits photons as it undergoes deceleration in matter. The radiation-stopping power is the average energy loss per propagation distance in the material by the charged particle due to photon or bremsstrahlung emission. The total stopping power of a material for a charged particle is, then, the sum of particle average energy loss per propagation distance from collisions and photon emission.

In samples and lenses, X-rays can eject atomic electrons as photoelectrons and Compton electrons. Gammas with energies above 1.02 MeV that occur in many thermal neutron capture reactions (e.g., 2.2-MeV gamma by thermal neutron capture by hydrogen) can produce positron-electron pairs in which the positrons eventually annihilate with other electrons in the material or its surroundings. In X-ray and neutron optics, the relatively low energy electrons and positrons lose almost all their energy by collisions and relatively little energy by photon emission. The anodes of X-ray tubes get very hot because most of the electron beam energy is lost through collisions with the anode atoms, with comparatively much smaller losses due to X-ray emission. However, the small percentage of stopped electron energy emitted as X-rays is very significant for the X-ray tube operator and the X-ray application.

The collision loss of energy per propagation length dT/dz along the z-direction of Compton-scattered electrons or pair-produced electrons and positrons is based on Coulomb scatter and is derived in this section as follows. Assume a charged particle, such as a proton, electron, positron, or alpha with mass m_c, is propagating through a material at velocity v, which has mass density ρ and is composed of atoms of average atomic weight A and number Z. Each atom has mass

$$m_A = Am_a, \tag{163}$$

where m_a is the mass of an atomic mass unit that is approximately the mass of a neutron or proton. The electron number density ρ_e in the material is approximately

$$\rho_e = \frac{\rho Z}{Am_a}. \tag{164}$$

The charged particle passes through a cylindrical channel of radius b in the material. Radius b is termed the impact parameter, and b is a distance measured perpendicular from the line of travel of the incident particle. The impact parameter can be considered the range of smallest distances between the straight line of flight of the incident-charged particle and the electron clouds of the atoms of material that surround the channel of radius b.

The minimum impact parameter distance b_{min} is assumed to be larger than the spatial extent Δx_c of the wave packet of the charged particle. The charged particle velocity v is assumed to be greater than 10% of the speed of light c, and thus, relativistic momentum p_c is calculated as follows:

$$p_c = \gamma m_c v. \tag{165}$$

The relativistic gamma factor γ is

$$\gamma = \frac{1}{\sqrt{1-(v/c)^2}} \tag{166}$$

By the Heisenberg uncertainty principle, the spatial extent Δx_c of the charged particle in terms of its relativistic momentum p_c can be approximated as the minimum impact parameter distance b_{min}, where

$$b_{\text{min}} = \Delta x_c = \frac{\hbar}{\Delta p} = \frac{\hbar}{\gamma m_c v}. \tag{167}$$

The maximum impact parameter distance b_{max} is determined by the ionization energy I of the bound atomic electrons of the material that surround the cylindrical channel of radius b. The incident line of charge particle propagation is the center axis of this channel. The maximum impact distance b_{max} of the charged particle from the atomic electron is sufficiently close to allow the charge of the propagating particle to interact with the charge of the bound electron with sufficient force to overcome the electron ionization energy I and ionize the atom. The charged particle thus loses energy I to the atomic electron due to the charged particle electric field performing work I on the atomic electron. The energy I received by the atomic electron from the charged particle is sufficient to eject the atomic electron from its bound state. The collision time is the time τ required for the charged particle electric field to perform work I on the atomic electron and cause its ejection from the atom; it is determined by the Heisenberg uncertainty principle, in which

$$\tau = \frac{\hbar}{I}. \tag{168}$$

The collision time τ is also approximated as the time required for the charged particle to travel axial distance b_{max} along its initial line of flight at its relativistic velocity v, in which b_{max} is also the maximum transverse distance between the bound atomic electron and the line of flight. Since the axial length along the particle line of flight is relativistically contracted, then the collision time is

$$\tau = \frac{b_{\text{max}}}{\gamma v}. \tag{169}$$

Equating the two expressions for the collision time τ, one solves for the maximum impact parameter b_{max}, where

$$b_{\mathrm{max}} = \frac{\gamma v \hbar}{I}. \tag{170}$$

The range of impact distances from the minimum b_{min} to the maximum b_{max} now allows the solution for the energy loss per unit distance traveled by the charged particle in the material due to Coulomb-based collisions. The Lorentz force equation is applied to the charged particle, which has momentum vector $\mathbf{p}$, charge state z_c, and charge $z_c e$ for elementary charge e. The charged particle travels through the electric field $\mathbf{E}$ of the bound atomic electrons and experiences force $d\mathbf{p}/dt$, where

$$\frac{d\mathbf{p}}{dt} = z_c e \mathbf{E}. \tag{171}$$

The magnetic field in the rest frame of the moving charged particle is ignored. Unit vector $\hat{\mathbf{r}}$ of the electric field $\mathbf{E}$ points from the bound electron to the moving charged particle. The distance of the closest approach is the perpendicular distance b from the line of charged particle flight to the electron.

Then the electric field of the bound electron, which has effective charge state z_e and charge $z_e e$ that acts on the charged particle of charge $z_c e$ at distance r, is

$$\mathbf{E} = \frac{1}{4\pi\varepsilon_0}\frac{z_e e}{r^2}\hat{\mathbf{r}}. \tag{172}$$

Distance r is

$$r = |\mathbf{r}| = \sqrt{b^2 + z^2}. \tag{173}$$

Unit vector $\hat{\mathbf{r}}$ is

$$\hat{\mathbf{r}} = \frac{\mathbf{r}}{|\mathbf{r}|} = \frac{\mathbf{r}}{\sqrt{b^2 + z^2}}. \tag{174}$$

The electric field has an axial component E_z along the z-direction of charged particle propagation, and a perpendicular component E_y, where

$$E_z = \frac{1}{4\pi\varepsilon_0}\frac{z_e e}{r^2}\left(\hat{\mathbf{r}}\cdot\hat{\mathbf{z}}\right) = \frac{1}{4\pi\varepsilon_0}\frac{z_e e}{r^2}\cos\theta = \frac{1}{4\pi\varepsilon_0}\frac{z_e e z}{r^3} = \frac{1}{4\pi\varepsilon_0}\frac{z_e e z}{\left(b^2 + z^2\right)^{3/2}} \tag{175}$$

and

$$E_y = \frac{1}{4\pi\varepsilon_0}\frac{z_e e}{r^2}\left(\hat{\mathbf{r}}\cdot\hat{\mathbf{y}}\right) = \frac{1}{4\pi\varepsilon_0}\frac{z_e e}{r^2}\cos\left(\frac{\pi}{2}+\theta\right)$$
$$= -\frac{1}{4\pi\varepsilon_0}\frac{z_e e y}{r^3} = -\frac{1}{4\pi\varepsilon_0}\frac{z_e e y}{\left(b^2+z^2\right)^{3/2}}, \tag{176}$$

where

$$\cos\theta = \frac{z}{r} \tag{177}$$

and

$$\cos\left(\frac{\pi}{2}+\theta\right) = -\sin\theta = -\frac{y}{r} \tag{178}$$

The change in the axial component momentum Δp_z of the charged particle with axial velocity v is found by integrating the Lorentz force equation in the axial z-direction over its entire time t of propagation through the material, where

$$\Delta p_z = z_c e\int_{-\infty}^{\infty} E_z dt \cong \frac{z_c z_e e^2}{4\pi\varepsilon_0}\int_{-\infty}^{\infty}\frac{z dt}{\left(b^2+z^2\right)^{3/2}}, \tag{179}$$

The electric field $\mathbf{E}$ at the charged particle positions outside the material slab for time $t = -\infty$ to the time of incidence t_{in} at the material slab, and for time of exit from the material slab t_{out} to time $t = \infty$, is approximated as zero. Therefore, the integrand is zero for the period of time the particle spends outside the material slab and at the slab entrance and exit boundaries. This integration over time is facilitated by approximating the axial propagation distance z by

$$z = vt. \tag{180}$$

Substitution of the expression for z of Eq. 180 into the right-most integrand of Δp_z in Eq. 179 yields

$$\Delta p_z = -\frac{z_c z_e e^2}{4\pi\varepsilon_0}\int_{-\infty}^{\infty}\frac{vt dt}{\left(b^2+(vt)^2\right)^{3/2}} = -\frac{z_c z_e e^2}{4\pi\varepsilon_0}\left[\frac{1}{\left(b^2+(vt)^2\right)^{1/2}}\right]_{t_{in}}^{t_{out}} = 0. \tag{181}$$

The change in the transverse component momentum Δp_y of the charged particle with axial velocity v is found by integrating the Lorentz force equation in the tranverse y-direction over its entire time t of propagation through the material, where

$$\Delta p_y = z_c e \int_{-\infty}^{\infty} E_y dt \cong -\frac{z_c z_e e^2}{4\pi\varepsilon_0} \int_{t_{in}}^{t_{out}} \frac{b dt}{\left(b^2 + z^2\right)^{3/2}}. \tag{182}$$

For integration purposes, one maintains the infinite limits of integration:

$$\Delta p_y = -\frac{z_c z_e e^2}{4\pi\varepsilon_0} \int_{-\infty}^{\infty} \frac{b dt}{\left(b^2 + (vt)^2\right)^{3/2}}. \tag{183}$$

To solve this integral, one substitutes the following values into the above Eq. 183:

$$vt = b \tan\phi \tag{184}$$

and

$$dt = \frac{b}{v} d(\tan\phi) = \frac{b}{v}\sec^2 d\phi. \tag{185}$$

The integral becomes

$$\begin{aligned} \Delta p_y &= -\frac{z_c z_e e^2}{4\pi\varepsilon_0}\frac{1}{bv} \int_{-\pi/2}^{\pi/2} \frac{\sec^2\phi d\phi}{\left(1 + \tan^2\phi\right)^{3/2}} \\ &= -\frac{z_c z_e e^2}{4\pi\varepsilon_0}\frac{1}{bv} \int_{-\pi/2}^{\pi/2} \cos\phi d\phi = -\frac{z_c z_e e^2}{4\pi\varepsilon_0}\frac{2}{bv}. \end{aligned} \tag{186}$$

No external forces act on the system of the material bound electrons, and these bound electron electric fields act on the charged particle moving through a cylindrical channel spanning the material. Thus, the system momentum is constant and conservation of momentum requires that the change in momentum of the charged particle Δp_y be equal and opposite to that of the momentum change of the bound electron Δp_e, where

$$\Delta p_y + \Delta p_e = 0. \tag{187}$$

Conservation of energy requires that the energy lost by the particle ΔE_c be the energy ΔE_e gained by the atomic electron, where

$$\Delta E_c + \Delta E_e = 0. \tag{188}$$

One assumes the acquired velocity of the ionized atomic electron is non-relativistic, so that the energy lost by the particle ΔE_c and gained by the atomic electron is

$$\Delta E_c = -\frac{1}{2m_c}\left(\Delta p_y\right)^2 = -\Delta E_e. \tag{189}$$

Substitution for Δp_y of Eq. 186 in Eq. 189 gives

$$\Delta E_c = -\frac{2}{b^2 v^2 m_c}\left(\frac{z_c z_e e^2}{4\pi\,\varepsilon_0}\right)^2 \tag{190}$$

If the charged particle moves differential distance dz along the center axis of its channel of radius b that spans the material slab. In distance dz, the charged particle encounters the electric fields of the atomic electrons that are within the bounds of the impact parameter

$$b_{\min} \leq b \leq b_{\max}. \tag{191}$$

Recall that the electron number density ρ_e in the material is approximately

$$\rho_e = \frac{\rho Z}{A m_a}. \tag{192}$$

The loss of energy ΔE of the charged particle to the electrons in axial propagation distance dz is the product of the energy loss per electron ΔE_c multiplied by the differential volume dV spanned by dz in the channel whose radius b varies from $b_{\min}$to $b_{\max}$, so that passing from the incremental ΔE to differential dE, one obtains

$$dE = dE_c \rho_e dV, \tag{193}$$

where

$$dV = 2\pi \bullet b db \bullet dz. \tag{194}$$

Substitution of dV in dE gives

$$dE = -4\pi \frac{\rho Z}{Am_a} \frac{1}{v^2 m_c} \left(\frac{z_c z_e e^2}{4\pi \varepsilon_0} \right)^2 \frac{db}{b} dz \tag{195}$$

Integration over the impact parameter over its range $b_{\min}$ to $b_{\max}$ gives the energy loss of the charged particle per unit propagation distance in the material slab or stopping power, where

$$\frac{dE}{dz} = -4\pi \frac{Z}{A} \frac{\rho}{m_a} \frac{z_c^2}{v^2 m_c} \left(\frac{e^2}{4\pi \varepsilon_0} \right)^2 Log_e \left(\frac{b_{\max}}{b_{\min}} \right). \tag{196}$$

The electron charge state is $z_e = 1$, and one replaces the ionization energy by the average ionization energy $< I >$ of the atomic electrons, where

$$\frac{b_{\max}}{b_{\min}} = \frac{\gamma^2 m_c v^2}{< I >}. \tag{197}$$

Multiplying the numerator and denominator of Eq. 196 by c^2 allows one to express the ratio in terms of the ratio of the rest mass energy of the charged particle E_{c0} divided by the average electron ionization energy of the material, where

$$\frac{b_{\max}}{b_{\min}} = \frac{E_{c0}}{< I >} \left(\gamma \frac{v}{c} \right)^2. \tag{198}$$

The classical electron radius r_e is

$$r_e = \frac{e^2}{4\pi \varepsilon_0 c^2} \frac{1}{m_e} \tag{199}$$

In Eq. 196 substitution and multiplication of the numerator and denominator by c^2 for each occurrence of a mass in the numerator or denominator allows the rest mass energies to be conveniently inserted:

$$\frac{dE}{dz} = -4\pi \frac{Z}{A} \frac{c^2 \rho}{(m_a c^2)} \frac{z_c^2 r_e^2 c^2}{v^2 (m_c c^2)} \left(m_e c^2 \right)^2 Log_e \left(\frac{b_{\max}}{b_{\min}} \right). \tag{200}$$

The ratio of the following rest mass energies is dimensionless:

$$\frac{\left(m_e c^2 \right)^2}{(m_c c^2)(m_a c^2)} \cong \frac{1}{1836} \frac{m_e}{m_c}. \tag{201}$$

The stopping power becomes

$$\frac{dE}{dz} = -4\pi \frac{Z}{A}\left(\frac{c}{v}\right)^2 \rho(z_c r_e c)^2 \left(\frac{1}{1836}\frac{m_e}{m_c}\right) Log_e\left(\frac{b_{\max}}{b_{\min}}\right). \qquad (202)$$

With substitution into Eq. 202 for $b_{\max}/b_{\min}$ of Eq. 198, the stopping power dE/dz becomes

$$\frac{dE}{dz} = -4\pi \frac{Z}{A}\left(\frac{c}{v}\right)^2 \rho(z_c r_e c)^2 \left(\frac{2}{1836}\frac{m_e}{m_c}\right) Log_e\left(\gamma \frac{v}{c}\sqrt{\frac{E_{c0}}{<I>}}\right) \qquad (203)$$

If the charged particle is an electron, then

$$m_c = m_e \quad \text{and} \quad z_c = -1. \qquad (204)$$

If the charged particle is a positron, then

$$m_c = m_e \quad \text{and} \quad z_c = 1. \qquad (205)$$

In either case, the stopping power of the electron or positron is

$$\frac{dE}{dz} = -\rho\left(\frac{4\pi}{1836}\right)\left(\frac{Z}{A}\right)\left(\frac{c}{v} r_e c\right)^2 Log_e\left(\gamma \frac{v}{c}\sqrt{\frac{E_{c0}}{<I>}}\right). \qquad (206)$$

The rate of particle energy loss with propagation distance becomes very large as the charge particle velocity v decreases toward the end of its trajectory or range of travel because of the $1/v^2$ term in dE/dx. That is, as the particle velocity v slows to zero, much of its energy is deposited within a short distance, forming the Bragg peak of energy deposition. Proton beam therapy for cancer exploits the linear decreasing energy of the proton beam as it travels toward the tumor site, which lies at the end of the particle range of travel in which the $1/v^2$ term quadratic increases the deposited energy within a Bragg peak occupying the tumor position.

16. BREMSSTRAHLUNG FROM CHARGED PARTICLE PASSAGE IN MATTER

Charged particles emit bremstahlung radiation as they decelerate via their decreasing speed and change of direction due to Coulomb scatter. These accelerated charges emit electromagnetic radiation (or bremsstrahlung) as they are slowed and scattered by the electric fields of the atomic electrons and nuclei. However, collision energy loss predominates over the energy loss by bremsstrahlung for the massive charged particles such as protons and alphas, and for the less energetic electrons and positrons. However, high-energy electrons and positrons are strong emitters of bremsstrahlung. The ratio of the energy loss ΔE_{rad} due to bremsstrahlung divided by the energy loss ΔE_{col} due to collisions is not derived here, but is stated as

$$\frac{\Delta E_{rad}}{\Delta E_{col}} = \frac{T(Z + 1.2)}{700}. \tag{207}$$

The incident charged particle kinetic energy is T [MeV], Z is the atomic number of the scattering material, and the 700 is given in MeV. For tungsten with $Z = 74$, an ion beam of protons or electrons accelerated to 10 MeV that strike tungsten lose about 50% of its energy by bremstrahlung and 50% by collisions, whereas a deuteron ion or electron accelerated to 100 keV loses 99% of its energy due to collisions and 1% by bremsstrahlung.

The number $N_{br}(E_0, E)$ of bremsstrahlung photons of energy E, per unit energy and per incident electron, emitted when an electron of incident energy E_0 is completely slowed down, can be approximated by the distribution

$$N_{br}\,(E_0, E) = 2kZ\left[\left(\frac{E_0}{E} - 1\right) - \frac{3}{4}LN\left(\frac{E_0}{E}\right)\right] \quad \text{for} \quad E < E_0. \tag{208}$$

The units of N_{br} are given in photons per electron per MeV. The normalization constant k is independent of the emitted photon energy E. The fraction of the incident electron's kinetic energy that is subsequently emitted as bremsstrahlung $Y(E_0)$ is found by integrating Eq. (208) with respect to emitted photon energy E to get

$$Y(E_0) = \frac{13}{16}kZE_0. \tag{209}$$

Solving for normalization constant k in the expression for $Y(E_0)$ yields

$$k = \frac{16}{13}\frac{Y(E_0)}{ZE_0}. \tag{210}$$

Substitution of the k expression in Eq. 210 into Eq. 208 for the bremsstrahlung photon energy distribution N_{br} gives

$$N_{br}(E_0, E) = \left(\frac{32Y(E_0)}{13E_0}\right)\left[\left(\frac{E_0}{E} - 1\right) - \frac{3}{4}LN\left(\frac{E_0}{E}\right)\right] \quad \text{for} \quad E < E_0. \tag{211}$$

The fraction $Y(E_0)$ of incident electron kinetic energy emitted as bremsstrahlung photons asymptotically increases from 0 to 1 as E_0 increases relative to the emitted photon energy E.

17. FORWARD-DIRECTED BREMSSTRAHLUNG FROM RELATIVISTIC ELECTRONS UNDERGOING ACCELERATION

Forward-directed electromagnetic radiation is emitted from a relativistic electron when the electron (1) changes direction by passing through a bending magnet or (2) changes (decreases in this case) speed and changes direction when striking a tungsten target. That is, the radiation is emitted primarily in the direction of the moving electron in which the electron has acceleration components perpendicular to the electron dominant velocity component. Electromagnetic radiation is emitted in all directions from an oscillating electron except in the direction of electron oscillation. The forward-directed electromagnetic radiation from the accelerating electron is a result of the Lorentz transformation, in which distances in the moving frame appear contracted or shrunk and time appears to the stationary observer in the laboratory frame to run slower. Polar angle θ in the frame of a moving electron is measured relative to the z-axis, which points in the direction of electron propagation. The z-axis in the stationary laboratory is coincident with the z-axis in the frame of the moving electron.

Consider a propagating wave in the two frames of reference in which the wave amplitude varies via its phase factor

$$\exp(i\phi) = \exp[i(wt - \mathbf{k}\cdot\mathbf{r})]. \tag{212}$$

The electron rest frame F' moves at velocity v along the positive z-direction relative to frame F of the stationary laboratory observer. The phase of the electromagnetic wave that is emitted from the moving electron must be the same in both frames F and F', where

$$\phi = \phi'. \tag{213}$$

That is, the relative amplitude (peak or trough) must be the same in both frames to have physical laws identically obeyed in both frames. From the exponent arguments in Eq. 212, substitution for ϕ and ϕ' in Eq. 213 gives the phase-invariant equation

$$wt - k_x x - k_y y - k_z z = w't' - k_x' x' - k_y' y' - k_z' z'. \tag{214}$$

The Lorentz transformation for time t and the z position in stationary frame F, and time t' and the z' position in moving frame F', are

$$t' = \gamma\left(t - \beta\frac{z}{c}\right) \tag{215}$$

and

$$z' = \gamma(z - \beta ct), \tag{216}$$

where

$$\beta = \frac{v}{c}, \tag{217}$$

Substitution for t' of Eq. 215 and z' of Eq. 216 in the phase-invariant expression of Eq. 214 gives

$$wt - k_x x - k_y y - k_z z = w'\gamma\left(t - \beta\frac{z}{c}\right) - k_x'x' - k_y'y' - k_z'\gamma(z - \beta ct). \tag{218}$$

The terms are regrouped to obtain

$$\begin{aligned} wt - k_x x - k_y y - k_z z &= t(w'\gamma + \beta ck_z'\gamma) \\ &\quad - z\left(w'\gamma\frac{\beta}{c} + k_z'\gamma\right) - k_x'x' - k_y'y'. \end{aligned} \tag{219}$$

Since this expression is for any arbitrary space-time coordinate, one can match the coefficients of each of the terms so that

$$w = \gamma(w' + \beta ck_z') \tag{220}$$

$$k_z = \gamma\left(w'\frac{\beta}{c} + k_z'\right) \tag{221}$$

$$k_x = k_x' \tag{222}$$

$$k_y = k_y'. \tag{223}$$

The photon or electromagnetic wave emitted from the accelerated electron has frequency w and wave vector $\mathbf{k}$, and propagates at speed c along the photon trajectory line that is oriented at angle θ relative to the z-direction of the electron propagation. In the lab frame, the transverse y-directed component k_y and axial z-directed component k_z of the wave vector $\mathbf{k}$ are

$$k_y = |\mathbf{k}| \sin\theta \tag{224}$$

and

$$k_z = |\mathbf{k}| \cos\theta\,, \tag{225}$$

where

$$w = ck = c\,|\mathbf{k}|\ . \tag{226}$$

Solving for $\sin\theta$ and $\cos\theta$, one obtains

$$\sin\theta = \frac{k_y}{|\mathbf{k}|} = c\frac{k_y}{w} \tag{227}$$

and

$$\cos\theta = \frac{k_z}{|\mathbf{k}|} = c\frac{k_z}{w}. \tag{228}$$

Substitution into Eqs. 227 and 228 of the moving F' frame expressions for k_y of Eq. 223, k_z of Eq. 221, and w of Eq. 220 yields

$$\begin{aligned}\sin\theta &= c\frac{k'_y}{\gamma\left(w' + \beta c k'_z\right)} = c\frac{k'\sin\theta'}{\gamma w'\left(1 + \beta c\frac{k'_z}{w'}\right)}\\ &= \frac{\sin\theta'}{\gamma\left(1 + \beta\frac{k'_z}{k'}\right)} = \frac{\sin\theta'}{\gamma(1 + \beta\cos\theta')}\end{aligned} \tag{229}$$

and

$$\cos\theta = c\frac{\gamma\left(w'\frac{\beta}{c} + k'_z\right)}{\gamma\left(w' + \beta c k'_z\right)} = \frac{k'\beta + k'_z}{k' + \beta k'_z} = \frac{\beta + \cos\theta'}{1 + \beta\cos\theta'}. \tag{230}$$

From these two equations, one obtains

$$\tan\theta = \frac{\sin\theta}{\cos\theta} = \frac{\sin\theta'}{\gamma(\beta + \cos\theta')} \tag{231}$$

If the electron is highly relativistic, where $\gamma \gg 1$ and $\beta \cong 1$, then for small photon emission angles θ' in the moving electron frame, the corresponding photon emission angle θ in the laboratory frame is also small. However, for large photon emission angles θ' in the electron frame, the corresponding photon emission angle θ in the laboratory frame is also small. Hence, the entire range of photon emission angles θ' in the electron frame is emitted in a forward-directed cone in the laboratory frame, in which the cone half-angle θ is

$$\tan\theta \cong \theta = \frac{1}{2\gamma}. \tag{232}$$

For relativistic electrons with incident energy E_0[MeV], the tungsten bremsstrahlung target is a strong source of forward-directed photons emitted primarily in a full cone angle $\theta[rad]$ given by

$$\theta[rad] = \frac{0.511}{E_0} \tag{233}$$

For example, 30-MeV electrons the bremsstrahlung is emitted within a very forward-directed cone with full cone angle $\theta = 0.017$ rad.

18. X-RAY WAVE PACKETS—PHASE VELOCITY AND GROUP VELOCITY, SPATIAL AND SPECTRAL WIDTHS

The *phase velocity* is defined as the velocity for a single-wavelength wave, whereas the group velocity is defined as the velocity for a packet of waves in which the waves vary in wavelength. First, the phase velocity v_p is calculated. If $\beta \ll \delta$, one can then approximate

$$k = (1 - \delta)\frac{w}{c} \quad \beta \ll \delta. \tag{234}$$

The X-ray phase velocity v_p is then real and given by

$$v_p = \frac{w}{k} = \frac{c}{n} = \frac{c}{1 - \delta} \tag{235}$$

The phase velocity v_p can be less than the speed of light. It can also be greater than the speed of light, but this does not violate special relativity because the single monochromatic wave or constant oscillation does not carry energy or information.

The group velocity for X-rays represents the velocity of a group of waves in which each wave has a different phase velocity v_p. That is, the group velocity v_g represents a modulation of a wave train packet, and as such, the wave packet carries information and energy and therefore must be less than the speed of light c, where

$$v_g \leq c. \tag{236}$$

A material is dispersive because waves of different wavelengths have different phase velocities; consequently, the individual waves of the wave packet travel at different speeds and change phase with each other as the wave packet propagates in time and space, and the wave packet can spread in space as it propagates. In a dispersive medium, the velocity of the energy flow may differ greatly from the phase velocity of the individual waves. Also in a dispersive medium, a pulse of X-rays will attenuate as it propagates, and the wave packet will change its pulse shape if the attenuation of the individual waves varies due to wavelength.

A simplistic introduction to group velocity is the summation of two individual sinusoids:

$$\psi(r,t) = \psi_1(r,t) + \psi_2(r,t), \tag{237}$$

where

$$\psi_1(r,t) = \psi_0 \sin(j2\pi(kr - wt)) \tag{238}$$

and

$$\psi_2(r,t) = \psi_0 \sin[j2\pi\left(r\left(k + \Delta k\right) - t\left(w + \Delta w\right)\right)]. \tag{239}$$

Assume

$$\Delta k \ll k \quad \text{and} \quad \Delta w \ll w. \tag{240}$$

The expressions in Eq. 238 and Eq. 239 are plugged into Eq. 237, and to the resulting sum-of-sines expression, one applies the trigonometric identity

$$\sin A + \sin B = 2\sin\left(\frac{A+B}{2}\right)\cos\left(\frac{A-B}{2}\right) \tag{241}$$

This yields the sinusoid

$$\psi(r,t) = 2\cos\left[j2\pi\left(r\frac{dk}{2} - t\frac{dw}{2}\right)\right]\sin\left[j2\pi(kr - wt)\right]. \tag{242}$$

The amplitude modulation of this wave form propagates at the group velocity v_g as follows:

$$v_g = \frac{r}{t} = \frac{dw}{dk}. \tag{243}$$

The center carrier wave propagates at phase velocity v_p:

$$v_p = \frac{r}{t} = \frac{w}{k}. \tag{244}$$

The wave packet of X-rays has a spatial pulse width Δz along the direction of propagation, which has a length inversely proportional to the packet's wave number width Δk, where $\Delta z \propto 1/\Delta k$, which is shown as follows. A packet of waves can be described by an integration or summation of the monochromatic waves, each with wave number k, angular frequency $w(k)$, which is dependent on k, and amplitude $E_0(k)$, in which each wave amplitude $E(k)$ is calculated as a function of its spatial position z along the z-axis and time t so that

$$E(k) = E_0(k)\exp(ikz - iwt). \quad (245)$$

The amplitude $u(z,t)$ of the wave packet as a function of its spatial position z along the z-axis, time t is the summation of waves of varying wave number $k = 2\pi/\lambda$, and the corresponding angular frequency $w = 2\pi f$, where

$$u(z,t) = \frac{1}{\sqrt{2\pi}} \int_{-\infty}^{\infty} E_o(k)\exp\left[i\left(kz - wt\right)\right]dk, \quad (246)$$

where $1/\sqrt{2\pi}$ allows the wave number amplitude $E_0(k)$ to represent the Fourier transform of the spatial amplitude $u(z,t)$ at $t = 0$ [namely $u(z,0)$], so that

$$E_0(k) = \frac{1}{\sqrt{2\pi}} \int_{-\infty}^{\infty} u(z,0)\exp(ikz)dk. \quad (247)$$

Suppose that $u\,(z,0)$ represents a wave of a single wavelength, λ_o, for all z, where

$$u(z,0) = \exp(ik_0z) = \exp(i\frac{2\pi}{\lambda_0}z). \quad (248)$$

Then

$$E_0(k) = \sqrt{2\pi}\,\delta(k - k_0). \quad (249)$$

Let the wave frequency $w(k_0)$ give rise to a traveling wave:

$$u(z,t) = \exp(ik_0z - iwt). \quad (250)$$

However, if $t = 0$, spatial wave amplitude $u(z,0)$ represents a wave packet, which consists of waves of varying wavelengths for all z, such that $u(z,0)$ is a wave packet of spatial root-mean-square (RMS) width Δz about center position z_0, then $E_0(k)$ is not a delta function, but rather a pulse in k-space with an RMS width of Δk, about center wave number k_0. Given intensities $|u(z,0)|^2$ and $|E_0(k)|^2$, the RMS spatial width Δz and wave number or spectral Δk width of the wave are related by

$$\Delta z \Delta k \geq \frac{1}{2}. \tag{251}$$

A sharp spatial pulse and its associated broad-wavelength spectrum is mathematically inherent in Fourier transforms, which do not have a Planck's constant but do have reciprocity between their spatial intervals and spatial frequency range or between their time intervals and temporal frequency range, and are observed in non-quantum domains such as radio or radar. This reciprocity of the standard deviation of the spatial packet width Δz and the corresponding standard deviation of the packet wave number or spectral width Δk is physically realized by Planck's constant h via the Heisenberg uncertainty principle, which is applied to electromagnetic and de Broglie matter waves as well as to matter, where

$$\Delta z \Delta k \geq \frac{h}{2\pi}. \tag{252}$$

19. X-RAY WAVE PACKETS—DISPERSION AND GROUP VELOCITY, AND RELATION TO REAL PART OF COMPLEX ATOMIC SCATTER FACTOR

We now consider the dispersion of the propagating packet of X-rays. If a wave train propagates in a highly dispersive medium in which the phase velocity of a wave changes rapidly with wavelength, and/or in a weakly dispersive medium, the wave train is a sharp spatial pulse with a broad wave number spectrum, the coherence or fixed phase relationship between all the waves will be lost to randomness and the pulse shape will become distorted. It is difficult to characterize such a pulse in space and time with simplicity. If the wave train has a narrow wave number spectrum and/or travels in a weakly dispersive medium, then the pulse maintains its shape and fixed phase relationships remain between the individual monochromatic wave constituents. That is, if the wave number distribution peaks about a center wave number k_0, then the angular frequency of the wave packet can be approximated by the linear expansion

$$w(k) = w_0 + (k - k_0)\frac{dw}{dk}(k = k_0) + \cdots \tag{253}$$

Summation of the individual waves in the packet yields the amplitude $u(z,t)$ of the wave packet in space z and time t, such that

$$u(z,t) = \frac{1}{\sqrt{2\pi}} \exp(ik_0 v_g t - iw_0 t) \int_{-\infty}^{\infty} E_0(k) \exp[ik\left(z - v_g t\right)] dk, \tag{254}$$

where

$$u(z - v_g t, 0) = \frac{1}{\sqrt{2\pi}} \int_{-\infty}^{\infty} E_0(k) \exp[ik\left(z - v_g t\right)] dk. \tag{255}$$

Here, the group velocity v_g is defined by

$$v_g = \frac{dw}{dk}(k = k_0). \tag{256}$$

The wave packet can be characterized as a pulse that has spatial width Δz and wave number width Δk, such that the pulse maintains its shape and coherence between its constituent member waves as it propagates through space with a packet or group velocity v_g, where

$$u(z,t) = u(z - v_g t, 0) \exp(ik_0 v_g t - iw_0 t). \tag{257}$$

Undistorted in shape, the packet amplitude distribution $u(z,t)$ propagates in space z and time t at the group velocity v_g, such that the position of the packet center or peak amplitude $u(0,0)$ propagates along z with time t as

$$z_0 = v_g t. \tag{258}$$

The energy density associated with the squared magnitude of the amplitude $|u\,(x,t)|^2$ of a wave, which has a narrow wave number spectrum and/or travels in a weakly dispersive medium, thus travels and conveys information at the group velocity v_g given by

$$v_g = \frac{dw}{dk} \tag{259}$$

We now derive an expression for the group velocity v_g of an X-ray packet for a narrow bandwidth of X-rays or a weakly dispersive medium as follows. If the wave train has a narrow wave number spectrum and/or travels in a weakly dispersive medium, then we can calculate the relevant group velocity of the packet, defined by $v_g = dw/dk$, from the wave phase velocity v_p:

$$v_p = \frac{c}{n} = \frac{w}{k}. \tag{260}$$

Rearrangement gives

$$w = \frac{kc}{n}. \tag{261}$$

Then one obtains

$$\frac{dw}{dk} = \frac{d}{dk}\left(\frac{kc}{n}\right) = \frac{c}{n} - \frac{c}{n^2}\frac{dn}{dk}. \tag{262}$$

The chain rule yields

$$\frac{dn}{dk} = \frac{dn}{dw}\frac{dw}{dk}. \tag{263}$$

Substituting of the Eq. 263 expression for dn/dk into Eq. 262, and solving dw/dk yields

$$\frac{dw}{dk} = \frac{c}{n + \frac{c}{n}\frac{dn}{dw}}, \tag{264}$$

where

$$n = 1 - \delta. \tag{265}$$

Rearrangement yields

$$\frac{dn}{dw} = -\frac{d\delta}{dw}. \tag{266}$$

Hence, the group velocity v_g becomes

$$v_g = \frac{dw}{dk} = \frac{c}{1 - \delta - \frac{c}{1-\delta}\frac{d\delta}{dw}}. \tag{267}$$

The refractive decrement $\delta(w)$ of a material medium as a function of the angular frequency w was derived earlier in the chapter, where,

$$\delta(w) = \frac{r_e c^2 2\pi}{w^2} \sum_{q=1}^{Q} \rho_q f_{1q}^0(w). \tag{268}$$

The material medium is composed of Q elements. For each sth electron of Z electrons in each atom of the qth element, g_s is the oscillator strength, $w_{s,q}$ is the resonant frequency, and γ_q is the damping constant for all the electrons in the atom. The real part $f_{1q}^0(w)$ of the complex scattering factor $f_q^0(w)$ is

$$f_{1q}^0(w) = \sum_{s=1}^{Z} \frac{g_{s,q} w^2 \left(w^2 - w_{s,q}^2\right)}{\left(w^2 - w_{s,q}^2\right)^2 + w^2 \gamma_q^2}. \tag{269}$$

Hence, the decrement $\delta(w)$ dependence on the X-ray angular frequency w is then determined to be

$$\delta(w) = r_e c^2 2\pi \sum_{q=1}^{Q} \rho_q \sum_{s=1}^{Z} \frac{g_{s,q} \left(w^2 - w_{s,q}^2\right)}{\left(w^2 - w_{s,q}^2\right)^2 + w^2 \gamma_q^2} \tag{270}$$

The variation of the decrement $\delta(w)$ with frequency w is then determined to be

$$\frac{d\delta(w)}{dw} = -2 r_e c^2 2\pi \sum_{q=1}^{Q} \rho_q \sum_{s=1}^{Z} \frac{\left[\left(w^2 - w_{s,q}^2\right)^2 - \gamma_q^2 w_{s,q}^2\right] \cdot g_{s,q} w}{\left[\left(w^2 - w_{s,q}^2\right)^2 + w^2 \gamma_q^2\right]^2}. \tag{271}$$

One then substitutes the expression for $\delta(w)$ of Eq. 270 and $d\delta/dw$ of Eq. 271 into Eq. 267 for the X-ray group velocity v_g in the material medium of refractive index $n(w) = 1 - \delta(w)$. The X-ray group velocity is thus expressed in terms of oscillator strength g_s, resonant frequency $w_{s,q}$, and damping constant γ_q of each sth electron for the Z electrons in each qth element of the Q elements of the material medium.

20. X-RAYS IN MATERIAL MEDIUM—PHASE AND GROUP VELOCITY AND PULSE BROADENING

A medium is characterized by electric permittivity ε and magnetic permeability μ, as well as the refractive index $n = 1 - \delta$, in which the electromagnetic group velocity v through the medium is less than the speed of light c, where

$$v_p = \frac{c}{n} = \frac{1}{n\sqrt{\mu_0 \varepsilon_0}} = \frac{1}{\sqrt{\mu\varepsilon}}. \tag{272}$$

In vacuum are electric permittivity ε_0 and magnetic permeability μ_0. The phase velocity v_p of light in terms of the decrement δ of the refractive index is

$$v_p = \frac{w}{k} = \frac{c}{n} = \frac{1}{(1-\delta)\sqrt{\mu_0\varepsilon_0}} \cong \frac{1+\delta}{\sqrt{\mu_0\varepsilon_0}} = (1+\delta)c. \tag{273}$$

Transposing this equation, one solves for frequency w in terms of wave number k, where

$$k = \frac{w}{c}(1-\delta). \tag{274}$$

The decrement δ is positive for electromagnetic waves and can be expressed in terms of plasma frequency w_p, where

$$\delta = \frac{1}{2}\frac{w_p^2}{w^2} > 0, \tag{275}$$

Taking the derivative of the decrement δ with respect to angular frequency w, via an excursion in recursion, gives

$$\frac{d\delta}{dw} = -2\frac{\delta}{w}. \tag{276}$$

For electron charge e and rest mass m_e with vacuum electric permittivity ε_0, the plasma frequency w_p depends on the electron density n_e, where

$$w_p = \sqrt{\frac{e^2 n_e}{\varepsilon_0 m_e}}. \tag{277}$$

The phase velocity v_p exceeds the speed of light $v_p > c$ for the positive decrements $\delta > 0$ of electromagnetic waves. However, the phase velocity

v_p of the electromagnetic wave is neither the speed of information (wave modulation) nor the speed of energy. Instead, information and energy of the electromagnetic wave propagate at group velocity v_g, which is less than or equal to the speed of light $v_g < c$, where

$$v_g = \frac{dw}{dk} = \left(\frac{dk}{dw}\right)^{-1} = \left[\frac{1}{c}\frac{d}{dw}\left(w\left(1-\delta\right)\right)\right]^{-1}$$
$$= c\left[1-\delta-w\frac{d\delta}{dw}\right]^{-1} = \frac{c}{1+\delta} \cong c(1-\delta). \tag{278}$$

We now derive the expressions for the pulse-broadening distance Δx and time Δt for electromagnetic waves, such as X-rays, in their passage through materials. If the refractive index is a function of frequency, then pulse broadening occurs. The amount of broadening in time Δt is determined by group velocity variation v_g over the pulse bandwidth, where

$$\frac{dv_g}{dw} = \frac{d^2w}{dk^2}\frac{dk}{dw}. \tag{279}$$

One approximates the phase velocity v_p by the speed of light c,

$$v_p = \frac{w}{k} \cong c. \tag{280}$$

With this approximation of the phase velocity v_p, the group velocity v_g is then approximated by the speed of light c, where

$$v_g = \frac{dw}{dk} = c. \tag{281}$$

Given the thickness L of a material or X-ray optic, such as a compound refractive lens, and given the bandwidth Δw of the pulse, the pulse broadens by a distance Δx, where

$$\Delta x = \frac{L}{c}\frac{dv_g}{dw}\Delta w. \tag{282}$$

The pulse broadens in time Δt given by

$$\Delta t = \frac{\Delta x}{c}. \tag{283}$$

Substitution in Eq. 282 for Δx from Eq. 283, dv_g/dw from Eq. 279, and dk/dw from the reciprocal of the expression in Eq. 278, where tiny δ is ignored, yields

$$\Delta t = \frac{L}{c^2}\frac{dv_g}{dw}\Delta w = \frac{L}{c^2}\frac{d^2w}{dk^2}\frac{dk}{dw}\Delta w = \frac{L}{c^3}\frac{d^2w}{dk^2}\Delta w. \tag{284}$$

One evaluates d^2w/dk^2 via substitution for $d\delta/dw$ of Eq. 276 and dw/dk of Eq. 278, where

$$\frac{d^2w}{dk^2} = \frac{d}{dk}\frac{dw}{dk} = \frac{d}{dk}c(1-\delta) = -c\frac{d\delta}{dk} = -c\frac{d\delta}{dw}\frac{dw}{dk} \cong 2c^2\frac{\delta}{w}. \tag{285}$$

The expression for the pulse broadening time Δt becomes

$$\Delta t = \frac{L}{c^3}\frac{d^2w}{dk^2}\Delta w = 2\delta\frac{L}{c}\frac{\Delta w}{w}. \tag{286}$$

Consider the case of pulse broadening of near-monochromatic X-rays $\Delta w/w = 10^{-4}$ in a 30-cm-long compound refractive lens with material decrement $\delta = 10^{-5}$. The maximum pulse broadening of $\Delta t = 10^{-18}$ sec is negligible.

21. REFLECTION AND TRANSMISSION OF X-RAYS AT A PLANAR INTERFACE OF TWO MEDIA

To determine surface boundary conditions, a rectangle contour is formed with a long side above the surface and the other long side below the surface, with both long sides parallel to the surface. The two very short sides of the rectangle are perpendicular to the surface. By arbitrarily making the two short sides approach zero so that the top and bottom long sides of the rectangle merge with the surface, the area of the rectangle goes to zero. However, a finite surface current can exist on the surface even as the rectangle area approaches zero. However, the time-varying finite flux density of a magnetic field multiplied by the rectangular area that is approaching zero as the two short sides go to zero yields a magnetic flux that approaches zero.

Ampere's law gives the boundary condition for the side-1 and side-2 tangential components H_{t1} and H_{t2} of the magnetizing field or magnetic field intensity $\mathbf{H}$ at the boundary surface. The tangential component boundary condition is stated in terms of the unit normal vector to the surface $\hat{\mathbf{n}}$ and the free surface current density $\mathbf{J}_f[A/m]$ on the boundary, where $\mathbf{J}_f \perp \hat{\mathbf{n}}$, and

$$\nabla \times \mathbf{H} = \mathbf{J}_f \Rightarrow \hat{\mathbf{n}} \times (\mathbf{H}_1 - \mathbf{H}_2) = \mathbf{J}_f. \tag{287}$$

Assuming there is no surface current $\mathbf{J}_f = 0$, then the tangential component of the magnetic field is continuous across the interface, where

$$H_{t1} = H_{t2}. \tag{288}$$

Faraday's law gives the boundary condition for the side-1 and side-2 tangential components E_{t1} and E_{t2} of the electric field $\mathbf{E}$ at the boundary surface. The tangential component boundary condition is stated in terms of the unit normal vector to the surface $\hat{\mathbf{n}}$. The magnetic field flux approaches zero because the rectangular area approaches zero as the rectangle top and bottom long sides merge with the surface, so

$$\nabla \times \mathbf{E} = -\frac{d\mathbf{B}}{dt} \Rightarrow \hat{\mathbf{n}} \times (\mathbf{E}_1 - \mathbf{E}_2) = 0. \tag{289}$$

Hence, the tangential component of the electric field is continuous across the interface, where

$$E_{t1} = E_{t2}. \tag{290}$$

The electric $\mathbf{E}$ and magnetic field vector $\mathbf{B}$ of a plane electromagnetic wave are perpendicular to each other. And the electric $\mathbf{E}$ and magnetic field vector $\mathbf{B}$ of a plane wave are each perpendicular to the direction of the plane wave vector $\mathbf{k}$ that point in direction of wave propagation. The wave number k is the magnitude of the wave vector $\mathbf{k}$. The direction electric field vector is used to denote the polarization of the electromagnetic plane wave. Consider an incident electromagnetic plane wave whose angle of incident θ_i relative to an interface is measured relative to the interface normal unit vector $\hat{\mathbf{n}}$. A portion of the wave is reflected at angle θ_r and a portion of the wave is refracted or transmitted at angle θ_t, both measured relative to the surface normal unit vector $\hat{\mathbf{n}}$. The tangential components of the electric and magnetic fields are continuous.

The magnetic field intensity $\mathbf{H}$(amp/m) and the electric field intensity $\mathbf{E}$(volt/m) are related by the intrinsic impedance, η(ohm), where for propagation in the positive z-direction,

$$E_x(z,t) = \eta H_y(z,t) \tag{291}$$

$$E_y(z,t) = -\eta H_x(z,t). \tag{292}$$

Given permeability μ and permittivity ε, the intrinsic impedance is

$$\eta = \sqrt{\frac{\mu}{\varepsilon}}. \tag{293}$$

In the case of the incident, reflected, and refracted plane electromagnetic waves, the perpendicular electric and magnetic fields are related by the intrinsic impedance η, where

$$E_i = \eta_i H_i \tag{294}$$

$$E_r = \eta_r H_r \tag{295}$$

$$E_t = \eta_t H_t. \tag{296}$$

The incidence and reflection media are the same, but they are different from the transmitted medium, where

$$\eta_i = \sqrt{\frac{\mu_i}{\varepsilon_i}} = \eta_r \tag{297}$$

$$\eta_t = \sqrt{\frac{\mu_t}{\varepsilon_t}}. \tag{298}$$

One assumes that the electromagnetic X-rays are plane harmonic waves in which (1) the incident wave $\mathbf{E}_i$ is specified by vector amplitude $\mathbf{E}_{0i}$ and phase is specified via frequency w_i and wave vector $\mathbf{k}_i$; (2) the reflected wave $\mathbf{E}_r$ is specified by vector amplitude $\mathbf{E}_{0r}$ and phase via frequency w_r and wave vector $\mathbf{k}_r$; and (3) the refracted or transmitted wave $\mathbf{E}_t$ is specified by vector amplitude $\mathbf{E}_{0t}$ and phase via frequency w_t and wave vector $\mathbf{k}_t$, where

$$\mathbf{E}_i = \mathbf{E}_{0i} \exp[i(\mathbf{k}_i \cdot \mathbf{r} - wt)] \tag{299}$$

$$\mathbf{E}_r = \mathbf{E}_{0r} \exp[i(\mathbf{k}_r \cdot \mathbf{r} - wt)] \tag{300}$$

$$\mathbf{E}_t = \mathbf{E}_{0t} \exp[i(\mathbf{k}_t \cdot \mathbf{r} - wt)]. \tag{301}$$

The incident, reflected, and refracted waves share the interface boundary at $\mathbf{r} = 0$, and all three waves exist simultaneously and exit at the same time t. The amplitudes $\mathbf{E}_{0i}$, $\mathbf{E}_{0r}$, and $\mathbf{E}_{0t}$ are expected to have a fixed relation, and for arbitrary boundary position $\mathbf{r}$ and time t, the phase terms for the incident, reflected, and refracted waves, which are dependent on $\mathbf{r}$ and t, should thus be equal in the boundary region of simultaneous occupation, where

$$\mathbf{k}_i \cdot \mathbf{r} - w_i t = \mathbf{k}_r \cdot \mathbf{r} - w_r t = \mathbf{k}_t \cdot \mathbf{r} - w_t t, \tag{302}$$

At the boundary point $\mathbf{r} = 0$ at time t, one finds

$$w_i t = w_r t = w_t t. \tag{303}$$

Hence, all frequencies are equal:

$$w_i = w_r = w_t = w. \tag{304}$$

Now consider time $t = 0$ points at vector position $\mathbf{r}$ within the boundary xy-plane at $z = 0$:

$$\mathbf{r} = x\hat{\mathbf{x}} + y\hat{\mathbf{y}}. \tag{305}$$

One finds

$$\mathbf{k}_i \cdot \mathbf{r} = \mathbf{k}_r \cdot \mathbf{r} = \mathbf{k}_t \cdot \mathbf{r}. \tag{306}$$

The grazing angles ϕ_i, ϕ_r, and ϕ_t are measured relative to the surface at position $\mathbf{r}$, and the dot product of vector position $\mathbf{r}$ in the surface plane with the incident $\mathbf{k}_i$, reflected $\mathbf{k}_r$, and refracted $\mathbf{k}_t$ wave vectors, which propagate in a plane perpendicular to the surface plane, gives

$$k_i r \cos\varphi_i = k_r r \cos\varphi_r = k_t r \cos\varphi_t. \tag{307}$$

The grazing angles ϕ_i, ϕ_r, and ϕ_t, measured relative to the surface tangent plane, are related to the normal-based angles θ_i, θ_r, and θ_t via

$$\theta_i = \frac{\pi}{2} - \phi_i \tag{308}$$

$$\theta_r = \frac{\pi}{2} - \phi_r \tag{309}$$

$$\theta_t = \frac{\pi}{2} - \phi_t. \tag{310}$$

This dot product relation is expressed equivalently in terms of normal-based angles:

$$k_i r \sin\theta_i = k_r r \sin\theta_r = k_t r \sin\theta_t. \tag{311}$$

The incident and reflected waves are in the same medium, and thus the magnitude of their wave vectors or wave numbers are equal, where

$$k_i = k_r = k. \tag{312}$$

Hence,

$$\sin\theta_i = \sin\theta_r = \frac{k_t}{k}\sin\theta_t. \tag{313}$$

One thus obtains the law of reflection, and the incident θ_i angle equals the reflected θ_r angle:

$$\theta_i = \theta_r. \tag{314}$$

The incident wave speed v_i and refracted (transmitted) wave speed v_t are related to the index of refraction of the medium of incidence and reflection n_i and the medium of the refracted (transmitted) wave n_t via common frequency w and light speed in vacuum c, where

$$k_i = \frac{w}{v_{i\,i}} = n_i \frac{w}{c} \tag{315}$$

$$k_t = \frac{w}{v_{t\,t}} = n_t \frac{w}{c}. \tag{316}$$

Substitution of Eqs. 315 and 316 into Eq. 311 gives Snell's law:

$$\sin\theta_i = \frac{k_t}{k_i}\sin\theta_t = \frac{n_t}{n_i}\sin\theta_t. \tag{317}$$

22. FRACTION OF TRANSMITTED AND REFLECTED POLARIZED X-RAYS WITH AN ELECTRIC FIELD COMPONENT PARALLEL TO THE PLANAR SURFACE INTERFACE

The transmission and reflection is determined for a polarized X-ray in which the electric field vector component is parallel to the planar surface interface.

The incident tangential electric E_i and magnetic H_i field components, reflected tangential electric E_r and magnetic H_r field components, and transmitted tangential electric E_t and magnetic H_t field components are continuous across the interface. The boundary conditions of tangential components of electric and magnetic fields for the case of the electric field parallel to the surface are

$$E_i + E_r = E_t \tag{318}$$

and

$$(H_i - H_r)\cos\theta_i = H_t\cos\theta_t. \tag{319}$$

The magnetic field lies in the plane of incidence. The plane of incidence is perpendicular to the surface interface, but in general, the magnetic field is not perpendicular to the surface and thus has components parallel to the surface via the cosine terms. The expressions for H_i, H_r, and H_t from Eqs. 294–296 are substituted into Eq. 319 for the tangential magnetic field continuity, and using $\eta_i = \eta_r$ produces the resulting expression for the tangential magnetic field continuity

$$\left(\frac{E_i}{\eta_i} - \frac{E_r}{\eta_i}\right) \cos\theta_i = \frac{E_t}{\eta_t} \cos\theta_t. \tag{320}$$

The equation for the continuity of the tangential magnetic field becomes

$$\frac{\eta_t}{\eta_i}\left(1 - \frac{E_r}{E_i}\right) \frac{\cos\theta_i}{\cos\theta_t} = \frac{E_t}{E_i}. \tag{321}$$

Rearrangement of the equation for continuity of the tangential electric field gives

$$1 + \frac{E_r}{E_i} = \frac{E_t}{E_i}. \tag{322}$$

The two expressions for E_t/E_i are equated, and one solves for the reflection, refraction, or reflectivity E_r/E_i, or $R_{E||}$, for the electric field parallel_to the surface, where

$$R_{E||} = \frac{E_r}{E_i} = \frac{\eta_t \cos\theta_i - \eta_i \cos\theta_t}{\eta_t \cos\theta_i + \eta_i \cos\theta_t}. \tag{323}$$

Substitution of the solved expression for reflectivity E_r/E_i of Eq. 323 into Eq. 322 yields the transmission fraction or transmittance E_r/E_i, or $T_{E||}$, for the electric field parallel to the surface interface, where

$$T_{E||} = \frac{E_t}{E_i} = \frac{2\eta_t \cos\theta_i}{\eta_t \cos\theta_i + \eta_i \cos\theta_t}. \tag{324}$$

The reflectivity $R_{E||}$ for the electric field parallel to the surface interface cannot be zero:

$$R_{E||} \neq 0. \tag{325}$$

If $R_{E||}$ were zero, then

$$\eta_t \cos\theta_i - \eta_i \cos\theta_t = 0 \Rightarrow \frac{\eta_t}{\eta_i} = \frac{\cos\theta_t}{\cos\theta_i}. \tag{326}$$

However, this result is in violation of Snell's law, given here:

$$\frac{n_t}{n_i} = \frac{\sin\theta_i}{\sin\theta_t}. \tag{327}$$

The consequences of violating Snell's law is demonstrated by equating the two expressions for n_t/n_i, where

$$\sin\theta_i \cos\theta_i \neq \sin\theta_t \cos\theta_t. \tag{328}$$

In general, the incidence angle θ_i is not equal to the transmitted angle θ_t, except in normal incidence at $\theta_i = 0$ and parallel incidence $\theta_i = \pi/2$. Thus, $\theta_i \neq \theta_t$, so that

$$\sin 2\theta_i \neq \sin 2\theta_t. \tag{329}$$

Thus, for the case of the electric field parallel to the surface with its associated perpendicular magnetic field in the plane of incidence, there is always some reflection and transmission except for incidence parallel to the surface $\theta_i = \pi/2$.

23. FRACTION OF TRANSMITTED AND REFLECTED POLARIZED X-RAYS WITH A MAGNETIC FIELD COMPONENT PARALLEL TO THE PLANAR SURFACE INTERFACE

The transmission and reflection is now calculated for a polarized X-ray in which its magnetic field vector component is parallel to the planar surface interface.

If the direction of the magnetic field vector in the plane wave is parallel to the surface, then the boundary conditions of tangential electric and magnetic fields are

$$H_i + H_r = H_t \tag{330}$$

and

$$(E_i - E_r)\cos\theta_i = E_t \cos\theta_t. \tag{331}$$

In this case, the electric field is in the plane of incidence; and in general, the electric field is not perpendicular to the surface and thus has components parallel to the surface via the cosine terms. The expressions for H_i, H_r, and H_t from Eqs. 294–296 are substituted into Eq. 330 for the tangential magnetic field continuity to yield the tangential magnetic field continuity expression

$$\frac{E_i}{\eta_i} + \frac{E_r}{\eta_i} = \frac{E_t}{\eta_t}. \tag{332}$$

The tangential magnetic field continuity equation becomes

$$\frac{\eta_t}{\eta_i}\left(1 + \frac{E_r}{E_i}\right) = \frac{E_t}{E_i}. \tag{333}$$

Rearrangement of the tangential electric field continuity equation gives

$$\left(1 - \frac{E_r}{E_i}\right)\frac{\cos\theta_i}{\cos\theta_t} = \frac{E_t}{E_i}. \tag{334}$$

The two expressions for E_t/E_i are equated, and one solves for the reflection, refraction, or reflectivity E_r/E_i, or $R_{B||}$, for the magnetic field parallel to the surface, where

$$R_{B||} = \frac{E_r}{E_i} = \frac{\eta_i\cos\theta_i - \eta_t\cos\theta_t}{\eta_i\cos\theta_i + \eta_t\cos\theta_t}. \tag{335}$$

Substitution of the solved expression for reflectivity E_r/E_i of Eq. 335 in the tangential electric field continuity expression of Eq. 334 yields the transmission fraction or transmittance E_r/E_i or $T_{B||}$ for the magnetic field parallel to the surface interface, where

$$T_{B||} = \frac{E_t}{E_i} = \frac{2\eta_t\cos\theta_i}{\eta_i\cos\theta_i + \eta_t\cos\theta_t}. \tag{336}$$

24. X-RAY BREWSTER ANGLE OCCURS ONLY FOR ELECTRIC FIELD PARALLEL TO PLANE OF INCIDENCE

Consider the case of the magnetic field parallel to the surface, where

$$\eta_i\cos\theta_i = \eta_t\cos\theta_t \Rightarrow R_{B||} = 0. \tag{337}$$

The reflectivity $R_{B||}$ for the parallel magnetic field is zero if this condition is met, which as is shown, is the Brewster angle condition for zero reflection of the magnetic field parallel to the surface and its associated perpendicular electric field in the plane of incidence. Thus, only the portion of the electromagnetic wave with the electric field parallel to the surface with its associated magnetic field, which is generally not perpendicular to the surface, is reflected when the Brewster angle condition is met. The Brewster angle for reflection is solved as follows. One invokes Snell's law:

$$n_i \sin\theta_i = n_t \sin\theta_t. \tag{338}$$

Dividing Snell's law of Eq. 338 by the boundary condition in Eq. 337 for continuity of the plane wave's magnetic field component parallel to a surface, which separates two different materials, gives

$$\tan\theta_i = \tan\theta_t. \tag{339}$$

The incidence angle θ_i does not equal the refracted angle θ_t, except for the trivial case of identical media $n_i = n_t$, or the nontrivial case in which the incident angle is at the Brewster angle θ_b:

$$\theta_i + \theta_t = \frac{\pi}{2} \quad \theta_i = \theta_b. \tag{340}$$

Into Snell's law, one inserts the following:

$$\theta_t = \frac{\pi}{2} - \theta_i \quad \theta_i = \theta_b. \tag{341}$$

One then obtains the Brewster incidence angle $\theta_i = \theta_b$ for complete polarized reflection. The Brewster incidence angle θ_b is the inverse tangent of the ratio of the refractive index n_t of the refractive (transmitting) medium, divided by the refractive index of the incident medium n_i, where

$$\tan\theta_b = \frac{n_t}{n_i}. \tag{342}$$

Bragg reflection at one polarization and not the other is explained as follows. One assumes that reflected X-rays are actually X-rays emitted from oscillating electrons driven by the incident X-ray electric field. At the planar interface, electrons are oscillated by the incident electric field of the X-ray in the direction of the incident electric field. If the electric field is in the plane of incidence, the electrons are oscillated in the plane of incidence. At the Brewster angle, the incident electric field vector points in the direction of the angle of reflection. The electrons oscillating in a direction along the line of reflection emit no X-ray electromagnetic field in the direction of their oscillation. An X-ray, with its magnetic field polarized parallel to the planar surface interface and its electric field polarized in the plane of incidence and tilted at the angle of reflection only for the Brewster angle, then reflects no X-rays because the electron oscillation is along the reflection angle. The reflection angle of the reflected X-ray wave vector is equal to its angle of the incident for all incidence angles. At the Brewster incidence angle, the direction of refraction and reflection is 90°, and the incident electric field oscillates electrons along the reflection angle. With a Brewster incident angle, if the observer looks at the corresponding equal reflection angle, there are no reflected X-rays.

In the case of the electric field parallel to the surface interface, the electrons oscillate parallel to the electric field. As before, electric fields are emitted from the oscillating electrons in all directions except the direction of oscillation. Hence, for every incidence angle, the corresponding equal reflection angle receives reflected X-rays that are produced by X-ray emission from the oscillating interface electrons driven by the incident X-ray electric field. Hence, there is always reflection of X-rays at all incident angles in the case of the electric field parallel to the interface and its associated magnetic field in the plane of incidence.

REFERENCES

Als-Nielsen, J., & McMorrow, D. (2011). *Elements of Modern X-ray Physics*. John Wiley & Sons, West Suffix, UK.

Attwood, D. (1999). *Soft X-rays and Extreme Ultraviolet Radiation*. (Chapter 2). Cambridge University Press, Cambridge, UK.

Eyges, L. (1972). *The Classical Electromagnetic Field*. (Chapter 14). Dover, New York.

Feynman, R. P. (1964). *The Feynman Lectures on Physics. Vol. II: Mainly Electromagnetism and Matter*. (Chapter 21). Addison Wesley, Reading, MA.

Henke, B. L., Gullickson, E. M., & Davis, J. C. (1993). X-ray interactions: Photoabsorption, scattering, transmission, and reflection. *Atomic and Nuclear Data, 54*, 181–342.

Hubbell, J. H., Veigele, W. J., Briggs, E. A., Brown, R. T., Cromer, D. T., & Howerton, R. J. (1975). Atomic form factors, incoherent scattering functions, and photon cross sections. *Journal of Physical and Chemical Reference Data, 4*, 471–616.

Jackson, J. D. (1975). *Classical Electrodynamics*. (Chapter 14). Wiley, New York.

Landau, L. M., & Lifshitz, E. M. (1975). *The Classical Theory of Fields*. (Chapter 21, 4th ed.). Pergamon Press, New York.

Marion, J. B. (1965). *Classical Electromagnetic Radiation*. (Chapter 7). Academic Press, New York.

Schwartz, M. (1972). *Principles of Electrodynamics*. (Chapter 6). Dover, New York.

Shadowitz, A. (1975). *The Electromagnetic Field*. (Chapter 19). Dover, New York.

Sommerfeld, A. (1964). *Electrodynamics*. (Chapters 29, 30). Academic Press, New York.

Stratton, J. A. (1941). *Electromagnetic Theory*. (Chapter 8). McGraw-Hill, New York.

Chapter 5

Neutron Optics

Jay Theodore Cremer, Jr.

Contents

1.	Neutron Phase and Group Velocity	561
2.	Derivation of the Schrödinger Wave Equation for Neutron Wave	566
3.	Derivation of the Schrödinger Wave Equation by the Electromagnetic Wave Equation Analogy	568
4.	Derivation of the Schrödinger Wave Equation by Assumption of the Neutron Plane Wave Solution	572
5.	Operator form of the Schrödinger Wave Equation for a Neutron Wave	574
6.	Reflection and Transmission of Neutrons at Media Interfaces	576
7.	Neutron Reflectometry	582
8.	Measurement of the Complex Refractive Index via Refraction	588
9.	X-ray and Neutron Interferometry	590
10.	Interferometry and the Fizeau Effect	592
11.	Pendellösung Oscillations and Anomalous Absorption in Perfect Crystals	596
12.	Measurement of Refractive Index via Interferometry	601
13.	The Maxwell–Boltzmann Distribution for the Reactor Source of Thermal Neutrons	603
	References	606

1. NEUTRON PHASE AND GROUP VELOCITY

In the particle view, the neutron energy E is related to its rest mass m_0 and momentum p by the Einstein relation

$$E = \gamma m_0 c^2. \tag{1}$$

Advances in Imaging and Electron Physics, Volume 172, ISSN 1076-5670, DOI: 10.1016/B978-0-12-394422-1.00005-0.
Copyright © 2012 Elsevier Inc. All rights reserved.

The velocity-dependent relativistic gamma factor γ is related to the neutron velocity v and the speed of light c by

$$\gamma = \left(1 - \frac{v^2}{c^2}\right)^{-1/2}. \tag{2}$$

The energy E of the neutron is related to the neutron momentum p by

$$E^2 = \left(m_0 c^2\right)^2 + p^2. \tag{3}$$

Also, the non-relativistic thermal neutron kinetic energy K is related to its velocity v, where the neutron mass m at velocity v is approximated by its rest mass $m \cong m_0$, where

$$K = \frac{mv^2}{2}. \tag{4}$$

The neutron momentum p is then

$$p = mv. \tag{5}$$

The non-relativistic kinetic energy K and momentum p are related by the neutron mass $m \cong m_0$:

$$K = \frac{p^2}{2m}. \tag{6}$$

A photon has rest mass m_0 because at the speed of light c, the photon would have infinite total energy E with a finite rest mass. Hence, the total energy E of a photon is related to its momentum p by

$$E = pc. \tag{7}$$

In a vacuum, the photon frequency f is related to its speed of light velocity c by the photon wavelength λ with the following equation:

$$f = \frac{c}{\lambda}. \tag{8}$$

The photon energy E is related to its wavelength λ by

$$E = \frac{hc}{\lambda}. \tag{9}$$

The Einstein photon energy cp is equated to the Planck photon energy expression hf, and one applies the wave relation $c = f\lambda$, where

$$E = cp \quad \text{and} \quad E = hf. \tag{10}$$

Equating the two expressions of Eq. 10 yields the de Broglie relation

$$p = \frac{h}{\lambda}. \tag{11}$$

Based on the work of Planck and Einstein, de Broglie was able to produce a wave representation of matter and energy. In the wave representation of matter, the de Broglie hypothesis relates to the momentum p of the neutron to the center wavelength λ of a localized wave packet. The neutron momentum p is related to the wave number $k = 2\pi/\lambda$ of the center wavelength of the representative wave packet by

$$p = \frac{h}{\lambda} = \hbar k. \tag{12}$$

The neutron momentum is the product of the neutron rest mass m_0 and velocity v, where

$$p = m_0 v. \tag{13}$$

The neutron wavelength λ is then inversely proportional to its group or particle velocity v as follows:

$$\lambda = \frac{h}{m_0 v}. \tag{14}$$

Alternatively, the neutron particle velocity v is related to the neutron wave number k by

$$v = \frac{\hbar k}{m_0}. \tag{15}$$

Given wavelength $\lambda\,[A]$, the neutron particle velocity in $v\,(\text{km/sec})$ is

$$v\,(\text{km/s}) = \frac{3.956}{\lambda\,[A]}. \tag{16}$$

In reciprocal space (K-space), the neutron energy E is

$$E = \frac{p^2}{2m_0} = \frac{\hbar^2 k^2}{2m_0}. \tag{17}$$

In direct space (R-space), the neutron energy E is

$$E = \frac{h^2}{2m_0\lambda^2}. \tag{18}$$

The neutron wave phase velocity, in terms of angular frequency $w/E = \hbar$ and wave number k, is

$$v_p = \frac{w}{k} = \frac{w}{E}\frac{E}{k} = \frac{1}{\hbar}\frac{1}{k}\frac{\hbar^2 k^2}{2m_0} = \frac{\hbar k}{2m_0}. \tag{19}$$

At non-relativistic speeds, the neutron wave phase velocity v_p of an unconfined neutron is half the neutron particle velocity v in a vacuum:

$$v_p = \frac{\hbar k}{2m_0} = \frac{v}{2}. \tag{20}$$

The neutron is confined to the region of the lens or detector, and thus the neutron is not a single sinusoid wave, but rather a superposition of waves. The amplitude modulation of the wave packet has group velocity

$$v_g = \frac{dw}{dk}. \tag{21}$$

Application of Eqs. 19 and 21 shows the neutron wave group velocity is equal to the neutron particle velocity:

$$v_g = \frac{dw}{dE}\frac{dE}{dk} = \frac{\hbar k}{m_0} = v. \tag{22}$$

The frequency distribution $w(k)$ of the wave packet is written in terms of expansion k about the packet center wave number $k = k_0$. The group velocity dw/dk is the linear term of the expansion, and it is evaluated at packet center peak $k = k_0$. Hence, rapid variation of w as a function of k invalidates the use of the group velocity v_g to represent the velocity of energy or information flow, because other terms, such as the quadratic term d^2w/dk^2, become important and the packet velocity can no longer be characterized by its center velocity, v_g. For more details, see Eisberg and Resnick (1974).

In a vacuum, which has unity refractive index $n = 1$, a thermal neutron with phase velocity $v_p = v_0$ (and group velocity $v_g = 2v_0$) travels length L_0 in time τ_0, where

$$\tau_0 = \frac{L}{v_0}. \tag{23}$$

However, in a slab with refractive index n with the same length L_0 as the vacuum, the thermal neutron wave propagates at phase velocity v_p over distance L_0 in time τ, where

$$v_p = \frac{v_0}{n} = \frac{v_0}{1-\delta} \tag{24}$$

and

$$\tau = \frac{L_0}{v_p} = \frac{nL_0}{v_0} n\tau_0 = (1-\delta)\tau_0. \tag{25}$$

For time interval τ_0, the neutron wave phase front propagates distance L_0 in a vacuum and distance L in the slab, where

$$L = v_p\tau_0 = \frac{v_0}{n}\frac{L_0}{v_0} = \frac{L_0}{n} = \frac{L_0}{1-\delta}. \tag{26}$$

A neutron wave packet with group velocity v_g is identical to the corresponding neutron particle velocity v, and they are twice the neutron wave phase velocity v_p, where

$$v = v_g = 2v_p. \tag{27}$$

For the identical time interval τ_0, a neutron wave packet with group velocity v_g or corresponding neutron particle velocity v, then travels distance L_g, where

$$L_g = v\tau_0 = v_g\tau_0 = 2v_p\frac{L_0}{v_0} = 2\frac{v_0}{n}\frac{L_0}{v_0} = 2\frac{L_0}{n} = 2L = 2\frac{L_0}{1-\delta}. \tag{28}$$

In a vacuum or material medium for the same time interval τ_0, the neutron wave packet (modulated sine wave), which is represented by total neutron wave function $\Psi(\mathbf{r},t)$ and the corresponding neutron particle, propagates at twice the velocity $v = v_g = 2v_p$ as phase velocity v_p of the individual sine wave, neutron wave function, and components $\psi_i(\mathbf{r},t)$, where $\Psi(\mathbf{r},t) = \sum_i \psi_i(\mathbf{r},t)$. Hence, in time interval τ_0, the neutron particle or equivalent wave packet $\Psi(\mathbf{r},t)$ then travels a distance $L_g = 2L$, and the individual sine wave, neutron wave function, and components $\psi_i(\mathbf{r},t)$ travel half the distance L.

2. DERIVATION OF THE SCHRÖDINGER WAVE EQUATION FOR NEUTRON WAVE

For the wave view of matter, Schrödinger derived the wave equation. In hindsight, the Schrödinger wave equation can be derived using the following five assumptions:

1. *The wave equation should be consistent with the Planck equation.*

   The Planck equation relates wave angular frequency w or period $T = 2\pi/w$ to wave energy E via Planck's proportionality constant h, where

$$w = \frac{E}{\hbar} \quad \text{(frequency)} \tag{29}$$

$$T = \frac{h}{E} \quad \text{(temporal)}. \tag{30}$$

2. *The wave equation should be consistent with the de Broglie relation.*

   The de Broglie relation between a particle's momentum p and the wavelength λ of the particle's wave representation is given by Planck's constant via

$$\lambda = \frac{h}{p} \quad \text{(direct space)} \tag{31}$$

$$k = \frac{p}{\hbar} \quad \text{(reciprocal space)}. \tag{32}$$

3. *The wave equation should be consistent with the Hamiltonian equation.*

   The Hamiltonian equation relates total particle energy E to its kinetic energy K and potential energy $V(\mathbf{r})$, where

$$E = K + V. \tag{33}$$

   The kinetic energy K of the particle is related to the particle rest mass m_0 and momentum p by

$$K = \frac{p^2}{2m_0}. \tag{34}$$

4. *The solution of the wave equation* $\Psi(\mathbf{r}, \mathrm{t})$ *as a function of position* $\mathbf{r}\left(\mathrm{x}, \mathrm{y}, \mathrm{z}\right)$ *and time* t *should be linear.*

   If $\Psi_1(\mathbf{r}, t)$ and $\Psi_2(\mathbf{r}, t)$ are solutions of the wave equation and c_1 and c_2 are constants, then their linear sum is also a solution:

$$\Psi(\mathbf{r}, t) = c_1\Psi_1(\mathbf{r}, t) + c_2\Psi_2(\mathbf{r}, t). \tag{35}$$

This linearity allows the addition of wave functions of one or more particles together and produce constructive and destructive interference. This wave interference causes the diffraction and refraction observed in experiments with particles, such as the Davisson–Germer experiments on electron and neutron diffraction.

5. *Assume that the total energy E of a particle is constant.*

   The total energy E is per the following equation:

$$E = K + V. \tag{36}$$

Taking the differential shows that the increase of kinetic energy is equal to the decrease in potential energy:

$$\Delta K = -\Delta V. \tag{37}$$

Given the work-energy theorem, relating force F and the differential distance Δx over which it acts give differential kinetic energy ΔK:

$$\Delta K = F\Delta x. \tag{38}$$

Substitution into Eq. 38 for the change of kinetic energy ΔK in terms of change in potential energy ΔV of Eq. 37 gives

$$F = -\frac{\Delta V}{\Delta x}. \tag{39}$$

The spatial gradient (or grad) of the potential is

$$\nabla V(\mathbf{r}, t) = \frac{\Delta V}{\Delta x}\hat{\mathbf{x}} + \frac{\Delta V}{\Delta y}\hat{\mathbf{y}} + \frac{\Delta V}{\Delta z}\hat{\mathbf{z}}. \tag{40}$$

The force $\mathbf{F}(\mathbf{r}, t)$ should be equal, but the force should act in the opposite direction of the gradient, where

$$\mathbf{F}(\mathbf{r}, t) = -\nabla V(\mathbf{r}, t). \tag{41}$$

That is, the force acts in a direction opposite to the spatial change of potential, similar to an induced current in a coil producing a magnetic field to counter the temporal change of the magnetic field linking the coil. This is an act of conservation of energy; without the minus sign, an increase in potential energy over a spatial interval would cause the kinetic energy to increase with increasing potential energy. In the case of a free particle, which has constant momentum and thus constant energy, one requires a zero gradient $\nabla V(\mathbf{r},t) = 0$, in which the potential is constant $V(\mathbf{r},t) = c$ throughout space, where constant c can also be zero.

The combination of assumptions 1–3 produces the following form of the conservation of energy expression or Hamiltonian expression H, where total energy E is

$$H = E = \frac{\hbar^2 k^2}{2m_0} + V(r,t) = \hbar w. \tag{42}$$

Rearrangement yields the required outcome of the matter wave equation that satisfies the conservation of energy or Hamiltonian expression:

$$\frac{\hbar}{2m_0}k^2 + \frac{1}{\hbar}V = w. \tag{43}$$

The neutron of rest mass $m_0 = 938\,\text{keV}$ is represented by a position r dependent wave function $\psi(r)$ of energy, $E = (h/2\pi)\,w$, and momentum, $p = h/\lambda$, and follows the time-independent Schrödinger wave equation. The Schrödinger wave equation with an average nuclear interaction potential for a neutron is

$$\left(\nabla^2 + \frac{8\pi^2 m_0 (E - V)}{h^2}\right)\psi(r) = 0. \tag{44}$$

The average potential $\langle V \rangle$ inside the scattering medium of volume U consists of N isotopes, with each jth isotope possessing scattering length b_j, and is

$$\langle V \rangle = \frac{h^2}{2\pi m_0 U}\sum_{j=1}^{N} b_j. \tag{45}$$

3. DERIVATION OF THE SCHRÖDINGER WAVE EQUATION BY THE ELECTROMAGNETIC WAVE EQUATION ANALOGY

One can look to the analogous wave equations to aid in the derivation of the Schrödinger wave equation. Consider the wave equation for an electromagnetic field propagating in the z-direction through a material zero charge, but constant conductivity σ, where

$$\nabla \cdot \mathbf{E}_\perp = 0. \tag{46}$$

In this case, a current density $\mathbf{J}_\perp$ is induced in the material by the incident electric field amplitude $E_\perp$, where

$$\mathbf{J}_\perp = \sigma \mathbf{E}_\perp \tag{47}$$

and

$$E_\perp = E_0 \exp\left[i\left(k_0 z - wt\right)\right]. \tag{48}$$

The electric and magnetic fields in CGS (centimeter-gram-second) units of an electromagnetic wave are equal, so the initial magnetic field is as follows:

$$B_\perp = E_\perp. \tag{49}$$

The vacuum space neutron wave number k_0 is

$$k_0 = \frac{w}{c}. \tag{50}$$

In CGS units, the resultant electric $E_\perp(z,t)$ and magnetic $B_\perp(z,t)$ fields in the material are identical, and so are their wave equations:

$$\nabla^2 E_\perp - \frac{1}{c^2}\frac{\partial^2 E_\perp}{\partial t^2} = k\sigma \frac{4\pi}{c}\frac{\partial E_\perp}{\partial t} \tag{51}$$

where

$$\nabla^2 B_\perp - \frac{1}{c^2}\frac{\partial^2 B_\perp}{\partial t^2} = k\sigma \frac{4\pi}{c}\frac{\partial B_\perp}{\partial t}. \tag{52}$$

In the material, a periodic solution is assumed for both the electric field $E_\perp$ and magnetic field $B_\perp$. In the material, the wave number is no longer the free space wave number k_0, but rather is a wave number k, which arises as a result of the interference of the incident electromagnetic wave and the electromagnetic wave induced in the material, so that

$$E_\perp = B_\perp = E_0 \exp\left[i\left(kz - wt\right)\right]. \tag{53}$$

Substitution of the resultant electric and magnetic field amplitudes of Eq. 53 into the two wave equation expressions of Eqs. 51 and 52 yields the solution of the wave number for the electromagnetic wave in the material:

$$\frac{c^2k^2}{4\pi\sigma} - \frac{w^2}{4\pi\sigma} = iw. \tag{54}$$

Solving for the neutron wave number k in the material gives

$$k = k_0\sqrt{1 + ik_0\sigma\frac{4\pi}{c}} = \frac{w}{c}\sqrt{1 + iw\sigma\frac{4\pi}{c^2}}. \tag{55}$$

This verifies the validity of the assumed exponential solution in a vacuum and the material, where

$$E_{\perp} = E_0 \exp\left[i\left(kz - wt\right)\right]. \tag{56}$$

Only the wave number changes when the electromagnetic wave moves from a vacuum into a material.

The electromagnetic wave moves at phase velocity c through a vacuum and phase velocity v_p through the material. The index of refraction of the material is

$$n = \frac{c}{v_p}. \tag{57}$$

In the material, the neutron wave number k is the ratio of the neutron wave angular frequency w divided by the neutron phase velocity v_p:

$$k = \frac{w}{v_p}. \tag{58}$$

The refractive index n is defined as the ratio of the neutron wave number in the material k divided by the neutron wave number k_0 in a vacuum, where

$$n = \frac{k}{k_0} \tag{59}$$

Substitution in Eq. 59 for k from Eq. 55 yields

$$n = \sqrt{1 + i\frac{4\pi\sigma}{k_0 c}}. \tag{60}$$

The conductivity σ of the material can be complex, where

$$\sigma = \sigma_r + i\sigma_i. \tag{61}$$

In this case, the real part σ_r contributes to wave attenuation, and the imaginary part σ_i contributes to the phase shift of the resultant wave in the material with respect to the incident wave from the vacuum.

The solution of the wave equation now provides a basis for determining the Schrödinger wave equation by analogy. The solution of the Schrödinger wave equation is assumed to be an exponential wave expression similar to that of the wave expression solution for the above electromagnetic plane wave, as well as the wave equations for scalar and vector potentials. Thus, the wave amplitude solution for the matter wave $\Psi(z,t)$ is assumed to have the same form as that of the above electromagnetic wave $E_{\perp}(z,t)$ solution, where

$$\Psi(z,t) = \exp\left[i\left(kz - wt\right)\right]. \tag{62}$$

Multiplication on both sides of this equation by matter wave amplitude Ψ yields

$$\frac{\hbar}{2m_0}k^2\Psi + \frac{1}{\hbar}V\Psi = w\Psi. \tag{63}$$

The imaginary argument of the exponential expression for the proposed matter wave amplitude $\Psi(z,t)$ is linear in time t (via iw) and position z (via ikx). The first derivative of $\Psi(z,t)$ taken with respect to time t or position z, yields the exponential term multiplied iw or ik. The second derivative of $\Psi(z,t)$ with respect to time t or position z multiplies the first derivative with respect to time t or position z by another factor of iw or ik. This suggests the following proposed form of the Schrödinger wave equation:

$$\frac{\hbar}{2m_0}\nabla^2\Psi + \frac{V}{\hbar}\Psi = \frac{\partial\Psi}{\partial t}. \tag{64}$$

Plugging the proposed wave amplitude solution $\Psi(z,t)$ into the proposed Schrödinger wave equation yields

$$-\frac{\hbar}{2m_0}k^2 + \frac{V}{\hbar} = -iw. \tag{65}$$

Comparison of the required outcome of the Schrödinger wave equation, the outcome of the proposed Schrödinger wave equation, and its proposed solution suggests a modification of the proposed equation. The form of the solution $\Psi(z,t)$ is unchanged, but the form of the proposed equation is slightly modified. The k^2 term of the proposed equation is multiplied by –1, and the $d\Psi/dt$ term is multiplied by $i = \sqrt{-1}$, so that the proposed, modified wave equation is

$$-\frac{\hbar}{2m_0}\nabla^2\Psi + \frac{V}{\hbar}\Psi = i\frac{\partial\Psi}{\partial t}. \tag{66}$$

Substitution of the unchanged, proposed wave amplitude solution $\Psi(z,t)$ of Eq. 62 thus yields the required outcome of the wave equation of Eq. 66 and its solution (namely, the conservation of energy E), where

$$\frac{\hbar}{2m_0}k^2\Psi + V\Psi = \hbar w = E \tag{67}$$

where

$$\Psi(z,t) = \exp\left[i\left(kz - wt\right)\right]. \tag{68}$$

The expression of the conservation of energy by this Schrödinger wave equation suggests that it can be represented by the Hamiltonian operator H, which was developed in classical mechanics to express total energy. In this case, the Hamiltonian operator H extracts its eigenvalue; namely, the total system energy E from its eigenfunction, the system state eigenfunction $\Psi(z,t)$, where

$$H\Psi = E\Psi \tag{69}$$

and the Hamiltonian operator H is then

$$H = \frac{\hbar^2}{2m_0}k^2 + V = E. \tag{70}$$

4. DERIVATION OF THE SCHRÖDINGER WAVE EQUATION BY ASSUMPTION OF THE NEUTRON PLANE WAVE SOLUTION

The simplest method to obtain the Schrödinger wave equation for a neutron is to consider a neutron plane wave propagating in the positive z-direction, and assume a neutron wave solution $\Psi(z,t)$, where

$$\Psi = \sin\left(wt - kz\right). \tag{71}$$

Differentiate $\Psi(z,t)$ twice with respect to distance z to obtain

$$\frac{\partial^2\Psi}{\partial z^2} = -k^2\Psi. \tag{72}$$

Recall that for a neutron particle of rest mass m_0 and particle and wave group velocity v, the neutron wave number k is

$$k = \frac{m_0 v}{h}. \tag{73}$$

Substitution of Eq. 73 into Eq. 72 yields

$$\frac{\partial^2 \Psi}{\partial z^2} = -\frac{\Psi}{\hbar^2} m_0^2 v^2 = -\frac{2m_0 \Psi}{\hbar^2} \frac{m_0 v^2}{2}. \tag{74}$$

The non-relativistic neutron kinetic energy K is

$$K = \frac{m_0 v^2}{2}. \tag{75}$$

The total energy E of the neutron is related to its kinetic energy K and potential energy V by

$$K = E - V. \tag{76}$$

One then obtains the Schrödinger wave equation, which describes the propagation of the neutron wave in a medium, whose refractive index varies with distance z, where

$$\frac{\partial^2 \Psi}{\partial z^2} + \frac{2m_0 \Psi}{\hbar^2} (E - V) = 0. \tag{77}$$

Similarly, we can derive the wave equation for an electromagnetic wave propagating in the positive z-direction, and assume a transverse electric field amplitude wave solution $E_\perp (z, t)$, where

$$E_\perp = \sin (wt - kz). \tag{78}$$

As before, differentiate $E_\perp (z, t)$ twice with respect to distance z to obtain

$$\frac{\partial^2 E_\perp}{\partial z^2} = -k^2 E_\perp. \tag{79}$$

The phase velocity v of an electromagnetic wave, which propagates through a material with index of refraction $n = c/v$ (with $n = 1$ for a vacuum), is related to its wave number $k = 2\pi/\lambda$, the vacuum speed of light c, and the refractive index n by

$$k = \frac{nw}{c}. \tag{80}$$

Substitution of Eq. 80 into Eq. 79 yields

$$\frac{\partial^2 E_\perp}{\partial z^2} = -\frac{n^2}{c^2} w^2 E_\perp. \tag{81}$$

Differentiation of $E_{\perp}(z,t)$ twice with respect to time t yields

$$\frac{\partial^2 E_{\perp}}{\partial t^2} = -w^2 E_{\perp}. \tag{82}$$

Comparing Eqs. 81 and 82 yields the electromagnetic wave equation propagating through a medium of refractive index n, where

$$\frac{\partial^2 E_{\perp}}{\partial z^2} - \frac{n^2}{c^2}\frac{\partial^2 E_{\perp}}{\partial t^2} = 0. \tag{83}$$

5. OPERATOR FORM OF THE SCHRÖDINGER WAVE EQUATION FOR A NEUTRON WAVE

The Schrödinger wave equation can be obtained by replacing the momentum p, potential energy V, and total energy E in the Hamiltonian expression H by operators p, V, and E, where

$$p \rightarrow i\hbar\nabla \tag{84}$$

$$V \rightarrow V \tag{85}$$

$$E \rightarrow i\hbar\frac{\partial}{\partial t}. \tag{86}$$

The resulting time-dependent Schrödinger wave equation is

$$\left(-\frac{\hbar^2}{2m_0}\nabla^2 + V\right)\Psi = i\hbar\frac{\partial \Psi}{\partial t}. \tag{87}$$

The Schrödinger wave equation describes the evolution in time and space of a particle of rest mass m_0, in which the particle and its trajectory is described by state function $\Psi(z,t)$, which has total energy E, which is the sum of the particle kinetic energy K and potential energy V, where

$$K = \frac{\hbar^2 k^2}{2m_0}. \tag{88}$$

In contrast, X-ray wave amplitudes are described by their electromagnetic field amplitudes via Maxwell's equations and the Lorentz force equation, which are both invariant under the Lorentz transformation. Maxwell's equations describe the generation and propagation of electromagnetic fields from accelerated charges via the scalar and vector Liénard–Wiechert potentials that are solutions of Maxwell's equations. However, in the Lorentz force equation, the electron rest mass m_0 must be replaced by the electron relativistic mass γm_0 to be invariant under the Lorentz transformation. The electron relativistic mass m is obtained by multiplying the relativistic gamma factor γ with the electron rest mass m_0, where

$$m = \gamma \cdot m_0 = \frac{1}{\sqrt{1-(v/c)^2}} \cdot m_0. \tag{89}$$

The Lorentz invariant force equation then describes the electric and magnetic forces acting on charge particles that lead them to accelerate and emit electromagnetic fields. There is tight coupling between Maxwell's equations and the Lorentz force equation, which are constrained by special relativity and boundary conditions of potential gradients. This tight coupling is evident in the undulators, wigglers, and bending magnets that change electron speed and/or electron direction—change in direction, change in speed, and a change in speed and direction are three ways to get acceleration and X-rays. Traditional X-ray sources in hindsight are relatively crude compared to the narrow-band undulator sources, in which traditional X-ray sources accelerate a beam of electrons, let them run into a tungsten anode and emit wide-bandwidth, bremsstrahlung X-rays.

Reflection, refraction, and diffraction occur at boundaries between two media in which the total energy E remains unchanged, but the potential energy V can vary in type (electric, magnetic, nuclear, gravity, etc.); in addition, the magnitude and spatial gradient of the potential energy can vary between media, and even within the media. As a result, the kinetic energy K and potential energy V change from one medium to the next, and even inside media, but this is provided that the magnitude $V(\mathbf{r})$ and gradient ∇V for the potential energy changed appropriately as well. From one medium to the next, the Schrödinger wave equation (in its differential equation form shown above or in its operator shown below) describes the time and space variation of the wave function amplitude $\Psi(z,t)$ with this very convenient constraint of total energy E, which makes simple solutions possible.

The above Schrödinger wave equation is for non-relativistic applications, and the above form of the Schrödinger wave equation is not invariant under the Lorentz transformation. The Schrödinger wave equation for relativistic particles must have the total energy E replaced by the special relativity identity between total energy E and momentum p, rest mass m_0, and speed of light c, where

$$E \rightarrow \sqrt{c^2p^2 + m_0^2c^2}. \tag{90}$$

6. REFLECTION AND TRANSMISSION OF NEUTRONS AT MEDIA INTERFACES

The neutron plane wave is described by its ray, a line in the direction of travel perpendicular to the wave front. The neutron ray also represents the particle trajectory of the neutron. Consider a neutron particle that propagates in a vacuum, where the potential energy is zero $V = 0$, such that its velocity is

$$v_0 = \sqrt{\frac{2E}{m_0}}. \tag{91}$$

If the neutron particle is incidental at a material surface at angle ϕ_0 with respect to the surface normal, then the particle refraction angle ϕ_1 is governed by the speed of the neutron particle and the continuity of its tangential velocity component across the interface. The potential energy V that the neutron particle encounters in the material yields the neutron particle velocity

$$v_1 = \sqrt{\frac{2\,(E - V)}{m_0}}. \tag{92}$$

The neutron particle tangential velocity remains unchanged across the boundary so that

$$v_0 \sin\phi_0 = v_1 \sin\phi_1. \tag{93}$$

Using Eq. 91 and substitution of Eq. 92 into Eq. 93 yields the description of the corpuscular or neutron particle refraction at an interface

$$\frac{\sin\phi_1}{\sin\phi_0} = \frac{v_0}{v_1} = \sqrt{\frac{E}{E - V}}. \tag{94}$$

In comparison, in the description of the same neutron refraction, but using the neutron wave approach, one applies Snell's law:

$$n_0 \sin\phi_0 = n_1 \sin\phi_1. \tag{95}$$

The index of refraction of the vacuum $n_0 = 1$ and the refractive index of the material are the ratio of the wave speed in the transmitting medium divided by the wave speed in the incident medium:

$$n_1 = \frac{v_1}{v_0}. \tag{96}$$

Hence, the wave description produces the same refraction result as this particle description.

Neutron reflection and transmission is exploited by neutron waveguides, mirrors, and curved surfaces. The Fresnel equations for neutron reflection and transmission at the surface interface between two media are derived as follows. As a function of vector position **r**, time t, frequency w, and wave vector **k**, a neutron plane wave with amplitude $\psi_i(\mathbf{r})$ is at incident angle θ_i measured relative to the surface tangent plane at the point of incidence. In general, a portion of the neutron wave is reflected and a portion transmitted.

The transmitted neutron plane wave is refracted by the interface to refracted or transmitted angle θ_t with respect to the surface tangent plane. The transmitted wave has the same frequency w incident wave; however, the transmitted wave amplitude E_t and wavelength λ_t (and associated wave vector $\mathbf{k}_t$ and wave number magnitude k_t) differ from the incident wave.

The remainder of the incident neutron wave is reflected from the surface at angle θ_r with respect to the surface tangent plane. The reflected wave has the same frequency w and wavelength λ_r (and associated wave vector $\mathbf{k}_r$ with wave number magnitude k_r) as the incident wave. However, the reflected wave amplitude differs from the incident wave. The incident wave and reflected wave each has the same wave number, wavelength, and thus phase velocity, as they are traveling in the same vacuum medium (where $\langle V \rangle = 0$). Hence, plugging in the incident and reflected wave amplitudes yields

$$k_i^2 = k_r^2 = \frac{8\pi^2 m_0 E}{h^2} \tag{97}$$

where

$$\langle V \rangle = 0. \tag{98}$$

The incident neutron total energy is also the neutron kinetic energy, where

$$E = \frac{m_0 v^2}{2}. \tag{99}$$

The neutron energy speed, v in terms of its wave number k, is

$$v = \frac{hk}{2\pi \cdot m_0}. \tag{100}$$

The index of refraction n is defined as the ratio of the wave speed v_t in the refracting medium, divided by the wave speed v_i in the incident medium.

The neutron velocity is proportional to its wave number ($v \propto k$), and thus, the index of refraction n is also the ratio of the respective wave number in the refracting medium k_t divided by the wave number in the incident medium k_i, where

$$n = \frac{v_t}{v_i} = \frac{k_t}{k_i}. \tag{101}$$

The incident, transmitted, and reflected neutron waves at the reflective surface are, respectively,

$$\psi_i = E_i \exp\left(wt - \mathbf{k}_i \cdot \mathbf{r}\right) \tag{102}$$

$$\psi_r = E_r \exp\left(iwt - i\mathbf{k}_r \cdot \mathbf{r}\right) \tag{103}$$

$$\psi_t = E_t \exp\left(iwt - i\mathbf{k}_t \cdot \mathbf{r}\right) \tag{104}$$

where

$$k_i = \frac{2\pi}{\lambda_i} \quad k_r = \frac{2\pi}{\lambda_r} \quad k_t = \frac{2\pi}{\lambda_t}. \tag{105}$$

The incident wave number k_i and reflected wave number k_r are equal:

$$k_i = k_r. \tag{106}$$

However, the incident number k_i and transmitted wave number k_r are not equal:

$$k_i \neq k_t. \tag{107}$$

The mirror or lens surface cannot change the tangential velocity component of the neutron wave. Hence, the tangential component of the incident wave $\mathbf{k}_i$ should be equal to the tangential component of the refracted (or transmitted) wave $\mathbf{k}_t$, where

$$k_i \cos\theta_i = k_t \cos\theta_t. \tag{108}$$

Snell's law for neutron refraction between two media is now expressed in terms of the grazing angles, which are measured relative to the tangent surface, is

$$\cos\theta_i = n \cos\theta_t. \tag{109}$$

In comparison, the prior expression of Snell's law that used angles measured relative to the surface normal—the 90° rotation of reference converts the sine to cosine.

Incidence angles θ_i less than or equal to the grazing incidence critical angle θ_c lead to the total reflection of incident neutrons, where $\theta_t = 0$ and

$$\cos\theta_c = n = 1 - \delta \quad \delta \ll 1. \tag{110}$$

The critical angle θ_c is a small except for ultracold neutrons so that

$$\cos\theta_c \cong 1 - \frac{\theta_c^2}{2} = 1 - \delta \quad \theta_c \ll 1 \quad \& \quad \delta \ll 1. \tag{111}$$

The critical angle of reflection for neutrons (and X-rays) is

$$\theta_c = \sqrt{2\delta}. \tag{112}$$

To calculate the neutron reflectivity of the surface, one can (for mathematical convenience) locate the surface at $z = 0$ in which the z-axis is perpendicular to the surface acting as the surface normal. Then, one invokes two continuity relationships. The first relationship is the continuity of the tangential components of the incident, reflected, and transmitted neutron waves at the surface, located at $z = 0$, where

$$\psi_i + \psi_r = \psi_t. \tag{113}$$

Rearranging gives

$$1 + \frac{\psi_r}{\psi_i} = \frac{\psi_t}{\psi_i}. \tag{114}$$

The second relationship is the continuity of their normal derivatives of the neutron waves with respect to the surface normal z-direction, where

$$\frac{d\psi_i}{dz} + \frac{d\psi_r}{dz} = \frac{d\psi_t}{dz}. \tag{115}$$

The normal derivative boundary condition evaluated at the surface at $z = 0$ yields

$$(-k_i \sin\theta_i)\psi_i + (k_r \sin\theta_r)\psi_r = (-k_t \sin\theta_t)\psi_t. \tag{116}$$

Only the components of the normal components of the wave vectors along the z-axis are involved in the differentiation of the arguments

$|\mathbf{k}_i \cdot \mathbf{r}|$, $|\mathbf{k}_r \cdot \mathbf{r}|$, and $|\mathbf{k}_t \cdot \mathbf{r}|$ of the exponentials. Note the sign convention of the z-components: The incident and transmitted wave z-components propagate in the same negative z-direction; whereas, the reflected wave z-components propagate in the opposite, positive z-direction. Since

$$k_i = k_r \quad \text{and} \quad \theta_i = \theta_r, \tag{117}$$

The normal derivative boundary condition yields

$$\left(\frac{\psi_r}{\psi_i} - 1\right) = -\frac{k_t \sin\theta_t}{k_i \sin\theta_i}\frac{\psi_t}{\psi_i}. \tag{118}$$

Substitution for ψ_t/ψ_i from Eq. 114 and solving for ψ_r/ψ_i gives the surface reflection fraction or reflectivity:

$$R = \frac{\psi_r}{\psi_i} = \frac{k_i \sin\theta_i - k_t \sin\theta_t}{k_i \sin\theta_i + k_t \sin\theta_t}. \tag{119}$$

The surface transmission fraction, or transmissivity T, is found by substitution of Eq. 119 into Eq. 118 or using, $R + T = 1$, where:

$$T = \frac{\psi_t}{\psi_i} = \frac{2k_i \sin\theta_i}{k_i \sin\theta_i + k_t \sin\theta_t}. \tag{120}$$

The probability that an incident neutron is reflected from a reflective surface P_R is the ratio of the square of the amplitude ψ_r^2 of the reflected neutron wave function divided by the square of the amplitude ψ_i^2 of the incident neutron wave function, or

$$P_R = |R|^2 = \left|\frac{\psi_r}{\psi_i}\right|^2 = \frac{|\psi_r|^2}{|\psi_i|^2} = \left|\frac{k_i \sin\theta_i - k_t \sin\theta_t}{k_i \sin\theta_i + k_t \sin\theta_t}\right|^2. \tag{121}$$

The probability of neutron transmission into the surface material P_T is the ratio of the square of the amplitude ψ_t^2 of the transmitted neutron wave function, divided by the square of the amplitude ψ_i^2 of the incident neutron wave function or $P_T = T^2$, where

$$P_T = |T|^2 = \left|\frac{\psi_t}{\psi_i}\right|^2 = \frac{|\psi_t|^2}{|\psi_i|^2} = \left|\frac{2k_i \sin\theta_i}{k_i \sin\theta_i + k_t \sin\theta_t}\right|^2. \tag{122}$$

With these results, the probabilities of reflection and transmission sum to unity, which is required for conservation of the number of neutrons at the mirror interface, where

$$P_R + P_T = 1. \tag{123}$$

Neutrons can be propagated from the reactor or spallation source to the user station over moderate distances (10–100 meters) via a neutron wave guide tube. The neutron wave guide tube, like a light wave guide, transmits neutrons of whose trajectory paths along the guide undergo 100% reflection for a grazing incident angle θ that is less than the neutron critical angle $\theta_c(\lambda)$ for neutrons of wavelength λ. The critical angle of neutron incidence $\theta_c(\lambda)$, at which or below which occur 100% reflection of neutrons with wavelength λ, is proportional to the neutron scatter length density $\rho b(\lambda)$ of the guide material, where ρ is the effective number density of the guide material scattering units (atoms or molecules) and $b(\lambda)$ is the effective scatter length (femtometer units) of the scatter unit, where

$$\theta_c = \lambda\sqrt{\frac{\rho b}{\pi}}. \tag{124}$$

The scatter length density $\rho b(\lambda)$ can be expressed in units of (#/cm^2) or (#/Å^2). There are some absorptive losses that attenuate the neutron beam as it undergoes grazing incidence reflection from one point to the next in the guide from the source to the user station.

Free neutrons have an average lifetime of 885.7 sec, and free ultracold neutrons, which have energies below 300 (neV corresponding to a velocity of 7.6 m/s and a wavelength of 52 nm, are reflected at all angles from most materials, including normal (perpendicular) incidence. The low density of the ultracold neutrons causes them to behave as an ideal gas with a temperature of 3.5° milli-Kelvin (mK). All neutrons with rest mass $m_n = 1.675 \times 10^{-27}$ kg are subject to gravity (acceleration $g = 9.8\,\mathrm{m/s^2}$), and 300-neV neutrons gain a potential energy U_g(neV) with an increase in height $h[m]$, where

$$U_g\,(\mathrm{neV}) = m_n g h = 102\,\mathrm{h}. \tag{125}$$

However, the downward fall of ultracold neutrons due to gravity is very significant. Beryllium and beryllium oxide are good materials for neutron waveguides due to their relatively large refractive index decrement δ and relatively low cross sections for absorption and incoherent scatter. A. I. Frank, a pioneer in ultracold neutron optics, measured refractive indices $n = 1 - \delta$ for 17 amino acids and their deuterated derivatives for 5 m/s, 20 m/s, and 50 m/s ultracold neutrons. The refractive n index was as low as 0.570 (for tryptophan) to 0.375 (for deuterated tryptophan); see Frank (1987).

The neutron spin produced a magnetic dipole moment μ (direction of μ points south S to north N), which interacts with the magnetic field **B** (**B** points north to south) to align the longitudinal component of the magnetic dipole moment point in the same direction of the applied magnetic field (lower-energy state) or in the opposite direction of the applied magnetic field (higher-energy state). The difference ΔE in the higher E_{hi} and lower E_{lo} energy states increases with applied magnetic field $\mathbf{B}[T]$ as

$$\Delta E = E_{hi} - E_{lo} = 60\,\mathrm{neV}/T. \tag{126}$$

7. NEUTRON REFLECTOMETRY

Reflection of neutrons is a simple, effective method of measuring the neutron scatter length of a material, and thus the material refractive index for neutrons. A collimated beam of neutrons is totally reflected at gazing angles relative to a planar mirrorlike surface of the material of interest until the critical angle is reached. Measurement of the critical angle θ_c with a narrow, finely collimated, monochromatic neutron beam with wavelength λ that is at grazing incidence to a very flat, planar surface composed of a single isotope with atom number density ρ then yields the neutron scatter length b via

$$\theta_c = \lambda\sqrt{\frac{\rho b}{\pi}}. \tag{127}$$

With measured critical angle θ_c, one can determine the atom density ρ of a single isotope material of known scatter length b, given a monochromatic neutron beam of wavelength λ. In the case of compounds, the critical angle θ_c yields the effective scatter length density ρb of the material. If the material is composed of N isotopes in which each nth isotope has scatter length b_n and atom number density ρ_n, then the scatter length density ρb of the material is

$$\rho b = \sum_{n=1}^{N} \rho_n b_n. \tag{128}$$

In neutron scatter, the Christiansen filter is used to measure the neutron scatter length b_p of a powdered material with a known atom, molecule, or particle density ρ_p that is dissolved (if soluble) or mixed in suspended state (insoluble) in a liquid of known scatter length b_l and atom or molecule number density ρ_l. The boundary between the liquid medium and the particle medium creates a refractive interface with a relative index of refraction n that is the ratio of the refractive index of the particle n_p divided by that of the liquid n_l, where

$$n = \frac{n_p}{n_l}. \tag{129}$$

A neutron beam wavelength λ, the refractive index of the particle n_p, and the liquid n_l are incident on the liquid with the suspended or dissolved powder in which the small angle scatter is measured as a function of the particle number density ρ_p, where

$$n_p \cong 1 - \frac{\lambda^2}{2\pi}\rho_p b_p \quad \frac{\lambda^2}{2\pi}\rho_p b_p \ll 1 \tag{130}$$

$$n_l \cong 1 - \frac{\lambda^2}{2\pi}\rho_l b_l \quad \frac{\lambda^2}{2\pi}\rho_l b_l \ll 1. \tag{131}$$

The relative refractive index n between the liquid and particle media is then defined by

$$n \cong \frac{1 - \frac{\lambda^2}{2\pi}\rho_p b_p}{1 - \frac{\lambda^2}{2\pi}\rho_l b_l}. \tag{132}$$

The binomial approximation is applied as follows:

$$n \cong \left(1 - \frac{\lambda^2}{2\pi}\rho_p b_p\right)\left(1 + \frac{\lambda^2}{2\pi}\rho_l b_l\right) \cong 1 - \frac{\lambda^2}{2\pi}\left(\rho_l b_l - \rho_p b_p\right). \tag{133}$$

When the scatter length density of the liquid and particle media are equal in which $n = 1$, then the minimum small-angle scatter angle is attained and measured, where

$$\rho_l b_l = \rho_{p0} b_p. \tag{134}$$

The particle number density ρ_{p0}, which corresponds to the minimum measured scatter angle, then allows the determination of the particle neutron scatter length b_p from the known neutron scatter length b_l and number density ρ_l of liquid atoms or molecules, where

$$b_p = \frac{\rho_l b_l}{\rho_{p0}}. \tag{135}$$

The accuracy of this method is about 0.1%. A large variety of liquids can be used, such as water, a combination of heavy water (D_2O) and water (H_2O), or organic liquids of known atom/molecule number density ρ_l and scatter length b_l. This allows measurement of the neutron scatter length b_p for a large variety of materials that can be dissolved, mixed, or suspended to create a particle-liquid refractive interface for each particle in the liquid.

Thermal neutron scatter is dominated by the hydrogen in water (H_2O, in which hydrogen has a scatter cross section of 82 barns. In D_2O, by contrast, the thermal neutron cross section is 7.6 barns. By varying the ratio of the number density $\rho_l^{H_2O}/\rho_l^{D_2O}$ of the D_2O and H_2O molecules in the liquid, and varying the density of the dissolved or mixed particle number density ρ_p, one can achieve equality between the liquid and particle scatter density in the neutron-interrogated sample, where $\rho_p = \rho_{p0}$ and the scatter angle is minimized that allows determination of the particle scatter length b_p.

For samples placed in water or on the water surface, contrast variation in small X-ray scatter can be accomplished by the replacement of H_2O by heavy water D_2O to alter the chemical composition of the sample. H_2O and D_2O have the same chemical properties so that the chemistry of the sample is not altered. In comparison, in X-ray diffraction and scatter contrast, variation must be done by altering the chemical composition of the sample by adding a heavy metal or altering the chemical composition of the liquid by adding a solute such as a salt or a sugar compound.

Gravity refractometry presents a 100-fold increase in accuracy of the measurement of the neutron scatter length b compared to the ordinary mirror method described previously. The gravity refractometry method is based on a horizontal beam of neutrons with initial horizontal velocity v_h that falls a vertical distance H at gravitational acceleration g onto a mirror surface of the material of interest (e.g., liquid mercury or molten lead). Conservation of energy states, the total neutron energy E, which is the sum of the initial horizontal, non-relativistic neutron kinetic energy K_h and neutron potential energy U_h, remains constant as the neutron falls from initial height H to the ground ($H = 0$). The neutron loses potential energy but gains vertical kinetic energy as it falls. However, the neutron horizontal kinetic energy remains constant ($K_{h1} = K_{h2}$) during its fall, where

$$E = K_{h1} + K_{v1} + U_{h1} = K_{h2} + K_{v2} + U_{h2}. \tag{136}$$

The initial neutron vertical kinetic energy is zero with $K_{h1} = 0$. However, as the neutron initial potential energy $U_{h1} = m_n g H$ decreases to zero potential energy $U_{h2} = 0$ at the ground, the neutron of rest mass m_n acquires vertical kinetic energy K_{v2}. The neutron strikes the mirror surface at terminal vertical velocity v_v. The energy conservation equation is, thus,

$$E = \frac{m_n v_h^2}{2} + m_n g H = \frac{m_n v_h^2}{2} + \frac{m_n v_v^2}{2} + 0. \tag{137}$$

Solving for vertical terminal velocity v_v at the mirror surface ($H = 0$) gives

$$v_v = \sqrt{2gH}. \tag{138}$$

The horizontal distance L that the neutron travels before hitting the mirror material is the horizontal velocity v_h multiplied by the time t_v required for the neutron to fall vertical distance H with gravitational acceleration g, where

$$\frac{1}{2}\,gt_v^2 = H \Rightarrow t_v = \sqrt{\frac{2H}{g}} \tag{139}$$

and

$$L = v_h t_v = v_h\sqrt{\frac{2H}{g}}. \tag{140}$$

The grazing angle of incidence θ_i of the neutron, which is measured relative to the mirror surface plane, is the inverse tangent of the final neutron vertical velocity v_v divided by its assumed constant horizontal velocity v_h. The experiment is set up for a small grazing incident angle θ_i so that the ratio of the velocities suffices, where

$$\theta_i = \tan^{-1}\frac{v_v}{v_h} \Rightarrow \theta_i \cong \frac{v_v}{v_h} = \frac{\sqrt{2gH}}{v_h}. \tag{141}$$

We now express the prior derived reflectivity magnitude $|R|$ in terms of the grazing incident angle θ_i and critical angle of incidence θ_c for total reflection at the mirror surface of the material of interest, where

$$|R| = \left|\frac{k_i \sin\theta_i - k_t \sin\theta_t}{k_i \sin\theta_i + k_t \sin\theta_t}\right|^2. \tag{142}$$

Equivalently,

$$|R| = \left|\frac{k_i \sin\theta_i - \sqrt{k_t^2 - k_t^2\cos^2\theta_t}}{k_i \sin\theta_i + \sqrt{k_t^2 - k_t^2\cos^2\theta_t}}\right|^2. \tag{143}$$

One recalls the boundary condition at the mirror (or lens surface), in which the tangential velocity component of the neutron wave cannot change upon crossing the interface, where

$$k_i \cos\theta_i = k_t \cos\theta_t. \tag{144}$$

The reflectivity magnitude $|R|$ becomes

$$|R| = \left| \frac{k_i \sin\theta_i - \sqrt{k_t^2 - k_i^2 \cos^2\theta_i}}{k_i \sin\theta_i + \sqrt{k_t^2 - k_i^2 \cos^2\theta_i}} \right|^2. \tag{145}$$

At the critical incident angle of reflection $\theta_i = \theta_c$, the refracted (transmitted) angle is zero, where $\theta_t = 0$. This allows one to obtain an expression for transmitted wave number k_t in terms of the incident wave number k_i and critical incident angle θ_c, where

$$k_i \cos\theta_c = k_t. \tag{146}$$

Substitution for k_t of Eq. 146 in the reflectivity magnitude $|R|$ of Eq. 145 gives

$$|R| = \left| \frac{\sin\theta_i - \sqrt{\cos^2\theta_c - \cos^2\theta_i}}{\sin\theta_i + \sqrt{\cos^2\theta_c - \cos^2\theta_i}} \right|^2. \tag{147}$$

The small-angle approximation applied to the expression for the reflectivity magnitude $|R|$ then gives

$$|R| = \left| \frac{\theta_i - \sqrt{\theta_i^2 - \theta_c^2}}{\theta_i + \sqrt{\theta_i^2 - \theta_c^2}} \right|^2 \tag{148}$$

where

$$\sin\theta \cong \theta \quad \text{and} \quad \cos^2\theta = \left(1 - \frac{\theta}{2}\right)^2 \cong 1 - \theta^2. \tag{149}$$

Simplifying, one obtains the reflectivity magnitude $|R|$ as a function of the ratio of the critical incidence angle for total reflection divided by the incident angle θ_c/θ_i, where

$$|R| = \left| \frac{1 - \sqrt{1 - (\theta_c/\theta_i)^2}}{1 + \sqrt{1 - (\theta_c/\theta_i)^2}} \right|^2. \tag{150}$$

From these derivations (measured relative to the mirror surface), the critical incident angle for total reflection θ_c and the neutron incidence angle at the mirror resulting from the fall of an initial horizontal neutron beam are

$$\theta_c = \lambda\sqrt{\frac{\rho b}{\pi}} \quad \text{and} \quad \theta_i \cong \frac{\sqrt{2gH}}{v_h} = \frac{m_n \lambda}{2\pi\hbar}\sqrt{2gH}. \tag{151}$$

The ratio of the critical angle for total reflection θ_c divided by the gravity fall incident angle is

$$\frac{\theta_c}{\theta_i} = \sqrt{\frac{2\pi\hbar^2\rho b}{m_n^2 gH}}. \tag{152}$$

One can thus define a particular height H_c for total reflection where

$$H_c = \frac{2\pi\hbar^2}{m_n^2 g}\rho b \tag{153}$$

and

$$\frac{\theta_c}{\theta_i} = \sqrt{\frac{H_c}{H}}. \tag{154}$$

The critical height H_c for neutrons is independent of the neutron beam initial horizontal velocity v_h and the wavelength λ. That is, H_c is dependent only on the mirror material scatter length density ρb, which is the product of the atom number density ρ and neutron scatter length b. The critical height H_c allows a horizontal beam of neutrons of any initial horizontal velocity v_h to fall onto a mirror surface at the critical grazing incidence angle θ_c for total neutron reflection.

The reflectivity magnitude $|R|$ as a function of the ratio of the critical height for total reflection divided by the chosen, initial height H_c/H is, then,

$$|R| = \left|\frac{1-\sqrt{1-(H_c/H)^2}}{1+\sqrt{1-(H_c/H)^2}}\right|^2. \tag{155}$$

Finally, the gravitational potential energy of the neutron at the critical height H_c is equal to the strong force nuclear scatter potential energy U of the mirror material, where

$$U = m_n g H_c. \tag{156}$$

The effective nuclear strong force nuclear potential for a material is thus measured by finding the critical height H_c at which 100% reflection occurs and in which an incident horizontal neutron beam falls onto the mirror surface by gravity. One measures the completely reflected neutron beam at the proper reflection angle θ_r, which is equal to the critical incidence angle of for total reflection θ_c, where $\theta_r = \theta_c$.

Frank and Nosov (2004) developed a one-dimensional magnetic gravitational trap for ultracold polarized neutrons in the higher-energy spin state using a Vladimirskiĭ magnetic mirror (Vladimirskiĭ, 1961). The Vladimirskiĭ mirror is a vertical (z-directed), one-dimensional magnetic field that varies sinusoidally along horizontal direction x. The sum of the magnetic and gravitational potentials has a minimum, and thus forms a potential well for the higher-energy spin-state neutrons, but no minimum and potential well forms for the lower-energy spin-state neutrons. Vertical (z-directed) magnetic field amplitudes that vary sinusoidally from –1.0 to 1.0 tesla on a planar surface in the horizontal x-direction with submillimeter periods allow the ultracold neutrons in the higher-energy spin state to be quantized in the magnetic-gravity potential well.

8. MEASUREMENT OF THE COMPLEX REFRACTIVE INDEX VIA REFRACTION

The complex refractive index for neutrons (or X-rays) in a material can be measured via the refraction of an incident neutron beam from a planar slab of known thickness. Alternatively, the slab thickness can be determined via the slab refractive index. One starts with a slab of thickness d with refractive index $n(\lambda)$ for neutrons (or X-rays) of wavelength λ given by

$$n = 1 - \delta + i\beta. \tag{157}$$

A beam of incident plane wave neutrons at incident angle θ_i is refracted to angle θ_t relative to their respective outward-pointing surface normals. By Snell's law:

$$n = \frac{\sin\theta_i}{\sin\theta_t}. \tag{158}$$

Absorption is not involved in the refraction angles, so one obtains the decrement δ (λ) for neutrons (X-rays) where

$$\delta = 1 - \frac{\sin\theta_i}{\sin\theta_t}. \tag{159}$$

On the downstream or back surface of the planar slab is a neutron (X-ray) position-sensitive detector (PSD) to measure the incident beam position without the slab and then with the slab. The change in the neutron (X-ray) beam position measured by the PSD allows the determination of the incident and refracted neutron (X-ray) angles relative to the reference z-axis in which the neutron beam incidence is perpendicular to the slab surfaces and the back detector plane.

A neutron (X-ray) PSD is placed on the front or upstream surface of the slab (facing the source plane) can measure the beam intensity at the front surface. And then removal of the front-surface PSD allows measurement of the refracted intensity of the beam that arrives at the downstream slab surface. The length of flight L of the neutron beam refracted at θ_t through the slab of thickness d is, then,

$$L = \frac{d}{\cos\theta_t}. \tag{160}$$

From Snell's law, with $n = 1 - \delta$:

$$\sin^2\theta_i^2\left(1 - \cos^2\theta_t\right). \tag{161}$$

One obtains

$$\cos\theta_t = 1 - \frac{\sin\theta_i}{n} \tag{162}$$

and

$$L = d\left(1 - \frac{\sin\theta_i}{1-\delta}\right)^{-1}. \tag{163}$$

The refracted neutron (X-ray) beam amplitude A_t at the downstream slab surface is related to the amplitude A_i at the upstream slab surface as follows:

$$A_t = A_i \exp\left[i\frac{2\pi}{\lambda}L\left(1 - \delta + i\beta\right)\right]. \tag{164}$$

The transmitted neutron (X-ray) intensity T is

$$T = \left|\frac{A_i}{A_t}\right|^2 = \exp\left(-\frac{4\pi}{\lambda}L\beta\right) = \exp\left(-\mu L\right). \tag{165}$$

The linear attenuation $\mu(\mathrm{cm}^{-1})$ coefficient for the neutrons (X-rays) is, then,

$$\mu = \frac{4\pi}{\lambda}\beta. \tag{166}$$

The slab linear attenuation coefficient $\mu(\lambda)$ or dimensionless, imaginary part of the slab refractive index $\beta(\lambda)$ for neutrons (X-rays) is then obtained from the transmission fraction T so that

$$\beta = -\frac{\lambda}{2\pi}\frac{\log_e T}{L}. \tag{167}$$

9. X-RAY AND NEUTRON INTERFEROMETRY

Visible light interferometry is widely used for metrology in visible light optics and increasingly with ultraviolet (UV) and soft X-rays. X-ray interferometry is an important potential tool in astronomy, X-ray diagnostics, and metrology in general. Neutron interferometry can be used to provide accurate measurement of neutron scatter lengths and is applied to research in quantum mechanics and gravitation. Two categories of interferometry devices are (1) division of wave fronts and (2) division of amplitude. In visible light optics, the division of wave fronts is accomplished via the Young double-slit interferometer, and the division of amplitudes is accomplished by the Michelson interferometer. Interferometry in optics in general, and X-rays and neutrons specifically, require splitting an incident ray at a single point into two rays that follow separate paths that then converge to a single point, in which the beams are then recombined with constructive and destructive summing of their amplitudes. The squared magnitude of the summed amplitudes produces a modulated intensity pattern that is characteristic of the phase shift differences incurred between the two ray paths from a single source point to a single image point.

Consider a monochromatic, small disk source of wave amplitude A_0, in which two very small diameter beams of equal amplitudes $A_0/2$ (conservation of neutrons) emerge at angles θ and $-\theta$ about the center z-axis. The beams travel a long object distance r_0 and intercept a neutron lens as plane waves. The lens refracts the two beams to converge to a locus of points in the image plane at image distance r_i. The transverse coherence length of each beam exceeds the radius of the image disk that conjugate to the source disk. The focal length f of the lens is half the object r_0 and r_i image distances so that $r_0 = r_i$ the magnification is unity $M = 1$.

The beams have traveled different total path lengths L_1 and L_2 from the source to image planes. Each beam is transverse coherent, such that the beams undergo constructive and destructive interference in the image plane. The amplitude A_r of the recombined pair of beams at the image plane is, then,

$$A_r = \frac{A_0}{2(r_0 + r_i)}\left\{\exp\left(i\frac{2\pi}{\lambda}L_1 + \phi_1\right) + \exp\left(i\frac{2\pi}{\lambda}L_2 + \phi_2\right)\right\} \tag{168}$$

where

$$r_0 + r_i \cong L_1, L_2. \tag{169}$$

One assumes that the beam path lengths L_1 and L_2 are approximately equal to the axial distance $r_0 + r_i$ between the source and image planes in the amplitude magnitude term. The intensity I_r of the recombining pair of beams at the image plane is, then,

$$I_r = |A_r|^2 = \frac{1}{2} I_0 \left\{ 1 + \cos\left[(w_1 - w_2) \cdot t + \frac{2\pi}{\lambda} (L_1 - L_2) + (\phi_1 - \phi_2) \right] \right\} \tag{170}$$

where

$$I_0 = \left[\frac{A_0}{(r_0 + r_i)} \right]^2. \tag{171}$$

At room temperature, the thermal neutron velocity $v = 2.2 \times 10^5$ cm/s and the wavelength is $\lambda = 1.8 \times 10^{-8}$ cm/s, so that

$$w = 7.7 \times 10^{13} \, \text{rad/sec}. \tag{172}$$

For a 10% bandwidth with $w_1 = w$ and $w_2 = w(1 \pm 0.05)$, the variation of the neutron wave frequency is 3.8 THz, where

$$w_1 - w_2 = w - w(1 \pm 0.1) = \pm 0.1\, w = 3.8 \times 10^{12} \, \text{rad/s}. \tag{173}$$

The maximum I_{max} and minimum I_{min} intensities of the recombined waves are

$$I_{\text{max}} = I_0 \quad \text{and} \quad I_{\text{min}} = I. \tag{174}$$

The contrast C is defined as

$$C = \frac{I_{\text{max}} - I_{\text{min}}}{I_{\text{max}} + I_{\text{min}}} = 1. \tag{175}$$

The interference pattern rapidly oscillates between I_{max} and I_{min} at 3.8 THz, in which the time-averaged interference pattern is then,

$$I_r = |A_r|^2 = \frac{1}{2} I_0 \left\{ 1 + \cos\left[\frac{2\pi}{\lambda} (L_1 - L_2) + (\varphi_1 - \varphi_2) \right] \right\}, \tag{176}$$

where

$$< \cos\left[(w_1 - w_2) \cdot t\right] >= \frac{1}{T}\int_0^T \cos\left[\Delta w \cdot t\right] = \frac{1}{T}\int_0^T \cos\left(2\pi\frac{t}{T}\right) = 0. \quad (177)$$

The spacing between adjacent maxima I_{max} and minima I_{min} is one-half period P, and the spacing between maxima pairs or between minima pairs is then

$$P = \lambda. \quad (178)$$

The interference pattern is shifted by phase Φ, where

$$\Phi = \phi_1 - \phi_2. \quad (179)$$

The phase shift Φ can arise if one path contains an object whose length and refractive index causes a phase shift. The optical length difference ΔL in the two paths can be varied so as to generate the interference pattern as a function of ΔL, where n_1 and n_2 are the refractive indexes of the mediums in paths 1 and 2 and

$$\Delta L = n_1 L_1 - n_2 L_2. \quad (180)$$

The intensity of the interference pattern of the combined waves is then

$$I_r = |A_r|^2 = \frac{1}{2}I_0\left\{1 + \cos\left[\frac{2\pi}{P}\Delta L + \Phi\right]\right\}. \quad (181)$$

10. INTERFEROMETRY AND THE FIZEAU EFFECT

Interferometry by division of the wave front, which used a Young double-slit interferometer, was done by Klein *et al.* (1981). Their purpose was to measure the Fizeau effect with neutrons. In the Fizeau effect, waves undergo a phase shift due to the movement of the medium in addition to the stationary medium. Given non-relativistic, thermal, or cold neutron vacuum velocities v_n and wavelength λ, and for moving medium velocity v, thickness D, and refractive index n, the neutron phase shift $\Delta\Phi$ incurred in a moving medium relative to a stationary medium is

$$\Delta\Phi = \frac{2\pi}{\lambda}D\frac{v}{v_n}\frac{n-1}{n}. \quad (182)$$

The Fizeau interferometer, which is used to measure the phase shift effects of visible light in a moving medium, is similar to a Fabry–Perot interferometer. The Fabry–Perot interferometer uses two mirrors, one of which is totally reflective and the other is partially reflective and transmissive, thereby allowing the splitting and then recombining of the wave amplitudes.

The Fizeau effect for neutron waves in a moving medium, which is measured by an interferometer, is described as follows. Assume that relative to the interferometer lab frame, the neutron velocity is $\mathbf{v}_i$, and the velocity of the phase-shifting medium velocity is $\mathbf{u}_i$. The velocity of the neutron $\mathbf{v}_m$ relative to the medium is

$$\mathbf{v}_m = \mathbf{v}_i - \mathbf{u}_i. \tag{183}$$

The neutron $\mathbf{v}_i$ and medium $\mathbf{v}_m$ velocities are assumed to be non-relativistic, and thus the Galilean transformations are applied.

The neutron particle has mass m and wave vector $\mathbf{k}_i$, and by the de Broglie relation, the particle and wave momentum $\mathbf{p}_i$, which is measured relative to the interferometer lab frame, is related via

$$\mathbf{p}_i = m\mathbf{v}_i = \hbar\mathbf{k}_i. \tag{184}$$

The neutron momentum $\mathbf{p}_m$ and wave vector $\mathbf{k}_m$, measured relative to the moving medium, is

$$\mathbf{p}_m = m\mathbf{v}_m = \hbar\mathbf{k}_m. \tag{185}$$

Taking the difference of these two equations yields

$$m\,(\mathbf{v}_i - \mathbf{v}_m) = \hbar\,(\mathbf{k}_i - \mathbf{k}_m). \tag{186}$$

Use of the medium velocity $\mathbf{u}_i$ in Eq. 186 for the $\mathbf{v}_i - \mathbf{v}_m$ term in Eq. 186 gives an equation for the wave vector $\mathbf{k}_m$ that is measured relative to the moving medium. The equation is expressed in terms of the incident neutron wave vector $\mathbf{k}_i$ in a vacuum and the medium velocity $\mathbf{u}_i$ that is measured relative to the interferometer lab frame, where

$$\mathbf{k}_m = \mathbf{k}_i - \frac{m}{\hbar}\mathbf{u}_i. \tag{187}$$

The phase shift χ is scalar and thus invariant in the Galilean transformation. As a result, the phase shift has the same value $\chi\,(\mathbf{k}_i)$ in the medium and the interferometer frames when the medium is at rest, relative to the interferometer lab frame. And when the medium is moving relative to the interferometer lab frame, then the phase shift changes to $\chi(\mathbf{k}_m)$, in which $\chi(\mathbf{k}_m)$ is the same in both frames. Thus, the phase shift $\Delta\chi$ between the moving medium and the stationary medium is

$$\Delta\chi = \chi(\mathbf{k}_m) - \chi(\mathbf{k}_i). \tag{188}$$

The moving medium is rectangular, with parallel input and output planar surfaces separated by medium thickness D. The neutron beam is incident at angle ϕ measured relative to the normal of the input surface of the moving medium. The path length L of the neutron beam through the moving or stationary medium is then

$$L = \frac{D}{\cos\phi}. \tag{189}$$

The z-component k_{iz} of the incident neutron wave vector $\mathbf{k}_i$ is perpendicular to the parallel input output faces of the medium, where

$$k_{iz} = |\mathbf{k}_i|\cos\phi. \tag{190}$$

The medium has refractive index n, expressed in terms of the nuclear potential U and the neutron total energy E. Outside the medium in the field-free vacuum, the neutron total energy E is equal to its kinetic energy W, where

$$n = 1 - \frac{U}{2E} = 1 - \frac{U}{2W} = 1 - \frac{U}{mv_i^2} = 1 - \frac{Um}{\hbar^2 k_i^2} \tag{191}$$

and

$$E = W = \frac{1}{2}mv_i^2 \quad \text{(Outside medium)} \tag{192}$$

$$E = U + K = W \quad \text{(Inside medium)}. \tag{193}$$

The neutron kinetic energy K inside the medium is decreased from its vacuum (and field-free) value W by the nuclear potential energy U that is acquired by the neutron inside the stationary or moving medium, where

$$K + U = E = W. \tag{194}$$

The total neutron energy E is constant throughout its trajectory, and the nuclear potential energy U of the medium that is experienced by the neutron is assumed to be constant.

The phase shift $\chi(\mathbf{k}_i)$, acquired by the neutron in its trajectory length L through the stationary medium, is

$$\chi(\mathbf{k}_i) = (n-1)\,k_i L = \frac{(n-1)\,k_i D}{\cos\phi} = -\frac{UmD}{\hbar^2 k_i \cos\phi} = -\frac{UmD}{\hbar^2 k_{iz}}. \tag{195}$$

The phase shift $\chi(\mathbf{k}_m)$, acquired by the neutron in its identical trajectory length L through the moving medium, is

$$\chi(\mathbf{k}_m) = (n-1)\,k_m L = \frac{(n-1)\,k_m D}{\cos\phi} = -\frac{UmD}{\hbar^2 k_m \cos\phi} = -\frac{UmD}{\hbar^2 k_{mz}}. \tag{196}$$

The difference $\Delta\chi$ between the phase shift for the neutron passing through a moving medium relative to a stationary medium is, then,

$$\Delta\chi = \chi(\mathbf{k}_i) - \chi(\mathbf{k}_m) = \frac{UmD}{\hbar^2}\left(\frac{1}{k_{mz}} - \frac{1}{k_{iz}}\right). \tag{197}$$

Next, one assumes that the z-component of the neutron wave vector in the moving medium and stationary medium are approximately the same, $k_{mz} \cong k_{iz}$. The neutron refractive index $n = 1 - \delta$ of the medium is assumed to be almost that of the vacuum $n = 1$, where $\delta \ll 1$. Hence,

$$\frac{1}{k_{mz}} - \frac{1}{k_{iz}} = \frac{k_{iz} - k_{mz}}{k_{iz}k_{mz}} = \frac{\Delta k_z}{k_{iz}^2} \quad \delta \ll 1. \tag{198}$$

Substitution of Eqs. 195 and 196 into the expression of Eq. 197 for the phase shift $\Delta\chi$ due to a moving medium, with use of Eq. 198, gives

$$\Delta\chi = \chi(\mathbf{k}_i) - \chi(\mathbf{k}_m) = \frac{UmD}{\hbar^2}\frac{\Delta k_z}{k_{iz}^2} = -\frac{\Delta k_z}{k_{iz}}\chi(\mathbf{k}_m). \tag{199}$$

Transposition yields

$$\frac{\Delta\chi}{\chi(\mathbf{k}_m)} = -\frac{\Delta k_z}{k_{iz}}. \tag{200}$$

The z-components of the difference between the neutron wave vectors measured relative to the moving medium $\mathbf{k}_{mz}$ and the stationary interferometer $\mathbf{k}_{iz}$ frames are

$$\Delta k_z = \mathbf{k}_{mz} - \mathbf{k}_{iz}. \tag{201}$$

One then recalls the previously derived expression:

$$k_{mz} = k_{iz} - \frac{m}{\hbar}u_{iz}. \tag{202}$$

Substitution of Eq. 202 into Eq. 201, and substitution of the result into Eq. 200 gives

$$\frac{\Delta\chi}{\chi(\mathbf{k}_m)} = \frac{mu_{iz}}{\hbar k_{iz}} = \frac{u_{iz}}{v_{iz}}. \tag{203}$$

The phase shift $\chi(\mathbf{k}_i)$, acquired by neutron trajectory length L through the stationary medium, differs by $\Delta\chi$ from that of a moving medium only when there is a nonzero component of the incident neutron velocity $\mathbf{v}_i$ (or wave vector $\mathbf{k}_i$) that is parallel to the direction of the moving medium velocity $\mathbf{u}_i$. To simplify matters, assume that the medium moves in the z-direction that is perpendicular to the two planar interfaces with the vacuum. The parallel, input, and output planar medium-vacuum interfaces are separated by the medium thickness D. And the neutron beam is incident on the input interface at angle ϕ that is measured relative to the surface normal. When the medium is moving, the neutron phase shift acquired by the neutron passage through the medium differs from that of a stationary medium by the fractional difference $\Delta\chi/\chi$, which is the ratio of the medium velocity $\mathbf{u}_i$ divided by the neutron velocity component $\mathbf{v}_i \cos\phi$ in the z-direction of the moving medium, where

$$\frac{\Delta\chi}{\chi(\mathbf{k}_m)} = \frac{u_{iz}}{v_{iz}} = \frac{\mathbf{u}_i}{\mathbf{v}_i \cos\phi}. \tag{204}$$

11. PENDELLÖSUNG OSCILLATIONS AND ANOMALOUS ABSORPTION IN PERFECT CRYSTALS

In visible light optics, one commonly uses the division of wave amplitude via partially reflecting and transmitting mirrors. In this way, the incident light wave can be divided into two coherent parts that are directed to travel different paths and finally to recombine. The different path lengths and phase-shifting obstacles or media in the paths cause the recombining coherent waves to undergo constructive and destructive interference that produces an interference pattern.

Mirrors for X-rays and neutrons work only for grazing incidence angles, and in the division of amplitude method, such as the Michelson interferometer, mirrors would produce very small beam separations, even for very long path lengths. Wide separations between the two beam paths are needed so that a phase-shifting material, magnetic field, or other phase-altering obstacle can be placed in one of the beam paths. The resulting amplitude summing between the recombined beams leads to an interference pattern that characterizes the obstacle.

An incident beam can be split into two widely separated beams by the combination of refraction and Bragg reflection in a large perfect crystal. With the development of pure, large, and almost perfect silicon crystals in the 1960s, the division of amplitude via Bragg reflection was developed for X-rays by Bonse and Hart (1965) and for neutrons by Rauch *et al.* (1974). With almost perfect silicon crystals, the two groups respectively developed an X-ray and neutron triple-Laue case (or LLL) interferometer. The LLL interferometer is fabricated from a single, large, perfect crystal. The rectangular crystal has two rectangular portions removed to form the letter "E." Each horizontal line of the letter "E" is a planar slab with parallel surfaces and Bragg reflection planes that are perpendicular the slab surfaces. As a result, there is also a set of perpendicular crystal planes that are parallel to the slab surface. The *z*-directed gaps between the first slab, S, and the second slab, M, is equal to the *z*-directed gap between the second slab, M, and the third slab, A.

In the kinematic theory of diffraction in crystals, the incident neutron (or X-ray) wave outside the crystal is assumed identical to the inside neutron (X-ray) waves incident on atomic planes in the crystal interior. Now consider an X-ray or neutron wave that is Bragg-reflected by a crystal plane in which the wave undergoes a second Bragg reflection by another parallel crystal plane. This double-Bragg-reflected X-ray or neutron wave now propagates in the same direction as the incident X-ray or neutron wave. The kinematic theory does not account for the double-, triple-, or further-Bragg-reflected beam inside the crystal. The dynamic theory, however, does account for the multiple Bragg reflections and interference between the incident beam and multiple Bragg-reflected beams.

In particular, we now consider the Borrmann effect or anomalous transmission for X-rays and the corresponding anomalous neutron transmission, which occurs only in perfect crystals. Consider the case of a perfect crystal in which the Bragg reflection planes are perpendicular to the planar crystal surface and thus parallel to the surface normals. This crystal configuration is termed the "symmetrical Laue case." In the symmetrical Laue case, the Bragg forward reflection (the Laue case) can occur not only from the horizontal crystal planes that are perpendicular to the surface, but also from the vertical crystal planes that are parallel to the surface. The transmitted wave can be an unrefracted wave that passes without scatter through the crystal. Alternatively, the transmitted rays can be refracted at the entrance crystal surface and pass unscattered to the crystal slab exit surface, where the ray is again refracted; or the transmitted wave can be the result of a multiple-Bragg-reflected X-ray or neutron ray that propagates in the same direction as the incident ray through the crystal.

As a result of the multiple Bragg reflections from the horizontal and vertical crystal planes, an incident X-ray or neutron wave is either transmitted directly through the crystal, or the wave is Bragg-reflected in the backward direction (the Bragg case) or Bragg-reflected in the forward direction (the Laue case). Each perfect crystal slab has a double-valued refractive index for refraction (i.e., transmission) and a double-valued refractive index for forward Bragg reflection (the Laue case) that are treated by the dynamical theory. The issue of perfect crystals for neutrons, dynamic theory of diffraction, and interferometers is covered by Sears (1989) and Squires (1996). X-ray interactions in perfect crystals are treated by Warren (1990).

The perfect crystal of the symmetrical Laue case has horizontal Bragg reflection planes perpendicular to the planar crystal input and output surfaces, as well as vertical Bragg reflection planes parallel to those surfaces. Each incident neutron (X-ray) wave on the crystal slab produces two refracted (i.e., transmitted) waves and two Bragg-reflected waves that can interact, leading to the effects of extinction, Pendellösung interference, and anomalous absorption. The coherent, oscillating transfer of neutron flux occurs between the pair of refracted (transmitted) and Bragg reflected waves of wavelength λ. This oscillation is analogous to the energy transfer between a pair of coupled pendulums.

Consider the case of the perfect crystal of thickness d, in which one ignores absorption because the crystal is optically thin for neutrons or X-rays. In this case, there is only the reflection and transmission of neutrons (X-rays). The neutron flux oscillation between the pair of refracted (transmitted) and Bragg-reflected waves of wavelength λ is observed by varying the crystal thickness d or the neutron wavelength λ. The fraction R of Bragg, forward-reflected neutrons (X-rays) and the fraction T of refracted (transmitted) neutrons (X-rays) are, respectively,

$$R = \sin^2 y \quad \text{and} \quad T = \cos^2 y \tag{205}$$

where

$$y = \frac{\pi}{\Delta_{hkl}} d. \tag{206}$$

The Pendellösung period Δ_{hkl} is inversely proportional to the neutron (X-ray) wavelength λ and to the cosine of the Bragg reflection angle for the hkl reflection planes, where

$$\Delta_{hkl} = \frac{\pi V_0 \cos \theta_{hkl}}{\lambda |F_{hkl}|}. \tag{207}$$

The Pendellösung period is also termed the extinction length Δ_{hkl}.

Each crystal unit cell has volume V_0, and θ_{hkl} is the angle of incidence of the neutron wave, which is measured relative to the normal of the crystal slab surface, in which Bragg reflection (the Laue case) occurs from the *hkl* crystal planes. A Laue symmetrical crystal is cut so that the *hkl* crystal slab planes (e.g., 220) are perpendicular to the planar slab faces. The structure factor $F(\boldsymbol{\tau}_{hkl})$ for each of the N crystal unit cells is evaluated for the neutron scatter vector $\boldsymbol{\kappa} = \boldsymbol{\tau}_{hkl}$ that leads to Bragg reflection from the set of *hkl* crystal planes, where

$$F(\boldsymbol{\tau}_{hkl}) = \sum_{d=1}^{r} b_d \exp(-W_{nd}) \exp(i\boldsymbol{\tau}_{hkl} \cdot \mathbf{d}_d). \tag{208}$$

where b_d is the neutron scatter length, which in the case of X-rays is replaced by the atomic form factor f_d; $\mathbf{d}_d$ is the position vector relative to the unit cell reference corner; and W_{nd} is the Debye–Waller factor of the dth atom of the r atoms in each of the N unit cells that comprise the perfect crystal. The incident wave vector $\mathbf{k}_0$ at incidence angle θ_{hkl} to the slab surface normal yields diffracted wave vector $\mathbf{k}$, in which scatter vector $\boldsymbol{\kappa}$ is equal to the reciprocal lattice vector $\boldsymbol{\tau}_{hkl}$ for the *hkl* set of Bragg reflection planes.

The Debye–Waller factor W_{nd} is the thermally induced average displacement of the dth atom in the nth unit cell, whose reference corner is located at position $\mathbf{R}_n$ relative to the crystal center, where

$$W_{nd} = \frac{1}{2} < (\boldsymbol{\tau}_{hkl} \cdot \mathbf{u}_{nd})^2 >. \tag{209}$$

The vibration amplitudes of the nth unit cell and the dth atom are expressed in terms of the set of $3Nr$ thermal vibration modes of the entire crystal in three dimensions.

We now consider absorption in the perfect crystal, in which A is the fraction of incident neutrons (X-rays) absorbed by the crystal slab. The incident number of neutrons must equal the sum of the neutrons transmitted, reflected, or absorbed. Hence, neutron (X-ray) conservation requires

$$R + T + A = 1. \tag{210}$$

For neutron path length D through the perfect crystal of thickness d and neutron linear attenuation coefficient μ for the crystal material, the absorption fraction A for anomalous absorption is given by

$$A = \frac{1}{2} - \frac{1}{2} \exp(-2\mu D). \tag{211}$$

That is, in a perfect crystal, the absorption independent of path length D and linear attenuation μ. The maximum fraction A of absorbed neutrons in even a very thick crystal is only 50% even as $\mu D \to \infty$. In comparison, in normal absorption in an imperfect crystal, the absorption fraction A is given by Lamberts's law, where

$$A = 1 - \exp(-2\mu D). \tag{212}$$

Thus, in a normal, imperfect crystal, the absorption fraction ranges from 0 to 1, and in a sufficiently thick or attenuating crystal, the transmission of neutrons goes to zero as μD increases.

Anomalous absorption of X-rays or the Borrmann effect in perfect crystals was discovered by Borrmann (1941) and interpreted by Laue (1949). The anomalous absorption of neutrons was observed by Knowles (1956). For neutron or X-ray incident angles at the Bragg angle for the set of planes perpendicular to the planar crystal surfaces, the nodes of the interior coherent X-ray or neutron waves at the Bragg angle occur at positions of the absorbing nuclei. Given the linear attenuation μ coefficient for X-rays or neutrons in the perfect crystal slab and propagation path length L, the fraction of transmitted X-rays or neutrons is $\exp(-\mu L)$.

Experimental observations showed that the X-rays or neutron rays were not extinguished in the crystal for path lengths of even $\mu L = 20$. The dynamic theory shows that at least 50% of the incident X-ray or neutron waves are transmitted by a perfect crystal, even for infinite thickness $L \to \infty$ or linear attenuation $\mu \to \infty$. Until the advent of perfect crystals, the observation of anomalous absorption was not possible. That is, the pair of transmitted waves ψ_t^+ and ψ_t^- interfere by amplitude summation $\psi_t^+ + \psi_t^-$, and the pair of Bragg-reflected waves ψ_b^+ and ψ_b^- also interfere by amplitude summation $\psi_b^+ + \psi_b^-$. However, in a perfect crystal, the zero-amplitude nodes occur for one of the transmitted wave functions ($\psi_t^- = 0$) at the absorbing crystal atoms when the other transmitted wave function ψ_t^+ is at its maximum. And zero-amplitude nodes also occur for one of the Bragg-reflected wave functions ($\psi_b^- = 0$) at the absorbing crystal atoms when the other Bragg-reflected wave function ψ_b^+ is at its maximum.

In mosaic crystals, primary extinction occurs as a result of multiple Bragg reflections in a single mosaic block, whereas secondary extinctions occur as a result of multiple Bragg reflections between different mosaic blocks. Each mosaic block is a perfect crystal in which the primary extinction is a dynamic scatter effect. The secondary extinction is a transport effect since the different mosaic blocks scatter incoherently. As a result, the mosaic crystal does not display the perfect crystal behavior of anomalous absorption, double refractive indexes for transmission and Bragg reflection, and the coupled interactions between the resulting four interior coherent waves produced by a single incident waves. Chapter 13 in the next volume will present the dynamic scatter of neutrons, which applies to X-rays as well, and provides derivations of neutron (X-ray) scatter in perfect and mosaic crystals.

12. MEASUREMENT OF REFRACTIVE INDEX VIA INTERFEROMETRY

In this section, the perfect crystal interferometer is described, and its application for measurement of the neutron or X-ray refractive index of sample materials is presented. The perfect crystal interferometer requires samples with very parallel surfaces providing a constant sample thickness D. The sample cross-sectional area must be sufficiently large that the beam, which is incident at angle ϕ relative to the sample surface normal, allows the beam trajectory length d through uniform density sample material to remain in the material during the entire sample trajectory length $d = D/\sin\phi$.

Assume that a perfect crystal in the symmetrical Laue case has a particular set of *hkl* Bragg reflection planes that are perpendicular to the planar crystal surface. The set of crystal vertical planes perpendicular to the *hkl* horizontal planes leads to a pair of Bragg forward-reflected X-rays or neutrons and a pair of refracted (transmitted) rays. An incident neutron beam at incidence angle θ relative to the z-directed surface normal is refracted (transmitted) and Bragg-reflected as it enters the crystal slab at point A. As a result, two refracted waves emerge as a combined, single-refracted wave at angle θ relative to the exit surface normal pointing in the positive z-direction at point A′. Entrance point A and exit point A′ are directly opposed to one another and separated by the planar crystal slab thickness D. The pair of transmitted and pair of Bragg forward-directed rays propagate in a narrow channel between the parallel planes that connect points A and A′. The two collinear Bragg-reflected waves emerge as a combined, single wave at angle $-\theta$ relative to the same positive z-directed normal.

The X-ray and neutron LLL interferometers are Mach-Zehnder interferometers. A single neutron or X-ray is incident at point A at angle θ relative to the backward-directed normal on the entrance surface of the first planar slab, S. On the direct opposite side of slab S at point A′ emerge two forward-directed rays at $\pm\theta$ relative to the forward-directed exit surface normal. The two forward-directed Bragg scatter rays in the crystal sum and emerge as a single Bragg forward-scattered ray (the Laue case), and the two transmitted rays in the crystal sum and emerge from the crystal as a single transmitted ray. The transmitted and forward Bragg-reflected beams diverge at angle 2θ from the point A′.

The second planar slab, M, intercepts the two divergent rays from slab S (splitter). Then slab M (as in "mirror") reflects the two incident beams in the forward direction (the Laue case) so the two beams converge to point X on the surface of a third planar slab A (analysis slab) with convergence angle 2θ. In this third (last) slab, A, the combined rays undergo transmission and Bragg forward reflection (the Laue case). The rays exit slab A at point X/ (opposite point X) as two diverging beams O and H with divergence angle 2θ. The two beams, O and H, that exit slab A are the result of interference between the recombined beams, and thus produce interference patterns. An object or rotating phase shifter is placed in one of the two beam paths between the second slab, M, and third slab, A. The rotating phase shifter can then vary the interference fringe pattern. And an object can be characterized by the interference patterns that it produces relative to the interference patterns without the object.

A sample whose refractive index is to be determined is placed in one of the beam paths between the second slab, M, and this slab, A. The neutron (X-ray) beam is incident at angle ϕ relative to the sample surface normal and travels path length d through the sample thickness D and refractive index n, where

$$d \sin \phi = D. \tag{213}$$

For accurate measurements, the sample surfaces must be extremely parallel to produce a constant sample thickness of D. For a particular neutron incidence angle ϕ, the phase shift $\Delta\chi$ produced by the sample with unknown refractive index n is

$$\chi = (n-1)\frac{2\pi}{\lambda}d = \delta\frac{2\pi}{\lambda}\frac{D}{\sin\phi}. \tag{214}$$

Variation of the neutron (X-ray) beam incidence angle ϕ via rotation of the sample by $\Delta\phi$ results in a change in phase shift $\Delta\chi$ between the recombined beams at the A slab of the interferometer:

$$\Delta\chi = -\delta\frac{2\pi}{\lambda}\frac{D\Delta\phi}{\sin^2\phi}\cos\phi = -\delta\frac{2\pi}{\lambda}\frac{D\Delta\phi}{\sin\phi}\cot\phi. \tag{215}$$

One varies the sample tilt angle $\Delta\phi$ until the interference fringes shift one full wavelength, which corresponds to phase shift $\Delta\chi = 2\pi$. This yields the measurement of the refractive index decrement δ, where

$$\delta = -\frac{\lambda \sin \phi \tan \phi}{D \Delta \phi}. \tag{216}$$

For 5-Å neutrons and sample thickness $D = 0.5$ cm with incident angle 45° and $\Delta \phi = -0.05$ rad, the decrement of the unknown material is then $\delta = 1.4 \times 10^{-6}$.

A more practical setup is to have a phase shifter between the second and third interferometer slabs, M and A. One then places the sample of thickness D on the same beam leg but between the first and second interferometer slabs, S and M. The beam is incident on the sample at angle ϕ and on the phase shifter at angle $-\phi$ relative to their flat surface normals. Both the phase shifter and the sample are slabs with very parallel surfaces. To produce a fringe shift $\Delta \chi = 2\pi$, the phase shifter is rotated angle $\Delta \phi_S$ with the sample in the beam, and then the phase shifter is rotated angle $\Delta \phi_0$ without the sample in the beam. The difference in the phase shifter angle with and without the sample yields the phase shift due to the sample alone, where

$$\Delta \phi = \Delta \phi_S - \Delta \phi_0. \tag{217}$$

The sample decrement δ for neutrons (X-rays) is, then,

$$\delta = -\frac{\lambda \sin \phi \tan \phi}{D \left(\Delta \phi_S - \Delta \phi_0 \right)}. \tag{218}$$

13. THE MAXWELL–BOLTZMANN DISTRIBUTION FOR THE REACTOR SOURCE OF THERMAL NEUTRONS

Reactor sources produce a wavelength-dependent λ Maxwell–Boltzmann neutron flux distribution $\Phi(v)$, which is derived from the velocity-dependent v flux distribution $\Phi(v)$ as follows. The Maxwell–Boltzmann neutron density $n(v)$ distribution for neutrons of rest mass m_n and velocities between v and $v + dv$ is given by

$$n(v) = \frac{4 n_0 v^2}{v_t^3 \sqrt{\pi}} \exp\left(-v_r^2\right). \tag{219}$$

The total, constant neutron density is n_0 for all neutron velocities v in the distribution. The ratio v_r is a particular neutron velocity v divided by the neutron peak velocity v_t in the distribution:

$$v_r = \frac{v}{v_t}. \tag{220}$$

The temperature of the Maxwell–Boltzmann distribution is T, and the peak neutron velocity v_t of the distribution is given by

$$v_t = \sqrt{\frac{2kT}{m_n}}. \tag{221}$$

Here, k is the Boltzmann constant, and the corresponding peak neutron de Broglie wavelength λ_t of the Maxwell–Boltzmann distribution is

$$\lambda_t = \frac{h}{\sqrt{2m_n kT}} = \frac{h}{m_n v_t}. \tag{222}$$

The Maxwell–Boltzmann distribution of beam neutrons of density $n(v)$ emerges from the reactor, and the neutron beam flux $\Phi(v)$ as a function of velocity v is the product of the neutron beam density $n(v)$ and neutron velocity v, where

$$\Phi(v) = nv = \frac{4n_0 v^3}{v_t^3\sqrt{\pi}} \exp\left(-v_r^2\right). \tag{223}$$

The increment of the velocity flux distribution $\Phi(v)dv$ and the corresponding increment of the wavelength flux distribution $\Phi(\lambda)\, d\lambda$ describe the same set of neutrons. As a result, the velocity $\Phi(v)\, dv$ and wavelength $\Phi(\lambda)d\lambda$ increments contain the same differential number dN of neutrons, where

$$dN = \left|\Phi(v)dv\right| = \left|\Phi(\lambda)d\lambda\right|. \tag{224}$$

The de Broglie relationship between the neutron particle and the neutron wave representation is

$$v = \frac{h}{m_n\lambda} \tag{225}$$

where

$$dv = -\frac{h}{m_n\lambda^2}d\lambda. \tag{226}$$

Substitution for the expressions in Eqs. 225 and 226 for v and dv as a function of wavelength λ in $\Phi(v)dv$ of Eq. 223 gives

$$\left|\Phi(v)dv\right| = \frac{4n_0}{v_t^3\sqrt{\pi}}\left(\frac{h}{m_n}\right)^4 \frac{d\lambda}{\lambda^5}\exp\left(-v_r^2\right). \tag{227}$$

Expression of v_t in terms of λ_t by the de Broglie relation gives

$$\left|\Phi(v)dv\right| = \frac{4n_0}{\sqrt{\pi}}\frac{2kT}{h}\frac{\lambda_t^5}{\lambda^5}\exp\left(-\frac{v^2}{v_t^2}\right)d\lambda. \tag{228}$$

Let λ_0 be the ratio of the distribution peak wavelength λ_t divided by the neutron wavelength λ, where

$$\lambda_0 = \frac{\lambda_t}{\lambda} = \frac{v}{v_t}. \tag{229}$$

Using the expression for λ_0 of Eq. 229 in the expression of Eq. 228 for $\left|\Phi(v)dv\right|$ gives

$$\left|\Phi(v)dv\right| = \left(\frac{8kT}{h\sqrt{\pi}}n_0\right)\lambda_0^5\exp\left(-\lambda_0^2\right)d\lambda. \tag{230}$$

Having determined $\left|\Phi(v)dv\right|$, one can equate to $\left|\Phi(\lambda)d\lambda\right|$ and then solve for the Maxwell–Boltzmann distribution neutron flux distribution $\Phi(\lambda)$ in terms of neutron wavelength λ. The neutron beam flux $\Phi(\lambda)$ for neutrons with wavelengths between λ and $\lambda + d\lambda$ is, then,

$$\Phi(\lambda) = \left(\frac{8kT}{h\sqrt{\pi}}n_0\right)\lambda_0^5\exp\left(-\lambda_0^2\right). \tag{231}$$

The peak neutron flux $\Phi_p(\lambda)$ occurs at wavelength $\lambda = \lambda_p$, where

$$\lambda_p = \lambda_t\sqrt{\frac{2}{5}} = \frac{h}{\sqrt{5m_nkT}}. \tag{232}$$

Solving for λ_t gives

$$\lambda_t = \sqrt{\frac{5}{2}}\lambda_p = \lambda_t\sqrt{\frac{2}{5}} = \frac{h}{\sqrt{5m_nkT}} = \frac{h}{\sqrt{2m_nkT}}. \tag{233}$$

The flux distribution $\Phi(\lambda)$ is normalized to unity at peak wavelength λ_p. With $\lambda_t^2 = 2.5\,\lambda_p^2$, the relative neutron flux distribution $\Phi\,(\lambda)$ becomes

$$\Phi(\lambda) = \frac{n_0 kT}{h}\sqrt{\frac{2 \cdot 5^5}{\pi}} \cdot \left(\frac{\lambda_p}{\lambda}\right)^5 \exp\left[2.5\left(1 - \frac{\lambda_p^2}{\lambda^2}\right)\right]. \tag{234}$$

REFERENCES

Bonse, U., & Hart, M. (1965). An x-ray interferometer. *Applied Physics Letters, 6*, 155–156.

Borrmann, G. (1941). Über Extinktionsdiagramme der Röntgenstrahlen (English translation: About the X-ray extinction diagram). *Zeitschrift für Physik, 42*, 157–162.

Eisberg, R., & Resnick, R. (1974). Chapter 3 De Broglie's Postulates - Wavelike Properties of Particles of *Quantum Physics of Atoms, Molecules, Solids, Nuclei, and Particles.* John Wiley and Sons, New York.

Frank, A. I. (1987). Optics of ultracold neutrons and the neutron microscope problem. *Soviet Physics Uspekhi, 30*(2), 110–132.

Frank, A. I., & Nosov, V. G. (2004). Quantum effects in a one-dimensional magnetic gravitational trap for ultracold neutrons. *JETP Letters, 79*(7), 313–315.

Klein, A. G., Kearney, P. D., Opat, G. I., Cimmino, A., & Gähler, R. (1981). Neutron interference by division of wavefront. *Physical Review Letters, 46*(14), 959–962.

Knowles, J. W. (1956). Anamolous absorption of slow neutrons and X-rays in nearly perfect single crystals. *Acta Crytallographica, 9*, 61–69.

Laue, M. V. (1949). Die Absorption der Röntgenstrahlen in Kristallen im Interferenzfall. *Acta Crystallographica, 2*, 106–113.

Rauch, H., Treimer, W., & Bonse, U. (1974). Test of a single crystal neutron interferometer. *Physics Letters, 47A*, 369–371.

Sears, V. F. (1989). *Neutron Optics.* Oxford University Press, Oxford, UK.

Squires, G. L. (1996). *Introduction to the Theory of Thermal Neutron Scattering.* Dover Publications, New York.

Vladimirskiĭ, V. V. (1961). Magnetic mirrors, channels, and bottles for cold neutrons. *Soviet Physics Journal of Experimental and Theoretical Physics, 12*, 740–746.

Warren, B. F. (1990). *X-ray Diffraction.* Dover Publications, New York.

Chapter 6

X-ray and Neutron Optics

Jay Theodore Cremer, Jr.

Contents

1. Derivation of the Index of Refraction for X-rays and Neutrons 607
2. Types of X-ray and Neutron Lenses and Significance of Delta/Mu Ratio 612
3. Coherent, Incoherent, Elastic, and Inelastic Scatter of X-rays and Neutrons 615
4. X-ray and Neutron Attenuation Comparison 619
5. Useful Formulas for X-rays and Neutrons 622
6. X-ray Multilayer Mirrors 624
7. Neutron Multilayer Mirrors 636
8. Capillary Optics and Kumakhov X-ray and Neutron Lenses 641
9. Bright-field, Dark-field, and Phase-Contrast Microscopy 646

References 653

1. DERIVATION OF THE INDEX OF REFRACTION FOR X-RAYS AND NEUTRONS

To derive the relation between the X-ray or neutron index of refraction n and the X-ray or neutron scattering length b of a material, one considers a thin circular disk of thickness t and radius R centered on the z-axis in a vacuum. The disk is situated halfway between a distant upstream point neutron source that emits neutrons of wavelength λ (and hence wave number $k = 2\pi/\lambda$) and a distant downstream detector. Both the source and detector are centered on the z-axis, each at a distance Z from the disk. The disk thickness t is negligible compared to the large distance Z, which allows a source of small radius R_s to be modeled as a point source, where $R_s \ll Z$.

Advances in Imaging and Electron Physics, Volume 172, ISSN 1076-5670, DOI: 10.1016/B978-0-12-394422-1.00006-2.
Copyright © 2012 Elsevier Inc. All rights reserved.

The disk consists of atoms of a single isotope with nuclei number density ρ. Each nucleus has a neutron scattering length of b. In terms of neutron ray optics, the disk material has an index of refraction of n, such that

$$n = 1 - \delta. \tag{1}$$

Here, δ is the decrement and represents the departure of the material index of refraction from that of a vacuum ($n = 1$). Typically, for most materials, the decrement is much less than 1, where $\delta \ll 1$.

A point source of amplitude $E_0[n/s]$ emits a spherical neutron wave, which passes through the thin disk at axial distance Z and arrives at the detector center point at a large distance 2Z with intensity $E_{\text{det}}(n/s)$, given by

$$E_{\text{det}} = \frac{E_0}{2Z} \exp\left(ik2Z\right) \exp[itk\,(n-1)]. \tag{2}$$

Compared to a vacuum, the neutron wave experiences scattering and an associated phase shift ϕ in the disk material, where

$$\phi = kt(n-1). \tag{3}$$

That is, the phase shift ϕ is the product of the dimensionless decrement δ multiplied by the ratio of the disk thickness t divided by the neutron wavelength λ, where

$$\phi = 2\pi \frac{t\delta}{\lambda}. \tag{4}$$

Since $\delta \ll 1$, a binomial expansion of the second exponential expression in Eq. 2 yields amplitude E_{det} at the point source detector:

$$E_{\text{det}} = \frac{E_0}{2Z} \left(1 - ikt\delta\right) \exp\left(ik2Z\right). \tag{5}$$

Next, one needs to evaluate the decrement δ in terms of the X-ray or neutron scattering length b, the nuclei number density of the material ρ, and the neutron wavelength λ by considering the scattering of neutrons from the individual nuclei of the material in the thin disk. In this case, we examine the portion of the wave incident at the disk on a particular nucleus situated at radius r from the center disk axis with amplitude

$E_{disk}[n/s]$. As before, the point source of amplitude $E_0[n/s]$ emits a spherical neutron wave that is incident on the thin disk at axial distance Z with amplitude $E_{disk}[n/s]$ at a nucleus situated in the disk, where

$$E_{disk} = \frac{E_0}{H} \exp\left(ikH\right). \tag{6}$$

The total or off-axis distance (hypotenuse) H from the point source to the disk nucleus at axial distance Z and radius r is

$$H = \sqrt{Z^2 + r^2}. \tag{7}$$

The axial distance Z between the point source and the disk is assumed to be much larger than the disk radius R, and thus the spherical wave appears to be planar to the scattering nuclei in the disk. The total distance from source point to scattering nucleus is, by binomial expansion,

$$H = Z\left(1 + \frac{r^2}{2Z^2}\right) \quad (Z \gg R). \tag{8}$$

This distance H is also the distance from the scattering disk nucleus to the detector center point.

At off-axis distance H from the source, an emitted incident neutron plane wave function of amplitude Ψ_0 coherently scatters from a disk nucleus with scattering length b. This plane wave is scattered from the disk nucleus to produce an outgoing spherical wave whose amplitude $\psi_s(d)$ at distance d from the scattering nucleus is given by

$$\psi_s(d) = -\frac{b\Psi_0}{d} \exp(ikd). \tag{9}$$

In this case, the incident amplitude is

$$\Psi_0 = E_{disk}. \tag{10}$$

The distance from the disk scattering nucleus to the detector center point is $d = H$, so the scattered amplitude $\Psi_{\text{det}}[n/s]$ is

$$\Psi_{\text{det}} = -\frac{E_0 b}{H^2} \exp(2ikH). \tag{11}$$

Substitution of the binomial approximation of the distance H of Eq. 8 into Eq. 11 gives

$$\Psi_{\text{det}} = -\frac{E_0 b}{Z^2} \exp(2ikZ) \exp\left(\frac{ikr^2}{Z}\right). \tag{12}$$

Next, we sum the scattering intensities from all the nuclei in the disk that arrive at the center point of the detector. The differential density of nuclei $\Delta\rho$ in an annular volume differential ΔV of thickness t at radius r is given by

$$\Delta V = 2\pi \cdot r \cdot t \cdot \Delta r. \tag{13}$$

The differential nuclei density $\Delta\rho$ is

$$\Delta\rho = \rho\Delta V. \tag{14}$$

The differential scattering intensity $\Delta\Psi_{\text{det}}$ received at the detector center point is

$$\Delta\Psi_{\text{det}} = \Psi_{\text{det}}\Delta\rho = \Psi_{\text{det}} \cdot t \cdot \rho \cdot 2\pi \cdot r \cdot \Delta r. \tag{15}$$

Integration over the disk radius yields the total scattered neutron intensity received at the center point of the detector $\Psi_{sc}[n/s]$, where

$$\Psi_{sc} = -\frac{I_0 bt\rho 2\pi}{Z^2}\exp(2ikZ)\int_0^R r\exp\left(\frac{ikr^2}{Z}\right)dr. \tag{16}$$

To solve the integral without loss of generality, one can replace the disk radius R upper limit of integration by infinity. That is, the contribution of the integrand from R to ∞ is zero, since $\rho = 0$ (no nuclei in this region). This allows one to evaluate the integral easily by using the following tabulated integral expression:

$$\int_0^\infty x\exp\left(-x^2\right)dx = \frac{1}{2}. \tag{17}$$

To solve our particular integral of Eq. 17, the following three substitutions are done:

$$-x^2 = \frac{ikr^2}{Z} \tag{18}$$

$$r = x\sqrt{\frac{iZ}{k}} \tag{19}$$

$$dr = dx\sqrt{\frac{iZ}{k}}. \tag{20}$$

At the detector center point, the total scattered intensity is obtained by performing the integration of Eq. 17, which yields

$$\Psi_{sc} = -\frac{iI_0 bt\rho\lambda}{2Z}\exp(2ikZ). \tag{21}$$

Arriving at the detector center point is also the spherical wave from the point source, which was unscattered by the disk and has amplitude $\Psi_{00}[n/s]$, where

$$\Psi_{00} = \frac{E_0}{2Z}\exp(ik2Z). \tag{22}$$

The scattered Ψ_{sc} and unscattered amplitude Ψ_{00} of the waves arriving at the center point of the detector sum to give the total amplitude Ψ_{tot}, where

$$\Psi_{tot} = \Psi_{sc} + \Psi_{00}. \tag{23}$$

Substitution for Ψ_{sc} and Ψ_{00} of Eqs. 21 and 22 into Eq. 23 gives

$$\Psi_{tot} = \frac{E_0}{2Z}\left(1 - ibt\rho\lambda\right)\exp(ik2Z). \tag{24}$$

Next, one equates the macroscopic neutron ray geometric optics expression for detected neutron amplitude E_{det} with the expression for detected neutron amplitude Ψ_{tot}, which considers the microscopic realm of neutron scattering from individual nuclei, where

$$E_{\text{det}} = \Psi_{tot}. \tag{25}$$

Substitution in Eq. 25 for E_{det} of Eq. 5 and Ψ_{tot} of Eq. 24 gives

$$\frac{E_0}{2Z}\left(1 - ikt\delta\right)\exp\left(ik2Z\right) = \frac{E_0}{2Z}\left(1 - ibt\rho\lambda\right)\exp(ik2Z). \tag{26}$$

Solving this expression, the decrement δ of the X-ray or neutron-refractive index of the disk material is then

$$\delta = \frac{b\rho\lambda^2}{2\pi}. \tag{27}$$

The decrement is proportional to the number density of the lens atoms ρ, the X-ray or neutron scatter length b, and the square of the X-ray or neutron wavelength λ. In the case of X-rays, the X-ray scatter length $b_{xray}(fm)$ of each element of the lens material is the product of the complex atomic scattering factor f^0 of the element multiplied by the classical radius of the electron r_e, where $r_e = 2.81\,fm$ and

$$b_{xray} = f^0 r_e = f^0 \frac{e^2}{4\pi\varepsilon_0 m_e c^2} = 2.81\left(f_r^0 - if_i^0\right) \quad \text{(X-rays).} \qquad (28)$$

where the real and imaginary parts of the atomic scatter factor are f_r^0 and f_i^0, e and m_e are the electron charge and mass, ε_0 is the permittivity of vacuum space, and c is the speed of light.

For low-Z atoms and long X-ray wavelengths, the real part of the atomic scatter factor is approximately equal to the atomic number, where $f_0^r = Z$. In the case of high-Z atoms, the electrons in some orbitals come very close to the nucleus at relativistic speeds. The increase of the electron mass due to relativistic speeds thereby decreases the electron scattering length compared to that of a free electron. As a result, the real part f_r^0 of the atomic scatter length is less than atomic number Z by corrections of approximately $(Z\alpha)^2$ (see Kissel and Pratt, 1985).

2. TYPES OF X-RAY AND NEUTRON LENSES AND SIGNIFICANCE OF DELTA/MU RATIO

Application of Snell's law determines whether the shape and material of a lens focuses or defocuses X-rays or neutrons. The incident angle is measured relative to the lens surface normal that the points toward the source or object plane in the opposite direction of the incident wave propagation. The refracted angle is measured relative to the lens surface normal that points in the same direction of incident wave propagation toward the image plane. Depending on the shape of the two lens surfaces and the lens material, the refracted ray angle, which is calculated from the incident ray angle and lens refractive index, determines whether the lens focuses or defocuses the X-rays or neutrons.

A biconcave lens focuses X-rays and neutrons if the decrement is positive ($\delta > 0$), with refractive index $n = 1 - \delta$ less than 1 ($n < 1$), which is the case for most materials for X-rays and neutrons. If the decrement of the biconcave material is negative ($\delta < 0$), with refractive index $n = 1 - \delta$ greater than 1 ($n > 1$), such as the case of visible light or titanium for neutrons, then the biconcave lens defocuses. Conversely, a biconvex lens defocuses X-rays and neutrons if the decrement is positive ($\delta > 0$), with refractive index $n = 1 - \delta$ less than 1 ($n < 1$), which is the case for most materials for X-rays and neutrons. If the decrement of the biconvex lens material is negative ($\delta < 0$), with refractive index $n = 1 - \delta$ greater than

TABLE 1 Focusing and Defocusing Lens by Shape and Refractive Index

Lens Shape	Decrement δ	Refractive Index $n = 1 - \delta$	X-ray and Neutron Refraction
Biconcave	Positive	Less than 1	Focus
Biconcave	Negative	Greater than 1	Defocus
Plano-concave	Positive	Less than 1	Focus
Plano-concave	Negative	Greater than 1	Defocus
Biconvex	Positive	Less than 1	Defocus
Biconvex	Negative	Greater than 1	Focus
Plano-convex	Positive	Less than 1	Defocus
Plano-convex	Negative	Greater than 1	Focus

1 ($n > 1$), such as the case of visible light or titanium for neutrons, then the biconcave lens focuses (Table 1).

The phase shift $\Delta\phi$ incurred by the neutron or X-ray passage through a distance Δl of a material or magnetic field with refractive index $n = 1 - \delta$, as compared to its passage through the same distance Δl in a mass-free or field-free vacuum (air) with $n = 1$, is given by

$$\Delta\phi = \frac{2\pi \cdot \Delta l \cdot \delta}{\lambda}. \tag{29}$$

The fractional phase shift of the neutron or X-ray wave $\Delta\phi$ with respect to 2π can be expressed as the fraction change of its wavelength given by

$$\frac{\Delta\lambda}{\lambda} = \frac{\Delta\phi}{2\pi}. \tag{30}$$

Hence, the distance Δl leads to a phase shift measured in terms of wavelength, or

$$\Delta\lambda = \delta \cdot \Delta l. \tag{31}$$

Thus, the decrement δ is the neutron wave phase shift (in units of its wavelength) per unit length of wave propagation through the material or magnetic field, where

$$\delta = \frac{\Delta\lambda}{\Delta l}. \tag{32}$$

In materials or gas-filled magnetic field regions, the incident X-ray or neutron wave amplitude Φ_0 is attenuated in distance Δl with linear attenuation coefficient μ_a, where

$$\Phi = \Phi_0 \exp\left(-\mu_a \Delta l\right). \tag{33}$$

For small distance Δl, the binomial series approximation yields the transmitted amplitude

$$\Phi = \Phi_0 \left(1 - \mu_a \Delta l\right). \tag{34}$$

Hence, the loss of amplitude $\Delta\Phi$ is

$$\Delta\Phi = \Phi_0 - \Phi = \mu_a \Delta l. \tag{35}$$

The linear attenuation μ_a is the decrease of X-ray or neutron wave amplitude $\Delta\Phi$ per unit length Δl of wave propagation through the lens material, where

$$\mu_a = \frac{\Delta\Phi}{\Delta l}. \tag{36}$$

The X-ray or neutron index of refraction n, which includes attenuation in a material compound refractive lens (CRL) or compound refractive prism (CRP) or through a gas-filled magnetic CRL or CRP, becomes

$$n = 1 - \delta + i\frac{\mu\lambda}{4\pi}. \tag{37}$$

The X-ray or neutron decrement-attenuation ratio δ/μ characterizes the ability of material to refract and yet be sufficiently transmitting. Materials for prism and lenses can be chosen on the basis of the largest δ/μ. The ratio δ/μ is the neutron or X-ray wave phase shift $\Delta\lambda$ in wavelengths per unit propagation length Δl in a material, divided by the decrease in the neutron wave amplitude $\Delta\Phi$ per unit propagation length Δl in the material, where

$$\frac{\delta}{\mu} = \frac{\frac{\Delta\lambda}{\Delta l}}{\frac{\Delta\Phi}{\Delta l}}. \tag{38}$$

Equivalently, the decrement/attenuation ratio δ/μ is the ratio of the neutron or X-ray wave phase shift $\Delta\lambda$ in wavelengths per decrease in the neutron wave amplitude $\Delta\Phi$, where

$$\frac{\delta}{\mu} = \frac{\Delta\lambda}{\Delta\Phi}. \tag{39}$$

3. COHERENT, INCOHERENT, ELASTIC, AND INELASTIC SCATTER OF X-RAYS AND NEUTRONS

The incoherent inelastic $\sigma_i^{ii}(\lambda)$, incoherent elastic scatter $\sigma_i^{ie}(\lambda)$, and coherent inelastic scatter $\sigma_i^{ci}(\lambda)$ have a complicated dependence on wavelength, temperature, and the physical form of the material (amorphous, single-crystal, polycrystal, porosity, density variation, etc.). The coherent elastic scatter cross section σ_i^{ce}, in which the neutrons are scattered without losing energy and maintain fixed phase relationships with other neutrons, is responsible for the refraction, reflection, and diffraction in optical materials and samples. The coherent elastic scatter cross section σ_i^{ce} for an isotope nucleus is related to its scattering length b by

$$\sigma_i^{ce} = 4\pi b^2. \tag{40}$$

And the scattering length b is used to calculate the nuclear strong force contribution to the decrement δ.

From these microscopic cross sections $\sigma(\text{cm}^2)$, one can determine the macroscopic cross section or linear attenuation $\mu(\text{cm}^{-1})$ by multiplication of $\sigma(\text{cm}^2)$ by the number density $\rho(\#/\text{cm}^3)$ of the atoms in the material in the case of X-rays and X-ray scatter, or the nuclei number density in the case of neutrons and neutron scatter cross sections, where

$$\mu(\text{cm}^{-1}) = \sigma(\text{cm}^2)\rho(\text{cm}^{-3}). \tag{41}$$

For a material composed of many elements in the case of X-rays, or composed of many isotopes in the case of neutrons, one can determine the macroscopic cross section or linear attenuation of the composite material. We determine the linear attenuation coefficient $\mu(\lambda)$ for X-ray or neutron scatter in optical or sample materials that are composed of Q elements for X-rays and Q isotopes for neutrons. Each ith element in the case of X-rays, or each ith isotope in the case of neutrons, has microscopic cross section $\sigma_i(\lambda)$ and nuclei number density ρ_i. Their product $\sigma_i\rho_i$ is the ith element or ith isotope macroscopic cross section (linear attenuation) $\mu_i(\lambda)$ contribution to the composite material, which is wavelength-dependent, where $\mu_i(\text{cm}^{-1})$ is

$$\mu_i(\text{cm}^{-1}) = \sigma_i\rho_i. \tag{42}$$

The wavelength-dependent total macroscopic cross section or linear attenuation $\mu[\lambda]$ for the composite material is the sum of its Q constituents

$\mu_i(\lambda)$, where $\mu(\text{cm}^{-1})$ is

$$\mu[\lambda] = \sum_{i=1}^{Q} \mu_i(\lambda) = \sum_{i=1}^{Q} \rho_i \sigma_i(\lambda). \tag{43}$$

To determine the number density of each *i*th element (for X-rays) or *i*th isotope (for neutrons), one must determine its weight fraction w_i or fraction of its weight (mass) in the mass density $\rho(\text{g/cm}^3)$ of the composite material, where $0 \leq w_i \leq 1$ for $i = 1, 2, 3, \ldots, Q$. The sum of the dimensionless weight fractions w_i of the Q elements (X-rays) or Q isotopes (neutrons) is unity, where

$$\sum_{i=1}^{Q} w_i = 1. \tag{44}$$

The product $w_i\rho(\text{g/cm}^3)$ represents the mass density of the *i*th element (for X-rays) or *i*th isotope (for neutrons), and *i*th mass density $w_i\rho$ is converted to its equivalent molar density $m_i(\text{mole/cm}^3)$ by dividing by gram atomic weight $A_i(\text{g/mole})$ of the *i*th element (isotope averaged) for X-rays or isotope, where the *i*th molar density $m_i(\text{mole/cm}^3)$ is

$$m_i = \frac{w_i\rho}{A_i}. \tag{45}$$

A *mole* is a quantity of 6.02×10^{23} of atoms, nuclei, neutrons, electrons, or molecules in which convenient counting units are millimoles or concentration units (millimoles per cubic centimeter), whereas molar quantities for larger entities (such as blood plasma platelets), the counts, and the count density are more conveniently described in units of picomoles and picomoles/cm^3 per or picomoles/liter, where, in SI units, 1.0 liter equals 1000 cm^3.

The mole is defined in terms of carbon-12, in which 12 grams or 1 mole of carbon-12 contains 6.02×10^{23} carbon atoms or nuclei, and the mole is represented by Avogadro's number N_A, where

$$N_A = 6.02 \times 10^{23}\ \text{units/mole}. \tag{46}$$

Hence, the atom number density $\rho_i(\text{atom/cm}^3)$ for the *i*th element (for X-rays) or isotope number density $\rho_i(\text{nuclei/cm}^3)$ for the *i*th isotope (for neutrons) is

$$\rho_i = N_A m_i = \frac{w_i\rho}{A_i}. \tag{47}$$

X-ray optical cross sections and associated atom number densities are less involved than those of neutrons when one must obtain and calculate macroscopic cross sections $\sigma_i \rho_i$ for the multiple isotopes of some elements. For example, beryllium (100% Be-9) or aluminum (100% Al-13) neutron or X-ray lenses require far less effort than that of MgF_2 (3 Mg isotopes, 1 F isotope) lenses.

Coherent X-ray or neutron scatter occurs when phase shift relationships between scattering centers are fixed so that constructive and destructive summing of scattered amplitudes occur, and one observes interference effects between the scatter neutron waves at an observation plane. Coherent elastic scatter of X-rays occurs in Rayleigh scatter. The X-ray scatters coherently from all the electrons of an atom in which the X-ray scatter recoil is shared by the atom as a whole rather than a single electron. The momentum transfer from the X-ray to the atom results in a negligible change in the velocity of the atom upon recoil. Thus, almost no X-ray energy is transferred to the atomic electron cloud, the X-ray scatter is elastic, and the scattered X-rays from the atoms have a fixed phase relationship that allows for interference. Coherent elastic scatter is responsible for neutron refraction, reflection, and diffraction. If in coherent scatter the energy of the neutron is unchanged, then the coherent scatter is elastic; otherwise, a change of neutron energy via the effects of temperature with its associated lattice vibrations and phonon exchange leads to coherent inelastic scatter.

Thermal vibrations in material lattices such as crystals can cause the coherent elastic scatter to become inelastic due to phonon exchange between the lattice and the incident X-rays or neutrons. Hence, the energy of the X-ray or neutron can slightly increase or decrease, depending on whether it receives or transfers a phonon of discrete packet lattice vibration energy to the lattice in coherent scatter. In magnetic structures, the coherent elastic scatter of neutrons or X-rays can become inelastic if there is an exchange of magnons, and discrete packets of energy arise from the precession of aligned, unpaired electron spins with their associated magnetic dipole moments. Absorption or emission of the magnon by the X-ray or neutron can cause it to inelastic scatter in coherent scatter with a magnetic material, such as helical spin lattice material such as Au_2Mn, antiferromagnetic material such as $KMnF_3$, or ferromagnetic material such as iron.

Neutron or X-ray incoherent scatter is random scatter in which the X-rays or neutrons are scattered in a random fashion from single scattering atoms in which there is no fixed phase relationship between the scattered waves of the X-rays or neutrons. If the X-ray or neutron energy is unchanged in this random scatter, then the scatter is termed *incoherent elastic scatter.* In Compton forward scatter, the X-rays are incoherently scattered by single electrons in the forward direction without any change in

X-ray energy and direction. Compton scatter only in the forward direction is an example of incoherent elastic scatter for X-rays. In all other directions, the Compton incoherent scatter of X-rays from a single electron changes the X-ray energy and direction and is thus incoherent, inelastic scatter.

X-rays have insufficient energy for pair production, so the absorption analog for X-rays is limited to photoelectric absorption. Absorption of X-rays by photoelectric absorption in the 0.5 – 2-Å wavelength range dominates over Compton incoherent scatter and Rayleigh coherent scatter. In photoelectric absorption, an X-ray is absorbed, usually by a *K*-shell electron in which the electron is ejected with energy that is equal to the difference between the incident X-ray energy and the electron binding energy. The electron vacancy is filled with another electron, leading to X-ray emission or ejection of Auger electrons. For neutrons, absorption occurs by thermal neutron capture.

The neutron absorption cross section $\sigma_i^a(\lambda)$ follows the $1/v$ rule, in which neutron absorption via capture by the nucleus increases with decreasing neutron velocity, where

$$\sigma_i^a(\lambda) \propto \lambda \quad \text{and} \quad \sigma_i^a(\lambda) \propto \frac{1}{v}. \tag{48}$$

Usually, neutron capture is followed by the emission of capture gammas that range from X-rays to 10.8-MeV gammas from nitrogen-14 thermal neutron capture. Neutron capture cross sections range from 0.0035 barn for carbon, 1.9×10^{-4} barn for oxygen, 9.6×10^{-3} barn for fluorine, 0.33 barn for hydrogen, 940 barn for 7.5% Li-6 in naturally occurring lithium, 3835 barn for 20% B-10 in naturally occurring boron, 20,600 barn for 12.22% Cd-113 in naturally occurring cadmium, 259,000 barn for 15.7% Gd-157, and 61,100 barn for 14.8% Gd-155 in naturally occurring gadolinium. Lithium has the distinction of emitting almost no gammas upon thermal neutron absorption. Upon thermal neutron absorption, Li-6 splits into a 2.72-MeV triton and 2.06-MeV alpha, with almost no emission of gammas.

This makes lithium a good choice for thermal neutron shielding around thermal neutron imaging detectors because the 1- to 2-MeV triton and alpha are easily absorbed by a millimeter of the lithium carbonate Li_2CO_3. With 30-cm-thick LiF $\left(2.63\,\text{g/cm}^3\right)$ or Li_2CO_3 $\left(2.1\,\text{g/cm}^3\right)$, one can reduce depressing thermal neutron images by the reduction of gamma background noise that would occur if one used boron-based shielding, hydrogen-based plastic or water shielding, cadmium, or gadolinium, which all create significant gamma background noise that reduces thermal neutron image contrast. One can press the lithium fluoride or lithium carbonate powder into bricks enclosed with thin Teflon ($-C_2F_4-$) sheet, tube, or ziplock plastic bags.

TABLE 2 Comparison of Scatter and Absorption Processes with X-rays and Neutrons

Neutron	X-ray	Comments
Coherent elastic scatter $\sigma_i^{ce}(\lambda)$	Rayleigh	Refraction, reflection, diffraction
Coherent inelastic scatter $\sigma_i^{ci}(\lambda)$	Rayleigh	Temperature effects—wavelength shift
Incoherent elastic scatter $\sigma_i^{ie}(\lambda)$	Compton	Forward scatter only—transmission
Incoherent inelastic scatter $\sigma_i^{ii}(\lambda)$	Compton	All other scatter directions—attenuation
Absorption by capture $\sigma_i^{ii}(\lambda)$	Photoabsorb	Absorption

Then, the attenuation cross section $\sigma_i^{att}(\lambda)$ for X-rays or neutrons, which excludes the coherent, elastic scatter cross section, is the sum of the coherent inelastic σ_i^{ci}, incoherent inelastic σ_i^{ii}, incoherent elastic σ_i^{ie}, and absorption cross sections σ_i^{a}, where

$$\sigma_i^{att}(\lambda) = \sigma_i^{ci} + \sigma_i^{ii} + \sigma_i^{ie} + \sigma_i^{a}. \tag{49}$$

Attenuation of X-rays or neutrons of wavelength λ in a particular direction occurs with inelastic scatter in which the X-ray neutron loses or gains energy, and attenuation occurs with incoherent scatter in which the X-ray or neutron changes direction randomly; and finally, attenuation occurs by absorption of the X-ray or neutron. A comparison of the X-ray and neutron coherent and incoherent elastic and inelastic processes is shown in Table 2.

4. X-RAY AND NEUTRON ATTENUATION COMPARISON

At the X-ray energy range of 1–2 Å, which is used with X-ray microscopes with CRLs, the attenuation cross sections is dominated by photoelectric absorption and to a much lesser extent by Compton scatter and coherent Rayleigh scatter of refraction. Table 3 displays the X-ray Rayleigh (coherent), Compton (incoherent), and absorption (photoelectric absorption) cross sections for 1.8-Å X-rays with comparison to the coherent scatter, incoherent scatter, and absorption cross sections of thermal neutrons of the same wavelength 1.8 Å that occurs at room temperature (293°K).

Table 3 also includes the commonly used thermal neutron-shielding elements of lithium, boron, cadmium, and gadolinium. Most elements, including isotope B11, have positive scatter lengths b, whereas Li-7, Ni-62,

TABLE 3 3 X-ray and Neutron Cross Sections (barns) and Scatter Factors at 1.8 .

	X-ray[1,2]					Neutron-isotope averaged[3]				
Element	Ray	Com	Photo	f_1	f_2	Coh	Incoh	Absorb	b_{coh}	b_{inc}
Hydrogen	0.08	0.6	0.02	1.0	1.6E–05	1.8	80.3	0.33	–3.74	25.3
Deuterium	0.08	0.6	0.02	1.0	1.6E–05	5.6	2.05	0.0005	6.67	4.04
Helium	0.7	0.9	0.8	2.0	5.8E–05	1.3	0	0.0075	3.26	–
Lithium	1.4	1.3	6.3	3.0	4.8E-04	0.4	0.9	70	–1.90	–
Beryllium	2	2	25	4.0	0.0022	7.6	0.0018	0.0076	7.79	0.12
Boron	3.3	2	69	5.0	0.0058	3.5	1.7	767	5.3-0.2i	–
Carbon	5	2	155	6.0	0.013	5.5	0.001	0.0035	6.65	0
Nitrogen	8	3	302	7.0	0.026	11	0.5	1.90	9.36	0
Oxygen	11	3	533	8.1	0.047	4.2	0.0008	0.0002	5.80	0
Fluorine	16	3	870	10.0	0.076	4.0	0.0008	0.0096	5.65	–0.08
Magnesium	33	4	2801	12.2	0.25	3.6	0.08	0.06	5.37	0.15
Aluminum	38	4	3841	13.3	0.33	1.5	0.0082	0.23	3.45	0.26
Silicon	44	4	5115	14.3	0.44	2.2	0.004	0.17	4.15	0
Titanium	109	6	26174	21.9	2.43	1.5	2.9	6.1	–3.44	0.26
Vanadium	122	6	30283	22.5	2.83	0.02	5.1	5.1	–0.38	6.33
Chromium	138	6	34345	23.0	3.13	1.7	1.8	3.0	3.65	0.65
Manganese	151	6	37273	22.7	3.60	1.7	0.4	13.3	–3.73	1.79
Iron	168	6	5362	22.9	0.51	11.2	0.4	2.6	9.45	0

Cobalt	189	6	6622	25.1	0.59	0.8	4.8	37.2	2.49	–6.2
Nickel	210	6	7860	26.5	0.70	13.3	5.2	4.5	10.3	0.04
Copper	235	6	9007	27.7	0.81	7.5	0.55	3.78	7.72	0.7
Zinc	255	6	10375	28.9	0.93	4.0	4.1	1.1	5.68	–0.06
Gallium	275	6	11890	30.1	1.06	6.7	0.2	2.7	7.23	–0.84
Germanium	295	7	13556	31.2	1.21	8.4	0.2	2.2	8.18	0.26
Tin	739	9	79569	49.7	7.15	4.9	0.022	0.63	6.22	0
Cadmium	678	9	67885	48.0	6.16	3.0	3.5	2520	4.9–0.7i	0
Gadolinium	1285	11	42693	54.4	4.42	29	151	49700	6.5–13.8i	–
Lead	2323	11	122103	78.5	11.37	11.1	0.003	0.171	9.40	0.03
Bismuth	2383	12	128250	79.5	11.86	9.2	0.0084	0.0338	8.53	0

[1]NIST XCOM X-ray cross sections (http://physics.nist.gov/PhysRefData/Xcom/html/xcom1.html).

[2]LBNL Center for X-ray Optics (http://henke.lbl.gov/optical_constants).

[3]NIST neutron cross section and scatter length tables (http://www.ncnr.nist.gov/resources/n-lengths/). $i = \sqrt{-1}$.

Ni-64, Ti-48, V-51, and 100% Mn have negative scattering lengths, and the thermal neutron-absorbing Li-6, B-10, Cd-113, Gd-155, and Gd-157 have complex scatter lengths. Finally, Table 3 lists for 1.8-Å wavelengths or energy 6882 eV the real f_1 and imaginary f_2 parts of the complex X-ray scatter factor f, where

$$f = f_1 + if_2. \tag{50}$$

And for 1.8 Å, the coherent b_{coh} and incoherent b_{inc} components of the neutron scatter length b are shown, where

$$b = b_{coh} + b_{inc}. \tag{51}$$

5. USEFUL FORMULAS FOR X-RAYS AND NEUTRONS

For X-rays, one converts between X-ray energy in keV and X-ray wavelength in Å, via Planck's constant $h = 4.14 \times 10^{-15}$ eV × s, where 6882 eV corresponds to 1.8 Å and

$$E(\text{eV}) = hf = \frac{hc}{\lambda} = \frac{1.24 \times 10^4}{\lambda[A]}. \tag{52}$$

One converts between neutron energy eV and neutron wavelength Å via $\lambda = h/p$. In the non-relativistic case of thermal neutrons, the neutron energy E is related to its momentum via

$$E = \frac{p^2}{2m_n}. \tag{53}$$

With rest mass of the neutron at 9.39×10^8 eV, the neutron energy E becomes

$$E[A] = \frac{1}{2m_n}\left(\frac{h}{\lambda}\right)^2 = \frac{1}{2m_n c^2}\left(\frac{hc}{\lambda}\right)^2 = \frac{0.082}{\lambda^2[A]}. \tag{54}$$

The incident neutron velocity v, in terms of the incident neutron energy E, is

$$v(\text{cm/s}) = \sqrt{\frac{2E}{m_n}} = 1.39 \times 10^6 \sqrt{E(\text{eV})}. \tag{55}$$

Thermal neutron energy is also described in terms of the material temperature. The neutron collisions with atoms in the material and the resulting elastic scatter energy exchanges cause the average random kinetic energy of the neutron $mv_n^2/2$ to attain the average thermal energy of the atoms. In elastic scatter, the neutron flux from a reactor has a velocity distribution $\Phi(v)$, which is described by a Maxwell–Boltzmann distribution, where

$$\Phi(v) \propto \frac{1}{v^3} \exp\left(-\frac{m_n v^2}{2k_B T}\right). \tag{56}$$

The Boltzmann constant k_B is

$$k_B = 8.6210^{-5} \frac{\text{eV}}{^\circ K}. \tag{57}$$

The velocity v_m at which the distribution $\Phi(v)$ is at its maximum is

$$v_m = \sqrt{\frac{3k_B T}{m_n}}. \tag{58}$$

The random neutron kinetic energy E, at the peak or most likely velocity v_m of the Maxwellian distribution, is

$$E = \frac{1}{2} m_n v_m^2 = \frac{3}{2} k_B T. \tag{59}$$

By convention, one expresses neutron energy E in terms of $k_B T$, where

$$E(\text{eV}) = k_B \times T = 8.6210^{-5}\, T(^\circ K). \tag{60}$$

A room temperature $T = 293$ K, a thermal neutron of energy 0.025 eV, wavelength 1.8 Å travels 2.2 mm/μs.

X-rays react to the electromagnetic potentials, fields, and forces (and gravitational fields of very large masses). However, neutrons react to (1) the strong force field of the nucleus; (2) the magnetic fields of unpaired atomic electrons; (3) the spin-orbit interaction of the nucleus electric field appearing as a magnetic field in the neutron rest frame due to relative velocity v; (4) the atomic electric field in the Foldy interaction due to the neutral, but slightly separated, positive and negative portions of the neutron that give rise to its magnetic dipole moment; (5) applied laboratory magnetic fields, electric fields (either an electric field by Foldy interaction or relative velocity of neutron in an electric field), and electromagnetic fields (powerful lasers in TEM_{00} mode); and (6) gravity. Neutrons have quite a variety of coherent and incoherent and elastic and inelastic scatter possibilities and applications.

At present, however, neutron optics is severely limited by the neutron sources. The spectral brightness, which is the number of photons or neutrons emitted unit source area per second into a unit solid angle within a 0.1% bandwidth, is the source advantage of X-rays over neutrons in terms of optics. That is, the existing reactor and spallation neutron sources of thermal and colder neutrons currently produce far less neutron flux for neutron optics than in the case of the powerful synchrotron radiation sources of X-rays and laser sources of light and microwaves.

On the imaging side, compared to X-rays, the smaller scatter and absorption neutron cross in optical materials allow more neutron transmission in the comparable wavelength ranges of 1 to 2 Å. This allows larger lens aperture diameters in compound refractive neutron lenses. The path length through the biconcave lens material for X-ray and neutrons increases with the aperture radii in compound refractive spherical and parabolic lenses. Hence, the smaller scatter and absorption cross section for neutrons in most materials, compared to that of X-rays, allows for larger aperture diameters in CRLs used with neutrons than is allowed with X-rays. Aluminum, copper, and beryllium have been used to focus both thermal neutrons and X-rays in the range of 1–2 Å. Compound refractive lenses in which the decrements δ are similar reflect their similar cross sections for coherent scatter.

The CRL depends on the number of lenses, the radius of curvature of the lens, and the complex decrement δ, whose real part is due to coherent scatter of refraction and whose imaginary part is due to attenuation by absorption and incoherent scatter. For similar complex decrements δ, the CRL yields the same optical effects for X-rays and neutrons. Presently, X-ray imaging optics provide resolutions below 0.1 μm, which is about 10-fold better than visible light, and present thermal neutron imaging provides resolutions of 10–100 μm, about 20- to 200-fold worse than visible light optics.

6. X-RAY MULTILAYER MIRRORS

A multilayer mirror for X-rays consist of alternating thin layers of high-Z and low-Z materials in which vacuum deposition technology can produce uniform layers with nanometer range thickness. With typically 10 to 20 periods or more of alternating low-Z and high-Z materials, the multilayers act as interference reflectors with layer thickness on the order of nanometers. The constructive interference from summing of tens to hundreds of reflected waves leads to narrow-band X-ray emission within narrow ranges of reflected angles. Also, multilayers can extend the range of X-ray incident angles that produce significant reflection.

Multilayer mirrors fabricated from alternating thin film layers of high-Z tungsten and low-Z carbon on a substrate of silicon or alternating film layers of high-Z molybdenum and low-Z silicon mounted on a silicon substrate are two examples of an X-ray multilayer mirror. A multilayer mirror with about 20 periods or 40 layers of alternating high-Z and low-Z thin film layers and individual layer thicknesses of 20 to 100 Å are readily fabricated and can provide larger ranges of X-ray incident angles with moderate to almost 100% reflectivity over certain ranges of incident angles.

The fraction of reflected and transmitted X-rays for an incident X-ray beam can be calculated via a transfer matrix method, which is derived as follows. One begins by application of the boundary condition for the electric and magnetic fields of the X-ray plane waves. The tangential components of the electric and magnetic fields are each continuous over each planar interface of the multilayer mirror. For nonpolarized X-rays, the electric field has components parallel and perpendicular to each planar interface surface. The case of the component X-ray electric field parallel to the surface with its associated perpendicular magnetic field in the plane of incidence, but generally not perpendicular to the surface interface, is derived first.

Consider the case of N layers of uniform thickness T that comprise $N/2$ periods, which is set on a thick $N + 1$ substrate layer. Each period consists of a high-Z layer and a low-Z layer, such that one produces a multilayer mirror of N layers alternating in high- and low-Z material. Incident electric field amplitude E^i_{q-1} of the $q - 1$ layer is incident on the planar surface interface between the $q - 1$ and q layers. Also incident at the same point on the first planar interface between the $q - 1$ and q layers are the following three fields: (1) the reflected electric field amplitude E^r_q from the layer q into the layer $q - 1$, (2) the electric field amplitude E^t_q transmitted into layer q from layer $q - 1$, and (3) incident electric field amplitude E^i_q produced by reflection from the $q + 1$ layer. Corresponding to the electric field amplitudes are their associated perpendicular magnetic field amplitudes B^i_{q-1}, B^r_q, B^t_q, and B^i_q.

For the case of the electric field polarized parallel to the surface interface, the boundary conditions for the tangential components of the electric and magnetic fields at the $q - 1$ and q layers interface are

$$E_q = E^i_{q-1} + E^r_q = E^t_q + E^i_q \quad (q = 1, 2, 3, \ldots, N). \tag{61}$$

The incident ray angle θ^i_{q-1} and the refracted (transmitted) ray angle θ^t_q are measured relative to the perpendicular or normal unit vector of the planar surface interfaces, where

$$B_q = B^i_{q-1}\cos\theta^i_{q-1} - B^r_q\cos\theta^i_{q-1} = B^t_q\cos\theta^t_q - B^i_q\cos\theta^t_q \quad (q = 1, 2, 3, \ldots, N). \tag{62}$$

Incident electric field amplitude E^i_{q+1} of the q layer is incident on the planar, surface interface between the q and $q+1$ layers. Incident at the same point on the second planar interface between the q and $q+1$ layers are the three following fields: (1) the reflected electric field amplitude E^r_{q+1} from the layer $q+1$ into the layer q, (2) the electric field amplitude E^t_{q+1} transmitted into layer $q+1$ from layer q, and (3) incident electric field amplitude E^i_{q+2} produced by reflection from the $q+2$ layer. Corresponding to the electric field amplitudes are their associated perpendicular magnetic field amplitudes B^i_{q+1}, B^r_{q+1}, B^t_{q+1}, and B^i_{q+2}.

For the case of the electric field polarized parallel to the surface interface, the boundary conditions for the tangential components of the electric and magnetic fields at the q and $q+1$ layers interface

$$E_{q+1} = E^i_{q+1} + E^r_{q+1} = E^t_{q+1} + E^i_{q+2} \quad (q = 1, 2, 3, \ldots, N). \tag{63}$$

The refracted (transmitted) ray angles θ^t_q, θ^t_{q+1}, and θ^t_{q+2} in the respective q, $q+1$, and $q+2$ layers are measured relative to the normal unit vector of the planar surface interfaces, where

$$B_{q+1} = B^i_{q+1}\cos\theta^t_q - B^r_{q+1}\cos\theta^t_q = B^t_{q+1}\cos\theta^t_{q+1} - B^i_{q+2}\cos\theta^t_{q+2} \quad (q = 1, 2, 3, \ldots, N). \tag{64}$$

In the plane wave X-ray, the magnetic field B can be expressed in terms of the electric field E divided by the wave propagation speed v through a medium of refractive index n, where

$$v = \frac{c}{n}. \tag{65}$$

In the propagation speed formula is light c, which is expressed in terms of the vacuum permittivity ε_0 and permeability μ_0, where

$$c = \frac{1}{\sqrt{\varepsilon_0\mu_0}}. \tag{66}$$

The magnetic field B is then

$$B = \frac{E}{v} = \left(\frac{n}{c}\right)E = n\sqrt{\varepsilon_0\mu_0}E. \tag{67}$$

The tangential electric and magnetic field component boundary conditions at the interface of the $q-1$ and q layers for electric field polarization parallel to the interface surfaces become

$$E_q = E^i_{q-1} + E^r_q = E^t_q + E^i_q \quad (q = 1, 2, 3, \ldots, N) \tag{68}$$

$$B_q = \gamma_{q-1}\left(E^i_{q-1} - E^r_q\right) = \gamma_q\left(E^t_q - E^i_q\right) \quad (q = 1, 2, 3, \ldots, N). \tag{69}$$

The tangential electric and magnetic field component boundary conditions at the interface of the q and $q+1$ layers for electric field polarization parallel to the interface surfaces become

$$E_{q+1} = E^i_{q+1} + E^r_{q+1} = E^t_{q+1} + E^i_{q+2} \quad (q = 1, 2, 3, \ldots, N) \tag{70}$$

$$B_{q+1} = \gamma_q\left(E^i_{q+1} - E^r_{q+1}\right) = \gamma_{q+1}E^t_{q+1} - \gamma_{q+2}E^i_{q+2} \quad (q = 1, 2, 3, \ldots, N). \tag{71}$$

The factors γ_{q-1}, γ_q, and γ_{q+1} with units of (sec /m) are given by

$$\gamma_{q-1} = n_{q-1}\sqrt{\varepsilon_0\mu_0}\cos\theta^i_{q-1} \quad (q = 1, 2, 3, \ldots, N) \tag{72}$$

$$\gamma_q = n_q\sqrt{\varepsilon_0\mu_0}\cos\theta^t_q \quad (q = 1, 2, 3, \ldots, N) \tag{73}$$

$$\gamma_{q+1} = n_{q+1}\sqrt{\varepsilon_0\mu_0}\cos\theta^t_{q+1} \quad (q = 1, 2, 3, \ldots, N) \tag{74}$$

$$\gamma_{q+2} = n_{q+2}\sqrt{\varepsilon_0\mu_0}\cos\theta^t_{q+2} \quad (q = 1, 2, 3, \ldots, N). \tag{75}$$

The phase shift ϕ between the reflected plane wave in layer $q-1$ from point A_q on the interface between the $q-1$ and q layers, and that of a reflected plane wave, also in layer $q-1$ from the corresponding point A_{q+1} directly below the interface between the q and $q+1$ layers, is

$$\phi = \frac{2\pi}{\lambda_{q-1}} n_q T \cos^t_q \quad (q = 1, 2, 3, \ldots, N). \tag{76}$$

The thickness T_q of the qth layer of the multilayer mirror can vary, but in the case of uniform layer thickness T,

$$T_q = T \quad (q = 1, 2, 3, \ldots, N). \tag{77}$$

The wavelength λ_{q-1} is expressed in terms of the wavelengths in a vacuum λ_0 and vacuum speed c. The frequency f of the electromagnetic wave is the same for a vacuum or material, but the wavelength λ_0 and phase velocity c of the electromagnetic wave changes in a material is divided by the refractive index n. In a vacuum,

$$c = f\lambda_0. \tag{78}$$

In a material with refractive index n, the phase velocity v is

$$v = \frac{c}{n} = f\frac{\lambda_0}{n}. \tag{79}$$

Hence,

$$\lambda_q = \frac{\lambda_0}{n_q} \quad (q = 1, 2, 3, \ldots, N). \tag{80}$$

The X-ray refractive index of the air or vacuum medium layer ($q = 0$) adjacent to the first $q = 1$ layer of the multilayer mirror is $n_0 = 1$. The X-ray index of refraction n_q for the qth layer is expressed in terms of the decrement $\delta_q$and absorption index β_q:

$$n_q = 1 - \delta_q + i\beta_q \quad (q = 1, 2, 3, \ldots, N). \tag{81}$$

The complex atomic scattering function f^0 is the sum of real f_1^0 and imaginary f_2^0 components, where

$$f^0 = f_1^0 + if_2^0. \tag{82}$$

The decrement real δ_q contribution from each element z of the M elements that comprise the qth layer is determined by the real component of the complex atomic scattering function $f_{1,z}^0$ and the atom number density $\rho_{q,z}$ of each element z, the classical electron radius r_e, and the vacuum space wavelength λ_0 of the X-ray, where

$$\delta_q = \frac{r_e}{2\pi}\lambda_0^2 \sum_{z=1}^{M} \rho_{q,z} f_{1,z}^0. \tag{83}$$

The absorption index β_q is obtained from the imaginary component of the complex atomic scattering function $f_{2,z}^0$, where

$$\beta_q = \frac{r_e}{2\pi}\lambda_0^2 \sum_{z=1}^{M} \rho_{q,z} f_{2,z}^0. \tag{84}$$

The phase shift ϕ_q between the reflected waves from two adjacent planar interfaces $q-1$ and q becomes

$$\phi_q = \left(2\pi \frac{n_{q-1}}{\lambda_0}\right) n_q T_q \cos_q^t \quad (q = 1, 2, 3, \ldots, N). \tag{85}$$

The phase shift ϕ_q, which occurs for the X-ray in its single pass through the qth layer of thickness T_q, is the same phase shift as in the Bragg diffraction between two layers, except that in refraction, the wave front is bent at the first interface twice, and in diffraction, the wave front is not bent. As a consequence, there are phase shifts as the electric field amplitudes pass through a qth layer of thickness T_q, in which

$$E_{q+1}^i = E_q^t \exp(-i\phi_q) \quad (q = 1, 2, 3, \ldots, N) \tag{86}$$

or

$$E_q^i = E_{q+1}^r \exp(-i\phi_q) \quad (q = 1, 2, 3, \ldots, N). \tag{87}$$

These two relations can be used to eliminate the fields E_{q+1}^i and E_{q+1}^r in the boundary condition between the q and $q+1$ layers so that, at the q and $q+1$ boundary interface,

$$E_{q+1} = E_q^t \exp(-i\phi_q) + E_q^i \exp(i\phi_q) = E_{q+1}^t + E_{q+2}^i \quad (q = 1, 2, 3, \ldots, N) \tag{88}$$

$$B_{q+1} = \gamma_q \left(E_q^t \exp(-i\phi_q) - E_q^i \exp(i\phi_q)\right) = \gamma_{q+1} E_{q+1}^t - \gamma_{q+2} E_{q+2}^i \quad (q = 1, 2, 3, \ldots, N). \tag{89}$$

We now examine the left pair of equations in these two equations, namely,

$$E_{q+1} = E_q^t \exp(-i\phi_q) + E_q^i \exp(i\phi_q) \quad (q = 1, 2, 3, \ldots, N) \tag{90}$$

$$B_{q+1} = \gamma_q \left(E_q^t \exp(-i\phi_q) - E_q^i \exp(i\phi_q)\right) \quad (q = 1, 2, 3, \ldots, N). \tag{91}$$

This pair of equations is solved for E_q^t and E_q^i in terms of E_{q+1} , B_{q+1}, γ_q, and ϕ:

$$E_q^t = \left(\frac{\gamma_q E_{q+1} + B_{q+1}}{2\gamma_q}\right) \exp(i\phi_q) \quad (q = 1, 2, 3, \ldots, N) \tag{92}$$

$$E_q^i = \left(\frac{\gamma_q E_{q+1} - B_{q+1}}{2\gamma_q}\right) \exp(-i\phi_q) \quad (q = 1, 2, 3, \ldots, N). \tag{93}$$

At the interface of the $q-1$ and q layers, the rightmost terms of the tangential electric and magnetic field expressions of E_q and B_q for the boundary condition equalities are

$$E_q = E_q^t + E_q^i \quad (q = 1, 2, 3, \ldots, N) \tag{94}$$

$$B_q = \gamma_q \left(E_q^t - E_q^i \right) \quad (q = 1, 2, 3, \ldots, N). \tag{95}$$

The following identities are used:

$$2\cos\phi = \exp(i\phi) + \exp(-i\phi) \tag{96}$$

$$2i\sin\phi = \exp(i\phi) - \exp(-i\phi). \tag{97}$$

These expressions for E_q^t and E_q^i of Eqs. 92 and 93 are substituted into the $q-1$ and q interface expressions of Eqs. 94 and 95 to obtain

$$E_q = E_{q+1}\cos\phi_q + B_{q+1}\left(\frac{i\sin\phi_q}{\gamma_q}\right) \quad (q = 1, 2, 3, \ldots, N) \tag{98}$$

$$B_q = E_{q+1}\left(i\gamma_q \sin\phi_q\right) + B_{q+1}\cos\phi_q \quad (q = 1, 2, 3, \ldots, N). \tag{99}$$

This pair of equations can be expressed by a 2×2 transfer matrix M_q which, for the $q-1$ and q interface, is

$$\begin{bmatrix} E_q \\ B_q \end{bmatrix} = M_q \begin{bmatrix} E_{q+1} \\ B_{q+1} \end{bmatrix} \quad (q = 1, 2, 3, \ldots, N). \tag{100}$$

The transfer matrix from the $q+1$ layer to the q layer is

$$M_q = \begin{bmatrix} \cos\phi_q & \frac{i\sin\phi_q}{\gamma_q} \\ i\gamma_q\sin\phi_q & \cos\phi_q \end{bmatrix} \quad (q = 1, 2, 3, \ldots, N). \tag{101}$$

One of these layers is of high-Z material and the other layer is low-Z material. The transfer matrix relates the fields of the q and $q+1$ layers. Hence, the terms for the $q+2$ layer are not involved.

For an N-layer multilayer mirror, the transfer matrix allows a simple mathematical representation and means of relating the tangential components of the electric and magnetic field at the surface of the Nth to that of the first layer, where

$$\begin{bmatrix} E_1 \\ B_1 \end{bmatrix} = M_1 M_2 M_3 \ldots M_q \ldots M_{N-1} M_N \begin{bmatrix} E_N \\ B_N \end{bmatrix}. \tag{102}$$

The overall transfer matrix M_T is, then,

$$M_T = M_1 M_2 M_3 \ldots M_q \ldots M_{N-1} M_N. \tag{103}$$

The transfer matrix relates the E_q and B_q of the q layer to E_{q+1}, and the B_{q+1} of the $q+1$ layer via transfer matrix M_q, where

$$\begin{bmatrix} E_q \\ B_q \end{bmatrix} = \begin{bmatrix} \cos\phi_q & \frac{i\sin\phi_q}{\gamma_q} \\ i\gamma_q\sin\phi_q & \cos\phi_q \end{bmatrix} \begin{bmatrix} E_{q+1} \\ B_{q+1} \end{bmatrix} \quad (q = 1, 2, 3, \ldots, N). \tag{104}$$

At the interface of the $q-1$ and q layers, E_q and B_q are expressed by

$$E_q = E^i_{q-1} + E^r_q \quad (q = 1, 2, 3, \ldots, N) \tag{105}$$

$$B_q = \gamma_{q-1}\left(E^i_{q-1} - E^r_q\right) \quad (q = 1, 2, 3, \ldots, N). \tag{106}$$

At the interface of the q and $q+1$ layers, E_{q+1} and B_{q+1} are expressed by

$$E_{q+1} = E^t_{q+1} + E^i_{q+2} \quad (q = 1, 2, 3, \ldots, N) \tag{107}$$

$$B_{q+1} = \gamma_{q+1}E^t_{q+1} - \gamma_{q+2}E^i_{q+2} \quad (q = 1, 2, 3, \ldots, N). \tag{108}$$

The expressions for E_q, B_q, E_{q+1}, and B_{q+1} are inserted into the matrix equation for the interface of the q and $q+1$ layers, where $q = 1, 2, 3, \ldots, N$ and

$$\begin{bmatrix} E^i_{q-1} + E^r_q \\ \gamma_{q-1}\left(E^i_{q-1} - E^r_q\right) \end{bmatrix} = \begin{bmatrix} \cos\phi_q & \frac{i\sin\phi_q}{\gamma_q} \\ i\gamma_q\sin\phi_q & \cos\phi_q \end{bmatrix} \begin{bmatrix} E^t_{q+1} + E^i_{q+2} \\ \gamma_{q+1}E^t_{q+1} - \gamma_{q+2}E^i_{q+2} \end{bmatrix}. \tag{109}$$

This equation is expressed by the transfer matrix M_q for the q and $q+1$ interface layer:

$$\begin{bmatrix} E^i_{q-1} + E^r_q \\ \gamma_{q-1}\left(E^i_{q-1} - E^r_q\right) \end{bmatrix} = M_q \begin{bmatrix} E^t_{q+1} + E^i_{q+2} \\ \gamma_{q+1}E^t_{q+1} - \gamma_{q+2}E^i_{q+2} \end{bmatrix} \quad (q = 1, 2, 3, \ldots, N). \tag{110}$$

The 2×2 matrix M_q has the four matrix elements m^q_{11}, m^q_{12}, m^q_{21}, and m^q_{22}, where

$$M_q = \begin{bmatrix} \cos\phi_q & \frac{i\sin\phi_q}{\gamma_q} \\ i\gamma_q\sin\phi_q & \cos\phi_q \end{bmatrix} = \begin{bmatrix} m^q_{11} & m^q_{12} \\ m^q_{21} & m^q_{22} \end{bmatrix} \quad (q = 1, 2, 3, \ldots, N). \tag{111}$$

The incident medium of air or a vacuum is the $q-1=0$ layer, which is adjacent to the $q=1$ first layer of the multilayer mirror; and the $q=N$ last layer of N, alternating high- and low-Z thin film layers, is followed by the thick $N+1$ layer of substrate to provide rigidity and support for mounting and orienting. Finally, there is a $N+2$ layer of medium composed of the air (vacuum) and/or the mounting device for the multilayer mirror. With the overall transfer matrix M_T, the multilayer mirror expression becomes

$$\begin{bmatrix} E_0^i + E_1^r \\ \gamma_0\left(E_0^i - E_1^r\right) \end{bmatrix} = M_T \begin{bmatrix} E_{N+1}^t + E_{N+2}^i \\ \gamma_{N+1}E_{N+1}^t - \gamma_{N+2}E_{N+2}^i \end{bmatrix}. \tag{112}$$

The 2×2 overall transfer matrix M_T is composed of matrix elements M_{11}, M_{12}, M_{21}, and M_{22}, where

$$M_T = M_1M_2M_3\ldots M_q\ldots M_{N-1}M_N = \begin{bmatrix} M_{11} & M_{12} \\ M_{21} & M_{22} \end{bmatrix}. \tag{113}$$

The $N+1$ layer of substrate is assumed to be sufficiently thick so that there are no reflected waves from the distant $N+2$ layer of mounting material and/or air (vacuum). Hence, one can drop the $N+2$ terms:

$$\begin{bmatrix} E_0^i + E_1^r \\ \gamma_0\left(E_0^i - E_1^r\right) \end{bmatrix} = \begin{bmatrix} M_{11} & M_{12} \\ M_{21} & M_{22} \end{bmatrix} \begin{bmatrix} E_{N+1}^t \\ \gamma_{N+1}E_{N+1}^t \end{bmatrix}. \tag{114}$$

On the left side of this equation are (1) the incident electric E_0^i and associated magnetic fields $\gamma_0E_0^i$ from the $q=0$ vacuum (air) layer adjacent to the surface of the $q=1$ first layer of the multilayer; and (2) the electric E_1^r and magnetic fields $\gamma_0E_1^r$ emerging from the first $q=1$ surface of the multilayer back into the vacuum (air) medium. The overall transfer matrix M_T relates the incident layer tangential electric and magnetic field amplitudes to the transmitted or refraction tangential electric field E_{N+1}^t and associated magnetic field $\gamma_{N+1}E_{N+1}^t$ in the $N+1$ substrate layer. The electric field is polarized parallel to the interface planar surfaces between the layers.

Dividing both sides of the above matrix representation of the multilayer mirror of Eq. 114 by the incident electric field E_0^i of the air (vacuum) $q=0$ layer, one obtains

$$\begin{bmatrix} 1+r \\ \gamma_0(1-r) \end{bmatrix} = \begin{bmatrix} M_{11} & M_{12} \\ M_{21} & M_{22} \end{bmatrix} \begin{bmatrix} t \\ t\gamma_{N+1} \end{bmatrix}. \tag{115}$$

The reflection r and transmission coefficients t are

$$r = \frac{E_1^r}{E_0^i} \tag{116}$$

and

$$t = \frac{E_{N+1}^t}{E_0^i}. \tag{117}$$

With two equations and two unknowns in this matrix expression, one solves the overall X-ray reflection r and transmission coefficients t for the X-ray multilayer mirror, where

$$r = \frac{\gamma_0 M_{11} + \gamma_0 \gamma_{N+1} M_{12} - M_{21} - \gamma_{N+1} M_{22}}{\gamma_0 M_{11} + \gamma_0 \gamma_{N+1} M_{12} + M_{21} + \gamma_{N+1} M_{22}} \tag{118}$$

and

$$t = \frac{2\gamma_0}{\gamma_0 M_{11} + \gamma_0 \gamma_{N+1} M_{12} + M_{21} + \gamma_{N+1} M_{22}} \tag{119}$$

where

$$M_q = \prod_{q=1}^{N} \begin{bmatrix} m_{11}^q & m_{12}^q \\ m_{21}^q & m_{22}^q \end{bmatrix} = \prod_{q=1}^{N} \begin{bmatrix} \cos\phi_q & \frac{i\sin\phi_q}{\gamma_q} \\ i\gamma_q \sin\phi_q & \cos\phi_q \end{bmatrix} \quad (q = 1, 2, 3, \ldots, N). \tag{120}$$

With vacuum wave speed c,

$$\gamma_q = n_q \sqrt{\varepsilon_0 \mu_0} \cos\theta_q^t = \frac{n_q}{c} \cos\theta_q^t \quad (q = 1, 2, 3, \ldots, N). \tag{121}$$

With vacuum wavelength λ_0,

$$\phi = \left(2\pi \frac{n_{q-1}}{\lambda_0}\right) n_q T_q \cos_q^t \quad (q = 1, 2, 3, \ldots, N). \tag{122}$$

For uniform layer thickness T,

$$T_q = T \quad (q = 1, 2, 3, \ldots, N). \tag{123}$$

Parameters γ_q and ϕ are calculated in an iterative fashion, starting with the first layer $q = 1$ in which Snell's law is used, where

$$n_{q-1} \sin\theta_{q-1}^i = n_q \sin\theta_q^t \quad (q = 1, 2, 3, \ldots, N). \tag{124}$$

Transposing

$$\theta_q^t = \sin^{-1}\left(\frac{n_{q-1}\sin\theta_{q-1}^i}{n_q}\right) \quad (q = 1, 2, 3, \ldots, N). \tag{125}$$

The refractive index n_q has two values, n_H for the high-Z layers and n_L for the low-Z layers, in which the low-Z material is at the top surface and the bottommost layer just above the substrate is high-Z material, so that

$$n_q = \begin{matrix} n_L & q = 1, 3, 5, \ldots \\ n_H & q = 2, 4, 6, \ldots \end{matrix}. \tag{126}$$

The X-ray reflectivity R and the transmittivity T are, respectively, the squared magnitudes of the reflection coefficient r and the transmission coefficient t, where

$$R = |r|^2 = \left|\frac{\gamma_0 M_{11} + \gamma_0\gamma_{N+1} M_{12} - M_{21} - \gamma_{N+1} M_{22}}{\gamma_0 M_{11} + \gamma_0\gamma_{N+1} M_{12} + M_{21} + \gamma_{N+1} M_{22}}\right|^2 \tag{127}$$

and

$$T = |t|^2 = \frac{2\gamma_0}{\gamma_0 M_{11} + \gamma_0\gamma_{N+1} M_{12} + M_{21} + \gamma_{N+1} M_{22}}. \tag{128}$$

Consider the case of a special high-reflectance multilayer mirror in which the low-Z material thickness T_H is twice as thick as the high-Z material thickness T_L, but the decrement of the high-Z material is approximately twice that of the low-Z material, where

$$n_q = \begin{matrix} n_L = 1 - \delta & q = 2, 4, 6, \ldots \\ n_H = 1 - 2\delta & q = 1, 3, 5, \ldots \end{matrix} \tag{129}$$

and

$$T_q = \begin{matrix} T_L = 2T & q = 2, 4, 6, \ldots \\ T_H = T & q = 1, 3, 5, \ldots \end{matrix}. \tag{130}$$

In the thicker, less-refractive, low-Z layer and the less thick, but more refractive, high-Z layer, a single passage across either layer causes a quarter-wavelength shift or a 90° phase shift:

$$\phi_q = \left(2\pi\frac{n_{q-1}}{\lambda_0}\right) n_q T_q \cos_q^t = \frac{\pi}{2} \quad (q = 1, 2, 3, \ldots, N) \tag{131}$$

and

$$\cos\phi_q = \cos\frac{\pi}{2} = 0 \quad \text{and} \quad \sin\phi = \sin\frac{\pi}{2} = 1 \tag{132}$$

$$M_q = \prod_{q=1}^{N} \begin{bmatrix} m_{11}^q & m_{12}^q \\ m_{21}^q & m_{22}^q \end{bmatrix} = \prod_{q=1}^{N} \begin{bmatrix} 0 & \frac{i1}{\gamma_q} \\ i\gamma_q 1 & 0 \end{bmatrix} \quad (q = 1, 2, 3, \ldots, N). \tag{133}$$

The reflection is zero $r = 0$, and only transmission occurs when the numerator of the reflection coefficient r is zero, where

$$\gamma_0 M_{11} + \gamma_0 \gamma_{N+1} M_{12} - M_{21} - \gamma_{N+1} M_{22} = 0 \Rightarrow r = 0. \tag{134}$$

So far, this analysis applies to the case of the electric field polarized parallel to the planar interface surfaces in which its associated perpendicular magnetic field lies in the plane of incidence, but in general, the magnetic field is not perpendicular to the planar interface surfaces.

Now consider the case of the magnetic field polarized parallel to the planar interface surfaces in which its associated perpendicular electric field lies in the plane of incidence, but in general, the electric field is not perpendicular to the planar interface surfaces. These results and equations for the electric field polarized parallel to the surface interfaces apply, except one must replace the parameter γ_q with the following expression:

$$\gamma_q = \frac{n_q\sqrt{\varepsilon_0\mu_0}}{\cos\theta_q^t} = \frac{n_q}{c}\frac{1}{\cos\theta_q^t} \quad (q = 1, 2, 3, \ldots, N). \tag{135}$$

If the X-ray has normal incidence to the multilayer, where $\theta_0^i = 0$, then

$$\theta_{q-1}^i = 0 \quad (q = 1, 2, 3, \ldots, N). \tag{136}$$

For normal incidence, the results for the electric field polarized parallel to the interface surfaces and the magnetic field to the interface surfaces are identical. In the case of general oblique incidence, if the incident X-rays are unpolarized with equal components of electric field parallel and the magnetic field parallel to the surface interfaces, then the total reflection coefficient r_T is the average of the reflection coefficient for electric field parallel $r_{E||}$ and magnetic field parallel $r_{B||}$, where

$$r_T = \frac{r_{E||} + r_{B||}}{2}. \tag{137}$$

7. NEUTRON MULTILAYER MIRRORS

Thin film multilayers are used as neutron mirrors and monochromators. As with X-rays, the neutron multilayers are fabricated with alternating layers of highly neutron-refractive material and less neutron-refractive material, such as iron and manganese or germanium and manganese, whose layer thickness is in the nanometer range. That is, one stacks bilayers or periods composed of one material whose neutron refractive index is significantly less than the other layer. About 10 to 100 periods, corresponding to 20 to 200 layers of two alternating high and low neutron refractive index material with 50-Å layer thickness, are readily fabricated for neutron mirrors and monochromators. If each bilayer or period is composed of a nonmagnetic material (e.g., germanium) and a magnetized material such as iron, then the multilayer mirror or monochromator can yield polarized neutron by reflection or transmission, given a nonpolarized incident neutron beam.

The equations for the multilayer X-ray mirror for the electric field polarized parallel to the planar interface boundaries can be applied to neutron multilayer mirrors. The only modification required is the expression for the complex index of refraction of the layers. The neutron refractive index of the air or vacuum medium layer ($q = 0$), adjacent to the first $q = 1$ layer of the multilayer mirror, is $n_0 = 1$. The neutron index of refraction n_q for the qth layer is expressed in terms of the decrement δ_q and absorption index β_q:

$$n_q = 1 - \delta_q + i\beta_q \quad (q = 1, 2, 3, \ldots, N). \tag{138}$$

The decrement real δ_q contribution from each isotope a of the M elements that comprise the qth layer is determined by the real component of the complex neutron scatter length b_a^{re} and the atom number density $\rho_{q,a}$ of each isotope a and the vacuum space wavelength λ_0 of the neutron, where

$$\delta_q = \frac{1}{2\pi}\lambda_0^2 \sum_{a=1}^{M} \rho_{q,a}.b_{q,a}^{re}. \tag{139}$$

In the vacuum (or air) layer prior to the multilayer, the neutron of mass m and magnetic dipole moment μ has wavelength λ_0, speed v_0, and total energy E given by kinetic energy K_0:

$$E = K_0 = \frac{mv_0^2}{2} = \frac{h^2}{2m\lambda_0^2}. \tag{140}$$

The qth layer decrement is calculated from the qth layer nuclear $< U_{q,a}^{nuc} >$ and magnetic $< U_{q,a}^{mag} >$ mean potential energies via Planck's constant h,

where

$$\delta_{q,a} = \frac{< U_{q,a}^{nuc} > + < U_{q,a}^{mag} >}{2E}, \tag{141}$$

$$< U_{q,a}^{nuc} > = \frac{h^2 \rho_{q,a} b_{q,a}^{re}}{2m\pi}, \tag{142}$$

and

$$< U_{q,a}^{mag} > = -\boldsymbol{\mu} \cdot \mathbf{B} = \pm \mu_{||} B. \tag{143}$$

The absorption index β_q is obtained from the imaginary component of the complex neutron scatter length b_a^{im}, where

$$\beta_q = \frac{1}{2\pi} \lambda_0^2 \sum_{a=1}^{M} \rho_{q,a} b_{q,a}^{im}. \tag{144}$$

The total, real part of the coherent scattering length $b_{q,a}^{re}$ of each ath ferromagnetic isotope of the qth layer is the sum of its material or nuclear bound, real, coherent scattering length $b_{q,a}^{re,nuc}$ and its real magnetic scattering length $b_{q,a}^{re,mag}$, where

$$b_{q,a}^{re} = b_{q,a}^{re,nuc} \pm \left(\frac{B}{B_s} \right) b_{q,a}^{re,mag}. \tag{145}$$

The complex magnetic scatter depends on the degree of magnetization of the ferromagnetic material. A ferromagnetic material with saturation magnetic field flux density B_s can have a magnetic field flux density B that is less than or equal to the saturation magnetic field flux density B_s, where

$$B \leq B_s. \tag{146}$$

The real part of the magnetic scattering length $b_{q,a}^{re,mag}$ is expressed in terms of the saturation magnetic field B_s, the neutron mass m, magnetic dipole moment μ, free space wavelength λ_0, and Planck's constant h, where

$$b_{q,a}^{re,mag} = \frac{2\pi \mu_{||} B_s m}{h^2 \rho_{q,a}}. \tag{147}$$

The positive sign indicates spin-up neutrons, and the minus sign indicates spin-down neutrons. That is, the magnetic scatter length component causes spin-up and spin-down neutrons to be refracted in opposite directions. Non-ferromagnetic isotopes cause only the nuclear scatter of neutrons:

$$b_{q,a}^{re} = b_{q,a}^{re,nuc}. \tag{148}$$

The total imaginary part of the coherent scattering length $b_{q,a}^{im}$ of each ath ferromagnetic isotope of the qth layer is the sum of its material or nuclear bound, imaginary, coherent scattering length $b_{q,a}^{im,nuc}$ and its imaginary magnetic scattering length $b_{q,a}^{im,mag}$, where

$$b_{q,a}^{im} = b_{q,a}^{im,nuc} \pm \left(\frac{B}{B_s}\right) b_{q,a}^{im,mag}. \tag{149}$$

Inelastic scatter of neutrons arises with magnon exchange between the neutron and scatter lattice, and thus produces an imaginary component to the magnetic scatter length. One usually ignores the small or zero imaginary part of the magnetic scatter length. Hence, for ferromagnetic isotopes, in which inelastic magnetic scatter is ignored, and for non-ferromagnetic isotopes, the imaginary part of the nuclear scatter length is due to inelastic nuclear scatter, where

$$b_{q,a}^{im} = b_{q,a}^{im,nuc}. \tag{150}$$

The magnetic scatter length b_q for the qth layer composed of M isotopes is, in general,

$$b_q = \sum_{a=1}^{M} \left[\left(b_{q,a}^{re,nuc} + i b_{q,a}^{im,nuc} \right) \pm \left(\frac{B}{B_s}\right) \left(b_{q,a}^{re,mag} + i b_{q,a}^{im,mag} \right) \right]. \tag{151}$$

If the qth layer contains no ferromagnetic isotopes, then the ferromagnetic terms are zero.

The critical angle of reflection ϕ_c from a single mirror is the maximum incident angle measured relative to the mirror surface plane in which total reflection form the mirror occurs, where

$$\phi_c = \sqrt{2\delta}. \tag{152}$$

If the mirror material includes both nuclear and magnetic scatter due to partly or fully magnetized ferromagnetic isotopes in the mirror, then the critical angle of reflection ϕ_c from the mirror that includes both nuclear and magnetic scatter is

$$\phi_c = \sqrt{\frac{1}{\pi} \lambda_0^2 \sum_{a=1}^{M} \rho_a \left[b_a^{re,nuc} \pm \left(\frac{B}{B_s}\right) \left(\frac{2\pi \mu_{||} B_s m}{h^2 \rho_a} \right) \right]}. \tag{153}$$

For grazing incident angles less than ϕ_c on a multilayer mirror, there is total reflection.

At angles greater than the critical angle of reflection ϕ_c, one generally has partial reflection and transmission from a single mirror or a multilayer mirror. The overall neutron reflection r and transmission coefficients t for the neutron multilayer mirror are, thus,

$$r = \frac{\gamma_0 M_{11} + \gamma_0 \gamma_{N+1} M_{12} - M_{21} - \gamma_{N+1} M_{22}}{\gamma_0 M_{11} + \gamma_0 \gamma_{N+1} M_{12} + M_{21} + \gamma_{N+1} M_{22}} \tag{154}$$

and

$$t = \frac{2\gamma_0}{\gamma_0 M_{11} + \gamma_0 \gamma_{N+1} M_{12} + M_{21} + \gamma_{N+1} M_{22}}, \tag{155}$$

where

$$M_q = \prod_{q=1}^{N} \begin{bmatrix} m_{11}^q & m_{12}^q \\ m_{21}^q & m_{22}^q \end{bmatrix} = \prod_{q=1}^{N} \begin{bmatrix} \cos\phi_q & \frac{i\sin\phi_q}{\gamma_q} \\ i\gamma_q \sin\phi_q & \cos\phi_q \end{bmatrix} \quad (q = 1, 2, 3, \ldots, N). \tag{156}$$

With vacuum wave speed c,

$$\gamma_q = n_q \sqrt{\varepsilon_0 \mu_0} \cos\theta_q^t = \frac{n_q}{c} \cos\theta_q^t \quad (q = 1, 2, 3, \ldots, N). \tag{157}$$

With vacuum wavelength λ_0,

$$\phi = \left(2\pi \frac{n_{q-1}}{\lambda_0}\right) n_q T_q \cos_q^t \quad (q = 1, 2, 3, \ldots, N). \tag{158}$$

For uniform layer thickness T,

$$T_q = T \quad (q = 1, 2, 3, \ldots, N). \tag{159}$$

Parameters γ_q and ϕ are calculated in an iterative fashion, starting with the first layer $q = 1$ in which Snell's law is used, where

$$\theta_q^t = \sin^{-1}\left(\frac{n_{q-1} \sin\theta_{q-1}^i}{n_q}\right) \quad (q = 1, 2, 3, \ldots, N). \tag{160}$$

The refractive index n_q has two values, n_H for the highly neutron-refractive layers and n_L for the less neutron-refractive layers, in which the less neutron-refractive material is at the top surface and the bottommost layer just above the substrate is more neutron-refractive material, so that

$$n_q = \begin{matrix} n_L & q = 1, 3, 5, \ldots \\ n_H & q = 2, 4, 6, \ldots \end{matrix}. \tag{161}$$

Strongly reflecting neutron multilayer mirrors are constructed as in the X-ray case, where

$$n_q = \begin{matrix} n_L = 1 - \delta & q = 2, 4, 6, \ldots \\ n_H = 1 - 2\delta & q = 1, 3, 5, \ldots \end{matrix} \tag{162}$$

and

$$T_q = \begin{matrix} T_L = 2T & q = 2, 4, 6, \ldots \\ T_H = T & q = 1, 3, 5, \ldots \end{matrix}. \tag{163}$$

where

$$\phi_q = \left(2\pi \frac{n_{q-1}}{\lambda_0}\right) n_q T_q \cos_q^t = \frac{\pi}{2} \quad (q = 1, 2, 3, \ldots, N) \tag{164}$$

and

$$M_q = \prod_{q=1}^{N} \begin{bmatrix} m_{11}^q & m_{12}^q \\ m_{21}^q & m_{22}^q \end{bmatrix} = \prod_{q=1}^{N} \begin{bmatrix} 0 & \frac{i1}{\gamma_q} \\ i\gamma_q 1 & 0 \end{bmatrix} \quad (q = 1, 2, 3, \ldots, N). \tag{165}$$

The neutron reflectivity R and the transmittivity T are, respectively, the squared magnitudes of the reflection coefficient r and the transmission coefficient t, where

$$R = |r|^2 = \left| \frac{\gamma_0 M_{11} + \gamma_0 \gamma_{N+1} M_{12} - M_{21} - \gamma_{N+1} M_{22}}{\gamma_0 M_{11} + \gamma_0 \gamma_{N+1} M_{12} + M_{21} + \gamma_{N+1} M_{22}} \right|^2 \tag{166}$$

and

$$T = |t|^2 = \frac{2\gamma_0}{\gamma_0 M_{11} + \gamma_0 \gamma_{N+1} M_{12} + M_{21} + \gamma_{N+1} M_{22}}. \tag{167}$$

The reflectivity R is zero if

$$\gamma_0 M_{11} + \gamma_0 \gamma_{N+1} M_{12} = M_{21} + \gamma_{N+1} M_{22}. \tag{168}$$

Neutron supermirrors are composed of multilayer mirrors in which the period P_p decreases with increasing period number p or depth into the multilayer. The thickness of period P_p is the sum thickness T_p and T_{p+1} of more-refractive and less-refractive layers. In neutron supermirrors, the period P_p decreases with increasing period number p or depth into the multilayer:

$$P_p = T_p + T_{p+1} \quad \left(p = 1, 2, 3, \ldots, \frac{N}{2}\right). \tag{169}$$

This increase of period P_p with period number p allows the Bragg peaks of individual layers overlap and the reflectivity R is large for incident angles θ_0 up to a maximum angle θ_{max}, which depends on the smallest layer thickness in the multilayer stack. For incident angles θ_{max} larger than θ_{max}, the reflectivity R is small.

8. CAPILLARY OPTICS AND KUMAKHOV X-RAY AND NEUTRON LENSES

Polycapillary optics for X-rays and neutrons act as an effective collection, collimator, focusing, and transfer optic. The polycapillary optic for X-rays and neutrons is very useful as a collection lens to collect photons or neutrons from a point source, or as a condenser lens with an extended source to focus X-ray or neutrons onto a sample. In an X-ray or neutron-refractive microscope, a polycapillary is used as a collection or condenser lens, and a CRL is used as the objective lens to magnify and image the sample. Reflective X-ray optics using a hollow glass tube or capillary to transmit X-rays was first discovered by Jentzch and Nahrung (1931).

A glass tube that was several meters long was used for analysis of resonance radiation of 14.4 KeV from the Fe-57 nucleus by Pound and Rebke (1959) at Harvard University, followed by measurements in X-ray transmission through a single capillary by Marton (1966) and Mosher and Stephanakis (1976). Pantell and Chung (1978) measured a $1/e$ transmission of 1.54-Å X-rays through a 51.5-cm-long, cylindrical, lead glass capillary with a 150-μm bore diameter, which was curved with a 20-meter radius of curvature. Then, Pantell and Chung (1979) also found that surface roughness on highly polished surfaces of glass have little effect on the propagation of X-rays through hollow glass capillaries.

Other investigations of the transmission of X-rays through glass monocapillaries and on the focusing of synchrotron radiation were performed at the Kurchatov Atomic Energy Insititute in Moscow, Russia, by Arkad'ev *et al.* (1986, 1988a,b, 1989), Chertov and Shakhparonov (1989), Kumakhov and Komarov (1990), and Kumakhov (1990), which led to the development of polycapillary optics by the Kumakhov group; see Arkad'ev *et al.* (1988a,b) and a U.S. patent developed by Kumakhov (1993). Also, note that Kumakhov (1990) investigated channeling radiation from charged particles propagating through crystals.

The first experiments on X-ray fluorescence using tapered monocapillaries for increasing the density of the X-ray beams on a sample were performed by Engström *et al.* (1989). Synchrotron radiation focusing with tapered monocapillaries was performed by Carpenter (1989) and by Thiel *et al.* (1992). For more on monocapillary and polycapillary X-ray and neutron optics (see the websites for the Institute for Roentgen Optics at http://www.xrayoptic.ru, for XOS at www.xos.com, and for the Center for X-ray Optics at the Albany University of the State University of New York at http://www.albany.edu/X-ray-optics).

In the 1980s, Kumakhov developed the curved polycapillary optic, which was a bundle of hollow capillaries that converged at both ends in the shape of a stretched biceps muscle. Also, the curved polycapillary optic could be converged or tapered at one end and parallel at the other end. Kumakhov found that the individual capillaries needed to have a small diameter, and required curving the capillary and modifying the capillary bore diameters. The process required drawing a single tube into a single capillary in a high-temperature furnace. The capillaries were then bundled together and placed into a glass tube, which is then placed in the high-temperature furnace and drawn again to produce a single polycapillary with even smaller individual capillary bore diameters. A multitude of these polycapillary arrays, each less than 1.0 mm in diameter, are then bundled again and placed in a glass tube exceeding 1.0 mm. Bundling these capillaries and heating and drawing can create even larger diameter polycapillary. The final assemblage of polycapillaries is heated, drawn, and shaped to produce convergence of the polycapillaries at one or both ends, thereby creating a tapered polycapillary whose center diameter exceeds 1 mm with individual capillary bore diameters of 10 to 20 μm.

Tapering the polycapillary optic at both ends could allow the polycapillary to accept a significant fraction of X-rays or neutrons emitted from a small spot source and transfer and focus the X-rays or neutrons upon a sample. The tapering of a polycapillary optic is matched to the acceptance angle that allows the entering X-rays or neutrons to have a grazing incidence angle ϕ less than the critical angle ϕ_C of the smooth capillary wall material for the X-ray or neutron bandwidth. A polycapillary optic tapered at the input end, but nontapered and parallel at the output end, can then serve as a collection and collimation optic. This polycapillary collection and collimation optic can then receive a significant fraction of the isotropic-emitted X-rays or neutrons from a small source and output them as a parallel rays in collimated beam of X-rays or neutrons. A single, nontapered capillary optic, or a capillary optic tapered at one or both ends, has a much smaller acceptance angle for a small point like source of X-rays or neutrons than a bundle of hollow capillaries that are converged or tapered on one or both ends.

The polycapillary optics can act as collection lenses or condenser lenses in X-ray and neutron optics. The first polycapillary neutron lens was developed and tested at the Kurchatov IR-8 reactor in Moscow, Russia, by the Kumakhov group in the 1980s and reported by Kumakhov and Sharov (1992). Kumakhov developed a polycapillary neutron lens, which was tested in experiments with collaborators at the Cold Neutron Research Facility reactor at the National Institute of Standards Technology (NIST), and reported by Chen *et al.* (1992). At NIST, the individual polycapillaries were shown to bend a cold neutron beam through 20° in a distance of 130 mm, with a transmission of 50% and a simple polycapillary array composed of a few polycapillaries demonstrated focusing of cold neutrons (wavelength 0.2–0.9 mm) down to a 1-mm^2 spot size.

In the Kumakhov and Sharov (1992) experiments at the Russian Kurchatov IR-8 reactor with the lead glass polycapillary neutron lens, the lens had parallel hollow capillaries on the 29-mm diameter polycapillary input end that converged or tapered to a 15-mm diameter output end. The polycapillary collection and focusing lens received the collimated (± 20 min divergence angle), a 2.4-Å quasi-monochromatic thermal neutron beam from a reactor source, and then focused the neutrons to a spot of about 1.0-mm diameter at 104-mm focal distance with a neutron flux density gain of 6 to 7. The measured reflection coefficient from the lead glass walls was 99%. The neutron lens polycapillary optic was constructed from 1,000 smaller hexagonal polycapillaries. Each smaller polycapillary was 0.4 mm in width, and composed of individual capillaries with 6-μm bore diameters. Note also, in general, capillary arrays tend to be mostly hollow, with thin glass walls so that the polycapillary optic can be about 80% hollow.

X-ray capillary optics use hollow, cylindrical tubes of lead or borosilicate glass, whereas neutrons use only non-borated glass, such as lead glass. A broad energy range of X-ray or neutron waves, which enter the hollow capillary at grazing incidences ϕ less than the energy-dependent critical angle of reflection ϕ_c, are propagated through the glass waveguide, where, given the energy-dependent refractive decrement $\delta(\lambda)$,

$$\phi < \phi_c = \sqrt{2\delta}. \tag{170}$$

The decrements $\delta(\lambda) \ll 1$ are very small for X-rays and neutrons, and give critical angles ϕ_c in the range of milliradians or less for wavelengths in the range of 0.1 to 10 Å. The grazing incidence angles ϕ of the X-rays and neutrons are measured with respect to the reflecting surface. For X-rays in borosilicate glass capillaries, the critical angle ϕ(mrad) in terms of the photon energy E(keV) is given by $\phi = 30/E$.

The X-rays or neutrons can be transmitted in a curved capillary with bore diameter d and radius of curvature R, provided the angle of incidence ϕ is less than the critical angle ϕ_c, where

$$(R + d) \cos \phi = R. \tag{171}$$

Dividing by $R + d$ and retaining the first two terms of the cosine series expansion gives

$$1 - \frac{\phi^2}{2} \cong \frac{R}{R + d} \cong 1 - \frac{d}{R}. \tag{172}$$

Rearrangement shows

$$\phi^2 \cong 2\frac{d}{R} < \phi_c^2. \tag{173}$$

Hence,

$$\frac{R\phi_c^2}{2d} > 1. \tag{174}$$

See Ulrich *et al.* (1993).

The maximum change in angle $\Delta\phi_{cap}$ upon each X-ray or neutron reflection from the inner wall surface of the hollow tube capillary, as well as the acceptance angle ϕ_a of the capillary, are twice the critical angle of reflection ϕ_c, where

$$\Delta\phi_{cap} = \phi_a = 2\phi_c = 2\sqrt{2\delta}. \tag{175}$$

We now compare the single-capillary optic with a CRL composed of parabolic biconcave lenses. The acceptance angle, as well as the maximum change in angle of a X-ray or neutron ray that passes through a single, biconcave, parabolic refractive lens, can be determined by the change in angle of the incident perimeter ray. The single biconcave parabolic lens has focal $f = R/2\delta$, an on-axis lens radius of curvature R, material decrement δ, and linear attenuation μ. The transmitted perimeter ray enters the lens at the effective aperture radius r_e, which is the minimum of the lens physical radius r_m and the biconcave, parabolic lens absorption aperture radius $r_{ap} = \sqrt{4f\delta/\mu}$, where $r_e = MIN(r_m, r_{ap})$.

The maximum change of the ray angle $\Delta\phi_{len}$ passing through a single parabolic lens at the lens perimeter (the effective aperture radius r_e) is equal to the change in angle $\Delta\phi$ of a parallel perimeter ray refracted to the focal point of the single lens, where, given the small angle approximation,

$$\tan \Delta\phi_{len} \cong \Delta\phi_{len} = \frac{r_e}{f} = \frac{r_e}{R} 2\delta. \tag{176}$$

One assumes the capillary and the parabolic CRL are made of the same material, such as beryllium, and thus have the same decrement δ. The ratio η of the maximum acceptance angle and change in angle per lens refraction in a CRL, divided by the maximum acceptance angle and change in angle per reflection in the capillary optic, is

$$\eta = \frac{\Delta\phi_{len}}{\Delta\phi_{cap}} = \frac{r_e}{R} \frac{2\delta}{2\sqrt{2\delta}} = \frac{r_e}{2R}\sqrt{2\delta} \ll 1. \tag{177}$$

Hence, the capillary optic is a much better collection of X-rays and neutrons since their common decrement $\delta \ll 1$ and the parabolic lens effective aperture radius r_e is less than the biconcave, parabolic lens on-axis radius R, where $r_e < R$.

The capillary and polycapillary optics are not true refractive lenses, which can provide high-resolution imaging and magnification of an object as the CRLs. Instead, the polycapillary optic provides a mosaic image of the object based on reflected light transmission of regions of the object through the capillary pixels. The light exits from the capillary pixels of the polycapillary optic, in which light from neighboring pixels overlaps to form a low-resolution, mosaic image of the object. For high-resolution X-ray or neutron imaging of an object, a CRL is used as an objective lens to image and magnify the object, whereas one can illuminate the object with a polycapillary condenser lens and collect X-rays or neutrons from a point source with a polycapillary collection lens.

For ϕ minus the critical angle $\phi_c(\lambda)$ for wavelength λ, the transmitted X-rays or neutrons propagate by multiple reflection from the smooth glass walls, which provides a high-reflection coefficient and allow excellent transmission efficiency. The X-ray and neutron capillary optics can be monocapillary or polycapillary, in which the X-rays or neutrons can be focused to spot diameters of a few microns and polycapillary optics can deliver high flux densities in spot diameters in the range of tens of microns. With synchrotron radiation sources, the use of the monocapillaries or polycapillaries in X-ray fluorescent studies can produce an intensity gain of several hundred-fold over pinhole collimators.

In X-ray spectroscopy with use of a capillary optic with a microfocused beam, the number of reflections in the capillary is minimized to obtain higher transmissions, and the distance between the end of the capillary and the sample should be sufficiently long to allow the illumination of arbitrarily shaped samples. With microspot X-ray tubes, a polycapillary optic tapered on its input end can receive a significant portion of the isotropic-emitted X-rays and deliver a collimated beam of X-rays on its non-tapered output end. If the polycapillary is tapered on both ends, then a significant solid angle of the microspot tube X-rays can be delivered and focused at high intensity on a tiny sample with applications in X-ray fluorescence and X-ray photoelectron spectroscopy (see Gibson and Havrilla, 2009; MacDonald, 2010).

A polycapillary optic, which is parallel on its input end and tapered on its output end, can be used as a neutron collection or condenser lens for receiving a collimated beam of neutrons and illuminating a specimen for refractive imaging or thermal neutron activation analysis. If the polycapillary is tapered on both ends to match a point source of thermal neutrons and a small sample, then the polycapillary can act as a collection and condensing neutron lens to provide focused intensity to an illuminated sample for neutron-refractive imaging or neutron activation analysis (see Gibson *et al.*, 2002a,b; Peurrung, 1996; Xiao *et al.*, 1994).

9. BRIGHT-FIELD, DARK-FIELD, AND PHASE-CONTRAST MICROSCOPY

The function of ordinary light microscopy, which creates images by light transmission through specimens, is to provide a magnified image of the specimen (magnification), separate the details in the image (resolution), and render the details visible to the eye or detector (contrast). The primary contrast mechanism in the light microscope is phase contrast and the secondary mechanism is amplitude contrast. In this section, we examine bright-field, dark-field, and phase-contrast microscopy for X-rays and thermal/cold neutrons. The evolution of light microscopy and electron microscopy provides an initial roadmap for the development of X-ray and neutron microscopy.

In X-ray microscopy, which creates an image of a sample by transmitting X-rays through it (bright-field X-ray microscopy), the primary interaction is photoelectric absorption and the minor interactions are coherent Rayleigh scattering and incoherent Compton scattering. Hence, in bright-field X-ray microscopy, the dominant contrast mechanism is absorption contrast. Given the atomic number Z of the scattering atom (that is, the number of scattering electrons per atom), and the X-ray photon energy E, the cross section $\sigma_{ph}(E)$ for photoelectric absorption of X-rays is calculated as follows:

$$\sigma_{ph} \propto \frac{Z^4}{E^3}. \tag{178}$$

The omission of X-rays via photoelectric absorption produces the image contrast, which hides the faint phase-contrast image produced by the interference of the Rayleigh-scattered coherent X-rays and the nonscattered X-rays that converge at points on the position-sensitive imaging detector. Also, the faint phase-contrast image is degraded in resolution by the overlay of a phase contrast image produced by incoherent scattering (Compton) of the incident X-rays by the specimen atom's individual electrons.

The Rayleigh-scattered X-rays are coherently scattered from the whole electron cloud of the specimen-scattering atom. Conservation of energy and momentum show the impact of the tiny X-ray momentum packet, with the large momentum of the electron cloud causing minimal energy transfer from the X-ray to the electron cloud. Hence, in Rayleigh scatter, the X-ray of energy E changes direction (momentum) but undergoes negligible energy change $\Delta E = 0$. Given the classical electron radius $r_e =$ 2.818 fm, the cross section σ_R dependence of Rayleigh X-ray scatter is

$$\sigma_R \propto \frac{r_e^2 Z^2}{2}. \tag{179}$$

In comparison, the cross section dependence for Compton X-ray scatter σ_c is

$$\sigma_c = \frac{r_e^2 Z}{2}. \tag{180}$$

The change of X-ray wavelength from incident wavelength λ to wavelength $\lambda - \Delta\lambda$ along the Compton scatter direction angle θ_s relative to the incident direction is as follows:

$$\frac{\lambda}{\lambda - \Delta\lambda} = \frac{\lambda}{\lambda + 1 - \cos\theta_s}. \tag{181}$$

In thermal and cold neutron microscopy, which creates an image of a sample by neutron transmission, the primary interaction is coherent scattering, followed closely by incoherent scattering and neutron absorption (capture). The thermal/cold neutron absorption cross section σ_{th} is proportional to the neutron wavelength, where

$$\sigma_{th} \propto \lambda. \tag{182}$$

Hydrogen is the exception, in that it primarily interacts with the cold and thermal neutrons by incoherent scattering. Inelastic scatter by hydrogen has an 80.26-barn cross section, in which neutron scatter by hydrogen-rich portions of a sample provide attenuation contrast. For materials in general, the neutron inelastic and incoherent scatter cross sections as a function of neutron wavelength λ and sample conditions are difficult to model and calculate theoretically, and thus the most direct approach is experimental measurement.

For bright-field X-ray and neutron microscopy, the omission of X-rays by photoelectric absorption is sufficiently dominant over the X-ray scattering that the image contrast is defined by the absorption contrast. In neutron microscopy, the absorption and hydrogen-atom scatter of thermal/cold neutrons by the sample creates an attenuation-contrast image. As a result, bright-field X-ray and neutron microscopy provides a magnified and inverted image of an object via a single refractive lens. A pinhole camera can provide a broadband, inverted, and magnified image of the object with the same magnification M as a refractive lens; that is, the ratio of the image distance r_i divided by the object distance r_0, where $M = r_i/r_0$. With a point source of X-rays or neutrons and a sample placed between the point source and position-sensitive detector (PSD), projection radiography provides a magnified, attenuation-contrast image of an object in which the magnification of the projection image is $M = (r_i + r_0)/r_0$.

A refractive lens, a pinhole camera, and projection radiography can produce a magnified, attenuation-contrast, two-dimensional image of a three-dimensional (3D) object. In comparison, contact radiography uses an extended source of X-rays or neutrons to provide a collimated beam at a sufficiently distant sample that is placed in tight contact with a PSD. Again, the contact radiography provides an attenuation-contrast, two-dimensional image of a 3D object. Tomography rotates the object or the detector so that a multitude of exposures are taken in different directions through the 3D object, which allows a reconstructed 3D, attenuation-contrast image.

Dark-field microscopy uses oblique illumination to enhance contrast in specimens that are not imaged well under normal bright-field illumination conditions. After the *0*th order, (direct) rays are blocked, with rays passing through or scattered by the object to form an image of the object superimposed onto a dark background. The dark-field mode is essentially a special form of the phase-contrast mode, in which the image contrast is proportional to the difference of the scattering strength between two neighboring resolution elements. Because of its low background, dark-field microscopy can emphasize features of the object, such as its boundary, which may be relatively invisible in a bright-field image.

Dark-field X-ray and neutron microscopy uses a compound refractive or polycapillary condenser lens that refracts incident rays through an object. An opaque disk is centered on the center optical axis of the microscope so as to block the direct rays from the source through the object from reaching the center portion of the objective lens. That is, the lens can no longer image the transmitted rays that pass directly in a line from the source to the objective lens. However, the center disk radius is sufficiently small to allow scattered rays to pass outside the disk perimeter onto the objective lens, which then images the scattered rays onto the PSD.

The absorption contrast between the absorbed rays and the transmitted (no absorption and no scatter) rays is eliminated from the PSD image and thus forms the dark background in the dark-field image. However, some of the coherent, Rayleigh-scattered X-rays or coherent-scattered neutrons, as well as incoherent and inelastic Compton-scattered X-rays or incoherent scattered neutrons, fall within the field of view of the objective lens. These scattered rays are magnified and imaged by the compound refractive objective lens onto the detector, thereby producing an illuminated image of the sample-scattered features in the dark image background. The random scatter by incoherent and inelastic scatter produces a blur in the sharp features formed by coherent scatter of the center wavelength of the bandwidth of X-rays, neutrons, or light.

We now consider phase-contrast microscopy for X-rays and neutrons as an evolution from phase-contrast light microscopy. In phase microscopy, the contrast mechanism is the coherently scattered rays from the specimen, which are phase-shifted but unchanged in wavelength. These scattered rays are converged to a PSD via a CRL. At the detector, the phase-shifted rays converge with the nonscattered (transmitted) rays that have not been phase-shifted. The different phase shift angles between the sample-scattered and nonscattered rays at the image plane results in an interference pattern or phase-contrast image of the sample via wave addition.

Phase-contrast microscope imaging records information from the bending or refraction of rays as they pass through samples. The refraction occurs because a small region A of the sample with a larger refractive decrement δ_A changes the speed of the rays more than a small adjoining sample region B with a smaller decrement δ_B, where $\delta_A > \delta_B$. The refraction and bending of the rays in regions A and B are different, which produce different path lengths through the objective lens to the imaging detector. The lens converges the rays from adjacent sample regions A and B onto the position-sensitive imaging detector, where the converged waves interfere constructively or destructively, depending on the phase shift difference incurred by the different ray paths from A and B to the detector.

The goal of the phase-contrast microscope is to convert this phase shift of the rays emitted from an illuminated sample into a sharp and magnified image that reveals subtle variations of sample structures. One must also prevent the rays directly transmitted from the sample from reaching the detector because the transmission ray intensity is much larger than the refracted ray intensity. That is, the intensity range of the attenuation contrast image is much larger than the intensity range of the phase-contrast image created by the interference of refracted rays. Hence, to see the much weaker phase-contrast image contribution, one must block the strong, overwhelming attenuation-contrast image contribution.

To penetrate a thicker sample requires sufficiently short X-ray or neutron wavelengths λ, but the higher-energy X-rays or neutrons are refractions and differential refractions of the rays due to the smaller energy-dependent decrement $\delta(\lambda)$, and thus a smaller sample decrement variation $\Delta\delta$, which decreases sample phase contrast. However, the thicker samples can provide a larger variation in the linear attenuation coefficient $\mu(\lambda)$, and thus larger sample attenuation variation $\Delta\mu$, which increases the sample attenuation contrast.

Features or structures in the sample can have large changes in their refractive index relative to the surrounding sample background. These large refractive gradient sample structures act in the same manner as a strongly refracting lens. The index of refraction of homogeneous sample background does not strongly refract the incident X-rays or neutrons. However, large decrement gradients $\Delta\delta$ arise that are directly related to the scatter lengths and number densities of the scatter units. The number of scatter units encountered by a ray (sample thickness) is directly related to the thickness of the sample features.

The change or gradient of the refractive index, with respect to its position along a line passing through the specimen but perpendicular to the incident X-rays, creates the relatively large refraction angles of coherent scattered X-rays or neutrons or light. In a specimen with refractive and reflective surfaces, one can see the increase in refraction and reflection angles by decreasing the X-ray or neutron energy. In summary, the shape of a sample and its features in the attenuation-contrast image is defined as the product variation $\Delta(t_z\mu)$ of the sample linear attenuation μ and thickness t_z. However, the shape of a sample and its features in the phase-contrast image are defined as the product variation $\Delta(t_z\delta)$ of the sample decrement δ and thickness t_z.

The change in refraction angle $\Delta\alpha_{x,y}(x, y, \lambda)$ is proportional to the gradient $\nabla_{x,y}\psi$ of the phase-shift angle $\psi(x, y)$. The phase-shift angle $\psi(x, y)$ of the X-ray, neutron, or light wave is a function of its wavelength λ and sample thickness $t(x, y)$ and decrement $\delta(x, y)$ at transverse sample position (x, y). The change of the refraction angle $\Delta\alpha_{x,y}(x, y, \lambda)$ per unit change in sample thickness Δt_z is the derivative $d\alpha(x, y)/dt_z$, given by

$$\frac{d\alpha_{x,y}}{dt_z} = \frac{\lambda}{2\pi}\left|\nabla_{x,y}\frac{d\psi}{dt_z}\right|. \tag{183}$$

The phase-shift angle $\psi(x,y,\lambda)$ for the incident waves propagating in the axial z-direction toward the detector, and passing perpendicularly though specimen thickness $t_z(x,y)$ at lateral sample coordinates (x,y), is

$$\psi = \frac{2\pi}{\lambda} t_z \delta. \tag{184}$$

For X-ray, neutron, and light wavelengths, which propagate through a range of sample thickness ranging from $t_z = 0$ to $t_z = t_\pi$, thereby cause the corresponding phase shifts of the refracted waves in the thin sample to range from $\psi = 0$ to $\psi = \pi$, with an average phase shift of $\bar{\psi}_{thin} = \pi/2$ or a quarter-wavelength shift of $\lambda/2$. For samples with a larger thickness range of $t_z = 0$ to $t_z = t_{2\pi}$, the corresponding phase shifts of the sample-refracted waves range from $\psi = 0$ to $\psi = 2\pi N$ with positive integer N. Effectively, the phase shifts of the refracted waves in the thick sample range from $\psi = 0$ to $\psi = 2\pi$ with average phase shift of $\bar{\psi}_{thik} = \pi$ or a half-wavelength shift of $\lambda/2$.

The change in the phase-shift angle per change in sample thickness measured in the z-direction, which is perpendicular to the sample plane (x,y), is

$$\frac{d\psi}{dt_z} = \frac{2\pi}{\lambda}\delta. \tag{185}$$

Substitution of Eq. 185 into Eq. 183 gives

$$\left|\nabla_{x,y}\frac{d\psi}{dt_z}\right| = \frac{2\pi}{\lambda}\left|\hat{\mathbf{x}}\frac{\partial\delta}{\partial x} + \hat{\mathbf{y}}\frac{\partial\delta}{\partial y}\right|. \tag{186}$$

The change in refraction angle per change in the sample thickness in the z-direction of the incident rays from the source through the sample to the detector $d\alpha_{x,y}/dt_z$ is equal to the change in the decrement $\Delta\delta$ along the z-direction, where

$$\frac{d\alpha_{x,y}}{dt_z} = \left|\nabla_{x,y}\delta\right|. \tag{187}$$

Here, the change in refraction angle per sample thickness $d\alpha_{x,y}/dt_z$ is measured with respect to the X-ray incident direction z, in which both $d\alpha_{x,y}(\lambda)/dt_z$ and $\delta(x,y,\lambda)$ depend on the wavelength λ of the X-rays, neutrons, or light. Also, complete reflection in the sample occurs when the grazing incidence angle ϕ_i is less than the critical angle ϕ_c. Grazing incidence angles that lead to complete forward-directed reflection toward the detector and large angles of refraction of incident rays each occur at steps and discontinuities in the thickness, shape, and composition of sample features.

We now examine the traditional phase-contrast microscope invented by Frits Zernike in 1934 (see the Nikon MicroscopyU website at http://www.microscopyu.com). The condenser optic of the phase-contrast microscope must illuminate the specimen, ideally with defocused, parallel wave fronts. The omnidirectional rays emitted are collected by a collector optic (mirror, polycapillary, or CRL) and are made parallel. These parallel rays from the collector optic pass through an annular aperture, which is aligned with the annular phase-shifting ring, and onto the condenser optic, which directs the defocused parallel rays as a hollow cone onto the sample. The parallel X-ray wave front illuminating the sample has a high spatial or lateral coherence. That is, the amplitude of the waves is highly correlated between different points transverse to the direction of ray propagation. The phase at any point on the wave front may be fluctuating randomly, but its phase difference with neighboring points on the wave front remains constant. Hence, points of the sample illuminated by an X-ray, neutron, or light wave can lead to fixed interference pattern at the position-sensitive imaging detector.

Absorption contrast in a thin sample is small. A sufficiently thin sample produces scattered rays with a phase shift of $\lambda/4$ (90°) relative to the nonscattered rays. As a result, the scattered waves with a $\lambda/4$ phase shift are not able to produce a phase-contrast image by destructive interference with the nonscattered waves. To use the phase information carried by the scattered rays to produce a visible intensity modulation (i.e., contrast) in the image plane, one must somehow phase-shift the nonscattered light by $-\lambda/4$ (–90°), thereby producing a $\lambda/2$ (180°) phase shift between the sample-scatter and sample-transmitted waves.

In the phase-contrast microscope, this is accomplished with a phase-shifting optic in the back focal plane of the objective. An annular hollow cone of rays is focused by the condenser optic onto the sample, and the rays that are transmitted through the sample diverge and intercept an annular zone of the objective lens. The annular portion of the objective lens refracts the divergent, nonscattered waves into parallel rays that are directed to the annular ring of a phase-shifting optic. The annular ring of the phase-shifting optic produces a phase shift of $3\lambda/4$ or $-\lambda/4$ (–90°) in the nonscattered (transmitted) rays from the sample.

Meanwhile, the other focused rays in the hollow cone that are scattered by the sample undergo a $\lambda/4$ (90°) phase shift by the sample. The sample scattered and 90° phase-shifted rays then diverge from the sample to the objective lens. However, the sample scattered rays are scattered out of the hollow cone of nonscattered rays and strike the center portion of the objective lens inside the outer annulus of sample transmitted rays. The sample scattered, phase-shifted rays are thus spatially separate from the non-scattered rays. The sample scattered rays fall within the center disk region of the objective lens, and the sample transmitted rays fall outside the center disk region into the surrounding outer annular zone of the objective lens. The divergent sample-scattered rays are refracted by the objective lens into parallel rays that pass through the air or a vacuum inside the annular zone of material of the phase-shifting optic. As a result, the sample scattered rays incur no phase shift by the annular phase-shift optic.

The sample scattered waves with phase shift $\lambda/4$ and the sample transmitted waves with phase shift $-\lambda/4$ are refracted by a second objective lens so the sample scattered and sample transmitted rays converge to the image plane and undergo destructive interference due to their $\lambda/2$ (180°) phase shift. That is, the 90° phase shift of the sample scattered rays and the −90° phase shift incurred by the sample transmitted rays in the annular zone of the phase-shift optic lead to amplitude contrast in the image via the destructive interference produced by the imposed 180° phase between sample scattered and transmitted waves.

Besides phase-shifting the nonscattered rays, the phase-shift optic attenuates the non-scattered rays. Since the nonscattered ray flux density typically dominates the scattered ray flux density, the attenuation helps equalize the amplitudes of the scattered and nonscattered rays. The annulus of the phase-shift optic is composed of a material with an elemental composition, density, and thickness that together produce the $\lambda/4$ phase shift and absorption in the sample-transmitted rays. By proper design of the phase shift optic, one can optimize the image contrast in a phase contrast or microscope using monochromatic or polychromatic X-rays, neutrons, or light. The complex phase shift $\Delta\varphi$, introduced to the nonscattered, sample transmitted waves by the phase-shift optic via its annulus of thickness t_{ps} and material decrement δ_{ps}, is given by

$$\Delta\varphi = \frac{2\pi}{\lambda}\delta_{ps}t_{ps} - j\mu_{ps}t_{ps}. \tag{188}$$

REFERENCES

Arkad'ev, V. A., Kolomiitsev, A. I., Kumakhov, M. A., Labuzov, V. V., & Ponomarev, I. Yu. (1988a). System using multiple reflections from curved surfaces. *Soviet Technical Physics Letters, 14*(1), 42.

Arkad'ev, V. A., Kolomiitsev, A. I., Kumakhov, M. A., Ponomarev, I. Yu., Khodeev, I. A., & Carpenter, D. A. (1989). Improved laboratory X-ray source for microfluorescence analysis. *X-ray Spectrometry, 18*(6), 253–257.

Arkad'ev, V. A., Kumakhov, M. A., & Faiazov, R. F. (1988b). Theoretical capabilities of grazing-incidence X-ray optics. *Soviet Technical Physics Letters, 44*(2), 101–102.

Arkad'ev, V. A., Kumakhov, M. A., & Ognev, L. I. (1986). Total reflection of γ rays from a surface. *Soviet Technical Physics Letters, 12*(11), 540–542.

Carpenter, D. A. (1989). Improved laboratory x-ray source for microfluorescence analysis. *X-ray Spectrometry 18*(6), 253–257.

Chen, H., Downing, R. G., Mildner, D., Gibson, W. M., Kumakhov, M. A., Ponomarev, I. Yu., & Gubarev, M. V. (1992). Guiding and focusing neutron beams using capillary optics. *Nature, 367*(6337), 391–393.
Chertov, Yu. P., & Shakhparonov, I. M. (1989). Wide band X-ray optics with a large angular aperture. *Soviet Physics Uspekhi, 32*(3), 271–276.
Engström, P., Larson, S., Rindby, A., & Stocklassa, B. (1989). A 200 μm X-ray microbeam spectrometer. *Nuclear Instruments and Methods in Physics Research, B36*(2), 222–226.
Gibson, W. M., Chen-Mayer, H. H., Mildner, D. F. R., Praskb, H. J., Schultz, A. J., Youngman, R., . . . Vitt, R. (2002a). Polycapillary optics based neutron focusing for small sample neutron crystallography. *JCPDS-International Centre for Diffraction Data 2002, Advances in X-ray Analysis, 45*, 306–312.
Gibson, W. A., & Havrilla, G. (2009). Requirements for X-ray fluorescence. In *Handbook of Optics, Third Edition Volume V: Atmospheric Optics, Modulators, Fiber Optics, X-Ray and Neutron Optics* (M. Bass, C. MacDonald, G. Li, C. DeCusatis, V. Mahajan, Eds.) (Chapter 29). McGraw Hill, New York.
Gibson, W. M., Huang, H. J., Nicolich, J., Klein, P., & MacDonald, C. A. (2002b). Polycapillary optics for angular filtering of X-rays and neutrons in two dimensions. *JCPDS-International Centre for Diffraction Data 2002, Advances in X-ray Analysis, 45*, 303–318.
Jentzch, F., & Nahrung, E. (1931). Die Fortleitung von Licht-und Röntgenstrahlen durch Röhren. *Zeitscgrift fur Technische Physik (Leipzig), 12*(3), 185–189.
Kissel, L., & Pratt, R. H. (1985). Rayleigh scattering: Elastic photon scattering by bound electrons. In B. Crasemann (Ed.), *Atomic Inner Shell Physics* (pp. 465–532). Plenum Press, New York.
Kumakhov, M. (March 9, 1993). Device for controlling beams of particles, X-ray, and gamma quanta. U.S. Patent No. 5,192,869.
Kumakhov, M. A. (1990). Channeling of photons and new x-ray optics. *Nuclear Instruments and Methods. B48*, 283–286.
Kumakhov, M. A., & Komarov, F. F. (1990). Multiple reflections from surface in X-ray optics. *Phys. Rep., 191*(5), 289–350.
Kumakhov, M. A., & Sharov, V. A. (1992). A neutron lens. *Nature, 357*, 390–391.
MacDonald, C. A. (2010). Focusing polycapillary optics and their applications. *X-Ray Optics and Instrumentation, 2010*, 1–17.
Marton, L. (1966). X ray fiber optics. *Applied Physics Letters, 9*(5), 194–195.
Mosher, D., & Stephanakis, S. J. (1976). X ray light pipes. *Applied Physics Letters, 29*(2), 105–107.
Pantell, R. H., & Chung, P. S. (1978). Transmission of X-rays through curved waveguides. *IEEE Journal of Quantum Electronics, 14*(9), 694–697.
Pantell, R. H., & Chung, P. S. (1979). Influence of surface roughness on the propagation of X-rays through capillaries. *Applied Optics, 18*(6), 897–899.
Peurrung, A. J. (1996). Capillary optics for neutron capture therapy. *Medical Physics, 23*(4), 487–494.
Pound, R. V., & Rebke, G. A. (1959). Resonant absorption of the 14.4 keV γ ray from 0.10-μsec Fe^{67}. *Physical Review Letters, 3*, 554–556.
Thiel, D. J., Bilderback, D. H., Lewis, A., Stern, E. A., & Rich, T. (1992). Guiding and concentrating hard X-rays by using flexible hollow-core tapered glass fiber. *Applied Optics, 31*(7), 987–992.
Ulrich, J. B., Kovantsev, V., & MacDonald, C. A. (1993). Measurements of polycapillary X-ray optics. *Journal of Applied Physics, 74*(10), 5933–5939.
Xiao, Q. F., Chen, H., Sharov, V. A., Mildner, D. F. R., Downing, R. G., Gao, N., & Gibson, D. M. (1994). Neutron focusing optic for submillimeter materials analysis. *Review of Scientific Instruments, 65*(11), 3399–3402.

Contents of Volumes 151–171

VOLUME 151[1]

Claas Bontus and Thomas Köhler, Reconstruction algorithms for computed tomography
Laurent Busin, Nicolas Vandenbroucke and Ludovic Macaire, Color spaces and image segmentation
Glenn R. Easley and Flavia Colonna, Generalized discrete Radon transforms and applications to image processing
Tomáš Radlička, Lie agebraic methods in charged particle optics
Valerie Randle, Recent developments in electron backscatter diffraction

VOLUME 152

Nina S.T. Hirata, Stack filters: from definition to design algorithms
Sameen Ahmed Khan, The Foldy–Wouthuysen transformation technique in optics
Saverio Morfu, Patrick Marquié, Brice Nofiélé and Dominique Ginhac, Nonlinear systems for image processing
Tohru Nitta, Complex-valued neural network and complex-valued backpropagation learning algorithm
Jérôme Bobin, Jean-Luc Starck, Y. Moudden and M.J. Fadili, Blind source separation: the sparsity revoloution
Ray L. Withers, "Disorder": structured diffuse scattering and local crystal chemistry

VOLUME 153

Aberration-corrected Electron Microscopy

Harald Rose, History of direct aberration correction

[1] Lists of the contents of volumes 100–149 are to be found in volume 150; the entire series can be searched on ScienceDirect.com

Maximilian Haider, Heiko Müller and Stephan Uhlemann, Present and future hexapole aberration correctors for high-resolution electron microscopy
Ondrej L. Krivanek, Niklas Dellby, Robert J. Kyse, Matthew F. Murfitt, Christopher S. Own and Zoltan S. Szilagyi, Advances in aberration-corrected scanning transmission electron microscopy and electron energy-loss spectroscopy
Philip E. Batson, First results using the Nion third-order scanning transmission electron microscope corrector
Andrew L. Bleloch, Scanning transmission electron microscopy and electron energy loss spectroscopy: mapping materials atom by atom
Florent Houdellier, Martin Hÿtch, Florian Hüe and Etienne Snoeck, Aberration correction with the SACTEM-Toulouse: from imaging to diffraction
Bernd Kabius and Harald Rose, Novel aberration correction concepts
Angus I. Kirkland, Peter D. Nellist, Lan-yun Chang and Sarah J. Haigh, Aberration-corrected imaging in conventional transmission electron microscopy and scanning transmission electron microscopy
Stephen J. Pennycook, M.F. Chisholm, A.R. Lupini, M. Varela, K. van Benthem, A.Y. Borisevich, M.P. Oxley, W. Luo and S.T. Pantelides, Materials applications of aberration-corrected scanning transmission electron microscopy
Nobuo Tanaka, Spherical aberration-corrected transmission electron microscopy for nanomaterials
Knut Urban, Lothar Houben, Chun-lin Jia, Markus Lentzen, Shao-bo Mi, Andreas Thust and Karsten Tillmann, Atomic-resolution aberration-corrected transmission electron microscopy
Yimei Zhu and Joe Wall, Aberration-corrected electron microscopes at Brookhaven National Laboratory

VOLUME 154

Henning F. Harmuth and Beate Meffert, Dirac's Difference Equation and the Physics of Finite Differences

VOLUME 155

Dmitry Greenfield and Mikhail Monastyrskiy, Selected Problems of Computational Charged Particle Optics

VOLUME 156

Vasileios Argyriou and Maria Petrou, Photometric stereo: an overview
Fred Brackx, Nele de Schepper and Frank Sommen, The Fourier transform in Clifford analysis

Niels de Jonge, Carbon nanotube electron sources for electron microscopes
Erasmo Recami and Michel Zamboni-Rached, Localized waves: a review

VOLUME 157

Mikhail I. Yavor, Optics of charged particle analyzers

VOLUME 158

Péter Dombi, Surface plasmon-enhanced photoemission and electron acceleration with ultrashort laser pulses
Brian J. Ford, Did physics matter to the pioneers of microscopy?
Jérôme Gilles, Image decomposition: theory, numerical schemes, and performance evaluation
Stina Svensson, The reverse fuzzy distance transform and its use when studying the shape of macromolecules from cryo-electron tomographic data
Marc van Droogenbroeck, Anchors of morphological operators and algebraic openings
Dong Yang, Shiva Kumar and Hao Wang, Temporal filtering technique using time lenses for optical transmission systems

VOLUME 159

Cold Field Emission and the Scanning Transmission Electron Microscope

Albert Victor Crewe, The work of Albert Victor Crewe on the scanning transmission electron microscope and related topics
Lyn W. Swanson and Gregory A. Schwind, A review of the cold-field electron cathode
Joseph S. Wall, Martha N. Simon and James F. Hainfeld, History of the STEM at Brookhaven National Laboratory
Hiromi Inada, Hiroshi Kakibayashi, Shigeto Isakozawa, Takahito Hashimoto, Toshie Yaguchi and Kuniyasu Nakamura, Hitachi's development of cold-field emission scanning transmission electron microscopes
Peter W. Hawkes, Two commercial STEMs: the Siemens ST100F and the AEI STEM-1
Ian R.M. Wardell and Peter E. Bovey, A history of Vacuum Generators' 100-kV STEM
H. Sebastian von Harrach, Development of the 300-kV Vacuum Generators STEM (1985–1996)
Bernard Jouffrey, On the high-voltage STEM project in Toulouse (MEBATH)

Andreas Engel, Scanning transmission electron microscopy: biological applications
Kenneth C.A. Smith, STEM at Cambridge University: reminiscences and reflections from the 1950s and 1960s

VOLUME 160

Zofia Baranczuk, Joachim Giesen, Klaus Simon and Peter Zolliker, Gamut mapping
Adrian N. Evans, Color area morphology scale-spaces
Ye Pu, Chia-lung Hsieh, Rachel Grange and Demetri Psaltis, Harmonic holography
Gerhard X. Ritter and Gonzalo Urcid, Lattice algebra approach to endmember determination in hyperspectral imagery
Reinhold Rüdenberg, Origin and background of the invention of the electron microscope
H. Gunther Rudenberg and Paul G. Rudenberg, Origin and background of the invention of the electron microscope: commentary and expanded notes on Memoir of Reinhold Rüdenberg

VOLUME 161

Marian Mankos, Vassil Spasov and Eric Munro, Principles of dual-beam low-energy electron microscopy
Jorge D. Mendiola-Santibañez, Iván R. Terol-Villalobos and Israel M. Santillán-Méndez, Determination of adequate parameters for connected morphological contrast mappings through morphological contrast measures
Ignacio Moreno and Carlos Ferreira, Fractional Fourier transforms and geometrical optics
Vladan Velisavlevic, Martin Vetterli, Baltasar Berufell-Lozano and Pier Luigi Dragotti, Sparse image representation by directionlets
Michael H.F. Wilkinson and Georgios K. Ouzounis, Advances in connectivity and connected attribute filters

VOLUME 162

Kiyotaka Asakura, Hironobu Niimi and Makoto Kato, Energy-filtered x-ray photoemission electron microscopy (EXPEEM)
Eireann C. Cosgriff, Peter D. Nellist, Adrian J. d'Alfonso, Scott D. Findlay, Gavin Behan, Peng Wang, Leslie J. Allen and Angus I. Kirkland, Image contrast in aberration-corrected scanning confocal electron microscopy

Christopher J. Edgcombe, New dimensions for field emission: effects of structure in the emitting surface
Archontis Giannakidis and Maria Petrou, Conductivity imaging and generalised Radon transform: a review
Olivier Losson, Ludovic Macaire and Yanqin Yang, Comparison of color demosaicing methods

VOLUME 163

Wolfgang S. Bacsa, Optical interference near surfaces and its application in subwavelength microscopy
Ruy H.A. Farias and Erasmo Recami, Introduction of a quantum of time ("chronon"), and its consequences for the electron in quantum and classical physics
Andrew Neice, Methods and limitations of subwavelength imaging
A. Sever Škapin and P. Ropret, Identification of historical pigments in wall layers by combination of optical and scanning electron microscopy coupled with energy-dispersive spectroscopy
Markus E. Testorf and Michael A. Fiddy, Superresolution imaging–revisited

VOLUME 164

Amos Bardea and Ron Naaman, Magnetolithography: from the bottom-up route to high throughput
Román Castañeda, The optics of spatial coherence wavelets
Junchang Li, Yanmei Wu and Yan Li, Common diffraction integral calculation based on a fast Fourier transform algorithm
Marcel Teschke and Stefan Sinzinger, A generalized approach to describe the interference contrast and phase contrast method
Dokkyu Yi and Booyong Choi, Nonlinear partial differential equations for noise problems
Henning F. Harmuth, Harmuth corrigenda

VOLUME 165

Natalie Baddour, Two-dimensional Fourier transforms in polar coordinates
Neil V. Budko, Superluminal, subluminal, and negative velocities in free-space electromagnetic propagation
Rowan Leary and Rik Brydson, Chromatic aberration correction: the next step in electron microscopy

Michele Marrocco, Methods for vectorial analysis and imaging in high-resolution laser microscopy
Tomoya Sakai, Masaki Narita, Takuto Komazaki, Haruhiko Nishiguchi and Atsushi Imiya, Image hierarchy in Gaussian scale space
Yusuf Ziya Umul, The theory of the boundary diffraction wave
Emil Wolf, History and solution of the phase problem in the theory of structure determination of crystals from x-ray diffraction measurements

VOLUME 166

Valeriy A. Syrovoy, Theory of Intense Beams of Charged Particles

VOLUME 167

Emmanuel de Chambost, A History of Cameca (1954–2009)
Johan Debayle and Jean-Charles Pinoli, Theory and Applications of General Adaptive Neighborhood Image Processing
Mohamed Ben Haj Rhouma, Mohamed Ali Khabou, and Lotfi Hermi, Shape Recognition Based on Eigenvalues of the Laplacian
Nicolas Loménie and Georges Stamon, Point Set Analysis
Leonid P. Yaroslavsky, Image Recovery from Sparse Samples, Discrete Sampling Theorem, and Sharply Bounded Band-Limited Discrete Signals

VOLUME 168

Luca Geretti and Antonio Abramo, The Synthesis of a Stochastic Artificial Neural Network Application Using a Genetic Algorithm Approach
Michel Jourlin, Josselin Breugnot, Frédéric Itthirad, Mohamed Bouabdellah, and Brigitte Closs, Logarithmic Image Processing for Color Images
Rainer A. Leitgeb, Current Technologies for High-Speed and Functional Imaging with Optical Coherence Tomography
Sergej A. Nepijko and Gerd Schönhense, Analysis of Optical Systems, Contrast Depth, and Measurement of Electric and Magnetic Field Distribution on the Object's Surface in Mirror Electron Microscopy
Chad M. Parish, Multivariate Statistics Applications in Scanning Transmission Electron Microscopy X-Ray Spectrum Imaging
Hidetaka Sawada, Fumio Hosokawa, Takeo Sasaki, Toshikatsu Kaneyama, Yukihito Kondo, and Kazutomo Suenaga, Aberration Correctors Developed Under the Triple C Project
Tobias Schulz, Martin Albrecht, and Klaus Irmscher, Spatially Resolved Thermoluminescence in a Scanning Electron Microscope

VOLUME 169

Erchan Aptoula and Sébastien Lefèvre, Morphological Texture Description of Grey-Scale and Color Images
Vera Guarrera and Herwig Ott, Electron Microscopy of Ultracold Gases
Konstantinos Konstantinidis, Ioannis Andreadis, and Georgios Ch. Sirakoulis, Application of Artificial Intelligence Methods to Content-Based Image Retrieval
Xingwei Yang, Daniel B. Szyld, and Longin Jan Latecki, Diffusion on a Tensor Product Graph for Semi-Supervised Learning and Interactive Image Segmentation
Sergej A. Nepijko and Gerd Schönhense, Electron Holography for Electric and Magnetic Field Measurements and Its Application for Nanophysics

VOLUME 170

Alexander S. Eggeman and Paul A. Midgley, Precession Electron Diffraction
Ray Hill, John A. Notte, and Larry Scipioni, Scanning Helium Ion Microscopy
Hone-Ene Hwang and Pin Han, Signal Reconstruction Algorithm Based on a Single Intensity in the Fresnel Domain
Kazuhisa Sato, Toyohiko J. Konno, and Yoshihiko Hirotsu, Electron Microscopy Studies on Magnetic $L1_0$-Type FePd Nanoparticles
D. A. Zanin, H. Cabrera, L. G. De Pietro, M. Pikulski, M. Goldmann, U. Ramsperger, D. Pescia, and John P. Xanthakis, Fundamental Aspects of Near-Field Emission Scanning Electron Microscopy

VOLUME 171

Gregor Esser, Wolfgang Becken, Werner Müller, Peter Baumbach, Josep Arasa, and Dietmar Uttenweiler, Derivation of the Reflection Equations for Higher-Order Aberrations of LocalWave Fronts by Oblique Incidence
Lila Iznita Izhar and Maria Petrou, Thermal Imaging in Medicine
Jean-Michel Tualle, Derivation of the Radiative Transfer Equation in a Medium with a Spatially Varying Refractive Index: A Review
Kamlesh Shrivas and Mitsutoshi Setou, Imaging Mass Spectrometry: Sample Preparation, Instrumentation, and Applications
Robert T. Thompson and Steven A. Cummer, Transformation Optics
Tobias Klein, Egbert Buhr, and Carl Georg Frase, TSEM: A Review of Scanning Electron Microscopy in Transmission Mode and Its Applications
Michel Jourlin, Maxime Carré, Josselin Breugnot, and Mohamed Bouabdellah, Logarithmic Image Processing: Additive Contrast, Multiplicative Contrast, and Associated Metrics

Index

Page numbers followed by "*f*" indicates figures and "*t*" indicates tables.

A

Abbe sine condition for coma aberration, 477–479
Abbe's theory, 439, 439*f*
Aberration equation, 468
Absorption aperture radius, 11, 12*f*
Absorption index, 510–513, 628, 637
Absorption-contrast mode, 222
Absorptive attenuation, X-ray, 527–528
Achromat lenses, 37, 38*f*
  for X-ray and neutrons, 31–34
Adelphi DD108 neutron generator, 243
Adelphi Technology, 70
  experiment, 108, 109*f*
  fabricated Fresnel lens, 164–165
  thermal neutron imaging experiments, 215
  X-ray line source, 168
Advanced Photon Source (APS), 128, 128*f*
Alligator X-ray lens, 136, 137*f*, 139*f*
Aluminum CRL, 49, 52*f*, 56*f*, 76, 105
  biconcave parabolic lens, 112, 115, 116*f*
  1D, 169
  photons of, 130
  transmission spectrum of, 78
AmBe neutron, 255
Ampere's law, 505, 550
Amplitude modulation FZP, 291, 292*f*
Amplitude-contrast X-ray/neutron microscope, 22–26, 24*f*
Analyzer, 302
Anger camera collimators, 238
Anger scintillation camera, 237, 238, 239*f*
Ångstrom-range neutron wavelength, 5
Angular beam divergence, 44
Anomalous absorption in crystals, 596–601
Aplanatic system, 285
Association Française de Normalisation (AFNOR), 202
Astigmatism, 446, 447*f*, 448, 481–483
  and curvature of field, 483–484
  Sturm's interval and remedy, 485–487
Atom number density, 628
Atomic electron oscillation, damped resonance for, 498–501
Atomic electrons, 505
Atomic scattering factor, complex, *see* Complex atomic scattering factor
Atoms, motion of, 59
Auger electrons, 618
Avogadro's number, 616
Axial magnification, CRL, 368–371
Axial-directed neutron beam, 99

B

Back-emitted neutrons, 261
Barrel distortion, 449*f*, 489
Beam purity indicator (BPI), 202, 203*f*
Bent crystals, 285
Bent double-Laue crystal, 169
Bent Laue pre-monochromator, 170
Berlin Neutron Scattering Center, 276
Beryllium CRL, 68–72, 106*f*
  biconcave parabolic lenses of, 115, 116*f*
  biconcave spherical lenses, 155–157
  gold test structure, 118
  imaging experiments, 117
  2D, 105
Beryllium lenses, 73
BFPP, *see* Blazed Fresnel phase plate
BGRR, *see* Brookhaven Graphite Research Reactor
Biconcave lenses, 338, 342, 343, 612
  chromatic aberration in spherical and parabolic, 430–434
  of CRL, 390*f*
  parabolic and spherical, 10–13
  stacked, 3*f*

Biconcave parabolic lens, 115, 340–344, 347*f*
  absorption aperture radius, 339
  focal length, 345–346
  1D, 121
  path length, 344–345
  transmission and aperture, 10–13
  2D, 148, 153, 157
  of various materials, 193
Biconcave refractive X-ray lenses, 100
Biconcave spherical lenses, 346–348, 347*f*
  CRL composed of, 101*f*
  MgF_2 lens, 44, 60*f*
  transmission and aperture, 10–13
  2D, 148, 153
Biconvex lens, 342, 612
Binomial approximation, 234, 609
Birefringent material, 299
Black garolite lens holder, 155
Blazed Fresnel phase plate (BFPP), 100, 291, 293*f*
Blur diameter, 431
BNCT, *see* Boron neutron capture therapy
Boltzmann constant, 623
Born approximation, 273
Boron neutron capture therapy (BNCT), 207, 216
Borosilicate glass capillaries, 644
Borrmann effect, 597
Bragg diffraction, 629
Bragg reflection, 558, 597–601
Bragg X-ray microscope, 288
Bragg–Fresnel diffracting lenses, 102
Bremsstrahlung
  from charged particle passage, 536–538
  forward-directed, 538–541
Brewster angle, X-ray, 557–559
Bright-field microscope, 20*f*, 22
  X-ray, 646–648
Broglie relation, 566, 605
Brookhaven Graphite Research Reactor (BGRR), 275
Bubble lenses, 143–144, 144*f*, 160–163, 161*f*, 176

C

Cadmium sheet, horizontal slit in, 91
Cadmium slit array, 66
Capillary optics, 44
  for X-rays and neutrons, 641–646
Carbon
  glassy, 121–123, 122*f*
  X-ray CRLs made of, 103*f*
Cartesian transverse detector coordinates, 211
Cauchy's principal value theorem, 513
CCD camera, *see* Charged-couple device camera
Centimeter-gram-second (CGS) units, 569
CFL, *see* Compound Fresnel lens
CGS units, *see* Centimeter-gram-second units
Charged-couple device (CCD) camera, 107
  based radiation imaging system, 254
  detector, 195
  dual-MCP night-vision camera, 263
  with MCP, 216
  software, 265
  Stanford Photonics, 261, 262
  for thermal neutrons, 202
Chief ray, 4*f*
Christiansen filter, 582
Chromatic aberration
  CRL image and object resolution, 438
  in parabolic and spherical lenses, 430–434, 453–456
Coddington shape factor, 479
Coded aperture
  imaging, 209–210, 210*f*
  plate function, 212
Coherent elastic scattering, 615–619
Coherent Gaussian beam X-ray sources, 144
Coherent inelastic scattering, 615–619
Coherent scatter, 519–521
Cold neutrons, 643, 648
  beams, 4, 278
  phase-contrast imaging for, 225*f*, 226
Collection lens, 641, 643
Collimators
  anger camera, 238, 239*f*
  geometric efficiency of, 239
  steel, 247
Collision
  energy, 536
  losses of charged particle passage, 528–536
Coma aberration, 448*f*, 449, 475–477
  Abbe sine condition for reduction of, 477–479
Comet's tail, 478
Complex atomic scattering factor
  complex refractive index and, 508–510
  differential cross section for, 517–519
  dispersion and group velocity relation to, 547

for mixture/compound, 503–504
for single element, 502–503
X-ray rayleigh scatter and, 514–517
Complex atomic scattering function, 628
Complex refractive index, 510–513
and complex atomic scatter factor, 508–510
neutrons, 4–10
X-ray, 4–10, 513–514
Compound Fresnel lens (CFL), 34–38
Compound magnetic refractive lens (CMRL), 82
Compound refractive lens (CRL), 336–338, 614, 624
aberrations, 444–450
absorption aperture depth and transmission in, 404–406
achromat lenses using, 31–34
aluminum, 49, 52*f*, 56*f*
chromatic aberration for, 430
cylindrical lens, 171
deflection of x-rays and neutrons in, 402–404
depth of field and depth of focus, 371–373, 372*f*
effective aperture, 147
FOV, 178, 180, 185, 380–382, 380*f*
gain *G* dependence, 46, 46*f*
image and object resolution, 438
imaging with synchrotron x-ray sources, 494–495
Institute of Microstructure Technology (IMT), 198, 198*f*
intensity, 361–367
lens misalignment on, 391–398
lithium, 157–159, 158*f*
MTF for, 373–377
calculation, 377–379
neutron and X-ray monochromator with, 38–39
numerical aperture, 145
planar arrays of, 167, 169*f*
resolution comparison with pinhole, 13
single-capillary optic comparing with, 644
surface roughness of, 390–391
thick lens, 382–389, 383*f*
transmission for x-rays/neutrons, 358–361
transverse and axial magnification, 368–371
X-ray, *see* X-ray compound refractive lens
Compound refractive neutron lenses, 272–279
Compound refractive prism (CRP), 80, 81*f*, 398–402, 614
to offset neutron gravity droop in SANS, 406–409
Compression molding technique, 157
Compton incoherent scatter, 618
Compton scatter, 100, 520, 647
of X-rays, 521–525
Concave lens, 37
center point of, 357, 358
Condenser lens, 641, 643
Confocal multilayer optics, 187, 189
Contact radiography, 205–208, 206*f*, 265
Contrast, 373
Conventional Fresnel lens, 291
Conventional thermal neutron radiography, 235, 236*f*
Convex lens, 37
Copper CRL, 68, 69, 71, 71*f*, 94, 97
Coulomb scatter, 536
CRL, *see* Compound refractive lens
Crossed polarizers, 302
Crown and flint method, 31
CRP, *see* Compound refractive prism
Crystals, 596–601
bent, 285
Curvature of field
astigmatism and, 483–484
Petzval condition and, 479–481
Curvature of lens, 445
Cusp magnetic lens, 99
Cylindrical neutron lens, 308*f*

D

D22 beam line, 50–52
D22 imaging detector, 51
Dark-field (DF) microscopy, 19, 22, 193–195, 299
X-ray, 297
Dark-field (DF) microscopy, 648–649
Davisson–Germer experiment, 567
DD109 fast neutron generator, 260*f*
de Broglie hypothesis, 284
de Broglie relation, 563, 593, 604
Debye–Waller factor, 273, 599
Decrement index, 510–513
Delta/Mu ratio, 612–614
Depth of field and focus, CRL, 371–373, 372*f*

Detector, 447
Detector noise, 189
Deuterium-based scintillator, 255
Deuterium-deuterium (DD), 200
  fusion reaction, 259
  generator, 260, 265
  neutrons, 258, 266, 270
Deuterium-tritium (DT), 200
  generator, 247, 265
  neutrons, 258
Diagonal ray, 337
Differential interference contrast (DIC) microscope, 299, 301–303
Differential nuclei density, 610
Differential phase-contrast (DPC) imaging, 303, 304, 305*f*
Differential scattering intensity, 610
Diffracted neutrons, 25
Diffraction grating, 286
Diffraction-limited image resolution, 438, 439*f*
Diffractive X-ray, 25
  development of, 289
  optics and microscopy, history of, 282
Direct space, 564
Dispersion velocity, X-ray wave packets, 544–547
Distortion
  aberration, 449, 487–489
  in compound refractive lens imaging, 494–495
  images without and with, 490–492
  prevention of, 492, 494
Divergent rays, 471, 472, 490
Doppler shift, 94
Doublet CRL, 433
Drift space matrix, 351
Dual-microchannel plate
  CCD night-vision camera, 263
  detectors, 217–218

E

Einstein relation, 561
Elastic scatter, 501, 519–521
Electric field
  amplitude, 17*f*
  component parallel, X-rays with, 554–556
  in plane of incidence, 557–559
Electromagnetic radiation, 538
Electromagnetic vector wave equation, 505–506
  contains refractive index, 506–508
Electromagnetic wave equation, 574
  Schrödinger wave equation derived by, 568–572
Electron acceleration, 500, 501
Electron backscatter diffraction (EBSD) image, 132
Electron microscopy, 307, 646
Electron oscillation, damped resonance for atomic, 498–501
Emission, damped resonance for X-ray, 498–501
Energy density, 545
Epithermal neutrons, refractive index of, 208
Euler's theorem, 454
European Synchrotron Radiation Facility (ESRF), 111
  synchrotron ring, 92
Extraordinary ray, 300

F

Fabrication, FZP, 296
Fabry–Perot interferometer, 593
Faraday's law, 505, 551
Fast neutrons (FNs), 237, 246*f*
  attenuation, 261
  false-color image, 244, 244*f*, 264*f*
  generator, 247, 258
  imaging detector, 243, 244*f*, 257*f*
  point sources of, 204
  radiography, 202, 255, 256*f*, 262*f*, 265
    contact and projection, 208
  refractive index of, 208
Fermats principle of least time, 414–417
Fermi's approach, 273
Ferromagnetic isotopes, 638
Field of view (FOV)
  compound refractive lens, 178, 180, 185, 380–382, 380*f*
  experiment, 56
  of microscope, 54
  X-ray tube, 187
Filtering, spatial, 252
Fizeau effect for neutron waves, 592, 596
Fizeau interferometer, 593
FLASH free-electron laser (FEL), 306
Fluorescent detection, 124
FNs, *see* Fast neutrons
Focal length, CRL, 430
Focal ray, 337
Forward-emitted neutrons, 261
FOV, *see* Filed of view
Free electron laser (FEL) X-ray source, 133

Frequency-dependent complex refractive index, 511
Fresnel CRL, 34
Fresnel equations for neutron reflection and transmission, 577
Fresnel lens, 34–38
  conventional, 291
  phase, 290
Fresnel zone plates (FZPs)
  achromat lenses using, 31–34
  amplitude modulation, 291, 292*f*
  efficiency, 294
  neutrons and X-ray, 29, 29*f*, 31
  phase, 290
  quaternary, 296*f*
Fresnel–Kirchhoff integrals, 148
FujiFilm detector, 65, 72, 215
Fundamental paraxial equation, 417–418
Fusion reactor, 43
FZPs, *see* Fresnel zone plates

G

Gabor zone plate, 31
Gabor's interferometric method, 31
Gadolinium neutron capture therapy, 208
Galilean transformations, 593
Gamma detector, 252
Gamma image intensity, 252
Gamma ratio, 250
Gamma rays, 240, 248, 249
  attenuation for, 239*f*
Gamma scan, 246*f*, 252, 253*f*
Gauss's formula for thin lens, 368
Gaussian beams
  of laser optics, 148
  propagation of, 147*f*
  X-ray sources, 144
Gaussian lens equation for thin lenses, 418–420
Gaussian-shaped X-ray transmission, 153
Geometric optics, 409
Green's function, 229
Group velocity, 561–565
  definition of, 541
  X-rays, 541
    in material medium, 548–550
    wave packets, 544–547

H

Hamiltonian equation, 566, 568, 572
Hard X-rays, 115, 127
Heisenberg uncertainty principle, 309, 530, 544
Helmholtz–Kirchhoff theorem, 228
High Flux Beam Reactor (HFBR), 275
High-purity germanium (HPGe) detector, 125
Higher-energy photons, 520
Holographic gratings, 289
Human ovarian adenocarcinoma (IGROV1) cell line, 129
Huygen's principle, 414–417, 509
  expression of, 13–17
Hydrogen, 201, 648
  rich materials, imaging of, 63
Hyperbolic tangent function, 377
Hypotenuse, 609

I

ID22 beam line, 111, 113*f*
Ideal chief ray, 471
Ideal lens, function of, 456
Image contrast, 373
Imperfect lens, 455
Imperfect spherical lens surface, aberrations from, 456–461
Incident angle, 612
Incident neutron energy, 622
Incident neutron velocity, 622
Incoherent elastic scattering, 615–619
Incoherent Gaussian beam X-ray sources, 144
Incoherent inelastic scattering, 615–619
Institut Laue-Lan (ILL)
  D22 beam line at, 50
  high flux reactor, 275
  refractive imaging of neutrons at, 49
  research reactor, 92
Intense Pulsed Neutron Source (IPNS), 276
Interference pattern, 591, 592, 596, 602
Interferometry, 592–596
IPNS, *see* Intense Pulsed Neutron Source
Isotope, refractive properties of, 47

J

Japan Atomic Energy Agency, 276

K

K-space, *see* Reciprocal space
Köhler illumination, 19, 20, 21*f*
Kapton compound refractive lens, 150*f*, 153–155
Kinematic theory of diffraction, 597

Kinetic energy of neutron, 562, 567, 577
Kirchhoff analysis, 67, 68, 436
Kirchhoff equation, 228, 230, 232, 373
  optical system analysis with, 13–17
Kirkpatrick–Baez microscopes, 304
Knife-edge images, line profile measurements of, 377–379
Knife-edge test, 376
Kramers–Kronig relation, 513–514
Kumakhov lens, 47, 641–646

L

Lamberts's law, 600
LANSCE, *see* Los Alamos Neutron Scattering Center
Laser, Vulcan, 42
Laser-produced plasma source, 285
Lateral aberration, 451
Laue diffraction, 307, 308*f*
Laue–Bragg equations, 273
Leith–Upatnieks hologram, 31
Lens aberrations, cause of, 471
Lens chord diameter, 482
Lens doublet, power and focal length of, 424–426
Lens surface
  curvature of, 409–411
  wave vergence change by, 413–414
Lensmaker's equation, 2, 337, 345, 369
  of Gaussian lens equation, 418
Lensmaker's formula, 368, 435, 441
Light microscope, 646
  archetypes for neutron and X-ray microscopy, 19–22
  DIC, 299
  zernike phase-contrast, 298
Line spread function (LSF), 375, 376, 378
Linear attenuation, 510–513
  coefficient, 336
Lithia water springs, 141
Lithium, 618
  biconcave parabolic lenses of, 157
  lens, 139–143, 139*f*
Lithography test mask, 135
LLL interferometer, 597
Lobster-eye X-ray telescope, 286
Long-wavelength X-rays, 4
Longitudinal aberration, 451, 489
Longitudinal chromatic aberration, 434
Longitudinal spatial coherence length, neutrons and X-ray, 26–28
Lorentz force equation, 531, 574–575
Lorentz transformation, 538
Los Alamos Neutron Scattering Center (LANSCE)
  experiment, 75
  LQD beam line, 77*f*
Lucite CRL, 108, 109*f*

M

Mach-Zehnder interferometers, 601
Magnesium fluoride (MgF_2)
  biconcave spherical lenses, 44
  CRL, 60, 76
    transmission spectrum of, 78
Magnetic CRP, 97
Magnetic defocusing effect, 84*f*
Magnetic dipole moment, 581
Magnetic field component parallel, X-rays with, 556–557
Magnetic field lenses, 98*f*
Magnetic prism, 398
Magnetic scatter length, 637, 638
Marginal rays, 4*f*, 440*f*
Material prisms, 398
Matrix expression for thin lens, 387
Maxwell's equations, 574–575
Maxwell–Boltzmann distribution, 47, 48, 214, 623
  for reactor source, 603–606
Maxwellian distribution, 623
Maxwells equations, 505–506
McClellan Nuclear Radiation Center (MNRC), 213–215
MCP, *see* Microchannel plate
Medipix2
  ASIC, 254
  imaging detector, 255
Michelson interferometer, 596
Micro–zone plate lens, 298*f*
Microbeam techniques, 123, 127
Microchannel plate (MCP), 215
  CCD camera with, 216
  neutron imaging detectors, 218–219, 219*f*
Microradiography, 222
Microspot X-ray tube, 187, 189
MLL, *see* Multilayer Laue lens
MNRC, *see* McClellan Nuclear Radiation Center
Modulation transfer function (MTF), 16, 66
  for CRL, 373–377
    calculation, 377–379
Monocapillary optics, for X-rays and neutrons, 641, 645
Monochromatic aberrations, 444
  in parabolic and spherical lenses, 453–456

Monochromatic non-paraxial rays, 450
Monochromatic Seidel aberrations, 454, 461–470
Monochromatic X-rays, 189
Monochromators, 636
PMF, 90, 91*f*
Monte Carlo simulation, 260
MTF, *see* Modulation transfer function
Multi-prism lens (MPL), 136, 137*f*
Multichannel collimators, 238
Multilayer Laue lens (MLL), 307
Multilayer mirrors, 2, 286, 287*f*, 289*f*
neutrons, 636–641
X-ray, 624
matrix representation of, 632
N-layer, 630
normal incidence to, 635
Mylar compound refractive lens, 150*f*, 153–155

N

N-layer multilayer mirror, 630
Neutron area detector, 44
Neutron compound refractive lens, 2–3
experiments with, 44
planar array of, 167, 169*f*
Neutron contact radiograph, 201*f*
Neutron flux, 266
emission of, 261
distribution, 603–606
Neutron Fresnel zone plates (NFZPs), 29–31, 29*f*
Neutron guide tubes, development of, 278
Neutron holography, 31
Neutron microscopy, 22–26, 60
experiment, 44, 50
MNRC Bay 4 setup, 71, 74*f*
setup, 56*f*
light microscopy archetypes for, 19–22
Neutron optics, 44, 414
history of, 272–279
matrix representation of, 350–353
Neutron Optics and Physics (NOP) group, 80
neutron camera, 94, 95, 96, 96*f*, 97
Neutron particle velocity, 563, 564
Neutron phase, 561–565
velocity, 415
Neutron plane wave equation Schrödinger wave equation derived by, 572–574
Neutron polarizers, 278
Neutron ratio, 249
Neutron refractometry, 582–588
Neutron resonance radiography (NRR), 265, 267, 268
Neutron stimulated emission computed tomography (NSECT), 208
Neutron wave
group velocity, 564
guide tube, 581
packet, 565
reflection and transmission of, 576–582
Schrödinger wave equation for, 566–568, 574–575
Neutrons
absorption aperture depth and transmission, 404–406
achromat lenses for, 31–34
attenuation comparison, 619–622
beam, 275
beam choppers, 434
bright-field microscopy for, 646–648
capillary optics, 641–646
complex refractive index, 4–10
CRL for, 2–3, 358–361
dark-field microscopy for, 648–649
deflection
in CRP, 402–404
by single prism, 398–402
diffracted, 25
focusing experiment, 86*f*
formulas for, 622–624
Fresnel and compound Fresnel lens, 34–38
guide tubes, 275
image, 49, 50, 95*f*
cadmium slits, 66, 66*f*
detector, 76
of rat paw, 58, 58*f*, 59*f*, 73*f*
with lens, 62*f*
interferometry, 590–592
lenses, 338
types of, 75*f*
light source intensities and images, 39–43
magnetic lenses, experiments with, 44
material lenses and prisms, 82*f*
monochromator with CRL and pinhole, 38–39
multilayer mirrors, 636–641
nuclear scatter of, 5, 6*f*
phase-contrast imaging, 223–226, 235
phase-contrast microscopy for, 649–653
polychromatic, 94
pulse, 437
ray, 248, 384, 433, 434
refraction of, 342*f*, 343, 399*f*
reflection of, 17–18, 278
refraction of, 385, 409

Neutrons (*continued*)
refractive imaging of, 49
refractive index derivation for, 607–612
scatter of, 272, 273, 615–619, 619*t*
length, 582–584, 590, 599
spatial coherence length, 26–28
spectrum, 259*f*
types of lenses, 612–614
wavelength, 43
wavelength of, 76, 433
Newton's lens equation, 337, 338, 337*f*
NFZPs, *see* Neutron Fresnel zone plates
NG-7 beam line, 59*f*
NG-7 SANS instrument, 54
No-lens coherent diffractive imaging, 306*f*
Nomarski-modified Wollaston prism, 300, 301*f*, 302, 303
Non-ferromagnetic isotopes, 638
Non-paraxial rays, 450
Nondiffracted neutrons, 25, 26
Nondiffracted X-rays, 26
Novel X-ray sources, 165
NRR, *see* Neutron resonance radiography
Nuclear scatter of neutrons, 5, 6*f*
Numerical aperture and thin lenses, 338–339

O

Object contrast, 493
Object resolution, 434–438
Object shape, 493
Offset-reference hologram, 31
Oil immersion, 19
Omnidirectional X-rays, 26
1D compound refractive X-ray lens, 103
Optical lenses and light imaging systems, 340
Optical path length difference, relation of, 470–473, 473
Optical system analysis with Kirchhoff equation, 13–17
Optical transfer function (OTF), 377
Optimized zone plate, 294, 295*f*
Ordinary ray, 300
Orthoscopic lens system, 492

P

Parabola, defined, 410
Parabolic aperture radius in spherical biconcave CRL, 349–350
Parabolic biconcave lenses, chromatic aberration in, 430–434
Parabolic compound refractive lens, 190, 191
Parabolic dies, 193
Parabolic lenses
chromatic aberration in, 453–456
curvature, measurement of, 411–413
Parabolic refractive X-ray lenses, 120
scanning electron micrograph of, 121*f*
Parabolic waves, converge to focal points, 450–453
Parallel-channel collimators, 238, 239*f*
Parametric X-ray radiation (PXR), 166
Paraxial approximation, 417–418
application of, 464
Paraxial rays, 440*f*
Pendellösung oscillations in crystals, 596–601
Perfect lens, 456
Perimeter ray, 4*f*
Periodic magnetic field (PMF), 82–86
spectrometer, 90, 91*f*
Petzval condition
curvature of field and, 479–481
prevention of, 492–494
Petzval surface, 487
Pf2 ILL beam line, 92
Pf2 UCN/VCN beam line, 92, 93*f*
Phase Fresnel lens, 290
Phase FZP, 290, 292*f*
Phase grating, 304*f*
Phase plate, 22
Phase shift effects of light, 593
Phase shifters, 602, 603
Phase velocity, 541
X-rays in material medium, 548–550
Phase-contrast FZP, 291
Phase-contrast imaging, 219, 227, 228*f*
amplitude at detector plane, 226–233
experimental results, 235–237
intensity of, 233–234
neutron, 223–226, 235
tungsten borate-boron interface, 119*f*
X-ray, 220–223, 224*f*
Phase-contrast light microscope, 23*f*
zernike, 298
Phase-contrast microscope, 22, 649–653
X-ray/neutron, 22–26, 24*f*
Phase-contrast mode, 222
Phase-shifting element, 25
Photodiodes, silicon, 248
Photoelectric absorption, 618, 646
X-ray attenuation by, 525, 527
Photon energy, 562, 563
Photon noise, 263
Photonics, 40

Photons, refraction of, 409
Phototubes, 237
PIN diode, *see* Positive-intrinsic-negative diode
Pincushion distortion, 197, 449*f*, 494
Pinhole aperture in CRL, 38
Pinhole camera imaging, 207*f*, 209
Pinhole collimation, 49
Pixel inversion, 215
Planar interface, X-rays at, 550–554
Planck equation, 563, 566
Planck's constant, 622, 636, 637
Plane surface, 410
Plano-concave lenses, 101
Plastic Fresnel lens, 43
Plastic scintillator, 245, 248, 255, 257, 258
PMF, *see* Periodic magnetic field
Point spread function (PSF), 17, 373
Polarized Neutrons (Williams), 279
Polarizer, 302
Polonium-210, 201
Polycapillary optics, for X-rays and neutrons, 641–646
Polychromatic neutrons, 94
Polycrystalline aluminum, biconcave parabolic lenses of, 110, 111*f*
Polyethylene CRL, 108
Polymethyl methylacrylate (PMMA), 136
Position-sensitive detector (PSD), 210, 588, 648, 649
Position-sensitive neutron detector, 254
Position-sensitive photomultiplier tube (PSPMT), 240–242, 244*f*, 245, 257, 258
Positive-intrinsic-negative (PIN) diode, 125, 129
Post-monochromator, 172
Potential energy of neutron, 567
Poynting vector, 515
Pre-monochromator, 172
Pressing tool, 194*f*
Principal curved surfaces, 478
Principal planes, 382, 384, 385, 388
Principal ray, 4*f*
Projection radiography, 204–205, 205*f*
Proof-of-principle experiment, 49
Protons, recoil, 261
Prototype scanner, 247, 252
PSD, *see* Position-sensitive detector
PSPMT, *see* Position-sensitive photomultiplier tube
Pulse broadening, 548–550
Pulsed neutron sources, 200
Pythagorean theorem, 412, 454

Q

Quadrupole lens and magnet, 84*f*
Quasi-monochromatic beam, 51
Quaternary FZP, 296*f*

R

R-space, *see* Direct space
Radiation
  electromagnetic, 538
  stopping power, 528, 529
Radiographic contrast, 373
Radiography, 237
Radius of curvature, 409–411, 445, 446
Ray, 456
  angle deviation in thin lens, 353–358
  bundle, 430
  origin of, 472
Rayleigh image resolution, 439*f*
Rayleigh resolution, 17, 373
Rayleigh scatter, 617, 647
  differential cross section for, 517–519
  Thomson, 519–521
  X-ray, 514–517
Reactor sources of thermal neutrons, 200
Reciprocal space, 563
Recoil protons, 261
Reference ray light beamlet, 302
Reflection, neutrons and X-ray, 17–18
Reflective X-ray optics, 641
  and microscopy, 282
Refracted X-ray, 434
Refraction, 597
Refraction angle, 650
Refractive Fresnel lens, 34
Refractive Fresnel X-ray lenses, 100
Refractive imaging experiment, 96*f*
Refractive index, 208, 414–417
  of air/vacuum medium layer, 628
  complex, *see* Complex refractive index
  derivation for X-rays and neutrons, 607–612
  electromagnetic vector wave equation, 506–508
  focusing and defocusing lens by, 613*t*
  gradient of, 650
  measurement via interferometry, 588–590, 601–603
  neutron, 4–10, 636
  X-ray, 4–10, 628
Refractive lens
  imaging, 206*f*, 208
  optics, 44
  for X-rays, 100

Region of interest (ROI), 123
Relativistic electrons, 538–541
Relief surfaces, 37
Resolution
  CRL, 13
  of microscope, 373
  Rayleigh, 17
Reversibility of optical path, 417
Risø experiment, 44, 45*f*
Rowland circle, 287, 288

S

Sag, length of, 485, 486
SANS, *see* Small-angle neutron scatter
Sapphire crystals, 214
Scanner, prototype, 247, 252
Scanning electron microscope, 307
Scanning near-field optical microscope (SNOM), 308
Scattering disk nucleus, 609
Schrödinger wave equation
  derivation
    by electromagnetic wave equation analogy, 568–572
    of neutron plane wave solution, 572–574
    for neutron wave, 566–568
  operator form of, 574–575
Scintillator, plastic, 245, 248, 255, 257, 258
Second-order differential equation, 499
Seidel monochromatic aberrations, 454
  of spherical biconcave lens, 461–470
Sensitivity indicator (SI), 202
Sextupole lens, 84*f*
Silicon dioxide layer, 120
Silicon photodiodes, 248
Silicon saw-tooth lens, 175
Silver–colored lithium, 141
Single biconcave parabolic lens, 110, 644
Single-capillary optics, 644
Single-crystal lens, 175
Single-crystal sapphire, 214
Small-angle neutron scatter (SANS), 44, 49, 406–409
  compound refractive neutron lenses for, 104
  refractive optics for, 50*f*
Smoothing filter, 252
Snell's law, 300, 414–417, 554, 612, 633, 639
  biconcave lens, 343
  linear form of, 356
  for neutron refraction, 578–579
SNOM, *see* Scanning near-field optical microscope
Spatial coherence length, neutron and X-ray, 26–28
Spatial filtering, 252
Spatial gradient of potential energy, 567
Spatial resolution, 373
Spherical aberration, 349, 473–475
  CRL image and object resolution, 438
  in spherical biconcave lenses, 434–438
Spherical biconcave CRL, parabolic aperture radius in, 349–350
Spherical biconcave lenses, 485
  chromatic aberration in, 430–434
  CRL, 67
  Seidel monochromatic aberrations of, 461–470
  spherical aberration in, 434–438
Spherical lens curvature, measurement of, 411–413
Spherical lens surface, aberrations from imperfect, 456–461
Spherical lenses
  chromatic aberration in, 453–456
  CRL, 436
Spherical neutron wave, 608
Spherical surface, 410
Spherical waves, 229, 455, 461
  converge to focal points, 450–453
  front, 470
Sphering
  CRL, 99
  magnetic lens, 97, 98*f*
Spin-down neutrons, 90, 637
Spin-up neutrons, 637
Stacked biconcave lenses, 3*f*
Stanford Photonics CCD camera, 261, 262
Stanford Synchrotron Radiation Lightsource (SSRL) beam line, 108
Steel collimators, 247
Stern–Gerlach effect, 278
Sturm's interval, 487
Submicron optical resolution, 299
Superconducting, 85*f*
Supermirrors, neutron, 641
Surface boundary conditions, 550
Surface roughness of CRL, 390–391
Surface roughness, cause of, 18
Symmetrical Laue case, 597–599, 601
Synchrotron X-ray
  imaging, 221
  sources, 182, 184, 494–495

Synchrotron-induced X-ray fluorescence (SXRF), 129
Synthetic multilayer X-ray mirrors, 286

T

Tabletop X-ray microscope, 181*f*, 184
  consist of, 188
  with multilayer optics, 186*f*
Teflon lenses, 73
Temporal spatial coherence length, neutrons and X-ray, 26–28
Thermal neutron absorption, 336
Thermal neutron energy, 623
Thermal neutron radiography
  application of, 200
  development, 202
  MNRC Bay 4 for, 215, 216*f*
Thermal neutrons, 43, 64, 564, 565, 648
  absorption and scatter, 200
  imaging, 66, 274
    experiments, 63
  Maxwell–Boltzmann-distributed, 214
  microscope setup, 64*f*
  radiographic imaging system, 262*f*
  reactor source of, 603–606
  scattering, 584
Thermal vibrations, in material lattices, 617
Thick lens
  CRL, 382–389, 383*f*
  power and focal length, 420–424
Thin lens
  approximation, 461
  Gaussian lens equation for, 418–420
  matrix and ray angle deviation in, 353–358
Thomson Rayleigh scatter, 519–521
Thomson, J. J., 341
3D lens array, 167, 168*f*
Time-of-flight (TOF)
  imaging detector, 70, 437
  neutron detector, 78
Titanium, 258
  neutron lenses, 338
TOF, *see* Time-of-flight
Tomographic techniques, 125
Transfer matrix, 625, 630–631
Transverse aberration, 451
Transverse chromatic aberration, 432
Transverse magnification, 3, 4*f*, 368–371
Transverse spatial coherence length, neutrons and X-ray, 26–28
Triple-Laue case interferometer, *see* LLL interferometer
TT neutron generator, 258
Tungsten borate-boron interface, 118, 119*f*
Two-dimensional (2D) neutron detector, 49
2D compound refractive X-ray lens, 101
Two-lens optical system, prevent distortion in, 492

U

UCN, *see* Ultracold neutron
ULDs, *see* Unit load devices
Ultracold neutron (UCN), 92, 93
  beams, 4
Ultracold neutrons, 579, 581
Ultrashort pulses, 307
Undulator X-ray source, 117
Unit load devices (ULDs), 246, 248, 254

V

Vacuum velocity, 415
VCNs, *see* Very cold neutrons
Vergence, wave, *see* Wave vergence
Very cold neutrons (VCNs), 92
Vinyl long-playing (LP) record, 137–138
Visible light optics, 2, 40
Vladimirskiĭ magnetic mirror, 588
Vulcan laser, 42

W

Wave fronts, 455
  tilted, 485
Wave packets, X-ray, 541–544
  dispersion and group velocity, 547
Wave surface, curvature of, 409–411
Wave vector, 501
Wave vergence, 413–414, 424–426
  change by lens surface, 413–414
  in thick lenses, 420–424
Wollaston prism, 299, 301*f*
  disadvantage of, 300
Wolter X-ray mirrors, 305
Work-energy theorem, 567

X

X-ray, 384, 433
  focusing and imaging, 289
  absorption aperture depth and transmission, 404–406
  absorptive attenuation by pair production, 528
  achromat lenses for, 31–34

X-ray (*continued*)
attenuation
comparison, 619–622
by photoelectric absorption, 525–527
Brewster angle, 557–559
bright-field microscopy for, 646–648
capillary optics, 641–646
complex index of refraction, 513–514
complex refractive index, 4–10
compton scatter of, 521–525
CRL transmission for, 358–361
dark-field microscopy for, 648–649
deflection
in CRP, 402–404
by single prism, 398–402
with electric field component parallel, 554–556
formulas for, 622–624
Fresnel and CFL, 34–38
Fresnel zone plates, 29–31, 29*f*
interferometry, 590–592
lenses, development of , 290
light source intensities and images, 39–43
lithography, 135
with magnetic field component parallel, 556
in material medium, 548–550
matrix representation of, 350–353
monochromator with CRL and pinhole, 38–39
multilayer mirror for, 286, 624
matrix representation of, 632
N-layer, 630
normal incidence to, 635
optics, 414
phase-contrast imaging, 220–223, 224*f*
phase-contrast microscopy for, 649–653
radiator, 166
rayleigh scatter and complex atomic scattering factor, 514–517
reflection and transmission of, 550–554
reflection of, 17–18
refraction of, 342*f*, 343, 385, 434
refractive index derivation for, 607–612
scatter of, 494, 615–619, 619*t*
sources, development of, 285
spatial coherence length, 26–28
transmission fraction, 178
types of lenses, 612–614
wave packets, 541–544
dispersion and group velocity, 544–547
X-ray absorption near-edge structure (XANES), 127, 129
X-ray compound refractive lens, 2–3
advantages of, 107–108
aluminum and beryllium parabolic, 131
bubble lenses, 143–144, 144*f*, 160–163, 161*f*, 176, 177*f*
diffraction-limited spot, 112
experimental demonstration of, 102–105, 105*f*
glassy carbon, 121–123, 122*f*
invention of, 100–102
lithium, 139–143, 142*f*, 152*f*
magnification of, 115
Mylar and Kapton, 150*f*, 153–155
planar array of, 167, 169*f*
X-ray contact radiograph, 201*f*
X-ray diffraction, 306*f*
gratings, 288, 289*f*
X-ray emission damped resonance for, 498–501
X-ray fluorescence, 642
X-ray fluorescence element microtomography (XFEMT), 123–124, 126*f*
X-ray grating, development of, 294
X-ray microscopy, 22–26, 307
dark-field, 297
light microscopy archetypes for, 19–22
phase-contrast, 297*f*
XM-1 bright-field soft, 298*f*
X-ray optics, 153, 166
development of, 282
elements, 304
X-ray photoelectron spectroscopy, 646
X-ray shearing interferometer, 305*f*
X-ray spectroscopy, 645
X-ray telescope, lobster-eye, 286
X-ray transmission, Gaussian-shaped, 153
X-ray tube
copper anode of, 185
FOV, 187
sources, 183, 184
XFEMT, *see* X-ray fluorescence element microtomography
XM-1 bright-field soft X-ray microscope, 298*f*

Z

Zernike phase-contrast light microscope, 298
Zernike phase-contrast microscope, 22, 298
Zone plate, 31, 290
doublet, 303
Gabor, 31
optimized, 294, 295*f*

FSC
www.fsc.org
MIX
Papier | Fördert gute Waldnutzung
FSC® C083411

Kolibri 360 GmbH
Ferdinand-Jühlke-Straße 7
99095 Erfurt, Deutschland
produktsicherheit@kolibri360.de